Mathematica® Navigator

Mathematica® Navigator

Mathematics, Statistics, and Graphics

SECOND EDITION

Heikki Ruskeepää

Department of Mathematics
University of Turku, Finland

ELSEVIER
ACADEMIC
PRESS

AMSTERDAM • BOSTON • HEIDELBERG • LONDON
NEW YORK • OXFORD • PARIS • SAN DIEGO
SAN FRANCISCO • SINGAPORE • SYDNEY • TOKYO

Senior Editor, Mathematics Barbara A. Holland
Associate Editor Tom Singer
Project Manager Justin Palmeiro
Cover Design Monty Lewis
Production Services Kolam USA
Composition Kolam
Printer The Maple-Vail Book Manufacturing Group

Camera ready copy was prepared by the author with *Mathematica*®.

Elsevier Academic Press
200 Wheeler Road, Burlington, MA 01803, USA
525 B Street, Suite 1900, San Diego, California 92101-4495, USA
84 Theobald's Road, London WC1X 8RR, UK

This book is printed on acid-free paper. ∞

Library of Congress Cataloging-in-Publication Data
Ruskeepää, Heikki.
 Mathematica navigator : mathematics, statistics, and graphics / Heikki Ruskeepää.—2nd ed.
  p. cm.
 Includes bibliographical references and index.
 ISBN 0-12-603642-X
  1. Mathematics—Data processing. 2. Mathematica (Computer file) I. Title.

 QA76.95.R87 2003
 510'.2855—dc22

                                                                         2003058330

British Library Cataloguing in Publication Data
A catalogue record for this book is available from the British Library

ISBN: 0-12-603642-X

For all information on all Academic Press publications
visit our website at www.academicpressbooks.com

Printed in the United States of America
03 04 05 06 07 08   9 8 7 6 5 4 3 2 1

To Marjatta

Contents

Preface

What is the difference between an applied mathematician and a pure mathematician?
An applied mathematician has a solution for every problem,
while a pure mathematician has a problem for every solution.

Welcome

The goals of this book, the second edition of *Mathematica Navigator: Mathematics, Statistics, and Graphics,* are as follows:

- to introduce the reader to *Mathematica*; and
- to emphasize mathematics (especially methods of applied mathematics), statistics, graphics, programming, and writing mathematical documents.

Accordingly, we navigate the reader through *Mathematica* and give an overall introduction. Often we slow down somewhat when an important or interesting topic of mathematics or statistics is encountered to investigate it in more detail. We then often use both graphics and symbolic and numerical methods.

Here and there we write small programs to make the use of some methods easier. Chapter 3 is devoted to *Mathematica* as an advanced environment of writing mathematical documents.

The online version of the book, which can be installed from the enclosed CD-ROM, makes the material easily available when working with *Mathematica*.

Changes in this second edition are numerous and are explained later on in the Preface.

■ Readership

The book may be useful in the following situations:

- for courses teaching *Mathematica*;
- for several mathematical and statistical courses (given in, for example, mathematics, engineering, physics, and statistics); and
- for self-study.

Indeed, the book may serve as a tutorial and as a reference or handbook of *Mathematica*, and it may also be useful as a companion in many mathematical and statistical courses, including the following:

differential and integral calculus • linear algebra • optimization • differential, partial differential, and difference equations • engineering mathematics • mathematical methods of physics • mathematical modelling • numerical methods • probability • stochastic processes • statistics • regression analysis • Bayesian statistics.

■ Previous Knowledge

No previous knowledge of *Mathematica* is assumed. In other words, this book may be used as an introduction to *Mathematica*.

On the other hand, we assume some knowledge of various topics in pure and applied mathematics. We study, for example, partial differential equations and statistics without giving detailed introductions to these topics. If you are not acquainted with a topic, you can simply skip the section of the book considering that topic (or you can browse through the section and try to get an idea of the subject, or even learn something about it).

Also, to understand the numerical algorithms, it is useful if the reader has some knowledge about the simplest numerical methods. Often we introduce briefly the basic ideas of a method (or they may become clear from the examples or other material presented), but usually we do not derive the methods. If a topic is unfamiliar to you, then consult a textbook about numerical analysis, such as Skeel and Keiper (1993).

■ Recommendations

If you are a newcomer to *Mathematica*, then Chapter 1, *Starting*, is mandatory, and Chapter 2, *Sightseeing*, is strongly recommended. You can also browse Chapter 3, *Notebooks*, and perhaps also Chapter 4, *Files*, so that you know where to go when you encounter the topics of these chapters. After that you can proceed more freely. However, read Section 12.1, "Basic Techniques", because it contains some very common concepts used constantly for expressions.

If you have some previous knowledge of *Mathematica*, you can probably go directly to the chapter or section you are interested in, with the risk, however, of having to go back to study some background material. Again, be sure to read Section 12.1.

Contents

An overview of the contents of the book is as follows:

As you can see, the 27 chapters of the book can be divided into six main parts: introduction, files, graphics, expressions, programs, and mathematics. The mathematics section is the largest and makes up about half of the book.

Most of the headings of the mathematical chapters resemble the names of some common mathematical and statistical courses. Exceptions are Chapters 19, *Equations*, 21, *Interpolation*, and 22, *Approximation*; these topics are often parts of courses in numerical methods (note that numerical methods are also used in all other mathematical chapters).

Dependencies between the chapters are generally quite low. If you read Chapter 2, *Sightseeing*, you will get a background that may serve you well when reading most other chapters; in some chapters you will also find references to earlier chapters, where you will find the needed background.

Next we describe the main parts of the book in a few words.

■ Introduction, Files, Graphics, Expressions, and Programs

The first two chapters introduce *Mathematica* and give a short overview.

The next two chapters consider files, in particular files created by *Mathematica*, which are called *notebooks*. We show how *Mathematica* can be used to write mathematical documents. We also explain how to load packages, how to export and import data and graphics into and from *Mathematica*, and how to manage memory and computing time. You may skip these two chapters until you need them.

Then we go on to graphics. One of the finest aspects of *Mathematica* is its high-quality graphics, and one of the strongest motivations for studying *Mathematica* is to learn to illustrate mathematics with figures. The material on graphics is divided into six chapters: two-dimensional graphics for functions (Chapters 5 and 6), three-dimensional graphics for functions (Chapters 7 and 8), and two- and three-dimensional graphics for data (Chapters 9 and 10).

In Chapters 11, 12, and 13, we study various types of expressions. We consider numbers in *Mathematica* and emphasize topics about floating-point numbers and precision of numerical routines. We explain how to manipulate several types of expressions. A chapter is devoted to lists, the very basic expressions of *Mathematica*.

Chapters 14 and 15 consider programs. First we consider special types of functions (such as composite, piecewise-defined, periodic, recursive, and implicit functions), so-called "pure" functions (very important in *Mathematica*), modules (functions consisting of several commands), and packages (collections of functions). Then we study programs. Three types of programming are introduced: procedural, functional, and rule-based.

■ Mathematics

In the remaining 12 chapters we study different areas of pure and applied mathematics. The chapters can be divided into 5 classes, each class containing chapters of more or less related topics. Descriptions of the 5 classes follow.

Topics of traditional *differential* and *integral calculus* include derivatives, Taylor series, limits, integrals, sums, and transforms.

Then we consider vectors and *matrices*; linear, polynomial, and transcendental *equations*; and unconstrained, constrained, and classical *optimization*.

In *interpolation* we have the usual interpolating polynomial, a piecewise-calculated interpolating polynomial, and splines. In *approximation* we distinguish the approximation of data and functions. For the former we can use the linear or nonlinear least-squares method, while for the latter we have, for example, minimax approximation.

Mathematica solves *differential equations* both symbolically and numerically. We can solve first- and higher-order equations, systems of equations, and initial and boundary value problems. For *partial differential equations*, we show how some equations can be solved symbolically, how to handle series solutions, and how to solve problems numerically with the method of lines or with the finite difference method. Then we consider *difference equations*. For linear difference equations, we can possibly find a solution in a closed form, but most nonlinear difference equations have to be investigated in other ways, such as studying trajectories and forming bifurcation diagrams.

Lastly we study *probability* and *statistics*. *Mathematica* contains information about most of the well-known probability distributions. Simulation of various random phenomena (like stochastic processes) is done nicely with random numbers. Statistical topics include descriptive statistics, frequencies, smoothing, confidence intervals, hypothesis testing, regression, and Bayesian statistics.

Special Aspects

The book explains a substantial part of the topics of *Mathematica*. However, some topics are emphasized, some are given less emphasis, and some are even excluded. We describe these special aspects of the book here.

■ Width

We have had the goal of studying important topics in some width and depth. This may mean detailed explanations, clarifying examples, programs, and applications. It may also mean introducing topics for which there is little or no material in the manuals by Wolfram (2003) and Wolfram Research (2003).

The headings of the chapters give a list of topics that are emphasized in this book and that are explained in some width. However, some emphasized topics cannot be seen from the chapter headings. One of them is numerical methods; they are used in every mathematical chapter. Another is methods relating to data. Indeed, we use several real life and artificial data sets in chapters about graphics for data, approximation, differential and difference equations, probability, and statistics.

■ Depth

To give an impression of the depth of various topics, we next describe some special topics in various chapters of the book.

- Chapter 3, *Notebooks:* An introduction to *Mathematica* as an environment for preparing technical documents; writing mathematical formulas
- Chapter 6, *Options for 2D Graphics:* Using texts, legends, and graphics primitives in graphics

- Chapter 7, *3D Graphics for Functions:* Stereographic figures; graphics for four-dimensional functions
- Chapter 9, *2D Graphics for Data:* Various types of plots for data; visualizations of several real-life data; dot plots; statistical plots
- Chapter 13, *Lists:* Tabulating data
- Chapter 14, *Functions:* Various types of functions
- Chapter 15, *Programs:* Three styles of programming (procedural, functional, and rule-based); emphasis on functional programming; lots of examples of programs
- Chapter 17, *Integral Calculus:* Programs for numerical quadrature
- Chapter 19, *Equations:* Iterative methods of solving linear equations; programs for nonlinear equations
- Chapter 20, *Optimization:* A program for numerical minimization; a program for classical optimization with equality and inequality constraints
- Chapter 22, *Approximation:* Graphical diagnostics of least-squares fits
- Chapter 23, *Differential Equations:* Analyzing and visualizing solutions of systems of nonlinear differential equations; study of a predator–prey model, a competing species model, and the Lorenz model; numerical solution of linear and nonlinear boundary value problems; estimation of nonlinear differential equations from data; solving integral equations
- Chapter 24, *Partial Differential Equations:* Series solutions for partial differential equations; solving parabolic and hyperbolic problems by the method of lines; solving elliptic problems by the finite difference method
- Chapter 25, *Difference Equations:* The logistic model as an example of nonlinear difference equations; bifurcation diagrams, periodic points, Lyapunov exponents; a discrete-time predator–prey model as an example of a system of nonlinear difference equations; estimation of nonlinear difference equations from data; fractal images
- Chapter 26, *Probability:* Simulation of several stochastic processes
- Chapter 27, *Statistics:* Visualizing confidence intervals and types of errors in statistical tests; confidence intervals and tests for probabilities; local regression; Bayesian statistics; Gibbs sampling; Markov chain Monte Carlo (MCMC)

■ Programs

Mathematica has many ready-to-use commands for symbolic and numerical calculations and for graphics. Nevertheless, in this book we also present more than 70 of our own programs. Indeed, programming is one of the strongest points of *Mathematica*. It is often amazing how concisely and efficiently we can write a program even for a somewhat complex problem. We think that our own programs can be of some value, despite the fact that they are not so fine and powerful as *Mathematica*'s built-in commands. We have included our own programs for the following reasons:

- First, a self-made implementation shows clearly how the algorithm works. You know (or should know) exactly what you are doing when you use your own implementation. The ready-made commands are often like black (or gray) boxes, because we do not know much about the methods.

- Second, writing our own implementations teaches us programming. We present short programs throughout the book (especially in the mathematical chapters). In this way, we hope that you will become steadily more familiar with programming and that you are encouraged to practice program writing.

- Third, a self-made implementation can be pedagogically worthwhile. For example, we implement Euler's method for differential equations. It has almost no practical value, but as the simplest numerical method for initial value problems, it has a certain pedagogical value. Also, programming a simple method first may help us to tackle a more demanding method later on.

■ Other Special Aspects

We have integrated the so-called "packages" tightly into the material covered in this book. Instead of presenting a separate chapter about packages, each package is explained in its proper context. In this way, it is hoped that your attention will be drawn more effectively to the many excellent packages.

We have tried to make the structure of the book such that finding a topic is easy. Usually a topic is considered in one and only one chapter or section so that you need not search in several places to find the whole story. Each numerical routine is also presented in the proper context after the corresponding symbolic methods. This helps you to find material for solving a given problem: it is usually best to try a symbolic method first and, if this fails, to then resort to a numerical method.

Some topics of a "pure" nature such as number theory, finite fields, quaternions, combinatorics, and graph theory are not considered in this book; *Mathematica* has many packages for these topics. Commands for string, box, and notebook manipulation are treated only briefly. We do not consider *MathLink* (a part of *Mathematica* that enables interaction between *Mathematica* and external programs), *J/Link* (a product that integrates *Mathematica* and Java), *XML* (a metamarkup language for the World Wide Web), or *MathML* (an XML-based markup language for representing mathematics). Also, we do not consider any of the many other *Mathematica*-related products like *webMathematica*, *gridMathematica*, *CalculationCenter*, or the *Applications Library* packages.

Changes in the Second Edition

The structure of and the broad topics covered in the second edition are very much the same as what was found in the first edition. However, the text has been revised throughout, the book now describes *Mathematica* 5 (the first edition was based on *Mathematica* 3), and new subjects are treated.

■ Changes in the Structure

The main change in the structure of the book is that we have two new chapters: Chapter 18, *Matrices*, and Chapter 25, *Difference Equations*. (In the first edition, matrices were considered in the chapter about lists and difference equations in the chapter about differential equations.)

The material about loading packages is now in Chapter 4, *Files*, instead of in Chapter 3, *Notebooks*. Mapping of lists is now considered in Chapter 13, *Lists*, instead of Chapter 15, *Programs*.

■ Major Changes in Contents

The most important changes in the contents are in the following five chapters:

- Chapter 3, *Notebooks:* Much more information about *Mathematica* as a writing tool: editing style sheets, using options, writing mathematical formulas, and using automatic numbering
- Chapter 9, *2D Graphics for Data:* Enhanced treatment of plotting one and several time series; coverage of `Histogram`, `BoxWhiskerPlot`, `PairwiseScatterPlot`, and `QuantilePlot`
- Chapter 20, *Optimization:* Reorganization of the material into three sections about unconstrained, constrained, and classical optimization; coverage of `Minimize` and `NMinimize`; includes a program for classical optimization
- Chapter 25, *Difference Equations:* Much broader coverage of both linear and nonlinear difference equations; includes the use of generating functions and the Z transform, a detailed analysis of the classical logistic model, an example of a system of nonlinear equations, and estimation of difference equation models
- Chapter 27, *Statistics:* New sections about smoothing (e.g., `ListCorrelate`), `ANOVA`, local regression, and Bayesian statistics; for local regression, a program for the loess method by Cleveland (1993) is presented; for Bayesian statistics, Gibbs sampling and Markov chain Monte Carlo (MCMC) are considered

■ Other Changes in Contents

Other changes in the contents include the following:

- Chapter 4, *Files:* Enhanced coverage of exporting and importing data and graphics
- Chapter 5, *2D Graphics for Functions:* Coverage of `InequalityPlot` and `Complex` `InequalityPlot`
- Chapter 6, *Options for 2D Graphics:* Enhanced treatment of texts and legends
- Chapter 7, *3D Graphics for Functions:* Coverage of the `RealTime3D`` environment for real time rotation
- Chapter 8, *Options for 3D Graphics:* More about the coloring of 3D graphics
- Chapter 12, *Expressions:* Treatment of logical expressions
- Chapter 13, *Lists:* Enhanced treatment of tabulating lists
- Chapter 15, *Programs:* A new section about simple programs
- Chapter 17, *Integral Calculus:* Coverage of the `Calculus`Integration`` package for advanced multiple integration
- Chapter 18, *Matrices:* Coverage of `SparseArray`, `MatrixPlot`, `Total`, `Norm`, `Matrix` `Rank`, and `SingularValueDecomposition`
- Chapter 19, *Equations:* Enhanced treatment of `Reduce`; solving inequalities; coverage of `FindInstance`
- Chapter 22, *Approximation:* Coverage of `FindFit` and trigonometric fits

Some Notes

■ New Properties of Version 5

Properties and commands of *Mathematica* available for the first time in version 5 are marked with (❀5). New commands, options, and constants include **BoxWhiskerPlot**, **PairwiseScatterPlot**, **QuantilePlot**, **MachinePrecision**, **EvaluationMonitor**, **Step↓ Monitor**, **Refine**, **Most**, **Reap**, **Sow**, **MatrixPlot**, **ArrayDepth**, **SparseArray**, **MatrixPlot**, **Total**, **Norm**, **MatrixRank**, **SingularValueList**, **SingularValueDecomposition**, **Find↓ Instance**, **FindMaximum**, **Maximize**, **NMaximize**, and **FindFit**.

Although we have explained the new commands of *Mathematica* 5 in their proper contexts and presented numerous examples of their use, note that in other places of the book, we have avoided the use of the new commands. The reason for this is that then users of *Mathematica* with versions earlier than 5 also have access to the examples and programs presented in the book. For example, in many places we could have used **Total** to calculate various sums or **Norm** to calculate norms of vectors and matrices. Instead, we have used **Apply** for sums and custom code for norms. Users of *Mathematica* 5 have the option to utilize the new commands in the examples and programs of this book.

Note that in version 5 we have several built-in commands that in earlier versions required loading a package; examples are **CholeskyDecomposition**, **Minimize**, **NMinimize**, **RSolve**, **Mean**, **Median**, **Quantile**, **Variance**, and **StandardDeviation**. Such commands are also marked with (❀5), and we mention the package that has to be loaded in earlier versions.

Note also that some new commands and some enhanced old commands in version 5 make obsolete some other old commands; examples are **SingularValueList** and **SingularValueDecomposition** pro **SingularValues**, **ArrayDepth** pro **TensorRank**, **Minimize** pro **ConstrainedMin**, **Maximize** pro **ConstrainedMax**, **FindFit** pro **Nonlinear↓ Fit**, **Reduce** pro **InequalitySolve**, and **Norm** pro **VectorNorm** and **MatrixNorm**.

Several commands are enhanced in version 5; examples are **Eigenvalues**, **Eigen↓ vectors**, **Reduce**, **FindRoot**, **FindMinimum**, **LinearProgramming**, **DSolve**, **NDSolve**, and **RSolve**.

Some results are slightly different in *Mathematica* 4.2 than in *Mathematica* 5. We have tried to denote such results by (❀4!), so that users of version 4 can prepare to see a somewhat different result.

■ Other Notes

Environment. During the work, I have used a Macintosh with MacOS X. *Mathematica* works in much the same way in various environments, but the keyboard shortcuts of menu commands vary among different environments. To some extent we mention the shortcuts for the Microsoft Windows and Macintosh environments.

Graphics. Most of the figures in the book are small in size to save space. When you produce graphics, use a larger size to make the figures clearer and more impressive (the default size is, in fact, much larger).

Options. Many commands of *Mathematica* have options for modifying the commands in some ways. All options have a default value, but we can input other values. When listing the options, we give either all possible values of them or some examples of possible values, but we do not explicitly mention what the default values are, to save space. In the context of this book, *the default value of an option is always the first value mentioned.* After that come other possible values or examples of other values.

Messages. In working with *Mathematica*, you will encounter warnings and error messages. Warnings about possible spelling errors are frequent, but often you can just go on with your calculations. If you work with the examples of the book, you will encounter messages about possible spelling errors, but please note that these messages have been removed from this text to save space. Some other messages have also been removed, but we often make note of such instances.

Simulations. In several places in the book, we simulate various random phenomena. Usually, each time a simulation is run, a slightly different result is obtained. However, in experimenting with the examples of the book, a reader may want to get exactly the same result as printed on the book. This can be achieved by using a seed to the random number generator with **SeedRandom[n]** for a given integer **n**. With the same seed, the result of a simulation remains the same in repeated executions. We will use **SeedRandom** now and then in this book. If you want to get other results of simulation than those of this book, give different seeds or do not execute **SeedRandom[n]** at all (in the latter case, the default seed is used).

CD-ROM. The entire book is contained on the CD-ROM that comes with it. With a few easy steps you can install the book into the *Help Browser* of *Mathematica* (the CD-ROM contains installation instructions). With the *Help Browser* you can easily find and read sections of the book, experiment with the commands, and copy material from the book to your document. You can see all of the figures of the book in color (color figures appear especially in Chapters 7, 8, 10, and 24) and show all animations explained in the book. In addition, the CD-ROM contains two packages, some data, and a notebook explaining the use of the **AuthorTools** package.

Notation. Throughout the book, the adjectives one-, two-, three-, and four-dimensional are abbreviated by 1D, 2D, 3D, and 4D. The symbol ⁀ is used as a hyphen for *Mathematica* commands.

Questions. If you have questions about the use of *Mathematica*, don't hesitate to contact me. I try to answer when I have the time. Also, please send comments and corrections.

Acknowledgments

In preparing this book, my main sources have been *The Mathematica Book* by Stephen Wolfram (2003) and the manual for the packages by Wolfram Research (2003). The technical support staff at Wolfram Research, Inc. has helped me a lot; I would like to especially thank Eric Bynum and Harry Calkins.

The whole book from Contents to Index has been written with *Mathematica*; each chapter was a *Mathematica* notebook. The notebooks were connected into a single project by the **AuthorTools** package of *Mathematica*. The package then automatically generated the index (after we had attached the index entries with the cells of the notebooks), and the package also prepared the *Help Browser* version of the book.

The anecdotes at the beginning of the chapters are from the wonderful book by Mac-Hale (1993) (the anecdotes are reproduced or adapted with the permission of the publishers, Boole Press, 26 Temple Lane, Dublin 2, Ireland).

I have been lucky enough to enjoy excellent working conditions at the Department of Mathematics of the University of Turku. For this my sincere thanks are due to Professor Emerita Ulla Pursiheimo and Professor Mats Gyllenberg.

For financial support I express my deep gratitude to Turun Yliopistosäätiö (Turku University Foundation) and Suomen Tietokirjailijat (The Association of Finnish Non-Fiction Writers).

For their review of the manuscript, I am very thankful to Donald Balenovich, Indiana University of Pennsylvania; Joaquin Carbonara, Buffalo State University; William Emerson, Metropolitan State University; Jim Guyker, Buffalo State University; Mike Mesterton-Gibbons, Florida State University; and Fred Szabo, Concordia University. Their valuable comments and suggestions greatly improved the book.

The editorial staff at Academic Press has done a fine work with the production of the book. Especially I would like to thank Barbara Holland, Tom Singer, and Justin Palmeiro. Christine Brandt at Kolam, Inc. supervised the final stages of the production with great care. I am very grateful to Jennifer Etling for doing an admirable job in correcting my English (however, even after an extensive correction, the text certainly reveals the poverty of my English). Thank you all!

Lastly, I would like to warmly thank my wife, Marjatta, for her encouragement and support during the long work.

Heikki Ruskeepää

Department of Mathematics
University of Turku
FIN-20014 Turku
Finland
ruskeepa@utu.fi

Mathematica® Navigator

Starting

Introduction

> *In 1903 at a meeting of the American Mathematical Society, F. N. Cole read a paper entitled "On the Factorization of Large Numbers." When called upon to speak, Cole walked to the board and, saying nothing, raised two to its sixty-seventh power and subtracted one from the answer. Then he multiplied, longhand, 193,707,721 by 761,838,257,287 and the answers agreed. Without having said a word, Cole sat down to a standing ovation. Afterwards he announced that it had taken him twenty years of Sunday afternoons to factorize the Mersenne number $2^{67} - 1$.*

This chapter is intended to give you an impression of *Mathematica* and to teach you some of its basic techniques and commands. A more complete insight is given in the next chapter, where we briefly present a selection of the most important commands of *Mathematica*.

Although this book puts some emphasis on the methods of applied mathematics, this chapter begins, in Section 1.1, with a "pure" example: factoring integers. We consider the problem mentioned in the anecdote above and show what we can do nowadays with such powerful systems as *Mathematica*. This example will enlighten you regarding some of the major aspects of the program. We emphasize that it is not intended that you do the calculations of this example, nor that you should understand the commands we use. The more direct instructions presented in this book begin in Section 1.2.

Following Section 1.1, we give a brief overview of some of *Mathematica*'s basic techniques and commands, beginning with the classical starting example of calculating $1 + 2$

and ending with calculus and graphics. Then we present and explain the important conventions of *Mathematica*, which often cause trouble for beginners.

In Sections 1.4 and 1.5, we tell how you can get help within *Mathematica* and how you can correct and edit what you have written. These two sections may give more information than you need now, but you can read the basic points and return to these sections later on, when getting help and editing become more relevant concerns.

Parts of this chapter depend on the computer you use. We will explain only the Windows and Macintosh environments, although some comments may be found about the basics of *Mathematica* in a Unix system.

1.1 What is *Mathematica*?

1.1.1 An Example

■ Verifying the Work of Cole

(*Note:* It is not intended that you do the calculations of Section 1.1.1. The example is only intended to be read and to demonstrate certain aspects of *Mathematica*. Your actual lessons begin in Section 1.2.)

Did you read the anecdote about F. N. Cole at the beginning of this chapter? Cole sacrificed every Sunday afternoon for 20 years to study the Mersenne number $M_{67} = 2^{67} - 1$:

```
2^67 - 1
```

```
147573952589676412927
```

The first line is the command entered to *Mathematica*, and the second line the answer given by *Mathematica*. At last he found that the number is the product of 193,707,721 and 761,838,257,287:

```
193707721 * 761838257287
```

```
147573952589676412927
```

M_{67} is thus not a prime. Cole's feat was admirable. Now, after 100 years, we have *Mathematica*, and the situation is totally different. It now takes only a fraction of a second to do the factorization:

```
FactorInteger[2^67 - 1] // Timing
```

```
{0.03 Second, {{193707721, 1}, {761838257287, 1}}}
```

Mathematica found that 193,707,721 and 761,838,257,287 are factors of multiplicity 1.

■ Difficult Factors

However, even today some problems can be surprisingly difficult. When *Mathematica*, in my computer (which is not very fast), factorized M_{227}, it needed about 1 hour and 45 minutes and about 78 megabytes of RAM:

```
FactorInteger[2^227 - 1] // Timing
```

```
{6326.68 Second, {{26986333437777017, 1},
    {799217777382059796264915069508677209535455660121688631, 1}}}
```

```
MaxMemoryUsed[]
```

```
77502848
```

However, note that M_{227} is very big:

```
2^227 - 1
```

```
215679573337205118357336120696157045389097155380324579848828881993727
```

and the two factors of this number also are big. Thus, factoring the number is obviously a difficult task. By the way, the difficulty of factoring large numbers is a key to some cryptographic methods.

However, for M_{227}, *Mathematica* can immediately tell that it is not a prime:

```
PrimeQ[2^227 - 1] // Timing
```

```
{0. Second, False}
```

Indeed, we can easily investigate Mersenne numbers for primality up to, say, index 521:

```
(mp = Table[{i, PrimeQ[2^i - 1]}, {i, 2, 521}];) // Timing
```

```
{1.02 Second, Null}
```

The indices for which the corresponding Mersenne number is a prime are as follows:

```
Transpose[Select[mp, #[[2]] == True &]][[1]]
```

```
{2, 3, 5, 7, 13, 17, 19, 31, 61, 89, 107, 127, 521}
```

■ A Demanding Computation

To further illustrate the use of *Mathematica*, we now factor the Mersenne numbers M_2 to M_{250}. (Note that you are not supposed to do the calculations in this example. Just cast an admiring glance at the commands. Later on in this book you will learn such commands as **Table**, **Apply**, and **Map**.) We do not show the factors themselves; we only count the number of factors:

```
(t = Table[Apply[Plus, Map[#[[2]] &, FactorInteger[2^i - 1]]],
    {i, 2, 250}]) // Timing
```

```
{11689.2 Second,
 {1, 1, 2, 1, 3, 1, 3, 2, 3, 2, 5, 1, 3, 3, 4, 1, 6, 1, 6, 4, 4, 2, 7, 3, 3, 3,
  6, 3, 7, 1, 5, 4, 3, 4, 10, 2, 3, 4, 8, 2, 8, 3, 7, 6, 4, 3, 10, 2, 7, 5,
  7, 3, 9, 6, 8, 4, 6, 2, 13, 1, 3, 7, 7, 3, 9, 2, 7, 4, 9, 3, 14, 3, 5, 7,
  7, 4, 8, 3, 10, 6, 5, 2, 14, 3, 5, 6, 10, 1, 13, 5, 9, 3, 6, 5, 13, 2, 5,
  8, 14, 2, 11, 2, 10, 11, 6, 1, 15, 2, 12, 6, 11, 5, 9, 6, 9, 9, 6, 6, 17,
  4, 3, 5, 8, 5, 14, 1, 9, 5, 9, 2, 15, 3, 5, 10, 11, 2, 9, 2, 16, 6, 6, 6,
  19, 5, 6, 7, 10, 2, 14, 5, 11, 8, 10, 8, 18, 4, 5, 8, 13, 7, 16, 5, 10,
  10, 8, 2, 19, 4, 7, 7, 10, 4, 11, 9, 14, 6, 5, 3, 24, 4, 11, 5, 11, 5, 8,
  5, 10, 10, 10, 5, 16, 3, 7, 8, 11, 2, 17, 2, 20, 6, 4, 7, 20, 6, 5, 9, 12,
  6, 22, 3, 10, 7, 4, 8, 21, 6, 5, 7, 19, 4, 13, 6, 16, 14, 10, 2, 17, 4,
  12, 10, 12, 4, 16, 7, 13, 8, 11, 6, 23, 2, 8, 10, 9, 7, 12, 6, 12, 5, 11}}
```

Thus, for example, M_4 has two factors, and M_{250} has 11 factors. The computations took about 3.2 hours and about 78 megabytes of RAM.

■ A Graphic Illustration

We continue studying Mersenne numbers and now form pairs from the indices and the numbers of factors:

```
s = Transpose[{Range[2, 250], t}];
```

Then we find a logarithmic least-squares fit for the number of factors as a function of the index of the Mersenne numbers:

```
lsq = Fit[s, {1, Log[i]}, i]
```

$$-3.05399 + 2.25251 \, \text{Log}[i]$$

We then plot the numbers of factors and the fit:

```
Plot[lsq, {i, 2, 250}, AspectRatio → 0.3, PlotRange → {-1, 25},
  TextStyle → {FontFamily -> "Arial", FontSize → 5},
  PlotStyle → {RGBColor[0, 0, 1], AbsoluteThickness[0.3]},
  Ticks → {Join[{2}, Range[10, 250, 10]], {1, 5, 10, 15, 20, 24}},
  Epilog → {AbsoluteThickness[0.3], Line[s],
    AbsolutePointSize[1.7], Map[Point, s]}];
```

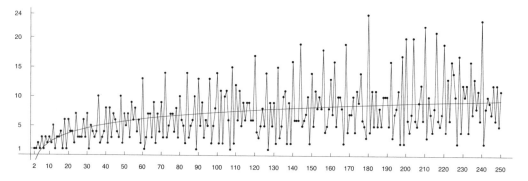

So, here are the numbers of factors of M_2 to M_{250}, together with the logarithmic fit. In the figure we see the prime Mersenne numbers with indices 2, 3, 5, 7, 13, 17, 19, 31, 61, 89, 107, and 127. These Mersenne primes had all been found by 1913. After the index 127, primes are not very common among Mersenne numbers. The next known primes occur with indices 521 (found in 1952); 607; 1279; 2203; 2281; 3217; 4253; 4423; 9689; 9941; 11,213; 19,937; 21,701; 23,209; 44,497; 86,243; 110,503; 132,049; 216,091; 756,839; 859,433; 1,257,787; 1,398,269; 2,976,221; 3,021,377; 6,972,593 (found in 1999); and 13,466,917 (found in 2001).

■ Lessons Learned

The preceding examples show how easy it now is to do long and complicated calculations and to visualize the results. *Mathematica* is one of the more popular systems for doing such calculations. However, even today, with powerful mathematical systems and machines, some problems remain very time consuming.

The example also illustrates some aspects of *Mathematica*, such as working with exact and approximate quantities, using graphics, and making programs. In general, *Mathematica* integrates symbolic calculation, numerical calculation, graphics, and programming into one system.

Mathematica contains still another aspect: a document-making environment (in versions of *Mathematica* that support the notebook interface). In this environment you can do symbolic and numerical calculations, produce graphics, and add text to explain what you have done. The result is a complete document of your work; in fact, this book has been written with *Mathematica*. In addition, the document is interactive. You can change parameters and functions, redo calculations, show animations, and continuously develop the document. The notebook interface is included in both the Windows and Macintosh versions of *Mathematica*. In plain Unix the interface is text-based and thus all features of *Mathematica* are not supported, but the X Window System supports notebooks.

1.1.2 The Structure of *Mathematica*

Mathematica is a big system: the total installation of *Mathematica* takes up about 300 megabytes (Mb) of space on the hard drive, and the effort spent directly on creating the source code for *Mathematica* is a substantial fraction of 1000 person-years. The online documentation of *Mathematica* is about 160 Mb and the system files about 60 Mb. The essential parts of *Mathematica* are the following three:

- the *kernel*, with the name **MathKernel** (about 12 Mb);
- the *front end*, with the name **Mathematica** (about 3 Mb); and
- the *packages*, with the name **StandardPackages** (about 4 Mb).

Next we consider each of these three components individually.

■ The Kernel

The main component of *Mathematica* is the kernel; it does all of the computations. It is written mainly with the C programming language and is one of the largest mathematical systems ever written, containing about 1.5 million lines of C code and 150,000 lines of *Mathematica* code. An important thing to understand is that the kernel is the same in all environments; this means that you get the same results in all environments (except possibly for the precision of floating-point calculations).

Mathematica commands are easy to use and quite versatile. However, behind the commands there is a huge amount of mathematical knowledge and a vast amount of work. For example, behind the single command **Integrate**, there are about 600 pages of C code and 500 pages of *Mathematica* code.

■ The Front End

The front end (about 700,000 lines of C code) is an environment for communicating with the kernel. When you open *Mathematica*, you open the front end. Commands are entered into the front end, and they are sent automatically to the kernel (the communication between the front end and the kernel is done with *MathLink*, a system that handles communication between parts of *Mathematica* and between *Mathematica* and other programs). The result of a calculation is then displayed by the front end. Figures come from the kernel in the form of PostScript code, and the front end then creates a screen image from this code.

There are two types of front ends: *notebook* and *text-based*.

There is a notebook front end in the Windows, Macintosh, and X Window versions of *Mathematica*. A notebook is an interactive document. It contains the commands you have entered and their results, graphics (animations can be shown), and comments you have added. You can do any kind of correcting, editing, and formatting in a notebook. In fact, you can make a notebook into a whole document of your work, even a book. (The chapters of this book were made as separate notebooks.)

The text-based front end is not nearly so handy and versatile as the notebook front end. This type of front end is found in the plain Unix version of *Mathematica*. With this kind of front end you can enter commands and see the results, but editing is very limited. Another disadvantage is that pictures are displayed in separate windows.

■ The Packages

The packages supplement the kernel. They are not normally loaded when *Mathematica* is loaded; you have to manually load the packages you want to use. Packages contain commands considered not to be as central and as common as the commands in the kernel. Packages are written using the programming capabilities of *Mathematica*. Wolfram Research (2003) has produced a separate manual for the packages.

To the packages that come with *Mathematica* you can also add the vast collection of packages found in MathSource (http://library.wolfram.com/infocenter/MathSource/). These freely available packages have been written by the users of *Mathematica* worldwide. You can copy interesting packages from this collection and use them in the same way as the ordinary packages. In addition, you can write your own packages and programs.

In addition, there is a *Mathematica Applications Library* that contains commercial *Mathematica* packages for special areas. At the time of this writing, the library contains such application packages as Dynamic Visualizer, Digital Image Processing, Experimental Data Analyst, Database Access Kit, Neural Networks, Time Series, Wavelet Explorer, Finance Essentials, Advanced Numerical Methods, Parallel Computing Toolkit, Signals and Systems, Control System Professional, Structural Mechanics, Mechanical Systems, Electrical Engineering Examples, Scientific Astronomer, Optica, and Fuzzy Logic.

1.2 First Calculations

1.2.1 Opening, Calculating, and Quitting

■ Opening Mathematica

To open *Mathematica*, do the following:

- In Windows and Macintosh environments: open *Mathematica* as other programs
- In Unix with X Window: type `mathematica`
- In plain Unix: type `math`

When you open *Mathematica* in a notebook environment, you get an empty notebook. A cursor appears once you have pressed the first key.

■ Calculating

Now, are you ready to begin? Let us start modestly and try to calculate $1 + 2$. Press these three keys:

1 + 2

The command is executed (that is, sent to the kernel for processing) by pressing a special key or key combination.

> To execute a command, press ENTER or SHIFT-RET

In Windows and Macintosh interfaces, both ways work (ENTER is the key at the bottom right corner of the keyboard; SHIFT-RET means that you are holding down the Shift key while at the same time pressing the Return key ↵). In Unix it may be that SHIFT-RET is the only possibility.

Note that, although we have spoken about "executing a command," this is not in keeping with the official terminology. We should speak about "*evaluating* an input." You should know the official term *evaluation*, but we ourselves feel free to use the more concrete term *execution*.

So, after typing the command or input **1 + 2**, press any of the executing or evaluating keys or key combinations. Now the kernel begins to evaluate the input. In most microcomputers, the kernel is not loaded until you ask it to do a calculation. This means that the first calculation takes some time, even if it is as simple as **1 + 2**. After you have entered the input, the label In[1]:= appears before the input, and the result has the label Out[1]=. You get

1 + 2

3

Note that here we do not show the In- and Out-labels (there is an option **Show In/Out Names** in the **Kernel** menu, and we have turned this option off). The input and the output can be distinguished by the font: the input is in boldface Courier, while the output is in plain Courier. Try the next command (note that, although you again cannot see a cursor in the notebook, just start typing; the new command appears where you see a horizontal line):

18 / 4

$$\frac{9}{2}$$

Fractions are automatically simplified, and they are written in a 2D form (instead of 9/2). If the expression contains a decimal number, then the result is also a decimal number:

10 / 3.

3.33333

■ Correcting

You may observe an error in a command when you write it, or after executing the command *Mathematica* may give you an error message telling you that there is something wrong with the command.

In a notebook environment (e.g., Windows, Macintosh, or X Window), you can do standard editing as with a word processor. For example, to delete an incorrect character, just move the cursor to the desired location by pressing the arrow keys or with the mouse and press the backspace key or the delete key to remove the character to the left or right, respectively. You can also highlight a portion of an input, cut ([CTRL]|x|; on a Macintosh, use ⌘ in place of [CTRL]) or copy ([CTRL]|c|) it, and then paste it ([CTRL]|v|) to a new location.

Once you have a corrected input, execute it. Note that after correcting a command you can leave the cursor where it is and then execute the command; the cursor need not be at the end of the command when you execute it. We consider editing in more detail in Section 1.5, p. 22. (In plain Unix possibly only the backspace key can be used to edit the input; arrow keys may not function. Using special editing commands such as **Edit** together with an editor like Emacs can help a lot in plain Unix.)

■ Quitting *Mathematica*

To quit *Mathematica*, do the following:

- In Windows, Macintosh, and X Window: quit *Mathematica* as other programs
- In plain Unix: execute the command **Quit**

Do not quit right now. Instead, continue reading and experimenting.

■ Aborting a Calculation

You may sometimes observe that a calculation is useless (perhaps because there was an error in the input or you do not have time to wait for the answer). You can then abort the calculation.

To abort a calculation, do the following:

- In Windows, Macintosh, and X Window: choose **Abort Evaluation** from the **Kernel** menu, or press [ALT]|.| (⌘|.| on a Macintosh)
- In plain Unix: press [CTRL]|c| [RET] a [RET]

It may happen that when you try to abort a calculation it seems that the calculation just goes on and on. You can then quit *Mathematica* (after possibly saving the notebook) and start a new session. In notebook environments, you can also quit only the kernel by choosing **Kernel ▷ Quit Kernel ▷ Local**. Start a new session by executing an input or by choosing **Kernel ▷ Start Kernel ▷ Local**.

1.2.2 Naming and Decimals

■ Before Continuing

Now we continue exploring *Mathematica*. Please note the following very important point:

- When you try these examples with your machine, write the commands exactly as they are printed here.

Mathematica will not forgive you even the smallest error in syntax. Be especially careful with small and capital letters: all *Mathematica* names such as **Sin** or **Integrate** begin with a capital letter. Also, you have to write all arguments in functions and commands in square brackets **[ ]**, for example, **Sin[x]** and **Integrate[a + b x, x]**; parentheses **( )** are not allowed. Parentheses are used only for grouping terms in expressions. Note that *Mathematica* automatically adds spaces in some places in your input (e.g., around **+** or **=**).

You probably will occasionally get error messages and wrong results because the syntax was not correct. Do not worry. This is normal. Getting used to *Mathematica* takes time, and only by working with the program can you learn to use it efficiently.

During a session, do not just enter one input after another in a hurry. Think carefully about each input and each result—how the input is written and what the result is. In this way you will learn more effectively.

Also, once you have tried some of the examples in this book, you can try other examples. This is recommended, because by trying out similar examples you strengthen your skills and get a clearer impression of each command and technique.

■ Referring to Earlier Results

`%`   Refers to the last result `%%`   Refers to the next-to-last result `%%...%`   Refers to the kth previous result if there are k `%` marks `Out[n]` or `%n`   Refers to the result in the output line `Out[n]`

Often you want to refer to earlier results. The percent mark `%` and the output names can be used for this. Try the following commands:

```
353^4
```
```
15527402881
```
```
30^4 + 120^4 + 272^4 + 315^4
```
```
15527402881
```
```
%% - %
```
```
0
```

Try also the output names by executing, say, `Out[4] + Out[5]` or `%4 + %5`.

Using `%` can be a problem if the command you execute contains errors to be corrected. Suppose you first calculate

```
16^2
```
```
256
```

Then you want to calculate $16^2 - 15^2$, but you write

```
% + 15^2
```
```
481
```

If you now correct the command to read `% - 15^2`, you do not get what you want, because `%` now refers to the result 481 of the last (wrong) command. So, you have to correct the command to read `%% - 15^2`. In general, if you correct several times a command that originally contained a `%`, you have to add one `%` each time. This may become awkward. It may be clearer to assign names to expressions, as explained below.

■ **Giving Names**

> **a = value** Assign **value** for **a**
> **a** Show the value of **a**
> **a =.** Clear the value of **a**

Another technique for referring to earlier results is to give names to results. Later on you can use the names as needed. For example, what is the probability of getting two 6s when tossing a die 6 times?

    a = Binomial[6, 2]

    15

    b = (1 / 6)^2 * (5 / 6)^4

    $\frac{625}{46656}$

    c = a * b

    $\frac{3125}{15552}$

You can always ask the value of a variable simply by entering the name of the variable:

    a

    15

When a symbol is no longer used, it is useful to clear the value of the variable so that it does not cause trouble later on:

    a =.

Now **a** has no value, as we see if we ask the value of **a**:

    a

    a

Mathematica printed only the name of the variable.

■ **Decimal Values**

> **expr//N** or **N[expr]** Calculate a decimal value for **expr**

You have perhaps noted that all calculations with integers and fractions are kept in an exact form; a decimal value is not automatically computed. A decimal value can be asked for with **N**. It can be used in two equivalent forms. The form **expr//N** may be easier to write than the form **N[expr]**. For example:

    b

    $\frac{625}{46656}$

    % // N

    0.0133959

```
c // N
```

0.200939

```
N[c]
```

0.200939

We can also ask directly for the decimal value (without first asking for a simplified fraction):

```
(1 / 6) ^ 2 * (5 / 6) ^ 4 // N
```

0.0133959

Note that, from now on, we here and there show the results of commands next to each command to save space.

1.2.3 Basic Calculations and Plotting

■ Basic Arithmetic

a + b   a − b   a b or a*b   a/b   a^b

The basic arithmetic operations plus, minus, division, and power are expressed in the usual manner, but multiplication is different. With *Mathematica*, multiplication is usually expressed by a space (press the space bar once). If you are more comfortable with the asterisk *, you can use it. For example:

```
a = 5      5

b = 3      3

a b      15
```

Note that, if you write **ab** without a space, *Mathematica* treats this as a single variable with the name **ab**:

```
ab      ab
```

We have not defined a value for the variable **ab**, so *Mathematica* just writes the name of the variable. This is a common error when using *Mathematica*. You have to write **a b** with a space or **a*b** with an asterisk if you want multiplication. More about multiplication is explained in Section 1.3, p. 15.

■ Basic Constants

Pi, E, I, Infinity

These well-known constants are usually denoted by π, e, i, and ∞ in mathematical texts. For example:

```
Pi // N      3.14159

E // N      2.71828

I ^ 2      − 1
```

```
1 / Infinity     0
```

Negative infinity is **-Infinity**. Note that *Mathematica* itself writes **Pi** as π, **E** as e, **I** as i, and **Infinity** as ∞:

```
{Pi, E, I, Infinity}
```

$\{\pi, e, i, \infty\}$

We can also write **Pi** as π, **Infinity** as ∞, and so on by using the Escape key, as follows:

> To write π, type $\boxed{\text{ESC}}$p$\boxed{\text{ESC}}$
> To write ∞, type $\boxed{\text{ESC}}$inf$\boxed{\text{ESC}}$

Just press the $\boxed{\text{ESC}}$, p, and $\boxed{\text{ESC}}$ keys in turn. From now on in this book, we will write **Pi** as π and **Infinity** as ∞.

■ Basic Functions

> **Sqrt[z]** or **z^(1/2)**   (square root)
> **Exp[z]** or **E^z**   (exponential function)
> **Log[z]**, **Log[b, z]**   (natural logarithm and logarithm to base **b**)
> **Abs[z]**   (absolute value)
>
> **Sin[z]**, **Cos[z]**, **Tan[z]**, **Cot[z]**, **Sec[z]**, **Csc[z]**
> **ArcSin[z]**, **ArcCos[z]**, **ArcTan[z]**, **ArcCot[z]**, **ArcSec[z]**, **ArcCsc[z]**
>
> **n!**, **Binomial[n, m]**
> **Max[{x, y, ...}]**, **Min[{x, y, ...}]**

Note that the natural logarithm is **Log[z]**. (In many mathematical texts, log(z) means a logarithm to base 10, while the natural logarithm is denoted by ln(z).) The arguments in trigonometric functions are in radians, and the values of the inverse trigonometric functions are in radians. **Binomial** gives the binomial coefficient. **Max** and **Min** give the maximum and the minimum of the arguments. For example:

```
Exp[Log[Sqrt[x]]]    √x
```

$$\mathbf{Cos[\pi / 4]} \qquad \frac{1}{\sqrt{2}}$$

$$\mathbf{ArcCos[1 / Sqrt[2]]} \qquad \frac{\pi}{4}$$

We throw a die 10 times and find the maximum of the results:

```
a = Table[Random[Integer, {1, 6}], {10}]
```

$\{5, 5, 3, 3, 5, 5, 2, 4, 1, 2\}$

```
Max[a]    5
```

With the **Table** command we get a *list*: an ordered collection of elements enclosed in curly braces { }.

■ Basic Calculus

> `D[expr, x]` Derivative of **expr** with respect to **x**
> `Integrate[expr, x]` Indefinite integral of **expr** with respect to **x**
> `Integrate[expr, {x, a, b}]` Definite integral of **expr** with respect to **x** from **a** to **b**
> `Simplify[expr]` or `expr//Simplify` Simplify the expression

Again a note about terminology. We have spoken about commands such as **D** or **Integrate**, but the official term is a *function*. However, we feel free to speak about commands and use the term *function* mainly for such expressions as **Sin[x]**, which are official mathematical functions. For example:

D[x Sin[x], x]

$x \, \text{Cos}[x] + \text{Sin}[x]$

Integrate[x^2 Exp[x], {x, 0, 1}]

$-2 + e$

Integrate[p x / (q + r x), x]

$p \left(\dfrac{x}{r} - \dfrac{q \, \text{Log}[q + r x]}{r^2} \right)$

We check the last integral by calculating the derivative of the result:

D[%, x]

$p \left(\dfrac{1}{r} - \dfrac{q}{r \, (q + r x)} \right)$

After simplification, we get the desired result:

% // Simplify

$\dfrac{p x}{q + r x}$

■ Basic Plotting

> `Plot[expr, {x, a, b}]` Plot **expr** when **x** takes on values from **a** to **b**

Mathematica has many plotting commands, but **Plot** is the basic one. An example:

Plot[Exp[-x] Sin[2 x], {x, 0, 2 π}]

- Graphics -

In notebooks, plotting is this easy. You can also change the size of a figure; click on it and then drag one of the handles. In plain Unix, things may be not so simple. Ask for more information from a person who knows your environment.

Congratulations! Now you have used *Mathematica* for some simple calculations and you have an impression of how *Mathematica* works. We will give you a better overview of *Mathematica* in Chapter 2. However, first, in Section 1.3, we will summarize the basic conventions of *Mathematica*. Then we explain, in Section 1.4, how you can get information about the commands of *Mathematica*. Section 1.5 considers writing, correcting, and editing in *Mathematica*.

1.3 Important Conventions

You have observed that all of the built-in *Mathematica* names we have presented have begun with a capital letter and that all arguments have been given in square brackets **[ ]**. These are the two most important conventions in *Mathematica*. Here are the six most important ones:

- All built-in *Mathematica* names begin with a capital letter.
- Multiplication must be expressed by a space or an asterisk (*****). (For numerical multipliers or complete expressions, nothing is needed).
- All arguments are given in square brackets **[ ]**.
- Parentheses **( )** are used only for grouping terms.
- Curly braces **{ }** are used for lists.
- Double square brackets **[[ ]]** are used to extract elements from lists.

It takes some time to get used to these conventions, and at the beginning you will often get error messages and wrong results because you have not remembered these rules. Later on you may see that these conventions have advantages. Let us consider the conventions in more detail.

■ **Names**

Mathematica is case sensitive. If a name is **Sin**, you cannot write **sin** or **SIN**; you must write **Sin** exactly. It is recommended that all names you introduce (like **a**, **b**, and **c** previously) begin with a small letter. If this convention is followed, then it is always clear which names are built-in and which are defined by the user. Such a distinction makes reading the *Mathematica* code easier; you need not remember whether a name is your own. Also, you cannot mistakenly define a symbol with the same name as a built-in command, thereby avoiding any confusion.

Many built-in names consist of several words run together, like **FindMinimum**, and in these cases each individual word begins with a capital letter. If you define a name consisting of several words, you can use capital letters in the middle of the name, as in **random‐Walk**; this makes reading the name easier.

Another convention is that all built-in names and words are written completely; abbreviations are not used. This can make some names long (e.g., **InverseLaplace‐Transform**, **BesselJPrimeYPrimeJPrimeYPrimeZeros**), but the advantage is that such complete names are often easier to remember than abbreviated names. Some abbreviations exist, though, such as **D** (derivative), **Det** (determinant), and **Tr** (trace). Names may be as long as you want. Names cannot begin with a number. User-defined names are also often written in full without abbreviations (the longest I have seen is in Shaw & Tigg (1994, p. 104): **NapoleonicMarchOnMoscowAndBackAgainPlot**).

Let us try out an example with the capital first letter. Instead of the correct form **Sin[π / 2]**, write

> **sin[π / 2]**
>
> General::spell1 : Possible spelling error: new symbol
>   name "sin" is similar to existing symbol "Sin". More…
>
> $\sin\left[\frac{\pi}{2}\right]$

A warning message tells you that **sin** possibly contains a spelling error, because **sin** resembles the built-in name **Sin** so closely. We do not get the expected answer 1 but rather $\sin[\pi/2]$. We correct the command:

> **Sin[π / 2]**   1

■ Multiplication

Multiplication was already considered in Section 1.2.3, p. 11, but let us still try some examples:

> **a = 3**   3

> **b = 4**   4

> **{a b, a ∗ b, ab}**
>
> {12, 12, ab}

Recall that you cannot write **ab** if you want **a** times **b**. If you write **ab**, *Mathematica* understands it as a variable with the name **ab**. Some more examples:

> **{5 a, a5, d (e + f), (d + e) (f + g), Sin[x] Cos[y]}**
>
> {15, a5, d (e + f), (d + e) (f + g), Cos[y] Sin[x]}

Note that, with a numeric multiplier, we do not need to write a multiplication indicator such as a space or an asterisk: we can write **5a**; *Mathematica* automatically adds a space between the terms. However, **a5** is interpreted as a name. No space or asterisk is needed with parentheses, either: we can write **c(d+e)** and **(c+d)(e+f)**, and *Mathematica* adds the space itself. A multiplication indicator is generally not needed between complete expressions. For example, we can write **Sin[x]Cos[y]**, and, again, *Mathematica* adds the space itself.

If a multiplication occurs at the end of a line, then it is safe to use the asterisk (∗). Place the asterisk either at the end of the first line or at the beginning of the next line. If you do not use the asterisk, *Mathematica* understands the two rows as separate commands, if they can be interpreted as complete commands.

■ Arguments

In traditional mathematical notation, parentheses are used for two purposes: for arguments and for grouping terms. *Mathematica* avoids this ambiguity by using different notation for these two purposes: square brackets for arguments and parentheses for grouping. For example, if we write, instead of the correct form `Sin[π/3]`, what you see below, we get a wrong result:

$$\texttt{Sin}\ (\pi\ /\ 3) \qquad \frac{\pi\,\text{Sin}}{3}$$

Mathematica interprets the expression according to its standard rules: `Sin` is a variable by which we want to multiply `Pi/3`. Here is the correct command:

$$\texttt{Sin}[\pi\ /\ 3] \qquad \frac{\sqrt{3}}{2}$$

■ Grouping

Be careful in entering expressions. Parentheses are sometimes easily forgotten, and the result will be incorrect. Special care is necessary with quotients and rational powers. Here are some examples of quotients:

$$\texttt{\{1 / 4 Sqrt[x] Log[x], 1 / (4 Sqrt[x]) Log[x], 1 / (4 Sqrt[x] Log[x])\}}$$

$$\left\{ \frac{1}{4}\ \sqrt{x}\ \text{Log}[x],\ \frac{\text{Log}[x]}{4\,\sqrt{x}},\ \frac{1}{4\,\sqrt{x}\ \text{Log}[x]} \right\}$$

Thus, `a/b*c` in interpreted as `(a/b)*c` and not as `a/(b*c)`. Here are some examples of powers:

$$\texttt{\{E\^{}-1, E\^{}-1 / 2, E\^{} (-1 / 2)\}}$$

$$\left\{ \frac{1}{e},\ \frac{1}{2\,e},\ \frac{1}{\sqrt{e}} \right\}$$

Thus `a^b/c` is interpreted as `(a^b)/c` and not as `a^(b/c)`. Remember to write the necessary parentheses, and if you are uncertain whether you should use parentheses or not, go ahead and use them, because unnecessary parentheses are harmless. Note that, if you want to square `Sin[x]`, you can simply write `Sin[x]^2`. If you want to calculate the value of `Sin` at `x^2`, write `Sin[x^2]`.

■ Lists

Lists are like vectors: a list is, mathematically, an ordered set of elements. Lists are used to store data and expressions. Here is an example of a list with three elements:

$$\texttt{c = \{6, 2 E, Sin[1.2 }\pi\texttt{]\}} \qquad \{6,\ 2\,e,\ -0.587785\}$$

Curly braces are reserved for lists. Another example:

$$\texttt{d = \{Cosh[3], Pi, 2\}} \qquad \{\text{Cosh}[3],\ \pi,\ 2\}$$

Calculations with lists are simple, because all operations are automatically done element by element:

$$\texttt{d\^{}2} \qquad \{\text{Cosh}[3]^2,\ \pi^2,\ 4\}$$

$$\texttt{c + d} \qquad \{6 + \text{Cosh}[3],\ 2\,e + \pi,\ 1.41221\}$$

Double square brackets are used to extract elements from lists. For example:

    `c[[2]]`   2 e

1.4 Getting Help

1.4.1 Palettes

Palettes can help you when you are entering input for *Mathematica* in notebook environments. We have nine standard palettes, and you can create your own palettes, too (see Section 3.1.3, p. 52). Palettes can be accessed by choosing **Palettes** from the **File** menu. Below we show four palettes: **AlgebraicManipulation**, **BasicInput**, **BasicCalculations**, and **BasicTypesetting**.

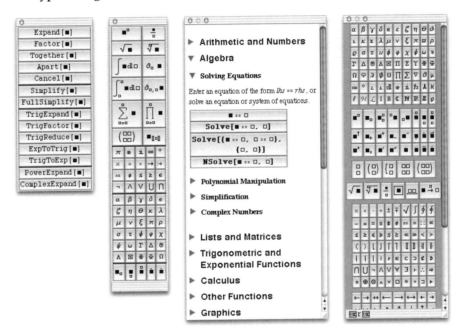

■ **AlgebraicManipulation**

The **AlgebraicManipulation** palette contains such commands as `Expand`, `Factor`, and `Simplify`. First, type the following:

    `(f + g) ^ 6`

Then select the whole expression with the mouse and click `Expand` in the palette. The expression is expanded. Then click `Factor` in the palette. The expression is now factored. In this way, whatever is currently selected in your notebook will be inserted into the position of the *selection placeholder*, ■.

■ BasicInput

The **BasicInput** palette contains buttons to perform some basic calculations and to input some basic symbols. Suppose you want to calculate the derivative of $x \sin(x) + \cos(x)$. First click the derivative button $\partial_\square \ \blacksquare$, then write **x**, press TAB, write **(x Sin[x] + Cos[x])**, and execute the resulting command:

∂_x **(x Sin[x] + Cos[x])**   x Cos [x]

You can also do the following: write **(x Sin[x] + Cos[x])**, select the whole expression, click the derivative button, press **x**, and execute.

For another example, suppose you want to calculate the definite integral of $x \sin(x) + \cos(x)$ on $(0, 2\pi)$. First click the integral button $\int_\square^\square \blacksquare \, d\square$, then write **0**, press TAB, write **2**, click π on the palette, press TAB, write **(x Sin[x] + Cos[x])**, press TAB, write **x**, and execute:

$$\int_0^{2\pi} (\textbf{x Sin[x] + Cos[x]}) \, \textbf{d} \textbf{x} \qquad - 2\pi$$

You can also do this the other way: write first **(x Sin[x] + Cos[x])**, click the integral button, and fill the limits and the integration variable with the help of TAB.

This palette also contains buttons for powers, fractions, roots, sums, products, 2×2 matrices, and part extraction. Also included are the four basic symbols π (= 3.14159...), e (= 2.71828...), i (= $\sqrt{-1}$), and ∞ (= infinity).

■ Other Palettes

The **BasicCalculations** palette contains about 140 buttons. The two palettes mentioned above can be considered as small collections from this larger palette. This palette contains a selection of the most important commands. You can find buttons for, among other things, solving equations and differential equations, numerical root finding, and plotting.

The **BasicTypesetting** palette contains a large collection of characters and constructions, while the **CompleteCharacters** palette contains all of the characters that can be entered into *Mathematica* (about 700).

1.4.2 The Help Browser

Mathematica used in notebook environments incorporates an excellent help system called *Help Browser*. Go to the **Help** menu and choose **Help Browser...** (or press SHIFT F1 in Windows or the Help key in Macintosh). A new *Help Browser* window appears.

Suppose we want to solve an equation. Select *Algebraic Computation* from the second column of subjects in *Help Browser* and *Equation Solving* from the third column. Among the commands in the fourth column, **Solve** may be the most promising for us. Select this command, and the window appears like so:

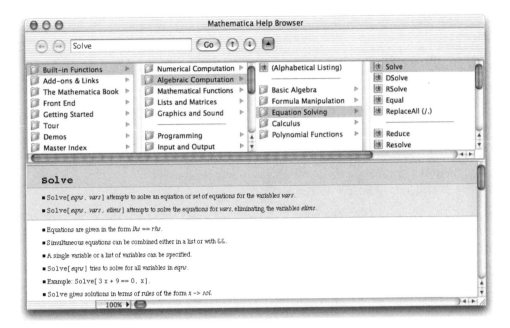

Note that in *Mathematica 5*, the interface of the *Help Browser* is somewhat new; in earlier versions, some of the subjects in the first column are denoted by buttons at the top part of the window.

You can now read much information about **Solve**. The text in *Help Browser* is mostly the same as that found in Appendix A.10 of the main manual by Wolfram (2003), but it often contains further examples. (Appendix A.10 explains all of the commands in *Mathematica*.)

Help Browser contains hyperlinks to various points in the main text of the manual and also to related packages in the manual of standard packages. Hyperlinks are blue and underlined. Clicking with the mouse on hyperlinks takes you to the relevant points in the manuals.

If you already know the name of the command but want information about it, then type, for example, **Solve** in the box at the top of the window and press RET (or click *Go*).

Help Browser can also be used as follows. In your notebook (not in *Help Browser*), type a command like **Solve**, leave the cursor at the end of the word, and press the F1 key in Windows or the Help key in Macintosh (or choose **Help ▷ Find Selected Function...**). The *Help Browser* window appears and shows the information about **Solve** directly.

■ Selecting the Source of Information

The default case is that the information is searched from among the built-in functions, but you can choose other sources by clicking one of the items of the first column of *Help Browser*. Here are the sources and some examples of their contents:

Built-in Functions: all built-in commands classified in logical groups; warning messages

Add-ons & Links: the standard packages and other add-ons; links

The Mathematica Book: the main manual

Front End: menu commands; keyboard shortcuts; 2D expression input; style sheets; front-end options (the ones accessible via **Format ▷ Option Inspector...**)

Getting Started: setting up; starting out; working with notebooks; system-specific information

Tour: a tour of *Mathematica*

Demos: formula, graphics, and sound galleries; notebook demos

Master Index: all index entries in alphabetical order

All printed material that comes with *Mathematica* can be read with the *Help Browser*, but some information shown by the *Help Browser* cannot be found printed anywhere. For example, explanations of the menu commands, listings of keyboard shortcuts, and information about style sheets and front-end options can only be read from the *Help Browser* under *Front End*.

Study the *Help Browser* in detail, and get used to using it continuously. It is an invaluable source of information.

1.4.3 Other Help

■ The Question Mark

With the question mark (?) and asterisk (*), we can get tables of names containing certain characters. Suppose you are interested in finding a minimum of a function. Perhaps such a command has `Min` in its name, and so we ask *Mathematica* to give a table of all names containing `Min`:

```
? *Min*
```

System`

ButtonMinHeight	Minimize	Minus	SpanMinSize
ConstrainedMin	Minors	NMinimize	$MinMachineNumber
FindMinimum	MinRecursion	RowMinHeight	$MinNumber
Min	MinSize	ScriptMinSize	$MinPrecision

Now we can ask for information about these commands by clicking the names of the commands. For example, if we click `FindMinimum`, we get the following short information (⌘4!):

```
FindMinimum[f, {x, x0}] searches for a local
    minimum in f, starting from the point x=x0. FindMinimum[
    f, {{x, x0}, {y, y0}, ... }] searches for a local
    minimum in a function of several variables. More...
```

By clicking the More link, we get *Help Browser* to open, and the information about **Find**‹ **Minimum** is displayed. Another way to use the question mark is to ask information about a particular command (⌘4!):

> **? FindMinimum**

> FindMinimum[f, {x, x0}] searches for a local
>   minimum in f, starting from the point x=x0. FindMinimum[
>   f, {{x, x0}, {y, y0}, ... }] searches for a local
>   minimum in a function of several variables. More...

?Abcd*   Give a table of all names beginning with **Abcd**

?*abcd   Give a table of all names ending with **abcd**

?*abcd*   Give a table of all names containing **abcd** somewhere

?Name   Give information about **Name**

??Name   Give also attributes and options of **Name**

The double question mark gives the same information as the single question mark and, in addition, information about attributes and options. Execute **?*** to give a list of all of the about 2000 *Mathematica* names.

■ Completing Names and Making Templates

Mathematica has quite long names for some commands, but in notebook environments you can let *Mathematica* do some of the typing work. Use palettes or the following technique.

Suppose we want to write the command **InterpolatingPolynomial**. We first write, say, **Interpo**, and then press ⌃k⌄ (or ⌘k⌄ on a Macintosh); this is the same as choosing **Complete Selection** in the **Input** menu. You get the following list:

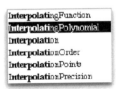

Here are all the commands beginning with **Interpo**. From the list you can choose the one you want by clicking with the mouse or by highlighting the appropriate command with the arrow keys and then pressing ⏎. **Interpo** is then automatically completed according to your choice.

Now that you have the complete name of the command **InterpolatingPolynomial**, press ⇧⌃k⌄ (or ⇧⌘k⌄ on a Macintosh); this is the same as choosing **Make Template** from the **Input** menu. You get a template for the command:

> **InterpolatingPolynomial[data, var]**

This is useful if you do not remember the syntax of a command. Now you can replace **data** with your data and **var** with your variable and then execute the command.

If the name of the command is uniquely determined by the first letters you have written, then ⌃k⌄ will complete the name directly, and ⇧⌃k⌄ will complete the name and make a template for the command. Try typing **InterpolatingP** and then pressing ⌃k⌄ or ⇧⌃k⌄.

■ Balancing Paired Characters

When writing commands and programs with *Mathematica*, you may use a lot of parentheses ( ), brackets [ ], and curly braces { }. It may be difficult to see whether they are correctly balanced. *Mathematica* helps with this as follows: unbalanced characters are shown in purple, and each time you write),], or }, *Mathematica* highlights the corresponding (, [, or { for a short time.

You can also use **Check Balance** from the **Edit** menu or press ⌜SHIFT⌐⌜CTRL⌐ b ⌉ (or ⌜SHIFT⌐⌜⌘⌐ b ⌉ on a Macintosh). Place the cursor somewhere in the command and then press the above-mentioned key combination. The smallest balanced part containing the cursor will be highlighted, or, if a correctly balanced part cannot be found, you will hear a beep.

■ Help from Wolfram Research

Help can also be found at the Web site http://support.wolfram.com of Wolfram Research. To display the version of *Mathematica* you are using (and thus find the appropriate information), execute the command $Version.

1.5 Editing

■ Correcting

In notebooks, you can use all the usual editing methods familiar from word processors. Look at the **Edit** menu for editing commands. You can edit the notebook very nicely with the mouse, but if you prefer to use key combinations to move the cursor and make deletions, you can see the appropriate key combinations by choosing **Motion** from the **Edit** menu.

Note especially that *you can edit all old inputs and execute them anew*. To recalculate an input, simply place the cursor anywhere in the command and then execute the command. In particular, if an input resulted in errors, you do not need to retype the input; simply correct the old input by means of standard editing and then execute the input again. Note also that the input and output numbers are assigned in the order of execution and not according to the physical order of the commands in the notebook.

■ Using Cells

You have probably noted the brackets at the right side of the *Mathematica* window. They indicate *cells*. An input is in a cell, the result is in another cell, and these two cells together form a higher-level cell.

A new input cell is automatically created when you start typing after a command is executed. A notebook is a structured document that is organized into a sequence of cells.

You can insert new cells between old cells. Simply place the cursor between two cells so that the cursor becomes horizontal, and then click with the mouse and start typing. In this way you can insert new calculations among old ones.

The cells are handy for moving a part of a notebook to another place: just click on the cell bracket so that it becomes black, then cut or copy the cell and paste it to a new location. You can select several cells by dragging with the mouse over the cell brackets. You can copy material from one notebook to another notebook with the usual copy-and-paste techniques. Copying is easy to do by selecting cells.

If you want to re-execute several cells, select the cell brackets and then execute them in the standard way.

Double-clicking a cell bracket closes the cells inside it so that only the first cell is visible. In this way you get a short outline of your notebook. Double-clicking the cell bracket of a closed cell opens the cell again.

■ Each Command into Its Own Cell

Mathematica is an interactive calculator that is designed such that we can easily proceed step by step: we execute one command and then proceed to the next command. A common bad habit is to write several commands in one cell and then execute all the commands at the same time. When this happens, the connection between inputs and the corresponding outputs becomes obscured. For example:

```
f = x Sin[x]
Integrate[f, x]
D[f, x]

x Sin[x]

-x Cos[x] + Sin[x]

x Cos[x] + Sin[x]
```

Here we put three commands into one cell. The outputs are shown one after another. This is not clear. In addition, if one of the commands contains an error, we have to execute all the commands anew instead of only re-executing the corrected command.

So, *write each command in its own cell*. As said earlier, a new input cell is automatically created when you start typing after a command is executed. If you want to write several commands before you execute any of them, create a new cell for each command as follows: write a command, press the down arrow key (↓), and start writing the next command. (Instead of the arrow key, you can also place the cursor below the last command so that the cursor becomes horizontal, click with the mouse, and start typing the next command.)

Note that, if a calculation takes some time, you need not be idle. You can write (but not execute) new commands so that they are ready when the present command has been executed. You can also edit the notebook in any way you like while you await the execution of a command.

■ Editing Inputs

Sometimes you want to slightly modify an old input and then execute it again. You have several possibilities. First, you can directly edit the old input and then re-execute it, as was explained above.

Second, you can select an old input or a part of it with the mouse, copy the selection, paste it to a new location, edit, and execute. A whole input can be selected by clicking the cell bracket, and the cell can then be copied and pasted.

Third, if the input you want to modify is the last input, you can get a copy of it by pressing CTRL[1] (or ⌘1 on a Macintosh); this is the same as choosing **Copy Input from Above** in the **Input** menu.

Mathematica divides long inputs automatically into several lines. However, you can also yourself press the Return key (RET) here and there to make long code easier to read.

■ Editing Outputs

You also have access to the results *Mathematica* writes. For example, you can copy a result or a part of it, paste the copy to a new cell, edit the new command, and then execute it.

If you want to edit the result of the last command, you can get a copy of the result by pressing SHIFT-CTRL[1] (or SHIFT-⌘1 on a Macintosh); this is the same as choosing **Copy Output from Above** in the **Input** menu.

You can also directly edit a result. For example, suppose the result of a calculation is 512. Now you want to divide this result by 2. You can type the new command **%/2**, but you can also type **/2** at the end of the result 512. If you do the latter, *Mathematica* immediately takes a copy of the result 512, changes the format to the input style (bold), and makes your changes to this copy. Then execute the new command. In this way, *Mathematica* automates copying and pasting outputs and prevents you from corrupting old outputs.

You can also execute a part of a command. Suppose you want to apply a transformation like **Factor** to a part of a result. Write the command to the output, select with the mouse the part you want to execute, and then press ALT-RET or ⌘-RET (this corresponds to **Kernel** ▷ **Evaluation** ▷ **Evaluate in Place**). Only the selected part is then executed.

■ Opening a Saved Notebook

Note that, *if you open a saved notebook and continue calculations, you cannot directly use any results in the saved document*. You have to recalculate all the commands with the results you need in your new session. Suppose, for example, that the saved notebook contains the result of **int = Integrate[Sqrt[x] Sin[x], x]**. When you open the notebook, **int** has no value! If you need the value of **int** in your new session, you have to execute anew the integrating command.

■ Writing a Document with *Mathematica*

You can write a whole document with *Mathematica* by calculating, plotting, and adding text comments to the results. To modify the look of the document, you can use *styles* and *style sheets*. These are considered in some detail in Chapter 3.

After you have written your document with *Mathematica*, check spelling by placing the cursor at the beginning of the document and then choosing **Edit** ▷ **Check Spelling...**. English is built into *Mathematica*, but you can buy dictionaries of other languages from Wolfram Research.

Sightseeing

Introduction

> *The relationship between pure and applied mathematics is based on trust and understanding:*
> *the pure mathematicians don't trust the applied mathematicians,*
> *and the applied mathematicians don't understand the pure mathematicians.*

Welcome all mathematicians, pure and applied, and everyone else, too, to a quick sightseeing tour through the vast and wonderful *Mathematica* factory, which produces graphics, eigenvalues, integrals, and so much more. We will visit the three main divisions of the factory: Graphics, Expressions, and Mathematics. In each division we will show you only the most important or most basic machines (a total of about 40, leaving more than 1000 that are not shown). We will introduce each machine only briefly, but we encourage you to spend more time investigating and experimenting (at your own risk, of course; use a helmet, because every erroneous input is thrown out of the machine). Later on we will explain all this and much more in detail, but the knowledge you get during this sightseeing tour may suffice for a while. In fact, this short tour gives you snapshots of Chapters 5 through 23. Please note that what you will see in this chapter is not anything spectacular but only some very basic facts. Later on we will show you some much more impressive results.

2.1 Graphics

2.1.1 2D Graphics for Functions

> **Plot[f, {x, a, b}]** Plot **f** when **x** takes on values from **a** to **b**
> **Plot[{f1, f2}, {x, a, b}]** Plot **f1** and **f2** in the same figure
>
> **Show[p1, p2]** Show figures **p1** and **p2** superimposed
> **Show[GraphicsArray[{p1, p2}]]** Show figures **p1** and **p2** side by side

We plot two functions in the same figure:

Plot[{Sin[x], Cos[x]}, {x, 0, 2 π}]

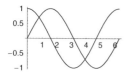

    - Graphics -

Recall that π can be written as **Pi** or as ⎡ESC⎤p⎡ESC⎤. From now on we end plotting commands in this book with a semicolon (**;**). This prevents the text "**- Graphics -**" from being displayed below the figure and saves space in the book.

We can also first plot the two figures separately:

p1 = Plot[Sin[x], {x, 0, 2 π}];

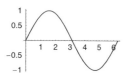

p2 = Plot[Cos[x], {x, 0, 2 π}];

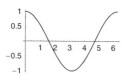

Then we can combine them:

Show[p1, p2];

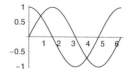

The figures can also be placed side by side (we have enlarged the figure after plotting):

> **Show[GraphicsArray[{p1, p2}]];**

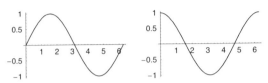

We can enter the expression to be plotted directly in the plotting command, as we have done thus far, but we can also first give a name to the expression:

> **f = x^2 Exp[-x] Sin[x]**

$e^{-x} x^2 \sin[x]$

> **Plot[f, {x, 0, 14}];**

In this example we do not see the whole function in the given interval. Indeed, sometimes *Mathematica* cuts a part of the figure out in order to give you a closer look at the more interesting parts of the curve. You can control the range of y values by using the **PlotRange** option. If you want to see the whole function, give the option **PlotRange → All** (write the arrow by pressing the hyphen and greater-than keys in turn [->]; *Mathematica* will then replace them with a genuine arrow):

> **Plot[f, {x, 0, 14}, PlotRange → All];**

> **f = .**

■ Suppressing Display

To save space, in this book we often use the **DisplayFunction** option to suppress the display of graphics, as follows:

> **p1 = Plot[Sin[x], {x, 0, 2 π}, DisplayFunction → Identity];**

> **p2 = Plot[Cos[x], {x, 0, 2 π}, DisplayFunction → Identity];**

The plots were prepared, but they were not shown. The plots can then be superimposed or shown side by side, as follows (for more about this technique, see Section 5.2.1, p. 117):

```
Show[p1, p2, DisplayFunction → $DisplayFunction];
```

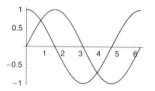

```
Show[GraphicsArray[{p1, p2}]];
```

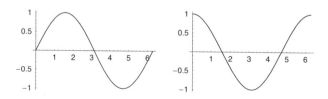

2.1.2 3D Graphics for Functions

> **Plot3D[f, {x, a, b}, {y, c, d}]** Plot **f** as a surface
> **ContourPlot[f, {x, a, b}, {y, c, d}, PlotPoints → n]** Plot **f** as contours

We plot a function of two variables first as a surface (※4!) and then as contours:

```
Plot3D[Sin[x^2] Cos[Sqrt[y]], {x, 0, 3}, {y, 0, 4}];
```

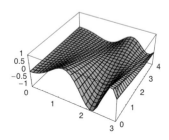

```
ContourPlot[Sin[x^2] Cos[Sqrt[y]], {x, 0, 3}, {y, 0, 4}, PlotPoints → 40];
```

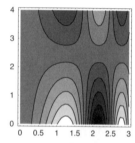

On the contours, the function takes on a constant value. Dark areas are lower than light areas. Give the **PlotPoints** option a large enough value so that the curves are smooth.

2.1.3 2D Graphics for Data

`ListPlot[data]`   Plot **data** as points
`ListPlot[data, PlotJoined → True]`   Plot **data** as joining lines
`ListPlot[data, PlotJoined → True, Epilog → {AbsolutePointSize[2],`
  `Map[Point, data]}]`   Plot **data** as points and joining lines

The data are given in either of the following forms:

  `{y1, y2, ..., yn}`

  `{{x1, y1}, {x2, y2}, ..., {xn, yn}}`

In the former case, the x values are automatically 1, 2, 3, …; in the latter case, we give explicit x values. If we want both points and joining lines, the data have to be given in the latter form. Here are some data and three plots:

  `data = {{0, 5}, {1, 7}, {2, 8}, {3, 7}, {4, 9}, {5, 8}, {6, 6},`
   `{7, 5}, {8, 5}, {9, 4}, {10, 5}, {11, 3}, {12, 2}, {13, 1}}`

  `{{0, 5}, {1, 7}, {2, 8}, {3, 7}, {4, 9}, {5, 8}, {6, 6},`
   `{7, 5}, {8, 5}, {9, 4}, {10, 5}, {11, 3}, {12, 2}, {13, 1}}`

  `p1 = ListPlot[data, DisplayFunction → Identity];`

  `p2 = ListPlot[data, PlotJoined → True, DisplayFunction → Identity];`

  `p3 = ListPlot[data, PlotJoined → True, DisplayFunction → Identity,`
    `Epilog → {AbsolutePointSize[2], Map[Point, data]}];`

  `Show[GraphicsArray[{p1, p2, p3}]];`

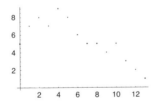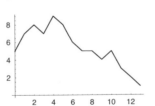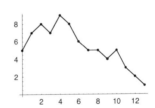

2.2 Expressions

2.2.1 Numbers and Expressions

■ **Numbers**

`N[expr]` or `expr//N` Calculate a decimal value of **expr**
`N[expr, n]` Calculate a decimal value of **expr** to **n**-digit precision

---

Here is a decimal value for π:

  `π // N`   `3.14159`

Now we ask for a decimal value to 30-digit precision:

> **N[π, 30]**   3.14159265358979323846264338328

Random[]   A random number from the continuous uniform distribution on (0, 1)
Random[Integer, {m, n}]   A random number from the discrete uniform distribution
on the integers from **m** to **n**

Random numbers can be used in simulating various phenomena. Here are some
uniform random numbers (**Table** is explained in Section 2.2.2):

> **Table[Random[], {7}]**
>
> {0.265706, 0.964275, 0.151301, 0.258857, 0.344144, 0.0321031, 0.0423591}

Then we simulate die tossing:

> **Table[Random[Integer, {1, 6}], {24}];**
>
> {1, 5, 6, 1, 6, 3, 3, 2, 4, 3, 1, 5, 1, 2, 5, 3, 4, 2, 3, 4, 3, 1, 1, 5}

Bad luck as usual: only two 6s, although four were expected.

■ Calculating the Value of an Expression

---

x = a   Give **x** the value **a**
expr   Show the value of **expr** when **x** has the value **a**

If you want to calculate the value of an expression for a certain value of a variable, one
possibility is to explicitly give the value for the variable and then ask the value of the
expression:

> **expr = Sin[x] Cos[x]**   Cos[x] Sin[x]
>
> **x = π / 6** $\frac{\pi}{6}$
>
> **expr** $\frac{\sqrt{3}}{4}$

This method has the drawback that from now on **x** has the value $\pi/6$ in all expressions,
and this may give unintended results. For example, if you now try to calculate the derivative of the expression, you get an error message:

> **D[expr, x]**
>
> General::ivar : $\frac{\pi}{6}$ is not a valid variable. More...
>
> $\partial_{\frac{\pi}{6}} \frac{\sqrt{3}}{4}$

Mathematica could not calculate the derivative with respect to a constant $\pi/6$. So, if you
give values for variables, remember to remove the values when you no longer need them:

> **x = .**

> **expr /. x → a** Replace **x** by **a** in **expr** (the arrow can be written as **->**)

This is the recommended method to calculate the value of an expression for a value of a variable. For example:

expr /. x → π / 6 $\dfrac{\sqrt{3}}{4}$

This is a very important technique. Here **x → π/6** is a transformation rule. It can be applied to any expression by preceding the transformation rule with **/.** . Note that now **x** has no value:

x   x

■ Manipulating Expressions

> **Simplify[expr]** Simplify **expr**
> **FullSimplify[expr]** Simplify **expr** thoroughly
> **Factor[expr]** Factor **expr**
> **Expand[expr]** Expand **expr**
> **Apart[expr]** Give the partial fraction expansion of **expr**

Note that these commands can be used also in the following way: **expr // Simplify**. For example:

a = (1 + x) ^ 2 + (1 + x) (2 + x)

$(1 + x)^2 + (1 + x)(2 + x)$

a // Simplify $3 + 5 x + 2 x^2$

a // Factor $(1 + x)(3 + 2 x)$

a // Expand $3 + 5 x + 2 x^2$

FullSimplify is often good for simplifying special functions. In the following example, **Simplify** does not work, but **FullSimplify** does:

n ! / (n − 1) ! // Simplify $\dfrac{n!}{(-1 + n)!}$

n ! / (n − 1) ! // FullSimplify   n

We calculate a partial fraction expansion:

(1 + x + x ^ 2 − x ^ 3) / (x + 2) ^ 2 // Apart

$5 - x + \dfrac{11}{(2 + x)^2} - \dfrac{15}{2 + x}$

■ Some Display Techniques

Sometimes we need not see the result of a computation. For example, we already know the result for certain, the result is so large an expression that it is useless to see it, or it takes too much time to have it displayed on the screen. We can prohibit displaying the result by ending the command with a semicolon (**;**).

> **expr;**   Calculate the value of **expr** but do not display the result

We do not want to see 100! (a number with 158 digits):

    a = 100!;

    a / 99!   100

We can execute several commands at the same time using the semicolon:

> **expr1; expr2; expr3**   Calculate the expressions; display the last result
> **expr1; expr2; expr3;**   Calculate the expressions; do not display anything

We calculate other factorials and display only the final result:

    b = 97!; c = 3!; a / (b c)   161700

If you want to display the values of all expressions, you can place the expressions in a list with curly braces (**{ }**):

> **{expr1, expr2, expr3}**   Calculate the expressions; display all values

    {Sin[π / 4], Sin[π / 5], Sin[π / 6], Sin[π / 7]}

$$\left\{ \frac{1}{\sqrt{2}}, \ \frac{1}{2} \sqrt{\frac{1}{2} (5 - \sqrt{5})}, \ \frac{1}{2}, \ \text{Sin}\left[\frac{\pi}{7}\right] \right\}$$

    % // N   {0.707107, 0.587785, 0.5, 0.433884}

For long expressions, it often suffices to see only some parts. This can be done with **Short**:

> **Short[expr]**   Give **expr** in a shortened form
> **Short[expr, c]**   Give **expr** in a shortened form having length **c**

We generate 50 uniform random numbers but show only a few of them:

    t = Table[Random[], {50}];

    Short[t]

    {0.0889886, ≪48≫, 0.652932}

    Short[t, 4]

    {0.0889886, 0.69749, 0.452841, 0.676295,
      ≪42≫, 0.803352, 0.169427, 0.755352, 0.652932}

In the first case 48 values and in the latter case 42 values are not shown. (To find an appropriate length, such as 4, for the expression may take some experimenting.)

Before we continue, we clear the values of **a**, **b**, **c**, and **t**. We could write **a=.; b=.; c=.; t=.**, but a more convenient way is the following:

    Clear[a, b, c, t]

2.2.2 Lists

■ Basic Operations

`{a, b, c}` A one-dimensional (1D) list

`{{a, b, c, d}, {e, f, g, h}}` A two-dimensional (2D) list

`list[[i]]` ith part of `list`

`list[[i, j]]` (i, j)th part of `list`

`Length[list]` Number of elements in `list`

`MatrixForm[list]` Display `list` in a 2D matrix form

`TableForm[list]` Display `list` in a 2D tabular form

`Sort[list]` Sort the elements of `list` into canonical order

Lists are very basic objects in *Mathematica*; you will use them all the time. Vectors and matrices are in fact lists, and in many, many other computations, you need lists. Lists can have as many elements as you want them to have (an empty list is `{}`). Lists with lists as elements are 2D, 3D, and higher-dimensional lists. We define two lists:

    a = {x, y, z, v};

    b = {{3, 2, 5, 4}, {4, 1, 6, 2}, {3, 1, 1, 6}};

Picking a part of a list is easy:

    a[[3]]   z

    b[[2]]   {4, 1, 6, 2}

    b[[2, 3]]   6

Here are the lengths of the lists:

    Length[a]   4

    Length[b]   3

Matrices can be displayed handily with `MatrixForm`:

    b // MatrixForm

$$\begin{pmatrix} 3 & 2 & 5 & 4 \\ 4 & 1 & 6 & 2 \\ 3 & 1 & 1 & 6 \end{pmatrix}$$

`TableForm` prints a table:

    b // TableForm

3	2	5	4
4	1	6	2
3	1	1	6

We try `Sort`:

    Sort[{r, 4, P, 2, q, p, 3}]

    {2, 3, 4, p, P, q, r}

■ **Forming Lists**

> **Range[m]** Form the list {1, 2, ..., **m**}
> **Range[m, n]** Form the list {**m**, **m** + 1, ..., **n**}
> **Range[m, n, d]** Form the list {**m**, **m** + **d**, **m** + 2 **d**, ..., **n**}

With **Range**, we can easily form equally spaced numbers:

> **Range[6]**   {1, 2, 3, 4, 5, 6}

> **Range[0, 6]**   {0, 1, 2, 3, 4, 5, 6}

> **Range[0, 7, 2]**   {0, 2, 4, 6}

> **Table[expr, {i, a, b}]** Form a list of values of **expr** when **i** takes on values from **a** to **b** (in steps of 1)
> **Table[expr, {i, a, b}, {j, c, d}]** Index **i** takes on values from **a** to **b** and, for each **i**, **j** takes on values from **c** to **d**

Table is one of the most useful commands in *Mathematica;* it forms a list from a general rule. Iteration specification of the form {**i, a, b**} is most common, but other forms can also be used:

> **{n}** Form a list from **n** values of **expr**
> **{i, b}** Index **i** has values from 1 to **b** (in steps of 1)
> **{i, a, b}** Index **i** has values from **a** to **b** (in steps of 1)
> **{i, a, b, d}** Index **i** has values from **a** to **b** in steps of **d**

As can be seen, if the starting value of **i** is 1, it can be left out (but it can also be written), and if the step size is 1, it, too, can be left out. For example:

> **Table[0, {10}]**

> {0, 0, 0, 0, 0, 0, 0, 0, 0, 0}

> **Table[n!, {n, 10}]**

> {1, 2, 6, 24, 120, 720, 5040, 40320, 362880, 3628800}

> **Table[Sin[n π / 6], {n, 0, 6}]**

> $\left\{0, \frac{1}{2}, \frac{\sqrt{3}}{2}, 1, \frac{\sqrt{3}}{2}, \frac{1}{2}, 0\right\}$

> **Table[Exp[x], {x, 0., 3., 0.5}]**

> {1., 1.64872, 2.71828, 4.48169, 7.38906, 12.1825, 20.0855}

> **Table[1 / (i + j - 1), {i, 4}, {j, 4}]**

> $\left\{\left\{1, \frac{1}{2}, \frac{1}{3}, \frac{1}{4}\right\}, \left\{\frac{1}{2}, \frac{1}{3}, \frac{1}{4}, \frac{1}{5}\right\}, \left\{\frac{1}{3}, \frac{1}{4}, \frac{1}{5}, \frac{1}{6}\right\}, \left\{\frac{1}{4}, \frac{1}{5}, \frac{1}{6}, \frac{1}{7}\right\}\right\}$

Often it is useful to make pairs of the value of the index and the corresponding value of the expression:

```
Table[{x, Exp[x]}, {x, 0., 3.}]
```

```
{{0., 1.}, {1., 2.71828}, {2., 7.38906}, {3., 20.0855}}
```

In the following example, we apply 200 times the recursion formula $x_{i+1} = 3.7\,x_i(1 - x_i)$, starting from $x_0 = 0.5$:

```
x = 0.5; t = Table[x = 3.7 x (1 - x), {200}];
```

```
ListPlot[t, AspectRatio → 0.4, PlotRange → {-0.05, 1.05}];
```

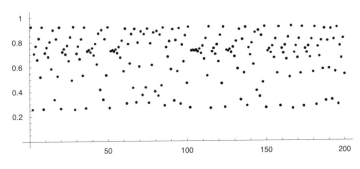

```
Clear[x, t]
```

■ Manipulating Lists

> **Total[list]** (⌘5) The sum of the elements of **list**
> **Mean[list], Median[list], Variance[list], StandardDeviation[list]** (⌘5)

We simulate the tossing of a die 20 times:

```
d = Table[Random[Integer, {1, 6}], {20}]
```

```
{3, 3, 6, 5, 6, 5, 3, 1, 1, 6, 4, 2, 3, 2, 2, 4, 2, 4, 1, 4}
```

The sum, mean, median, (unbiased) variance, and standard deviation are:

```
{Total[d], Mean[d], Median[d], Variance[d], StandardDeviation[d]} // N
```

```
{67., 3.35, 3., 2.76579, 1.66307}
```

Users of *Mathematica* with versions earlier than 5 have to use the command **Apply[Plus, list]** in place of **Total[list]** and have to load a package with **<<Statistics`DescriptiveStatistics`** before calculating means, medians, variances, or standard deviations:

```
Apply[Plus, d]      67
```

```
<< Statistics`DescriptiveStatistics`
```

```
{Mean[d], Median[d], Variance[d], StandardDeviation[d]} // N
```

```
{3.35, 3., 2.76579, 1.66307}
```

> `Map[f[#]&, {a, b, c}]` Calculate `{f[a], f[b], f[c]}`

Map is one of the most useful commands for manipulating lists; with it we can map each element of a given list with a given function. The effect of **Map** is that each element of the list is substituted in turn for **#**, and a list is formed from the results. The function is given in a special form having the name *pure function*. In such a function, the argument is expressed as **#**, and at the end of the function we have **&** (we consider pure functions in Section 14.1.5, p. 359). Here is an example:

> `Map[Sin[# π / 6] &, {1, 2, 4, 6}]`
>
> $\{ \frac{1}{2}, \frac{\sqrt{3}}{2}, \frac{\sqrt{3}}{2}, 0 \}$

The result is the value of $\sin(n\pi/6)$ for $n = 1, 2, 4, 6$. Recall then the **b** we defined earlier:

> **b**   `{{3, 2, 5, 4}, {4, 1, 6, 2}, {3, 1, 1, 6}}`

Now we find the maximum element of each list of **b**:

> `Map[Max[#] &, b]`   `{5, 6, 6}`

If the function to be mapped is a single built-in command (like **Max**), it suffices to write the name of the function (that is, we need not write **#** and **&**). So, we can simply write the following:

> `Map[Max, b]`   `{5, 6, 6}`

In plotting data, we often need to form points from given pairs of numbers (see Section 9.1.1, p. 230). For example:

> `Map[Point, {{2, 1}, {4, 7}, {9, 3}}]`
>
> `{Point[{2, 1}], Point[{4, 7}], Point[{9, 3}]}`
>
> `Clear[a, b, d]`

2.2.3 Functions and Programs

■ Functions

> `f[x_] := expr` Define a function

When defining a function, it is important to remember the underscore (_) after each argument. The underscore makes the function capable of calculating the value of the function for *any* value of the argument. The colon (**:**) before the equals sign results in the value of **expr** not being calculated in the definition; the value is calculated only when using the function.

In everyday use of *Mathematica*, we rarely need to define functions for expressions to be, say, differentiated, integrated, or plotted. Mostly we can use the expressions as such or we can give a name to the expression and then use that name.

For example, consider integration. First we enter the expression of the integrand as shown:

```
Integrate[x / (a + x), x]      x - a Log[a + x]
```

Then we give a name for the expression and use that name:

```
f = x / (a + x);
```

```
Integrate[f, x]      x - a Log[a + x]
```

We can also define a function, and then use it:

```
g[x_] := x / (a + x)
```

```
Integrate[g[x], x]      x - a Log[a + x]
```

However, this is unnecessarily complicated. The first two methods are the most useful. Giving a name is especially handy if we do several calculations with the same expression.

Function definitions are mostly used to form more complicated functions and programs. For example, here is a function for calculating the characteristic polynomial of a matrix:

```
charPoly[m_, x_] := Det[m - x IdentityMatrix[Length[m]]]
```

An example of using the function:

```
charPoly[{{2, 5}, {3, 1}}, x]      - 13 - 3 x + x²
```

■ **Programs**

`f[x_] := Module[{local variables}, body]` Define a function as a module

More complicated programs are often written as modules. In a module, we can define local variables that are used only within the module and that have no values outside of the module.

As an example, we develop a program to simulate random walk, in which the object starts at zero and moves one step up or down, each with a probability of 0.5. In the following way, we can generate the steps:

```
Table[2 Random[Integer, {0, 1}] - 1, {20}]
```

```
{1, -1, 1, -1, -1, 1, 1, 1, 1, -1, 1, 1, 1, -1, 1, 1, -1, 1, -1, 1}
```

The random walk is the cumulative sum of the steps. Cumulative sums can easily be calculated with **FoldList**:

```
FoldList[Plus, x0, {x1, x2, x3}]
```

```
{x0, x0 + x1, x0 + x1 + x2, x0 + x1 + x2 + x3}
```

In our example, the walk is shown in this way:

```
FoldList[Plus, 0, %%]
```

```
{0, 1, 0, 1, 0, -1, 0, 1, 2, 3, 2, 3, 4, 5, 4, 5, 6, 5, 6, 5, 6}
```

So, a program for a random walk with *n* steps could be written as follows:

```
randomWalk[n_] := Module[{steps, walk},
   steps = Table[2 Random[Integer, {0, 1}] - 1, {n}];
   walk = FoldList[Plus, 0, steps];
   ListPlot[walk, PlotJoined → True]]
```

The variables **steps** and **walk** in the program are local and thus they have no value outside the module. We simulate 200 steps of the random walk:

randomWalk[200];

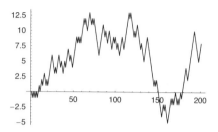

2.3 Mathematics

2.3.1 Differential and Integral Calculus

■ Differential Calculus

D[f, x]   Derivative of **f** with respect to **x**
D[f, x, x]   Second order derivative of **f** with respect to **x**
D[f, x, y]   Mixed second order derivative of **f** with respect to **x** and **y**
Series[f, {x, a, n}]   nth order Taylor series of **f** with respect to **x** at **a**
Limit[f, x → a]   Limit of **f** as **x** approaches **a**

Here are some examples:

a = x Sin[y];

{D[a, x], D[a, y], D[a, x, x], D[a, x, y], D[a, y, y]}

$\{ Sin[y], x Cos[y], 0, Cos[y], -x Sin[y] \}$

Series[Sin[Sqrt[x]], {x, 0, 4}]

$$\sqrt{x} - \frac{x^{3/2}}{6} + \frac{x^{5/2}}{120} - \frac{x^{7/2}}{5040} + O[x]^{9/2}$$

Limit[(1 + c / x) ^ x, x → ∞] e^c

(Recall that ∞ can be written as **Infinity** or as ⎡ESC⎤inf⎡ESC⎤.)

■ Integral Calculus

Integrate[f, x]   Indefinite integral of **f** with respect to **x**
Integrate[f, {x, a, b}]   Definite integral of **f** when **x** varies from **a** to **b**
NIntegrate[f, {x, a, b}]   Calculate the definite integral by numerical methods
Sum[f, {i, a, b}]   Sum of the values of **f** when **i** varies from **a** to **b**

Prepare to see special functions when you integrate functions that are not easy:

a = Integrate[Sin[Exp[x]], x] $SinIntegral[e^x]$

This is one of the many special functions in *Mathematica*. Do not worry! You can do the same with the special functions as you do with the more usual functions. For example, you can check the result by differentiation:

D[a, x] $Sin[e^x]$

You can ask for a value:

a /. x → 1.   **1.82104**

You can ask for a plot:

Plot[a, {x, 0, 3}];

a = .

Sometimes even *Mathematica* does not know an integral:

Integrate[Sin[Sin[x]], {x, 0, 1}] $\int_0^1 Sin[Sin[x]]\, dx$

Mathematica just writes the command as such. You can then resort to numerical integration (Gaussian quadrature):

NIntegrate[Sin[Sin[x]], {x, 0, 1}]   **0.430606**

Sums are calculated like integrals:

Sum[1 / 2^n, {n, 1, 10}] $\dfrac{1023}{1024}$

Sum[1 / n^2, {n, 1, ∞}] $\dfrac{\pi^2}{6}$

Sum[r^n, {n, 1, m}] $\dfrac{r\,(-1 + r^m)}{-1 + r}$

2.3.2 Matrices, Equations, and Optimization

■ Matrices

{a, b, c} A vector
{{a, b, c, d}, {e, f, g, h}} A matrix with two rows

A 1D list is also a vector; a 2D list is a matrix. Vectors and matrices can have as many elements as you want. *Mathematica* does not distinguish column and row vectors but, nevertheless, it does calculations with matrices and vectors so that the results are, almost always, what you intended. We define a vector:

a = {2, 5};

A useful fact to know is that *Mathematica* automatically does all operations with vectors element by element:

> **{a^2, Sqrt[a], a / {3, 6}}**

$$\{\{4, 25\}, \{\sqrt{2}, \sqrt{5}\}, \{\tfrac{2}{3}, \tfrac{5}{6}\}\}$$

Here is a matrix:

> **MatrixForm[m = {{2, 1}, {3, 2}}]**

$$\begin{pmatrix} 2 & 1 \\ 3 & 2 \end{pmatrix}$$

Note that **MatrixForm** is used only in displaying matrices. You cannot do any calculations with such a form. If you would write **m = {{2, 1}, {3, 2}} // MatrixForm**, then the value of **m** would be the *matrix form* of the given matrix, and with such an **m** we cannot calculate. However, you could write **(m = {{2, 1}, {3, 2}}) // MatrixForm**.

> **a m**   The product of a scalar **a** and a vector or matrix **m**
>
> **m + n**   The sum of two vectors or matrices **m** and **n**
>
> **m.n**   The product of two vectors or matrices **m** and **n**
>
> **Transpose[m]**   The transpose of a matrix **m**
>
> **Det[m]**   The determinant of a square matrix **m**
>
> **Inverse[m]**   The inverse of a square matrix **m**
>
> **Eigenvalues[m]**   The eigenvalues of a square matrix **m**

Note that the point (**.**) has to be used when calculating products of vectors and matrices; the space and the asterisk do not work properly. You cannot use powers, either. So, to calculate the second power of a matrix **m**, you have to write **m.m**; you cannot write **m^2**. Also, to calculate the inverse of a matrix **m**, you have to write **Inverse[m]**; you cannot write **m^-1**.

In the following example, **a** is interpreted to be a row vector:

> **a.m**   {19, 12}

Here, **a** is a column vector:

> **m.a**   {9, 16}

We calculate the square, transpose, determinant, inverse, and eigenvalues of **m**:

> **m.m**   {{7, 4}, {12, 7}}
>
> **Transpose[m]**   {{2, 3}, {1, 2}}
>
> **Det[m]**   1
>
> **Inverse[m]**   {{2, -1}, {-3, 2}}
>
> **Eigenvalues[m]**   {2 + $\sqrt{3}$, 2 - $\sqrt{3}$}
>
> **Clear[a, m]**

■ Equations: Symbolic Solutions

> **expr1 == expr2** An equation (== can be written as ==)
> **Solve[eqn, x]** Solve a (polynomial) equation with respect to **x**
> **Solve[{eqn1, eqn2}, {x, y}]** Solve two (polynomial) equations with respect to **x**
> and **y**

Equations are formed with two equal signs (==) but *Mathematica* replaces them with the special symbol ==. Forgetting the second = is a common error; remember that = is used only to assign values for variables.

Here is a polynomial equation familiar to you (we give the name **eqn** to this equation):

> **eqn = a x^2 + b x + c == 0**
>
> $c + b x + a x^2 == 0$
>
> **sol = Solve[eqn, x]**
>
> $\left\{\left\{x \rightarrow \dfrac{-b - \sqrt{b^2 - 4\, a\, c}}{2\, a}\right\}, \left\{x \rightarrow \dfrac{-b + \sqrt{b^2 - 4\, a\, c}}{2\, a}\right\}\right\}$

The result is in the form of transformation rules. If you want only the values of **x**, apply the transformation rule to **x** (see Section 2.2.1, p. 31):

> **x /. sol**
>
> $\left\{\dfrac{-b - \sqrt{b^2 - 4\, a\, c}}{2\, a}, \dfrac{-b + \sqrt{b^2 - 4\, a\, c}}{2\, a}\right\}$

We can also check that the solution is correct by inserting the solution into the equation:

> **eqn /. sol // Simplify**
>
> {True, True}

Then we solve two linear equations (larger systems are solved similarly). Enclose a system of equations and the variables within curly braces (**{ }**):

> **Solve[{2 x + 5 y == 4, x - 3 y == 3}, {x, y}]**
>
> $\left\{\left\{x \rightarrow \dfrac{27}{11}, y \rightarrow -\dfrac{2}{11}\right\}\right\}$

■ Equations: Numerical Solutions

> **NSolve[eqn, x]** Solve a (polynomial) equation with numerical methods

Polynomial equations of a degree higher than four can rarely be solved:

> **eqn2 = x^5 - x^3 + x^2 - 2 == 0;**
>
> **Solve[eqn2, x]**
>
> $\{\{x \rightarrow \text{Root}[-2 + \#1^2 - \#1^3 + \#1^5 \,\&,\, 1]\},$
> $\{x \rightarrow \text{Root}[-2 + \#1^2 - \#1^3 + \#1^5 \,\&,\, 2]\}, \{x \rightarrow \text{Root}[-2 + \#1^2 - \#1^3 + \#1^5 \,\&,\, 3]\},$
> $\{x \rightarrow \text{Root}[-2 + \#1^2 - \#1^3 + \#1^5 \,\&,\, 4]\}, \{x \rightarrow \text{Root}[-2 + \#1^2 - \#1^3 + \#1^5 \,\&,\, 5]\}\}$

We did not obtain the solution in an explicit form (*Mathematica* only gives a symbolic list representing the five roots). So we resort to numerical methods:

```
NSolve[eqn2, x]
```

$\{\{x \to -1.09595 - 0.361002\,\mathbb{i}\}, \{x \to -1.09595 + 0.361002\,\mathbb{i}\},$
$\{x \to 0.508323 - 1.00984\,\mathbb{i}\}, \{x \to 0.508323 + 1.00984\,\mathbb{i}\}, \{x \to 1.17525\}\}$

Solve can solve some transcendental equations:

```
Solve[Exp[a x] == b, x]
```

Solve::ifun :
   Inverse functions are being used by Solve, so some solutions may not
      be found; use Reduce for complete solution information. More…

$\left\{\left\{x \to \dfrac{\text{Log}[b]}{a}\right\}\right\}$

However, **FindRoot** is the general-purpose command for such equations. It calculates a zero iteratively by Newton's method and other methods.

> **FindRoot[eqn, {x, x0}]**   Solve an equation with numerical methods, starting from **x0**

To find a zero for the following function, we first plot it:

```
f = Exp[-x] - 0.5 x;
```

```
Plot[f, {x, 0, 3}];
```

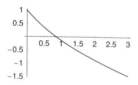

The zero seems to be near one, and so we start from this point:

```
x0 = FindRoot[f == 0, {x, 1}]      {x → 0.852606}
```

The value of the function at this point is zero, with a high degree of accuracy (⌘4!):

```
f /. x0        - 5.55112 × 10⁻¹⁷
```

■ **Optimization**

> **FindMinimum[f, {x, x0}]**   Find a local minimum of **f** starting from **x0**
> **FindMaximum[f, {x, x0}]** (⌘5)   Find a local maximum of **f** starting from **x0**

FindMinimum uses Brent's iterative method. To find a local maximum for the following function, we first plot it:

```
f = Cos[x] + Log[1 + x];
```

```
Plot[f, {x, 0, 2}];
```

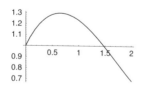

The maximum seems to be near 0.5 and so we start from this point:

```
x0 = FindMaximum[f, {x, 0.5}]      {1.29686, {x → 0.650752}}
```

Users of *Mathematica* 4 can do as follows:

```
FindMinimum[-f, {x, 0.5}]      {-1.29686, {x → 0.650752}}
```

Thus, the local maximum is at the point 0.650752, and the maximum value is 1.29686. The derivative of the function is, indeed, zero, with a high degree of accuracy at the given point (✥4!):

```
D[f, x] /. x0[[2]]      6.83054 × 10⁻¹²
```

2.3.3 Interpolation and Approximation

■ Interpolation

> `Interpolation[data]` Find a piecewise third-degree interpolating polynomial for
> data

Let us first generate some data:

```
points = Table[{x, Cos[Exp[x]]}, {x, 0., 2., 0.2}]
```

```
{{0., 0.540302}, {0.2, 0.342328}, {0.4, 0.0788896}, {0.6, -0.248685},
 {0.8, -0.608957}, {1., -0.911734}, {1.2, -0.984107},
 {1.4, -0.610894}, {1.6, 0.238328}, {1.8, 0.972854}, {2., 0.448356}}
```

```
ListPlot[points];
```

We calculate a piecewise third-degree interpolating polynomial through the points:

```
int = Interpolation[points]
```

```
InterpolatingFunction[{{0., 2.}}, <>]
```

Mathematica calls the result an interpolating function. We do not see the actual function, only the interval where it is defined. However, we can calculate values of the interpolating function:

```
int[1.5]      -0.208865
```

We can plot it:

```
Plot[int[x], {x, 0, 2}];
```

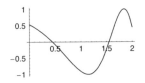

The function goes exactly through all the points and is of degree three between each pair of points. The result is a good representation of the data. In Section 2.3.4, when we solve differential equations numerically, we encounter these functions; the numerical solution is expressed as an interpolating function.

■ Approximation

Fit[data, basis, var]   Fit **data** by a linear combination of functions of **var** in **basis**

Fit calculates a least-squares function to smoothly represent data containing errors. Consider the following data:

```
data =
  {{0, 0.185}, {1, 0.935}, {2, 0.649}, {3, 1.231}, {4, 2.279}, {5, 3.913},
   {6, 4.670}, {7, 5.620}, {8, 6.767}, {9, 9.044}, {10, 11.045}};
```

```
p1 = ListPlot[data];
```

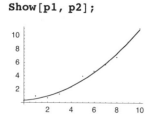

The points seem to follow roughly a quadratic pattern, and so we try a quadratic fit:

```
lsq = Fit[data, {1, x, x^2}, x]
```

$0.293972 + 0.146282\, x + 0.0910618\, x^2$

```
p2 = Plot[lsq, {x, 0, 10}];
```

To see how close the fit is to the data, we can show both:

```
Show[p1, p2];
```

The fit seems to be good.

2.3.4 Differential Equations

■ Symbolic Solutions

> `sol = y[t] /. DSolve[eqn, y[t], t]` Give the general solution of a differential equation
>
> `sol = y[t] /. DSolve[{eqn, y[a] == α}, y[t], t]` Solve a first-order initial value problem
>
> `sol = y[t] /. DSolve[{eqn, y[a] == α, y'[a] == β}, y[t], t]` Solve a second-order initial value problem
>
> `Plot[sol, {t, a, b}]` Plot the solution

A differential equation is like a usual equation containing ==, but now the equation contains an unknown function such as `y[t]` and its derivatives such as `y'[t]` and `y''[t]`. Note that initial conditions must also be written as equations (containing ==).

DSolve can solve a large number of differential equations. We first ask for a general solution to the following linear equation with constant coefficients:

`eqn1 = y'[t] == a y[t] + b`

$y'[t] == b + a y[t]$

`DSolve[eqn1, y[t], t]`

$$\left\{ \left\{ y[t] \rightarrow -\frac{b}{a} + e^{a t} \, C[1] \right\} \right\}$$

The solution is in the familiar form of a transformation rule. The `C[1]` is an undetermined constant. Often it is useful to directly ask the value of `y[t]`:

`y[t] /. DSolve[eqn1, y[t], t][[1]]`

$$-\frac{b}{a} + e^{a t} \, C[1]$$

Then we solve a first-order initial value problem:

`eqn2 = y'[t] + 2 t y[t] - t == 0`

$-t + 2 t y[t] + y'[t] == 0$

`sol2 = y[t] /. DSolve[{eqn2, y[0] == 0}, y[t], t][[1]]`

$$\frac{1}{2} e^{-t^2} \left(-1 + e^{t^2} \right)$$

We plot the solution:

`Plot[sol2, {t, 0, 3}];`

Next, we solve a second-order initial value problem and plot the solution:

```
eqn3 = y''[t] + y[t] == -t
```

$y[t] + y''[t] == -t$

```
sol3 = y[t] /. DSolve[{eqn3, y[0] == 0, y'[0] == 1}, y[t], t][[1]]
```

$-t + 2 \sin[t]$

```
Plot[sol3, {t, 0, 8}];
```

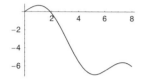

■ Numerical Solutions

If *Mathematica* is not able to solve a differential equation symbolically with **DSolve**, then we can resort to numerical methods with **NDSolve**. We can simply replace **t** with **{t, a, b}** defining the interval (a, b) where the solution is calculated.

> **sol = y[t] /. NDSolve[{eqn, y[a] == α}, y[t], {t, a, b}]** Solve a first-order equation numerically
> **sol = y[t] /. NDSolve[{eqn, y[a] == α, y'[a] == β}, y[t], {t, a, b}]** Solve a second-order equation numerically
> **Plot[sol, {t, a, b}]** Plot the solution

Mathematica cannot solve the following nonlinear initial value problem:

```
eqn4 = y''[t] == -y[t]^2 + t
```

$y''[t] == t - y[t]^2$

```
sol4 = DSolve[{eqn4, y[0] == 0, y'[0] == 1}, y[t], t]
```

$\mathrm{DSolve}[\{y''[t] == t - y[t]^2, y[0] == 0, y'[0] == 1\}, y[t], t]$

However, we can use numerical methods and obtain an approximate solution:

```
sol4 = y[t] /. NDSolve[{eqn4, y[0] == 0, y'[0] == 1}, y[t], {t, 0, 7}]
```

$\{\mathrm{InterpolatingFunction}[\{\{0., 7.\}\}, <>][t]\}$

The solution is expressed as an interpolating function, that is, as a piecewise third-degree interpolating polynomial (we considered these in Section 2.3.3, p. 43). We plot the solution:

```
Plot[sol4, {t, 0, 7}];
```

Notebooks

Introduction

During his stay in Berlin, Euler developed the habit of writing memoir after memoir, placing each when finished on top of a pile of manuscripts. Whenever material was needed to fill the journal of the Academy, the printers would help themselves to a few papers from the top of the pile. This meant that papers at the bottom remained there a long time and earlier papers often contained developments and improvements on later papers.

Note that you may skip this chapter if you, right now, do not need to do the following:

- get more information about notebooks (saving, opening, printing, styles, style sheets, palettes, hyperlinks);
- edit the outlook of notebooks;
- write special characters and 2D formulas with the keyboard; or
- write a mathematical document with display and inline formulas.

Regardless of your level of comfort with *Mathematica*, it may be helpful for you to read Sections 3.1.1, 3.1.2, and 3.3.1.

This chapter emphasizes *Mathematica* as a writing tool. Indeed, you can use *Mathematica* to write all kinds of mathematical and other documents. *Mathematica* is a strong alternative to traditional writing tools, because it has the remarkable advantages that the whole document can be done with the same high-quality application (for example, graphics or formulas need not be prepared with other applications) and that *Mathematica*'s computing power gives you excellent possibilities to do all kinds of calculations needed for the preparation of your document.

Adding notebook material into the *Help Browser* and other advanced matters about notebooks are explained in a separate document, **Using Author Tools.nb**, that can be found on the CD-ROM that is included with this book.

3.1 Working with Notebooks

3.1.1 Saving, Opening, and Printing

■ Saving

Mathematica documents are called *notebooks*. To save a notebook for the first time, choose **Save As...** from the **File** menu; later on choose **Save** to save the modified notebook. It is customary to give the notebook a name ending with **.nb**. The handling of file names depends somewhat on the system used. For example, in Windows, the name of a notebook will automatically end with **.nb**, and without this extension, the system does not automatically know that the file should be opened with *Mathematica*. On the other hand, in MacOS X *Mathematica* does not suggest a name ending with **.nb**; indeed, the name can be without that extension.

You can arrange for the notebook to be automatically saved after each command (you must, however, do the first saving). Choose **Option Inspector...** from the **Format** menu, choose *Show option values for* to be *notebook*, and then go to **Notebook Options ▷ File Options** and click the box for **NotebookAutoSave** (for more information about *Option Inspector*, see Section 3.2.2, p. 57).

The notebook can also be saved in certain special formats like TeX or HTML (see Section 4.3.2, p. 101) by choosing **Save As Special...** from the **File** menu.

■ Opening

A saved notebook can be opened by double-clicking the document or by choosing **Open...** from the **File** menu. Recently used notebooks are listed in the **File ▷ Open Recent** menu (in version 4 they are listed at the end of the **File** menu). When you open a notebook, you may observe that it no longer has the In- and Out-labels such as In[1] and Out[1].

After opening a notebook, you can continue working with it by adding new calculations, deleting old calculations, and doing other kinds of editing and modifying (this topic was already shortly considered in Section 1.5, p. 22).

One important thing to know is that, when you open an old notebook, *you cannot directly use any of its results in the new session*. For example, suppose that in the old notebook you have defined **a = 5**. When you open the notebook, the variable **a** has no value. If you want **a** to have the value 5 in the new session, you have to execute the command **a = 5** anew (place the cursor anywhere in the relevant cell and execute). Similarly, all the results you want to use in the new session have to be recalculated. In other words, *you can use only the results that have been calculated in the current session.*

A straightforward way to continue working with an old notebook is to re-execute all its commands by choosing **Kernel ▷ Evaluation ▷ Evaluate Notebook**. However, some commands may require substantial time for execution. A better method in such a situation may be to use the old outputs directly. For example, if you have calculated an integral, place the cursor at the beginning of the *output* and write **int =**. *Mathematica* now automatically takes a copy of the output and transforms it to an input. You can then execute the resulting command. Now you have the value of the integral in **int** without having to recalculate the integral. You can also save results to files and in later sessions, load them (see Section 4.3.1, p. 98).

■ Checking Spelling

One of the last steps in preparing a document is the checking of the spelling. Choose **Edit ▷ Check Spelling...**. *Mathematica* comes with English spell checking, but you can buy spell checking for many other languages; contact Wolfram Research. A **Spelling Language** menu command then appears either in the **Edit** or **Format** menu so that you may choose the language used. Open the *Option Inspector* by choosing **Format ▷ Option Inspector...**, choose *Show option values for* to be *notebook*, and go into **Formatting Options ▷ Text Content Options ▷ SpellingOptions**; you will find some options that control the spell checker (for *Option Inspector*, see Section 3.2.2, p. 57).

Note that, if you want it to, *Mathematica* will automatically hyphenate words according to the rules of English. To turn hyphenation on or off, open *Option Inspector*, choose *Show option values for* to be *notebook*, go into **Formatting Options ▷ Text Layout Options**, and click the box for **Hyphenation**. The international spell-checking products also contain hyphenation rules for other languages.

■ Adjusting Printing Settings

Before printing the notebook, we can check, from **File ▷ Printing Settings**, the page setup, the margins, and the headers and footers. Headers and footers can be adjusted in the following window:

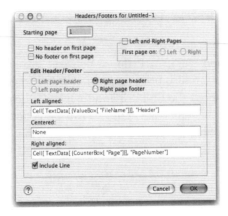

Here we can adjust the starting page number, whether the notebook has separate left and right pages, and what to include in the left, center, and right of the headers and footers. The default header of the default notebook defines that, at the left, center, and right, respectively, we have the following:

    Cell[TextData[{ValueBox["FileName"]}], "Header"]
    None
    Cell[TextData[{CounterBox["Page"]}], "PageNumber"]

This means that at the left we have the name of the file, at the center we have nothing, and at the right we have the page number. We can modify the header and footer items. For example, we can put other text into the left:

    Cell[TextData["Chapter 3 • Notebooks"], "Header"]

If you want page numbers like *i, ii, iii, iv*, etc., define them as follows:

    CounterBox["Page", CounterFunction -> RomanNumeral]

The styles "Header" and "PageNumber" can be adjusted by the style sheet (see Section 3.2.4, p. 63) and the margin above the header by an option (see Section 3.2.3, p. 63).

■ Adjusting Page Breaks

Before printing, it may also be useful to see the page breaks. Choose **Format ▷ Show Page Breaks**. A page break between two cells is shown by a dashed line, and a page break within a cell is shown by a short horizontal line at the right of the window. The current page number can be seen in the bottom left corner of the window. The appearance of the notebook on the screen changes to reflect the printed output.

   In some special cases, you may want to manually adjust the automatically set page breaks. Suppose you want a page break above a certain cell. Select the cell bracket of this cell with the mouse, choose **Format ▷ Option Inspector...**, go to **Cell Options ▷ Page Breaking**, and choose **True** as the value for `PageBreakAbove`. Now there is a forced page break above the selected cell. This kind of page break has a different dashed line than a normal page break. Page break options are considered in Section 3.2.3, p. 63.

   The following is another method for adding a page break. Place the cursor somewhere in the cell and choose **Format ▷ Show Expression**. The cell is then converted into the code *Mathematica* internally uses for the cell. Before the closing square bracket, write `,Page BreakAbove -> True`, and then again choose **Format ▷ Show Expression**.

■ Printing

To print your notebook, choose **File ▷ Print...**. You can print the whole notebook or a selected range of pages. You can also print selected cells by dragging over their cell brackets (or clicking the cell bracket of the first cell and then shift-clicking the cell bracket of the last cell) to select them and then choosing **File ▷ Print Selection...**. If you have problems with fonts when printing graphics, you may find useful information in Section 6.3.2, p. 161.

3.1.2 Cell Styles and Style Sheets

■ *Mathematica* as an Advanced Writing Tool

An important thing to realize is that *Mathematica is an advanced environment for technical writing*. Indeed, with *Mathematica* we can write a complete structured document with title, headings, texts, formulas, tables, graphics, inputs, and outputs. For example, each chapter of this book is a *Mathematica* notebook.

When using *Mathematica* as a writing environment, two things are important: cell styles and style sheets. These are considered next (they are also explained in the *Help Browser* under **Front End ▷ Style Sheets**). *Mathematica* as a writing tool is considered in Section 3.4.

■ Cell Styles

In *Mathematica*, each cell has a *style*. The names of the styles can be seen by choosing **Format ▷ Style** (the styles can also be seen from a toolbar; choose **Format ▷ Show Toolbar**). For example, when we write a new command, the style is automatically **Input**, and the style of the result is **Output**. We can add text into the notebook with the **Text** style. Titles and headings of sections can be written with styles like **Title**, **Section**, and **Subsection**.

The default style of a new cell is **Input**. However, we can change the style of a cell by highlighting its bracket and then choosing an appropriate style from **Format ▷ Style**. Or we can, before we write anything, choose the style of the next cell and then type the text. Within a cell of a certain type, we can apply other styles for smaller parts. If we want to change the appearance of all cells of a certain type in our notebook, we can just change the *style definition cell* of the *style sheet* we have used (see below).

One tip is that, if you intend to write a new cell with the same style as the current cell (**Text**, for example), choose **Input ▷ Start New Cell Below** or press $\boxed{\text{ALT}}$-$\boxed{\text{ENTER}}$ in Windows or $\boxed{\text{OPTION}}$-$\boxed{\text{RET}}$ in Macintosh (normally the style of a new cell is **Input**).

■ Style Sheets

The cell styles have default settings (font, face, size, color, etc.), but we can adjust these if we are not satisfied with them. One way to adjust the styles is to choose an appropriate *style sheet* from the several ready-to-use options that come with *Mathematica*. Style sheets are special notebooks that define styles for normal notebooks. The style sheets can be seen by choosing **Format ▷ Style Sheet**. The default style sheet is **Default**, but there are several others, such as **ArticleClassic**, **Classroom**, **Demo**, and **Textbook**. If you have prepared a notebook, try out these style sheets and choose the one you prefer.

Note that different style sheets define different sets of cell styles. For example, the style **NumberedEquation** is only available with certain style sheets.

After you have chosen an appropriate style sheet, it may be that you would still like to change the styles of some cells. This can easily be done with the style sheet, as is shown in Section 3.2.4, p. 63. We can even create new style sheets.

■ Style Environments

In addition to style sheets, there are *style environments*, like **Working**, **Presentation**, **Condensed**, and **Printout**. These allow you to modify the appearance of your notebook according to how you plan use it. When writing the notebook, you can use the **Working** environment (with large fonts); for presentations, the **Presentation** environment (with still larger fonts); for small screens, the **Condensed** environment (with small fonts and a condensed style); and before printing, the **Printout** environment (to see how the notebook will look when printed).

Furthermore, the style environment can be defined separately for the screen and for printing. Choose the environments from **Format ▷ Screen Style Environment** and **Format ▷ Printing Style Environment**.

■ Magnification

A simple way to change the size of the text on the screen is to choose an appropriate magnification percentage from the bottom of the notebook window or from **Format ▷ Magnification** (the default percentage is 100%).

3.1.3 Palettes

■ Pasting Palettes

Palettes, which are available from **File ▷ Palettes**, help when you are writing inputs. The built-in palettes cover a significant part of all characters and commands of *Mathematica*, but you may find other things you want to do with a palette. Create your own palette by doing the following:

> - Select **Input ▷ Create Table/Matrix/Palette....**
> - In the dialog box, click the **Palette** button, specify the number of rows and columns you want, and click **OK**. A form of a palette, with blank buttons, appears in your notebook at the cursor location.
> - Write in the buttons whatever text you want. When you then click a button in the prepared palette, the corresponding text is pasted at the cursor location.
> - Let the cursor remain in the palette, and choose **File ▷ Generate Palette from Selection**. The palette then appears as a new notebook; the original form of the palette remains in your notebook (you can delete it or let it stay in the notebook).
> - You can already use the palette that appeared on the screen, but to get the palette under **File ▷ Palettes**, save the palette as follows. Click the close button of the palette; you are then prompted to save it. Save the palette in **SystemFiles ▷ Front End ▷ Palettes** in **$UserBaseDirectory** (in **$UserAddOnsDirectory** in *Mathematica* 4) (the name of the palette must end with **.nb**).
> - Quit and then re-open *Mathematica*, and choose the new palette from **File ▷ Palettes**.

For example, on my computer with MacOs X, the correct directory is as follows:

```
$UserBaseDirectory

/Users/ruskeepaa/Library/Mathematica
```

If you want to modify a palette, select it and then choose **File ▷ Generate Notebook from Palette**. A copy of the palette appears in a separate notebook, where you can modify it. If you want to add a row somewhere in the palette, place the cursor in the preceding row and press CTRL-RET. To add a new column, place the cursor in the preceding column and press CTRL,. Once you have made the changes, choose **File ▷ Generate Palette from Selection**, click the **Close** button, and save.

■ An Example

We have created the following palette containing some plotting commands to plot data. Clicking any button in this palette pastes the corresponding text at the cursor location.

```
                    DataPlotting.nb
                   ListPlot[■];
            ListPlot[■, PlotJoined → True];
        Map[ListPlot[#, PlotJoined → True,
             Epilog → {AbsolutePointSize[3],
                 Map[Point, #]}] &, {■}];
   Map[ListPlot[#, PlotStyle →
        AbsolutePointSize[3], Prolog →
        Map[Line[{{#[[1]], 0}, #}] &, #]] &, {■}];
```

In this palette, we have used placeholders (■). Creating a palette that has buttons with placeholders is handy, particularly if the buttons create whole commands to be executed. We can then complete the command by using the TAB key to go from one placeholder to the next. Write a selection placeholder ■ as ESC spl ESC and a placeholder □ as ESC pl ESC. A selection placeholder is automatically selected after clicking a button (so that what we then write is entered in place of it), whereas we can go to regular placeholders with the TAB key.

To try the palette, first generate some data:

```
data = Table[{i, i + 15 Random[]}, {i, 30}];
```

Then click the last button of the palette, type **data** (it comes in place of the selection placeholder), and execute:

```
Map[ListPlot[#, PlotStyle →
    AbsolutePointSize[3], Prolog →
    Map[Line[{{#[[1]], 0}, #}] &, #]] &, {data}];
```

■ Evaluating Palettes

Thus far we have created palettes that paste text into the notebook. The result of clicking a button can also be set to be automatically executed.

Suppose you want to create a button that, when clicked after you have selected a list, will automatically plot the list. Fill in the buttons in the palette using the selection place-holder ■ in the place where you want to substitute the list. Then select the whole palette, choose **Input ▷ Edit Button...**, choose **EvaluateCell** as the **Button Style**, and click **OK**. Then choose **File ▷ Generate Palette from Selection**, and save the palette.

■ Other Kinds of Palettes

As an example of other kinds of palettes, we will create a palette with which we can select all cells of a certain type in the current notebook. With the palette we can then select, for example, all **Graphics** cells (in order to, say, convert them into bitmaps). Macintosh users do not need this palette, because on a Macintosh, all cells of a certain type are selected by pressing OPTION and then selecting one cell of the desired type.

Create a palette with 13 rows and 1 column. Type the following texts in the buttons: `Title`, `Subtitle`, `Subsubtitle`, `Section`, `Subsection`, `Subsubsection`, `Text`, `SmallText`, `Input`, `Output`, `Message`, `Graphics`, and `Print`. Select the palette, choose **Input ▷ Edit Button...**, clear the text in the box under **Button Function**, type the new text `Notebook-Find[SelectedNotebook[], #, All, CellStyle] &`, and click **OK**. Choose **File ▷ Generate Palette from Selection**, and save the palette with the name **SelectAllCellsOfType.nb**.

3.1.4 Hyperlinks

In the *Help Browser*, you have probably seen hyperlinks. These are special buttons that consist of underlined words in blue type. When a hyperlink is clicked, *Mathematica* jumps to another part of the notebook. To create a hyperlink, do the following:

> • Select the destination cell.
> • Choose **Find ▷ Add/Remove Cell Tags...**. A dialog box appears in which you write a word or phrase as the cell tag. It identifies the destination cell. Click **Add** and **Close**.
> • Select the words in the notebook that will represent the hyperlink, that is, the words you want to be able to click to jump to the destination cell.
> • Choose **Input ▷ Create Hyperlink...**. In the dialog box, select the tag you created earlier, and then click **OK**.

The text of the hyperlink button can be edited by clicking the text near the button and then moving the cursor with the arrow keys into the button.

Hyperlinks to *Help Browser* documentation can be created simply. As examples, we will create hyperlinks to the command **Integrate** in the *Built-in Functions*, to the standard package **Statistics`HypothesisTests`** in the *Add-ons & Links*, and to Section 3.4.5 in *The Mathematica Book*. Just select the topic (the name of the command, the name of the package, or the number of the section) with the mouse, choose **Input ▷ Create Button**, and then select **RefGuideLink**, **AddOnsLink**, or **MainBookLink**, respectively. Similarly, if you have installed *Mathematica Navigator* into the *Help Browser* and you want to create a hyperlink to Section MN:27.7.1, simply select MN:27.7.1 and then create an **AddOnsLink**. When creating these links, note that the word or phrase from which the hyperlink is created has to be in the same form as shown in the top left box of the *Help Browser* when the corresponding information is shown in the *Help Browser*.

3.2 Editing Notebooks

3.2.1 Basic Editing

■ Steps

When creating a document with *Mathematica,* you want to be aware of how the document looks on screen and when printed out. The look can be adjusted in many ways. Proceed with the following steps:

- *Select a style sheet:* try several style sheets (Section 3.1.2, p. 51), and choose the one you like most.
- *Modify the style sheet:* if the style sheet does not satisfy you in all respects, modify the style sheet.
- *Adjust the notebook:* if you still find some small parts of the document needing adjustment, modify these parts directly in the document.

The key to modifying a notebook is the use of the style sheet. The notebook should be modified directly only in exceptional cases. Why? To save your work and to maintain the consistency of the look of the document. Suppose you want to change the font size of all text cells. You could manually change the style of each text cell, but if you then continue writing your document, you have to modify all new text cells as well. If you instead modify the style sheet, all text cells immediately change accordingly, and all new text cells also have the correct style.

Thus, resist direct modification of the notebook, and modify the style sheet instead. The editing of style sheets is described in Section 3.2.4, p. 63.

■ Methods

Whether you modify the style sheet or the document, you can use several methods in the modification:

- Use the **Format** menu.
- Use the *Option Inspector.*
- Use options directly.

Below we consider the use of the **Format** menu and the direct use of the options, while the *Option Inspector* is considered in Section 3.2.2, p. 57. For each method, we set values of some *front-end options.* All options can be set with the second and third methods, whereas only a small number of options (although important ones) can be set with the **Format** menu.

The *Option Inspector* is a tool that is used to view and modify various options of cells and notebooks, among other things. The options can also be adjusted directly, without the *Option Inspector,* as is shown below. When we use the **Format** menu, we actually also adjust options. With the *Option Inspector,* we can do all the things we can do with the **Format** menu, but the *Option Inspector* has a large number of additional options. Note that options are explained in the *Help Browser* under **Front End ▷ Front End Options.**

■ Using the Format Menu

The editing tool used most often may be the **Format** menu. With it we can set various font options: **Font**, **Face**, **Size**, **Text Color**, and **Background Color**. We can also adjust **Text Alignment**, **Text Justification**, **Word Wrapping**, **Cell Dingbats**, and **Horizontal Lines**. Furthermore, with **Format** we can choose the **Style** of each cell and use style sheets by **Style Sheet** and **Edit Style Sheet**.

To use the **Format** menu, first select with the mouse the part of the document you want to modify. The part is typically a cell (click its cell bracket), but it can also be a part of a cell (drag over the part), several cells (drag over their cell brackets), or a group of cells (click its bracket). Then, choose a suitable command from the **Format** menu. As an example, write an **EmphasizedText** cell:

Here is some text.

Increase the size of the font to 14, so that the cell becomes

Here is some text.

To show that the **Format** menu actually uses options, we next show how to look at the internal code *Mathematica* uses for cells.

■ Looking at and Modifying the Code of Cells

Front-end options are normally not visible; we only see the effect of the options. However, with a special menu command, we can see the internal code behind a cell:

- Put the cursor somewhere in the cell, or select the bracket of the cell.
- Open the code of the cell by choosing **Format ▷ Show Expression** or pressing
  ⌷SHIFT⌷-⌷CTRL⌷-[e] (⌷SHIFT⌷-⌘-[e] on a Macintosh).
- Modify the code, if you so choose.
- Close the code by choosing the same menu command again; the cell is then formatted according to the code.

As examples, here are the codes of the two text cells we considered above:

```
Cell["Here is some text.", "EmphasizedText"]
```

```
Cell["Here is some text.", "EmphasizedText", FontSize->14]
```

In both cases the text is in an **EmphasizedText** style cell and that the latter cell has an option for the font size. So, we have shown that the **Format** menu actually uses options.

When the code is open, we also have the possibility to modify it. This is the third method of modification mentioned earlier, "Use options directly." We can directly write new options, delete options, and change values of options. For example, we could have opened the code of the cell of "Here is some text.", written the option FontSize -> 14, and then closed the code. It may be that the more you use options and learn their names, the more you will like the direct modification of the code of the cells. In Section 3.2.3, p. 59, we list some useful options, many of which are not available with the **Format** menu.

■ The Structure of Expressions and Notebooks

It may be interesting to look at the code of a mathematical result. For example:

```
a = Sqrt[8] + Sin[Pi / 6]
```

$$\frac{1}{2} + 2\sqrt{2}$$

The code of the output cell is as follows (put the cursor in the result and press $\boxed{\text{SHIFT}}\boxed{\text{CTRL}}\boxed{\text{e}}$):

```
Cell[BoxData[
    RowBox[{
        FractionBox["1", "2"], "+",
        RowBox[{"2", " ",
            SqrtBox["2"]}]}]], "Output",
    CellLabel->"Out[1]="]
```

We see that the formula is built up from various box constructs.

A whole notebook consists of a list of cells, together with possible options. The general form of a notebook is thus `Notebook[{cell1, cell2, ...}, options]`. In this way, *Mathematica* represents cells and notebooks as text expressions containing only 7-bit ASCII characters. From this, it follows that notebooks work independently of the platform they are opened with and thus can be used unchanged with any computer system. We shall not go into the details of cell and notebook expressions here; see Sections 2.9 and 2.11 in Wolfram (2003). We only study the use of the front-end options.

3.2.2 Using the Option Inspector

■ Setting Values of Options

The *Option Inspector* is a special window where we can view and modify the options of the front end. The window can be opened by choosing **Format ▷ Option Inspector...**, and it looks like this (here we have opened some groups of options):

The options are grouped into seven categories: Global, Notebook, Cell, Editing, Formatting, Graphics, and Button. There are a total of more than 500 options, but it may be relaxing to know that most of us will never need the majority of them. Clicking the triangle before a category opens the category into a list of subcategories, and there we can find the options and their current values. This categorical listing is the default way that the *Option Inspector* shows the options.

To use the *Option Inspector* to modify your document, do the following:

> - Select the part of your document that you want to modify with options (typically the part is a cell, but it can also be a part of a cell or several cells).
> - Open the *Option Inspector* by choosing **Format ▷ Option Inspector...** or pressing SHIFT CTRL o (SHIFT ⌘ o on a Macintosh).
> - Set the values of the options you want. The values of many options can be set by checkboxes, pop-up menus, or dialog boxes, but if you type the value, you have to *press the return key* RET after typing for your setting to go into effect.
> - Go back to your document (you can leave the *Option Inspector* open, in case you need it again).

An option for which we have set a nondefault value has a bullet or another symbol before it. Clicking the symbol changes the value of the option to the default value and removes the symbol.

After experimenting with some options of a cell, we may find that we would like to return to the default option values for this cell. Select the cell bracket and choose **Format ▷ Remove Options...**.

The *Option Inspector* can also be accessed from **Edit ▷ Preferences...** (**Mathematica ▷ Preferences...** in MacOs X).

■ More about the Option Inspector

At the top of the window, it reads *Show option values for,* and then we have some choices:

Show option values for	global	by category
	notebook	alphabetically
	selection	as text

As can be seen, the options can also be seen alphabetically and as text (the last way shows the nondefault options as text like, "FontSize→12"). A key property of *Option Inspector* is that it allows us to specify the level at which we want to set the value of an option. Possible levels are *global, notebook,* and *selection.*

> - Option values set at the *global* level affect the whole *Mathematica* application: the present session and all future sessions, all currently open notebooks, and all notebooks opened or created in the future.
> - Option values set at the *notebook* level only affect the current notebook.
> - Option values set at the *selection* level only affect the part of the current notebook that is selected with the mouse (the selected part may be a cell, several cells, or a part of a cell).

When working with *Option Inspector,* it is important to choose the suitable level; otherwise you can easily generate unwanted effects. Note also that some options cannot be set at all levels; options that cannot be set at the currently selected level are dimmed.

■ Finding Suitable Options

Finding suitable options from among the more than 500 available may be a problem, but we have some help. The categorical listing may help you with finding appropriate options (many but not all font options are in **Formatting Options** ▷ **Font Options**). The alphabetical listing may also be helpful (many font options probably begin with **Font**...).

The inspector window also has a **Lookup** field where we can type a word or a few words describing the options we are looking for. After doing this, clicking the **Lookup** button repeatedly or pressing the ⌷RET⌷ key flashes each match in turn. For example, if we are interested in font options, we can type the word "font" in the **Lookup** field and then press ⌷RET⌷ until we find a suitable option.

Note that all of the front-end options are explained in the *Help Browser:* go to **Front End** ▷ **Front End Options** (**Other Information** ▷ **Front End Options** in version 4). Many of the options are also explained in Sections 2.11.8 to 2.11.14 of Wolfram (2003). In the next section, we give some lists of front-end options that may be useful when you are writing documents with *Mathematica*. The lists contain some 40 options, and elsewhere in the book we explain about 20 additional options.

3.2.3 Useful Options

■ Font Options

Most options concerning fonts can be set with the **Format** menu. These options can also be found from the *Option Inspector*, and they can be written directly into the code of the cells.

> **FontFamily** Examples of values: `"Times"`, `"Arial"`, `"Courier"`
> **FontSize** Examples of values: `12`, `10`, `9`
> **FontWeight** Examples of values: `"Plain"`, `"Bold"`
> **FontSlant** Examples of values: `"Plain"`, `"Italic"`
> **FontTracking** Examples of values: `"Plain"`, `"Condensed"`, `"Extended"`
> **FontColor** Examples of values: `Automatic`, `RGBColor[1,0,0]`
> **Background** Examples of values: `Automatic`, `RGBColor[0,0,1]`
> **Magnification** Examples of values: `1`, `0.75`, `1.5`
> **FontVariations** An example: `{"Underline" → True, "Outline" → True}`; possible variations: `"Underline"`, `"Outline"`, `"Shadow"`, `"StrikeThrough"`, `"Masked"`, `"CompatibilityType"`, and `"RotationAngle"`; each can be set to `False` or `True`, except that `"CompatibilityType"` can be set to `"Normal"`, `"Superscript"`, or `"Subscript"`, and `"RotationAngle"` (available only in Windows) to an angle

For color definitions, see Section 5.4.7, p. 144. In the following example, we have used several font options (to see the options, open the code of the cell in the *Help Browser* version of the book):

An example of font options.

■ **Cell Options**

The look of a document is greatly affected by various margins and spacings; we will present lists of options that can be used to adjust these. The value of many of these options are given in *printer's points*. One printer point is about 1/72 of an inch (this unit is also used with **AbsolutePointSize** and **AbsoluteThickness**, see Sections 5.4.3, p. 135, and 5.4.4, p. 136). The value of some options are given in *ems* or *x-heights*. An em is about the width of an m, and x-height is the height of an x. The values of many options may be given in the form **{{left, right}, {bottom, top}}**, meaning that numerical values are given that control the left, right, bottom, and top parts. This value is written below in the short form **{{l, r}, {b, t}}**.

CellMargins   Margins, in printer's points, around a cell; value is of the form **{{l, r}, {b, t}}**, where each number is the size of a part of the margin (note that left and right margins can be set with the ruler, available from **Format ▷ Show Ruler**; its unit can be set with **RulerUnits**)

CellDingbat   Dingbat to be used to emphasize a cell; examples of values: **None**, "**■**", "**\[FilledSmallSquare]**"

Background   Background color of a cell; examples of values: **Automatic**, **RGBColor[1,0,0]**

Magnification   Magnification factor for the cell; examples of values: **1**, **0.75**, **1.5**

CellMargins determines, in addition to the white space to the left and right of the content of the cell, the white space between cells. Usually each paragraph of text is written into its own cell, and then the cell margins determine the spacing between paragraphs.

One application of **Magnification** may be as follows. When using magnification values of more than 100% from **Format ▷ Magnification**, all text remains in the window, but it may be that graphics can no longer be seen completely. Setting **Magnification → 1** keeps the size of graphics the same, irrespective of the magnification percentage used for the notebook.

■ **Text Layout Options**

LineSpacing   The spacing between successive lines of text; value is of the form **{c, n}**, meaning that the space is **c** times the height of the contents of the line plus **n** printer's points

ParagraphSpacing   The extra space between two paragraphs (a new paragraph begins after each explicit ⏎ character); value is of the form **{c, n}**, meaning that the extra space is **c** times the height of the font plus **n** printer's points

ParagraphIndent   The indentation, in printer's points, of the first line of a paragraph (a new paragraph starts at the beginning of a cell and after each explicit ⏎); a negative value causes all but the first line be indented

TabSpacings   The number of spaces, in ems, that the cursor advances (at most) when ⎇TAB is pressed in a text cell; examples of values: **4**, **{10, 15, 12, 7}**

TabFilling   Determines how a ⎇TAB is represented; examples of values: **None**, "**.**", **Underline**, **GrayUnderline**

LineSpacing and **CellMargins**, together with fonts and their sizes, are important options that affect the overall look of the pages of your document. Usually each paragraph is in its own cell, and then **CellMargins** determines the spacing between paragraphs. However, if you write several paragraphs into the same cell, then **ParagraphSpacing** determines the space between the paragraphs in such cells. With **LineSpacing**, either **c** or **n** can also be zero, and then the spacing is determined only with the height of the contents or only with the given printer's points. A typical value of this option is **{1, 3}**.

With a negative **ParagraphIndent**, we get paragraphs like this one. Usually we need both indented and nonindented paragraphs. A normal paragraph is indented, but the first paragraph in a section is nonindented. Also, if the same paragraph continues after a formula, we also need a nonindented paragraph. One way to use both indented and nonindented paragraphs is to let a text cell have a zero indentation and to press TAB if an indentation is needed; we can give **TabSpacings** a suitable value.

If one value, say, m, is given for **TabSpacings**, then the width of the space between two tab stops is m; that is, tab stops are at positions m, $2\,m$, $3\,m$, etc. However, if the value is a list like $\{m, n, k\}$, then the widths of the spaces between tab stops are m, n, k, k, k, etc. The default is that a tab is represented as white space, but we can use any character, like a period " **.** " and also the values **Underline** and **GrayUnderline**.

TextAlignment Examples of values: **Left**, **Right**, **Center**

TextJustification Examples of values: **0** (natural spacing) **0.5**, **1** (full justification)

PageWidth Examples of values: **WindowWidth**, **PaperWidth**, ∞

Hyphenation Possible values: **False**, **True**

HyphenationOptions → {"HyphenationMinLengths" → {m, n}} A minimum of **m** and n characters can be split off the start and end of a word, respectively; default value: **{3, 3}** (this option is most easily set with the *Help Browser*)

■ Cell Frame Options

CellFrame Whether a frame is drawn around a cell; value may be **False**, **True**, **f**, or **{{l, r}, {b, t}}**, where each number is the absolute thickness of the frame or a part of it in printer's points (the value **True** implies the value 0.25)

CellFrameMargins Margins, in printer's points, inside a frame; value may be **m** or **{{l, r}, {b, t}}** (typical default value is 8)

CellFrameColor Color of the frame; examples of values: **GrayLevel[0]**, **RGBColor[1,0,0]**

CellFrameLabels Labels of the frame; value is of the form **{{left text, right text}, {bottom text, top text}}** (the style of frame labels can be adjusted by using the style sheet)

CellFrameLabelMargins Margins, in printer's points, between a cell frame and the labels; value may be **m** or **{{l, r}, {b, t}}** (typical default value is 6)

With **CellFrame**, together with **CellFrameMargins**, we can form various frames, like the following ones:

CellFrame → True

CellFrame → {{0, 0}, {0, 0.25}}

CellFrame → {{0, 0}, {0.25, 0}}

CellFrame → {{5, 0}, {0, 1}}

With **Format ▷ Background Color ▷ Cell Grey Box**, we get a result like this:

CellFrame → True, Background → GrayLevel[0.85]

In the following example, we have cell frame labels:

top

left ⎪ A text with a frame and frame labels. ⎪ right

bottom

■ Frame Box Options

FrameBoxOptions → {BoxFrame → value, BoxMargins → value} Options controlling the frame around a box (to get a frame, use **Edit ▷ Expression Input ▷ Add Frame**):

BoxFrame Whether to draw lines around a frame box. Value may be **True**, **False**, **f**, or **{{l, r}, {b, t}}**, where each number is the thickness of the frame or a part of it (the value **True** implies the value 1; each numerical value implies an absolute thickness of about 0.8 times the value). The color of the frame is determined by the color of the font.

BoxMargins Margins between the contents of a grid box and the surrounding frame. Left and right margins are given in ems, bottom and top margins in x-heights. Value may be **True**, **False**, **m**, or **{{l, r}, {b, t}}** (the default value is about {{0.25, 0.25}, {0.35, 0.35}}, and the value **True** implies the value {{1, 1}, {1, 1}}).

If we want a frame around a smaller part than a cell, then select that part and choose **Edit ▷ Expression Input ▷ Add Frame**. An example is $\boxed{\sin(x)^2 + \cos(x)^2 = 1}$. As another example, we emphasize a display formula:

$$\boxed{\lim_{x \to \infty} \left(1 + \frac{a}{x}\right)^x = e^a}$$

Such frames can be controlled with the two **FrameBoxOptions**. With **BoxFrame**, we can set the thickness of the frame, and with **BoxMargins**, we set the margins inside the frame (these options can be set by the *Option Inspector* or by opening the code of the cell).

■ Printing Options

> **PageBreakAbove** Whether a page break should be made above a cell; possible values: **Automatic**, **True**, **False**
>
> **PageBreakBelow** Whether a page break should be made below a cell; possible values: **Automatic**, **True**, **False**
>
> **PageBreakWithin** Whether a page break should be allowed within a cell; possible values: **Automatic**, **True**, **False**
>
> **GroupPageBreakWithin** Whether a page break should be allowed within a group of cells; possible values: **Automatic**, **True**, **False**
>
> **PrintingOptions → {"PageHeaderMargins" → {left, right}}** Vertical margins, in printer's points, above the header of the left and right facing pages, respectively (recall that the margins of the pages can be set with **File ▷ Printing Settings ▷ Printing Options...**)

■ Miscellaneous Options

Additional useful options are now described. With **StartupSound**, we can turn off the startup sound of *Mathematica*. **InputAutoReplacements** tells the sequences of characters that are automatically replaced with other characters. For example, in input cells, -> is automatically replaced with →. With **DragAndDrop**, we can turn on the property that allows us to drag a selection to a new location with the mouse. **RulerUnits** determines the units in the ruler toolbar.

Moreover, several options are explained elsewhere in the book: **NotebookAutoSave** (Section 3.1.1, p. 48), **Hyphenation** (Section 3.1.1, p. 49), **SpellingOptions** (Section 3.1.1, p. 49), **ScriptSizeMultipliers** and **ScriptMinSize** (Section 3.3.1, p. 67; Section 3.4.2, p. 78), **EvaluationCompletionAction** (Section 4.4.1, p. 103), **ImageSize** (Section 5.1.3, p. 116), **"GraphicsPrintingFormat"** (Section 6.3.2, p. 161), **PrintPrecision** (Section 11.1.2, p. 282), and **RowAlignments**, **ColumnAlignments**, etc. (Section 13.1.3, p. 331).

3.2.4 Modifying Style Sheets

As noted in Section 3.2.1, p. 55, notebooks should be edited mainly with style sheets, and they should be directly edited only in exceptional cases. In this section, we show how to modify style sheets.

An important thing to know is that there are two types of style sheets: *private* and *shared*. A private style sheet is used only for a particular notebook; a shared style sheet can be used for many notebooks. All the shared style sheets can be seen from **Format ▷ Style Sheet**.

Choose a shared style sheet that is as near to your needs as possible. If such a style sheet does not fully satisfy you, you may want to make some modifications to the styles. You have three possibilities (they are discussed below):

- create (import) a *private* copy of the shared style sheet and modify the private copy;
- modify the *shared* style sheet; or
- create a *new* style sheet.

■ **Editing Existing Style Sheets and Cell Styles**

Choose **Edit Style Sheet...** from the **Format** menu. You are given two possibilities: **Import Private Copy** or **Edit Shared Style Sheet**.

If you choose **Import Private Copy**, a private copy of the shared style sheet is created for just the current notebook, and you will then change the styles of this private copy. With this method, the changes of the styles affect only the current notebook. This is the normal way to modify styles of a notebook.

If you choose **Edit Shared Style Sheet**, you will change the shared style sheet that you are using for your current notebook. Your modifications will affect all old, current, and future notebooks based on this style sheet, and you may not want this happen. Indeed, a better method may be to leave the standard shared style sheets untouched and to create a new style sheet. This is explained in the next section.

Once you have made your choice (import a private copy of the style sheet or edit the shared style sheet), a style sheet appears, a portion of which is shown here:

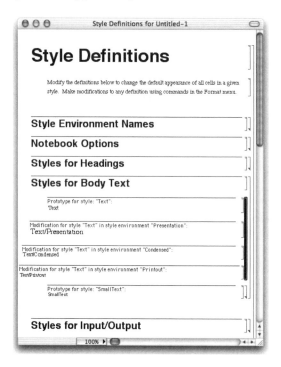

In the style sheet, we can see *style definition cells* with titles such as **Styles for Headings** and **Styles for Body Text**. When we open a closed group of cells by double-clicking the cell bracket, a more detailed view appears. Generally there are style definitions for each of the four normal screen and printing style environments (Working, Presentation, Condensed, and Printout) we considered in Section 3.1.2, p. 52.

Choose the environment that you want to change by selecting its cell bracket. Now you can use all three editing methods mentioned in Section 3.2.1, p. 55, to modify the style of the selected cell in the style sheet. The cells in your document using this style change accordingly.

The main modification method is to use the **Format** menu (see Section 3.2.1, p. 56). Another method is the use of the *Option Inspector* (Sections 3.2.2, p. 57, and 3.2.3, p. 59). A third method (Section 3.2.1, p. 56) is to open the cell code of the selected cell by ⌷SHIFT⌷CTRL⌷e⌷, modify the code, and then close the code with ⌷SHIFT⌷CTRL⌷e⌷.

After your modifications, close the style sheet. Now whatever you have written in your notebook and whatever you write later on will follow the modified styles of the style sheet. If you edited the shared style sheet, save it before closing it. All notebooks based on this shared style sheet will follow the modified styles.

■ Creating New Style Sheets and Cell Styles

To create a new style sheet, open a new notebook, choose a style sheet from **Format ▷ Style Sheet** that is close to the style sheet you intend to create, choose **Format ▷ Edit Style Sheet...**, select **Import Private Copy**, make your changes to the style sheet, choose **File ▷ Save As...**, give a suitable name to the style sheet (it has to end with **.nb**), and save the new style sheet to **SystemFiles ▷ FrontEnd ▷ StyleSheets** in **$UserBaseDirectory** (in **$UserAddOnsDirectory** in *Mathematica* 4). The new style sheet then appears in **Format ▷ Style Sheet** (after restarting *Mathematica*).

To create a new cell style for a style sheet, choose the style sheet from **Format ▷ Style Sheet**, choose **Format ▷ Edit Style Sheet...**, select **Import Private Copy** or **Edit Shared Style Sheet**, copy (from the present style sheet or from another style sheet) a style definition cell that is close to the cell style you intend to create, paste the cell, select the new cell, choose **Format ▷ Show Expression**, write a new representative name into Style Data["name"], choose **Format ▷ Show Expression** again, change the style of the cell (using, for example, the **Format** menu), and close the style sheet. The new style then appears in **Format ▷ Style** (after restarting *Mathematica*).

Note that you can copy cell styles from one style sheet to another. For example, if your style sheet does not have the **NumberedEquation** style, choose a style sheet that has this style, open the style sheet, copy the style definition cell, and paste it into your style sheet. Note also that if you find yourself now and then manually modifying a new cell of your document, this may be an indication that you should define a new cell style to your style sheet.

3.3 Inputs and Outputs

3.3.1 Forms of Input and Output

Here in Section 3.3, we show that *Mathematica* can give the results of computations in various forms. Likewise, *Mathematica* accepts inputs of various forms. In particular, we can write inputs in a linear or 1D form (**a/b**), but we can also give the inputs in a 2D form ($\frac{a}{b}$).

2D inputs and special characters can be given with the help of palettes, but they can also be directly entered with the keyboard. The techniques of this section may be useful when calculating with *Mathematica*, but they may even be more useful when writing mathematical documents with *Mathematica*.

■ Forms of Output

Mathematica can show results in several forms. The forms can be seen by choosing **Default Output FormatType** from the **Cell** menu; these forms are **InputForm**, **OutputForm**, **StandardForm**, and **TraditionalForm**. The default form is **StandardForm**. As an example, we calculate an integral and show the result with all four format types:

```
Integrate[Sqrt[x] / (a^2 + b^2 x), x] // InputForm

(2*Sqrt[x])/b^2 - (2*a*ArcTan[(b*Sqrt[x])/a])/b^3

Integrate[Sqrt[x] / (a^2 + b^2 x), x] // OutputForm
```

$$\frac{2\ \mathrm{Sqrt}[x]}{b^2} - \frac{2\ a\ \mathrm{ArcTan}[\dfrac{b\ \mathrm{Sqrt}[x]}{a}]}{b^3}$$

```
Integrate[Sqrt[x] / (a^2 + b^2 x), x] // StandardForm
```

$$\frac{2\sqrt{x}}{b^2} - \frac{2\,a\,\mathrm{ArcTan}\left[\frac{b\sqrt{x}}{a}\right]}{b^3}$$

```
Integrate[Sqrt[x] / (a^2 + b^2 x), x] // TraditionalForm
```

$$\frac{2\sqrt{x}}{b^2} - \frac{2\,a\,\tan^{-1}\left(\frac{b\sqrt{x}}{a}\right)}{b^3}$$

InputForm is a linear or 1D form that uses only standard characters. **OutputForm** is a 2D form that also uses only standard characters. Both of these forms are only seldomly used. One application of the input form is to ask for all of the 16 decimals that *Mathematica* uses in its internal calculations:

```
π // N      3.14159
```

```
% // InputForm      3.141592653589793
```

StandardForm is a 2D form and uses special characters like $\sqrt{\ }$, special spacings, and special character sizes. This is the default form of output (so writing **//StandardForm** was unnecessary).

TraditionalForm imitates all aspects of traditional mathematical notation. For example, variables are in italics, arguments of functions are in parentheses ( ) and not in square brackets [ ], and other traditional notations like $\tan^{-1}$ (instead of ArcTan) are used. This form is mainly used in preparing mathematical documents.

■ Some Notes

Usually **StandardForm** is the best form of output. Results are then in neat 2D forms that can also be used for input. This means that you can copy a 2D output or a part of it and use it as a 2D input. For example, we can copy the second part of the standard form output above and take its derivative:

$$\mathbf{D}\!\left[\frac{2\,\mathbf{a}\,\mathbf{ArcTan}\left[\frac{b\sqrt{x}}{a}\right]}{\mathbf{b^3}},\ \mathbf{x}\right] \qquad \frac{1}{b^2\,\sqrt{x}\,\left(1 + \frac{b^2\,x}{a^2}\right)}$$

In Section 3.3.3, p. 70, you will learn how you can write 2D inputs yourself.

The form of an output can be set with commands like **TraditionalForm** (as we did above), but we can also select the cell of the output and then use **Cell ▷ Convert To ▷ TraditionalForm**.

There are many options for controlling 2D formulas. Select a formula, choose **Format ▷ Option Inspector…**, and go to **Formatting Options**. The **Expression Formatting** section in particular may contain options that enable you to put the formula in the form you like. For example, *Mathematica* reduces the size of higher-level subscripts and superscripts by multiplying the previous sizes by numbers in **ScriptSizeMultipliers** (default is 0.71) until size 5 is obtained (this holds true for the default printing style environment; in the default screen style environment, the minimum size is 9). If you want to stop at a larger point size, go to **Expression Formatting ▷ Display Options** and change the size of **ScriptMinSize**.

■ Forms of Input

From **Cell ▷ Default Input FormatType**, we see that inputs can be written in three different forms: **InputForm**, **StandardForm**, and **TraditionalForm**. The default is again **StandardForm**, and other forms are seldom used. If the default input format type is **StandardForm**, we can write inputs both in a 1D and in a 2D form. For example:

```
Integrate[7 / 2 (x^2 + a Pi / x) Sqrt[x], {x, 0, 1}]
```

$1 + 7\,a\,\pi$

$$\int_0^1 \frac{7}{2}\left(x^2 + \frac{a\,\pi}{x}\right)\sqrt{x}\ \mathrm{d}x$$

$1 + 7\,a\,\pi$

In the latter form, we used the **BasicInput** palette; see Section 1.4.1, p. 17, about the use of palettes to write 2D inputs. In Section 3.3.3, p. 70, we learn how to write 2D inputs without palettes using special key combinations.

Choosing between the 1D and 2D input forms may mainly be a matter of habit. I personally use 1D inputs, because they are so straightforward to write with the keyboard and because I then do not need the mouse, the palettes, or special key combinations. However, 2D inputs are very clear and illustrative, as they resemble traditional mathematical notation. Using palettes to form inputs may be handy at least in the early stages of learning the use of *Mathematica*, because the user then does not need to remember the form of, say, an integrating command. A compromise may be to write in 2D form only the parts of the input that can easily be written with the keyboard:

$$\mathtt{Integrate}\left[\frac{7}{2}\left(x^2 + \frac{a\,\pi}{x}\right)\sqrt{x},\ \{x,\ 0,\ 1\}\right]$$

$1 + 7\,a\,\pi$

Still another way is to write a 1D input and then convert it to 2D form with **Cell ▷ Convert To ▷ StandardForm**.

In Sections 3.3.2 and 3.3.3, we learn how to write special characters and 2D inputs.

3.3.2 Writing Special Characters

Mathematica has a large set of special characters like π, ∞, $\in$, and $\sum$. To write these characters, we have several possibilities. As an example we consider various ways to write ∞.

1. In an input, we can write **Infinity**.
2. From the keyboard, we can possibly find ∞ (on a Macintosh, press ⟨OPTION⟩⟨5⟩).
3. All characters have a *full name*; for ∞, it is \[Infinity]. After we have typed], the sequence of characters transforms to ∞.
4. Many characters also have an *alias*. For ∞, it is ⟨ESC⟩inf⟨ESC⟩ (⟨ESC⟩ means the escape key; it appears as ∶ on the screen). After pressing the second ⟨ESC⟩, the sequence of characters transforms to ∞. This is very handy way to type special characters, although we have to remember the aliases.
5. Many characters can be found in the **BasicInput** palette; more are found in the **BasicTypesetting** palette, the **InternationalCharacters** palette, and the **Complete-Characters** palette. At the bottom of the **CompleteCharacters** palette, we can see both the alias and the full name of a character over which the mouse pointer is hovering. In this way we can learn the aliases and full names so that we possibly more rarely need to use the palette.

In addition, characters can be typed using TeX aliases like ⟨ESC⟩\infty⟨ESC⟩ or HTML and SGML aliases.

During the early stages of learning to use *Mathematica*, it may be handy to use the palettes. In time, it may be more useful to learn the aliases of the characters you need often. When we next study in some detail the special characters of *Mathematica*, we will give the aliases of the characters.

■ Greek and Other Letters

char.	α	β	γ	δ	ϵ	ε	ζ	η	θ	ϑ	ι	κ	χ	λ	μ	ν	ξ	o	π	ϖ	ρ	ϱ	σ	τ	υ	ϕ	φ	χ	ψ	ω
alias	a	b	g	d	e	ce	z	h	q	cq	i	k	ck	l	m	n	x	om	p	cp	r	crs	s	t	u	f	j	c	y	o
char.	A	B	Γ	Δ	E		Z	H	Θ		I	K		Λ	M	N	Ξ	O	Π		P		Σ	T	Υ	Φ		X	Ψ	Ω
alias	A	B	G	D	E		Z	H	Q		I	K		L	M	N	X	Om	P		R		S	T	U	cU	F		C	Y	O

The second and fourth rows tell how the alias of a Greek letter is formed: for example, to write α, press ⟨ESC⟩a⟨ESC⟩. In general, these aliases are easy to remember: the alias is formed by the corresponding usual letter (exceptions are q for θ and y for ψ, among others). The "c" letter in some aliases comes from "curly." If you do not remember this kind of short alias, you can use a longer alias containing the whole name of the Greek letter, for example, ⟨ESC⟩alpha⟨ESC⟩ or ⟨ESC⟩Gamma⟨ESC⟩.

In addition to Greek letters, *Mathematica* has script letters like a, b, c, d, e, f, g, $\mathcal{A}$, $\mathcal{B}$, $\mathcal{C}$, $\mathcal{D}$, $\mathcal{E}$, $\mathcal{F}$, $\mathcal{G}$, Gothic letters like $\mathfrak{a}$, $\mathfrak{b}$, $\mathfrak{c}$, $\mathfrak{d}$, $\mathfrak{e}$, $\mathfrak{f}$, $\mathfrak{g}$, $\mathfrak{A}$, $\mathfrak{B}$, $\mathfrak{C}$, $\mathfrak{D}$, $\mathfrak{E}$, $\mathfrak{F}$, $\mathfrak{G}$, and double-struck letters like $\mathbb{a}$, $\mathbb{b}$, $\mathbb{c}$, $\mathbb{d}$, $\mathbb{e}$, $\mathbb{f}$, $\mathbb{g}$, $\mathbb{A}$, $\mathbb{B}$, $\mathbb{C}$, $\mathbb{D}$, $\mathbb{E}$, $\mathbb{F}$, $\mathbb{G}$. Their aliases are formed by enclosing, for example, sca, scA, goa, goA, dsa, or dsA with ⟨ESC⟩ keystrokes.

In the following table, we have collected some special symbols:

char.	π	e	i	∞	$\mathbb{C}$	$\mathbb{R}$	$\mathbb{Q}$	$\mathbb{Z}$	
alias	p	ee	ii	inf	dsC	dsR	dsQ	dsZ	a.

To write π, you can type ESC p ESC. Note that, among the Greek k
meaning in *Mathematica*. $\mathbb{C}$, $\mathbb{R}$, $\mathbb{Q}$, $\mathbb{Z}$, and $\mathbb{N}$ are often used to denote
rational, integer, and natural numbers, respectively. In outputs, *Math* has a special
four symbols given in the table for Pi (= 3.141), E (= 2.718), I (= $\sqrt{-1}$... mplex, real,
respectively:
finity,

    {Pi, E, I, Infinity}   {π, e, i, ∞}

■ Mathematical Symbols

char.	$\pm$	$\mp$	$\times$	$\times$	$\div$	$\sqrt{}$	∂	∇	d	$\int$	$\oint$	$\sum$	$\prod$
alias	+-	-+	*	cross	div	sqrt	pd	del	dd	int	cint	sum	prod
char.	$\in$	$\notin$	$\subset$	$\subseteq$	$\cup$	$\cap$	$\emptyset$	$\equiv$	$\approx$	$\simeq$	$\cong$	$\neq$	$\propto$
alias	elem	!elem	sub	sub=	un	inter	es	===	~~	~=	~==	!=	prop
char.	$\Longrightarrow$	$\Longleftrightarrow$	$\rightarrow$	$\longrightarrow$	$\longleftrightarrow$	$\forall$	$\exists$	$\nexists$	$\wedge$	$\vee$	$\neg$	$\langle$	$\rangle$
alias	=>	<=>	->	--->	<->	fa	ex	!ex	and	or	not	<	>

To write $\pm$, type ESC +- ESC (note that the arrows are normally not so long as those shown
in the table: $\Rightarrow$, $\Leftrightarrow$). The symbol $\times$ means multiplication, $\times$ the cross product of two vectors,
∂ partial differentiation, ∇ a gradient or a backward difference, d the differential in an
integral, $\oint$ a contour integral, and $\emptyset$ an empty set.

Among the symbols in the table, many have special mathematical meanings in *Mathematica*. Such symbols include $\times$, $\times$, $\div$, $\sqrt{}$, ∂, $\int$, $\sum$, $\prod$, $\in$, $\cup$, $\cap$, $\wedge$, $\vee$, and $\neg$. For example:

$$\{2 \times 4, \ \{1, 2, 3\} \times \{a, b, c\}, \ 2 \div 4, \ \sqrt{4}, \ \partial_x \text{Sin}[x],$$

$$\int \text{Sin}[x] \, dx, \ \sum_{i=1}^{4} i, \ \{a, b, c\} \cap \{a, c, d\}, \ \neg \ (\text{True} \wedge \text{False})\}$$

$$\{8, \ \{-3b + 2c, \ 3a - c, \ -2a + b\}, \ \frac{1}{2}, \ 2, \ \text{Cos}[x], \ -\text{Cos}[x], \ 10, \ \{a, c\}, \ \text{True}\}$$

■ Characters for Fine Tuning

\[InvisibleSpace]	ESC is ESC	\[InvisibleComma]	ESC , ESC
\[VeryThinSpace]	ESC _ ESC	\[NonBreakingSpace]	ESC nbs ESC
\[ThinSpace]	ESC __ ESC	\[NoBreak]	ESC nb ESC
\[MediumSpace]	ESC ___ ESC	\[IndentingNewLine]	ESC nl ESC
\[ThickSpace]	ESC ____ ESC	\[AlignmentMarker]	ESC am ESC

Mathematica has some special characters that are useful for fine tuning text and expressions. One of them is an invisible comma. An example is x_{ij}, where we used an invisible

ve write *i* and *j* one after another, the indices are not italicized:
between *i* and *j*, the indices may be too far from each other: x_{ij}.
racter is an invisible space. An example is xy, where we used an
x_{ij}, and if en *x* and *y*. Such an xy may be used as the product of *x* and *y* if we
etween *x* and *y* is not aesthetically what we want: $x\ y$.
spaces of various widths; note that ‿ means the space key. Not shown in
are the corresponding negative spaces from \[NegativeVeryThinSpace] to
the hickSpace], which allow us to move characters nearer to each other. We also
\[ace that does not allow a break (that is, the words on both sides of the nonbreak-
e will always be on the same text line) and a character that does not allow a break.
that the special spaces are of fixed width and thus do not stretch like a normal space.

An indenting character inserts a line break while always maintaining the correct indent-
ng level (a usual new line character (↵) sets the indenting level at the time the new line is
started). With an alignment marker, we can tell how to align rows, for example, in a
mathematical formula containing several rows (see Example 4 in Section 3.4.2, p. 76).

■ Keys on a Keyboard

char.	SPACE	‿	TAB	↵	RET	ENTER	SHIFT	CTRL	ALT	⋮	ESC	CMD	⌘	OPTION	DEL	{	]	∎
alias	spc	space	tab	ret	ret	ent	sh	ctrl	alt	esc	esc	cmd	cl	opt	del	[	]	kb

These characters are useful in describing the use of the keyboard.

3.3.3 Writing 2D Expressions

Recall that we can write inputs in a 2D form by using palettes. Now we learn how to write
such inputs with the keyboard. These techniques are useful also when writing formulas in
a mathematical document (see Section 3.4).

2D expressions are written with the control key. Here is a list of mathematical construc-
tions made in that manner (these can also be found by choosing **Expression Input** from
the **Edit** menu, and they are explained in the *Help Browser* under **Front End ▷ 2D Expres-
sion Input**).

CTRL /	Go to the numerator of a *fraction* to be written or to the denominator of a fraction whose numerator is already written
CTRL @ or CTRL 2	Go into a *square root*
CTRL ^ or CTRL 6	Go to the *superscript* position (e.g., of a power)
CTRL _ or CTRL −	Go to the *subscript* position (e.g., of a variable or derivative)
CTRL & or CTRL 7	Go to the *overscript* position (e.g., of a sum, product, or integral)
CTRL + or CTRL =	Go to the *underscript* position (e.g., of a sum, product, integral, or limit)
CTRL % or CTRL 5	Go from superscript position to subscript position or vice versa, or from overscript position to underscript position or vice versa, or from the root position to the exponent position or vice versa

[CTRL][,]	Add a *column* to a table or matrix
[CTRL][↵]	Add a *row* to a table or matrix
[CTRL][␣]	*Return* from a special position
[CTRL][.]	*Select* the next larger subexpression

The symbols [CTRL][/] mean that we hold down the [CTRL] key while pressing /. The symbol ↵ means the return key, and ␣ means the space key. Some constructions can be made in two ways. The first way mentioned should be possible with most keyboards, while the second should work with any keyboard. If both work with your keyboard, choose the way you feel more comfortable with or that is easier for you to remember.

In formulas, we can move the cursor with the arrow keys (←, →, ↑, and ↓). To go to the next selection placeholder (■), press [TAB].

To form derivatives, integrals, sums, and products, we need the characters ∂, $\int$, d, Σ, and $\prod$ (d is a special differential used in integrals). They can be written by enclosing pd, int, dd, sum, and prod between two [ESC] keystrokes (see Section 3.3.2, p. 69).

Use [CTRL][␣] to get out of the denominator of a fraction, to get out of a square root, to get down from a power, to get to the baseline from an index, and so on.

■ Examples

Here we have some simple examples of writing 2D formulas with the keyboard:

To write	in the form	type
a / b	$\frac{a}{b}$	a [CTRL][/] b [CTRL][␣]
Sqrt[a]	$\sqrt{a}$	[CTRL][2] a [CTRL][␣]
x^2	x^2	x [CTRL][^] 2 [CTRL][␣]
x[n]	x_n	x [CTRL][␣] n [CTRL][␣]
xbar	$\bar{x}$	x [CTRL][&] ␣ [CTRL][␣]

Next we show some more elaborate constructs:

To write	in the form	type
D[a + b x, x]	$\partial_x\,(a + b\,x)$	[ESC]pd[ESC] [CTRL][␣] x [CTRL][␣] (a + b x)
Integrate[a + b x, {x, c, d}]	$\int_c^d (a + b\,x)\,\mathrm{d}x$	[ESC]int[ESC] [CTRL][+] c [CTRL][%] d [CTRL][␣] (a + b x) [ESC]dd[ESC] x
Sum[x^i, {i, 0, n}]	$\sum_{i=0}^{n} x^i$	[ESC]sum[ESC] [CTRL][+] i = 0 [CTRL][%] n [CTRL][␣] x [CTRL][^] i [CTRL][␣]
{{a, b, c}, {d, e, f}}	$\begin{pmatrix} a & b & c \\ d & e & f \end{pmatrix}$	(a [CTRL][,] b [CTRL][,] c [CTRL][↵] d [TAB] e [TAB] f [CTRL][␣])

Note that parentheses are needed around the expression to be differentiated, integrated, or summed if the expression is in the form of a sum.

Writing 2D formulas is easier than it seems to be when looking at the ⌈CTRL⌉ and ⌈ESC⌉ key sequences above. Indeed, *Mathematica*'s way of writing formulas is one of the best and easiest in the market, if not **the** best and easiest. Try it.

■ Some Notes

Remember that matrices of size at most 2×2 can be formed with palettes. A handy way to create matrices is to use **Input ▷ Create Table/Matrix/Palette…**. In the appearing dialog box, ask for a matrix, and then enter the number of rows and columns. We get a blank matrix:

$$\begin{pmatrix} \square & \square & \square \\ \square & \square & \square \end{pmatrix}$$

By pressing ⌈TAB⌉, we can then go through the matrix and enter the elements. If we ask for a table, the result does not have the parentheses.

2D input can also be typed by using only standard printable characters. For example, a fraction can be typed as follows:

$$\backslash!\backslash(a\backslash/b\backslash) \qquad \frac{a}{b}$$

If you want to see the result without executing the command, select the expression `\!\(a\/b\)` with the mouse and then choose **Make 2D** from the **Edit** menu. We will not explain this type of notation in more detail here; see Section 1.10.2 of Wolfram (2003).

3.4 Writing Mathematical Documents

3.4.1 Introduction

As noted in Section 3.1.2, p. 51, *Mathematica is an advanced environment for technical writing.* One of the advantages of using *Mathematica* as a writing tool is that the whole document can be prepared with just one application.

For example, graphics can be created with *Mathematica,* and *Mathematica* writes the results of mathematical computations in a form ready to be printed. Writing and computation take place in the same application so that they have a fruitful interaction; when writing the text, you may observe new needs for mathematical computation, and, from the results, you may observe new things to be reported.

In addition, the properties of *Mathematica* needed for using *Mathematica* as a writing environment are of high quality. For example, with *Mathematica* we get first-rate

- *text layout* by using styles and style sheets (Section 3.1.2, p. 51);
- *formulas* by using the traditional form (Sections 3.4.2, p. 74, and 3.4.3, p. 79);
- *graphics* of PostScript form (Chapters 5 through 10);
- *tables* with many kinds of fine tuning (Sections 13.1.2, p. 326, and 13.1.3, p. 330);
- *indexes* with hyperlinks (see Author Tools, p. 73, below); and
- *Help Browser material* with hyperlinks (see Author Tools, p. 73, below).

Furthermore, *Mathematica,* as one of the most powerful mathematical systems available, helps you with all kinds of computations needed to prepare your document. *Mathematica* documents can also be converted into TeX and HTML documents from **File ▷ Save As**

Special.... A disadvantage of *Mathematica* is that scientific journals prefer TeX documents and may not accept *Mathematica* documents.

While text layout, graphics, and tables are considered elsewhere, in Sections 3.4.2 through 3.4.4 we present some other material that is helpful to know when using *Mathematica* as a writing tool. In particular, we show how to write mathematical formulas and how to number formulas and sections automatically.

■ Selecting a Style Sheet

The key to writing with *Mathematica* is the use of *styles* (Section 3.1.2, p. 51). With styles, you can easily write titles, headings, text, and formulas. Styles help in getting a consistent look throughout the document.

Remember that the styles of cells vary according to the *style sheet* you use (see Section 3.1.2). You may start the writing work with the **Default** style sheet, but at some point you should consider the style sheet in more detail. In particular, if you want to write numbered formulas, note that only the style sheets **ArticleClassic**, **ArticleModern**, **Classic**, **Classroom**, **HelpBrowser**, and **Report** have the **NumberedEquation** style. You can also try the style sheet **MathematicaNavigator** which is provided on the CD-ROM that accompanies this book. After choosing a style sheet, you may want to modify it (see Section 3.2.4, p. 63).

■ Main and Working Documents

When writing a mathematical document with *Mathematica*, it may be useful to work simultaneously with two documents: a *main document* and a *working document*. The main document is aimed to grow into the final publication, while all computations are done in the working document. Mathematical results, tables, and graphics are copied from the working document into the main document. This division into two documents may be needed because the main document may not contain the *Mathematica* commands but only the results. The working document contains all used *Mathematica* commands so that all computations can easily be done again.

The working document should include the same sections as are in the main document so that you can easily find the computations of a certain section. Add into the working document comments about the computations, such as any difficulties that may arise; they may be valuable later on if you need to do similar computations again.

When you have completed the writing project, you will then have the main document ready to be printed and the working document that will enable you to redo and modify computations as needed.

■ Author Tools

In preparing a long, possibly book-sized document, and adding documents into the *Help Browser*, the add-on package **AuthorTools** may be very valuable. With the author tools, we can do the following:

- create a table of contents with hyperlinks;
- add index entries into the document;
- create an index with hyperlinks;
- create a Browser Categories file (for adding information into the *Help Browser*); and

- create a Browser Index file (to add the index entries into the Master Index of the *Help Browser*).

With this package, we can also compare differences among notebooks (✿5); fix corrupted notebooks (✿5); create bilaterally formatted cells (for displaying examples of *Mathematica* calculations); extract all cells of a particular type and save them in a desired format; insert objects to display the current values of variables such as the date, time, and the file name; and set printing options such as headers and footers.

Each of these operations can be done either on a single notebook or on a set of notebooks. For example, we can generate a unified index or table of contents for a book consisting of several notebooks. We used the package in the writing of this book to add the index entries, to create an index, and to create the Browser Categories and Browser Index files.

The author tools can be applied either by command lines or by using palettes. The palettes can be accessed from **File ▷ Palettes ▷ OpenAuthorTools**. Documentation of the package can be found with the *Help Browser* from **Add-ons & Links ▷ AuthorTools**. The documentation as a notebook **AuthorToolsGuide.nb** (about 50 pages) can be found from **Mathematica ▷ AddOns ▷ Applications ▷ AuthorTools ▷ Documentation ▷ English**. The CD-ROM included with this book also contains a guide to the package as the notebook **Using Author Tools.nb**.

3.4.2 Display Formulas

■ Writing Display Formulas

To write a *display formula*, which is a mathematical formula in a separate line, complete the following simple steps:

> - Write the formula in the form of a *Mathematica* command, bearing in mind that you can, if you want, execute any parts of the command.
> - Execute the parts of the command you want by selecting each part in turn with the mouse and then choosing **Kernel ▷ Evaluation ▷ Evaluate in Place** or by pressing [ALT]-[RET] (⌘-[RET] on a Macintosh).
> - Convert the cell into **Text** style by choosing **Format ▷ Style ▷ Text** or by pressing [ALT]-7 (⌘-7 on a Macintosh).
> - Convert the cell into **TraditionalForm** by choosing **Cell ▷ Convert To ▷ Traditional-Form** or by pressing [SHIFT]-[CTRL]-t ([SHIFT]-⌘-t on a Macintosh).

After this, the formula is displayed in a nice traditional form. Note that in the first step we can use any of the ways to write *Mathematica* commands: we can write usual commands like **Integrate[...]**, we can use palettes, and we can write 2D formulas directly with the keyboard (see Section 3.3.3, p. 70). In this book, instead of converting formulas into **Text** cells, we convert them into **DisplayEquation** cells. Writing an *inline formula* (that is, a mathematical formula among text) is considered in Section 3.4.3, p. 79.

■ Example 1

Suppose you want to write the following formula:

$$f_n = \int_0^\pi \frac{\sin(x)}{n+x}\,dx, \; n = 1, 2, \ldots$$

First, write the formula in the form of a *Mathematica* command:

```
Subscript[f, n] = Integrate[Sin[x] / (n + x), {x, 0, π}]
```

The lefthand side could also be written directly as f_n with a palette or a key combination (see Section 3.3.3, p. 70). We do not execute any parts of the formula. After converting the cell into **Text** form, we have the following:

Subscript[f,n]=Integrate[Sin[x]/(n+x),{x,0,π}]

After converting into **TraditionalForm**, the formula appears like this:

$$f_n = \int_0^\pi \frac{\sin(x)}{n+x}\,dx$$

Then write "$n = 1, 2, \ldots$" at the end. A final touch is to move the formula somewhat right by adding a tab or to center the formula by choosing **Format ▷ Text Alignment ▷ Align Center**. Also, insert a period at the end of the formula. (Note that here we showed the formula in three separate forms in three cells to show how the formula proceeds, but normally the conversions are done in only one cell. The formula in the command form may, however, be useful to save in the working document, in case you want to modify the formula later on.)

■ Example 2

Next we want to write the following formula:

$$f_n = \mathrm{Ci}(n)\sin(n) - \mathrm{Ci}(n+\pi)\sin(n) - \cos(n)\,\mathrm{Si}(n) + \cos(n)\,\mathrm{Si}(n+\pi).$$

Copy the command we wrote in Example 1:

```
Subscript[f, n] = Integrate[Sin[x] / (n + x), {x, 0, π}]
```

Execute the righthand side by selecting it with the mouse and pressing ALT-RET. Here is the result:

```
Subscript[f, n] = CosIntegral[n] Sin[n] - CosIntegral[n + π] Sin[n] -
    Cos[n] SinIntegral[n] + Cos[n] SinIntegral[n + π]
```

Then convert this into text and traditional form:

$$f_n = \mathrm{Ci}(n)\sin(n) - \mathrm{Ci}(n+\pi)\sin(n) - \cos(n)\,\mathrm{Si}(n) + \cos(n)\,\mathrm{Si}(n+\pi)$$

■ Example 3

Now we want to write this formula:

$$\int_0^\pi \frac{\sin(x)}{n+x}\,dx = \mathrm{Ci}(n)\sin(n) - \mathrm{Ci}(n+\pi)\sin(n) - \cos(n)\,\mathrm{Si}(n) + \cos(n)\,\mathrm{Si}(n+\pi).$$

Write the same integrating command on both sides of the equation:

```
Integrate[Sin[x] / (n + x), {x, 0, π}] = Integrate[Sin[x] / (n + x), {x, 0, π}]
```

Execute the righthand side by pressing ⌐ALT╕RET⌐:

```
Integrate[Sin[x] / (n + x), {x, 0, π}] =
  CosIntegral[n] Sin[n] - CosIntegral[n + π] Sin[n] -
  Cos[n] SinIntegral[n] + Cos[n] SinIntegral[n + π]
```

Convert this into text and traditional form:

$$\int_0^\pi \frac{\sin(x)}{n+x}\,dx = \mathrm{Ci}(n)\sin(n) - \mathrm{Ci}(n+\pi)\sin(n) - \cos(n)\,\mathrm{Si}(n) + \cos(n)\,\mathrm{Si}(n+\pi)$$

■ Example 4

To write the following formula,

$$d = \int_0^1 \frac{\log^7(x)}{\sqrt{1-x^2}}\,dx$$

$$= -\frac{1}{768}\,\pi$$

$$\{275\,\pi^6 \log(4) + 133\,\pi^4\,[\log^3(4) + 12\,\zeta(3)] + 21\,\pi^2\,[\log^5(4) + 120\,\zeta(3)\log^2(4) + 720\,\zeta(5)] +$$
$$3\,[\log^7(4) + 420\,\zeta(3)\log^4(4) + 15120\,\zeta(5)\log^2(4) + 10080\,\zeta(3)^2\log(4) + 90720\,\zeta(7)]\}$$

$$= -5040.39$$

so that the equal signs are aligned, write:

```
d = Integrate[Log[x]^7 / Sqrt[1 - x^2], {x, 0, 1}] =
  Integrate[Log[x]^7 / Sqrt[1 - x^2], {x, 0, 1}] =
  Integrate[Log[x]^7 / Sqrt[1 - x^2], {x, 0, 1}] // N
```

Execute separately the second and the third integral, and convert to text and traditional form:

$$d = \int_0^1 \frac{\log^7(x)}{\sqrt{1-x^2}}\,dx = -\frac{1}{768}\,\pi$$

$$(275\,\pi^6 \log(4) + 133\,\pi^4\,(\log^3(4) + 12\,\zeta(3)) + 21\,\pi^2\,(\log^5(4) + 120\,\zeta(3)\log^2(4) + 720\,\zeta(5)) +$$
$$3\,(\log^7(4) + 420\,\zeta(3)\log^4(4) + 15120\,\zeta(5)\log^2(4) +$$
$$10080\,\zeta(3)^2\log(4) + 90720\,\zeta(7))) = -5040.39$$

Add a return character before the second and third equal sign, write an alignment marker (⌐ESC╕am⌐ESC╕) before each equal sign, write a no break sign (⌐ESC╕nb⌐ESC╕) after the third equal sign, select the cell bracket, and choose **Format ▷ Text Alignment ▷ On AlignmentMarker**. Lastly, replace some usual parentheses with brackets ([ ]) and curly braces ({ }) to make the formula more readable.

■ Example 5

To write the following formula,

$$\int x^a\,dx = \begin{cases} \frac{x^{a+1}}{a+1} & a \in \mathbb{R},\ a \neq -1 \\ \log(x) & a = -1 \end{cases}$$

first write the same integrating command twice:

```
Integrate[x^a, x] = Integrate[x^a, x]
```

Execute the righthand side by pressing ALT RET, and convert the cell into text and traditional form:

$$\int x^a \, dx = \frac{x^{a+1}}{a+1}$$

Place the cursor after = and choose $\left\{ \begin{smallmatrix} \square \\ \square \end{smallmatrix} \right.$ from the **BasicTypesetting** palette. Replace the two placeholders with four placeholders by selecting the two placeholders and clicking $\begin{smallmatrix} \square & \square \\ \square & \square \end{smallmatrix}$ in the same palette. Then fill the four boxes by cutting the value of the integral and pasting into the first box and writing "log(x)" and the conditions into the other boxes (see Section 3.3.2, p. 68, to see how to write special characters with the keyboard):

$$\int x^a \, dx = \left\{ \begin{matrix} \frac{x^{a+1}}{a+1} & a \in \mathbb{R}, \, a \neq -1 \\ \log(x) & a = -1 \end{matrix} \right.$$

To expand the {, select it with the mouse and choose **Edit ▷ Expression Input ▷ Spanning Characters ▷ Expand Indefinitely**:

$$\int x^a \, dx = \left\{ \begin{matrix} \frac{x^{a+1}}{a+1} & a \in \mathbb{R}, \, a \neq -1 \\ \log(x) & a = -1 \end{matrix} \right.$$

To get the items left aligned, put the cursor somewhere in the formula and choose **Format ▷ Show Expression**, or press SHIFT CTRL e. The code of the cell appears. Add the option `ColumnAlignments->{Left}` (see Section 13.1.3, p. 331) so that the end of the code looks like this:

```
          },ColumnAlignments->{Left}],
        ShowAutoStyles->False]}],
      (#&)]}]}],TraditionalForm]], "Text"]
```

Press SHIFT CTRL e again to get back to the usual formula.

To fine tune the formula, select the cell of the formula and then choose **Edit ▷ Expression Input ▷ Add Frame**. Widen the margins of the frame by adding the option `FrameBoxOptions` (see Section 3.2.3, p. 62) so that the end of the code reads `"Text"`, `FrameBoxOptions->{BoxMargins->True}]`. Select $\frac{x^{a+1}}{a+1}$, and press CTRL ← twice to nudge this part a little to the left (another way to nudge part of a formula left, right, down, or up is the use of the menu commands, found in **Edit ▷ Expression Input**).

■ Example 6

To write the following formula,

$$\lim_{x \to \infty} \left(\frac{a}{x} + 1 \right)^x = e^a$$

type:

```
Limit[(1 + a / x)^x, x → ∞] = Limit[(1 + a / x)^x, x → ∞]
```

Execute the righthand side, and convert the cell to text and traditional form.

To write this one,

$$\underbrace{111\cdots1}_{k}\underbrace{000\cdots0}_{n}$$

click ■ two times from the **BasicTypesetting** palette to get three placeholders. Write "1 1 1
··· 1" into the first box (the three dots can be found from the same palette), ‿ from the
same palette in the second box, and k in the last box. Copy this whole expression, paste it
next to the expression, and replace the 1s with 0s and k with n. Convert to text and traditional form.

■ Example 7

The following expressions can be found in the **BasicTypesetting** palette:

$$x^{+} \quad x_{+} \quad x^{-} \quad x_{-} \quad x^{*} \quad x_{*} \quad x^{\dagger} \quad x' \quad x''$$

The expressions x' and x'' are often used to denote the first and second derivative of a
function x. These expressions are easy to write with $\boxed{\text{CTRL}}\{\wedge\}$ and $\boxed{\text{CTRL}}\{\_\}$ (see Section 3.3.3, p.
70). For example, x^{+} is written with $x\boxed{\text{CTRL}}\{\wedge\}+\boxed{\text{CTRL}}\{\_\}$. The characters †, ′, and ″ can be written
as \[Dagger], \[Prime], and \[DoublePrime], respectively.

The same palette also contains the following expressions:

$$\bar{x} \quad \underline{x} \quad \vec{x} \quad \tilde{x} \quad \hat{x} \quad \dot{x} \quad \ddot{x}$$

The expressions $\dot{x}$ and $\ddot{x}$ are often used to denote the first and second derivative of a
function x. These expressions are easy to write with $\boxed{\text{CTRL}}\{\&\}$ and $\boxed{\text{CTRL}}\{+\}$. For example, $\bar{x}$ is
written with $x\boxed{\text{CTRL}}\{\&\}\_\boxed{\text{CTRL}}\{\_\}$. The characters →, ~, and ·· can be written as $\boxed{\text{ESC}}$vec$\boxed{\text{ESC}}$, \[Tilde],
and \[DoubleDot].

Many of these expressions can also be written as commands such as **SuperPlus[x]**,
OverBar[x], **OverDot[x]**; remember also **Subscript[x, n]**. For example, to write the
following formula,

$$\dot{y}(x) = a + b\,x$$

write **ẏ[x] = a + b x** or

    OverDot[y][x] = a + b x

Convert this into text and traditional form.

■ Example 8

If you have superscripts or subscripts at many levels, such as is shown here,

$$x^{2^{a^{3}}}$$

and you find the scripts to be too small, consider using the options **ScriptSizeMultipli⹂
ers** (default value is 0.71) and **ScriptMinSize** (default value is 5) (see Section 3.3.1, p. 67).
When going to higher-level scripts, the size of the font is each time multiplied by **Script⹂
SizeMultipliers**, but the size will not become smaller than **ScriptMinSize**. To adjust
the options, place the cursor in the formula and choose **Format ▷ Show Expression**. After
"Text", write an option like **ScriptSizeMultipliers → 0.85** or write both options, and
again choose **Format ▷ Show Expression**:

$$x^{2^{a^{3}}}$$

3.4.3 Inline Formulas

An inline formula is among the text of a **Text**-style cell. Such formulas are often small, like i, a_i, x, $f(x)$, $f'(x)$, $\partial f / \partial x$, $\frac{\partial f}{\partial x}$, $\sqrt{x}$, $\sin(x)$, a/b, $\frac{a}{b}$, $\lim_{x \to \infty} (1 + \frac{a}{x})^x = e^x$, $\int \sin(x)\,dx$, $\int_o^\pi \cos(x)\,dx$, or $\sum_{i=1}^\infty 1/i^2$. If the formula is larger, it is often more useful to put it in a separate line as a display formula.

Note that *Mathematica* formats inline formulas lower as display formulas (as is usual in traditional mathematical notation). For example, in a fraction the font size is smaller, in a limit the variable and limiting value are written as a subscript next to "lim," and in a definite integral or sum the lower and upper limits are written as sub- and superscripts (and not as under- and overscripts).

To write an inline formula, we begin and end the formula with the special key combinations [CTRL]$($ and [CTRL]$)$. The simple steps are as follows:

> - Begin the formula by typing [CTRL]$($ or [CTRL]9.
> - Write the formula.
> - End the formula by typing [CTRL]$)$ or [CTRL]0.

The formula can be written in any way: linearly, in a 2D form with the keyboard (see Section 3.3.3, p. 70), or in a 2D form with the palettes. I often write the simplest formulas (e.g., variables, special characters, subscripts, powers, functions, square roots, fractions) with the keyboard and less simple formulas (e.g., integrals, sums) with palettes.

Note that the default style of inline formulas is traditional form, as can be seen from **Cell ▷ Default Inline FormatType**. This means that if you write the formula in a 2D form like $\int \frac{a}{1+x}\,dx$ by using palettes or with special key combinations, the formula is automatically in traditional form (and no conversion is needed).

On the other hand, if you write the formula *linearly*, like Integrate[$a/(1 + x)$, x], a conversion to traditional form is needed. This is done by typing at least one character after the formula and then selecting the formula with the mouse and choosing **Cell ▷ Convert To ▷ TraditionalForm** or pressing [SHIFT][CTRL]t ([SHIFT]$\mathbb{B}$$t$ on a Macintosh).

Note that a part of an inline formula can also be executed by selecting that part and then choosing **Kernel ▷ Evaluation ▷ Evaluate in Place** or by pressing [ALT][RET] ($\mathbb{B}$[RET] on a Macintosh). However, this may seldom be needed, as inline formulas are, as a rule, simple.

■ **Example 1**

We want to write a simple text cell:

> Assume that x is positive. Then...

Just begin a **Text** cell by pressing [ALT]7, write the text before x, then write [CTRL]$($ x [CTRL]$)$, and continue the text. Note that you need not worry about italicizing the variable; *Mathematica* does it for you. Another way to write x is to press [CTRL]i x [CTRL]i to italicize the variable. Choose the way you find easiest. I consistently use [CTRL]$($ ... [CTRL]$)$ for all inline formulas, even formulas as simple as x.

■ **Example 2**

Next we want to write the following text cell:

Consider the function $f(x) = a \sin(x + \pi)$. Let…

To write the formula, press [CTRL]([(], write "f(x)=a_sin(x+[ESC]p[ESC])", press [CTRL])[)], and continue the text. Note that when writing the formula you should not press the space key in any other place than in multiplications; *Mathematica* adds suitable spaces when needed (e.g., before and after = or +).

■ **Example 3**

Now we write the text cell below:

Let $f(x) = \frac{a}{x}$ and assume that…

To write the formula, do one of the following:

- Use the keyboard: [CTRL]([(] f(x)=a [CTRL][/] x [CTRL])[)] (see Section 3.3.3, p. 70)
- Use the **BasicInput** palette to write the fraction: [CTRL]([(] f(x)=$\frac{a}{x}$ [CTRL])[)].
- Write the formula linearly as [CTRL]([(] f(x)=a/x [CTRL])[)], select a/x, and press [SHIFT][CTRL][t].

■ **Example 4**

Let us now write the below formula in the text cell:

If the integral $\int_0^\infty f(x)\,dx$ converges, then…

Do one of the following:

- Use the keyboard to write the lower and upper bounds of the integral as sub- and superscripts: [CTRL]([(] [ESC]int[ESC] [CTRL][_] 0 [CTRL][%] [ESC]inf[ESC] [CTRL][␣] f(x) [ESC]dd[ESC] x [CTRL])[)]. (This is easier than it looks; try it.)
- Use the **BasicInput** palette to write the integral: [CTRL]([(]$\int_0^\infty f(x)\,dx$[CTRL])[)] (go with [TAB] from one selection placeholder to the next one).
- Write the formula linearly as [CTRL]([(]Integrate[f[x],{x,0,Infinity}][CTRL])[)], write a space after the formula, select the formula, and press [SHIFT][CTRL][t].

■ **Example 5**

Now we want to write the following:

Because $\int \frac{1}{a+x^2}\,dx = \frac{1}{\sqrt{a}} \tan^{-1}\left(\frac{x}{\sqrt{a}}\right)$, then…

First write the following, with a palette:

Because $\int \frac{1}{a+x^2}\,dx = \int \frac{1}{a+x^2}\,dx$, then…

Select the latter integral and execute it with [ALT][RET]. Here is the result:

Because $\int \frac{1}{a+x^2}\,dx = \frac{\tan^{-1}\left(\frac{x}{\sqrt{a}}\right)}{\sqrt{a}}$, then…

To make the formula lower, cut the denominator, insert a fraction after the equal sign, write 1 into the numerator, and paste the cut square root into the denominator:

Because $\int \frac{1}{a+x^2}\,dx = \frac{1}{\sqrt{a}} \tan^{-1}\left(\frac{x}{\sqrt{a}}\right)$, then...

In this way, you can also modify the results *Mathematica* gives.

■ Example 6

Another way to write the same formula is to write the following:

Because $\int \frac{1}{a+x^2}\,dx =$, then...

Then execute the integral:

```
Integrate[1 / (a + x^2), x]
```

$$\frac{\text{ArcTan}\left[\frac{x}{\sqrt{a}}\right]}{\sqrt{a}}$$

Copy the result into the formula:

Because $\int \frac{1}{a+x^2}\,dx = \frac{\text{ArcTan}\left[\frac{x}{\sqrt{a}}\right]}{\sqrt{a}}$, then...

Select the value of the integral, convert it into traditional form, and then move the denominator as shown above. In this way, results from calculations can be inserted into inline formulas.

3.4.4 Automatic Numbering

Sections and some formulas in a mathematical document often are numbered, so that we can easily refer to them. Often we also want to refer to some pages numbers of our document. With *Mathematica,* we can give the numbers manually, but we can also let *Mathematica* automatically choose the appropriate numbers.

In a small document, the numbers of sections and references to page numbers can easily be written manually, and to some extent the same is true for formulas. However, in a larger document the automatic numbering becomes more tempting, because chances grow that sections and formulas are moved to other places and new sections and new formulas are written among old ones. After such a modification, automatic numbers and references to them are again correct without you having to manipulate them in any way, while manually given numbers and references to them have to be manually corrected. However, using the automatic numbering system of *Mathematica* requires some work in the form of handling the cell tags. Next we consider automatic numbering of pages, formulas, and sections.

■ Referring to Page Numbers

Suppose you want to write "According to Theorem 2.3 (see p. XXX) ..." where XXX should be replaced with the correct page number. First, we have to assign a *cell tag* to the cell that contains Theorem 2.3. Select the cell bracket of the theorem, and do the following:

- Choose **Find ▷ Add/Remove Cell Tags....**
- Write a name (like "Theorem 2.3") for the cell into the **Cell tag** field.
- Click **Add** and **Close**.

(By the way, the cell tags can be seen in the document by choosing **Find** ▷ **Show Cell Tags**. To go to a cell with a tag, choose the tag from **Find** ▷ **Cell Tags**.)

To create a reference to the page number of Theorem 2.3, put the cursor in the place where you want to refer to the theorem (i.e., after "see p. "), and do the following:

- Choose **Input** ▷ **Create Automatic Numbering Object…**.
- Set **Counter type** to **Page**.
- From the list of cell tags, click the cell tag of the theorem, and click **OK**.

The page number now appears, at the location of the cursor, in the form XXX. When the document is printed, XXX is automatically replaced with the correct page number. Before printing, you can see the correct page number by choosing **Format** ▷ **Show Page Breaks**. The page number also has the useful property that when it is clicked on, the document is scrolled to the cell to which the page number refers (to go back to your current location in the document, choose **Find** ▷ **Go Back**).

■ Manual Numbering of Formulas

Before explaining the automatic numbering of equations, we show how to manually add a number to a formula. First, write the formula (see Section 3.4.2, p. 74):

$$f = b + a x$$

Open the code behind this cell by selecting the formula and then choosing **Format** ▷ **Show Expression**:

```
Cell[BoxData[
    FormBox[
        RowBox[{"f", "=",
            RowBox[{"b", "+",
                RowBox[{"a", " ", "x"}]}]}]], TraditionalForm]], "Text"]
```

After "Text", add the following option (see Section 3.2.3, p. 61):

```
CellFrameLabels → {{None, "(3.1)"}, {None, None}}
```

After choosing **Format** ▷ **Show Expression** and inserting a tab before the formula, the result is as follows:

$$f = b + a x \tag{3.1}$$

■ Automatic Numbering of Formulas

As mentioned in Section 3.4.1, p. 73, automatic numbering of formulas can only be used with the style sheets **ArticleClassic**, **ArticleModern**, **Classic**, **Classroom**, **HelpBrowser**, and **Report**; these style sheets have the **NumberedEquation** style. You can also try the **MathematicaNavigator** style sheet provided on the CD-ROM of this book.

Giving an automatic number for a formula is easy. Write the formula:

$$f = b + a x$$

Select its cell bracket, then choose **Format** ▷ **Style** ▷ **NumberedEquation**. Here is the result:

$$f = b + a x \tag{1}$$

If we write a second numbered equation, such as the following,

$$g = c + d x \tag{2}$$

it automatically gets the correct running number. If we would change the order of these formulas, the numbers would change accordingly, so that g would get number (1) and f would get number (2). If we would remove formula (1), which is currently f, the number of g would change to (1). If we would add a numbered formula between formulas (1) and (2), the new formula would get the number (2), and the number of g would change to (3). Basically, with this method, the numbers will always be in running order.

If you would like to remove the number of an equation, just choose **Text** as the style of the cell. Dot-numbered equations having numbers like (4.3) are considered after we have explained how to refer to numbered equations.

■ Referring to Numbered Equations

To be able to refer to a numbered equation, we have to first assign a cell tag to the equation. Select the cell bracket of the numbered equation, and do the following:

- Choose **Find ▷ Add/Remove Cell Tags…**.
- Write a name for the equation into the **Cell tag** field.
- Click **Add** and **Close**.

To create a reference to a numbered equation, put the cursor in the place where you want to refer to the numbered equation, and do the following:

- Choose **Input ▷ Create Automatic Numbering Object…**.
- Set **Counter type** to **NumberedEquation**.
- Click the cell tag of the equation, and click **OK**.

The number of the equation now appears at the location of the cursor. Add parentheses around the number to get a reference like "(2)." This reference changes automatically if the number of the equation changes. By clicking the number, the notebook is scrolled to where the equation is located.

■ Dot-Numbered Equations

Equations often have dotted numbers like (4.3), with the first number referring to the current section and the latter to the number of the equation within this section. The style sheets that come with *Mathematica* do not have a style for this kind of numbered equation, but try the **DotNumberedEquation** style of the **MathematicaNavigator** style sheet that is included on the CD-ROM that comes with this book. For example:

$$h = p + q\,x \tag{4.3}$$

To refer to a dot-numbered equation, assign a cell tag to both the equation and to the cell of the **Section** style that corresponds to the section where the equation is. Then do the following:

- Type the opening parenthesis, choose **Input ▷ Create Automatic Numbering Object…**, set **Counter type** to **Section**, click the cell tag of the section, click **OK**, and type the period.
- Choose again **Input ▷ Create Automatic Numbering Object…**, set **Counter type** to **NumberedEquation**, click the cell tag of the equation, click **OK**, and finally type the closing parenthesis.

■ Automatic Numbering of Sections

Manual numbering of sections is, of course, very easy: just write a suitable number at the beginning of the heading of the section. With automatic numbering, however, we have the advantage that the numbers of sections and references to them are correct after modifications in the order of the sections. To give an automatic number to a section, place the cursor at the beginning of the section and do the following:

- Choose **Input ▷ Create Automatic Numbering Object....**
- Set **Counter type** to **Section** and click **OK**.

If you want an automatic number for a subsection, like 2.4, first create the number 2 of the section in the way we showed above, type a period, and then create the number 4 of the subsection in the same way (just set **Counter type** to **Subsection**).

■ Referring to Numbered Sections

To write a reference to a section having an automatic number, first assign a cell tag to the corresponding **Section** style cell in the same way that tags are assigned for numbered equations: select the cell bracket, choose **Find ▷ Add/Remove Cell Tags...**, write a name for the cell, and then click **Add** and **Close**.

A reference to the section is then written in the same way a reference is written for a numbered equation: choose **Input ▷ Create Automatic Numbering Object...**, set **Counter type** to **Section**, click the cell tag of the **Section** cell, and click **OK**. In this way we can write, for example, "In Section 3 we will show that...", and the number 3 will change if the automatic number of the section changes.

To write a reference to a subsection that has an automatic number, first assign a cell tag to the corresponding **SubSection** style cell. To write a reference like "In Subsection 2.4 we will show that...", first create the number 2 of the section by choosing **Input ▷ Create Automatic Numbering Object...**, type a period, and then, in the same way, create the number 4 of the subsection.

Assigning cell tags to sections and subsections also has the advantage that by using **Find ▷ Cell Tags** we can quickly go to each section and subsection of our notebook.

Files

Introduction

> *The world's greatest and most powerful computer was constructed, so mathematicians decided to test it out by seeing if it could make any impression on some classical unsolved problems. They decided on Fermat's last theorem (this happened prior to the work of Andrew Wiles), namely that $x^n + y^n = z^n$ has no solutions over the natural numbers for $n \geq 3$. For days they fed it with every known piece of information, conjecture, and partial result, and at last they set it to work. After a few minutes it printed out: "I have a wonderful proof of this result, but my memory is too small to store it."*

Note that you may skip this chapter if you right now do not need (or already know how) to do the following:

- use the packages of *Mathematica;*
- read data or graphics from a file;
- write data or graphics into a file;
- save results for later use;
- convert results and notebooks for TeX, HTML, C, or Fortran;
- speed up calculations; or
- save memory consumption.

With packages we can give added functionality to *Mathematica*. Using packages is straightforward, unless you forget to load them; we show how to solve this problem.

Read Section 4.2.1 if you have data in a file and you want to visualize or analyze it with *Mathematica* by, for example, using some plotting commands or some statistical methods. You then need to import the data in such a way that *Mathematica* understands it. This means that you must form lists that contain the elements of the data. We consider here some simple examples, but real-life data are considered, e.g., in Chapters 9 and 10. Data can also be exported for use in other applications.

You may want to export a *Mathematica* plot to another application or to import into *Mathematica* a plot made by another application. Section 4.2.2 is devoted to these topics.

In Section 4.3, we consider saving and loading results. This may be useful if we want to continue calculations in later sessions. We also consider converting *Mathematica* notebooks and results to the forms required by TeX, HTML, C, and Fortran.

In Section 4.4, we show some ways to manage and save computing time and computer memory.

4.1 Loading Packages

4.1.1 Basic Loading of Packages

■ About Packages

Mathematica packages supplement the kernel by providing more commands (see Section 1.1.2, p. 6). The standard packages of *Mathematica* are divided into groups according to the subject area of the packages. For example, we have several numerical mathematics packages, like **NumericalMath`BesselZeros`** and **NumericalMath`InterpolateRoot`**.

The names of the packages contain two times the backquote or grave accent character `` ` ``. This is actually a context mark in *Mathematica*. Note that after the first `` ` `` you sometimes have to press the space key to avoid the creation of a symbol like **à** or **ì**. Also, after the second `` ` `` you always have to press the space key so that the accent character appears.

The packages have a separate manual, but the same information can also be found with the *Help Browser*. Go to **Help ▷ Help Browser…** and then to **Add-ons and Links ▷ Standard Packages**. You can even open a package to look at the code. Just choose **Open** from the **File** menu and select a package from **AddOns ▷ StandardPackages**. Looking at the code of a package may be interesting if you are wondering how the package works. You can then perhaps even learn something about programming with *Mathematica*.

In addition to the standard packages, you may want to buy additional packages or create your own. See Section 4.1.3, p. 89, for information about where to put such add-ons.

■ The Basic Loading Command

The packages are not normally loaded when *Mathematica* is loaded. Instead, we have to take care of loading each package we intend to use. The basic loading command is **<<**. We use the **NumericalMath`InterpolateRoot`** package as an example:

```
<< NumericalMath`InterpolateRoot`    Load the package
```

First we load the package:

```
<< NumericalMath`InterpolateRoot`
```

Now we can use its commands. In this case we can find the root of an equation with **InterpolateRoot**:

```
InterpolateRoot[Exp[x] - 2 == 0, {x, 0, 2}]
```

```
{x → 0.69314718055994530941723}
```

Using packages is this easy. (Note that **<<** is equivalent to the **Get** command: **Get["Numeri⌐ calMath`InterpolateRoot`"].**)

■ If a Package Does not Seem to Work

Now and then it seems as if a package does not work. In such a situation, you have, with high probability, first used a command of the package, observed that you have not yet loaded the package, loaded the package, and then used the command again, only to observe that the command still does not work. To get the package working, do the following:

- remove with **Remove[name]** all the names of the package you have used and then execute the nonworking command again, or
- choose **Kernel ▷ Quit Kernel ▷ Local**, then choose **Kernel ▷ Start Kernel ▷ Local**, load the package, and (after executing all preparing commands) execute the nonworking command again.

As an example, we try to use **BesselJZeros**:

```
BesselJZeros[2, 5]
```

```
BesselJZeros[2, 5]
```

The command did not work, because we have not loaded the corresponding package. We now load the package:

```
<< NumericalMath`BesselZeros`
```

```
BesselJZeros::shdw : Symbol BesselJZeros appears in
   multiple contexts {NumericalMath`BesselZeros`, Global`};
   definitions in context NumericalMath`BesselZeros` may
   shadow or be shadowed by other definitions. More…
```

We get a warning that **BesselJZeros** appears in multiple contexts. This is explained in more detail in Section 14.3.2, p. 371, but the warning means, briefly, that now we have two **BesselJZeros** names: the one we implicitly defined when we tried to use the **Bessel⌐ JZeros** command and the one that is in the package. We try to use **BesselJZeros** again:

```
BesselJZeros[2, 5]
```

```
BesselJZeros[2, 5]
```

We still do not get the result. Indeed, *Mathematica* uses our own **BesselJZeros** and not the one in the package. A solution is to remove our definition:

```
Remove[BesselJZeros]
```

Now only one **BesselJZeros** remains—the one in the package—and now it works:

```
BesselJZeros[2, 5]
```

{5.13562, 8.41724, 11.6198, 14.796, 17.9598}

4.1.2 More about Loading Packages

■ Forgetting that a Package Is Already Loaded

If we have forgotten that we have already loaded a package and try to load it again, we may get some warnings, because we are trying to redefine some protected names:

```
<< NumericalMath`InterpolateRoot`
```

```
Set::write :
  Tag InterpolateRoot in Options[InterpolateRoot] is Protected. More…
SetDelayed::write :
  Tag InterpolateRoot in InterpolateRoot[f_ == g_, {x_, a_, b_},
      options___] is Protected. More…
SetDelayed::write : Tag InterpolateRoot in
      InterpolateRoot[fg_, {x_, a_, b_}, options___] is Protected. More…
```

However, these are only warnings, and we can continue using the package in the usual way. Also, warnings like the above currently only appear when loading one of about a dozen packages.

Nevertheless, *Mathematica* also has a special loading command that only loads the package if it has not been already loaded. This command is considered next.

■ Loading a Package if it Is not Already Loaded

Needs["NumericalMath`InterpolateRoot`"]   Load the package if not already loaded

---

If we use **Needs**, then no warnings appear:

Needs["NumericalMath`InterpolateRoot`"]

Needs finds that the package is already loaded and does not load it again.

■ Autoloading Packages

<< NumericalMath`   Load numerical math packages when needed

This is a very handy way of using the standard packages. In that way, we only declare the group of packages in which we are interested. When we then use a command defined in one of these packages, *Mathematica* automatically loads the appropriate package (if it is not already loaded). Use this method if you use several packages from a given subject area.

We use this method to load the numerical math packages:

<< NumericalMath`

Although we have not loaded the package **NumericalMath`NLimit`**, which defines **ND** (it calculates a numerical derivative), we can use **ND**:

```
ND[Sin[x], x, 1]
```

```
0.540302
```

Mathematica found that **ND** is defined in one of the numerical math packages, loaded the correct package, and then executed our command.

■ Automating Autoloading Packages

An **init.m** file is located in **Mathematica** ▷ **Configuration** ▷ **Kernel**. The commands in this file are executed each time *Mathematica* is started. So, if you frequently use, for example, graphics packages, write **<<Graphics`Graphics`** at the end of the **init.m** file. Then you can always use graphics packages without loading them.

Another way to load standard packages is to use the **Autoload** folder in **$UserBase** **Directory** (**$UserAddOnsDirectory** in *Mathematica* 4). For example, if you frequently use several graphics packages, copy the **Graphics** folder of the **StandardPackages** folder into the **Autoload** folder (or create, in the **Autoload** folder, a folder named **Graphics** and then copy the folder **AddOns** ▷ **StandardPackages** ▷ **Graphics** ▷ **Kernel** into **Graphics**). Now you need not bother about loading graphics packages: each package is automatically loaded when *Mathematica* is started.

4.1.3 User Add-Ons

■ Packages and Applications

In addition to the standard packages, you may have additional packages and applications. If you want them to be available for all users, put them in the **Applications** folder that is located in the **$BaseDirectory** (**$AddOnsDirectory** in *Mathematica* 4). For example, on my Macintosh with MacOS X, here is the location:

```
$BaseDirectory
```

```
/Library/Mathematica
```

On a Windows machine, the location may be **C:\Documents and Settings\All Users\Application Data\Mathematica**. On the other hand, if you want the packages to be available only for you, put them in the **Applications** folder that is located in the **$UserBaseDirectory** (**$User** **AddOnsDirectory** in *Mathematica* 4). For example, on my Macintosh, here is the location:

```
$UserBaseDirectory
```

```
/Users/ruskeepaa/Library/Mathematica
```

On a Windows machine, the location may be of the form **C:\Documents and Settings** *username***\Application Data\Mathematica**.

■ Style Sheets and Palettes

Mathematica comes with several style sheets and palettes. Additional style sheets and palettes can be put in the **SystemFiles** folder, which is located in the same directories as the **Application** folder we considered above. Put style sheets in **SystemFiles** ▷ **FrontEnd** ▷ **StyleSheets** and palettes in **SystemFiles** ▷ **FrontEnd** ▷ **Palettes**.

4.2 Exporting and Importing Data and Graphics

4.2.1 Exporting and Importing Data

In Sections 4.2.1 and 4.2.2, we consider **Export** and **Import**. These are versatile commands to use for the writing and reading of data and graphics (and other material). A mathematician or statistician may mainly be interested in reading existing data files into *Mathematica* and writing *Mathematica* graphics into files that some other programs can use. Note that, in place of **Export**, we can also use the menu command **Edit ▷ Save Selection As** to save material in several different forms. To learn how to locate files in various folders, read Section 4.2.3, p. 96.

Export["file", data, "format"]   Write **data** into **file** in **format**
!! file   Look at **file**
data = Import["file", "format"]   Read **file** into **data** in **format**

Examples of formats:

List   A 1D table. Writing: put each item of a 1D list in its own row. Reading: form a 1D list of all of the items (of a 1D or 2D table).

Table   A 2D table (items in a row separated by spaces; **.dat**). Writing: put each sublist of a 2D list in its own row. Reading: form a sublist of the items of each row to form a 2D list.

Text   A single string of ordinary characters (**.txt**). Writing: put the whole text into the file. Reading: form a single string from the text.

Lines   Each line is a string. Writing: put the whole text into the file. Reading: form a string from each row to form a 1D list of strings.

Words   Each word is a string. Writing: put the whole text into the file. Reading: form a string from each word to form a 1D list of strings.

To see all available formats:

$ExportFormats, **$ImportFormats**

List and **Table** may be the most important formats. They can handle both numeric and textual items. A textual item can be a string like **"Donkey"**, which has the quotation marks, or a word like **Donkey**, which does not have quotation marks. When exporting a string, the text is written without quotation marks, but when exporting a textual item that is not a string, its *value* (if any) is written. When importing a textual item like **Donkey**, the item is converted to a string (**"Donkey"**).

Note that the **Table** and **Text** formats can also be indicated by the extensions **.dat** and **.txt** of the file. For example, instead of **Export["file", data, "Table"]**, we can also write **Export["file.dat", data]**; instead of **Import["file.dat", "Table"]**, we can also simply write **Import["file.dat"]**. For spreadsheet compatibility, we have the **CSV** format, where items in a row are separated with commas. To see, for example, all export formats, execute **$ExportFormats**.

With **Import** we can use various options if the format is **List** or **Table** (see *Additional Information* for **Import** in the *Help Browser*).

For easy reference, here are the two most important commands to use to read data files:

`data = Import["file", "List"]`	Read a 1D table of **file** into **data**
`data = Import["file", "Table"]`	Read a 2D table of **file** into **data**

Note that by default, **Export** writes the file into the current working directory; the command to view this directory is **Directory[]**. Also, by default, **Import** searches for a file only from certain directories; these directories can be seen by asking the value of **$Path**. If you want to write a file into or read a file from a nondefault directory, you have to specify the full name of the file or modify the default directories (see Section 4.2.3, p. 96).

■ A 1D List

Define the data:

```
data1 = {23, 41.7, 39.5, 143, 8.4 10^-7};
```

Write the list into a file **listdata**, putting each item in its own row:

```
Export["listdata1", data1, "List"]

listdata1
```

Look at the file:

```
!! listdata1
```

```
23
41.7
39.5
143
8.4e-7
```

Note that small and large numbers are written in a C- or Fortran-like e-form.

Suppose then that we have data in **listdata**. The file in this example is written by *Mathematica*, but it could be done with a text editor by saving the file in a plain text format. Now we read the file, forming a 1D list:

```
data1a = Import["listdata1", "List"]
```

$$\{23, 41.7, 39.5, 143, 8.4 \times 10^{-7}\}$$

We could also form a 2D list:

```
data1b = Import["listdata1", "Table"]
```

$$\{\{23\}, \{41.7\}, \{39.5\}, \{143\}, \{8.4 \times 10^{-7}\}\}$$

We could also form strings from the numbers:

```
data1c = Import["listdata1", "Words"]
```

$$\{23, 41.7, 39.5, 143, 8.4e-7\}$$

The quotations marks of the strings can be seen by asking for the **InputForm** of the result:

```
% // InputForm
```

```
{"23", "41.7", "39.5", "143", "8.4e-7"}
```

■ A 2D List

Consider the 2D list:

```
data2 = {{23, 41.7, 39.5}, {143, 8, 56}, {28.8, 74, 13}};
```

Export it in a table form:

```
Export["tabledata2", data2, "Table"];
```

Look at the resulting file:

```
!! tabledata2
```

```
23     41.7    39.5
143    8       56
28.8   74      13
```

Next we read the file, forming sublists from the rows:

```
data2a = Import["tabledata2", "Table"]
```

```
{{23, 41.7, 39.5}, {143, 8, 56}, {28.8, 74, 13}}
```

We could also form a 1D list:

```
data2b = Import["tabledata2", "List"]
```

```
{23, 41.7, 39.5, 143, 8, 56, 28.8, 74, 13}
```

■ A 2D List with Textual Items

Define the data:

```
data3 = {{"Results from an experiment"}, {Individual, Measurement},
    {a, 23}, {b, 41.7}, {c, 39.5}, {d, 143}, {e, 8}};
```

Write it into a file:

```
Export["alphanumericdata3", data3, "Table"];
```

```
!! alphanumericdata3
```

```
Results from an experiment
Individual    Measurement
a    23
b    41.7
c    39.5
d    143
e    8
```

Then read the file:

```
data3a = Import["alphanumericdata3", "Table"]
```

```
{{Results, from, an, experiment}, {Individual, Measurement},
    {a, 23}, {b, 41.7}, {c, 39.5}, {d, 143}, {e, 8}}
```

All textual items were converted into strings:

```
% // InputForm
```

```
{{"Results", "from", "an", "experiment"},
  {"Individual", "Measurement"}, {"a", 23},
  {"b", 41.7}, {"c", 39.5}, {"d", 143}, {"e", 8}}
```

We can drop the first two rows:

```
data3b = Drop[data3a, 2]
```

{{a, 23}, {b, 41.7}, {c, 39.5}, {d, 143}, {e, 8}}

If we begin to analyze the measurements, we can assign them to a variable **meas**:

```
{ind, meas} = Transpose[data3b]
```

{{a, b, c, d, e}, {23, 41.7, 39.5, 143, 8}}

■ Text

Consider the following text:

```
data4 = "A mnemonic for the digits of pi = 3.1415926535:\n
    May I have a large container of coffee - sugar and cream?"
```

A mnemonic for the digits of pi = 3.1415926535:
May I have a large container of coffee - sugar and cream?

Here **\n** is a new line character, which inserts a line break. Write the text into a file (in the **Text** format; with **Lines** and **Words** we get the same result):

```
Export["textdata4", data4, "Text"];
```

```
!! textdata4
```

A mnemonic for the digits of pi = 3.1415926535:
May I have a large container of coffee - sugar and cream?

If we read the file in **Text** form, we get a single string:

```
data4a = Import["textdata4", "Text"]
```

A mnemonic for the digits of pi = 3.1415926535:
May I have a large container of coffee - sugar and cream?

```
% // InputForm
```

"A mnemonic for the digits of pi = 3.1415926535:\nMay I have \
a large container of coffee - sugar and cream?"

Using the **Lines** format gives a list of two strings, one for each row:

```
data4b = Import["textdata4", "Lines"]
```

{A mnemonic for the digits of pi = 3.1415926535:,
    May I have a large container of coffee - sugar and cream?}

With **Words**, each word is transformed into a string:

```
data4c = Import["textdata4", "Words"]
```

{A, mnemonic, for, the, digits, of, pi, =, 3.1415926535:, May,
    I, have, a, large, container, of, coffee, -, sugar, and, cream?}

■ Other Commands

For reading files, *Mathematica* also has **ReadList**, which allows, for example, detailed declaration of the types of items of the files. Files can also be read item by item with **OpenRead**, **Read**, **Skip**, and **Close**. For item-specific writing, we have **OpenWrite**, **Open** **Append**, and **Write**. In the **Utilities`BinaryFiles`** package, there are commands for reading from and writing to binary files.

The contents of files can be searched with **FindList**:

> **FindList["file", "text"]**   Get a list of all lines in **file** containing **text**

> **FindList["alphanumericdata3", "c"]**
>
> {c   39.5}

4.2.2 Exporting and Importing Graphics

Export["file", fig, "format"]   Write graphics **fig** into **file** in **format**
fig = Import["file", "format"]   Read **file** into graphics **fig** in **format**

Options for **Export**:

ImageSize   Absolute size of the image in printer's points (1/72 inch); default value:
  Automatic

ImageResolution   Resolution of the image in dpi (dots per inch); default value:
  Automatic

ImageRotated   Whether to rotate the image to get an image in the landscape form;
  possible values: **False, True**

Examples of formats:

AI   Adobe Illustrator format (**.ai**) (only with **Export**)

EPS   Encapsulated PostScript format (**.eps**)

EPSI   EPS format with a device-independent preview (**.epsi**)

EPSTIFF   EPS format with a TIFF preview

PDF   Adobe Acrobat portable document format (**.pdf**) (only with **Export**)

PICT   Macintosh PICT format (only with **Export**)

WMF   Microsoft Windows metafile format (**.wmf**) (only with **Export**)

BMP   Microsoft bitmap format (**.bmp**)

GIF   GIF format (**.gif**)

JPEG   JPEG format (**.jpg, .jpeg**)

TIFF   TIFF format (**.tif, .tiff**)

To see all available formats:

$ExportFormats, $ImportFormats

Remember that in place of **Export** we can also use the menu command **Edit ▷ Save Selection As** to save material in many different formats.

Note that most formats can be indicated by the extensions of the file names; the extensions are given in parentheses above. For example, instead of **Export["file", fig, "WMF"]**, we can also write **Export["file.wmf", fig]**; instead of **Import["file.gif", "GIF"]**, we can also simply write **Import["file.gif"]**.

When exporting, the first group of formats from **AI** to **WMF** (except **EPSI** and **EPSTIFF**) are independent of the setting for **ImageResolution**, while the formats in the second group, together with **EPSI** and **EPSTIFF**, depend on the value of this option.

When importing **EPS**, **EPSI**, and **EPSTIFF** files, the result is of the form **Graphics[data, opts]** (containing graphics primitives and directives), while the result of importing **BMP**, **GIF**, **JPEG**, and **TIFF** files is of the form **Graphics[Raster[data], opts]** (**Raster** gives a bitmap plot).

With **Export** and **Import**, we can use various options (see *Additional Information* for **Export** and **Import** in the *Help Browser*). From the formats mentioned above, options exists, particularly for **GIF**, **JPEG**, and **TIFF**.

Note that by default, **Export** writes the file into the current working directory; the command to view this directory is **Directory[]**. Also, by default, **Import** searches for a file only from certain directories; these directories can be seen by asking the value of **$Path**. If you want to write a file into or read a file from a nondefault directory, you have to specify the full name of the file or modify the default directories (see Section 4.2.3, p. 96).

■ **Example**

We first make a plot:

```
fig = Plot[Sin[x], {x, 0, 2 π}];
```

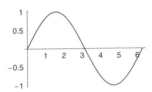

Then we export it into a file:

```
Export["fig.eps", fig, ImageSize → 100]
```

```
fig.eps
```

Next we import the file back into *Mathematica*:

```
fig1 = Import["fig.eps"]
```

```
- Graphics -
```

We show the resulting plot:

```
Show[fig1];
```

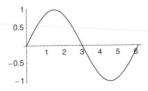

■ **Exporting Graphics into Microsoft Word**

If you are preparing a mathematical document with Microsoft Word, you may want to produce some plots with *Mathematica* and insert them into your Word document. First, make the plot:

```
fig = Plot[Sin[x], {x, 0, 2 π}];
```

To export the plot into Word, the best method may vary from system to system, but the following method should work on a PC or Macintosh. Adjust the size of the plot to be suitable for your Word document, click on the plot, choose **Edit ▷ Save Selection As ▷ EPS**, save the plot in a suitable place, go to your Word document, place the cursor at a suitable point, choose **Insert ▷ Picture ▷ From File…**, locate and click the file you saved from the dialog box, and click **Insert**. In Word, do not adjust the size of the plot, as the size of all text in the plot then also changes, which should be avoided; any changes in size should be done with *Mathematica*, which does not change the size of text.

On some PC systems, **Save Selection As ▷ Metafile** may also yield a good result. On Macintosh systems prior to MacOS X, a simple method is to click the plot, choose **Edit ▷ Copy As ▷ PICT with Embedded PostScript**, and then choose **Edit ▷ Paste** in the Word document. **Export** and **Display** can also be tried:

```
Export["fig.wmf", fig, ImageSize → s]

Display["fig.wmf", fig, "Metafile", ImageSize → s]
```

■ Exporting Graphics into LaTeX

To export a plot into a LaTeX document, save the plot in EPS format, either with **Edit ▷ Save Selection As ▷ EPS…** or with **Export["fig.eps", fig]**. Save the file in the same folder where you have saved your LaTeX document.

In the starting rows of the document, add **\usepackage[dvips]{graphicx}**. At a suitable place in the document, add **\includegraphics{fig.eps}**. If you want to modify the plot, write **\includegraphics[key=value]{fig.eps}**, where key is **width**, **height**, **angle**, or **scale**. For example, write **\includegraphics[width=8cm]{fig.eps}**. This kind of command adds the plot in the place you have chosen.

A more recommended method is to use a **figure** environment, as follows:

```
\begin{figure}[htbp]
\includegraphics[width=8cm]{fig.eps}
\caption{caption to be added}
\label{name used to refer to the plot}
\end{figure}
```

Now LaTeX places the plot in the best position, a caption can be added, and we can refer to the plot with **\ref**. The placement of the plot in the example is directed with **[htbp]**. This asks to place the plot at the present point (**h** = here), if possible, but if not possible, then at the top of the page (**t** = top), then at the bottom of the page (**b** = bottom), and then in a separate page (**p** = page), in this order of preference.

4.2.3 Locating the File

■ The Default Locations

As noted earlier, using short names of files like **data1.dat** with **Export** and **Import** has some restrictions. With **Export**, *Mathematica* saves the file in a default location; the location can be seen by using the command **Directory[]**. With **Import**, *Mathematica* only searches the file from a list of default locations; this list can be seen by using the command **$Path**.

> **Directory[]**   The default directory where **Export** writes a file
> **$Path**   The default list of directories from which **Import** searches a file

More generally, **Directory[]** gives the *current working directory,* and **$Path** the *search path* (i.e., the default list of directories to search in attempting to find an external file). On my Macintosh with MacOS X, the directories are as follows:

Directory[]

/Users/ruskeepaa

Short[$Path, 3]

{/Applications/Mathematica 5.0.app/AddOns/JLink, ≪15≫,
 /Applications/Mathematica 5.0.app/Configuration/Kernel}

If it suits to you that **Export** writes files into **Directory[]** and **Import** searches files only from **$Path**, then you can use short names for the files. Otherwise we have two possibilities. Either use the full names of files or modify the values of **Directory[]** and **$Path**. First, we consider an example of using full names.

■ Using Full Names of Files

We have a folder **MNData** in a **Documents** folder, and now we write a data file into this folder. Because this is not the default location, we have to give the full name of the file:

data5 = {46, 71, 22, 38, 52};

Export["/Users/ruskeepaa/Documents/MNData/listdata5", data5, "List"]

/Users/ruskeepaa/Documents/MNData/listdata5

This location is not among the default search directories of **$Path**, so if we read the file, we have to again use the full name of the file:

data5 = Import["/Users/ruskeepaa/Documents/MNData/listdata5", "List"]

{46, 71, 22, 38, 52}

Note that you need not write the full names by yourself; let *Mathematica* do it. Simply choose **Input ▷ Get File Path...** and then in the dialog box select the file in which you are interested. The full name of this file is now pasted at the current location of the cursor. Note also that the full names are different in form with various computer systems.

■ Modifying the Current Working Directory

The current working directory can be changed by **SetDirectory**:

SetDirectory["/Users/ruskeepaa/Documents/MNData"]

/Users/ruskeepaa/Documents/MNData

This is now the current working directory:

Directory[]

/Users/ruskeepaa/Documents/MNData

Export now writes files into this default directory (without having to use full names of files). So, exporting a file into **MNData** is easy:

```
Export["listdata5", data5, "List"]
```

listdata5

We can go back to the original directory by **ResetDirectory[]**:

```
ResetDirectory[]
```

/Users/ruskeepaa

This is now the current working directory:

```
Directory[]
```

/Users/ruskeepaa

■ Modifying the Search Path

A nondefault folder can easily be added into the search path:

```
$Path = Append[$Path, "/Users/ruskeepaa/Documents/MNData"];
```

Now we can easily read files from this folder:

```
data5 = Import["listdata5", "List"]
```

{46, 71, 22, 38, 52}

   If it seems that you will in several sessions read data from a certain folder, you may consider adding a command **$Path = Append[$Path, "…"]** in the **init.m** file, which can be found in the folder **Mathematica ▷ Configuration ▷ Kernel**. If you modify **$Path** in this way, then every time you open *Mathematica*, the **$Path** variable has the appropriate value and you can easily read the data.

4.3 Saving for Other Purposes

4.3.1 Continuing Work in Later Sessions

As noted in Section 3.1.1, p. 48, when we open a notebook to continue calculations, we cannot directly use any results that are already in the notebook. Suppose a notebook contains the following calculation:

```
int = Integrate[2 x Sin[x] Exp[x], x] // Simplify
```

e^x (Cos[x] − x Cos[x] + x Sin[x])

Now we open this notebook and want to continue by differentiating the integral. Note that opening a notebook only shows the notebook on the screen; opening it does not execute any commands. Thus, after opening the notebook, we cannot write **D[int, x]**, because *Mathematica* does not yet know the value of **int**. We have to tell *Mathematica* what the value is. We have at least three ways to do this:

- execute anew the command defining **int**;
- write **int =** at the beginning of the *value* of **int** and execute the resulting command; or
- save the value of **int** into a file and, in a new session, load the file to get the value of **int**.

The first method is very straightforward if the execution of the command does not take much time.

The second method also is very handy and quick: we get the value of **int** without doing the calculation anew. To illustrate this, we paste a copy of the command above and then write **int =** at the beginning of the result:

> **int = Integrate[2 x Sin[x] Exp[x], x] // Simplify**
>
> e^x (Cos[x] − x Cos[x] + x Sin[x])
>
> **int = e^x (Cos[x] − x Cos[x] + x Sin[x])**

Mathematica immediately makes a copy of the result, and our modification uses this copy. Then we simply execute the resulting command, and we have the value of **int**. Note that this method is not available if we have ended the original command with the semicolon (;) because then the notebook does not have the result.

The third method may be worth considering if you have a very large expression that is clumsy to keep and handle in the notebook but with which you want to continue calculations in later sessions. The expression may be, for example, a large symbolic expression, a complicated plot, or a large set of generated random numbers.

■ Saving and Loading an Expression

> **a >> file** Save the value of a variable **a** into **file** (clearing the file if it already exists)
> **!! file** View **file**
> **a = << file** Load **file** and assign the content as the value of **a**

In place of **>>** and **<<**, we can also use **Put** and **Get**; remember that we already discussed **<<** in section 4.1.1 as it pertains to loading packages. As an example, we save the value of **int** into a file:

> **int >> intfile**

We can check that everything is all right by viewing the file:

> **!! intfile**
>
> E^x*(Cos[x] − x*Cos[x] + x*Sin[x])

In a new session, we can then load the file:

> **int = << intfile**
>
> e^x (Cos[x] − x Cos[x] + x Sin[x])

Now **int** has the desired value.

■ Saving and Loading Several Expressions

> **Save["file", {a, b, …}]** Save definitions of variables **a, b,** … into **file**
> **<< file** Load **file** (variables **a, b,** … then have the saved values)

The value of a single variable can nicely be saved with **>>**. With **Save**, we can save several values in the same file. Note that **Save** appends the values to the file if it already exists (remember that **>>** clears an existing file; **>>>** or **PutAppend** can also be used to append expressions to an existing file). As an example, we generate some random numbers:

```
r = Table[Random[Integer, {1, 6}], {10}]
```

```
{4, 3, 6, 6, 3, 5, 5, 5, 1, 5}
```

We then save both **int** and **r** in a file:

```
Save["intfile2", {int, r}]
```

We look at the file:

```
!! intfile2
```

```
int = E^x*(Cos[x] - x*Cos[x] + x*Sin[x])
```

```
r = {4, 3, 6, 6, 3, 5, 5, 5, 1, 5}
```

We now observe that each expression has the name we have used (note that **>>** does not save the name of the saved expression). Then, in a new session, we load the file:

```
<< intfile2;
```

The variables **int** and **r** now have the saved values:

```
{int, r}
```

$\{ e^x \, (\mathrm{Cos}\,[\mathrm{x}] - \mathrm{x}\,\mathrm{Cos}\,[\mathrm{x}] + \mathrm{x}\,\mathrm{Sin}\,[\mathrm{x}]) , \; \{4, \, 3, \, 6, \, 6, \, 3, \, 5, \, 5, \, 5, \, 1, \, 5\}\}$

Another saving command is **DumpSave**. It saves expressions in a binary format and may be advantageous for very large and complicated expressions. These files can be read with **<<**.

■ Saving and Loading Plots

Now and then you may want to manipulate existing plots in your notebooks. Manipulating with the mouse is straightforward (see Section 5.1.3, p. 115). However, the plots cannot be otherwise manipulated. For example, new options cannot be added, new values of options cannot be used, and plots cannot be combined without replotting the old plots. However, we may want to avoid replotting, because it may take some time if the plots are very complex. Saving and loading plots is a solution. As an example, we plot a function and save it into a file:

```
p1 = ParametricPlot[{t Cos[2 t], t Sin[t]},
    {t, 0, 8 Pi}, AspectRatio → Automatic];
```

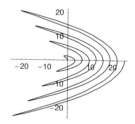

```
p1 >> plot1file
```

(Remember also **DumpSave**, which may be more efficient for very large and complex plots.) In a new session, we can load the file:

```
p1 = << plot1file
```

- Graphics -

With **Show**, we can see the pl

```
Show[p1, Frame → True];
```

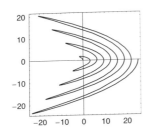

...tions or change old ones:

4.3.2 Exporting to TeX, C, Fortran, and HTML

■ Exporting to TeX

TeXSave["file.tex"]	Write the present notebook into a file in LaTeX form
TeXSave["file.tex", "source.nb"]	Write the given notebook into a file in LaTeX form
File ▷ Save As Special... ▷ TeX	Write the present notebook into a file in LaTeX form
TeXForm[expr]	Show the (plain) TeX form of the expression

The *Help Browser* has much information about **TeXSave**. The TeX material of *Mathematica* is in **SystemFiles ▷ IncludeFiles ▷ TeX**. For exporting graphics into LaTeX, see Section 4.2.2, p. 96. Here is an example of **TeXForm**:

```
i = Integrate[a / (x^2 + 2 b x + c^2), x]
```

$$\frac{a\, \text{ArcTan}\left[\frac{b+x}{\sqrt{-b^2+c^2}}\right]}{\sqrt{-b^2+c^2}}$$

```
TeXForm[i]
\frac{a\,\arctan (\frac{b + x}{{\sqrt{-b^2 + c^2}}})}
   {{\sqrt{-b^2 + c^2}}}
```

■ Exporting to C and Fortran

CForm[expr]	Show the C form of the expression
FortranForm[expr]	Show the Fortran form of the expression

These commands help when you want to export *Mathematica* results into a C or Fortran program. For example:

```
CForm[i]
(a*ArcTan((b + x)/Sqrt(-Power(b,2) + Power(c,2))))/
    Sqrt(-Power(b,2) + Power(c,2))
```

102

)))/Sqrt(-b**2 + c**2)

FortranForm

(a*ArcTan(

, you may also be interested in **Splice**. Suppose
ve of a function. In the C code, write **<*D[…,…]*>**,
between **<*** and ***>**. Give the file a name ending with
command, and you get a file where the C code is as it was
command has been executed and written in a C form.

If you write
your C prog
including a
.mc. Ther
input, a

■ E**x** **file.tex"]** Write the present notebook into a file in HTML form

ve["file.tex", ConversionOptions → {"MathOutput" → "MathML"}]** Write
.e present notebook into a file in HTML form, converting all typeset expressions to
MathML

HTMLSave["file.tex", "source.nb"] Write the given notebook into a file in HTML
form

File ▷ Save As Special… ▷ HTML Write the present notebook into a file in HTML
form

File ▷ Save As Special… ▷ XML (XHTML+MathML) (❀4!) Write the present notebook into a file in XHTML form, converting all typeset expressions to MathML

MathMLForm[expr] Show the MathML form of the expression

The *Help Browser* has much information about **HTMLSave**.

4.4 Managing Time and Memory

4.4.1 Managing Time Consumption

■ **Information about Time**

Timing[expr] Evaluate **expr**; give the time the kernel has used, together with the
result obtained

AbsoluteTiming[expr] (❀5) Evaluate **expr**; give the absolute elapsed time, together
with the result obtained

TimeConstrained[expr, t] Stop evaluating **expr** after **t** seconds

TimeUsed[] Show the used central processing unit time in the current session

SessionTime[] Show the elapsed time in the current session

Date[] Current time: {year, month, day, hour, minute, second}

The time is measured in steps of **$TimeUnit** which, in many systems, has the default
value of 1/60 second. Note that doing the same calculation again or doing a similar
calculation may take much less time, because *Mathematica* may have already loaded some
files or stored results. An example of **Timing**:

```
FactorInteger[2^255 - 1] // Timing
```

{20.04 Second, {{7, 1}, {31, 1}, {103, 1}, {151, 1},
   {2143, 1}, {11119, 1}, {106591, 1}, {131071, 1}, {949111, 1},
   {9520972806333758431, 1}, {5702451577639775545838643151, 1}}}

If we only want to see the time—not the result—we can use the semicolon. In place of the result, we then have Null:

```
(ran = Table[Random[], {100000}];) // Timing
```

{0.07 Second, Null}

The execution time can be seen in the lower lefthand corner of the window by choosing **Format ▷ Option Inspector...**, setting *Show option values for* to be *notebook*, going to **Notebook Options ▷ Evaluation Options**, and then selecting **ShowTiming** as the value of **EvaluationCompletionAction**. This time is somewhat larger than the time given by **Timing**, because it includes the time to format and show the result.

If we load the **Utilities`ShowTime`** package, we automatically get the **Timing** time for each command.

■ Time-Saving Tips

• It is good to have plenty of random access memory (RAM). To some extent, the more memory *Mathematica* has available, the speedier it is. Also, remember that virtual memory on the hard disk is much slower than actual physical RAM.

• If you work with a heavy and time-consuming problem in several sessions, you perhaps need not start from scratch each time. Instead, try to continue the work from where you left off (see Section 4.3.1, p. 98).

• Avoid using so-called "arbitrary-precision" numbers. Calculations with these numbers are done by software in *Mathematica* and are much slower than calculations done with the hardware-implemented, machine-precision numbers. This point is explained in more detail in Sections 11.2.2, p. 287, and 11.3.1, p. 292. The precision used in calculations can often be set with the **WorkingPrecision** option. Resist changing the default value **MachinePrecision** (※5) (16 in older versions of *Mathematica*) of this option. Of course, if the problem is ill-conditioned and round-off errors have an effect, the option mentioned should be used by giving it a large enough value, like 20. In this way the problem may be solved without difficulties.

• In numerical routines such as **NIntegrate**, **FindRoot**, **FindMinimum**, and **NDSolve**, the default precision may sometimes be more than you actually need (e.g., if the result is used in plotting). Lowering the precision requirements **PrecisionGoal** and **Accuracy-Goal** reduces the calculation time (for these options, see Section 11.3.1, p. 292).

• When plotting largely varying functions, you can sometimes get better results if you increase the value of the **PlotPoints** option from the default value of 25. However, after a certain value you cannot see any difference in the plot, and you merely waste time, because the larger the value of **PlotPoints** is, the longer it takes to produce the plot. When plotting 3D graphics, you can also increase the default value of 25 of the **PlotPoints** option, and then the surface will be smoother. However, the plotting time increases very rapidly as **PlotPoints** increases. Also, the higher the value is, the more grid lines you get, and the result may be a dark and unattractive plot.

• The execution time of a program may significantly depend on the design of the program; see Wagner (1996) for detailed experiments and recommendations. Some points are as follows. Use built-in commands if possible. If the result of the program consists of decimal numbers, use decimal numbers as early as possible (i.e., avoid calculating with exact numbers). Use the functional programming style (see Section 15.2, p. 392). Avoid using **Append**, **AppendTo**, **Prepend**, and **PrependTo** (see Section 15.1.3, p. 391). Compile your functions (see Section 14.2.2, p. 365).

• While awaiting the result of a time-consuming command, you can edit the notebook by adding, for example, text to explain what you have done and what the results mean. You can also write new commands so that they are ready to be executed when the time-consuming command has printed its result.

• You can calculate with *Mathematica* while waiting for the result of a time-consuming command. This can be done by using a subsession. Once you have entered a subsession, the execution of the time-consuming command is interrupted, and you can do some shorter calculations during the subsession. After you exit the subsession, the execution of the time-consuming command continues. To enter a subsession, choose **Kernel ▷ Evalua-tion ▷ Enter Subsession**. Now you can do some shorter calculations. To exit the subsession and continue the long calculation, choose **Kernel ▷ Evaluation ▷ Exit Subsession**.

• A good way to save time may be to use a remote kernel. If your own machine is not powerful enough, you can save time by using the greater power of a remote machine. You can also use several kernels simultaneously. Let a remote kernel (or even several remote kernels) do some of your long and tedious jobs. Meanwhile, you can do more interesting and shorter calculations with the kernel in your machine. In this way you can do several calculations at the same time. To get information about using a remote kernel, open the *Help Browser* and go to **Front End ▷ Menu Commands ▷ Kernel Menu ▷ Kernel Configu-ration Options**.

4.4.2 Managing Memory Consumption

■ Information about Memory

In some systems the memory consumption of the front end of *Mathematica* can be seen with the memory monitor located in the bottom lefthand corner of the notebook window. The black portion is the part used at present, and the whole bar is the complete amount of memory allocated to the front end. The window elements can be set with the *Option Inspector*; go to **Notebook Options ▷ Window Properties ▷ WindowElements**.

In some systems the memory consumption of the kernel of *Mathematica* can be seen by choosing **MathKernel** from the list of open applications and then choosing **Show Memory Usage** from the **File** menu. A separate small window appears and shows the total and used memory of the kernel.

Statistics of the cells in a notebook can be seen from **Cell ▷ Cell Size Statistics…**.

In the following box, we list commands that relate to the consumption of kernel memory. The kernel needs memory to store both the code of *Mathematica* and the results of computations. The latter material is called "data."

> **ByteCount[expr]**   Bytes used by **expr** if sharing (see Memory-Saving Tips below) is
>   not used
> **MemoryConstrained[expr, b]**   Stop evaluating **expr** if more than **b** bytes are needed
>   for the evaluation
> **MemoryInUse[]**   Memory in bytes currently being used to store data
> **MaxMemoryUsed[]**   Maximum memory in bytes used to store data thus far

Note that the kernel of *Mathematica* keeps in RAM all the results of the current session, and this means that the memory consumption of the kernel steadily grows as your session continues. However, *Mathematica* uses kernel memory sparingly and deletes all intermediate results that are no longer needed. In any case, you may run out of memory. If this happens, you get a message telling you the bad news, and you are given only one possibility, which is quitting the kernel. This means that you have to restart the kernel and will probably have to redo some calculations. Note, however, that the front end does not quit, so you do not lose unsaved work in your notebook.

If the front end runs out of memory, you do not always get a message, and the front end may just suddenly disappear, along with your valuable notebook. Unsaved work will be lost, so be sure to save frequently (see also Section 3.1.1, p. 48, for autosaving).

It may be wise to divide long sessions into shorter ones by quitting *Mathematica* now and then and starting a new session with a clean sheet, maximum memory, and maximum speed. In notebook environments, we can also quit the kernel only by choosing **Kernel ▷ Quit Kernel ▷ Local**.

■ Memory-Saving Tips

• Ending a command with the semicolon (**;**) prevents the result from being displayed. This saves memory, especially for large expressions (it also saves your screen area by not cluttering it with uninteresting formulas and so helps you to manage the flow of computation).

  • Use the option **DisplayFunction → Identity** for plots you do not need to show (see Section 5.2.1, p. 117). This technique saves the memory of the front end and the kernel, reduces the size of the notebook, and also saves your screen space.

  • You can delete all output cells by choosing **Kernel ▷ Delete All Output**. This releases front-end memory and makes the notebook smaller. The memory saving may be very substantial if the notebook contains many plots. You can recalculate the commands later on if needed.

  • When you have finished a plot and no longer want to change or print it, you can greatly save front-end memory and the size of the notebook by converting the plot to bitmap form. Select the plot with the mouse and choose **Cell ▷ Convert To ▷ Bitmap**. The plot does not change in any way on the screen, but the PostScript code behind the plot is discarded. Converting graphics to bitmaps is particularly useful in animation because animation is done on the screen, and you do not need the high resolution of PostScript to print the plots. Before converting, make sure the size of the plots is the final size you need.

  • After copying a large expression (e.g., a complex plot), do not use **Edit ▷ Paste** but **Edit ▷ Paste As ▷ Paste and Discard**. This clears the clipboard and releases memory.

• Execute **Share[]** to reduce the amount of kernel memory. This command shares the storage of common subexpressions between different parts of an expression or between different expressions. The output of the command is the memory saved. If we load the **Utilities`MemoryConserve`** package, then **Share** is used automatically whenever the amount of memory used grows by a certain amount.

• A good method of saving memory is to run only the front end on your machine and to run the kernel in another machine. Then only the front end takes memory from your machine, and you can take advantage of the possibly larger RAM and speedier processor of the remote kernel. To get information about using a remote kernel, open the *Help Browser* and go to **Front End** ▷ **Menu Commands** ▷ **Kernel Menu** ▷ **Kernel Configuration Options**.

• Sometimes you do not need the kernel at all. If you do not execute any kernel commands but only read, edit, and print notebooks or run animations, the kernel is not necessary. If you do not run the kernel, the front end will have more memory available.

• If you are doing a very memory-demanding calculation, you can also consider using the kernel only (this may be possible in some systems). When you do this, the front end does not take up valuable RAM, and the kernel has more memory available. To open only the kernel, double-click the **MathKernel** icon ⊛ (and not the front-end icon ⊛). You will get a window similar to the window that is seen in plain Unix.

2D Graphics for Functions

Introduction

> *Straight line — the shortest way between two points. — Euclid*
> *Cycloid — the fastest way between two points. — Johann Bernoulli*
> *Curve — the loveliest way between two points. — Mae West*

The plotting capabilities of *Mathematica* are impressive. There are many ready-to-use commands such as **Plot**, **Plot3D**, **ParametricPlot**, **ParametricPlot3D**, **ListPlot**, **ListPlot3D**, and **ContourPlot**, which often give good results. In case we want to modify the plots, we have many options at our disposal that may help us to obtain just the result we want. And if there is not a suitable plotting command, we can build the plot from graphics primitives, or we can write a program.

We consider plotting in six chapters: 2D graphics for functions in this chapter and the next, 3D graphics for functions in Chapters 7 and 8, and graphics for data in Chapters 9 and 10. In this chapter we study the various 2D plotting commands. The whole next chapter is then devoted to a detailed study of the options for 2D graphics.

This chapter explains the basic techniques used with graphics, several special 2D plotting commands, and graphics primitives and directives. All plots in *Mathematica* are made up of a few *graphics primitives*, and the style of these primitives is controlled by *graphics directives*. The importance of the primitives and directives is explained in Section 5.4.1, p. 132. During your first reading, you may wish to skip the rest of Section 5.4. However, you will soon need knowledge about the primitives and directives, because they are used by many graphics options (e.g., **PlotStyle** and **Epilog**).

Several plotting commands are defined in packages. You may want to read Section 4.1.1, p. 86, about loading packages and about what to do if you mistakenly use a command from a package before you have loaded it. A simple method for using all graphics packages is to execute **<<Graphics`**. The correct graphics package is then automatically loaded if you use a command from a graphics package. In this book, however, we load each package explicitly.

Exporting figures to other applications is explained in Section 4.2.2, p. 94. Working with a complex plot in several sessions is explained in Section 4.3.1, p. 100. Methods for saving computing time and computer memory are considered in Sections 4.4.1, p. 103, and 4.4.2, p. 105; those methods also contain some tips for working with graphics.

Graphics is one of the central parts of *Mathematica* and contains a wealth of material. You may first want to read only the topics you are interested in now and go on to other topics later. More about *Mathematica* graphics can be found in Smith and Blachman (1995) and Wickham-Jones (1994). The latter text contains much information about interesting and important applications of graphics.

5.1 Basic Plotting

5.1.1 Plotting One Curve

Plot[expr, {x, a, b}]   Plot **expr** when **x** takes on values from **a** to **b**

The **expr** can be an explicit expression, the name of an expression, or the name of a function (for functions, see Section 2.2.3, p. 36). So, to plot the density function of the standard normal distribution, we can write the following:

    Plot[Exp[-x^2/2]/Sqrt[2 π], {x, -4, 4}]

Or:

    f = Exp[-x^2/2]/Sqrt[2 π]

    Plot[f, {x, -4, 4}]

Or:

```
g[x_] := Exp[-x^2 / 2] / Sqrt[2 π]
```

```
Plot[g[x], {x, -4, 4}]
```

The second method is often handy, because from the output of **f = …** we can first check that the expression is correct and because the plotting command then becomes simpler and shorter. We try the second method:

```
f = Exp[-x^2 / 2] / Sqrt[2 π]
```

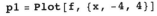

```
p1 = Plot[f, {x, -4, 4}]
```

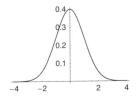

- Graphics -

Below a plot, *Mathematica* prints - Graphics -. It shows the type of the plot (in later chapters we encounter, for example, - SurfaceGraphics - and - Graphics3D -). The type of graphic is unimportant in this chapter, and so from this point on we end all plotting commands with the semicolon (;) so that - Graphics - is not printed.

■ Other Examples

The expression to be plotted can consist of several definitions:

```
mu = 1; sigma = 0.7; c = 1 / (sigma Sqrt[2 π]);
g = Exp[-0.5 ((x - mu) / sigma) ^2];
```

```
p2 = Plot[c g, {x, -4, 4}];
```

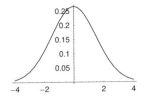

Next we try a function:

```
h[x_, mu_, sigma_] := 1 / (sigma Sqrt[2 π]) Exp[-0.5 ((x - mu) / sigma) ^2]
```

```
p3 = Plot[h[x, 0, 1.5], {x, -4, 4}];
```

We can plot almost any kind of expression. Here is a discontinuous function (the syntax of **If** is **If[condition, then, else]**):

```
Plot[If[x < 3 π / 4, Sin[x], Cos[x]], {x, 0, π}];
```

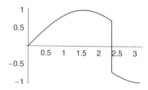

(Note that *Mathematica* plots vertical connecting lines at discontinuous points; actually such a vertical line does not belong to the graph of the function. A better plot for discontinuous functions is obtained by plotting the function in several pieces and then combining the plots [see Section 5.2.1, p. 118].)

Next we plot a function defined with an integral:

```
Plot[NIntegrate[Exp[-t^2], {t, 0, x}], {x, 0, 2}];
```

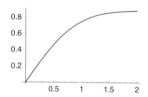

Plot works by first sampling the function to be plotted at 25 (almost) equally spaced points. If the function changes rapidly somewhere in the interval, then more points are sampled automatically in such regions. **Plot** is in this sense adaptive. The sampled points are then joined by straight lines. For more about the algorithm behind **Plot**, see Section 6.4.2, p. 163.

■ Using Options

Mathematica often draws very nice plots. Now and then we may, however, want to make some adjustments. When this is the case, many options are available. Each option has a default value that is used if another value is not given. One of the options is **AspectRatio**. The default value of this option is **1/GoldenRatio** = 0.618. With the default value, we get the following plot:

```
p4 = Plot[Sin[x], {x, 0, 2 π}];
```

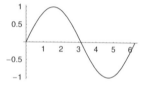

If we want to plot in such a way that one unit on the *x* axis has the same length as one unit on the *y* axis, then we can give **AspectRatio** the value **Automatic**:

```
Plot[Sin[x], {x, 0, 2 π}, AspectRatio → Automatic];
```

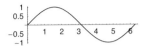

We could also use **Show** and define the option there:

```
Show[p4, AspectRatio → Automatic];
```

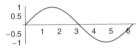

The standard way when plotting is to first use the default values for the options (i.e., writing no options in the plotting command). If the result is not satisfactory, change the value of some of the options. To make changes, we have two approaches. First, we can write the options into the original plotting command and then execute the command anew. Second, we can write the options into a **Show** command. Note that **Show** does not redo the computations; only the way the figure is shown is changed. This means that computer time is saved, at least for complex figures.

■ Important Options

We consider in detail all the options of **Plot** in the next chapter. However, here are some of the most important options. For each option we mention several examples of values. Throughout this book, we use the convention that the default value of each option is always mentioned first.

AspectRatio Ratio of height to width of the plot; examples of values: **1/GoldenRatio** (= 0.618), **Automatic** (one unit on both axes has the same length), **0.4**

PlotRange Range for y in the plot; examples of values: **Automatic**, **All**, **{-1, 1}**

Ticks Ticks on the axes; examples of values: **Automatic**, **{{ π, 2 π, 3 π}, Automatic}**, **{{1, 2, 3}, {-2, -1, 1, 2}}**

Sometimes **Plot** cuts off low or high parts of the function in order to plot the remaining parts more accurately. If you want to see the whole function in the given interval, give **PlotRange** the value **All**. If the ticks of a plot do not satisfy you, define them with **Ticks**. You can specify the ticks on the x or y axis and let **Plot** choose the ticks on the other axis, or you can specify the ticks on both axes. Options can be written in any order, but they must be the last entries in the command, that is, they must be written after the expression and the plotting interval. For example:

```
Plot[Sin[x], {x, 0, 2 π}, AspectRatio → Automatic,
    PlotLabel → "sin(x)", Ticks → {{π, 2 π}, {-1, 1}}];
```

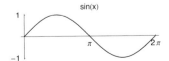

5.1.2 Plotting Several Curves

Show[p1, p2, …]   Combine several plots in the same figure

Plot[{expr1, expr2, … }, {x, a, b}]   Plot the given expressions in the same figure

exprs = {expr1, expr2, … }   Plot the expressions of the list **exprs** in the same figure
Plot[Evaluate[exprs], {x, a, b}]

Table[Plot[expr, {x, a, b}], {i, 1, n}]   Plot several expressions, each separately

Show[GraphicsArray[{p1, p2, … }]]   Show several plots side by side

To get several curves in the same plot, we have two methods. First, plot each curve separately, giving them names, and then combine the plots with **Show**. Second, plot the curves in one command, giving the expressions in a list. With this second method, if the list of expressions **exprs** is formed outside **Plot** (the third case in the table above), then **Evaluate** is needed around the name of the list (otherwise **Plot** does not work). Likewise, **Evaluate** is needed if the list is formed by **Table** inside **Plot**:

   Plot[Evaluate[Table[expr, {i, 1, n}]], {x, a, b}]

We combine the plots **p1** and **p3** we created in Section 5.1.1, p. 108:

   Show[p1, p3];

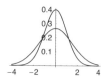

We could also plot the curves in one command:

   h[x_, mu_, sigma_] := 1 / (sigma Sqrt[2 π]) Exp[-0.5 ((x - mu) / sigma) ^ 2]

   Plot[{f, h[x, 0, 1.5]}, {x, -4, 4}];

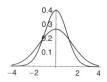

Plots can be arranged side by side or in other ways with **GraphicsArray**. We have already used this command in Section 2.1.1, p. 26, and we will consider it in more detail in Section 5.2.2, p. 120. Here is an example:

   Show[GraphicsArray[{p1, p2}]];

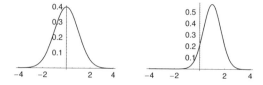

We have here enlarged the figure with the mouse (see Section 5.1.3, p. 115). The enlargement has to be done almost always for figures produced by **GraphicsArray**.

Next we consider conflicting settings for options and the use of **Evaluate**. If this is your first time through this chapter, you can skip these topics.

■ How Show Handles Conflicting Settings

When **Show** is used to combine plots, *Mathematica* has to decide what to do if the plots have conflicting settings. Indeed, the plotting intervals may be different, and the values of some options may be different. *Mathematica* uses the following rules:

- use the *union* of the plotting intervals;
- use the values of the options of the *first* plot; and
- options given in **Show** override those of the first plot.

If the union of the intervals is not suitable, we can give in **Show** a suitable value of **PlotRange** to define a better interval. Indeed, this option can also be used to define the *x* range, as in **{{0, 2 π}, {-1, 1}}**.

The second rule, as inevitable it is for *Mathematica*, may cause trouble for us. It means that all option settings other than those in the first plot are simply discarded! We have perhaps carefully set various options in all plots, but when the plots are combined, only the options from the first plot are used. To correct the situation, in **Show**, we can give some of the options we have used in the plots. If you know in advance that you do not need the separate plots as such but will combine the plots into one figure, then fine tune only the combined plot by giving suitable options in **Show**; do not fine tune each plot.

As an example, consider the following plots (we use **Block** to prevent the showing of the figures [see Section 5.2.1, p. 118]):

```
Block[{$DisplayFunction = Identity},
  p1 = Plot[Sin[x], {x, -π, π}];
  p2 = Plot[Cos[x], {x, 0, 2 π}];
  p3 = Plot[Sin[x], {x, 0, 2 π}, PlotRange → {0, 1}];
  p4 = Plot[Cos[x], {x, 0, 2 π}, PlotRange → {-1, 0}];
  p5 = Plot[Cos[x], {x, 0, 2 π}, Ticks → {{π, 2 π}, {1}}];]
```

Plots **p1** and **p2** have different plotting ranges, so the combined plot uses their union:

```
Show[p1, p2];
```

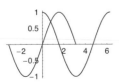

Plots **p3** and **p4** have conflicting plot ranges, so the plot range of the first plot is used:

```
Show[p3, p4];
```

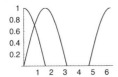

Plots **p1** and **p5** have conflicting ticks (the first plot uses the default ticks), so the ticks of the first plot are used:

    Show[p1, p5];

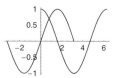

We can adjust the options in **Show**:

    Show[p1, p5, PlotRange → {{0, 2 π}, {0, 1}}, Ticks → {{ π, 2 π}, {1}}];

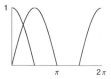

■ Using Evaluate

In the box above, we encountered **Evaluate** for the plotting of a previously defined *list of expressions*:

    exprs = {expr1, expr2, … }
    Plot[Evaluate[exprs], {x, a, b}]

Plot simply does not work without **Evaluate** in this case. Also, **Evaluate** may be useful when plotting *a named expression* or *an expression defined as a function*, because in these cases **Evaluate** speeds up the plotting. In simple plots the saving of time is not considerable, but for complex, time-consuming plots, the saved time may be noticeable. Thus, consider using **Evaluate** in such expressions as the following:

    f = expr
    Plot[Evaluate[f], {x, a, b}]

Another example:

    g[x_] := expr
    Plot[Evaluate[g[x]], {x, a, b}]

In these cases, plotting without **Evaluate** may take extra time.

The explanation for the use of **Evaluate** can be given with the attributes of **Plot**:

    Attributes[Plot]

    {HoldAll, Protected}

The attribute **HoldAll** means that **Plot** does not evaluate its argument before plotting. It only substitutes several values of x into the expression and tries to get a numerical value. In general, this is a valuable feature, but for plotting a list of expressions either created with **Table** in the plotting command or defined earlier, the list must first be evaluated so that **Plot** gets the desired list in an explicit form into which numerical values of x can be inserted.

If we plot a named expression or a function, **Plot** is not able—without **Evaluate**—to use compiling (for compiling, see Section 14.2.2, p. 365). **Evaluate** enables the compiling of the expression, and the values of a compiled expression can be calculated speedily.

5.1.3 Manipulating and Animating

■ Manipulating

In notebook environments we can change a plot in several ways using the mouse. The plot must first be selected by clicking on it so that the selection rectangle with eight handles becomes visible.

- To *move* the plot on the screen, place the cursor on the plot, and drag with the mouse. The position (given in printer's points; 72 printer's points equals one inch) of the top lefthand corner of the plot can be read from the bottom lefthand corner of the window.
- To *resize* the plot, click on the plot and drag with the mouse by one of the handles. The width and height of the plot (in printer's points) can be read from the bottom lefthand corner of the window. Note that the size of all text remains unchanged when resizing the plot. Also, the form of the plot remains the same (use the **AspectRatio** option to change the form of the plot). Changing the default size of figures is considered shortly.
- To *crop* the plot, click on the plot, hold down the CTRL key in Windows or the ⌘ key on Macintosh, and drag by one of the handles. The size of the plot can be read from the bottom lefthand corner of the window. By cropping you can either make more room around the plot or cut off some parts of the plot (croppings are not permanent).
- To *go back* to the standard position, size, and cropping, click on the plot and choose **Make Standard Size** from the **Cell** menu.
- To *read coordinates* from the plot, click on the plot, hold down the CTRL key in Windows or the ⌘ key on Macintosh, and move the cursor on the plot without pressing the button on the mouse. The coordinates appear in the lower lefthand corner of the window.
- To *copy coordinates* from the plot, click on the plot, hold down the CTRL or ⌘ key, and click or drag the mouse on the plot. If you then choose **Edit ▷ Copy**, place the cursor somewhere, and choose **Edit ▷ Paste**, the coordinates of the point or points are pasted. (On a Macintosh, the selected points also appear on the plot, but they disappear when you click on the plot.)
- To *align* several plots according to the first plot, select the plots by dragging over the cell brackets so that all plots to be aligned become selected (do not worry if some non-graphics cells also become selected), and then choose **Cell ▷ Align Selected Graphics....** Plots can be aligned with respect to any side of the first plot. Plots can also be made the same size as the first plot.
- To *convert to bitmap*, select a plot or several plots and choose **Cell ▷ Convert To ▷ Bitmap**. This method allows you to save memory; see Section 4.4.2, p. 105, for more details.
- To *render* graphics anew, select a plot or several plots and choose **Cell ▷ Render Graphics**. Indeed, after you have changed the magnification percentage from **Format ▷ Magnification**, the quality of all existing graphics is poor on the screen. To again get high-quality graphics for the existing plots on the screen, render all graphics anew.

To change the default size of all future plots in the current notebook or in all current and future notebooks, choose **Format ▷ Option Inspector...**, select *Show option values for* to be *notebook* (if you want to change the size in the current notebook only) or *global* (if you want to change the size in all notebooks), go to **Graphics Options ▷ Image Bounding Box**, and change the width and height values for **ImageSize**. The default value is 288 printer's points. Divide the value by 72 to get the width of figures in inches. For example, with the default value, the width of the figure is 4 inches. Once you have typed a new value, remember to press RET. Note that the sizes of old plots do not change according to the new standard size unless you replot them.

■ Animating

Animation is a useful way to illustrate the development of a series of plots. Of course, we can draw and show (with **GraphicsArray**) a sequence of plots separately, but this is not so illustrative as if we had showed all the plots in one place, one right after another. In the notebook environment, animations can be made as follows:

- Generate a sequence of plots by executing, for example, a command of the type **Do[Plot[expr, {x, a, b}], {c, c0, c1, dc}]**, where a parameter **c** (appearing in **expr**) gets values from **c0** to **c1** in steps of **dc**.
- Double-click any of the plots (or select all of the plots by clicking their cell bracket and choose **Cell ▷ Animate Selected Graphics**); animation starts.
- Control the animation from the buttons at the bottom lefthand corner of the window.

To hide all but the first plot before animating, double-click the cell bracket of the plot sequence. When the animation has started, the following buttons appear at the bottom of the window:

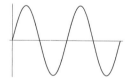

By pressing these buttons, we can adjust the speed and direction of the animation (the buttons mean the following: backward, back and forth, forward, stop/continue, slower, and faster, respectively). The animation can also be stopped by clicking with the mouse.

As an example, we plot 20 figures but show here only the first of them (all of the figures are shown on the CD-ROM version of the book). The series of plots can then be animated:

Do[Plot[Sin[x - c], {x, 0, 4 π}, Ticks → None], {c, 0, 2 π - π / 10, π / 10}]

Note that, if the height of the plots varies in the series, we must specify for each plot the same **PlotRange** (see Section 5.1.1, p. 111) so that the animation works properly. Try the following example:

```
Do[Plot[Exp[-x^2] x^n, {x, 0, 2},
    PlotRange → {0, 1}, Ticks → None], {n, 0, 5, 0.5}];
```

Also, if the plots are not properly aligned, the animation looks bad. Select the plots and choose **Cell ▷ Align Selected Graphics**....

Although the sequence of plots used in an animation can be easily generated with **Do**, we also have **Animate** in the **Graphics`Animation`** package.

■ Sound

With the command **Play**, we can hear sound. For example, the following command plays a pure tone with a frequency of 400 hertz for two seconds:

> **Play[Sin[2 π 400 t], {t, 0, 2}]**

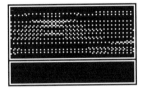

- Sound -

The figure contains an approximation of the sound's wave form. Double-clicking the speaker in the cell bracket plays the sound again. Double-clicking the sound figure plays the sound until you click the mouse.

5.2 Arranging Plots

5.2.1 Suppressing Display

Figures can be arranged in various ways with **GraphicsArray** and **Rectangle**. Before we explain these commands, we consider ways of not displaying a figure at all. The key point is that in books, to save space, usually only the final, combined plot is displayed; the intermediate plots are not shown. This technique can also be used in routine interactive use of *Mathematica* to save screen space, minimize scrolling of the notebook, and reduce RAM consumption.

■ Using DisplayFunction

All plotting commands have the option **DisplayFunction**, which is used to tell whether or not to display the plot on the screen. Essentially only two values are used for this option: **Identity** and **$DisplayFunction**. These values are used as follows:

```
p1 = Plot[expr1, {x, a, b}, DisplayFunction → Identity];
p2 = Plot[expr2, {x, c, d}, DisplayFunction → Identity];

Show[p1, p2, DisplayFunction → $DisplayFunction];
Show[GraphicsArray[{p1, p2}]];
```

The value **Identity** is first used in some plotting commands to suppress the display of the plots. The value **$DisplayFunction** is then used in **Show** to display all the plots in the same figure. If the plots are shown with **GraphicsArray**, then the **DisplayFunction** option need not be used (this is because **GraphicsArray** has its own display function). The actual value of **$DisplayFunction** is the following:

`$DisplayFunction`

`Display[$Display, #1] &`

As an example, we first plot two figures without showing the results:

```
p1 = Plot[Sin[x], {x, 0, 2 π}, DisplayFunction → Identity];
p2 = Plot[Cos[x], {x, 0, 2 π}, DisplayFunction → Identity];
```

Then we combine the plots:

`Show[p1, p2, DisplayFunction → $DisplayFunction];`

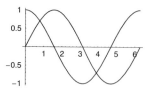

We then show them side by side:

`Show[GraphicsArray[{p1, p2}]];`

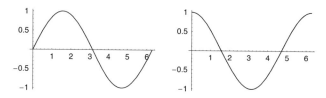

If you are doing the computations of this book with your computer, you can also drop the option `DisplayFunction → Identity` so that you can see the intermediate plots.

■ Using Block

```
Block[{$DisplayFunction = Identity},
  p1 = Plot[expr1, {x, a, b}];
  p2 = Plot[expr2, {x, c, d}];
  p3 = Plot[expr3, {x, e, f}];]
Show[p1, p2, p3];
Show[GraphicsArray[{p1, p2, p3}]];
```

If we plot several functions and want to show only the final combined figure, then it may be handy to use the **Block** construct, where we define the local value **Identity** for `$DisplayFunction`. Then we do not need to write `DisplayFunction → Identity` in each plotting command. Also, when using **Block**, we do not need to define `DisplayFunction →` `$DisplayFunction` in **Show**. **Block** is used frequently in the graphics chapters of this book (for more information about **Block**, see Section 14.1.5, p. 362).

As an example, consider the following plot:

```
Plot[Tan[x], {x, 0, 2 π}, PlotRange → {-16, 16}];
```

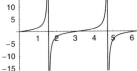

You can see two vertical lines (like asymptotes), which you may want to avoid. You can plot the function separately in each interval of continuity and then combine the plots:

```
Block[{$DisplayFunction = Identity},
 p1 = Plot[Tan[x], {x, 0, π / 2}];
 p2 = Plot[Tan[x], {x, π / 2, 3 π / 2}];
 p3 = Plot[Tan[x], {x, 3 π / 2, 2 π}];]

Show[p1, p2, p3, PlotRange → {-16, 16}];
```

■ Using a Package

> *In the* **Graphics`Graphics`** *package:*
>
> **DisplayTogether[p1, p2, …, opts]**
> **DisplayTogetherArray[plotarray, opts]**

These commands can be used to plot several figures at a time without displaying each figure separately. With **DisplayTogether**, we can use options that are suitable for a combined plot. **DisplayTogetherArray** accepts options suitable for **GraphicsArray** (like **GraphicsSpacing**; see Section 5.2.2, p. 120). Here are two examples:

```
<< Graphics`Graphics`

DisplayTogether[
  Plot[Sin[x], {x, 0, 2 π}],
  Plot[Cos[x], {x, 0, 2 π}],
  AspectRatio → Automatic];
```

```
DisplayTogetherArray[
  {Plot[Sin[x], {x, 0, 2 π}, AspectRatio → Automatic],
   Plot[Cos[x], {x, 0, 2 π}, AspectRatio → Automatic]},
  GraphicsSpacing → 0.2];
```

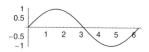 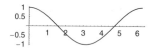

5.2.2 Arranging Plots

■ Regular Arrangements

With **GraphicsArray**, we get regular arrangements of plots; the plots are of the same size and form a matrix-like arrangement. Here are some simple examples (generalizations are obvious):

Show[GraphicsArray[{p1, p2}]]   The arrangement is $p_1\ p_2$

Show[GraphicsArray[{{p1}, {p2}}]]   The arrangement is $\begin{matrix} p_1 \\ p_2 \end{matrix}$

Show[GraphicsArray[{{p1, p2}, {p3, p4}}]]   The arrangement is $\begin{matrix} p_1 & p_2 \\ p_3 & p_4 \end{matrix}$

A special option of **GraphicsArray**:

GraphicsSpacing   Defines the horizontal and vertical space between plots as fractions of the width and height of the plots; examples of values: **0.1, {0.15, 0.1}, 0, −0.1**

The value of the **GraphicsSpacing** option can be either a single number, which then defines both the horizontal and the vertical spaces, or a list of two numbers, which define separately the horizontal and the vertical spaces. For example, the default value **0.1** means the same as **{0.1, 0.1}**. Negative values cause the figures to overlap. In addition to **GraphicsSpacing**, **GraphicsArray** has all the same options as **Graphics** (see Section 6.1.2, p. 153).

Shown here is an example in which we also draw a frame around the combined plot (**Frame** and other options are considered in Chapter 6):

```
Block[{$DisplayFunction = Identity},
 p1 = Plot[Sin[x], {x, 0, 2 π}];
 p2 = Plot[Cos[x], {x, 0, 2 π}];
 p3 = Plot[Tan[x], {x, 0, 2 π}, PlotRange → {-16, 16}];
 p4 = Plot[Cot[x], {x, 0, 2 π}, PlotRange → {-16, 16}];]

Show[GraphicsArray[{{p1, p2}, {p3, p4}},
  GraphicsSpacing → 0.05, Frame → True]];
```

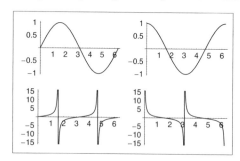

   The normal size of a graphics array is the same as the size of a usual plot. This means that we must almost always enlarge graphics arrays. Another possibility is to use the **ImageSize** option, either in the command containing **GraphicsArray** or in a **SetOptions** command such as **SetOptions[GraphicsArray, ImageSize → 300];**.

■ A Problem with AspectRatio

There may be some problems with **GraphicsArray** if you use your own setting for **Aspect・Ratio**. With the default setting, the result is neat:

```
Block[{$DisplayFunction = Identity},
 p1 = Plot[Sin[x], {x, 0, 4 π}];
 p2 = Plot[Tan[x], {x, 0, 2 π}, PlotRange → {-3.2, 3.2}];]

Show[GraphicsArray[{p1, p2}]];
```

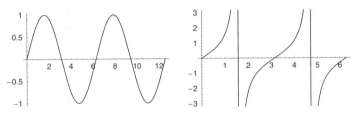

However, if you define another **AspectRatio**, then the result may be as follows:

```
Block[{$DisplayFunction = Identity},
 p3 = Plot[Sin[x], {x, 0, 4 π}, AspectRatio → Automatic];
 p4 = Plot[Tan[x], {x, 0, 2 π},
   PlotRange → {-3.2, 3.2}, AspectRatio → Automatic];]

Show[GraphicsArray[{p3, p4}, GraphicsSpacing → 0]];
```

The latter component of the figure seems to be too small, and there is too much space between the components. Setting a suitable **AspectRatio** for **GraphicsArray** helps:

```
Show[GraphicsArray[{p3, p4}, GraphicsSpacing → 0, AspectRatio → 0.25]];
```

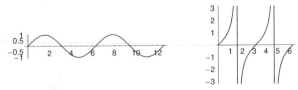

Hardly any problems arise if we use the same vertical **PlotRange** for all components.

■ Stacking Graphics

With **ParametricPlot3D**, we can arrange 2D plots in a 3D parallelepiped. As an example, we plot $\sin(x)^n$ for various values of n. First we form a table of the functions to be plotted. The functions are represented in a parametric form, and we have added a fourth argument to specify the color of the corresponding curve (we can also use three arguments):

```
t1 = Table[{x, n, Sin[x]^n, Hue[n / 10]}, {n, 10}];
```

When plotting curves, it is advantageous to change the form of the surrounding box with **BoxRatios** to make room for the curves and also to change the viewpoint with **ViewPoint**, so that the curves do not overlap very much (such options are explained in Chapter 8):

```
ParametricPlot3D[Evaluate[t1], {x, 0, π}, ViewPoint → {3, -1.5, 0.5},
    BoxRatios → {1, 4, 0.6}, Ticks → {{0, 1, 2, 3}, Range[10], {0, 1}}];
```

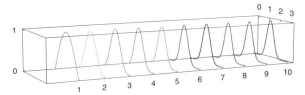

We could also use **StackGraphics** from the **Graphics`Graphics3D`** package.

■ Irregular Arrangements

```
Show[Graphics[{ Rectangle[{x1, y1}, {x2, y2}, p1],
                Rectangle[{x3, y3}, {x4, y4}, p2] }]];
```

In this example we arrange two plots, **p1** and **p2**, such that **p1** is placed in a rectangle with the lower lefthand coordinate of **{x1, y1}** and the upper righthand coordinate of **{x2, y2}** and **p2** is placed in a rectangle with the corresponding coordinates of **{x3, y3}** and **{x4, y4}**. The relative values of **xi** and **yi** determine the positions of the plots. Often it is easiest to let the x and y coordinates of the combined plot run from 0 to 1. More generally, **Rectangle** can be used to arrange as many plots as we want.

With **Rectangle**, we can place the plots so that they overlap and are of different sizes:

```
Show[Graphics[{Rectangle[{0, 0}, {0.8, 1}, p1],
      Rectangle[{0.6, 0.5}, {1, 1}, p2]}], AspectRatio → 0.4];
```

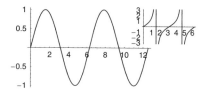

5.3 Other Plots for 2D Functions

5.3.1 Parametric Plots

```
ParametricPlot[{fx, fy}, {t, a, b}]   Make a parametric plot
```

We can define a function parametrically by giving a list of two expressions **{fx, fy}**. The expressions define the x and y coordinates of the function and are functions of a parameter, often denoted by **t**. The usual plotting command **Plot[expr, {x, a, b}]** can be seen as the special case `ParametricPlot[{x, expr}, {x, a, b}]` of the parametric plotting command. `ParametricPlot` has the same options and default values as `Plot` (the options are considered in Chapter 6).

With parametric plots, it is usually advantageous to define `AspectRatio → Automatic`, because then, for example, a circle looks like a circle (see `p1` below) and not like an ellipse. Here we use `SetOptions` to set the values of `AspectRatio` and `Axes`:

```
SetOptions[ParametricPlot, AspectRatio → Automatic, Axes → False];

Block[{$DisplayFunction = Identity},
 p1 = ParametricPlot[{Cos[t], Sin[t]}, {t, 0, 2 π}, Axes → True];
 p2 = ParametricPlot[{2 Cos[t], Sin[t]}, {t, 0, 2 π}];
 p3 = ParametricPlot[
   {(2 Cos[t] - 1) Cos[t], (2 Cos[t] - 1) Sin[t]}, {t, 0, 2 π}, Axes → True];
 p4 = ParametricPlot[{t Cos[t], t Sin[t]}, {t, 0, 12 π}];
 p5 = ParametricPlot[{t Cos[t] Sin[t], t Sin[t]^2}, {t, 0, 8 π}];
 p6 = ParametricPlot[{Sin[2 t] + Sin[5 t], Cos[2 t] + Cos[5 t]}, {t, 0, 2 π}];
 p7 = ParametricPlot[{Sin[2 t] Sin[5 t], Cos[2 t] Sin[5 t]}, {t, 0, 2 π}];
 p8 = ParametricPlot[{Cos[t] + 1 / 2 Cos[7 t] + 1 / 3 Cos[-17 t + π / 2],
   Sin[t] + 1 / 2 Sin[7 t] + 1 / 3 Sin[-17 t + π / 2]}, {t, 0, 2 π}];]

Show[GraphicsArray[{{p1, p2, p3, p4}, {p5, p6, p7, p8}}]];
```

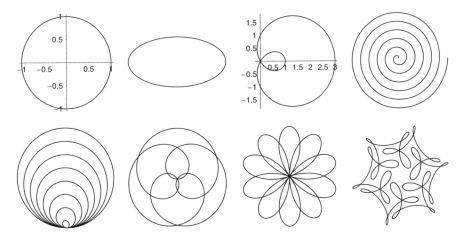

The `Graphics`Graphics`` package defines `PolarPlot[r, {th, a, b}]`, thereby plotting the radius `r` as a function of angle `th`.

5.3.2 Implicit Plots

■ Plotting an Implicit Function

An implicit function is defined by an equation `expr1 == expr2` (note the two equals signs needed in equations) in which the expressions are functions of two variables, often denoted by `x` and `y`. From the equation we can, at least in principle, solve `y` for each `x`, and so we get a function `y` of `x`. With `ImplicitPlot`, we can plot implicit functions.

> *In the* **Graphics`ImplicitPlot`** *package:*
>
> **ImplicitPlot[expr1 == expr2, {x, a, b}]**   Plot by using **Solve**
> **ImplicitPlot[expr1 == expr2, {x, a, b}, {y, c, d}]**   Plot by using **ContourPlot**

Regarding the options of **ImplicitPlot**, the default value of **PlotPoints** is 39 and that of **AspectRatio** is **Automatic**.

We have two methods for plotting implicit functions. The first method is to solve the equation for **y** by using several values for **x**. This method is used if we supply only the **x** range. **Solve** is used to solve the **y** values, and this means that essentially only algebraic implicit functions can be plotted with this method (an expression is algebraic if it contains the basic arithmetic operations and rational powers). With this method, we can use the same options as with **Plot**.

Another method is to plot a contour plot for the function **expr1 - expr2** by asking for only one contour: the one in which the function takes on the value 0. This method is used if we also supply the **y** range. The method actually uses **ContourPlot** (to be considered in Sections 7.1.1, p. 180, and 8.2.2, p. 222) so that the options of this command can be used. **ContourPlot** is also easy to use directly:

> **ContourPlot[expr1 - expr2, {x, a, b}, {y, c, d}, Contours → {0},**
>   **ContourShading → False, AspectRatio → Automatic]**

ContourPlot (either with **ImplicitPlot** or directly) is a faster plotting method and can also handle nonalgebraic or transcendental expressions, but the result may not be as good as with the other method, especially around singularities or intersections of the curve.

ImplicitPlot also has the following form:

> **ImplicitPlot[expr1 == expr2, {x, a, m1, m2, …, b}]**

In this case, the (possibly problematic) points **m1**, **m2**, and so on are avoided in the generation of the plot. In addition, we can plot several implicit functions at a time by giving a list of equations.

Note that, while **ImplicitPlot** plots the points satisfying an equality, **InequalityPlot** (see Section 5.3.3, p. 127) plots the points satisfying one or more inequalities.

■ Examples

```
<< Graphics`ImplicitPlot`

Block[{$DisplayFunction = Identity},
  p1 = ImplicitPlot[x^3 - 2 x y + y^3 == 0, {x, -1, 1.1}];
  p2 = ImplicitPlot[x^3 - 2 x y + y^3 == 0, {x, -1, 1.1}, {y, -1.7, 1.2}];
  p3 = ImplicitPlot[x^3 - 2 x y + y^3 == 0,
    {x, -1, 1.1}, {y, -1.7, 1.2}, PlotPoints → 100];
  p4 = ContourPlot[x^3 - 2 x y + y^3, {x, -1, 1.1}, {y, -1.7, 1.2},
    Contours → {0}, ContourShading → False, AspectRatio → Automatic,
    Frame → False, Axes → True, AxesOrigin → {0, 0}, PlotPoints → 100]];
```

```
Show[GraphicsArray[{p1, p2, p3, p4}]];
```

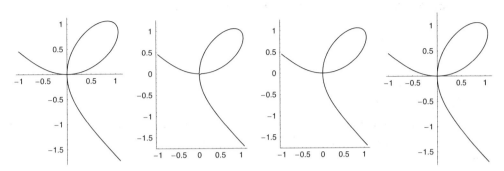

The first figure was generated with **Solve**, and the result is good. The second figure was generated with **ContourPlot**, and the result is not so good (as would be seen if the plot were larger). However, if we increase the default value of **PlotPoints** from 39 to, say, 100, then the result is good, as can be seen from the third figure. In the fourth figure, we have used **ContourPlot** directly.

■ Plotting a Function of an Implicit Function

We may have the situation in which **y** is defined implicitly as a function of **x** by an equation **expr1 == expr2** but we want to plot, by using **x** as the independent variable, not the plain **y** but an expression **expr3** containing **y** (and possibly also **x**). In this way, **expr3** is a function of the implicit function **y**. Is this too complicated? Not for *Mathematica*. Try the following:

```
Plot[y = y /. FindRoot[expr1 == expr2, {y, y0}]; expr3, {x, a, b}]
```

Here we have used the fact that with **Plot** we can enter some commands before the expression to be plotted. The command we have used finds, for each **x** used by **Plot**, the corresponding **y**, so that the value of **expr3** can then be calculated. In **FindRoot**, we have specified only one starting value, **y0**, so that Newton's method is used. If derivatives cannot be calculated, supply two starting values.

■ An Example

Consider the salmon model of Mesterton-Gibbons (1989, p. 62). In studying the stability of a period-2 equilibrium cycle, we have the following equation:

```
eqn = y Exp[r (1 - y)] == 2 - y;
```

This implicitly defines **y** for each **r**, and we want to plot the following expression as a function of **r**:

```
delta = Abs[(1 - r y) (1 - 2 r + r y)];
```

First we plot the implicit function defined by **eqn**:

```
ImplicitPlot[eqn, {r, 1.9, 2.8},
  {y, 0, 2}, AspectRatio → 0.6, PlotPoints → 50];
```

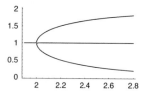

So, **y** as a function of **r** has multiple values. When plotting **delta** for **r** in (2, 2.8), we are interested in the smallest **y** value. Plot **delta** as follows (we also plot the line 1 with the **Epilog** option):

```
Plot[y = y /. FindRoot[y Exp[r (1 - y)] == 2 - y, {y, 0.1}];
  delta, {r, 2, 2.8}, Epilog → Line[{{0, 1}, {2.8, 1}}]]; y =.
```

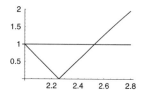

We used the starting value 0.1, for which we get the smallest **y** value (as can be verified).

The stability of the cycle requires that **delta** be smaller than 1. From the plot above, we see that **r** should be at most about 2.5. To get a more accurate value, we solve a system of two equations:

```
FindRoot[{delta == 1, eqn}, {r, 1, 2}, {y, 0, 0.1}]
```

$\{r \to 2.52647, y \to 0.277704\}$

We have given two starting values for **r** and **y** so that **FindRoot** does not use Newton's method (this method requires derivatives, and *Mathematica* is not able to calculate the derivative of the **Abs** function in **delta**).

5.3.3 Filled and Inequality Plots

■ **Filled Plots**

In the **Graphics`FilledPlot`** *package:*

FilledPlot[expr, {x, a, b}] Fill the space between the curve and the *x* axis
FilledPlot[{expr1, expr2, … }, {x, a, b}] Fill the space between several curves

Special options:

Fills The colors used in the fills; examples of values: **Automatic, GrayLevel[0.9],**
 RGBColor[0,0,1], {{Axis, 1}, GrayLevel[0.9]}, {{1, 2}, GrayLevel[0.8]}}
Curves Whether each curve is on the same layer as the corresponding fill (**Back,**
 causing later fills to hide parts of earlier curves), all curves are on top of fills (**Front,**
 causing all curves to be wholly visible), or the curves are not drawn at all (**None**)
AxesFront Whether axes are on top of the fills (**True,** causing axes to be wholly
 visible) or behind the fills (**False,** causing fills to possibly hide parts of the axes)

FilledPlot has the three options mentioned, in addition to the options of **Plot**. For example:

```
<< Graphics`FilledPlot`

Block[{$DisplayFunction = Identity},
  p1 = FilledPlot[Cos[x], {x, 0, 2 π}];
  p2 = FilledPlot[{Sin[x], If[π / 4 < x < 3 π / 4, 0, Sin[x]]},
    {x, 0, π}, Fills → GrayLevel[0.8]];]

Show[GraphicsArray[{p1, p2}]];
```

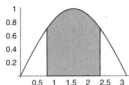

In the following plot, we define the colors to be used between succeeding curves. The *x* axis can be denoted by **Axis**:

```
FilledPlot[{Tanh[x], 2 Tanh[x], 3 Tanh[x], 4 Tanh[x]}, {x, 0, 2 π},
  Fills → {{{Axis, 1}, GrayLevel[0.9]}, {{1, 2}, GrayLevel[0.8]},
    {{2, 3}, GrayLevel[0.7]}, {{3, 4}, GrayLevel[0.6]}}];
```

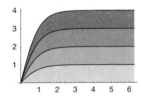

■ Inequality Plots

In the **Graphics`InequalityGraphics`** *package:*

InequalityPlot[ineqs, {x, a, b}, {y, c, d}]   Show by a filled plot the region
  satisfying the given logical combination of nontranscendental inequalities

Special options:
Fills, Curves, AxesFront, BoundaryStyle

Inequalities can be combined with logical operations like **&&** (AND), **||** (OR), **!** (NOT), or **Xor** (exclusive OR); these are explained in Section 12.3.5, p. 314. If the region is bounded, we can specify the variables simply with **{x}** and **{y}**. **InequalityPlot** has the same three special options as **FilledPlot** plus **BoundaryStyle**. This latter option defines the style used to show the boundaries of the region (the default value is **{}**, meaning that the boundaries are black). The default value of **AspectRatio** is **Automatic**. For example:

```
<< Graphics`InequalityGraphics`
```

```
Block[{$DisplayFunction = Identity},
  p1 = InequalityPlot[x^2 < y < Sqrt[x], {x}, {y},
    BoundaryStyle → {AbsoluteThickness[2], RGBColor[1, 0, 0]}];
  p2 = InequalityPlot[1 < x^2 - 1.5 x y + y^2 < 2, {x}, {y}];
  p3 = InequalityPlot[x^3 - 2 x y + y^3 < 0, {x, -1, 1.1}, {y, -1.7, 1.2}];
  p4 = InequalityPlot[Xor[(x + 1)^2 + y^2 < 2, (x - 1)^2 + y^2 < 3], {x}, {y}];]

Show[GraphicsArray[{p1, p2, p3, p4}]];
```

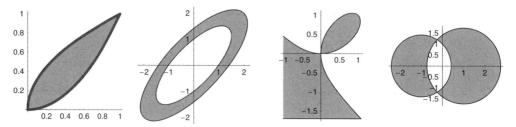

5.3.4 Complex Plots

■ **Using Re, Im, and Abs**

The function to be plotted with **Plot** must have real values. Where the function has complex values, nothing is plotted, and warnings are printed (in this example, we have deleted some copies of the same warning):

```
Plot[Sqrt[Cos[x]] - 0.5, {x, 0, 2 π}];
```

Plot::plnr : $\sqrt{Cos[x]}$ - 0.5 is not a machine-
    size real number at x = 1.5854770701281846`. **More…**

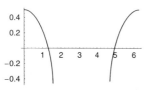

One possibility to plot functions that have complex values is to use **Re**, **Im**, or **Abs** to plot only the real or imaginary part or the absolute value:

```
Block[{$DisplayFunction = Identity},
  p1 = Plot[Re[Sqrt[Cos[x]] - 0.5], {x, 0, 2 π}];
  p2 = Plot[Im[Sqrt[Cos[x]] - 0.5], {x, 0, 2 π}];
  p3 = Plot[Abs[Sqrt[Cos[x]] - 0.5], {x, 0, 2 π}];]

Show[GraphicsArray[{p1, p2, p3}]];
```

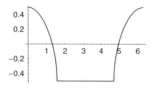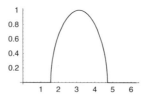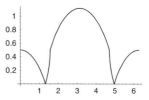

For other methods of illustrating complex-valued functions, see the packages `Graphics`ComplexMap``, `Graphics`ArgColors``, and `Graphics`PlotField`` (the last package defines `PlotPolyaField`).

■ Avoiding Complex Powers

Consider the function $x^{1/3}$. For a given x like -2.0, it takes on three values, which can be obtained by solving the equation $y^3 = -2.0$:

```
Solve[y^3 == -2.0]
```

$\{\{y \rightarrow -1.25992\}, \{y \rightarrow 0.629961 - 1.09112\,i\}, \{y \rightarrow 0.629961 + 1.09112\,i\}\}$

Let us look at what value *Mathematica* gives us for $(-2)^{1/3}$:

```
(-2.) ^ (1 / 3)
```

$0.629961 + 1.09112\,i$

We have a complex value rather than the real value -1.25992. In general, *Mathematica* gives the value with the smallest positive argument. To plot such a function, one possibility is to use an **If** construct:

```
Plot[If[x ≥ 0, x^(1/3), -(-x)^(1/3)], {x, -2, 2}];
```

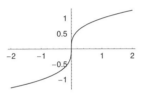

We can also load a package that automates the task of picking a real value:

```
<< Miscellaneous`RealOnly`
```

```
(-2.) ^ (1 / 3)
```

-1.25992

```
Plot[x^(1/3), {x, -2, 2}];
```

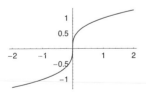

The package redefines the power function so that odd roots of negative numbers are negative instead of complex. More exactly, $b^{m/n}$ gives $(-(-b)^{1/n})^m$ if b is negative and m and n are integers with n being odd.

■ Complex Inequalities

In the `Graphics`InequalityGraphics`` *package:*

`ComplexInequalityPlot[ineqs, {z, zmin, zmax}]` Show by a filled plot the region satisfying the given logical combination of inequalities of the complex variable **z**

The expressions in **ineqs** must be made to have real values with **Re**, **Im**, or **Abs**. For example:

```
<< Graphics`InequalityGraphics`

Block[{$DisplayFunction = Identity},
  p1 = ComplexInequalityPlot[Abs[z] ≤ 1, {z}];
  p2 = ComplexInequalityPlot[Abs[1 - z^3] ≤ 1, {z}];
  p3 = ComplexInequalityPlot[
    Abs[1 - z^2] < Abs[1 - z + z^2], {z, -0.5 - 2 I, 2 + 2 I}];]

Show[GraphicsArray[{p1, p2, p3}]];
```

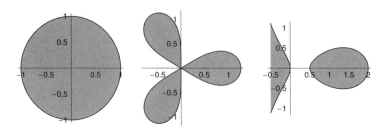

5.3.5 Miscellaneous Plots

■ Logarithmic Plots

In the **Graphics`Graphics`** *package:*

LinearLogPlot[expr, {x, a, b}]   A plot of **Log[expr]** as a function of **x**
    = **LogPlot[expr, {x, a, b}]**
LogLinearPlot[expr, {x, a, b}]   A plot of **expr** as a function of **Log[x]**
  LogLogPlot[expr, {x, a, b}]   A plot of **Log[expr]** as a function of **Log[x]**

```
<< Graphics`Graphics`

Block[{$DisplayFunction = Identity},
  p1 = Plot[Exp[x], {x, 0, 4}];
  p2 = LogPlot[Exp[x], {x, 0, 4}, GridLines → Automatic]];

Show[GraphicsArray[{p1, p2}]];
```

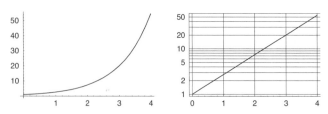

The exponential function becomes a straight line in a logarithmic scale. Logarithmic plots have the same options as **Plot**.

■ Regular Polygons

With the graphics primitive **Line** (see Section 5.4.4, p. 136), we can easily create regular polygons. Here is a function for doing this (for information about **Map**, see Section 2.2.2, p. 36):

```
regularPolygon[n_, x0_, y0_, r_, th_, opts___] := Graphics[
  Line[Map[{r Cos[#] + x0, r Sin[#] + y0} &, Range[th, th + 2 π, 2 π / n]]],
  AspectRatio → Automatic, opts]
```

The arguments of this function are the order **n** of the polygon, the center point (**x0**, **y0**), the radius **r** (from the center point to a vertex), an angle **th** (one of the vertices is at this angle), and zero or more options. To show, for example, a triangle, type **Show[regpol[3, 0, 0, 1, π/2]]**. Here are some examples:

```
g1 = regularPolygon[3, 0, 0, 1, π / 2]; g2 = regularPolygon[3, 0, 0, 1, -π / 2];
g3 = regularPolygon[3, 0, 0, 1, -π]; g4 = regularPolygon[3, 0, 0, 1, 0];
g5 = regularPolygon[4, 0, 0, 1, π / 4]; g6 = regularPolygon[4, 0, 0, 1, 0];

Show[GraphicsArray[{g1, g2, g3, g4, g5, g6}]];
```

In the **Graphics`MultipleListPlot`** package, we have ready-to-use commands for regular polygons. Load this package and try, for example, **Show[Graphics[Regular﹀ Polygon[5], AspectRatio → Automatic]]**. With the **Geometry`Polytopes`** package, we can get information about regular polygons, from **Digon** to **Dodecagon**.

■ Maps

With the **Miscellaneous`WorldPlot`** package, we can plot the world, the continents, the countries, and other regions:

```
<< Miscellaneous`WorldPlot`

WorldPlot[Africa];
```

■ Other Plots

The **DiscreteMath`Combinatorica`** package has several commands to illustrate graphs, and with the **DiscreteMath`ComputationalGeometry`** package, we can compute and show, for example, Delaunay triangulations (see Section 10.1.2, p. 272) and Voronoi diagrams. Wickham-Jones (1994) has a whole chapter about 2D geometry.

5.4 Graphics Primitives and Directives

5.4.1 Introduction

All plots in *Mathematica* are made up of a few basic components called *graphics primitives*. We have such primitives as **Point**, **Line**, **Circle**, **Rectangle**, and **Text**. The style of these primitives is controlled with *graphics directives* like **PointSize**, **Thickness**, **Dashing**, **GrayLevel**, or **RGBColor**. The importance of the primitives and directives derives from three facts:

- *Mathematica* uses the primitives and directives in the construction of all plots.
- We can construct a plot directly from the primitives and directives.
- The primitives and directives can be used to modify plots with options.

In practice, the third fact is the most important. Indeed, all plotting commands have several options, such as **PlotStyle**, to modify the plot with directives (e.g., **PlotStyle →** **Thickness[0.02]**). In addition, with the **Prolog** and **Epilog** options, we can add primitives to plots (e.g., **Epilog → {Thickness[0.02], Line[{{0, 1}, {2, 1}}]}**). Options are considered in Chapter 6. Examples 1 and 2 below illustrate the first and second facts.

■ Example 1

Here is a very simple plot:

```
p = Plot[1, {x, 0, 4}, Ticks → {{4}, None}, PlotPoints → 3]
```

```
- Graphics -
```

With **InputForm**, we can see how *Mathematica* made the plot:

```
Short[InputForm[p], 2]
```

```
Graphics[{{Line[{{2.*^-6, 1.},
    {1.9472155954999577, 1.}, {3.999998, 1.}}]}}, {<<25>>}]
```

(We showed only the basic part of **p**; options [a total of 25] are not shown.) We see that the line is drawn with the **Line** primitive. It goes through three points. **Graphics** creates a graphics object from the primitive. With **FullGraphics**, we can see all the graphics primitives of a plot, including the primitives used for axes and ticks, among other things:

```
InputForm[FullGraphics[p]]

Graphics[{{{Line[{{2.*^-6, 1.}, {1.9472155954999577, 1.},
      {3.999998, 1.}}]}},
  {{GrayLevel[0.], AbsoluteThickness[0.25],
    Line[{{4., 0.}, {4., 0.02123669610234237}}]}],
  Text[4, {4., -0.04247339220468474}, {0., 1.}],
  {GrayLevel[0.], AbsoluteThickness[0.25],
    Line[{{-0.09999790000000001, 0.}, {4.0999979, 0.}}]}],
  {GrayLevel[0.], AbsoluteThickness[0.25],
    Line[{{0., -0.05}, {0., 2.05}}]}}}]
```

Here we first see the line. Then we have the tick mark and the tick label 4 on the x axis, and lastly we can see the two lines forming the axes.

■ Example 2

Usually it is most practical to use commands like **Plot**, but if we so choose, we can collect a plot from graphics primitives. As an example, we make the same plot as above, with slight simplification:

```
Graphics[{Line[{{0, 1}, {4, 1}}],
  Line[{{4, 0}, {4, 0.02}}], Text[4, {4, -0.04}, {0, 1}],
  Line[{{-0.1, 0}, {4.1, 0}}], Line[{{0, -0.05}, {0, 2.05}}]}]

- Graphics -
```

With **Show**, we can see the result:

```
Show[%, PlotRange → All]
```

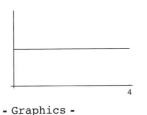

```
- Graphics -
```

■ Building Plots from Primitives and Directives

The examples show that we get a graphics object from the primitives with **Graphics** and that we can show an object with **Show**.

Graphics[{directives and primitives}]   Create a graphics object from primitives
Show[graphics object]   Show a graphics object

All primitives have their default styles. If these are used, we need not define directives; we simply write a list of primitives (like in Example 2). We can modify each primitive with one or more directives. They are written before the corresponding primitive and thus work like adjectives. However, note that a directive affects all primitives after the directive. Thus, if you want to apply a directive or several directives only to the next primitive, enclose the directives and the primitive in curly braces (**{ }**), as seen in the following example:

> `Graphics[{{dir`$_{11}$`, dir`$_{12}$`, …, prim`$_1$`}, {dir`$_{21}$`, dir`$_{22}$`, …, prim`$_2$`}, … }]`

Graphics has mostly the same options as **Plot**. One exception is the default value **False** of the option **Axes** for **Graphics** (for **Plot**, the default value is **True**). **Graphics** does not have certain options that **Plot** does, which control the sampling algorithm.

With the **Prolog** and **Epilog** options, we can add graphics primitives to plots produced by usual plotting commands like **Plot** (see Section 6.6.2, p. 170):

> `Plot[expr, {x, a, b}, Epilog →` {directives and primitives}`]`   Add graphics
> primitives

5.4.2 Summary

■ Graphics Primitives

Below are all the built-in graphics primitives (the primitive **PostScript** is lacking, though) and some primitives from packages. With **p**, **p1**, **p2**, and **pn** we denote a point, made up of the x and y coordinates, such as {**1, 4**}. For a mathematician, the primitives **Point**, **Line**, **Text**, and **Arrow** may be the most important ones.

> `Point[p]`   Point at **p**
> `Line[{p1, …, pn}]`   Line through points **p1**, …, **pn**
> `Spline[{p1, …, pn}, type, opts]`   Spline through points **p1**, …, **pn**
> `Rectangle[p1, p2]`   Filled rectangle with two opposite corners at **p1** and **p2**
> `Polygon[{p1, …, pn}]`   Filled polygon with vertices **p1**, …, **pn**
> `Circle[p, r]`   Circle with center **p** and radius **r** (also ellipse)
> `Disk[p, r]`   Filled circle with center **p** and radius **r** (also filled ellipse)
> `Ellipsoid[p, {r1, r2}, d1]`   Ellipse with center **p** and radii **r1** and **r2**, one in direction **d1** (with norm 1), the other perpendicular
> `Text[expr, p, opts]`   Text **expr** centered at **p**
> `Arrow[p1, p2, opts]`   Arrow from **p1** to **p2**
> `Raster[graylevels]`   Raster image in gray levels
> `RasterArray[colors]`   Raster image in colors

Spline is in the **Graphics`Spline`** package, **Ellipsoid** in the **Statistics`Multi‑DescriptiveStatistics`** package, and **Arrow** in the **Graphics`Arrow`** package.

In Sections 5.4.3 through 5.4.6, we consider all the primitives, except for the last two; section 5.4.7 contains some examples of **Raster** and **RasterArray**. In Section 6.6.2, p. 170, and in numerous other sections, we present other examples of using graphics primitives.

■ Graphics Directives

Here are all the built-in graphics directives:

> `PointSize[d], AbsolutePointSize[d]`
> `Thickness[d], AbsoluteThickness[d]`
> `Dashing[{d1, d2, … }], AbsoluteDashing[{d1, d2 ,… }]`

> **GrayLevel[g]** Gray **g**
> **Hue[h]** Hue **h** (with maximum saturation and brightness)
> **Hue[h, s, b]** Hue **h** with saturation **s** and brightness **b**
> **RGBColor[r, g, b]** Red **r**, green **g**, and blue **b**
> **CMYKColor[c, m, y, k]** Cyan **c**, magenta **m**, yellow **y**, and black **k**

Each primitive can be modified with certain directives. The table above gives all the built-in directives. A summary is as follows. With **Raster** and **RasterArray**, we can use no directives. The color of all other graphics primitives can be defined with one of **Gray**‹ **Level**, **Hue**, **RGBColor**, and **CMYKColor**; colors are considered in Section 5.4.7, p. 144. Color is the only directive that can be used with **Rectangle**, **Polygon**, **Disk**, and **Text**. In addition to color, the style of a point can be controlled with **PointSize** or **AbsolutePointSize** (see Section 5.4.3) and the style of a line, spline, circle, ellipsoid, or arrow with **Thickness** or **AbsoluteThickness** and **Dashing** or **AbsoluteDashing** (see Section 5.4.4, p. 136). The **Spline**, **Text**, and **Arrow** primitives also have their own options for controlling the style. Controlling the style of a **Text** primitive is considered in Section 6.3.2, p. 159.

5.4.3 Point

> **Point[p]** A point at **p**
> **Map[Point, {p1, …, pn}]** Points at **p1**, …, **pn**

With **Point** and **Line**, we can easily get the same result as with **ListPlot**:

```
data = Table[{i, Random[]}, {i, 1, 20}];
```

```
Show[Graphics[
    {Line[data], AbsolutePointSize[2], Map[Point, data]}, Axes → True]];
```

This is, indeed, a noteworthy method to use to plot data, in particular in some more complex situations, and we will use it again in Chapters 9 and 10. To compare, we could get the same plot as above by writing the following:

```
ListPlot[data, PlotJoined → True,
    Epilog → {AbsolutePointSize[2], Map[Point, data]}];
```

■ PointSize

> **PointSize[d]** The diameter of a point is a fraction **d** of the width of the graph
> **AbsolutePointSize[d]** The diameter of a point is a multiple **d** of 1/72 inch

The diameter of a point can be defined in two ways: either as a fraction of the width of the graph or as a multiple of the printer's point, which is about 1/72 inch. For those of us more familiar with millimeters, we note that 3 printer points is about 1 millimeter. The default argument of **PointSize** is 0.008 for 2D graphics and 0.01 for 3D graphics.

The size of a point defined by **PointSize** depends on the size of the plot: the larger the plot, the larger the points. On the other hand, the size of a point defined with **Absolute**⸗ **PointSize** does not depend on the size of the plot. In the points below, the absolute diameter varies from 0.5 to 5:

```
t = Table[Graphics[{AbsolutePointSize[0.5 i], Point[{i, 0}]}], {i, 10}];

Show[t, Axes → {True, False}, AxesStyle → GrayLevel[1],
   Ticks → {Transpose[{Range[10], Range[0.5, 5, 0.5], Table[0, {10}]}],
      None}, AspectRatio → 0.07];
```

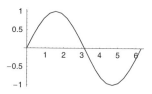

I use absolute point sizes, because I have found that usually I do not want the size of the points to grow if I enlarge the plot. An absolute size between 1 and 3 is often suitable. Likewise, I use absolute thickness and absolute dashing.

5.4.4 Line and Spline

■ Line

> **Line[{p1, p2}]**   A straight line between points **p1** and **p2**
> **Line[{p1, …, pn}]**   A broken line through points **p1**, …, **pn**

With **Line**, we can easily mimic **Plot**:

```
t = Table[{x, Sin[x]}, {x, 0, 2 π, π / 10}];

Show[Graphics[Line[t]], Axes → True];
```

However, we recommend that you do not use this method, because **Plot** chooses the points *adaptively* and gives a smoother plot.

■ Thickness

> **Thickness[d]**   The thickness of a line is a fraction **d** of the width of the graph
> **AbsoluteThickness[d]**   The thickness of a line is a multiple **d** of 1/72 inch

The default thickness of 2D curves is, according to Wolfram (2003), **Thickness[0.004]**, but it seems actually to be **AbsoluteThickness[0.5]**. Thickness defined by **Thickness** depends on the size of the plot: the larger the plot, the thicker the curves. Thickness defined with **AbsoluteThickness** does not depend on the size of the plot. In the plot below, the absolute thickness varies from 0 to 3:

```
t = Table[Graphics[{AbsoluteThickness[d], Line[{{0, d}, {1, d}}]}],
    {d, 0, 3, 0.5}];

Show[t, Axes → {False, True},
   AxesOrigin → {0, -0.001}, AspectRatio → 0.4, PlotRange → All];
```

As can be seen, the ends of a line float outside of the intended interval (the ends are right-angled on the screen and half circles on paper). The floating is clearly visible with thicker lines. Note also that the thickness depends somewhat on the resolution of the printer: the larger the resolution, the thinner the lines.

■ Dashing

Dashing[{d1, d2, … }]   The length of a segment is a fraction **di** of the width of the graph
AbsoluteDashing[{d1, d2, … }]   The length of a segment is a multiple **di** of 1/72 inch

In the dashing style, a line consists of small segments, and they alternate between black and white. The lengths of the segments are defined with the dashing directive. Usually only a few segments are defined, and they are used cyclically. For example:

Dashing[{}]   No dashing is used; lines are solid (this is the default)
Dashing[{d}]   Black and white segments, each of length **d**, alternate
Dashing[{d1, d2}]   Black and white segments of lengths **d1** and **d2**, respectively, alternate

Dashing defined with **Dashing** depends on the size of the figure: the larger the figure, the longer the segments. Dashing defined with **AbsoluteDashing** does not depend on the size of the figure. Next we show some absolute dashings between 1 and 3:

```
t = Table[Graphics[{AbsoluteDashing[{d}], Line[{{0, d}, {1, d}}]}],
    {d, 1, 3, 0.5}];

Show[t, Axes → {False, True},
   AxesOrigin → {0, -0.001}, AspectRatio → 0.35];
```

Note that the dashing is probably not exactly what we would expect: the white segments are smaller than they should be. The thicker the dashed curve is, the smaller the white segments are. This is a consequence of the fact that the ends of lines float outside of the desired interval, as we saw earlier when we considered thickness. By the way, this problem with dashing may actually be a useful feature: the dashing looks better when the white parts are shorter than the black ones.

■ Spline

> *In the* **Graphics`Spline`** *package:*
>
> **Spline[{p1, p2, … }, type]**   Spline of type **type** through points **p1**, **p2**, …
>
> *An option:*
>
> **SplineDots**   The style used for the given points; examples of values: **None** (points are not plotted), **Automatic** (points are red and of size 0.03), **{AbsolutePointSize[3], RGBColor[0, 0, 1]}**

Splines are considered in some detail in Section 21.3, p. 561. Here we only note that a spline is a smooth curve through all or some of the points and that three types of splines can be used: **Cubic** (goes through all points), **Bezier** (goes through the end points), and **CompositeBezier** (goes through every other point). The default is that the given points are not shown. The splines are drawn by an adaptive method, which can be controlled with the additional options **SplinePoints**, **MaxBend**, and **SplineDivision**. Splines may be useful, say, for drawing some arcs:

```
<< Graphics`Spline`
```

```
Show[Graphics[{Circle[{0, 0}, 1], Circle[{3, 0}, 1], Circle[{6, 0}, 1],
    Spline[{{0, 1}, {0.7, 1.7}, {2.3, 1.7}, {3, 1}}, Cubic],
    Spline[{{3, 1}, {3.7, 1.7}, {5.3, 1.7}, {6, 1}}, Cubic],
    Spline[{{0, -1}, {0.7, -1.7}, {2.3, -1.7}, {3, -1}}, Cubic],
    Spline[{{3, -1}, {3.7, -1.7}, {5.3, -1.7}, {6, -1}}, Cubic],
    Text[StyleForm[1, FontSize -> 11], {0, 0}],
    Text[StyleForm[2, FontSize -> 11], {3, 0}],
    Text[StyleForm[3, FontSize -> 11], {6, 0}]}], AspectRatio -> Automatic];
```

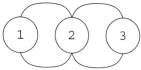

5.4.5 Rectangle, Polygon, Circle, Disk, and Ellipsoid

■ Rectangle and Polygon

> **Line[{p1, p2, p3, p4, p1}]**   Rectangle with vertices **p1**, …, **p4**
> **Rectangle[p1, p2]**   Filled rectangle with two opposite corners at **p1** and **p2**
>
> **Line[{p1, …, pn, p1}]**   Polygon with vertices **p1**, …, **pn**
> **Polygon[{p1, …, pn}]**   Filled polygon with vertices **p1**, …, **pn**

When working with geometric objects like rectangles, polygons, circles, or ellipses, it is useful to define **AspectRatio → Automatic**, because then a square really looks like a square and not like a rectangle, and so on. The default fill color of a rectangle, polygon, circle, or ellipse is black. We can change the color with color directives (see Section 5.4.7, p. 144). Here is an example, where we use a yellow filled rectangle, a black rectangle, and white text on a red background:

```
Show[Graphics[{RGBColor[1, 1, 0], Rectangle[{0, 0}, {2, 1}], GrayLevel[0],
    AbsoluteThickness[4], Line[{{0, 0}, {2, 0}, {2, 1}, {0, 1}, {0, 0}}],
    Text["STOP", {1, 0.5}, Background → RGBColor[1, 0, 0],
     TextStyle → {FontSize → 30, FontWeight → "Bold",
      FontColor → GrayLevel[1]}]}], AspectRatio → Automatic];
```

■ Circle, Disk, and Ellipsoid

`Circle[p, r]`	Circle with center **p** and radius **r**
`Circle[p, Offset[{r, r}]]`	Circle with center **p** and radius **r**/72 inch
`Disk[p, r]`	Filled circle with center **p** and radius **r**
`Circle[p, {rx, ry}]`	Ellipse with center **p** and radii **rx** and **ry**
`Disk[p, {rx, ry}]`	Filled ellipse with center **p** and radii **rx** and **ry**
`Ellipsoid[p, {r1, r2}, d1]`	Ellipse with center **p** and radii **r1** and **r2**, one in direction **d1** (with norm 1), the other perpendicular

`Ellipsoid` is in the `Statistics`MultiDescriptiveStatistics`` package. This graphics primitive can be used with `Graphics` but not with `Epilog`.

One method to get true circles is to set `AspectRatio → Automatic`. However, this form of the plot may not be suitable, and the size of the circles changes as we change the size of the plot. Another method is to define the radii of the circles with `Offset`. In the second line of the box above, we define the radius of the circle to be **r** printer's points. Such a radius is absolute; the size or form of the circle does not change when the size or form of the figure is changed. This ensures that we get exact circles of exactly the size we have specified. The same method can be used with `Disk`, and it also works for ellipses.

A useful application of `Circle` is to plot data. While `ListPlot` uses filled points, with `Circle` we get open points:

```
data = Table[{i, Random[]}, {i, 1, 40}];

Show[Graphics[{Line[data],
    GrayLevel[1], Map[Disk[#, Offset[{1, 1}]] &, data],
    GrayLevel[0], Map[Circle[#, Offset[{1, 1}]] &, data]},
   Axes → True], AspectRatio → 0.25];
```

Here we first plotted a line connecting the points. Then we plotted white disks to hide the lines inside the circles to be drawn. Then the black circles are plotted.

With circles, we can plot networks.

```
networkPlot[points_, labels_, lines_, r_, opts___] := Show[Graphics[
   {Map[Line, Map[{points[[#[[1]]]], points[[#[[2]]]]} &, lines]],
    {GrayLevel[1], Map[Disk[#, Offset[{r, r}]] &, points]},
    Map[Circle[#, Offset[{r, r}]] &, points],
    Map[Text[#[[1]], #[[2]]] &, Transpose[{labels, points}]]}, opts]]
```

The program draws circles with labels and connects some of the circles with lines. Circles of absolute radius **r** are drawn at **points**, and **labels** are written inside them. The list **lines** gives the pairs of point numbers between which lines are drawn; point numbers are 1, 2, and so on in the order they are given in **points**. For example:

```
points = {{0, 2}, {2, 4}, {2, 3}, {2, 2}, {2, 1}, {2, 0}, {4, 3}, {4, 1}};
lines = {{1, 2}, {1, 3}, {1, 4}, {1, 5}, {1, 6}, {2, 7}, {2, 8},
   {3, 7}, {3, 8}, {4, 7}, {4, 8}, {5, 7}, {5, 8}, {6, 7}, {6, 8}};

networkPlot[points, Range[8],
   lines, 7, PlotRange → {{-0.5, 4.5}, {-0.5, 4.5}}];
```

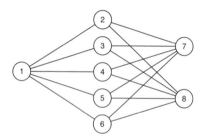

An example of an ellipse:

```
<< Statistics`MultiDescriptiveStatistics`

Show[Graphics[{Ellipsoid[{4, 1}, {3, 1}, {2, 1} / Sqrt[5]], Point[{4, 1}]},
   AspectRatio → Automatic, Ticks → {Range[7], {1, 2}},
   PlotRange → {{-0.2, 7.2}, Automatic}], Axes → True];
```

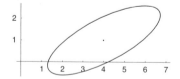

With the primitives **Circle** and **Disk**, we can also define circular and elliptical arcs. We show the forms for **Circle**:

```
Circle[p, r, {t1, t2}]   Circular arc with center p, radius r, and angle (t1, t2)
Circle[p, {rx, ry}, {t1, t2}]   Elliptical arc with center p, radii rx and ry, and angle
   (t1, t2)
```

5.4.6 Text, Arrow, and Coordinates

■ Text

> **Text[expr, {x, y}]** The center of the text **expr** is at the point **{x, y}**
>
> **Text[expr, {x, y}, {u, v}]** The point **{u, v}**, expressed in text coordinates, of the text is at the point **{x, y}**

The expression to be printed can be a mathematical expression such as **Sin[x]** or a string such as **"Here is text"**. The expression in printed, by default, in **OutputForm**. The default is that the text is centered at the given point. A text element has, however, its own coordinate system ranging from −1 to 1 in both the x and the y directions. For example, the point **{-1, -1}** in text coordinates represents the bottom lefthand corner of the text element and **{1, 1}** the top righthand corner. With the help of the text coordinates, we can place the text element in other ways. For example:

> **Text[expr, {3, 4}, {-1, 0}]**

The vertical middle of the lefthand end of **expr** is at the point **{3, 4}**, that is, the text starts from **{3, 4}**. The following table explains the most frequently used text coordinates:

Text[t,{x,y},{-1,1}] The top left corner of **t** is at **{x,y}**	**Text[t,{x,y},{0,1}]** **t** is centered below **{x,y}**	**Text[t,{x,y},{1,1}]** The top right corner of **t** is at **{x,y}**
Text[t,{x,y},{-1,0}] **t** starts from **{x,y}**	**Text[t,{x,y},{0,0}]** **t** is centered at **{x,y}**	**Text[t,{x,y},{1,0}]** **t** ends at **{x,y}**
Text[t,{x,y},{-1,-1}] The bottom left corner of **t** is at **{x,y}**	**Text[t,{x,y},{0,-1}]** **t** is centered above **{x,y}**	**Text[t,{x,y},{1,-1}]** The bottom right corner of **t** is at **{x,y}**

An example (note that axes labels can be set more easily with the **AxesLabel** option):

```
Plot[Sin[x], {x, 0, 2 π}, PlotRange → All, Ticks → {{π, 2 π}, {-1, 1}},
   Epilog → {AbsolutePointSize[3], Point[{π / 2, 1}], Point[{3 π / 2, -1}],
      Text["x", {2 π + 0.4, 0}, {-1, 0}],
      Text["sin(x)", {0, 1.2}, {0, -1}],
      Text["maximum", {π / 2, 1}, {-1, -1}],
      Text["minimum", {3 π / 2, -1}, {-1, 1}]}];
```

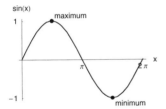

If a part of a text goes outside the ordinary plot region, that part is not shown in the plot. Give the option **PlotRange → All** to see all of the text. Another way to see all of the text is to enlarge the figure with the mouse (this solves the problem, because while the size of the plot grows, the size of all text remains the same).

For fonts, formatting, and colors in **Text**, see Sections 6.3.1, p. 157, and 6.3.2, p. 159.

Text[expr, {x, y}, {u, v}, {r, s}]   Text is rotated to have slope **s/r**

The following plot shows some slope definitions **{r, s}**:

```
p1 = Graphics[Text["bottom to top: {0,1}", {-1, 0}, {0, 0}, {0, 1}]];
p2 = Graphics[Text["left to right: {1,0}", {0, 1}, {0, 0}, {1, 0}]];
p3 = Graphics[Text["top to bottom: {0,-1}", {1, 0}, {0, 0}, {0, -1}]];
p4 = Graphics[Text["right to left: {-1,0}", {0, -1}, {0, 0}, {-1, 0}]];
p5 = Graphics[Text["ascending: {2,1}", {-.4, .3}, {0, 0}, {2, 1}]];
p6 = Graphics[Text["descending: {2,-1}", {.3, -.4}, {0, 0}, {2, -1}]];

Show[p1, p2, p3, p4, p5, p6, PlotRange → All];
```

Note that on the screen of a Macintosh, rotated text does not appear correctly. With versions of *Mathematica* that are older than 5.0, the text should, however, print correctly. If you use *Mathematica* 5.0 with MacOS X, you have to set **GraphicsPrintingFormat** to **"DownloadPostScript"** with the *Option Inspector* to get correctly rotated text.

■ Arrow

In the **Graphics`Arrow`** *package:*

Arrow[p1, p2]   Arrow from point **p1** to point **p2**

Options:

HeadLength   Length of the arrowhead (as a fraction of the width of the whole plot); examples of values: **Automatic, 0.03**

HeadWidth   Width of the arrowhead (expressed as a factor of **HeadLength**); examples of values: **0.5, 0.7**

HeadCenter   Location of the center of the base of the arrowhead (expressed as a factor of **HeadLength**); examples of values: **1, 0**

In addition, we have the more seldom-used options **HeadScaling**, **HeadShape**, and **ZeroShape**. For example:

```
<< Graphics`Arrow`
```

```
Plot[Sin[x], {x, 0, 2 π}, Ticks → {{π, 2 π}, {-1, 1}},
  Epilog → {Text["point of inflection", {3.7, 0.4}, {-1, -1}],
    Arrow[{3.7, 0.4}, {π + 0.05, 0.05}]}];
```

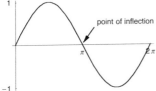

I find the default arrow somewhat heavy with a filled triangle as the head. An example of a more open head is in Section 6.6.2, p. 171.

■ Coordinates

> {x, y} A point in the original coordinates of the plot
> Scaled[{sx, sy}] A point in the scaled coordinates

Thus far we have defined the positions of graphics primitives with the normal coordinates used in the plot. Another way is to use scaled coordinates that run from 0 to 1 in both directions. An example:

```
Plot[Sin[x], {x, 0, 2 π}, Ticks → {{π, 2 π}, {-1, 1}},
  Epilog → {AbsolutePointSize[2], Point[{π / 2, 1}], Point[{3 π / 2, -1}],
    Text["A function with\na minimum and\na maximum",
    Scaled[{1, 1}], {1, 1}]}];
```

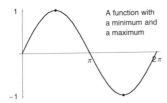

Here we told *Mathematica* that the top right corner of the text is at the point **Scaled[{1, 1}]**, which is at the top right corner of the plot (note that in the text string, **\n** defines a new line). Using scaled coordinates for primitives may be useful, for example, when the original coordinates vary from figure to figure but we want a primitive at the same position in each figure.

> Scaled[{sdx, sdy}, {x, y}] Scaled offset from {x, y}
> Offset[{adx, ady}, {x, y}] Absolute offset from {x, y}
> Offset[{adx, ady}, Scaled[{sx, sy}]] Absolute offset from Scaled[{sx, sy}]

The position of a primitive can also be expressed as an offset from a given point. In this way we can define a position relative to another position. The offset can be set either in scaled coordinates or in absolute units. The absolute unit is one printer's point (1/72 inch). We already used **Offset** in a special way when we considered circles and disks: for these primitives, we can give the radii in absolute units with **Offset[{adx, ady}]**.

5.4.7 Colors

> **GrayLevel[g]**   Gray level **g** between 0 (black) and 1 (white)
> **Hue[h]**   Color with hue **h** between 0 and 1 (with maximum saturation and brightness)
> **Hue[h, s, b]**   Color with hue **h**, saturation **s**, and brightness **b**, each between 0 and 1
> **RGBColor[r, g, b]**   Color with specified red **r**, green **g**, and blue **b** components, each
>   between 0 and 1
> **CMYKColor[c, m, y, k]**   Color with specified cyan **c**, magenta **m**, yellow **y**, and black **k**
>   components, each between 0 and 1 (used in four-color printing)

■ Gray Level

Gray level 0 corresponds to black and 1 to white (contrary perhaps to what you might expect). Here is a sequence of grays from 0 to 1 in steps of 0.05 (the last rectangle, which is white, is not visible):

```
t = Table[Graphics[
    {GrayLevel[0.05 i], Rectangle[{i, 0}, {i + 1, 1}]}], {i, 0, 20}];
Show[t, Axes → {True, False}, AxesStyle → GrayLevel[1], Ticks →
    {Transpose[{Range[0.5, 20.5, 1], Range[0, 1, 0.05], Table[0, {21}]}],
    None}, AspectRatio → Automatic];
```

```
0   0.05  0.1  0.15  0.2  0.25  0.3  0.35  0.4  0.45  0.5  0.55  0.6  0.65  0.7  0.75  0.8  0.85  0.9  0.95  1.
```

The graphics primitive **Raster** is designed to show a table of gray levels:

```
Show[Graphics[Raster[Transpose[Table[{1 - g, g}, {g, 0, 1, 0.05}]]]],
    AspectRatio → 0.1];
```

■ Hue

When the argument in the hue specification ranges from 0 to 1/6, 2/6, 3/6, 4/6, 5/6, and 1, the color changes from red to yellow, green, cyan, blue, magenta, and back to red. The default saturation and brightness is the maximum value 1 (**Hue[h]** is equivalent to **Hue[h, 1, 1]**). In the following three rows, we have several hues. The first row has maximum saturation and brightness. In the middle row, the saturation is lower, and in the last row, the brightness is lower.

```
rec = Rectangle[{0, 0}, {1, 1}];
t1 = Table[Graphics[{Hue[h], rec}], {h, 0, 1, 1 / 12}];
t2 = Table[Graphics[{Hue[h, 0.5, 1], rec}], {h, 0, 1, 1 / 12}];
t3 = Table[Graphics[{Hue[h, 1, 0.5], rec}], {h, 0, 1, 1 / 12}];
```

```
Show[GraphicsArray[{t1, t2, t3}, GraphicsSpacing → -0.1],
    Frame → {True, False, False, False},
    FrameStyle → GrayLevel[1], FrameTicks →
        {Transpose[{Range[0.5, 12.5, 1], Range[0, 1, 1 / 12], Table[0, {13}]}]},
        None, None, None}];
```

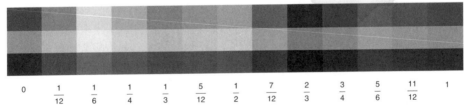

| 0 | $\frac{1}{12}$ | $\frac{1}{6}$ | $\frac{1}{4}$ | $\frac{1}{3}$ | $\frac{5}{12}$ | $\frac{1}{2}$ | $\frac{7}{12}$ | $\frac{2}{3}$ | $\frac{3}{4}$ | $\frac{5}{6}$ | $\frac{11}{12}$ | 1 |

Next we show a color wheel showing how the hue changes (the wheel is implemented by plotting 300 narrow sectors of a disk):

```
d = π / 150;
Show[Graphics[Map[{Hue[# / (2 π - d)], Disk[{0, 0}, 1, {#, # + d}]} &,
    Range[0, 2 π - d, d]], AspectRatio → Automatic]];
```

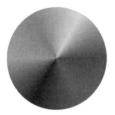

■ **RGB Color**

In the RGB system, we specify the intensity of red, green, and blue. For example, **RGBColor[1, 0, 0]** is red, **RGBColor[0, 1, 0]** is green, and **RGBColor[0, 0, 1]** is blue. This color system corresponds to the one used in color monitors. In the following four tables, we have various RGB colors when the blue component changes from 0 to $1/3$, $2/3$, and 1. **RasterArray** is designed to show tables of colors.

```
tic = {{0.5, 0}, {1.5, "1/3"}, {2.5, "2/3"}, {3.5, 1}};
t = Table[Graphics[
        RasterArray[Table[RGBColor[r, g, b], {g, 0, 1, 1 / 3}, {r, 0, 1, 1 / 3}]],
        Frame → True, FrameLabel → {"r", "g"},
        RotateLabel → False, FrameTicks → {tic, tic, None, None},
        PlotLabel → SequenceForm["b = ", InputForm[b]]], {b, 0, 1, 1 / 3}];

Show[GraphicsArray[t], GraphicsSpacing → 0, AspectRatio → 0.18];
```

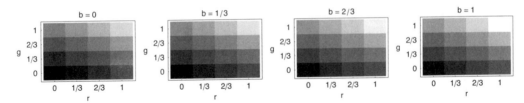

■ **Summary**

Here is a summary of how the basic colors can be obtained with the four color systems:

		GrayLevel	Hue	RGBColor	CMYKColor
Black	■	0	0, 0, 0	0, 0, 0	0, 0, 0, 1
Grey	■	.5	0, 0, .5	.5, .5, .5	0, 0, 0, .5
White	□	1	0, 0, 1	1, 1, 1	0, 0, 0, 0
Red	■		0	1, 0, 0	0, 1, 1, 0
Yellow	□		1/6	1, 1, 0	0, 0, 1, 0
Green	■		2/6	0, 1, 0	1, 0, 1, 0
Cyan	■		3/6	0, 1, 1	1, 0, 0, 0
Blue	■		4/6	0, 0, 1	1, 1, 0, 0
Magenta	■		5/6	1, 0, 1	0, 1, 0, 0

The **Graphics`Colors`** package defines other color systems, namely **CMYColor**, **YIQ**, **Color**, and **HLSColor**. In addition, the package defines 193 ready-to-use colors with illustrative names like **ForestGreen**, **Orange**, or **Ultramarine**. For a list of all names, type **AllColors**. For example, **Orange** is defined in the package as **RGBColor[1, 0.5, 0]**.

■ **Interactive Choice of Colors**

For help defining a color, there is a **Color Selector...** item in the **Input** menu. This item gives you various ways to select colors interactively. On a Macintosh, some of the ways are as follows:

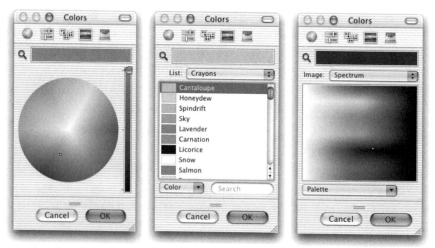

Choose a color from a window by clicking with the mouse and then clicking **OK**. The chosen color definition appears at the location of the cursor as an RGB color definition. For example, type:

```
Plot[Sin[x], {x, 0, 2 π}, PlotStyle →];
```

Place the cursor after →, then select a color by the color selector.

Options for 2D Graphics

6

Introduction

> *Isaac Newton, it seems, was one of the original absentminded mathematicians.*
> *He once cut a hole in the bottom of the door of an outhouse to allow his favorite cat*
> *easy access. When the cat had kittens, he added a small hole next to the big one.*

We shall now explore in detail all the options of **Plot**, **ParametricPlot**, and **Graphics** (the option **DisplayFunction** has, however, already been considered in Section 5.2.1, p. 117). The options of these three commands are mostly the same. In Section 5.3, we mentioned how the options of other plotting commands differ from those of **Plot**.

We go somewhat beyond options in that we consider fonts, formatting, and add-ons more widely. In fact, font and formatting can be defined by global variables as well as by options, and add-ons can also be made using separate graphics.

In Section 6.1.2, p. 150, we present a summary of the options. The options are classified into global options, which modify the plot as a whole, and local options, which modify separate components of the figure, such as the curve, the axes, the frame, and the add-ons. The varying importance of the options is also shown. In Sections 6.2 through 6.6, we then consider the options in the same order as in the summary.

Many options use graphics directives, and the options **Prolog** and **Epilog** also use graphics primitives. You may want to study them in Section 5.4.

6.1 Introduction to Options

6.1.1 Using Options

In Section 5.1.1, p. 110, we already explained how options can be used to modify a plot. Below we summarize four methods of using options. We use **Plot** as an example of a plotting command.

(a) Setting options in **Plot**:

    Plot[expr, {x, a, b}, opt1 → val1, opt2 → val2, …]

(b) Setting options in **Show**:

    p = Plot[expr, {x, a, b}]
    Show[p, opt1 → val1, opt2 → val2, …]

(c) Giving a name to the options:

    opts = Sequence[opt1 → val1, opt2 → val2, …]
    Plot[expr, {x, a, b}, Evaluate[opts]]

(d) Setting options with **SetOptions**:

    SetOptions[Plot, opt1 → val1, opt2 → val2, …]
    Plot[expr, {x, a, b}]

(a) The first method may be the most convenient in a notebook environment (as in Windows and on a Macintosh). You can first plot the function without any options. If the result is not satisfactory, add an option or several options to the original plotting command and then execute the command anew. Continue adding and modifying the options and executing the plotting command until you are satisfied with the result.

(b) The second method has a certain advantage. Indeed, **Show** does not execute the plotting command anew; only the appearance of the figure is changed. Thus, if the plotting command is time consuming, it is better to use **Show** to avoid executing the plotting command anew every time.

(c) The third method may be useful if you use the same options (**opts**) for several plots: just write the options once and use the name of the set of options in the subsequent plots (see Example 1 in Section 6.1.3, p. 153).

(d) The fourth method is useful if, during a session, you continuously use certain values for some options. Before using this method, it may be useful to look at the default values with **Options[Plot]** for the case you want to return to the default values. The default values can then be set with another **SetOptions** command.

Note that if you combine two or more figures with **Show**, then you should be aware of the fact that if the figures have different values for the same option, then **Show** takes the value given in the *first* figure; the values given in later figures for this option are disregarded (see Section 5.1.2, p. 113).

Note also that, with **Show**, we can adjust most of the options, *but not all of them*. Indeed, the options **Compiled**, **MaxBend**, **PlotDivision**, **PlotPoints**, and **PlotStyle** controlling the plotting algorithm cannot be adjusted with **Show**. If you want to use new values for any of these options, you have to write the options in the original plotting command and execute the command anew.

■ **Information about Options**

> **Options[comm]**   Give the options and their default values of a command **comm**
> **Options[comm, opt]**   Give the default value of an option **opt** of a command **comm**
> **AbsoluteOptions[p]**   Give the detailed values of the options used in a plot **p**, even if a
>   value is **Automatic** or **All**
> **AbsoluteOptions[p, opt]**   Give the detailed value of an option **opt** used in a plot **p**

Here are all the 30 options and their default values for **Plot** (**DefaultFont** is obsolete):

Options[Plot]

$\Big\{$AspectRatio $\rightarrow \dfrac{1}{\text{GoldenRatio}}$, Axes $\rightarrow$ Automatic, AxesLabel $\rightarrow$ None,
AxesOrigin $\rightarrow$ Automatic, AxesStyle $\rightarrow$ Automatic, Background $\rightarrow$ Automatic,
ColorOutput $\rightarrow$ Automatic, Compiled $\rightarrow$ True, DefaultColor $\rightarrow$ Automatic,
DefaultFont $\rightarrow:$ $DefaultFont, DisplayFunction $\rightarrow:$ $DisplayFunction,
Epilog $\rightarrow$ {}, FormatType $\rightarrow:$ $FormatType, Frame $\rightarrow$ False, FrameLabel $\rightarrow$ None,
FrameStyle $\rightarrow$ Automatic, FrameTicks $\rightarrow$ Automatic, GridLines $\rightarrow$ None,
ImageSize $\rightarrow$ Automatic, MaxBend $\rightarrow$ 10., PlotDivision $\rightarrow$ 30.,
PlotLabel $\rightarrow$ None, PlotPoints $\rightarrow$ 25, PlotRange $\rightarrow$ Automatic,
PlotRegion $\rightarrow$ Automatic, PlotStyle $\rightarrow$ Automatic, Prolog $\rightarrow$ {},
RotateLabel $\rightarrow$ True, TextStyle $\rightarrow:$ $TextStyle, Ticks $\rightarrow$ Automatic$\Big\}$

The default value of a certain option or a list of options can also be displayed:

Options[Plot, PlotRange]

{PlotRange $\rightarrow$ Automatic}

Remember also that we can ask for information about an option by typing, for example, **?PlotRange** or using the *Help Browser* (see Sections 1.4.2, p. 18, and 1.4.3, p. 20).

Let us explain some typical values of options. Most default values of the options of **Plot** are **Automatic**. This value means that there is a special value chosen by **Plot** according to certain rules. Some options, like **AxesLabel**, have the value **None**, meaning that the plot does not have the corresponding components. Some options, like **Frame**, have the value **True** or **False**, which tells whether or not something is present. **PlotRange** can have the value **All**, meaning that the whole plot is shown.

We can also ask for the options of a given plot:

```
p = Plot[Sin[x], {x, 0, 2 π}];
```

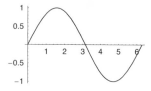

```
AbsoluteOptions[p, PlotRange]
```

{PlotRange → {{-0.15708, 6.44026}, {-1.05, 1.05}}}

Note that **AbsoluteOptions** does not show the values of **Compiled**, **MaxBend**, **Plot⁑Division**, **PlotPoints**, and **PlotStyle** nor the values behind the **Automatic** values of **Background**, **ColorOutput**, **DefaultColor**, **ImageSize**, and **PlotRegion**.

■ Comparing Options

We can easily compare the options of various plotting commands. For example, **Paramet⁑ricPlot** has exactly the same options and default values as **Plot**:

```
Options[Plot] == Options[ParametricPlot]
```

True

The options of **Plot** and **Graphics** have some differences:

```
p = Options[Plot]; g = Options[Graphics];
```

```
Complement[p, g]
```

{Axes → Automatic, Compiled → True, MaxBend → 10.,
 PlotDivision → 30., PlotPoints → 25, PlotStyle → Automatic}

```
Complement[g, p]
```

{Axes → False}

Complement[p, g] gives the elements of **p** that are not in **g**. Thus, the default value of **Axes** is different (**Automatic** for **Plot** and **False** for **Graphics**), and **Graphics** does not have the options **Compiled**, **MaxBend**, **PlotDivision**, **PlotPoints**, and **PlotStyle** that **Plot** has.

6.1.2 Summary

Below is a sorted list of all of the options for **Plot** (see Section 5.1.1, p. 108), **Parametric⁑Plot** (see Section 5.3.1, p. 122), and **Graphics** (see Section 5.4.1, p. 133), with short descriptions and some common values. *The default value of an option is mentioned first,* and after that we mention either *all other possible values* or *some examples of other values* (the examples are simple; more advanced forms may exist).

The six options assumed to be the most important and most often used are marked with two asterisks (**). Nine options assumed to be less important are marked with one asterisk (*). With parentheses we have shown seven options that most of us will almost never need. The remaining options, without any special markings, may sometimes be useful. Of course, the given classification of the options according to importance reflects the personal impression of the author. You may well have a different classification.

Most of the options are common to all three commands. Options and their default values applicable only to certain commands are expressed in the lists below by a superscript after the option name or value:

> ^{c}: applicable to **Plot** (shorthand for *curve*)
> ^{p}: applicable to **ParametricPlot** (shorthand for *parametric*)
> ^{g}: applicable to **Graphics** (shorthand for *graphics*)

For example, **PlotStyle**^{cp} means that this option applies to **Plot** and **ParametricPlot** but not to **Graphics**, and **False**^{g} means that the default value of the option in question (**Axes**) is **False** for **Graphics**.

The options are divided into global and local options. *Global options* modify some global aspects of the plot: whether to display the plot at all and how the plot looks in general (form, ranges of x and y, margins, size, colors, font, and formatting). *Local options* modify local components of a plot: the curve, axes, frame, and add-ons.

Global Options of **Plot**, **ParametricPlot**, *and* **Graphics**

Option for display:

** **DisplayFunction** Function used to display the plot; values used: **$Display‹ Function** (plot is shown), **Identity** (plot is not shown)

Options for form, ranges, margins, and size:

** **AspectRatio** Ratio of height to width of the plotting rectangle; examples of values: **1/GoldenRatio, Automatic, 0.4**

** **PlotRange** Ranges for x and y in the plot; examples of values:

Automatic,	**{{xmin, xmax}, Automatic},**
All,	**{{xmin, xmax}, All},**
{ymin, Automatic},	**{{xmin, xmax}, {ymin, Automatic}},**
{ymin, ymax},	**{{xmin, xmax}, {ymin, ymax}}**

PlotRegion Region of the final display area that the plot should fill; examples of values: **Automatic** (means **{{0, 1}, {0, 1}}**), **{{xmin, xmax}, {ymin, ymax}}**

ImageSize The absolute size of the plot; examples of values: **Automatic** (usually means 288), **width, {width, height}**

Options for colors:

Background Color of the background; possible values: **Automatic** (usually means **GrayLevel[1]**), **GrayLevel[g], Hue[h,s,b], RGBColor[r,g,b], CMYKColor[c,m,y,k]**

(**DefaultColor** Color of the curve, axes, labels, etc; possible values: **Automatic** [usually means **GrayLevel[0]**], **GrayLevel[g], Hue[h,s,b], RGBColor[r,g,b], CMYKColor[c,m,y,k]**)

(**ColorOutput** Type of color output to produce; examples of values: **Automatic, None, GrayLevel, RGBColor, CMYKColor**)

Options for font and formatting:

TextStyle   Font used in a plot; examples of values: **$TextStyle**, **{FontFamily →**
  "Times", FontSize → 8}

FormatType   Format type of text used in a plot; examples of values: **$FormatType**,
  StandardForm, TraditionalForm, InputForm, OutputForm

DisplayFunction is marked above with two asterisks. This reflects the importance of this option, at least in this book. The option may also be important for you if you are making a document with *Mathematica*. For interactive use of *Mathematica*, the option is probably not so central.

Local Options of **Plot**, **ParametricPlot**, *and* **Graphics**

Options for curve:

** **PlotStyle**[cp]   Style(s) of the curve(s); examples of values: **Automatic, Absolute**⟩
  Thickness[2], {AbsoluteThickness[2], RGBColor[1,0,0]}
* **PlotPoints**[cp]   Number of initial sampling points; default value: **25**
(**PlotDivision**[cp]   Maximum number of subdivisions of an interval; default value: **30**)
(**MaxBend**[cp]   Maximum bend angle between successive line segments; default value: **10**)
(**Compiled**[cp]   Whether to compile the expression to be plotted; possible values: **True,**
  False)

Options for axes and ticks:

* **Axes**   Whether to draw the axes; default values: **Automatic**[cp] (usually means **True**),
  False[g]; examples of other values: **{True, False}, {False, True}**
* **AxesOrigin**   Point where the axes cross; examples of values: **Automatic, {0, 0}**
* **AxesLabel**   Labels for the axes; examples of values: **None, "y", {"x", None}, {"x",**
  "y"}
(**AxesStyle**   Style of the axes; examples of values: **Automatic, {Absolute**⟩
  Thickness[1], GrayLevel[0.5]})
(**AxesFront**[g]   Whether axes are behind or on top of graphics elements; possible values:
  False, True [in the **Graphics`FilledPlot`** package])
** **Ticks**   Ticks on the axes; simple examples of values: **Automatic, None, {{0,1,2},**
  Automatic}, {Automatic, {-1,0,1}}, {{0,1,2}, {-1,0,1}}

Options for frame, frame ticks, and grid lines

* **Frame**   Whether to draw a frame; examples of values: **False, True, {True, True,**
  False, False}
* **FrameLabel**   Labels for the frame; examples of values: **None, "y", {"x", None}, {"x",**
  "y"}, {"bottom", "left", "top", "right"}
* **RotateLabel**   Whether to rotate the label for the vertical edges; possible values:
  True, False
FrameStyle   Style of the frame; examples of values: **Automatic, {GrayLevel[0.5],**
  AbsoluteThickness[1]}
* **FrameTicks**   Ticks on the frame; simple examples of values: **Automatic, None, {{0,**
  π, 2π}, {0, π}}, {{0, π, 2π}, {0, π}, None, None}

GridLines How the grid lines are drawn; simple examples of values: **None**, **Auto**-
  matic, **{None, Automatic}**, **{None, {-1,0,1}}**, **{{0,1,2}, {-1,0,1}}**

Options for plot label, primitives, and legend:
* **PlotLabel** Label of the plot; examples of values: **None**, **Sin[x/2]**, **"sin(x/2)"**
Prolog Graphics primitives to be plotted before the main plot; examples of values: **{}**,
  {{AbsolutePointSize[3], RGBColor[1,0,0], Point[{0.4, 0.7}]}}
** **Epilog** Graphics primitives to be plotted after the main plot; examples of values:
  {}, **{AbsolutePointSize[3], RGBColor[1,0,0], Point[{0.4, 0.7}]}**
PlotLegend, **LegendPosition**, **LegendSize**, etc. (in the **Graphics`Legend`** package)

Graphics does not have the options for the curve (indeed, **Graphics** is mainly used to
form a plot from graphics primitives, and these options are therefore of no use). The two
asterisks of **Epilog** reflect the importance of this option at least in this book; for you this
option may well be less important. Graphics primitives like **Point**, **Line**, and **Text** and
graphics directives like **AbsolutePointSize**, **AbsoluteThickness**, **AbsoluteDashing**,
GrayLevel, **Hue**, and **RGBColor** are explained in Section 5.4.

Note that, with **GraphicsArray** (see Section 5.2.2, p. 120), we can use the special option
GraphicsSpacing and all the same options as with **Graphics**. However, many of the
options of **Graphics** are not relevant for **GraphicsArray**; relevant are many global options
and the local options **PlotLabel**, **Frame**, and **FrameStyle**. The default values of **Aspect**-
Ratio, **FrameTicks**, and **Ticks** are **Automatic**, **None**, and **None** for **GraphicsArray** and
1/GoldenRatio, **Automatic**, and **Automatic** for **Graphics**, respectively.

All of the options are explained in detail in the remaining sections of this chapter.

6.1.3 Examples

■ Example 1

With the help of the options we can get interesting results, such as the following. The plot
is pretty impressive! (OK, the plot is overdone, but our aim is just to show what can be
done with options.)

```
t = "Times";
opts1 = Sequence[FontFamily → t, FontWeight → "Bold", FontSize → 12];
opts2 = Sequence[FontFamily → t, FontWeight → "Bold", FontSize → 9];
opts3 = Sequence[FontFamily → t, FontWeight → "Bold", FontSize → 6];

<< Graphics`Arrow`

Plot[Sin[x], {x, 0, 2 π}, PlotRange → {{-0.7, 2 π + 0.7}, {-1.6, 1.6}},
    PlotRegion → {{-0.06, 1}, {0.04, 0.92}}, ImageSize → {200, 130},
    Background → GrayLevel[0], DefaultColor → GrayLevel[1],
    TextStyle → {FontFamily → "Times", FontSize → 9},
    FormatType → TraditionalForm, PlotStyle → AbsoluteThickness[2],
    PlotLabel → StyleForm["An interesting wave", opts1], Frame → True,
    FrameLabel → {StyleForm[x, opts2], StyleForm[Sin[x], opts2]},
    RotateLabel → False,
    FrameStyle → {GrayLevel[0.6], AbsoluteThickness[1.5]}, FrameTicks →
      {{0, {π / 2, "π/2"}, π, {3 π / 2, "3π/2"}, 2 π}, {-1, 0, 1}, None, None},
```

```
Epilog → {Text[StyleForm["Maximum point", opts3], {π / 2, 1}, {-1, -1}],
   Text[StyleForm["Minimum point", opts3], {3 π / 2, -1}, {1, 1}],
   Text[StyleForm["Point of inflection", opts3], {3.5, 0.4}, {-1, -1}],
   AbsoluteThickness[0.8], Arrow[{3.55, 0.4}, {π + 0.1, 0.1},
     HeadLength → 0.03, HeadWidth → 0.7, HeadCenter → 0.2],
   AbsolutePointSize[5], GrayLevel[0.4],
   Map[Point, {{π / 2, 1}, {π, 0}, {3 π / 2, -1}}]}];
```

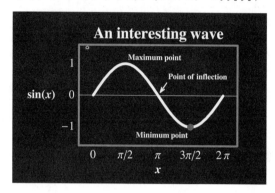

■ Example 2

Here we use some of the options of **GraphicsArray**:

```
Block[{$DisplayFunction = Identity, a = GrayLevel[1]},
 p1 = Plot[Sin[x], {x, 0, 2 π}, PlotLabel → "sin", Background → a];
 p2 = Plot[Cos[x], {x, 0, 2 π}, PlotLabel → "cos", Background → a];
 p3 = Plot[Tan[x], {x, 0, 2 π}, PlotLabel → "tan", Background → a,
   PlotRange → {-16, 16}];
 p4 = Plot[Cot[x], {x, 0, 2 π}, PlotLabel → "cot", Background → a,
   PlotRange → {-16, 16}];]

Show[GraphicsArray[{{p1, p2}, {p3, p4}}, GraphicsSpacing → {0.035, 0.05},
   AspectRatio → 0.7, PlotRegion → {{0, 1}, {0.05, 0.96}},
   ImageSize → {220, 160}, Background → GrayLevel[0.7], TextStyle →
   {FontFamily → "Helvetica", FontWeight → "Bold", FontSize → 8},
   PlotLabel → "Some trigonometric functions", Frame → True,
   FrameStyle → {AbsoluteThickness[1], GrayLevel[0]}]];
```

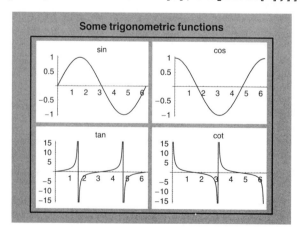

6.2 Options for Form, Ranges, and Colors

6.2.1 Form, Ranges, Margins, and Size

> **** AspectRatio** Ratio of height to width of the plotting rectangle; examples of values:
> **1/GoldenRatio, Automatic, 0.4**

The default value **1/GoldenRatio** = 0.618 gives an aesthetically pleasing form. How-ever, sometimes you may want one unit on the x axis to have the same length as one unit on the y axis. This can be obtained with the value **Automatic**; an example is in Section 5.1.1, p. 111. In principle, the aspect ratio can be any positive real number.

Functions taking on very large positive or negative values can be problematic if plotted with the **Automatic** setting of the aspect ratio, because the plot becomes very narrow. Constraining the plot range with **PlotRange** often helps in such situations. The problem with the first plot below is corrected in this way:

```
Block[{$DisplayFunction = Identity},
  p1 = Plot[Tan[x], {x, 0, 2 π}, AspectRatio → Automatic];
  p2 = Plot[Tan[x], {x, 0, 2 π},
    AspectRatio → Automatic, PlotRange → {-3.5, 3.5}];]

Show[GraphicsArray[{p1, p2}], AspectRatio → 0.3];
```

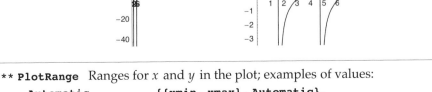

> **** PlotRange** Ranges for x and y in the plot; examples of values:
>
> | | |
> |---|---|
> | **Automatic,** | **{{xmin, xmax}, Automatic},** |
> | **All,** | **{{xmin, xmax}, All},** |
> | **{ymin, Automatic},** | **{{xmin, xmax}, {ymin, Automatic}},** |
> | **{ymin, ymax},** | **{{xmin, xmax}, {ymin, ymax}}** |

When showing a plot, *Mathematica* normally displays all values of the function in the given interval. However, if the function takes on very small or very large values on a small interval, *Mathematica* may decide to cut such values away from the plot so that the remain-ing parts of the function can be seen more clearly; this may happen if the default value **Automatic** is used for **PlotRange**. Sometimes this feature gives good plots, but perhaps more often we would like to see more of the curve. To see the whole function, give the value **All** to **PlotRange**, or specify a suitable range **{ymin, ymax}**.

The values in the lefthand column above specify the range for y only; the values in the righthand column specify the ranges for x and y. There is seldom the need to also specify the x range. A situation where this is relevant, however, may arise when using the

AxesOrigin option (see Section 6.5.1, p. 165). The x range can also be of the form {**xmin**, **Automatic**}.

PlotRegion   Region of the final display area that the plot should fill; examples of
   values: **Automatic** (means **{{0, 1}, {0, 1}}**), **{{xmin, xmax}, {ymin, ymax}}**

Each plot has a *display area* that can be seen by clicking the plot: a rectangle around the plot appears. The curve normally fills this area, but with **PlotRegion** we can define other ways for the curve to be placed in the display area. The plot region is given in scaled coordinates ranging from 0 to 1 in each direction. The default setting **Automatic** is the same as **{{0, 1}, {0, 1}}**, which is the whole display area. By specifying other values (values less than 0 and greater than 1 are allowed), we can adjust the margins around the curve in the display area. For example, plots with gray or colored backgrounds often look better with a somewhat reduced plot region (that is, with larger margins); see the examples in Sections 6.1.3, p. 153, and 6.2.2, p. 156.

ImageSize   The absolute size of the plot; examples of values: **Automatic** (usually
   means 288), **width**, **{width, height}**

The size of the plot is easy to change with the mouse, but we can also use the **Image- Size** option. It determines the absolute size of the plot in units of printer's points (1/72 inch). The default value of 288 is equal to 4 inches. One number as the value of the option defines the width and a list of two numbers both the width and the height. Note that if both the width and the height are specified, the plot fills this area only if the aspect ratio is exactly **height**/**width**. The default size can be changed with the *Option Inspector* (see Section 5.1.3, p. 116). **ImageSize** is sometimes needed, together with **PlotRegion**, to adjust the margins around the plot (see examples in Section 6.1.3, p. 153).

6.2.2 Colors

Background   Color of the background; possible values: **Automatic** (usually means
   GrayLevel[1]), **GrayLevel[g]**, **Hue[h,s,b]**, **RGBColor[r,g,b]**,
   CMYKColor[c,m,y,k]
(**DefaultColor**   Color of the curve, axes, labels, etc; possible values: **Automatic**
   (usually means **GrayLevel[0]**), **GrayLevel[g]**, **Hue[h,s,b]**, **RGBColor[r,g,b]**,
   CMYKColor[c,m,y,k])
(**ColorOutput**   Type of color output to produce; examples of values: **Automatic**, **None**,
   GrayLevel, **RGBColor**, **CMYKColor**)

Color systems were considered in Section 5.4.7, p. 144. The default value **Automatic** of **Background** leaves the background white, and then the curve is black. If the gray level of the background is less than 0.5, the figure is drawn in white. Note that if you use a non-white background, you may want to make a little more space around the plot with **Plot- Region** so that the plot looks good:

```
p1 = Plot[Sin[x], {x, 0, 2 π},
    Background → GrayLevel[0.8], DisplayFunction → Identity];
p2 = Show[p1, PlotRegion → {{0.03, 0.97}, {0.03, 0.97}}];
p3 = Show[p2, Background → GrayLevel[0.4]];

Show[GraphicsArray[{p1, p2, p3}]];
```

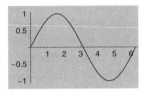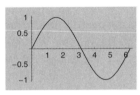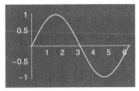

DefaultColor defines the color used for all components of the plot, except the color of the background and the color of components having their own color definition. The default value **Automatic** chooses a color complementary to the background used, as seen in the third plot above.

The default value **Automatic** of **ColorOutput** leaves the colors unchanged; the value **None** converts colors to black and white; **GrayLevel** converts colors to gray levels; and **RGBColor** and **CMYKColor** convert colors to these systems. If you write a document to be printed with no color plots, then it may be useful to set **ColorOutput** to **GrayLevel**, because then you see on the screen how the figures will look when printed.

6.3 Font and Formatting

6.3.1 Font and Formatting in a Session and in a Plot

■ **Introduction to Fonts in Graphics**

The default is that all text in all graphics is written in the Courier font at size of 10 points. This is probably far from being the most pleasing font. By defining a new font and a new size, you may get more attractive plots. I prefer the Helvetica font, and because I often scale the figures to a small size, I use a small font size like 7 or even 6 (size 6 is used in this book). Defining a new font is easy, and we shall give an example right now:

```
$TextStyle = {FontFamily → "Times", FontSize → 8};
```

If you give this command at the beginning of a session, then Times size 8 is used in all text in all graphics in that session (unless you define otherwise).

Font and formatting in graphics can be defined at three levels: for the current session (as we did in the example above), for a particular plot, and for a piece of text. The first two levels are considered here in Section 6.3.1, while the third level is considered in Section 6.3.2.

Remember that a style sheet is used to define fonts in other places than in graphics, such as in inputs, outputs, text, headings, and so on (see Section 3.1.2, p. 51).

■ Font and Formatting in a Session

Variables used to define font and formatting in a session:

** **$TextStyle** Font used in all graphics
$FormatType Format type of text used in all graphics

All text in graphics is written with the default font, which is Courier size 10. Further, all text is formatted as **$FormatType**, with the default value of **StandardForm** (the formatting types are explained in Section 3.3.1, p. 65). We ask for the values of these variables:

    {$TextStyle, $FormatType}

    {{FontFamily → Helvetica, FontSize → 6}, StandardForm}

So, in this book we have changed the default font in graphics: we use Helvetica size 6.

Changing the font and format is easy. We can write the following:

    $TextStyle = {FontFamily → "Times", FontSize → 8};
    $FormatType = TraditionalForm;

These values of **$TextStyle** and **$FormatType** are then applied in all figures plotted during the rest of the current session. With the given definitions, we get the following result:

    f = Exp[x] ArcTan[x] / Sqrt[x];

    Plot[f, {x, 0, 5}, AxesLabel → {x, y}, PlotLabel → f];

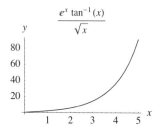

Now the font is Times, and all variables and formulas are formatted according to the traditional form (e.g., with italic variables).

The style of font can be defined with various options, two (and the most important) of them being **FontFamily** and **FontSize** (the size is given in the same absolute units as **AbsolutePointSize** [see Section 5.4.3, p. 135]). Here we list options that can be used for **$TextStyle** and values of **$FormatType**:

Font options and examples of their values:

** **FontFamily:** "Courier", "Times", "Helvetica"
** **FontSize:** 10, 8
FontWeight: "Plain", "Bold"
FontSlant: "Plain", "Italic", "Oblique"
FontColor: GrayLevel[0], RGBColor[1, 0, 0]
Background: GrayLevel[1], RGBColor[0, 0.5, 0.5]

> *Format types:*
> **InputForm, OutputForm, StandardForm, TraditionalForm**

Use **$TextStyle**. By defining a more attractive font, you will get better text in all plots. Note that you can also define the font and formatting in the **init.m** file; your definitions then hold automatically for all sessions, unless you change the definitions (see Section 4.1.2, p. 89).

■ Font and Formatting in a Plot

> *Options used to define font and formatting in a plot:*
>
> **TextStyle** Font used in a plot; examples of values: **$TextStyle, {FontFamily →**
>   **"Times", FontSize → 8}**
> **FormatType** Format type of text used in a plot; examples of values: **$FormatType,**
>   **StandardForm, TraditionalForm, InputForm, OutputForm**

If we define font and formatting with **$TextStyle** and **$FormatType**, then we need not carry these matters for each plot separately. However, the options **TextStyle** and **Format﹀ Type** are available for a particular plot.

The default values of these options are **$TextStyle** and **$FormatType**. The options affect all text in the plot: the plot label, axes labels, frame labels, tick labels, frame tick labels, and text primitives. The value of the **TextStyle** option is a list of options. Available options like **FontFamily** and **FontSize** were mentioned above. As an example, we plot the same figure as earlier:

```
Plot[f, {x, 0, 5}, AxesLabel → {x, y}, PlotLabel → f,
  TextStyle → {FontFamily → "Times", FontSize → 8},
  FormatType → TraditionalForm];
```

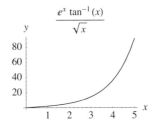

6.3.2 Font and Formatting in a Piece of Text

Sometimes we want to use several fonts or several sizes of a font in the same figure. We can then define the font and format used most often with the **$TextStyle** and **$Format﹀ Type** variables or with the **TextStyle** and **FormatType** options and use **StyleForm** in places where we want to use special fonts, sizes, and formats.

> *Commands used to define font and formatting in a piece of text:*
>
> **StyleForm[expr, fontOptions]** Define the font in a text
> **formatType[StyleForm[expr, fontOptions]]** Define the font and format in a text

In **fontOptions** we can use the options mentioned in Section 6.3.1, p. 158, and **format**
Type can be one of the four formatting types mentioned there. As examples we consider text in a plot label, text in axes labels, and text in a **Text** primitive within the **Epilog** option.

■ Plot Label

In the first example, we define the style of the plot label:

```
f = Exp[x] ArcTan[x] / Sqrt[x];
```

```
Plot[f, {x, 0, 5}, AxesLabel → {x, y}, PlotLabel →
    TraditionalForm[StyleForm[f, FontFamily → "Times", FontSize → 8]]];
```

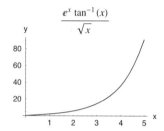

■ Axes Label

If we define styles for axes labels, then it may be convenient to use **Map** to apply the same definitions to both labels (for **Map**, see Section 2.2.2, p. 36):

```
Plot[f, {x, 0, 5}, AxesLabel → Map[TraditionalForm[
    StyleForm[#, FontFamily → "Times", FontSize → 7]] &, {x, f}]];
```

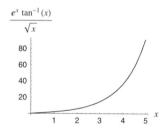

■ Text

```
Text[StyleForm[expr, fontOptions], {x, y}, {p, q}]
Text[formatType[StyleForm[expr, fontOptions]], {x, y}, {p, q}]
```

The usual form of a text primitive is as follows (see Section 5.4.6, p. 141):

```
Text[expr, {x, y}, {u, v}]
```

This defines a text so that the point {**u, v**}—expressed in text coordinates—of the text is at the point {**x, y**} (if {**u, v**} is omitted, the text is centered at {**x, y**}). To define the font and formatting in a **Text** primitive, simply replace **expr** with one of the forms declared for **StyleForm** above. For example:

```
Plot[f, {x, 0, 5},
  Epilog → Text[TraditionalForm[StyleForm[f, FontFamily → "Times",
    FontSize → 8, Background → GrayLevel[0.85]]], {2.7, 73}]];
```

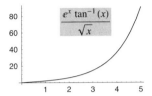

Another method to define font and formatting in **Text** is to use the options **TextStyle**, **FormatType**, and **Background** inside the **Text** primitive:

> **Text[expr, {x, y}, {u, v}, TextStyle → {fontOptions}, FormatType → type,**
> **Background → color]**

■ Some Notes

Mathematica normally uses its own fonts to display formulas in graphics (even when printing parentheses and brackets). This may cause problems if the plots are to be transferred outside *Mathematica* (for exporting, see Section 4.2.2, p. 94), for example, into an EPS file. To avoid the use of *Mathematica*'s special fonts, write the option **FormatType → OutputForm**.

If you have problems printing graphics, try downloading the fonts you use in graphics. Fonts can be downloaded, for example, with PCSEND.EXE or Downloader from Adobe. You may also need to download the fonts that come with *Mathematica*.

Another method to use to solve printing problems is to render the document in the front end (and not in the printer). To do this, choose **Format ▷ Option Inspector...**, set *Show option values for* to be *notebook*, then go to **Notebook Options ▷ Printing Options ▷ PrintingOptions** and change the value of **GraphicsPrintingFormat** from **Automatic** to **RenderInFrontEnd**. Now you do not need to download the fonts. When you print a document, it is rendered in the front end.

On a Macintosh with MacOS X and *Mathematica* 5, to get rotated text correctly printed, set **GraphicsPrintingFormat** to **DownloadPostScript**.

6.4 Options for Curve

6.4.1 Plot Style

> **** PlotStyle^{cp}** Style(s) of the curve(s); examples of values: **Automatic, Absolute**↘
> **Thickness[2], {AbsoluteThickness[2],RGBColor[1,0,0]}**

When plotting 2D curves, **PlotStyle** defines the thickness, dashing, and color of the curves. The default value **Automatic** means a thin, nondashed, black curve. The style can be defined with *graphics directives* (see Section 5.4.1, p. 132). We repeat the directives here:

```
PointSize[d], AbsolutePointSize[d]
Thickness[d], AbsoluteThickness[d]
Dashing[{d1, d2, … }], AbsoluteDashing[{d1, d2 ,… }]
GrayLevel[g], Hue[h], Hue[h, s, b], RGBColor[r, g, b], CMYKColor[c, m, y, k]
```

For thickness and dashing, see Section 5.4.4, p. 136, and for colors, see Section 5.4.7, p. 144, (remember that `GrayLevel[0]` is black and `GrayLevel[1]` is white). The size of points is not relevant for curves; the default thickness is `AbsoluteThickness[0.5]`.

`PlotStyle` is handy for distinguishing different curves in the same plot. We can also add a legend telling which style belongs to which curve (see Section 6.6.3, p. 173). The `PlotStyle` option can be written in the following forms:

Style definitions for one curve:

`PlotStyle → s` One style for the curve

`PlotStyle → {s1, s2, … }` Several styles for the curve

Style definitions for several curves:

`PlotStyle → s` Same style for each curve

`PlotStyle → {{s1, s2, … }}` Same styles for each curve

`PlotStyle → {s1, s2, … }` Different style for each curve

`PlotStyle → {{s11, s12, … }, {s21, s22, … }, … }` Different styles for each curve

Note that the default style can be denoted by the empty list `{}`. This is useful when plotting several curves. In the first example, we plot one curve:

```
p1 = Plot[Sin[x], {x, 0, 2 π},
    PlotStyle → {RGBColor[1, 0, 0], AbsoluteThickness[1.5]}];
```

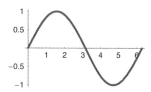

Next we plot two curves, each with its own style:

```
Plot[{Sin[x], Cos[x]}, {x, 0, 2 π},
    PlotStyle → {{RGBColor[1, 0, 0], AbsoluteThickness[1.5]},
      {RGBColor[0, 1, 0], AbsoluteThickness[1.5]}}];
```

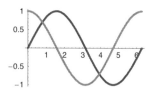

Here we use the default style for the first curve:

```
Plot[{Sin[x], Cos[x]}, {x, 0, 2 π},
  PlotStyle → {{}, {RGBColor[0, 0, 1], AbsoluteThickness[1.5]}}];
```

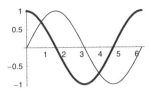

Note that the value of **PlotStyle** cannot be changed by **Show**. The figure must be replotted with the new value for **PlotStyle**.

6.4.2 Plotting Algorithm

> * **PlotPoints**^{cp} Number of initial sampling points; default value: **25**
> (**PlotDivision**^{cp} Maximum number of subdivisions of an interval; default value: **30**)
> (**MaxBend**^{cp} Maximum bend angle between successive line segments; default value: **10**)
> (**Compiled**^{cp} Whether to compile the expression to be plotted; possible values: **True**,
>   **False**)

The quality of the plot is controlled by the options **PlotPoints**, **PlotDivision**, and **MaxBend**. Compiling an expression speeds up the computations.

The default values for these options are mostly sufficient, and so we seldom have to adjust the values, except possibly the value of **PlotPoints**. If we want to give our own values, note that the values cannot be changed by **Show**. We must replot the figure with the options we want.

■ PlotPoints

Plot works by sampling the function initially at some (almost) equally spaced points and then joining the points by straight lines. If the function changes rapidly somewhere in the interval, then more points are sampled automatically. **Plot** is in this sense adaptive. Accordingly, the resulting plot is usually sufficiently smooth and accurate.

The initial number of sample points is controlled by the option **PlotPoints**. Its default value is 25, which means that the function is first sampled at 25 (almost) equally spaced points; however, if this seems not to suffice, then more points are sampled. As an example, we investigate the following figure:

```
p = Plot[Sin[x], {x, 0, 2 π},
  AspectRatio → 0.25, DisplayFunction → Identity];
```

Here are the final sampled points:

```
Short[sample = p[[1, 1, 1, 1]], 3]
```

$$\{\{2.61799 \times 10^{-7}, 2.61799 \times 10^{-7}\}, \ll 80 \gg, \{6.28319, -2.61799 \times 10^{-7}\}\}$$

These 82 points could be plotted by **ListPlot[sample]**, but to get vertical lines instead of points, we do the following:

```
Show[Graphics[
   {AbsoluteThickness[0.1], Map[Line[{{#[[1]], 0}, #}] &, sample]}],
   Axes → True, AspectRatio → 0.25, Ticks → {{π, 2 π}, {-1, 1}}];
```

We see that, where the function changes rapidly, more points are calculated than where the function behaves almost linearly.

As another example, consider the following two plots:

```
Block[{$DisplayFunction = Identity},
  p1 = Plot[Sin[1 / x], {x, 0.1, 0}, AspectRatio → 0.4];
  p2 = Plot[Sin[1 / x], {x, 0.1, 0}, PlotPoints → 500, AspectRatio → 0.4];]
```

```
Show[GraphicsArray[{p1, p2}]];
```

The numbers of sampled points are as follows:

```
{Length[p1[[1, 1, 1, 1]]], Length[p2[[1, 1, 1, 1]]]}
```

```
{432, 2602}
```

The function behaves wildly near the origin. In the first figure, **Plot** gradually increased the number of sampled points from 25 to 432, but by comparing the two figures, we see that even this is not sufficient. In the second figure, **Plot** started from 500 points and ended with 2602 points, and now the result is good.

■ Other Options

When deciding whether a plot is smooth enough, *Mathematica* investigates two successive joining lines. If they form an angle (bend angle) larger than the threshold value, **MaxBend** degrees, then an additional point is chosen midway in the larger of these two intervals. However, the same interval is not divided more than **PlotDivision** times (default value is 30). When plotting $\sin(\frac{1}{x})$, we get a good result by using the default value of **Plot-Points** but defining **PlotDivision → 500** (**Plot** then samples the function at 2219 points).

The option **Compiled** has the default value **True**, and this means that the expression to be plotted is *compiled* before the sampling of the function begins (for compiling, see Section 14.2.2, p. 365). The idea of compiling is to speed up computations by transforming the expression to a special compiled code that is very effective in numerical computations. Read, in Section 5.1.2, p. 114, a note about the use of **Evaluate** to enable compiling when plotting named expressions and functions. When using high-precision numbers, compiling has to be turned off (see an example in Section 11.2.3, p. 291).

■ How Plots Are Displayed

Once the sample points are ready, the plotting proceeds in three steps. First the figure is presented in *Mathematica* form by using graphics primitives. For example, in a **Plot** figure, the lines joining the sample points are formed with the **Line** primitive. If the name of the figure is **p**, we can see the code by writing the command **?p**. An even more complete code can be seen by typing **InputForm[FullGraphics[p]]** (see Section 5.4.1, p. 132). The graphic is then transformed to the device-independent PostScript form. We can see this code by clicking a plot and choosing **Format ▷ Show Expression**. Lastly, the figure is rendered on the screen (the front end of *Mathematica* does this).

6.5 Options for Axes, Ticks, Frames, and Grid Lines

6.5.1 Axes

> * **Axes**   Whether to draw the axes; default values: **Automatic**[cp] (usually means **True**), **False**[g] ; examples of other values: **{True, False}**, **{False, True}**
> * **AxesOrigin**   Point where the axes cross; examples of values: **Automatic**, **{0, 0}**
> * **AxesLabel**   Labels for the axes; examples of values: **None**, **"y"**, **{"x", None}**, **{"x", "y"}**
> (**AxesStyle**   Style of the axes; examples of values: **Automatic**, **{Absolute ⸲ Thickness[1], GrayLevel[0.5]}**)
> (**AxesFront**[g]   Whether axes are behind or on top of graphics elements; possible values: **False**, **True** [in the **Graphics`FilledPlot`** package])

The default value **Automatic** of **Axes** for **Plot** and **ParametricPlot** means that axes are drawn. By default, **Graphics** does not produce axes. The point where the axes cross is **AxesOrigin**. Its default value is determined by an algorithm; this method usually chooses the point **{0, 0}** if it is in the region defined by **PlotRange**.

If you want to move the axes origin, the result is often a gap in the axes, as can be seen in the second plot below. In fact, the axes are drawn only within the plot range. The solution is to define a value for **PlotRange** that contains the axes origin; see the third plot below.

```
p1 = Plot[2 + Sin[x], {x, π / 2, 2 π},
    AxesLabel → {"x", "y"}, DisplayFunction → Identity];
p2 = Show[p1, AxesOrigin → {0, 0}];
p3 = Show[p1, PlotRange → {{-0.1, 2 π + 0.1}, {-0.1, 3.1}}];

Show[GraphicsArray[{p1, p2, p3}]];
```

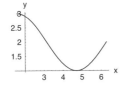  

Axes labels are placed at the ends of the axes (frame labels, instead, are in the middle of the frame edges). If the axes labels are long, consider using a frame instead of axes (see Section 6.5.3, p. 167). See Section 6.3.2, p. 160, for an example of defining font and formatting for axes labels.

The default style for axes is **Automatic**, which means that they are drawn with thin, nondashed, black lines. The default thickness is **AbsoluteThickness[0.25]**. You can define your own styles with style directives as done for **PlotStyle** (see Section 6.4.1, p. 161).

Axes are normally behind graphics elements, but if we load the **Graphics`Filled‹ Plot`** package, we can use the option **AxesFront → True**, and then axes are on top of the elements. This option can be used, for example, with **Graphics** and **Show**.

6.5.2 Ticks

■ Defining Positions

> **** Ticks**   Ticks on the axes; simple examples of values: **Automatic**, **None**, **{{π,2π},**
> **Automatic}**, **{Automatic, {-1,0,1}}**, **{{π,2π}, {-1,0,1}}**

The general form of the value of **Ticks** is **{xticks, yticks}**. We can define ticks on both axes or let *Mathematica* choose the ticks on one of the axes. The automatic algorithm often uses appropriate ticks but frequently also too many ticks, especially if the plots are scaled to be small (as in this book). In particular, the minor ticks (without labels) between the major ticks (with labels) are often unnecessary. A few carefully selected ticks often suffice. Indeed, **Ticks** is perhaps the option I use most often. For example:

```
Block[{$DisplayFunction = Identity},
 p1 = Plot[Sin[x], {x, 0, 2 π}];
 p2 = Plot[Sin[x], {x, 0, 2 π}, Ticks → {{π, 2 π}, {-1, 0, 1}}]];]

Show[GraphicsArray[{p1, p2}]];
```

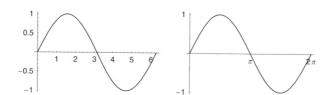

In the second plot, we asked for only five ticks. Note, however, that we got only four tick labels: the label on the *y* axis at 0 was not drawn (probably because the small extension of the *x* axis to the left takes up the space); for a solution, see below.

■ Defining Labels, Lengths, and Styles

Ticks has more advanced forms, where we can define—in addition to the positions of the ticks—the labels and the length and style of the tick marks. In the next box we give possible forms of a *single* tick and examples where each form is used to define two ticks on the *x* axis (on the *y* axis, the default ticks are used).

```
position
Example: {{π, 2π}, Automatic}

{position, label}
Example: {{{π, a}, {2π, b}}, Automatic}

{position, label, {poslength, neglength}}
Example: {{{π, a, {0.02, 0}}, {2π, b, {0.02, 0}}}, Automatic}

{position, label, {poslength, neglength}, {style}}
Example: {{{π ,a, {0.02, 0}, {Hue[0]}}, {2π, b, {0.02, 0}, {Hue[0]}}},
  Automatic}
```

Consider the following example:

```
Plot[Sin[x], {x, 0, 2 π}, Ticks → {{{π / 2, "π/2"}, π, {3 π / 2, "3π/2"}, 2 π},
    {{-1, "min = -1"}, {0.001, 0}, {1, "max = 1"}}}];
```

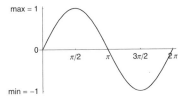

Here we use the tick labels "π/2" and "3π/2" in string form at positions $\pi/2$ and $3\pi/2$. This is done because the nonstring labels $\frac{\pi}{2}$ and $\frac{3\pi}{2}$ look somewhat large in a small plot. At positions −1 and 1 on the y axis, we have placed some text. At the height 0.001 on the y axis, we have placed 0; this solves the problem mentioned earlier.

We can also define the length and style of tick marks. The length can be defined separately in the positive and negative direction (for example, above and below the x axis). Lengths are expressed as a fraction of the width of the plot. The length can also be defined with a single number, and then this value is the total length of the tick mark.

6.5.3 Frames

* **Frame** Whether to draw a frame; examples of values: **False**, **True**, **{True, True, False, False}**
* **FrameLabel** Labels for the frame; examples of values: **None**, **"y"**, **{"x", None}**, **{"x", "y"}**, **{"bottom", "left", "top", "right"}**
* **RotateLabel** Whether to rotate the label for the vertical edges; possible values: **True, False**

  FrameStyle Style of the frame; examples of values: **Automatic**, **{GrayLevel[0.5], AbsoluteThickness[1]}**
* **FrameTicks** Ticks on the frame; simple examples of values: **Automatic**, **None**, **{{0, π, 2π}, {0, π}}**, **{{0, π, 2π}, {0, π}, None, None}**

The default setting is that a frame is not drawn. We can define for each edge of the frame separately whether or not it is to be drawn (the edges are considered in the order bottom, left, top, right). Frame labels are placed midway on the edges (axes labels are placed at the ends of the axes). The default value of **RotateLabel** is **True**, which means that the frame labels on the vertical parts of the frame are rotated so that they read from bottom to top. A short label like y looks better and is easier to read when not rotated. If we want a frame, then the figure often looks better if there is somewhat more space around the curve. This can be done with **PlotRange**:

```
Plot[Sin[x], {x, -π/2, 5π/2}, PlotRange → {-1.3, 1.3},
    Frame → True, FrameLabel → {"x", "y"}, RotateLabel → False];
```

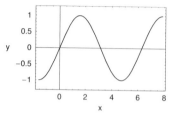

The default is that there are tick marks on all edges but tick labels only on the bottom and left edges. If we define our own x and y ticks, they are used on all four edges, but with **None** we can remove the ticks from the edges on which we do not want them. Frame ticks are defined in the same way as axes ticks (see Section 6.5.2, p. 166):

```
Plot[Sin[x], {x, -π/2, 5π/2}, PlotRange → {-1.3, 1.3}, Frame → True,
    FrameTicks → {Range[-π/2, 5π/2, π/2], {-1, 0, 1}, None, None}];
```

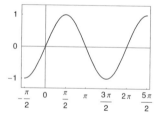

Note that a plot with a frame also has axes, and the options of axes can be used. If you do not want the axes, define **Axes → False**. A plot can have either axes ticks or frame ticks. By defining **FrameTicks → None**, axes ticks are automatically drawn, and a simple box is drawn around the plot:

```
Plot[Sin[x], {x, -π/2, 5π/2},
    Frame → True, FrameTicks → None, AxesLabel → {"x", "y"},
    PlotRange → {{-π/2 - 0.4, 5π/2 + 0.4}, {-1.3, 1.3}}];
```

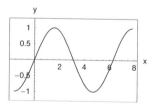

6.5.4 Grid Lines

> **GridLines** How the grid lines are drawn; simple examples of values: **None**, **Auto**
> **matic**, **{None, Automatic}**, **{None, {-1,0,1}}**, **{{0,1,2}, {-1,0,1}}**

Ticks often suffice to give information about the values of the coordinates. However, if we want to read approximate coordinates from a figure, then grid lines may help us. By default, grid lines are not drawn. The value **Automatic** chooses some lines and colors them blue. For example:

```
p1 = Plot[Sin[x], {x, 0, 2 π},
    GridLines → Automatic, DisplayFunction → Identity];
p2 = Show[p1, GridLines → {None, Automatic}];
p3 = Show[p1, GridLines → {{π / 2, π, 3 π / 2, 2 π}, {-1, -0.5, 0.5, 1}}];

Show[GraphicsArray[{p1, p2, p3}]];
```

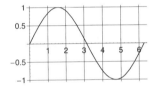 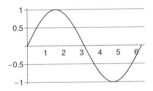 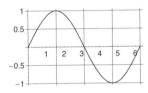

The style of a grid line can be defined with **{location, {style}}**. For example:

```
xgrids = Table[{x, {GrayLevel[0.6]}}, {x, 0, 2 π, π / 2}];
ygrids = Table[{y, {GrayLevel[0.6]}}, {y, -1, 1, 0.5}];

Plot[Sin[x], {x, 0, 2 π}, GridLines → {xgrids, ygrids}];
```

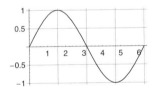

6.6 Options for Plot Labels, Primitives, and Legends

6.6.1 Plot Labels

> *** PlotLabel** Label of the plot; examples of values: **None**, **Sin[x/2]**, **"sin(x/2)"**

A label can be defined as a usual formula (without the quotation marks) or as a string (enclosed by the quotation marks " "). If the label is, say, **Sin[x/2]**, then remove the possible value of **x** before plotting. See Section 6.3.2, p. 160, for an example of defining font and formatting in a plot label. For another example:

```
Plot[Sin[x/2], {x, 0, 4 π}, PlotLabel → Sin[x/2],
  TextStyle → {FontFamily → "Times", FontSize → 7},
  FormatType → TraditionalForm];
```

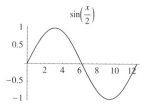

Sometimes we may want to vary a parameter in the label:

```
t = Table[Plot[Sin[n x], {x, 0, 2 π}, DisplayFunction → Identity,
    PlotLabel → Sin[n x], TextStyle → {FontFamily → "Times", FontSize → 7},
    FormatType → TraditionalForm], {n, 3}];

Show[GraphicsArray[t]];
```

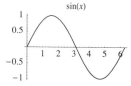 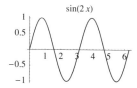 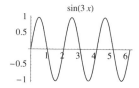

If the label is written as a string, then give **PlotLabel** the value **StringForm["sin(``x)"**, **n]** (here the marks `` are replaced by the current value of **n**) or **SequenceForm["sin(", n, "x)"]**.

6.6.2 Primitives

■ Adding Primitives with Prolog and Epilog

Now and then we may want to add some components to a plot. We may want to have clarifying text, an important point, an arrow, or a line. Such small additions may considerably improve the quality of a plot: they guide the eyes of the reader to important aspects of the plot. The options **Prolog** and **Epilog** are the tools for making such additions.

Prolog Graphics primitives to be plotted before the main plot; examples of values: **{}**,
   {{AbsolutePointSize[3], RGBColor[1,0,0], Point[{0.4, 0.7}]}}
**** Epilog** Graphics primitives to be plotted after the main plot; examples of values:
   {}, {AbsolutePointSize[3], RGBColor[1,0,0], Point[{0.4, 0.7}]}

The values of **Prolog** and **Epilog** are lists of graphics directives and primitives (their default values are empty lists). (Note that **Prolog** needs double curly braces; with single curly braces, the directives in **Prolog** also affect the main plot.) Directives and primitives were explained in Section 5.4. For easy reference, we list here the four primitives that are most useful when using **Prolog** and **Epilog**. We also list all directives.

```
Point[p]    Point at p
Line[{p1, …, pn}]   Line through points p1, …, pn
Text[expr, p, q]    Text expr placed so that the point q, expressed in text coordinates,
    of expr is at the point p
Arrow[p1, p2, opts]   Arrow from p1 to p2 (in the Graphics`Arrow` package)
```

```
PointSize[d], AbsolutePointSize[d]
Thickness[d], AbsoluteThickness[d]
Dashing[{d1, d2, …}], AbsoluteDashing[{d1, d2 ,…}]
GrayLevel[g], Hue[h], Hue[h, s, b], RGBColor[r, g, b], CMYKColor[c, m, y, k]
```

We have already used some primitives with the **Epilog** option in a few examples; see Sections 1.1.1, p. 4 (**Point**, **Line**), 5.4.6, p. 141 (**Point**, **Text**, **Arrow**), 6.1.3, p. 153 (**Point**, **Text**, **Arrow**), and 6.3.2, p. 160 (**Text**). Many more examples are in the forthcoming chapters. In the next example, we use the primitives **Point**, **Line**, **Text**, and **Arrow**:

```
<< Graphics`Arrow`

Plot[{Log[x] + 1, Sqrt[x]}, {x, 0, 1.5}, FormatType → TraditionalForm,
  TextStyle → {FontFamily -> "Times", FontSize → 7},
  PlotRange → {{0, 2.05}, {0, Automatic}},
  Ticks → {{0.5, 1, 1.5, 2}, {0.5, 1, 1.5}},
  Epilog → {Text[Log[x] + 1, {1.55, 1.42}, {-1, 0}],
    Text[Sqrt[x], {1.55, 1.22}, {-1, 0}], {AbsoluteThickness[0.3],
     Arrow[{1.19, 0.82}, {1.05, 0.95}, HeadLength → 0.03,
      HeadWidth → 0.7, HeadCenter → 0.3]}, RGBColor[0, 0, 1],
    Text[Log[x] + 1 == Sqrt[x], {1.2, 0.85}, {-1, 1}], RGBColor[0, 0.7, 0],
    AbsoluteDashing[{2}], Line[{{0, 1}, {1, 1}, {1, 0}}],
    RGBColor[1, 0, 0], AbsolutePointSize[3.5], Point[{1, 1}]}];
```

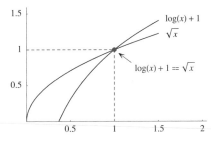

With **Prolog** and **Epilog**, before each primitive, we write the directives with which we want to modify the primitive. However, remember that a directive affects all primitives that come after it. To avoid this, we often have to use additional curly braces to restrict the effect of some directives to only the primitives inside the curly braces. For example, in the plot above, the thickness directive is restricted to the arrow.

An application of **Prolog** is in Section 9.1.1, p. 230. There we plot data as points with vertical lines: the points are plotted with **ListPlot** and the vertical lines with **Prolog**. **Prolog** is used to first draw the lines; **ListPlot** then plots the points on top of the top parts of the lines. The result is then good, even if the points and the lines have different colors (with **Epilog**, the lines would appear on top of the points).

■ Nonappearing Primitives

Sometimes a text primitive is not shown wholly in the figure. We can then add the option **PlotRange → All** or enlarge the figure with the mouse. For other primitives, we are not so lucky. If, for example, a point or a line defined in **Epilog** is not shown in the figure, defining **PlotRange → All** does not help us. In such a situation, we have two possibilities: either define explicitly a large enough plot range with **PlotRange** so that the primitives fall inside this range, or plot the primitives separately with **Graphics**. The latter method is considered after we have considered disappearing primitives.

■ Disappearing Primitives

We may also have the problem of disappearing primitives. Remember what we said in Section 5.1.2, p. 113, about options with **Show**: if plots to be combined with **Show** have different values for some options, the values of the first plot are applied. Thus, if we combine two or more plots with **Show**—each having an **Epilog** option—then **Show** takes on the value of **Epilog** given in the first figure. This means that all of the graphics primitives in the other figures are left out. You may think the problem is solved by writing an **Epilog** in **Show** to add the lacking primitives. In reality, the primitives of the first figure are now lacking.

Thus, if you intend to combine several figures with **Show**, prepare to use **Epilog** also in **Show**: write in the **Epilog** option *all* the primitives of the component figures. Or, if you do not need the intermediate figures as such, do not add primitives into them but only into the final combined figure. However, if you plot the primitives with **Graphics**, then the problem does not occur; this method is considered next.

■ Adding Primitives with Graphics

Prolog and **Epilog** are not the only ways of adding graphics primitives; we can also use **Graphics** (see Section 5.4.1, p. 133). In fact, we can separately plot the main figure with **Plot** (or with another command) and the primitives with **Graphics** and combine the two plots with **Show**. For example:

```
p1 = Plot[{Log[x] + 1, Sqrt[x]}, {x, 0, 1.5}, FormatType → TraditionalForm,
    TextStyle → {FontFamily -> "Times", FontSize → 7},
    PlotRange → {{0, 2.05}, {0, Automatic}},
    Ticks → {{0.5, 1, 1.5, 2}, {0.5, 1, 1.5}}];
```

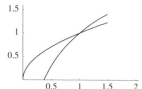

```
p2 = Show[Graphics[{Text[Log[x] + 1, {1.55, 1.42}, {-1, 0}],
    Text[Sqrt[x], {1.55, 1.22}, {-1, 0}], {AbsoluteThickness[0.3],
      Arrow[{1.19, 0.82}, {1.05, 0.95}, HeadLength → 0.03,
        HeadWidth → 0.7, HeadCenter → 0.3]}, RGBColor[0, 0, 1],
    Text[Log[x] + 1 == Sqrt[x], {1.2, 0.85}, {-1, 1}], RGBColor[0, 0.7, 0],
    AbsoluteDashing[{2}], Line[{{0, 1}, {1, 1}, {1, 0}}],
    RGBColor[1, 0, 0], AbsolutePointSize[3.5], Point[{1, 1}]}],
  PlotRange → All];
```

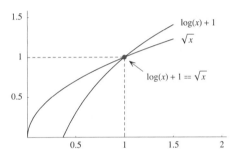

```
Show[p1, p2];
```

This method of adding primitives does not have the problems of nonappearing or disappearing primitives.

6.6.3 Legends

■ Legends with Primitives

If there are several curves in the same plot, it might be useful to identify the curves in some way. This can be done with graphics primitives in an **Epilog** option or with a package. First we consider the use of **Epilog**.

A good and simple way for identifying the curves is to place a suitable **Text** primitive near each curve. See an example in Section 6.6.2, p. 171. This method has several advantages. First, it is simple to implement. Second, we need not look at a separate legend to see the style of each curve. Third, we can use the same style for each curve (as was done in the example in Section 6.6.2).

We can also make a legend that associates each line style with an explanation. To this end we plot, with **Epilog**, short lines of the types used and place text primitives next to them:

```
Plot[{Sin[x], Cos[x]}, {x, 0, 2 π}, AspectRatio → 0.4,
   PlotRange → {{-0.1, 9.6}, All}, Ticks → {{π, 2 π}, {-1, 1}},
   PlotStyle → {{}, AbsoluteDashing[{1.3}]}, AxesStyle → GrayLevel[1],
   Epilog → {Line[{{-0.1, 0}, {2 π + 0.1, 0}}],
     Line[{{0, -1.1}, {0, 1.1}}], Line[{{7.0, 0.8}, {8.0, 0.8}}],
     AbsoluteDashing[{1.3}], Line[{{7.0, 0.4}, {8.0, 0.4}}],
     Text[Sin[x], {8.2, 0.8}, {-1, 0}], Text[Cos[x], {8.2, 0.4}, {-1, 0}]}];
```

If the legend is placed outside the *x* range of the curves as above, then **PlotRange** must be used to extend the *x* range so that the legend is within it (**PlotRange → All** does not help, because the lines in the legend do not appear, although the text does appear; see **Nonappearing Primitives** in Section 6.6.2, p. 172). The extended *x* range causes that the *x* axis continues from 2π to 9.6. To stop the *x* axis at 2π, we have defined white as the style of the axes and drawn the axes ourselves with **Line** primitives.

■ Legends with a Package

The **Graphics`Legend`** package adds some new options for **Plot** and **MultipleListPlot** enabling the use of legends. Note that *this holds only for these two commands* and not for any other plotting commands. For other commands, the package defines the **ShowLegend** command, which is considered at the end of this section.

Here we first consider legends for **Plot**; legends for **MultipleListPlot** are considered in Section 9.2.2, p. 244. The basic option is **PlotLegend**:

> **Plot[{expr1, expr2 ,… }, {x, a, b}, PlotLegend → {text1, text2, … }]** Plot the expressions and add a legend having **text1**, **text2**, … as explanations for the curves

When using a legend, it is important to plot the curves with styles having *absolute* sizes—**AbsoluteThickness** and **AbsoluteDashing**—because then the styles shown in the legend look the same as they do in the main plot. For example:

```
<< Graphics`Legend`
```

```
Plot[{Sin[x], Cos[x]}, {x, 0, 2 π}, PlotStyle →
   {{}, AbsoluteDashing[{1.3}]}, PlotLegend → {Sin[x], Cos[x]}];
```

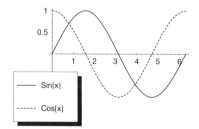

The default position, size, and style of the legend may not satisfy you (in my opinion, a better position is at the right, outside of the plot, the legend could be smaller, and it is too heavy with its border and shadow). The legend can, however, be modified with many options, as explained below. Using the options may not be easy due to the various coordinate systems and units and because changing the value of an option may have unexpected side effects. However, with a few options, we can create appealing legends.

■ Options

Before going to the options, we explain some terms. The legend is in a *legend box*. The box has a *border* (a thin, black line in the example above) and a *background* (white in the example). The legend box may have a *shadow* (black in the example). Inside the box we have the *key boxes* that contain samples of the styles of the curves and the *texts* explaining the styles. Above the explanations of the curves (but inside the box), we can have a *legend label* (there is not one in the example).

In the following list we have marked with two asterisks (**) the three most important options and with one asterisk (*) three not-so-important options (the classification reflects my personal opinion).

Options for the position, size, and orientation of the legend box:

** **LegendPosition** Position of the lower-left corner of the legend box (in a coordinate system where the center of the figure is {0, 0} and the longest side of the plot runs from −1 to 1); default value: **{-1.2, -0.82}** (means somewhat below and left of the bottom-left corner of the plot); another example: **{1.1, 0}** (means outside the plot at the right)

** **LegendSize** Size of the legend box (in the same units as **LegendPosition**: 2 units means the length of the longest side of the plot); examples of values: **Automatic** (usually means **0.8**), **size** (the longest side), **{width, height}**, **{0.8, 0.5}**

LegendOrientation Direction in which the key boxes are laid out; possible values: **Vertical** (top to bottom), **Horizontal** (left to right)

Options for the style of the legend box:

LegendBorder Style of the border line; examples of values: **Automatic** (usually means **{AbsoluteThickness[0.25], GrayLevel[0]}**), **GrayLevel[0.5]**

LegendBorderSpace Space between the border and the entire set of legend label, key boxes, and texts (1 unit means the width of the key box; sets mainly the lefthand space); default value: **Automatic** (usually means **0.1**)

LegendBackground Color of the background; examples of values: **Automatic** (usually means **GrayLevel[1]**), **GrayLevel[0.9]**

Options for the shadow of the legend box:

** **LegendShadow** Offset of the shadow from the legend box (in the same units as **LegendPosition**: 2 units means the length of the longest side of the plot); examples of values: **Automatic** (usually means **{0.05, -0.05}**), **None** (no border and no shadow), **{0, 0}** (no shadow), **{xoffset, yoffset}**

ShadowBackground Color of the shadow; examples of values: **GrayLevel[0]**, **RGBColor[0,0,1]**

An option for the key boxes:

LegendSpacing Space around each key box (1 unit is the width of the key box); default
   value: **Automatic** (usually means **0.08**)

Options for the texts:

* **LegendTextSpace** The width of the space allocated for each text (1 unit is the width
   of a key box); default value: **Automatic** (usually means **2**)

LegendTextOffset Offset of a text from a key box: if the value of the option is **{u, v}**,
   then the point **{u, v}**—expressed in text coordinates (ranging from –1 to 1 in both
   directions of the text)—of the text is at the right middle end of the key box in vertical
   legends and at the top middle of the key box in horizontal legends. Examples of
   values: **Automatic** (usually means **{-1, 0}** so that the text starts where the key box
   ends), **{-1.5, 0}** (to make a larger offset).

LegendTextDirection Direction of text: if the value of the option is **{r, s}**, then the
   text has a slope **s/r** (may be useful if **LegendOrientation → Horizontal**, to avoid
   overlapping texts). Examples of values: **Automatic** (usually means **{1, 0}** so that text
   is horizontal), **{3, 1}** (for ascending text).

Options for the legend label:

* **LegendLabel** Label of the legend; examples of values: **None**, **"Functions"**

* **LegendLabelSpace** Space above and below the label (1 unit is about the space
   between key boxes); default value: **Automatic** (usually means **0**, if the legend does
   not have a label, and **1**, if the legend has a label)

To summarize, **LegendPosition**, **LegendSize**, and **LegendShadow** use the coordinate
system where the longest side of the plot runs from –1 to 1 so that the longest side has a
length of 2. **LegendBorderSpace**, **LegendSpacing**, and **LegendTextSpace** use the width of
the key box as the unit; in addition, the unit of **LegendLabelSpace** seems to be about the
space between key boxes. **LegendTextOffset** and **LegendTextDirection** use the same
systems as the **Text** primitive (see Section 5.4.6, p. 141).

Note that the form of the legend box changes somewhat unexpectedly if we change the
value of **LegendSpacing**, **LegendTextSpace**, or **LegendLabelSpace** unless we have
defined a fixed size with **LegendSize → {w,h}** (the larger **LegendSpacing** or **LegendLabel**
Space is, the narrower is the legend box; the larger **LegendTextSpace** is, the lower the
legend box).

■ **Examples**

Using **LegendPosition**, **LegendSize**, and **LegendShadow**, we can move the legend to a
better place, give a smaller size for the legend, and get rid of the border and shadow. The
result is then more aesthetically pleasing:

```
Plot[{Sin[x], Cos[x]}, {x, 0, 2 π},
  PlotStyle → {{}, AbsoluteDashing[{1.3}]}, PlotLegend → {Sin[x], Cos[x]},
  LegendPosition → {1.1, 0}, LegendSize → {0.8, 0.5}, LegendShadow → None];
```

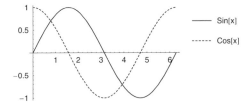

Next we plot three curves and use a legend label:

```
Plot[{Sin[x], Sin[2 x], Sin[3 x]}, {x, 0, 2 π}, PlotStyle →
  {{}, {AbsoluteDashing[{1.5}]}, {AbsoluteDashing[{3, 2, 0.5, 2}]}},
  PlotLegend → {"n = 1", "n = 2", "n = 3"}, LegendPosition → {1.05, -0.1},
  LegendSize → {0.7, 0.6}, LegendShadow → None, LegendTextSpace → 0.8,
  LegendLabel → StyleForm[Sin[n x], FontSize → 7], LegendLabelSpace → 1.5];
```

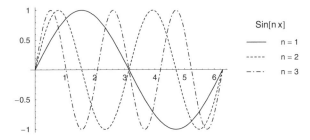

■ Legends for Other Plots

The options for the legend can only be used with **Plot** and **MultipleListPlot**. However, the same package also contains the command **ShowLegend**, with which we can also add a legend to other plots.

ShowLegend[graphic, {{{key1, text1}, {key2, text2}, … }, opts}] Show **graphic** with a legend containing the key graphics or key colors **key1**, **key2**, … and texts **text1**, **text2**, …; modify the legend with the legend options **opts**

When using **ShowLegend**, we must explicitly give the styles used in the plot (with **Plot**, the correct styles appear automatically in the legend). As an example, we plot two parametric curves and add a legend to the combined plot:

```
st1 = AbsoluteThickness[1];
st2 = Sequence[AbsoluteThickness[1], GrayLevel[0.5]];

Block[{$DisplayFunction = Identity},
  p1 = ParametricPlot[{Cos[t], Sin[t]}, {t, 0, 2 π}, PlotStyle → {st1}];
  p2 = ParametricPlot[{Sin[t], t}, {t, -π, π}, PlotStyle → {st2}];
  p3 = Show[{p1, p2}]];
```

```
<< Graphics`Legend`

ShowLegend[p3,
   {{{Graphics[{st1, Line[{{0, 0}, {1, 0}}]}], {Cos[t], Sin[t]}},
     {Graphics[{st2, Line[{{0, 0}, {1, 0}}]}], {Sin[t], t}}},
    LegendPosition → {1.1, 0}, LegendSize → {1.1, 0.5},
    LegendShadow → None}];
```

For another application of **ShowLegend**, see the bar charts in Section 9.3.2, p. 249. With contour plots (see Section 8.2.2, p. 224) and density plots (see Section 8.2.3, p. 228), we use special forms of **ShowLegend**. As mentioned earlier, with **Plot** and **MultipleListPlot** (see Section 9.2.2, p. 244), we do not need **ShowLegend**.

3D Graphics for Functions

Introduction

> *A man went in to the post office to post a fishing rod which was five feet long and all in one piece. He was upset to find that the maximum length of parcel that the post office would accept was four feet. He solved the problem by placing his fishing rod diagonally in a rectangular box of length four feet and width three feet.*

A function of two variables x and y is of the form $z = f(x, y)$. For each pair (x, y), we get a value z so that the result is a surface in three dimensions. Illustrating such a function on the 2D surface of a screen or a sheet of paper involves problems, but by using some guides for the eye (e.g., mesh lines, shading, a bounding box), we get quite a realistic illusion of the 3D reality. There are techniques for improving the illusion, such as rotations and stereograms (see Section 7.3). There are even some ways to illustrate 4D functions of the form $v = f(x, y, z)$ (see Section 7.4).

Options for the main 3D plotting commands **Plot3D**, **ParametricPlot3D**, **Graphics3D**, **ContourPlot**, and **DensityPlot** are considered in Chapter 8; in the present chapter we introduce only the most important options. Special options for special 3D plotting commands are explained in Section 7.2, together with the commands.

7.1 Surface, Contour, and Density Plots

7.1.1 Three Types of Plots

`Plot3D[expr, {x, a, b}, {y, c, d}]` Plot **expr** as a surface plot
`Plot3D[{expr, shading}, {x, a, b}, {y, c, d}]` Color the surface with **shading**
`ContourPlot[expr, {x, a, b}, {y, c, d}]` Plot **expr** as a contour plot
`DensityPlot[expr, {x, a, b}, {y, c, d}]` Plot **expr** as a density plot

Surface, contour, and density plots are the three main plot types to illustrate 3D functions. For example:

```
f = Cos[x y] Cos[x];
```

```
Block[{$DisplayFunction = Identity},
  p1 = Plot3D[f, {x, 0, 3}, {y, 0, 4}, AxesLabel → {x, y, None}];
  p2 = ContourPlot[f, {x, 0, 3}, {y, 0, 4},
     FrameLabel → {x, y}, RotateLabel → False];
  p3 = DensityPlot[f, {x, 0, 3}, {y, 0, 4},
     FrameLabel → {x, y}, RotateLabel → False]];
```

```
Show[GraphicsArray[{p1, p2, p3}], GraphicsSpacing → 0];
```

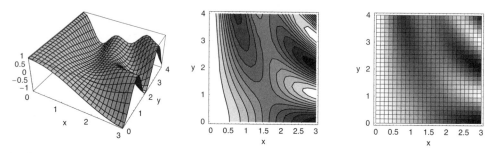

With **Plot3D**, we can give a shading function that tells how the surface is shaded. The function must use the **GrayLevel**, **Hue**, or **RGBColor** directive (see Section 5.4.7, p. 144). An example is in Section 8.1.7, p. 217.

■ The Plotting Method

First 25×25 or 625 equally spaced points (15×15 or 225 points in *Mathematica* 4 and earlier versions) are calculated from the plotting region in the (x, y) plane. The function is then sampled at these points to get the corresponding z values. From this point on, the three commands proceed differently.

Plot3D joins the (x, y, z) points with lines, producing 24×24 polygons. The resulting surface is colored.

ContourPlot produces curves that are *contours of constant value*, that is, along each of the curves the value of the function is a certain constant. The default is that contours are drawn to correspond to 10 equally spaced values of the function between the minimum and maximum values. Indeed, let **zmin** and **zmax** be the minimum and maximum values of the function at the 25×25 sample points, and let **d = (zmax − zmin)/10**. The constants to be used in the contour plot are approximately **Table[zmin + d/2 + n d, {n, 0, 9}]**. The curves are then determined by interpolation between grid points. The areas between the curves are colored so that the highest parts of the function are white and the lowest parts black.

DensityPlot simply produces a small rectangle around each of the sample points (there are thus 25×25 rectangles). Each rectangle is a shade of gray, with the highest parts of the function being white and the lowest parts black.

Remember that **Plot** and **ParametricPlot** are adaptive in that they sample more points if the function changes rapidly. **Plot3D**, **ContourPlot**, and **DensityPlot** *are not adaptive*, which means that they sample the function at a fixed number of points and try to plot the function with this information without having the possibility of taking more sample points. If the plot is not smooth enough, we can increase the value of the **Plot‹ Points** option (see Section 7.1.2).

■ Converting among Types of Plots

Each of the three plotting commands produces its own type of graphics: **‑ Surface‹ Graphics ‑**, **‑ ContourGraphics ‑**, and **‑ DensityGraphics ‑**. (The type can be seen below the plot if the plotting command does not end with a semicolon.) Remember that the type of graphics produced by **Plot** is **‑ Graphics ‑**. Realizing the fact that all three commands use the same sampled points, *Mathematica* has the commands **Surface‹ Graphics**, **ContourGraphics**, and **DensityGraphics** to transform a plot to another type. For example, if you have produced a surface plot **p**, you can get a contour plot by **Show[ContourGraphics[p]]**. Such a conversion does not resample the function, and this saves some computing time.

■ Combining Plots

As in the case with 2D graphics, we can combine 3D plots with **Show**:

```
p1 = Plot3D[Sin[x y], {x, 0, π}, {y, 0, π / 2}, DisplayFunction → Identity];
p2 = Plot3D[Cos[x y], {x, 0, π}, {y, 0, π / 2}, DisplayFunction → Identity];

Show[p1, p2, DisplayFunction → $DisplayFunction, Ticks → None];
```

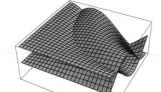

7.1.2 Important Options

We consider all of the options of the main 3D graphics commands in Chapter 8, but here we give some of the most important options. The default value of all options is given first and, after that, examples of other values.

Some common options for **Plot3D**, **ContourPlot**, *and* **DensityPlot**:

PlotPoints Number of sampling points; examples of values: **25, 40, {15, 25}**
PlotRange Range for z in the plot; examples of values: **Automatic, All, {zmin, zmax}**

If we consider the plot not to be smooth enough, we can increase the value of the **PlotPoints** option and replot the function. The default value **25** (**15** in *Mathematica* 4) is often quite good for a surface or density plot. However, for a contour plot, the default value is almost always too small. So, *to get smooth contours in a contour plot, use* **PlotPoints** *with a value of about 40 to 50*:

```
ContourPlot[Cos[x y] Cos[x], {x, 0, 3}, {y, 0, 4}, PlotPoints → 50];
```

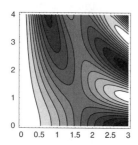

The default value **Automatic** for **PlotRange** has the effect that some low and high parts of some functions may be clipped, and this results in plateaus at the top and bottom of the surface. To ensure that no parts of the function are clipped, we can set **PlotRange → All**.

Some options for **Plot3D**:

BoxRatios Ratios of side lengths of the bounding box; examples of values:
  {1, 1, 0.4}, Automatic
ViewPoint Point from which the surface is viewed; default value: **{1.3, -2.4, 2}**
Ticks Ticks on the axes; examples of values: **Automatic, {{0, 1, 2}, {0, 1},**
  Automatic}, {{0, 1, 2}, {0, 1}, {-1, 0, 1}}

The default value **{1, 1, 0.4}** of **BoxRatios** means that the x and y axes have the same length and that the length of the z axis is 0.4 times the length of the other axes. If we set **BoxRatios → Automatic**, then one unit on each of the x, y, and z axes has the same length in the surface plot.

With **ViewPoint**, we can view a surface from different points. Note that the coordinates of the viewpoint are not expressed in the coordinate system of the surface: the viewpoint has its own coordinate system in which the origin is at the center of the bounding box and the longest side of the box runs from –0.5 to 0.5. The viewpoint can be selected interactively by using **3D ViewPoint Selector...** from the **Input** menu (see Section 8.1.3, p. 209).

Some common options for **ContourPlot** *and* **DensityPlot**:

AspectRatio Ratio of height to width of the plotting rectangle; examples of values: **1,**
  Automatic
FrameTicks Ticks on the frame; examples of values: **Automatic, {{0, π, 2 π},**
  {0, π}, None, None}

> *An option for* **ContourPlot**:
>
> **Contours** The number of *z* values for which contour lines are to be drawn, or a list of
> *z* values; examples of values: **10**, **{-2, -1, 0, 1, 2}**

The default value of **AspectRatio** is **1**, which gives a square. If we set **AspectRatio →**
Automatic, then one unit on both the *x* and *y* axes has the same length. Note that contour
and density plots have a *frame* (and no axes), and so ticks on the frame can be set with
FrameTicks.

For 3D plots, we can perform similar manipulations with the mouse as we did for 2D
plots (see Section 5.1.3, p. 115). For example, we can move, resize, and crop the plot with
the mouse. To go back to the standard position, size, and cropping, choose **Cell ▷ Make**
Standard Size.

7.1.3 Animating

3D plots can be animated in the usual way by plotting a sequence of plots and double-
clicking one of the plots. Animation in three dimensions can be used for two main
purposes.

First, we can animate a sequence of *different* graphics objects having (in most cases) the
same viewpoint. Such an animation shows how the objects evolve step by step. This type
of animation is used in Section 24.2.3, p. 653, to illustrate how the solution of a hyperbolic
partial differential equation evolves with time. Another example is shown next.

Second, we can animate *one* graphics object, plotted with different viewpoints. Such an
animation can be used to rotate a graphics object. This gives a good impression of the 3D
structure of the object. For an example, see Section 7.3.1, p. 194.

Note that animating 3D objects takes a lot of computer memory.

■ Example

We present a program **sputnik** to animate the movement of a satellite around the earth.
The program has five arguments:

number: the number of separate plots to be generated
distance: the distance of the satellite from the surface of the earth, expressed as a
multiple of the radius of the earth
diameter: the diameter of the satellite, given as a fraction of the width of the whole plot
viewheight: the height of the viewpoint (i.e., the *z* coordinate of the **ViewPoint** option)
animation: 0 if only one plot is wanted where we can see all the positions of the satellite, and 1 if we want to see all the separate plots and actually do the animation (the
value 0 can be used to first ensure that the parameters of **sputnik** are OK)

```
sputnik[number_, distance_, diameter_, viewheight_, animation_] :=
  Module[{earth, satellite, h = distance + 1, d = N[2 π / number]},
    earth = ParametricPlot3D[{Cos[t] Cos[u], Sin[t] Cos[u], Sin[u]},
      {t, 0, 2 π}, {u, -π / 2, π / 2}, Axes → None, Boxed → False,
        ViewPoint → {1.3, -2.4, viewheight}, DisplayFunction → Identity];
    satellite = Table[Point[h {Cos[t], Sin[t], 0}], {t, 0, 2 π - d, d}];
```

```
If[animation == 0,
  Show[earth, Map[Graphics3D[{PointSize[diameter], #}] &, satellite],
    PlotRange → {{-h, h}, {-h, h}, {-1, 1}},
    DisplayFunction → $DisplayFunction],
  Map[Show[earth, Graphics3D[{PointSize[diameter], #}],
      PlotRange → {{-h, h}, {-h, h}, {-1, 1}},
      DisplayFunction → $DisplayFunction] &, satellite];]]
```

We show 50 positions of the satellite. The distance of the satellite from the surface of the earth is the radius of the earth, the diameter of the satellite is 0.04, and the height of the viewpoint is 0.8:

```
sputnik[50, 1, 0.04, 0.8, 0];
```

The corresponding animation could be done with the following:

```
sputnik[50, 1, 0.04, 0.8, 1];
```

Then, we could double-click one of the 50 plots. The CD-ROM that comes with this book contains a ready-to-animate sequence of 50 plots.

7.2 Other Plots for 3D Functions

7.2.1 Parametric Curves and Surfaces

`ParametricPlot3D[{fx, fy, fz}, {t, a, b}]` Plot a parametric curve
`ParametricPlot3D[{fx, fy, fz, styles}, {t, a, b}]` Plot a parametric curve;
   styles is a graphics directive or a list of graphics directives
`ParametricPlot3D[{fx, fy, fz}, {s, a, b}, {t, c, d}]` Plot a parametric surface
`ParametricPlot3D[{fx, fy, fz, styles}, {s, a, b}, {t, c, d}, Lighting → False]`
   Plot a parametric surface

The expressions **fx**, **fy**, and **fz** give the x, y, and z coordinates of the curve or surface. We can add a style definition as a fourth expression. For parametric curves, the style may consist of thickness, dashing, and color directives. For parametric surfaces, the style may consist of an **EdgeForm** directive to define the style of the edges (see Section 7.2.4, p. 191) and a color definition used in the shading of the polygons. The fourth argument is used in **p2** and **p7** below (more examples of the fourth argument are in Sections 8.1.6, p. 215, and 8.1.7, p. 218; to see the fine colors of **p2** and **p7**, look up the plots in the *Help Browser* if you have installed the CD-ROM of the book):

```
SetOptions[ParametricPlot3D, Axes → False, Boxed → False];
```

```
Block[{$DisplayFunction = Identity}, p1 =
   ParametricPlot3D[{t, Cos[9 t], Sin[9 t]}, {t, 0, π}, PlotPoints → 200];
  p2 = ParametricPlot3D[{Cos[2 t] + Cos[10 t], Cos[2 t] - Sin[10 t], t,
      {AbsoluteThickness[1.5], Hue[t / π]}}, {t, 0, π}, PlotPoints → 200];
  p3 = ParametricPlot3D[{Sin[s] Cos[t], Sin[s] Sin[t], 1 + Cos[s]},
      {s, 0, π}, {t, 0, 2 π}, PlotPoints → 20];
  p4 = ParametricPlot3D[2 {Sin[s] Cos[t], Sin[s] Sin[t], Cos[s]},
      {s, π / 2, π}, {t, -π, π}, PlotPoints → {10, 20}];
  p5 = Show[{p3, p4}];
  p6 = ParametricPlot3D[{Cos[s] (3 + Cos[t]), Sin[t], Sin[s] (3 + Cos[t])},
      {s, 0, 2 π}, {t, 0, 2 π}, PlotPoints → {30, 15}];
  p7 = ParametricPlot3D[{s Cos[t] Sin[s], s Cos[s] Cos[t], -s Sin[t],
      Hue[s t / (2 π^2)]}, {s, 0, 2 π}, {t, 0, π}, Lighting → False];
  p8 = ParametricPlot3D[1.2^t {Sin[s] ^2 Sin[t], Sin[s] ^2 Cos[t],
      Sin[s] Cos[s]}, {s, 0, π}, {t, -π / 4, 5 π / 2}, PlotRange → All];]

Show[GraphicsArray[
   {{p1, p2, p3, p4}, {p5, p6, p7, p8}}, GraphicsSpacing → -0.2]];
```

ParametricPlot3D is not adaptive (and in this respect resembles the commands of
Section 7.1). For space curves, the default value of **PlotPoints** is 75, meaning that the
function is sampled at 75 equally spaced points in the parameter interval. For surfaces, the
default value of **PlotPoints** is 30 (20 in *Mathematica* 4), meaning that the function is
sampled at a 30×30 grid for the parameters.

   Note that the default value of **BoxRatios** for **ParametricPlot3D** is **Automatic** (as
compared with {1, 1, 0.4} for **Plot3D**). This means that a parametric plot is shown in the
natural scaling, where one unit on the *x*, *y*, and *z* axes has the same length.

   ParametricPlot3D produces a graphics object of the type - Graphics3D -. This type
includes all kinds of surfaces, such as multiple-valued and self-crossing. Instead, **Plot3D**
produces an object of the type - SurfaceGraphics -, which is a more restricted type that
includes only the usual one-valued functions. Everything you can plot with **Plot3D[f[x,
y], {x, a, b}, {y, c, d}]** you can also plot with **ParametricPlot3D**:

   ParametricPlot3D[{x, y, f[x, y]}, {x, a, b}, {y, c, d}]

However, use **Plot3D** if you can, because this command is designed to plot normal surfaces and is fast.

The commands we present next are found in packages. You may want to read Section 4.1.1, p. 86, about packages and about what to do if you accidentally use a command from a package before you have loaded it. A simple method for using all graphics packages is to execute **<<Graphics`**. The correct graphics package is then automatically loaded if you use a command from a graphics package. In the following examples, however, we load each package explicitly.

■ Spherical and Cylindrical Coordinates

> *In the* **Graphics`ParametricPlot3D`** *package:*
>
> **SphericalPlot3D[r, {θ, a, b}, {φ, c, d}]**
> **CylindricalPlot3D[z, {r, a, b}, {θ, c, d}]**

SphericalPlot3D and **CylindricalPlot3D** use **ParametricPlot3D** with arguments **r{Sin[θ] Cos[φ], Sin[θ] Sin[φ], Cos[θ]}** and **{r Cos[φ], r Sin[φ], z}**, respectively.

```
<< Graphics`ParametricPlot3D`

SetOptions[SphericalPlot3D, Axes → False, Boxed → False];
SetOptions[CylindricalPlot3D, Axes → False, Boxed → False];

Block[{$DisplayFunction = Identity},
 p1 = SphericalPlot3D[1, {θ, 0, π}, {φ, 0, 2 π}];
 p2 = SphericalPlot3D[{θ - 1, Hue[0.08 θ φ]}, {θ, 0, π},
   {φ, 0, π}, PlotPoints → {20, 10}, Lighting → False];
 p3 = SphericalPlot3D[φ, {θ, π, 2 π}, {φ, 0, 5 π / 2}, PlotPoints → 20];
 p4 = CylindricalPlot3D[1, {r, 0, 3}, {φ, 0, 2 π}, PlotPoints → {6, 30}];
 p5 = CylindricalPlot3D[r^2, {r, 0, 2}, {φ, 0, 2 π}];
 p6 = CylindricalPlot3D[φ, {r, 0, 3}, {φ, 0, 2 π}, PlotPoints → {6, 30}];]

Show[GraphicsArray[{p1, p2, p3, p4, p5, p6}, GraphicsSpacing → -0.2]];
```

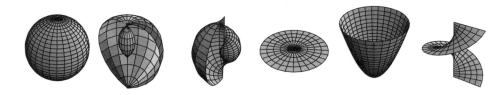

7.2.2 Special Surface Plots

■ Well-Known Surfaces

> *In the* **Graphics`Shapes`** *package:*
>
> **Show[Graphics3D[shape]]**
>
> *Shapes:*
>
> **Cylinder[r, h, n], Cone[r, h, n], MoebiusStrip[r1, r2, n], Helix[r, h, m, n],**
> **DoubleHelix[r, h, m, n], Sphere[r, n, m], Torus[r1, r2, n, m]**

Here **r**, **r1**, and **r2** are radii, **h** is half-height, and **m** and **n** specify the numbers of polygons used to draw the shapes. The shapes can also be rotated and translated. For example:

```
<< Graphics`Shapes`

Show[GraphicsArray[Map[Graphics3D[#, Boxed → False] &,
    {Cylinder[1, 0.5, 30], Cone[1, 1, 20], MoebiusStrip[2, 1, 25],
     Helix[1, 1.1, 2, 20], DoubleHelix[1, 1.6, 2, 20],
     Sphere[1, 20, 20], Torus[2, 1, 25, 15]}], GraphicsSpacing → -0.3]];
```

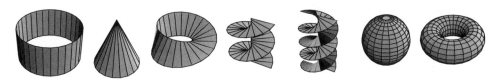

■ Surfaces of Revolution

In the `Graphics`SurfaceOfRevolution`` *package:*

`SurfaceOfRevolution[f, {x, a, b}]` Rotate around *z* axis the curve **f** in the (*x*, *z*) plane

`SurfaceOfRevolution[{fx, fy}, {t, a, b}]` Rotate around *z* axis the parametrically defined curve in the (*x*, *z*) plane

`SurfaceOfRevolution[{fx, fy, fz}, {t, a, b}]` Rotate around *z* axis the parametrically defined space curve

We can add a third argument **{u, c, d}** to these commands so that the curve is rotated only from angle **c** to angle **d**. With the **RevolutionAxis** option, we can define the axis around which the curve is rotated. A value **{x, z}** revolves around an axis obtained by connecting the origin to the point **{x, z}** (the default point is **{0, 1}**). The revolution axis can also be defined by a 3D point.

```
<< Graphics`SurfaceOfRevolution`

SetOptions[SurfaceOfRevolution, Axes → False, Boxed → False];

Block[{$DisplayFunction = Identity},
  p1 = SurfaceOfRevolution[Cos[x], {x, 0, 2 π}, PlotPoints → {10, 20}];
  p2 = SurfaceOfRevolution[{t Cos[t], t Sin[t]},
     {t, 0, π}, ViewPoint → {.7, -3.2, -.5}, PlotPoints → 20];
  p3 = SurfaceOfRevolution[{t, Cos[2 t], Sin[2 t]},
     {t, 0, π / 4}, {u, 0, 3 π / 2}, PlotPoints → {10, 30}];
  p4 = SurfaceOfRevolution[Cos[x], {x, 0, 2 π},
     RevolutionAxis → {1, 0}, PlotPoints → 20];
  p5 = SurfaceOfRevolution[Cos[x], {x, 0, 2 π},
     PlotPoints → {8, 20}, RevolutionAxis → {1, 1}];]
```

```
Show[GraphicsArray[{p1, p2, p3, p4, p5}, GraphicsSpacing → -0.1]];
```

■ Surfaces with Contours

Wickham-Jones (1994) presents the following method for showing contours on a surface:

```
f[x_, y_] := Cos[x y] Cos[x]

Block[{$DisplayFunction = Identity},
   s = Plot3D[f[x, y] - 0.03, {x, 0, π},
     {y, 0, 5 π / 4}, PlotPoints → 30, Mesh → False];
   c = ContourPlot[f[x, y], {x, 0, π}, {y, 0, 5 π / 4},
     PlotPoints → 30, ContourShading → False];
   c3d = Graphics3D[Graphics[c][[1]] /. Line[pts_] :→
     (val = Apply[f, First[pts]]; Line[Map[Append[#, val] &, pts]])]];

Show[s, c3d];
```

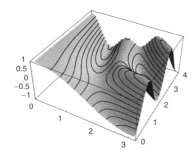

First the surface plot is drawn. We must lower the function somewhat by subtracting a small constant from the function (0.03 turned out to be a suitable constant in this example); otherwise the surface and the contours intersect at some points, and so parts of some contours go under the surface and are invisible. Then the contour plot is drawn. Lastly the contour plot is transformed into a 3D plot by inserting the value of the function at each point. Then the surface plot and the 3D contour plot can be combined with **Show**.

■ Inequality and Shadow Plots

In the **Graphics`InequalityGraphics`** *package:*

InequalityPlot3D[ineqs, {x, a, b}, {y, c, d}, {z, e, f}]   Show the region that satisfies the given logical combination of rational inequalities

```
<< Graphics`InequalityGraphics`
```

`InequalityPlot3D[ x^2 + y^2 + z^2 ≤ 1 && -x + y + z ≥ 0.05, {x}, {y}, {z}];`

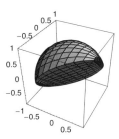

The `Graphics`Graphics3D`` package defines `ShadowPlot3D`, which produces both the surface and the density plot in the same figure. The same package also contains `Shadow`, which projects shadows of a given surface onto the coordinate planes, and `Project`, which shows various projections of a 3D graphics object.

■ Polyhedra

In the `Graphics`Polyhedra`` *package:*

`Show[Polyhedron[name]]`
`Show[Polyhedron[name, point, scale]]`

Convex polyhedra:
`Tetrahedron`, `Cube = Hexahedron`, `Octahedron`, `Dodecahedron`, `Icosahedron`

We can specify the center point (`point`) and the size (`scale`). The package also defines some nonconvex polyhedra. The `Geometry`Polytopes`` package has information about polyhedra (e.g., number of vertices, edges, and faces).

`<< Graphics`Polyhedra``

`Show[GraphicsArray[Map[Polyhedron[#, Boxed → False] &,`
`    {Tetrahedron, Cube, Octahedron, Dodecahedron, Icosahedron}]]];`

7.2.3 Gradient Fields

Plotting a gradient field is a way to describe a 3D function, and vector fields are useful in describing solutions of differential and difference equations.

In the `Graphics`PlotField`` *package:*

`PlotVectorField[{fx, fy}, {x, a, b}, {y, c, d}]` Plot the vector field of the vector-valued function `{fx, fy}`

`PlotGradientField[f, {x, a, b}, {y, c, d}]` Plot the gradient vector field of the scalar-valued function `f`

PlotHamiltonianField[f, {x, a, b}, {y, c, d}]   Plot the Hamiltonian vector field
of the scalar-valued function **f**

Options:

PlotPoints   Number of evaluation points; examples of values: **15**, **{10,15}**

ColorFunction   Function to define the style of the vectors; examples of values: **None**,
(**RGBColor[1, 0, 0] &**), (**AbsoluteThickness[0.25] &**), (**{RGBColor[#, 0, 1 - #]**,
AbsoluteThickness[0.25]} &)

ScaleFunction   Function to use for rescaling the magnitude of the vectors; examples
of values: **None**, **(0.2 # &)**

MaxArrowLength   Eliminates vectors that are longer than the specified value (applied
after **ScaleFunction**); default value: **None** (no vectors are removed)

ScaleFactor   Lengths of vectors are linearly scaled so that the length of the longest
vector is equal to the specified value (applied after **MaxArrowLength**); examples of
values: **Automatic** (fits the vectors in the mesh), **None** (no rescaling is used, use this
value with **ScaleFunction**)

PlotVectorField plots the vectors **{fx, fy}** on a 15×15 grid. **PlotGradientField**
plots the vectors **{D[f, x], D[f, y]}**; gradient points to the direction of maximum
increase of the function. **PlotHamiltonianField** plots the vectors **{D[f, y], -D[f, x]}**.
Gradient and Hamiltonian fields are orthogonal. For all three commands, **AspectRatio**
has the default value **Automatic**.

With **ColorFunction** we can define the color, thickness, and dashing of the arrows.
The default is that arrows are black and have **Thickness[0.0001]**, which means that the
arrows are very thin (too thin, in my opinion). The color function must be a pure function.
For example, color function (**RGBColor[#, 0, 1 - #] &**) makes long arrows red and short
arrows blue.

All options of the **Arrow** primitive (see Section 5.4.6, p. XXX) can also be used to control
appearance of the arrows; these options include **HeadLength** (default is 0.02), **HeadWidth**
(default is 0.5), and **HeadCenter** (default is 1, meaning that the head hides small vectors; if
the value is 0, we can see the magnitude of even small vectors).

The default length of the arrows is such that the arrows do not overlap. If you find the
arrows to be too small, define a suitable **ScaleFunction** (e.g., **(0.2 # &)**) and set **Scale**
Factor → None.

PlotVectorField is handy in describing differential equations (see Sections 23.1.1, p.
594; 23.1.3, p. 601; and 23.3.2, p. 616) and difference equations (see Sections 25.1.1, p. 672;
25.1.2, p. 675; and 25.2.1, p. 681). Here we plot a function both as a surface and as a gradient field:

```
<< Graphics`PlotField`

p1 = Plot3D[Cos[x y] Cos[x], {x, 0, 3 π / 4},
    {y, 0, π / 2}, PlotPoints → 20, DisplayFunction → Identity];

p2 = PlotGradientField[Cos[x y] Cos[x], {x, 0, 3 π / 4},
    {y, 0, π / 2}, PlotPoints → 10, HeadCenter → 0, ColorFunction →
    (AbsoluteThickness[0.25] &), DisplayFunction → Identity];
```

```
Show[GraphicsArray[{p1, p2}]];
```

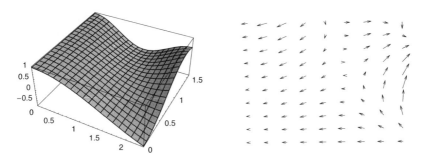

7.2.4 Graphics Primitives and Directives

■ 3D Graphics Primitives

Below are all five 3D graphics primitives. With **p**, **p1**, **p2**, and **pn**, we denote a point that consists of the x, y, and z coordinates; an example is {**1**, **4**, **3**}.

Point[p]   Point at **p**
Line[{p1, …, pn}]   Line through points **p1**, …, **pn**
Cuboid[p]   Unit cube with opposite corners **p** and **p + 1**
Cuboid[p1, p2]   Rectangular parallelepiped with opposite corners **p1** and **p2**
Polygon[{p1, …, pn}]   Polygon with vertices **p1**, …, **pn**
Text[expr, p, {u, v}]   Text **expr** placed so that the point {**u**, **v**}, expressed in text coordinates, of **expr** is at the point **p**

Four of these primitives—**Point**, **Line**, **Polygon**, and **Text**—have 2D counterparts with the same names (see Section 5.4); however, the 2D counterpart of **Cuboid** is **Rectangle**.

■ 3D Graphics Directives

For 3D primitives, we have all of the directives of point size, thickness, dashing, and color as we do for 2D primitives. The default point size in three dimensions is **Point･ Size[0.01]** (the default is **PointSize[0.008]** for 2D graphics). The default thickness in three dimensions is **AbsoluteThickness[0.5]** (which is the same as for 2D graphics).

PointSize[d], AbsolutePointSize[d]
Thickness[d], AbsoluteThickness[d]
Dashing[{d1, d2, … }], AbsoluteDashing[{d1, d2 ,… }]

GrayLevel[g], Hue[h], Hue[h, s, b], RGBColor[r, g, b], CMYKColor[c, m, y, k]

EdgeForm[{styles}]   Styles of the edges of polygons (no edges if **EdgeForm[]**)
FaceForm[color1, color2]   Colors of the front and back faces of polygons (requires **Lighting → False**)
SurfaceColor[diff]   Surface with diffuse intrinsic color (requires **Lighting → True**)
SurfaceColor[diff, spec]   Surface with diffuse and specular intrinsic colors
SurfaceColor[diff, spec, n]   Surface with specular exponent **n**

In addition, we have three directives that control the style of polygons (and hence also the style of cuboids). **EdgeForm** controls the style of the edges of polygons (dashing cannot be used). The faces of polygons are normally (with **Lighting → True**) colored by a simulated illumination. In this case, the way in which polygons reflect light can be controlled with **SurfaceColor** (the default is **SurfaceColor[GrayLevel[1]]**; see Section 8.1.7, p. 218). If simulated illumination is not used (**Lighting → False**), the colors of the faces are set with **FaceForm** (the front face of a polygon is defined to be the one for which the corners, as you specify them, are in counterclockwise order).

■ Using the Primitives and Directives

The primitives and directives are important because *Mathematica* uses them in 3D graphics and also because we can use them ourselves. Indeed, we can build a whole plot directly from the primitives in the same way that we did in two dimensions. However, now we do not use **Graphics** but rather **Graphics3D**:

Graphics3D[{directives and primitives**}]**   Create a graphics object from primitives
Show[graphics object**]**   Show a graphics object

Note that the 3D primitives cannot be used in the options **Prolog** and **Epilog**. In these options, we can only use 2D primitives. Style directives are also important because they are used with many options to modify the appearance of plots.

Graphics3D has mostly the same options as **Plot3D**, but the default values of **Axes** and **BoxRatios** are **False** and **Automatic**, while the default values for **Plot3D** are **True** and **{1, 1, 0.4}**.

Here is an example of how to build a plot from graphics primitives:

```
p1 = {{0, 0, 0}, {1, 0, 0}, {1, 1, 0}, {0, 1, 0}, {0, 0, 0}};
p2 = {{0, 0, 1}, {1, 0, 1}, {1, 1, 1}, {0, 1, 1}};

Show[Graphics3D[{RGBColor[0, 0, 1], AbsoluteThickness[2], Line[p1],
    RGBColor[1, 0, 0], AbsolutePointSize[6], Map[Point, p1],
    EdgeForm[{RGBColor[0, 1, 0], AbsoluteThickness[2]}],
    RGBColor[0, 0, 1], Polygon[p2], GrayLevel[0],
    Text["3D primitives", {0.8, 0.9, 0.5}, {-1, 0}]},
    Boxed → False, Lighting → False]];
```

3D prir

Note that text falling outside of the selection rectangle (shown when the plot is clicked on with the mouse) is not printed (even with **PlotRange → All**). Thus, text should be placed so that it is wholly inside the rectangle, or else the plot should be enlarged with the mouse so that the text comes inside the rectangle.

7.3 Improving the Illusion of Space

7.3.1 Rotations

A surface plot gives quite a good impression of 3D reality. Mesh lines, shading, and a bounding box guide the eye. However, there are techniques for obtaining an even better impression. We now consider real-time and animated rotations and two-image and single-image stereograms.

■ Real-Time Rotations

`<< RealTime3D``	Begin rotating new 3D plots in real time with the mouse
`<< Default3D``	Go back to normal 3D plots

After loading the context `RealTime3D``, the properties of all new 3D plots change:

- The plot can be rotated in real time with the mouse.
- The coloring of the surface differs from the usual one.
- Ticks are not shown.
- The plot is a bitmap plot (it looks good on the screen, but the printing quality is poor).
- Only the following options can be used: **DisplayFunction**, **ViewPoint**, **PlotPoints**, **Compiled**, **Mesh**, **Boxed**, **BoxRatios**.

To give an example, we load the context and plot the Klein bottle:

```
<< RealTime3D`

a = 6 Cos[s] (1 + Sin[s]); b = 16 Sin[s]; c = 4 (1 - Cos[s] / 2);
fx = If[π < s ≤ 2 π, a + c Cos[t + π], a + c Cos[s] Cos[t]];
fy = If[π < s ≤ 2 π, b, b + c Sin[s] Cos[t]];
fz = c Sin[t];

ParametricPlot3D[{fx, fy, fz},
   {s, 0, 2 π}, {t, 0, 2 π}, PlotPoints → {40, 20}];
```

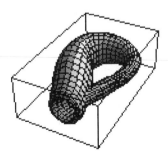

The produced plot can be rotated with the mouse (try on your screen) so that we can get a good impression of the surface. We will not plot any more rotatable surfaces, and hence we go back to the usual plots:

```
<< Default3D`
```

■ Animated Rotations

In the `Graphics`Animation`` *package:*

`SpinShow[p]`   Generate a sequence of plots by showing the 3D plot **p** from various
  viewpoints

Options:
`Frames` Number of separate plots; default value: `24`
`SpinRange` Range of angles; default value: `{0 Degree, 360 Degree}`

The options `SpinOrigin` (default value is `{0, 0, 1.5}`) and `SpinDistance` (default
value is `2`) can be used to define the viewpoint, `SpinTilt` (default value is `{0, 0}`) the
angles used to tilt the rotation, and `RotateLights` (default value is `False`) whether the
light sources should rotate with the object.

As an example, we first plot a surface:

```
p = ParametricPlot3D[{s Cos[t] Sin[s], s Cos[s] Cos[t], -s Sin[t]},
    {s, 0, 2 π}, {t, 0, π}, Axes → False, Boxed → False];
```

Then we use `SpinShow` to generate 30 plots with different viewpoints. We can reduce the
range of angles to (0 °, 180 °), because the function is symmetric:

```
<< Graphics`Animation`
```

```
SpinShow[p, Frames → 30, SpinRange → {0 Degree, 180 Degree}];
```

We do not show the 30 plots here; they can be seen on the CD-ROM version of this book. If
the sizes of the plots are not suitable, resize the first plot, select the cell bracket of the
group of plots, choose **Cell ▷ Align Selected Graphics...**, and click **OK**.

7.3.2 Stereograms

■ Two-Image Stereograms

An excellent 3D impression (from one viewpoint) can be obtained by making only two
plots with slightly different viewpoints and placing the plots on top of each other using
your eyes. Many of you are probably able to do this. For most people it is easiest to try to
focus beyond the paper surface. First you get four plots, and the goal is to relax the eyes so
much that two of the four plots are superimposed. The result is three plots, with the one in
the middle giving the stereo view of the plot. For example:

```
p1 = ParametricPlot3D[
    {s Cos[t] Sin[s], s Cos[s] Cos[t], -s Sin[t]}, {s, 0, 2 π}, {t, 0, π},
    Axes → False, Boxed → False, DisplayFunction → Identity];
```

```
p2 = Show[p1, ViewPoint → {1.4, -2.3, 2.0}];

Show[GraphicsArray[{p1, p2}, GraphicsSpacing → -0.4]];
```

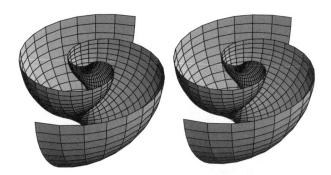

Here the first plot has the default viewpoint {1.3, -2.4, 2.0}. The other viewpoint {1.4, -2.3, 2.0} is at the same height (2.0) but the x and y coordinates are somewhat different. To show the two figures very near to each other, we used the **GraphicsSpacing** option with a negative value. For the stereographic method to be possible to the eyes, the distance of two corresponding points in the two plots should be at most about 4 to 5 cm.

The pair of figures we have considered gives the correct result if you focus your eyes beyond the paper. For some people it is easier to focus on the front of the paper surface. In this case, the order of the figures has to be changed to get the correct result:

```
Show[GraphicsArray[{p2, p1}, GraphicsSpacing → d]]
```

For surfaces, the improvement in the illusion of three dimensions is clear with two-image stereograms but not so remarkable as for curves, dots, and arrows in space. For these latter graphics objects, if plotted in the usual way, the eye has too few guides to obtain an adequate impression of the positions of the objects, but the stereo view makes a big improvement. For examples of space curves, dots, and arrows, see Sections 7.4.1, p. 198 (gradient field of a 4D function), 10.1.2, p. 271 (irregular 3D data), 10.2.2, p. 278 (an illustration of a galaxy), and 23.3.3, p. 626 (solution of a system of three differential equations).

■ Single-Image Stereograms

Single-image (or random-dot) stereograms are explained in detail in Maeder (1995a). The package of this article can be downloaded from the address http://www.mathematica-journal.com/backissues. From there, click "Vol. 5 (1)" and download the electronic supplement of this issue. See Section 4.1.3, p. 89, to learn about the correct place to put the **SIS.m** package. The **SIS-EX.MA** notebook contains some examples.

In the **SIS.m** *package:*

SIS[expr, {x, a, b}, {y, c, d}]   Produce a single-image stereogram from the given function by looking from the positive z axis

Options:

PlotRange   Expected range of the function; default value: {0, 1}

PlotPoints Number of random initial points; default value: **100**
PlaneDistance Distance from the back plane to the viewing plane; default value: **2**
EyeDistance Distance from the viewing plane to the eyes; default value: **2**
Guides Whether guide dots at the top of the plot are drawn; possible values: **True**,
   False
EyeSeparation Separation of the guide dots measured as a fraction of half of the
   horizontal width of the plot; default value: **1/4**
Object Objects used in the plot; default value: **Point**
PlotStyle Style of the objects; default value: **{PointSize[0.01]}**

As an example, we produce a single-image stereogram of the following function:

```
Plot3D[Cos[Sqrt[x^2+y^2]], {x, -7, 7}, {y, -7, 7}];
```

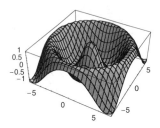

```
<< SIS.m
```

```
SIS[Cos[Sqrt[x^2+y^2]], {x, -7, 7}, {y, -7, 7}, PlotRange → {-1, 1},
    PlotPoints → 150, PlaneDistance → 3, EyeDistance → 3];
```

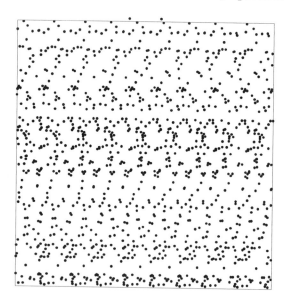

   The top of the plot contains two dots as guides for the eyes. You get the stereographic
view if you can get from the two dots—by focusing beyond the paper surface—first four
dots and then three dots (two of the four dots are then superimposed). We have presented
the plot in a small size; enlarge the plot to have a more impressive experience.

7.4 Plots for 4D Functions

7.4.1 Simple Methods

2D, 3D, and 4D functions are of the forms $y = f(x)$, $z = f(x, y)$, and $v = f(x, y, z)$, respectively. For 2D and 3D functions, we can use 2D and 3D graphics. Graphical illustration of 4D functions involves difficulties, but something can be done with 3D and even 2D graphics.

■ Plotting Values on Curves and Surfaces

Consider the following function:

```
v[x_, y_, z_] := Cos[x^2 + y^2 + z^2]
```

With 2D graphics, we can show the values of the function along parametrically defined curves. For example, along the curves $x = t$, $y = z = 0$; $x = y = t$, $z = 0$; $x = y = z = t$; and $x = \cos(t)$, $y = \sin(t)$, $z = 0$, the value of the function is as follows:

```
Block[{$DisplayFunction = Identity},
 p1 = Plot[v[t, 0, 0], {t, -1.5, 1.5}, PlotRange → {-1, 1}];
 p2 = Plot[v[t, t, 0], {t, -1.5, 1.5}, PlotRange → {-1, 1}];
 p3 = Plot[v[t, t, t], {t, -1.5, 1.5}, PlotRange → {-1, 1}];
 p4 = Plot[v[Cos[t], Sin[t], 0], {t, -1.5, 1.5}, PlotRange → {-1, 1}];]

Show[GraphicsArray[{p1, p2, p3, p4}], GraphicsSpacing → 0];
```

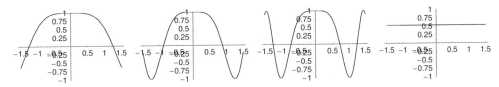

With 3D graphics, we can show the values of the function on parametrically defined surfaces. For example, on the surface $x = s$, $y = t$, $z = 0$ (that is, on the (x, y) plane), the value of the function is as follows:

```
Plot3D[v[s, t, 0], {s, -1.5, 1.5}, {t, -1.5, 1.5}];
```

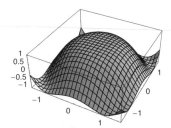

The properties of the first, second, and fourth 2D plots can also be seen (or at least guessed) from this single plot.

■ **Gradient Fields**

In the `Graphics`PlotField3D`` *package:*

`PlotVectorField3D[{gx, gy, gz}, {x, a, b}, {y, c, d}, {z, e, f}]` Plot the vector
  field of the vector valued function `{gx, gy, gz}`

`PlotGradientField3D[g, {x, a, b}, {y, c, d}, {z, e, f}]` Plot the gradient vector
  field of the scalar valued function `g`

Options:

`PlotPoints, ColorFunction, ScaleFunction, MaxArrowLength, ScaleFactor,`
  `VectorHeads`

These commands have the same five options as the corresponding 2D commands (see Section 7.2.3, p. 189) plus the option `VectorHeads`. This new option can be used to tell whether the vectors should have a head (`True`) or not (`False`, the default). `PlotPoints` now has the default value 7 (it was 15 in Section 7.2.3); the value must now be a single number.

One method for illustrating a 4D function is to plot the gradient field. `PlotGradient‹ Field3D` plots the vector field `{D[expr, x], D[expr, y], D[expr, z]}`. The gradient points in the direction of maximum increase of the function. The longer the arrow, the stronger the growth. The heads or the arrows show the direction of growth. As an example, consider the same function that we studied earlier:

```
<< Graphics`PlotField3D`

p1 = PlotGradientField3D[Cos[x^2 + y^2 + z^2],
    {x, -1.5, 1.5}, {y, -1.5, 1.5}, {z, -1.5, 1.5}, PlotPoints → 5,
    VectorHeads → True, DisplayFunction → Identity];

p2 = Show[p1, ViewPoint → {1.4, -2.3, 2.0}];

Show[GraphicsArray[{p1, p2}, GraphicsSpacing → 0]];
```

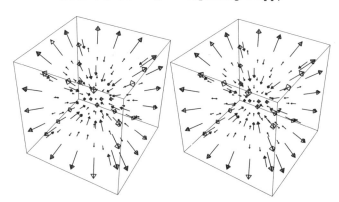

Near the boundary of the box, the function grows toward the outside; in the middle of the box, it grows toward the inside. This means that, where the directions of the arrows change, the function has a surface of local minimum points.

7.4.2 Surfaces of Constant Value

With **ContourPlot**, we can plot *curves* of constant value for a 3D function $z = g(x, y)$. With **ContourPlot3D**, we can plot *surfaces* of constant value for a 4D function $v = g(x, y, z)$.

In the **Graphics`ContourPlot3D`** *package:*

ContourPlot3D[g, {x, a, b}, {y, c, d}, {z, e, f}] Plot some surfaces of constant value of **g**

Options:

Contours Constants for which surfaces are drawn; examples of values: **{0.}, {0, 1}**

ContourStyle Style of the surfaces (requires **Lighting → False**); examples of values: **{}, {{Hue[0]}, {Hue[1/3]}}**

PlotPoints Number of evaluation points in each direction in the first and later recursions; examples of values: **{3, 5}, {4, 5}**

MaxRecursion Maximum number of recursions used in each cube; examples of values: **1, 2**

Compiled Whether to compile the expression to be plotted; possible values: **True, False**

The command has the same options as **Graphics3D** plus the five options mentioned in the box.

The default value of **Contours** is **{0.}**, meaning that only the surfaces are plotted where the function has the value 0. A list of a few values can be given, but another possibility (often a better one) is to plot separate figures for each value. **ContourStyle** can be useful to identify the surfaces if several are plotted at the same time.

The command works by first dividing the plotting region into a small number of cubes and then deciding whether the surface sought intersects each cube. If the surface intersects a cube, this cube is subdivided further, and so on.

The number of cubes used at each recursion is determined by the option **PlotPoints**. Its default value is **{3, 5}**, which means that, in the first division into cubes, 3 evaluation points are used in *each* direction (we interpret this to mean that first the plot region is divided into $2 \times 2 \times 2 = 8$ smaller cubes). If the surface intersects any of these cubes, that cube is divided into even smaller cubes by the use of 5 evaluation points in each direction (again, we interpret this to mean that the cube in question is divided into $4 \times 4 \times 4 = 64$ smaller cubes). This method of division is continued but not more than **MaxRecursion** times. The default value of this option is 1, so that one subdivision at most is made.

■ Example

We consider the same function we did in Section 7.4.1, p. 197. We ask for the surfaces where the function has the value 0:

```
<< Graphics`ContourPlot3D`

g = Cos[x^2 + y^2 + z^2];
```

```
ContourPlot3D[g, {x, -1.5, 1.5}, {y, -1.5, 1.5}, {z, -1.5, 1.5}];
```

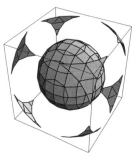

A series of plots with different values of the constant gives a good description of the function. For symmetric functions, we can reduce the plotting region. We plot the same function, but now y only goes from 0 to 1.5:

```
Block[{$DisplayFunction = Identity},
  p1 = ContourPlot3D[g, {x, -1.5, 1.5}, {y, 0, 1.5}, {z, -1.5, 1.5},
    Contours → {-0.5}, PlotLabel → "Surfaces with value -0.5"];
  p2 = ContourPlot3D[g, {x, -1.5, 1.5}, {y, 0, 1.5}, {z, -1.5, 1.5},
    Contours → {0.0}, PlotLabel → "Surfaces with value 0.0"];
  p3 = ContourPlot3D[g, {x, -1.5, 1.5}, {y, 0, 1.5}, {z, -1.5, 1.5},
    Contours → {0.5}, PlotLabel → "Surfaces with value 0.5"];
  p4 = ContourPlot3D[g, {x, -1.5, 1.5}, {y, 0, 1.5},
    {z, -1.5, 1.5}, Contours → {0.9}, PlotPoints → {7, 5},
    PlotLabel → "Surfaces with value 0.9"]];

Show[GraphicsArray[{p1, p2, p3, p4}]];
```

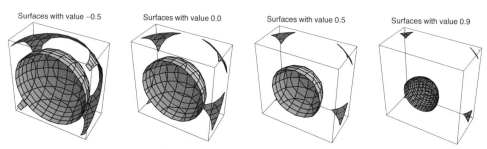

Surfaces with value −0.5   Surfaces with value 0.0   Surfaces with value 0.5   Surfaces with value 0.9

The function seems to grow when the argument approaches the origin but also when the argument moves near the corners of the bounding box. In the fourth plot, we had to increase the first element of **PlotPoints** from 3 to 7 to get the correct plot.

For another example of surfaces of constant value, see Section 24.2.6, p. 660, where we plot the solution of a 3D elliptic partial differential equation.

Options for 3D Graphics

Introduction

> *When Albert Einstein first arrived in the United States, he was invited to speak at Princeton University. The large hall was packed to capacity with people who had come out of curiosity to hear the world-famous scientist. Einstein was amazed at the huge crowd and remarked to the chairman, "I never realized that so many Americans were interested in tensor analysis."*

In Section 8.1, we consider the options of **Plot3D**, **ParametricPlot3D**, and **Graphics3D**. These commands produce a genuine 3D surface plot (or a space curve), and they have mostly the same options. In Section 8.2, we study the options of **ContourPlot** and **DensityPlot**. These two commands describe a 3D surface, but the plots they produce are similar to 2D plots. Most of their options are the same. Many of them are also the same as the options of **Plot**.

8.1 Options for Surface Plots

8.1.1 Summary

This section includes sorted lists of all of the options of **Plot3D**, **ParametricPlot3D**, and **Graphics3D**, with short descriptions and with some common values (**Graphics3D** was explained in Section 7.2.4, p. 192).

The default value of each option is mentioned first, and after that we mention either all other possible values or some examples of other values (the examples are simple; more advanced forms may exist).

Among all of the 38 options of **Plot3D**, the four options considered to be the most important and most often used are marked with two asterisks (**). Seven options assumed to be less important are marked with one asterisk (*). With parentheses we have shown 11 options that most of us will almost never need. The remaining 16 options without any notation may sometimes be useful.

Most of the options are common to all three commands. Options and their default values applicable only to certain commands are expressed in the lists below by a superscript after the option name or value:

^{s}: applicable to **Plot3D** (shorthand for *surface*)
^{p}: applicable to **ParametricPlot3D** (shorthand for *parametric*)
^{g}: applicable to **Graphics3D** (shorthand for *graphics*)

For example, **PlotPoints**^{sp} means that this option applies to **Plot3D** and **Parametric‹ Plot3D** but not to **Graphics3D**.

The options are divided into global and local options. *Global options* modify some global aspects of the plot: whether to display the plot at all and how the plot looks in general (form; ranges of x, y, and z; margins; viewpoint; colors; font; and formatting). *Local options* modify local components of a plot (the surface, box, axes, plot label, and add-ons). The local options are presented in two parts: local options common to all three commands and local options specific to each command. (**BoxRatios** is mentioned in both global and local options: it controls the global form of the surface, but it is also one of the options that controls the bounding box.)

Style directives like **AbsoluteThickness**, **GrayLevel**, and **RGBColor** are explained in Section 5.4.

Global Options of **Plot3D**, **ParametricPlot3D**, *and* **Graphics3D**

Option for display:

DisplayFunction Function used to display the plot; values used: **$DisplayFunction** (plot is shown), **Identity** (plot is not shown)

Options for form, ranges, margins, and size:

* **BoxRatios** Ratios of side lengths of the bounding box; default values: **{1, 1, 0.4}**^{s}, **Automatic**^{pg}

(**AspectRatio** Ratio of height to width of the plotting rectangle; default value: **Auto‹ matic**)

* **PlotRange** Ranges for x, y, and z in the plot; examples of values:

Automatic,	{{xmin, xmax}, {ymin, ymax}, Automatic},
All,	{{xmin, xmax}, {ymin, ymax}, All},
{zmin,Automatic},	{{xmin, xmax}, {ymin, ymax}, {zmin, Automatic}},
{zmin,zmax},	{{xmin, xmax}, {ymin, ymax}, {zmin, zmax}}

ClipFill^s How to draw clipped parts of the surface; examples of values: **Automatic**, **None**, **RGBColor[1,0,0]**, **{RGBColor[0,1,0]**, **RGBColor[1,0,0]}**

PlotRegion Region of the final display area that the plot should fill; examples of values: **Automatic** (means **{{0, 1}, {0, 1}}**), **{{xmin, xmax}, {ymin, ymax}}**

(**SphericalRegion** Whether to leave room in the plotting rectangle for a sphere enclosing the bounding box; possible values: **False**, **True**)

ImageSize The absolute size of the plot; examples of values: **Automatic** (usually means 288 printer's points), **width**, **{width, height}**

Options for viewpoint:

** **ViewPoint** Point from which the surface is viewed; default value: **{1.3, -2.4, 2}**

(**ViewCenter** Point of the surface placed in the center of the display area; default value: **Automatic**)

(**ViewVertical** Vertical direction; default value: **{0, 0, 1}**)

Options for colors:

Background Color of the background; possible values: **Automatic** (usually means **GrayLevel[1]**), **GrayLevel[g]**, **Hue[h,s,b]**, **RGBColor[r,g,b]**, **CMYKColor[c,m,y,k]**

(**DefaultColor** Color of the box, axes, labels, mesh, etc.; possible values: **Automatic** (usually means **GrayLevel[0]**), **GrayLevel[g]**, **Hue[h,s,b]**, **RGBColor[r,g,b]**, **CMYKColor[c,m,y,k]**)

(**ColorOutput** Type of color output to produce; examples of values: **Automatic**, **GrayLevel**, **RGBColor**, **CMYKColor**)

Options for font and formatting:

TextStyle Font used in the plot; examples of values: **$TextStyle**, **{FontFamily → "Times", FontSize → 8}**

FormatType Format type of text used in the plot; examples of values: **$FormatType**, **StandardForm, TraditionalForm, InputForm, OutputForm**

The main difference in the options above for the three commands is in the default value of **BoxRatios**: **{1, 1, 0.4}** for **Plot3D** and **Automatic** for the other two commands. *Later on in this chapter, we do not consider the options controlling colors, font, or formatting;* we refer to Sections 6.2.2, p. 156; 6.3.1, p. 157; and 6.3.2, p. 159.

Next we consider the local options. First we list local options that are *common* to all three commands. These options control local components other than the surface.

Common Local Options of **Plot3D**, **ParametricPlot3D**, *and* **Graphics3D**

Options for axes and ticks:

* **Axes** Whether to draw the axes; default values: **True**[sp], **False**[g]; examples of other values: **{True, True, False}, {False, False, True}**

AxesEdge Where to draw the axes; examples of values: **Automatic, {{1,1}, {-1,1}, {1,1}}, { Automatic, {-1,1}, {1,1}}, {{1,1}, {-1,1}, None}**

** **AxesLabel** Labels for the axes; examples of values: **None**, **"z"**, **{"x", "y", None}**, **{"x", "y", "z"}**

AxesStyle Style of the axes; examples of values: **Automatic**, **{Absolute⸴
Thickness[1], GrayLevel[0.5]}**

** **Ticks** Ticks on the axes; simple examples of values: **Automatic**, **None**, **{{0,1,2},
{0,1}, Automatic}**, **{{0,1,2}, {0,1}, {-1,0,1}}**

Options for box and face grids:

* **Boxed** Whether to draw a box around the surface; possible values: **True**, **False**

* **BoxRatios** Ratios of side lengths of the bounding box; default values: **{1, 1, 0.4}**[s],
Automatic[p 8]

BoxStyle Style of the box; examples of values: **Automatic**, **{AbsoluteThickness[1],
GrayLevel[0.8]}**

FaceGrids Grid lines drawn on the faces of the bounding box; examples of values:
None, **All**, **{{-1,0,0}, {0,1,0}, {0,0,-1}}**, **{{{-1,0,0}, {ygrid, zgrid}},
{{0,1,0}, {xgrid, zgrid}}, {{0,0,-1}, {xgrid, ygrid}}}**

Options for plot label and primitives:

* **PlotLabel** Label of the plot; examples of values: **None**, **Sin[x y]**, **"A surface plot"**

(**Prolog** 2D graphics primitives to be plotted before the main plot; examples of values:
{}, **{{AbsolutePointSize[3], RGBColor[1,0,0], Point[{0.4, 0.7}]}}**)

Epilog 2D graphics primitives to be plotted after the main plot; examples of values:
{}, **{AbsolutePointSize[3], RGBColor[1,0,0], Point[{0.4, 0.7}]}**

Lastly we list the *special* local options of each of the three commands. These options control the properties of the surface. Although many of these options are the same for two or all three of the commands, there are so many differences that we considered it better to show the options for each command separately.

Special Local Options of **Plot3D**

Options for function sampling:

** **PlotPoints**[sp] Number of sampling points; examples of values: **25**, **30**, **{15, 20}**

(**Compiled**[sp] Whether to compile the expression to be plotted; possible values: **True**,
False)

Options for mesh lines:

* **Mesh**[s] Whether to draw the mesh lines; possible values: **True**, **False**

MeshStyle[s] Style of the mesh lines; examples of values: **Automatic**, **Absolute⸴
Thickness[1]**, **{AbsoluteThickness[1], GrayLevel[1]}**

Options for coloring the surface:

Shading Whether to shade the surface; possible values: **True**, **False**

HiddenSurface[s] Whether to eliminate hidden parts of the surface; possible values:
True, **False**

* **Lighting** Whether to use simulated illumination; possible values: **True**, **False**

(**LightSources** Direction and color of point light sources for simulated illumination;
default value: **{{{1,0,1}, RGBColor[1,0,0]}, {{1,1,1}, RGBColor[0,1,0]},
{{0,1,1}, RGBColor[0,0,1]}}**)

(**AmbientLight** Isotropic ambient light for simulated illumination; examples of values: **GrayLevel[0]**, **GrayLevel[0.2]**, **RGBColor[0,0,1]**)

ColorFunction[s] How the surface is colored if simulated illumination is not used; examples of values: **Automatic** (means **GrayLevel**), **(Hue[1 - #] &)**, **(RGBColor[#, 1 - #, 0] &)**

(**ColorFunctionScaling**[s] Whether values provided for a color function should be scaled to lie between 0 and 1; possible values: **True**, **False**)

Special Local Options of **ParametricPlot3D**

Options for function sampling:

** **PlotPoints**[s p] Number of sampling points; examples of values: one parameter: **75**, **100**; two parameters: **30**, **40**, **{15, 25}**

(**Compiled**)

Options for polygons of the surface:

(**RenderAll**[p s] Whether to draw all polygons (**True**) or only the polygons or parts of polygons visible in the final plot (**False**))

(**PolygonIntersections**[p s] Whether intersecting polygons are left as such (**True**) or broken into nonintersecting polygons (**False**))

Options for coloring the surface:

Shading, * **Lighting**, (**LightSources**), (**AmbientLight**)

Special Local Options of **Graphics3D**

Options for polygons of the surface:

(**RenderAll**), (**PolygonIntersections**)

Options for coloring the surface:

Shading, * **Lighting**, (**LightSources**), (**AmbientLight**)

The most important differences here are in the default value of **PlotPoints**: **25** (**15** in *Mathematica* 4) for **Plot3D**, **75** for parametric space curves, and **30** (**20** in *Mathematica* 4) for parametric surfaces. (**Graphics3D** does not have this option at all, because this command does not sample a function but rather forms a graphics object from primitives. Note also that the default value of **PlotPoints** for **ParametricPlot3D** is **Automatic**, but we have shown what this actually means for parametric curves and parametric surfaces.)

The mesh lines can be controlled by options only for graphics produced by **Plot3D**. Some coloring options, too, are applicable only for this command.

Note that there are no options to control the style of a parametric space curve. The style of such a curve can be defined by entering a fourth expression into the plotting command in addition to the expressions defining the *x*, *y*, and *z* coordinates (see Section 7.2.1, p. 184). From the special local options of **ParametricPlot3D** above, only **PlotPoints** and **Compiled** are relevant for a parametric space curve.

Options for 3D graphics can be changed with **Show**, as is the case for 2D plots (see Section 6.1.1, p. 148). However, **PlotPoints** cannot be changed with **Show**. The **MeshRange** option can be used with **Show** to modify the result of **Plot3D** (see Section 8.2.3, p. 227). With **MeshRange**, we can redefine the ranges of x and y for ticks.

■ Example

First we define the grid lines for x, y and z:

```
xg = {π / 4, π / 2, 3 π / 4}; yg = {π / 4, π / 2, 3 π / 4}; zg = {0};
```

In the plot we color the surface with the **ColorFunction** option according to the height (to see the wonderful colors, look at the plot in the *Help Browser* if you have installed the CD-ROM of this book):

```
p = Plot3D[Cos[x y] Cos[x], {x, 0, π}, {y, 0, π},
    DisplayFunction → Identity, BoxRatios → Automatic,
    PlotRegion → {{0, 1}, {0.03, 0.95}}, Background → GrayLevel[0],
    TextStyle → {FontFamily → "Times", FontSize → 8},
    FormatType → TraditionalForm, PlotPoints → 60, Mesh → False,
    Lighting → False, ColorFunction → (Hue[1 - #] &),
    AxesEdge → {Automatic, {1, -1}, Automatic},
    AxesLabel → Map[StyleForm[#, FontWeight → "Bold"] &, {x, y, z}],
    Ticks → {{0, {π / 2, "π/2"}, π}, {0, {π / 2, "π/2"}, π}, {-1, 0, 1}},
    BoxStyle → AbsoluteThickness[0.5], FaceGrids → {{{-1, 0, 0}, {yg, zg}},
      {{0, 1, 0}, {xg, zg}}, {{0, 0, -1}, {xg, yg}}}, PlotLabel →
     StyleForm[Cos[x y] Cos[x], FontWeight → "Bold", FontSize → 10]];

Show[GraphicsArray[{p}, Frame → True,
    FrameStyle → {AbsoluteThickness[3], GrayLevel[0.5]}]];
```

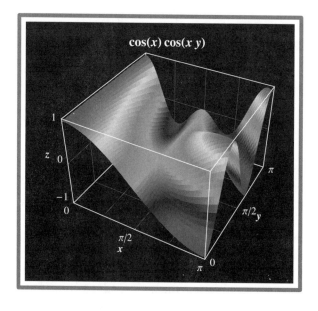

8.1.2 Form, Ranges, Margins, and Size

> * **BoxRatios**   Ratios of side lengths of the bounding box; default values: **{1, 1, 0.4}**[s],
>   **Automatic**[p g]
> (**AspectRatio**   Ratio of height to width of the plotting rectangle; default value:
>   **Automatic**)

(Remember that options and values denoted by s, p, and g apply for **Plot3D**, **Paramet**·
ricPlot3D, and **Graphics3D**, respectively.)

BoxRatios determines the form of the 3D bounding box. The default value **{1, 1, 0.4}**
for **Plot3D** means that the x and y axes have the same length and that the z axis is 0.4
times this length. The default value **Automatic** for **ParametricPlot3D** and **Graphics3D**
tells us that one unit in all three axes has the same length. Thus, the setting **BoxRatios →**
Automatic for 3D graphics corresponds to the setting **AspectRatio → Automatic** for 2D
graphics. Before the examples we change the default value of **PlotPoints** from **25** to **12**,
because our plots are small in size:

```
SetOptions[Plot3D, PlotPoints → 12];

p1 = Plot3D[Sin[x y], {x, 0, π}, {y, 0, π}, DisplayFunction → Identity];
p2 = Show[p1, BoxRatios → Automatic];
p3 = Show[p1, BoxRatios → {1, 1, 1}];

Show[GraphicsArray[{p1, p2, p3}]];
```

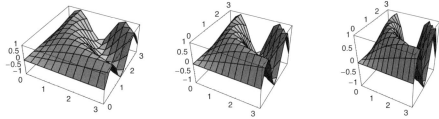

For surfaces produced by **Plot3D**, it may be worth setting the **BoxRatios** such that one
unit on the x and y axes has the same length to get the correct impression of the form of
the (x, y) region and setting the third component such that we get a good impression of
the form of the surface.

AspectRatio defines the 2D form of the plot. The default value **Automatic** leaves
unchanged the form of the graphics determined by the **BoxRatios** option. **AspectRatio** is
seldom used for 3D graphics. Instead, **BoxRatios** is the option for defining the 3D form of
the surface.

> * **PlotRange**   Ranges for x, y, and z in the plot; examples of values:
>
> | Automatic, | {{xmin, xmax}, {ymin, ymax}, Automatic}, |
> |---|---|
> | All, | {{xmin, xmax}, {ymin, ymax}, All}, |
> | {zmin,Automatic}, | {{xmin, xmax}, {ymin, ymax}, {zmin, Automatic}}, |
> | {zmin,zmax}, | {{xmin, xmax}, {ymin, ymax}, {zmin, zmax}} |

The values in the lefthand column specify the range for z only. The values in the righthand column also specify the ranges for x and y. The value **Automatic** determines the z range with an algorithm that may clip some high or low parts of the surface. Giving the value **All** ensures that the surface is wholly plotted, without clipping. For example:

```
p1 = Plot3D[Exp[x y], {x, 0, 3}, {y, 0, 3}, DisplayFunction → Identity];
p2 = Show[p1, PlotRange → All];

Show[GraphicsArray[{p1, p2}]];
```

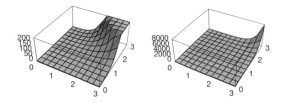

In the first plot, high values near the point (3, 3) are clipped. The clipping can normally be observed from plateaus at the top or bottom of the surface, but the **ClipFill** option can be used to specify other ways to show the clipped parts.

ClipFill[s]   How to draw clipped parts of the surface; examples of values: **Automatic**, **None**, **RGBColor[1,0,0]**, **{RGBColor[0,1,0], RGBColor[1,0,0]}**

The default value **Automatic** for **ClipFill** produces plateaus at the top and bottom of the surface, and they are colored in the usual way. The result is that the clipped parts may not be very clearly shown; see the first plot below (there we have simulated clipping of the surface by specifying a suitable plot range). The value **None** leaves the clipped parts empty, and this reveals the clipped parts very clearly; see the second plot. The clipped parts can also be shown with colors; see the third plot, where the bottom clipped parts are green and the top clipped parts red:

```
p1 = Plot3D[Sin[x y], {x, 0, π}, {y, 0, π}, Ticks → None,
      PlotRange → {-0.7, 0.85}, PlotPoints → 20, DisplayFunction → Identity];
p2 = Show[p1, ClipFill → None];
p3 = Show[p1, ClipFill → {RGBColor[0, 1, 0], RGBColor[1, 0, 0]}];

Show[GraphicsArray[{p1, p2, p3}]];
```

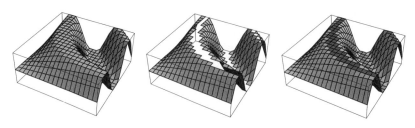

> **PlotRegion**   Region of the final display area that the plot should fill; examples of
>   values: **Automatic** (means **{{0, 1}, {0, 1}}**), **{{xmin, xmax}, {ymin, ymax}}**
> (**SphericalRegion**   Whether to leave room in the plotting rectangle for a sphere
>   enclosing the bounding box; possible values: **False**, **True**)
> **ImageSize**   The absolute size of the plot; examples of values: **Automatic** (usually
>   means 288 printer's points), **width**, **{width, height}**

   PlotRegion determines the 2D margins around the plot in the usual way, as it does for
2D graphics (see Section 6.2.1, p. 156). The default value **Automatic** means **{{0, 1}, {0, 1}}**: the plot fills the whole display area. A value such as **{{0.05, 0.95}, {0.05, 0.95}}**
leaves wider margins; this may be useful when using a colored background (see the
example of Section 8.1.1, p. 206).

   SphericalRegion determines whether to leave room in the plotting rectangle for a
sphere enclosing the bounding box. The value **True** is useful when rotating a graphic with
animation, because then the graphic has the same size from all viewpoints (this is important in animation). If we use **SpinShow** from the **Graphics`Animation`** package, then we
need not worry about this option (see Section 7.3.1, p. 194).

   For **ImageSize**, see Section 6.2.1, p. 156.

8.1.3 Viewpoint

> **∗∗ ViewPoint**   Point from which the surface is viewed; default value: **{1.3, -2.4, 2}**
> (**ViewCenter**   Point of the surface placed in the center of the display area; default
>   value: **Automatic**)
> (**ViewVertical**   Vertical direction; default value: **{0, 0, 1}**)

   We plot the same surface from three different viewpoints:

```
p1 = Plot3D[Sin[x y], {x, 0, π},
    {y, 0, π}, Ticks → None, DisplayFunction → Identity];
p2 = Show[p1, ViewPoint → {1.3, -2.4, 0}];
p3 = Show[p1, ViewPoint → {1.3, -2.4, -1}];

Show[GraphicsArray[{p1, p2, p3}]];
```

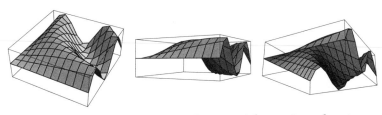

   The viewpoint coordinates are not coordinates of the surface; the viewpoint has its own
coordinate system. The origin of this system is at the center of the bounding box, and the
longest side of the box runs from −0.5 to 0.5. The other sides then run so that the lengths of
the sides satisfy the proportions expressed in **BoxRatios**. Thus, if we have **BoxRatios →**
{1, 1, 0.4}, then the viewpoint coordinates of the box run from −0.5 to 0.5 for x and y
and from −0.2 to 0.2 for z.

Recall from Section 7.3.1, p. 193, that we can rotate a surface in real time with the mouse so that we can easily see the surface from various viewpoints. The viewpoint can also be selected interactively, as explained next.

■ Interactive Selection of the Viewpoint

The viewpoint coordinates are not easy to use, but the viewpoint can also be selected interactively. This helps a lot. Suppose you have a plot like the following:

```
p = Plot3D[Sin[x y], {x, 0, π}, {y, 0, π}, Ticks → None];
```

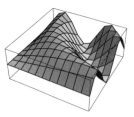

Now you want to look at it from another viewpoint. Write the following:

```
Show[p,];
```

Place the cursor before the closing bracket. Choose **Input ▷ 3D ViewPoint Selector...**, and you get the following window:

Rotate the cube with the mouse or use the scroll bars until you have the desired viewpoint. Then click **Paste** and go back to the notebook window. The chosen viewpoint has appeared in the **Show** command. Then execute this command:

```
Show[p, ViewPoint → {3.146, -0.685, 1.040}];
```

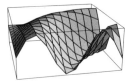

If the result is not exactly what you want, select **ViewPoint → {3.146, -0.685, 1.040}** in the **Show** command with the mouse (so that this part becomes black), choose another viewpoint interactively, and click **Paste**. The new viewpoint is now in the **show** command; execute the command again. Repeat this until the viewpoint is suitable.

Note that the coordinates in the viewpoint selector are shown by default in spherical coordinates. Do not worry: when pasted, the viewpoint is in Cartesian coordinates. Note also that in the spherical coordinate system you can rotate the cube with the mouse, but if you want to move closer to or farther away from the surface, the lowest scroll bar must be used. To get the default viewpoint, click **Defaults**.

8.1.4 Axes, Ticks, Boxes, and Face Grids

> ∗ **Axes** Whether to draw the axes; default values: **True**[sp], **False**[g]; examples of other values: **{True, True, False}, {False, False, True}**
>
> **AxesEdge** Where to draw the axes; examples of values: **Automatic, {{1,1}, {-1,1}, {1,1}}, { Automatic, {-1,1}, {1,1}}, {{1,1}, {-1,1}, None}**
>
> ∗∗ **AxesLabel** Labels for the axes; examples of values: **None, "z", {"x", "y", None}, {"x", "y", "z"}**
>
> **AxesStyle** Style of the axes; examples of values: **Automatic, {Absolute Thickness[1], GrayLevel[0.5]}**
>
> ∗∗ **Ticks** Ticks on the axes; simple examples of values: **Automatic, None, {{0,1,2}, {0,1}, Automatic}, {{0,1,2}, {0,1}, {-1,0,1}}**

The default value of **Axes** is **True** for **Plot3D** and **ParametricPlot3D** and **False** for **Graphics3D**. If the plot has the bounding box, some edges of the box are selected as the axes. Note, however, that axes and the box can be drawn independently: you can have axes without the box, the box without the axes, or neither the axes nor the box; see examples of the box below.

The edges of the box that are used as axes are, by default, determined automatically. Each axis can, however, be drawn on any of four edges with **AxesEdge**. In the example below, the first **{-1, -1}** defines the position of the x axis as the edge where y and z have the minimum values. **{1, -1}** defines the position of the y axis as the edge where x has the maximum value and z the minimum value. The last **{-1, -1}** defines the position of the z axis as the edge where x and y have the minimum values. (By the way, these definitions are unnecessary in this example, because they happen to be the default definitions.) Any definition can also be **Automatic** (that axis is chosen automatically) or **None** (that axis is not drawn).

```
Plot3D[Sin[x y], {x, 0, π}, {y, 0, π},
  AxesEdge → {{-1, -1}, {1, -1}, {-1, -1}}, AxesLabel → {x, y, z},
  AxesStyle -> AbsoluteThickness[1], Ticks → {{0, π}, {0, π}, {-1, 1}}];
```

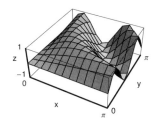

AxesLabel, **AxesStyle**, and **Ticks** are used as they are with 2D plots (see Sections 6.5.1, p. 165, and 6.5.2, p. 166). We have marked **AxeLabel** with two asterisks because, in 3D plots, the labels are useful: it is not obvious which one of the two horizontal axes is the x axis and which the y axis.

---

* **Boxed** Whether to draw a box around the surface; possible values: **True**, **False**
* **BoxRatios** Ratios of side lengths of the bounding box; default values: **{1, 1, 0.4}**^{s}, **Automatic**^{pg}
  BoxStyle Style of the box; examples of values: **Automatic**, **{AbsoluteThickness[1], GrayLevel[0.8]}**

BoxRatios was already considered in Section 8.1.2, p. 207. The default thickness for the edges of the box is **AbsoluteThickness[0.25]**. For example:

```
p1 = Plot3D[Sin[x y], {x, 0, π}, {y, 0, π},
    Ticks → {{0, π}, {0, π}, {-1, 1}}, DisplayFunction → Identity];
p2 = Show[p1, Boxed → False];
p3 = Show[p1, Axes → False,
    BoxStyle → {AbsoluteThickness[1], GrayLevel[0.8]}];
p4 = Show[p1, Axes → False, Boxed → False];

Show[GraphicsArray[{p1, p2, p3, p4}]];
```

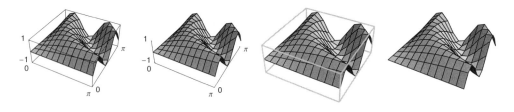

FaceGrids Grid lines drawn on the faces of the bounding box; examples of values:
  None, **All**, **{{-1,0,0}, {0,1,0}, {0,0,-1}}**, **{{{-1,0,0}, {ygrid, zgrid}}**, **{{0,1,0}, {xgrid, zgrid}}, {{0,0,-1}, {xgrid, ygrid}}}**

The value **All** draws grids on all six faces. We can define the faces where we want the grids. For example, **{-1,0,0}** is the face where x has the smallest value, **{0,1,0}** the face where y has the largest value, and **{0,0,-1}** the face where z has the smallest value. In addition, we can define the grid lines and even their styles (styles are defined in the same way as are the styles for grid lines in 2D graphics [see Section 6.5.4, p. 169]). For example, grid lines for x could be the following:

```
xgrid = {π / 4, π / 2, 3 π / 4};
```

If we want the grid lines to be white, we write the following:

```
xgrid = Transpose[{{π / 4, π / 2, 3 π / 4}, Table[{GrayLevel[1]}, {3}]}]
```

$$\left\{\left\{\frac{\pi}{4}, \{\text{GrayLevel}[1]\}\right\}, \left\{\frac{\pi}{2}, \{\text{GrayLevel}[1]\}\right\}, \left\{\frac{3\pi}{4}, \{\text{GrayLevel}[1]\}\right\}\right\}$$

(The example in Section 8.1.1, p. 206, also contains grid lines.)

8.1.5 Plot Label and Primitives

* **PlotLabel** Label of the plot; examples of values: **None, Sin[x y]**, "**A surface plot**"
(**Prolog** 2D graphics primitives to be plotted before the main plot; examples of values:
  {}, {{AbsolutePointSize[3], RGBColor[1,0,0], Point[{0.4, 0.7}]}})
Epilog 2D graphics primitives to be plotted after the main plot; examples of values:
  {}, {AbsolutePointSize[3], RGBColor[1,0,0], Point[{0.4, 0.7}]}

PlotLabel was considered in Section 6.6.1, p. 169; there is also an example in Section 8.1.1, p. 206. For **Prolog** and **Epilog**, we refer to Section 6.6.2, p. 170. These options can be used to add 2D graphics primitives to the plot. Note that the primitives cannot be 3D. In addition, the coordinates used in the primitives have to be given with scaled coordinates, which run from 0 to 1 in both the horizontal and vertical directions.

Usually **Epilog** is the better option, because it draws the primitives on top of the main figure (primitives drawn with **Prolog** can remain behind the main plot). Note that all primitives must be wholly inside the selection rectangle of the plot, because otherwise they are not plotted wholly. For example:

```
Plot3D[Sin[x y], {x, 0, π}, {y, 0, π}, Ticks → {{π}, {π}, {-1, 1}},
    Epilog → Text[StyleForm[Sin[x y], FontFamily -> "Times", FontSize → 8],
      {0.2, 0.1}, FormatType → TraditionalForm]];
```

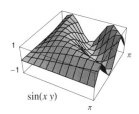

If you want to add 3D primitives (see Section 7.2.4, p. 191), plot the primitives with **Graphics3D** and then use **Show** to combine the main plot and the primitives. For example:

```
p1 = ParametricPlot3D[{Cos[s] (3 + Cos[t]), Sin[t], Sin[s] (3 + Cos[t])},
    {s, 0, 2 π}, {t, 0, 2 π}, Boxed → False, Axes → False,
    PlotPoints → {30, 15}, DisplayFunction → Identity];

p2 = Graphics3D[{AbsoluteThickness[1.5], Line[{{-5, 0, 0}, {5, 0, 0}}],
    Line[{{0, -5, 0}, {0, 6, 0}}], Line[{{0, 0, -5.5}, {0, 0, 5}}],
    AbsolutePointSize[4], RGBColor[1, 0, 0], Point[{0, 0, 0}]}];

Show[p1, p2, PlotRange → All, PlotRegion → {{-.2, 1.2}, {-.2, 1.2}},
    DisplayFunction → $DisplayFunction];
```

8.1.6 Sampling, Mesh Lines, and Polygons

> ** **PlotPoints**^{sp} Number of sampling points; examples of values: **Plot3D**: 25, 30,
> {15, 20}; **ParametricPlot3D**, one parameter: 75, 100; **ParametricPlot3D**, two
> parameters: **30, 40, {15, 25}**
> (**Compiled**^{sp} Whether to compile the expression to be plotted; possible values: **True**,
> **False**)

The quality of the plot is controlled by **PlotPoints** (for **Compiled**, see Section 6.4.2, p. 163). A single value of **PlotPoints** such as the default value **25** (the default value is **15** in *Mathematica* 4) means that the function is sampled on a 25×25 grid. A pair of values such as **{15, 20}** means that a 15×20 grid is used.

The basic thing to know is that 3D plotting commands are not adaptive: no more points are sampled, even if the function seems to change rapidly. If the figure is not good enough, we have to replot the function using a higher value of **PlotPoints**. Note that this option cannot be changed with **Show**; we have to redo the whole plot.

However, do not use unnecessarily high values of **PlotPoints**. The computing time can be quite long, the resulting plot takes much memory, and the plot has a lot of mesh lines, thereby making it dark and unpleasant. If you use a large value of **PlotPoints**, it may be better not to draw the mesh lines at all:

```
Block[{$DisplayFunction = Identity},
  p1 = Plot3D[Sin[x y], {x, 0, 2 π}, {y, 0, 3 π / 2}, Ticks → None];
  p2 = Plot3D[Sin[x y], {x, 0, 2 π}, {y, 0, 3 π / 2},
    Ticks → None, PlotPoints → 50];
  p3 = Plot3D[Sin[x y], {x, 0, 2 π}, {y, 0, 3 π / 2},
    Ticks → None, PlotPoints → 50, Mesh → False]];

Show[GraphicsArray[{p1, p2, p3}]];
```

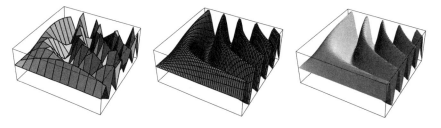

> * **Mesh**^{s} Whether to draw the mesh lines; possible values: **True, False**
> **MeshStyle**^{s} Style of the mesh lines; examples of values: **Automatic, Absolute**‹
> **Thickness[1], {AbsoluteThickness[1], GrayLevel[1]}**

A high value of **PlotPoints** is a good reason to consider not drawing the mesh lines (see the preceding example and the example of Section 8.1.1, p. 206). With **MeshStyle** we can set the thickness, dashing, and color of the mesh lines. The default value **Automatic** draws thin (**AbsoluteThickness[0.5]**), nondashed, and usually black lines.

The options for mesh are not available for **ParametricPlot3D** or **Graphics3D**. With **Graphics3D** we can, however, use the **EdgeForm** directive (see Section 7.2.4, p. 191). In **ParametricPlot3D**, we can add **EdgeForm** as a fourth element in the list of coordinates to define the style of the edges (remember that **EdgeForm[ ]** means no edges at all):

```
p1 =
   ParametricPlot3D[{Sin[s] Cos[t], Sin[s] Sin[t], 1 + Cos[s], EdgeForm[]},
      {s, 0, π}, {t, 0, 2 π}, Axes → False, DisplayFunction → Identity];

p2 = ParametricPlot3D[{Sin[s] Cos[t], Sin[s] Sin[t], 1 + Cos[s],
      EdgeForm[AbsoluteThickness[1], GrayLevel[1]]}, {s, 0, π}, {t, 0, 2 π},
      Axes → False, PlotPoints → 20, DisplayFunction → Identity];

Show[GraphicsArray[{p1, p2}]];
```

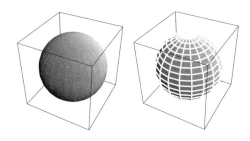

(**RenderAll**[p s]   Whether to draw all polygons (**True**) or only the polygons or parts of polygons visible in the final plot (**False**))
(**PolygonIntersections**[p s]   Whether intersecting polygons are left as such (**True**) or broken into nonintersecting polygons (**False**))

8.1.7 Coloring the Surface

■ Coloring or No Coloring

Shading   Whether to shade the surface; possible values: **True, False**
HiddenSurface[s]   Whether to eliminate hidden parts of the surface; possible values: **True, False**

Normally a surface is shaded. If the value of **Shading** is **False**, then the surface is white but has the mesh lines.

Normally the parts of a surface not visible from the used viewpoint are not shown. If the value of **HiddenSurface** is **False**, then only the mesh lines are drawn, and no parts of the surface are eliminated: parts not visible are also drawn. (This option is not available for **ParametricPlot3D** or **Graphics3D**.) For example:

```
p1 = Plot3D[Sin[x y], {x, 0, π}, {y, 0, π}, Ticks → None,
      ViewPoint → {1.4, -2.8, 1}, DisplayFunction → Identity];
p2 = Show[p1, Shading → False];
p3 = Show[p1, HiddenSurface → False];
```

```
Show[GraphicsArray[{p1, p2, p3}]];
```

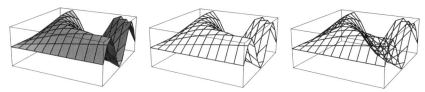

If shading is used, then there are two ways to color the surface. One is simulated illumination (the default), and the other is to color the function according to the height of the surface. First we consider the simulated illumination. (To see the fine colors of the plots that are coming, look at the plots in the *Help Browser* if you have installed the CD-ROM of this book.)

■ Coloring with Simulated Illumination

> ∗ **Lighting** Whether to use simulated illumination; possible values: **True, False**
> (**LightSources** Direction and color of point light sources for simulated illumination;
>   default value: **{{{1,0,1}, RGBColor[1,0,0]}, {{1,1,1}, RGBColor[0,1,0]},**
>   **{{0,1,1}, RGBColor[0,0,1]}}**)
> (**AmbientLight** Isotropic ambient light for simulated illumination; examples of values:
>   **GrayLevel[0], GrayLevel[0.2], RGBColor[0,0,1]**)

Simulated illumination is normally used. If the value of **Lighting** is **False**, then the surface is by default colored with shades of gray according to the height of the surface.

The default value of **LightSources** indicates that there are red, green, and blue point light sources (the distances of the light sources from the surface can be considered to be infinity, so that the light rays are parallel). The default value of **AmbientLight** is **Gray‹ Level[0]** so that there is no ambient light. A value such as **GrayLevel[0.2]** can be used to make the colors of the surface lighter. For example:

```
p1 = Plot3D[Sin[x y], {x, 0, π},
    {y, 0, π}, Ticks → None, DisplayFunction → Identity];
p2 = Show[p1, LightSources → {{{2, 3, 4}, RGBColor[1, 0, 0]},
      {{1, -5, 6}, RGBColor[0, 1, 0]}, {{2, 6, -12}, RGBColor[0, 0, 1]}}];
p3 = Show[p1, AmbientLight → GrayLevel[0.2]];

Show[GraphicsArray[{p1, p2, p3}]];
```

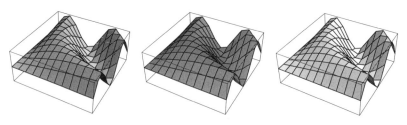

■ Coloring Usual Surfaces According to z or (x, y)

> **ColorFunction**[s]   What colors are used to color the surface according to the height, if
> simulated illumination is not used; examples of values: **Automatic** (means **Gray**⸱
> **Level**), (**Hue[1 - #] &**), (**RGBColor[#, 1 - #, 0] &**)
> (**ColorFunctionScaling**[s]   Whether values provided for a color function are scaled to
> lie between 0 and 1 (**True**) or left as such (**False**))

If simulated illumination is not used (**Lighting → False**), then the surface is colored according to the height *z* of the function. The default is that gray levels are used, but with **ColorFunction** we can define other coloring systems (if **ColorFunction** is used, then **Lighting** is **False** by default). The color function is expressed as a *pure function* where the argument of the function is expressed by the pound sign (**#**) and at the end of the function there is an ampersand (**&**) (we encountered pure functions in Section 2.2.2, p. 36, in studying **Map**). Examples:

```
p1 = Plot3D[Sin[x y], {x, 0, π}, {y, 0, π},
    Lighting → False, Ticks → None, DisplayFunction → Identity];
p2 = Show[p1, ColorFunction → (Hue[1 - #] &)];
p3 = Show[p1, ColorFunction → (Hue[0.6 (1 - #)] &)];
p4 = Show[p1, ColorFunction → (RGBColor[#, 1 - #, 0] &)];
p5 = Show[p1,
    ColorFunction → (If[# > 0.7, RGBColor[1, 0, 0], RGBColor[0, 1, 0]] &)];
p6 = Plot3D[{Sin[x y], Hue[x y / π^2]}, {x, 0, π}, {y, 0, π},
    Ticks → None, DisplayFunction → Identity];

Show[GraphicsArray[{{p1, p2, p3}, {p4, p5, p6}}]];
```

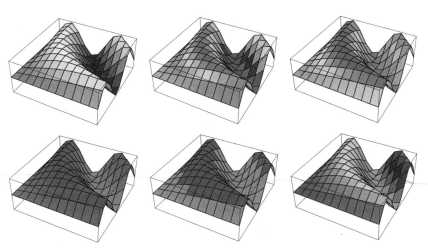

In the first plot we have the default gray levels, in the second we use **Hue** (another example is in Section 8.1.1, p. 206), in the third a modified **Hue**, in the fourth the **RGBColor** system, and in the fifth a two-color system (polygons above 0.7 are red and polygons below 0.7 green). In the sixth plot we colored the surface not according to the *z* coordinate but according to the *x* and *y* coordinates (see Section 7.1.1, p. 180).

■ Coloring Parametric Surfaces According to (s, t)

`ColorFunction` cannot be used with `ParametricPlot3D`. With this command we can define the color of the polygons and the style of the edges as a fourth element in the list of coordinates:

```
Block[{$DisplayFunction = Identity},
  p1 = ParametricPlot3D[{s Cos[t] Sin[s], s Cos[s] Cos[t],
     -s Sin[t], {EdgeForm[], Hue[s t / (2 π^2)]}}, {s, 0, 2 π},
    {t, 0, π}, Lighting → False, Boxed → False, Axes → False];
  p2 = ParametricPlot3D[{s Cos[t] Sin[s], s Cos[s] Cos[t], -s Sin[t],
     SurfaceColor[GrayLevel[0.3], GrayLevel[1], 2]}, {s, 0, 2 π},
    {t, 0, π}, Boxed → False, Axes → False, PlotPoints → 20];
  p3 = ParametricPlot3D[{x, y, Sin[x y],
     SurfaceColor[GrayLevel[0.5], GrayLevel[1], 1]}, {x, 0, π},
    {y, 0, π}, Boxed → False, Axes → False, PlotPoints → 15];]

Show[GraphicsArray[{p1, p2, p3}, GraphicsSpacing → -0.1]];
```

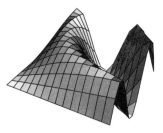

In the first plot we use no edges and color the surface with **Hue** (**Lighting → False** is required in this type of color definition). In the second plot we use **SurfaceColor** (see Section 7.2.4, p. 191) to produce somewhat stronger colors (**Lighting → True** is required with this directive). In the third plot we use **ParametricPlot3D** to plot a usual surface, because then we can use **SurfaceColor** to get stronger colors. Parametric space curves can also have style definitions (see Section 7.2.1, p. 184).

8.2 Options for Contour and Density Plots

8.2.1 Summary

We present here a summary of the options of **ContourPlot** and **DensityPlot**. The same principles are used as in Section 8.1.1, p. 201. The default values of the options are mentioned first. For **ContourPlot** we have marked three options with two asterisks, and for **DensityPlot**, only two.

Global Options of **ContourPlot** *and* **DensityPlot**

Option for display:

DisplayFunction Function used to display the plot; values used: **$DisplayFunction** (plot is shown), **Identity** (plot is not shown)

Options for form, ranges, margins, and size:

* **AspectRatio** Ratio of height to width of the plotting rectangle; examples of values: **1**, **Automatic**

* **PlotRange** Ranges for *x*, *y*, and *z* in the plot; examples of values:

Automatic,	**{{xmin, xmax}, {ymin, ymax}, Automatic},**
All,	**{{xmin, xmax}, {ymin, ymax}, All},**
{zmin,Automatic},	**{{xmin, xmax}, {ymin, ymax}, {zmin, Automatic}},**
{zmin,zmax},	**{{xmin, xmax}, {ymin, ymax}, {zmin, zmax}}**

PlotRegion Region of the final display area that the plot should fill; examples of values: **Automatic** (means **{{0, 1}, {0, 1}}**), **{{xmin, xmax}, {ymin, ymax}}**

ImageSize The absolute size of the plot; examples of values: **Automatic** (usually means 288 printer's points), **width**, **{width, height}**

Options for colors:

Background Color of the background; possible values: **Automatic** (usually means **GrayLevel[1]**), **GrayLevel[g]**, **Hue[h,s,b]**, **RGBColor[r,g,b]**, **CMYKColor[c,m,y,k]**

(**DefaultColor** Color of the curve, axes, labels, etc.; possible values: **Automatic** (usually means **GrayLevel[0]**), **GrayLevel[g]**, **Hue[h,s,b]**, **RGBColor[r,g,b]**, **CMYKColor[c,m,y,k]**)

(**ColorOutput** Type of color output to produce; examples of values: **Automatic**, **None**, **GrayLevel**, **RGBColor**, **CMYKColor**)

Options for font and formatting:

TextStyle Font used in the plot; examples of values: **$TextStyle**, **{FontFamily →** **"Times", FontSize → 8}**

FormatType Format type of text used in the plot; examples of values: **$FormatType**, **StandardForm, TraditionalForm, InputForm, OutputForm**

The default value of **AspectRatio** for contour and density plots is **1**, meaning that the form of the plot is a square. It may be worth considering the use of the value **Automatic**, because it sets one unit in the *x* and *y* axes to have the same length so that we get a good impression of the proportions of the (*x*, *y*) region. Prepare to use the value **All** for **Plot-Range**, because otherwise the *z* values may be clipped. Fonts and formatting are considered in Section 6.3.

Common Local Options of **ContourPlot** *and* **DensityPlot**

Options for function sampling:
** **PlotPoints** Number of sampling points; examples of values: **25**, **40**, **{25, 35}**
(**Compiled** Whether to compile the expression to be plotted; possible values: **True**,
  False)

Options for axes and ticks:
Axes Whether to draw the axes; examples of values: **False**, **True**, **{True, False}**,
  {False, True}
AxesOrigin Point where the axes cross; examples of values: **Automatic**, **{0, 0}**
AxesLabel Labels for the axes; examples of values: **None**, **"y"**, **{"x", None}**, **{"x",
  "y"}**
(**AxesStyle** Style of the axes; examples of values: **Automatic**, **{Absolute⋅
  Thickness[1], GrayLevel[0.5]}**)
Ticks Ticks on the axes; simple examples of values: **Automatic**, **None**, **{{0,1,2},
  Automatic}**, **{Automatic, {-1,0,1}}**, **{{0,1,2}, {-1,0,1}}**

Options for frame and frame ticks:
Frame Whether to draw a frame; examples of values: **True**, **False**, **{True, True,
  False, False}**
* **FrameLabel** Labels for the frame; examples of values: **None**, **"y"**, **{"x", None}**, **{"x",
  "y"}**, **{"bottom", "left", "top", "right"}**
* **RotateLabel** Whether to rotate the label for the vertical edges; possible values:
  True, **False**
FrameStyle Style of the frame; examples of values: **Automatic**, **{GrayLevel[0.5],
  AbsoluteThickness[1]}**
** **FrameTicks** Ticks on the frame; simple examples of values: **Automatic**, **None**, **{{0,
  π, 2π}, {0, π}}**, **{{0, π, 2π}, {0, π}, None, None}**

Options for plot label and primitives:
* **PlotLabel** Label of the plot; examples of values: **None**, **Sin[x y]**, **"A contour
  plot"**
(**Prolog** Graphics primitives to be plotted before the main plot; examples of values:
  {}, **{{AbsolutePointSize[3], RGBColor[1,0,0], Point[{0.4, 0.7}]}}**)
Epilog Graphics primitives to be plotted after the main plot; examples of values: **{}**,
  {AbsolutePointSize[3], RGBColor[1,0,0], Point[{0.4, 0.7}]}

Note that the default value of **PlotPoints** is **25** as it is for **Plot3D**. In a density plot this value may suffice, but in a contour plot the result is often quite rough, because the contours are based on the 25×25 points (15×15 points in *Mathematica* 4), and more points are not sampled (like other 3D plotting commands, **ContourPlot** is not adaptive). Thus, in contour plots, you should almost always increase the value of **PlotPoints**; a value of about 40 to 50 is generally sufficient. Note that **PlotPoints** cannot be used in **Show**; the plot has to be prepared wholly anew. (Note also that the options **PlotStyle**, **Plot⋅ Division**, **MaxBend**, and **GridLines**, which are familiar for **Plot**, are absent.)

One of the basic points to note about a contour or density plot is that the plot is like a 2D plot *with a frame* (and without axes). You can modify the frame as you can for 2D plots. For example, ticks are modified with **FrameTicks** (and not with **Ticks**). Axes are usually not used with contour and density plots. For a contour plot, however, axes can sometimes be useful if you turn shading off.

Special Local Options of **ContourPlot**

Options for contours and fills:

** **Contours** Number of z values for which contour lines are to be drawn or a list of z values; examples of values: **10**, **15**, **{-2, -1, 0, 1, 2}**

* **ContourLines** Whether to draw the contour lines; possible values: **True, False**

ContourStyle Style of the contour lines; examples of values: **Automatic, Gray\ Level[0.5]**, **{AbsoluteThickness[0.25], GrayLevel[0.5]}**, **Table[{Gray\ Level[gr]}, {gr, 1, 0, -1/9}]**

* **ContourShading** Whether to shade the regions between the contours; possible values: **True, False**

ColorFunction How the regions between the contour lines are colored; examples of values: **Automatic** (means **GrayLevel**), **(Hue[0.85(1 - #)] &)**, **(RGBColor[#, 0, 1 - #] &)**

(**ColorFunctionScaling** Whether values provided for a color function should be scaled to lie between 0 and 1; possible values: **True, False**)

Special Local Options of **DensityPlot**

Options for mesh lines and fills:

* **Mesh** Whether to draw the mesh lines; possible values: **True, False**

MeshStyle Style of the mesh lines; examples of values: **Automatic, {Gray\ Level[0.5]}**, **{AbsoluteThickness[0.5], GrayLevel[0.5]}**

ColorFunction How the regions between the mesh lines are colored; examples of values: **Automatic** (means **GrayLevel**), **(Hue[0.85(1 - #)] &)**, **(RGBColor[#, 0, 1 - #] &)**

(**ColorFunctionScaling** Whether values provided for a color function should be scaled to lie between 0 and 1; possible values: **True, False**)

These local options control the contour and mesh lines and the coloring of the fills between the lines. They are considered in detail in Sections 8.2.2, p. 222, and 8.2.3, p. 226.

Note that, in addition, the **MeshRange** option can be used with **Show** to modify the result of **ContourPlot** and **DensityPlot** (see Section 8.2.3, p. 227). **MeshRange** can be used to redefine the ranges of x and y for ticks.

Note also that the **Graphics`Legend`** package adds several options that are used to show and manipulate a legend for a plot. A legend can be useful for a contour or density plot: it shows what z values correspond to the gray levels of the plot (see Sections 8.2.2, p. 224, and 8.2.3, p. 228).

■ **Example**

In the plot below, we use a nonstandard package (explained in Section 8.2.2, p. 225) to add an enhanced legend:

```
p = ContourPlot[-x^4 - 3 x^2 y - 5 y^2 - x - y, {x, -1.3, 0.1}, {y, -0.8, 0.1},
    DisplayFunction → Identity, AspectRatio → Automatic, PlotRange → All,
    PlotRegion → {{-0.17, 1}, {0.02, 0.96}}, Background → GrayLevel[0],
    TextStyle → {FontFamily → "Times", FontSize → 9},
    FormatType → TraditionalForm, PlotPoints → 50,
    FrameStyle → {AbsoluteThickness[1.3], GrayLevel[1]},
    FrameTicks → {{-0.89, 0}, {-0.34, 0}, None, None},
    PlotLabel → StyleForm[-x^4 - 3 x^2 y - 5 y^2 - x - y,
      FontWeight → "Bold", FontSize → 11], Epilog →
    {{AbsolutePointSize[4], RGBColor[1, 0, 0], Point[{-0.89, -0.34}]},
      Text[StyleForm["Maximum point", FontColor → RGBColor[0, 0, 1]],
        {-0.86, -0.34}, {-1, -1}]},
    Contours → Range[-2.2, 0.8, 0.3], ContourStyle → GrayLevel[0.4]];
```

```
<< ExtendGraphics`LabelContour`
```

```
LabelContourLegend[p,
    LegendPosition → {0.800, -0.523}, LegendSize → {0.325, 1.000}];
```

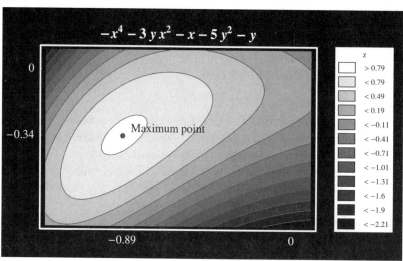

8.2.2 Options for Contour Plots

■ **AspectRatio, PlotRange, PlotPoints, and Frame**

Before going to the special local options of **ContourPlot**, we comment on some other important options.

As mentioned above, the default value of **AspectRatio** for contour and density plots is **1**, which means that the form of the plot is a square. Consider using the value **Automatic** (as in the preceding example) to get a better impression about the proportions of the (x, y) region.

As for **Plot3D**, the z values may be clipped. In a contour plot, the clipped parts appear as black and white and may easily remain unobserved. Use **PlotRange → All** to see all z values.

The default value **25** (**15** in *Mathematica* 4) of **PlotPoints** is almost always too low a value: the contours are not smooth enough. So, *the default value of* **PlotPoints** *should almost always be increased when using* **ContourPlot**. A value of about 40 to 50 generally provides smooth enough contours.

A contour or density plot is basically a 2D plot *with a frame*. This means that ticks are set with **FrameTicks** and not with **Ticks** and x and y labels with **FrameLabel** and not with **AxesLabel**. If the y label is short, the rotating of the label should be avoided by setting **RotateLabel → False**. Axes are usually not used with contour and density plots. For a contour plot, however, axes can sometimes be useful if shading is turned off:

```
ContourPlot[-x^4 - 3 x^2 y - 5 y^2 - x - y, {x, -1.6, 0.6},
    {y, -1.1, 0.6}, AspectRatio → Automatic, PlotRange → All,
    PlotPoints → 40, Frame → False, Axes → True, AxesOrigin → {0, 0},
    Ticks → {Range[-1.5, 0.5, 0.5], Range[-1, 0.5, 0.5]},
    Contours → 24, ContourShading → False];
```

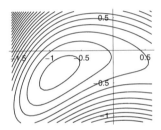

■ Special Local Options of ContourPlot

** Contours Number of z values for which contour lines are to be drawn or a list of z values; examples of values: **10**, **15**, **{-2, -1, 0, 1, 2}**

* ContourLines Whether to draw the contour lines; possible values: **True**, **False**

ContourStyle Style of the contour lines; examples of values: **Automatic**, **GrayLevel[0.5]**, **{AbsoluteThickness[0.25], GrayLevel[0.5]}**, **Table[{GrayLevel[gr]}, {gr, 1, 0, -1/9}]**

* ContourShading Whether to shade the regions between the contours; possible values: **True**, **False**

ColorFunction How the regions between the contour lines are colored; examples of values: **Automatic** (means **GrayLevel**), (**Hue[0.85(1 - #)] &**), (**RGBColor[#, 0, 1 - #] &**)

(**ColorFunctionScaling** Whether values provided for a color function should be scaled to lie between 0 and 1; possible values: **True**, **False**)

The default value **10** of **Contours** is often suitable, but often we also want more contours. With the default value we get ten contours corresponding to ten z values evenly spaced between the minimum and maximum values of the function in the region considered. One drawback is that we do not know the z values of the contours. If we want to know these values, we can define them ourselves in a list: **Contours → {-2, -1, 0, 1, 2}**.

By default, contour lines are drawn. The default is that the lines are black with **Abso**ᵇ
luteThickness[0.5]. Gray contours are more easily visible around the black regions (see
the example in Section 8.2.1, p. 222). Each contour can also have its own style (see the last
example for **ContourStyle** in the box above).

The regions between the contour lines are, by default, shaded. Turn shading off if you
want to use axes instead of the frame and the axes are inside the main plot. The shading is
done, by default, with **GrayLevel** so that the regions are various shades of gray, but
ColorFunction can be used to color the regions with **Hue**, **RGBColor**, or **CMYKColor**.

Some examples follow (to see the fine colors of the plots, look at the plots in the *Help
Browser* if you have installed the CD-ROM of this book):

```
Block[{$DisplayFunction = Identity},
  p1 = ContourPlot[Cos[x y] Cos[x], {x, 0, π},
    {y, 0, π}, PlotPoints → 40, ContourLines → False];
  p2 = ContourPlot[Cos[x y] Cos[x], {x, 0, π}, {y, 0, π},
    PlotPoints → 40, ColorFunction → (RGBColor[#, 0, 1 - #] &)];
  p3 = ContourPlot[Cos[x y] Cos[x], {x, 0, π}, {y, 0, π},
    Contours → 60, ContourLines → False, PlotPoints → 40,
    ColorFunction → (Hue[0.85 (1 - #)] &)];]

Show[GraphicsArray[{p1, p2, p3}]];
```

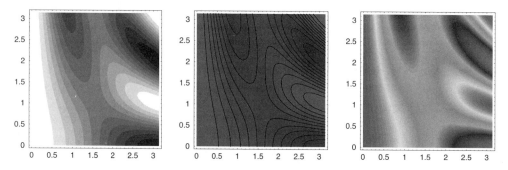

In the first plot, we do not have contour lines at all. In the second plot, the high values
are red and the low values blue. In the third plot, high values are red and low values
magenta, and values in between are various shades of orange, yellow, green, cyan, and
blue. The third plot also has a large number of contours without contour lines.

■ Legends and Labels

> *In the* **Graphics`Legend`** *package:*
>
> **ShowLegend[p, {GrayLevel[1 - #] &, n, "max-value", "min-value", opts}]**

Legends were considered in Section 6.6.3, p. 173. With the **ShowLegend** command, we
specify the graphic **p** for which we want to add a legend, the color function (e.g., **Gray**ᵇ
Level), the number **n** of different colors we want to show in the legend, and what the
maximum and minimum values of the function are. Options can be added (see Section
6.6.3).

Note that, if we use the default color function **(GrayLevel[#] &)** in the plot, we have to use the color function **(GrayLevel[1 - #] &)** in the legend. Similarly, if we use the color function **(RGBColor[#, 0, 1 - #] &)** in the plot, we have to use the color function **(RGBColor[1 - #, 0, #] &)** in the legend.

We plot a function using 8 contours (so that there are 9 regions between the contours):

```
p = ContourPlot[Cos[x y] Cos[x], {x, 0, π}, {y, 0, π},
    Contours → 8, PlotPoints → 40, DisplayFunction → Identity];
```

```
<< Graphics`Legend`
```

```
ShowLegend[p, {GrayLevel[1 - #] &, 9, "   1", " -1", LegendShadow → None,
    LegendPosition → {1.2, -0.78}, LegendSize → {0.6, 1.76}}];
```

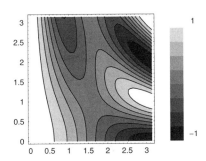

In the **ExtendGraphics`LabelContour`** *package:*

LabelContourLines[g, opts]   Place z values on the contour lines

LabelContourLegend[g, opts]   Form a legend with z values

With the above-mentioned nonstandard package, we can place the z values on the contour lines, or we can form an enhanced legend that shows the z values. This package is not a standard package but comes with Wickham-Jones (1994) (the book contains a whole chapter on labeling contour plots). All of the packages in that book can be downloaded from http://library.wolfram.com/database/Books/3753/. See Section 4.1.3, p. 89, to see where the packages can be placed. An example of **LabelContourLegend** is in Section 8.2.1, p. 222. Below is an example of **LabelContourLines** (note that, at least on a Macintosh, a warning message results when loading the package; the message is not shown here):

```
<< ExtendGraphics`LabelContour`
```

```
p = ContourPlot[Cos[x y] Cos[x], {x, 0, π}, {y, 0, π},
    ContourShading → False, Contours → Range[-0.9, 0.9, 0.2],
    PlotPoints → 40, FrameTicks → {{0, π}, {0, π}, None, None},
    DisplayFunction → Identity];
```

```
LabelContourLines[p,
   LabelPlacement → Automatic, LabelFont → {"Courier", 5}];
```

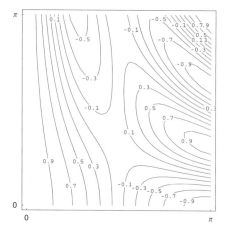

8.2.3 Options for Density Plots

■ **AspectRatio, PlotRange, PlotPoints, and Frame**

Before going to the special local options of **DensityPlot**, we comment on some other important options. Consider the use of the value **Automatic** for **AspectRatio** to get one unit in the x and y axes to have the same length. Clipped parts appear as black and white areas and may easily remain unobserved. Use **PlotRange → All** to see all z values. The default value **25** (**15** in *Mathematica* 4) of **PlotPoints** is often suitable for **DensityPlot**. A density plot is a 2D plot *with a frame*. Thus, set ticks with **FrameTicks** and x and y labels with **FrameLabel**. Axes are not usually used with density plots.

■ **Special Local Options of DensityPlot**

> * **Mesh** Whether to draw the mesh lines; possible values: **True, False**
> **MeshStyle** Style of the mesh lines; examples of values: **Automatic**, **{Gray** **Level[0.5]}**, **{AbsoluteThickness[0.5], GrayLevel[0.5]}**
> **ColorFunction** How the regions between the mesh lines are colored; examples of values: **Automatic** (means **GrayLevel**), (**Hue[0.85(1 - #)]** &), (**RGBColor[#, 0, 1 - #]** &)
> (**ColorFunctionScaling** Whether values provided for a color function should be scaled to lie between 0 and 1; possible values: **True, False**)

The default thickness for the mesh is **AbsoluteThickness[0.25]**. The default is that gray levels are used in the shading, but with **ColorFunction**, we can define different ways of coloring. Some examples follow (to see the fine colors of the plots, look at the plots in the *Help Browser* if you have installed the CD-ROM of this book):

```
Block[{$DisplayFunction = Identity},
  p1 = DensityPlot[Cos[x y] Cos[x],
     {x, 0, π}, {y, 0, π}, PlotPoints → 20, Mesh → False];
  p2 = DensityPlot[Cos[x y] Cos[x], {x, 0, π}, {y, 0, π},
     ColorFunction → (RGBColor[#, 0, 1 - #] &), PlotPoints → 15];
  p3 = DensityPlot[Cos[x y] Cos[x], {x, 0, π}, {y, 0, π}, PlotPoints → 60,
     Mesh → False, ColorFunction → (Hue[0.85 (1 - #)] &)];]

Show[GraphicsArray[{p1, p2, p3}]];
```

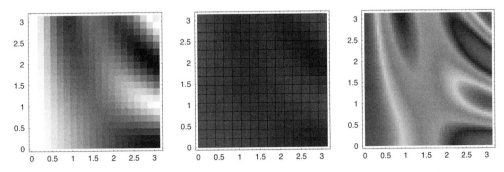

In the first example, we do not use mesh lines at all. In the second plot, the rectangles are colored in red and blue: high values are red and low values blue. In the third plot, we use **Hue** and a large value of **PlotPoints**.

■ MeshRange

The ticks in a density plot are somewhat biased. For example, if you make a density plot of a function for x in (0, 4) and y in (0, 3), then the ticks should indicate that the point (0, 0) is at the *center* of the rectangle in the lower lefthand corner of the plot (as in the second plot below) instead of at the lower lefthand corner of the rectangle (as in the first plot below), because this rectangle is drawn in such a way that (0, 0) is at the center of the rectangle. Similarly, the ticks should indicate that the point (4, 3) is at the center of the rectangle in the upper righthand corner of the figure instead of at the upper righthand corner of the rectangle.

If you want the correct ticks, then you can use the **MeshRange** option, available with **Show**. **MeshRange** redefines the x and y ranges used for ticks. Proceed as follows:

```
p = DensityPlot[expr, {x, a, b}, {y, c, d}, PlotPoints → {m, n}];
dx = (b - a) / (2 (m - 1));
dy = (d - c) / (2 (n - 1));
Show[p, MeshRange → {{a - dx, b + dx}, {c - dy, d + dy}}];
```

For example:

```
p1 = DensityPlot[Cos[x y] Cos[x], {x, 0, 4}, {y, 0, 3}, PlotPoints → {5, 4},
     FrameTicks → {{0, 1, 2, 3, 4}, {0, 1, 2, 3}, None, None},
     AspectRatio → Automatic, DisplayFunction → Identity];
dx = 4 / 8; dy = 3 / 6;
p2 = Show[p1, MeshRange → {{-dx, 4 + dx}, {-dy, 3 + dy}}];
```

```
Show[GraphicsArray[{p1, p2}]];
```

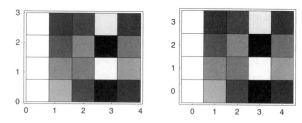

The ticks in the first plot are biased, but the aspect ratio is also somewhat distorted (the small rectangles are not true squares, as they should be). Both problems are solved in the second plot by using **MeshRange**. In this way, **MeshRange** can be used with **Show** to modify the ticks. It can also used for plots produced by **Plot3D** and **ContourPlot** (although the ticks of these plots are unbiased).

■ Legends

A legend can be added to a density plot in the same way as it is added to a contour plot (see Section 8.2.2, p. 224):

```
p = DensityPlot[Cos[x y] Cos[x], {x, 0, π},
    {y, 0, π}, Mesh → False, DisplayFunction → Identity];
```

```
<< Graphics`Legend`
```

```
ShowLegend[p, {GrayLevel[1 - #] &, 10, "  1", " -1", LegendShadow → None,
    LegendPosition → {1.2, -0.78}, LegendSize → {0.6, 1.76}}];
```

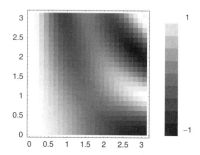

9

2D Graphics for Data

Introduction

> *"Data! data! data!" he cried impatiently. "I can't make bricks without clay."* — *Sherlock Holmes*

In this chapter we present ways to illustrate data with 2D graphics. Chapter 10 is devoted to 3D graphics of data having two independent variables and one dependent variable.

A basic situation encountered when plotting data is a time series: a data set with one dependent variable and time as the independent variable. Plotting such data is considered in Section 9.1. The basic command is **ListPlot**, but we can also easily use **Graphics**. Plotting several time series is considered in Section 9.2. The command is **MultipleList‑Plot**, but, again, **Graphics** is handy.

When the independent variable is not time, useful plotting methods include bar and pie charts, dot plots, and box-and-whisker plots. These are considered in Sections 9.3 and 9.4.

When we want to study the relationships between two or more dependent variables, the methods of Section 9.5 may be useful. The basic approach is to plot one variable against another variable, thereby yielding a scatter plot or, more generally, a scatter plot matrix or correlation plot. This method of pairing observations is also used in labeled plots and quantile-quantile plots.

9.1 Plotting One Time Series

9.1.1 Built-in Plotting

```
ListPlot[data]   Plot data as points

ListPlot[data, PlotJoined → True]   Plot data as joining lines

ListPlot[data, PlotJoined → True, Epilog →
  {AbsolutePointSize[d], Map[Point, data]}]   Plot data as points and joining lines

ListPlot[data, PlotStyle → AbsolutePointSize[d], Prolog →
  Map[Line[{{#[[1]], 0}, #}] &, data]]   Plot data as points and vertical lines
```

Data can be given in either of the following forms:

$\{y_1, y_2, \dots\}$ Plot the points $\{1, y_1\}, \{2, y_2\}, \dots$
$\{\{x_1, y_1\}, \{x_2, y_2\}, \dots\}$ Plot the given points (this form is required for the **Epilog** and **Prolog** options)

The basic form of **ListPlot** shows the data as points, but with some options we can also show the data as broken lines going from datum to datum, or as a plot containing both the points and the joining lines, or as vertical lines with points at the top. As an example, we plot 30 random numbers in all four ways:

```
data = Table[{x, 0.2 x + 2 Random[]}, {x, 1, 30}];

Block[{$DisplayFunction = Identity},
 p1 = ListPlot[data];
 p2 = ListPlot[data, PlotJoined → True];
 p3 = ListPlot[data, PlotJoined → True,
    Epilog → {AbsolutePointSize[2], Map[Point, data]}];
 p4 = ListPlot[data, PlotStyle → AbsolutePointSize[2],
    Prolog → Map[Line[{{#[[1]], 0}, #}] &, data]];]

Show[GraphicsArray[{{p1, p2}}]];
```

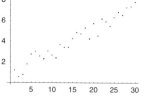 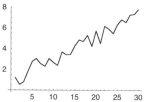

```
Show[GraphicsArray[{{p3, p4}}]];
```

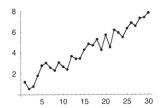 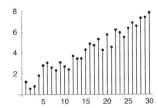

The first plot shows clearly the points and the second the path, but the third shows both the points and the path. The fourth type of plot, which makes the data somewhat more concrete with the lines and clearly reveals possible lacking observations, may also be considered.

In the last two plots, we used the graphics primitives **Point** and **Line** (see Sections 5.4.3, p. 135, and 5.4.4, p. 136), and **Map** was used to form lists of points and lines; **Map** was considered in Section 2.2.2, p. 36. In our examples, **Map** works as follows:

```
Short[Map[Point, data], 3]
```

```
{Point[{1, 1.19948}], Point[{2, 0.54398}],
  ≪26≫, Point[{29, 7.36487}], Point[{30, 7.80679}]}
```

```
Short[Map[Line[{{#[[1]], 0}, #}] &, data], 3]
```

```
{Line[{{1, 0}, {1, 1.19948}}], Line[{{2, 0}, {2, 0.54398}}],
  ≪27≫, Line[{{30, 0}, {30, 7.80679}}]}
```

■ Options

The options and their default values for **ListPlot** are basically the same as for **Plot**. **ListPlot** has the new option **PlotJoined**, which has the default value **False** (i.e., the dots are not joined). On the other hand, **ListPlot** does not have the options **Compiled**, **MaxBend**, **PlotDivision**, and **PlotPoints** of **Plot**.

Thus, we can mainly refer to Chapter 6 for the options. However, one point deserving attention is the **PlotStyle** option.

> **** PlotStyle** Style of the points or joining lines; examples of values: **Automatic**,
> **AbsolutePointSize[2]**, **{AbsoluteThickness[2], RGBColor[1,0,0]}**

PlotStyle controls the style of points or lines plotted with the main command but not the style of points or lines given with **Epilog** or **Prolog**. Thus, if we plot only the points or the joining lines, then **PlotStyle** controls the appearance of the points or the lines. If we plot both the points and the joining lines, then **PlotStyle** controls the style of the lines, and the style of the points is controlled with **Epilog**. If we plot both the points and the vertical lines, then **PlotStyle** controls the style of the points, and the style of the lines is controlled with **Prolog**. Also, recall the following from Sections 5.4.3, p. 135; 5.4.4, p. 136; and 5.4.7, p. 144:

- The style of points can be controlled with `PointSize[d]`, `AbsolutePointSize[d]`, and colors (the default point size is `PointSize[0.008]`).
- The style of lines can be controlled with `Thickness[d]`, `AbsoluteThickness[d]`, `Dashing[{d1, d2, … }]`, `AbsoluteDashing[{d1, d2, … }]`, and colors (the default thickness is `AbsoluteThickness[0.5]`).
- The color can be defined with `GrayLevel[g]`, `Hue[h]`, or `RGBColor[r, g, b]`.

Next we give four examples of style definitions:

```
Block[{$DisplayFunction = Identity},
  p1 =
   ListPlot[data, PlotStyle → {AbsolutePointSize[3], RGBColor[1, 0, 0]}];
  p2 = ListPlot[data, PlotJoined → True,
    PlotStyle → {AbsoluteThickness[1.2], RGBColor[0, 0, 1]}];
  p3 = ListPlot[data, PlotJoined → True,
    PlotStyle → {AbsoluteThickness[1.2], RGBColor[0, 0, 1]},
    Epilog → {AbsolutePointSize[3], RGBColor[1, 0, 0], Map[Point, data]}];
  p4 = ListPlot[data, PlotStyle → {AbsolutePointSize[3],
      RGBColor[1, 0, 0]}, Prolog → {{AbsoluteThickness[0.75],
      RGBColor[0, 0.3, 0.2], Map[Line[{{#[[1]], 0}, #}] &, data]}}];]
```

```
Show[GraphicsArray[{{p1, p2}, {p3, p4}}]];
```

9.1.2 Self-Made Plotting

It is very easy to plot data directly with the graphics primitives **Point** and **Line** (see Sections 5.4.3, p. 135, and 5.4.4, p. 136). In the plotting, we use **Graphics** and **Show** (see Section 5.4.1, p. 133). For example, to plot the points, we can simply write the following:

```
Show[Graphics[Map[Point, data], Axes → True]];
```

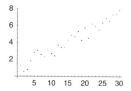

(We could also write `Map[Point[#] &, data]`.) Similarly, to plot the joining lines we could write the following:

```
Show[Graphics[Line[data], Axes → True]]
```

To plot both the points and the joining lines, we could write this:

```
Show[Graphics[
    {Line[data], AbsolutePointSize[3], Map[Point, data]}, Axes → True]]
```

To plot both the points and the vertical lines, we could write this:

```
Show[Graphics[{Map[Line[{{#[[1]], 0}, #}] &, data],
    AbsolutePointSize[3], Map[Point, data]}, Axes → True]]
```

In these examples we only used the **AbsolutePointSize** directive, but other directives can be added if suitable.

This method of using **Point** and **Line** is simple and can be considered to be a strong alternative to **ListPlot**, particularly when plotting several sets of data (see Section 9.2.1, p. 238). We will use the method now and then.

■ A Program

The following program adds four more types of plots to the four thus far considered. Of note is that with it we can also use circles (with or without joining or vertical lines). Using circles for plotting data was introduced in Section 5.4.5, p. 139. Circles are popular in scientific publications; for example, Cleveland (1993) uses mainly circles.

```
<< Utilities`FilterOptions`

Options[dataPlot] = {plotType → "p", lineThickness → 0.5,
    lineDashing → {}, lineColor → GrayLevel[0], pointDiameter → 3,
    pointColor → GrayLevel[0], circleDiameter → 3, circleThickness → 0.5,
    circleColor → GrayLevel[0], circleInsideColor → GrayLevel[1]};

dataPlot[data_, opts___] :=
 Module[{pt, lth, ld, lc, pd, pc, cd, cth, cc, cic},
  {pt, lth, ld, lc, pd, pc, cd, cth, cc, cic} =
   {plotType, lineThickness, lineDashing, lineColor, pointDiameter,
       pointColor, circleDiameter, circleThickness, circleColor,
       circleInsideColor} /. {opts} /. Options[dataPlot];
  Show[Graphics[{{AbsoluteThickness[lth], AbsoluteDashing[ld], lc,
      If[pt == "jl" || pt == "pjl" || pt == "cjl", Line[data], {}],
      If[pt == "vl" || pt == "pvl" || pt == "cvl",
       Map[Line[{{#[[1]], 0}, #}] &, data], {}]},
     If[pt == "p" || pt == "pjl" || pt == "pvl",
      {AbsolutePointSize[pd], pc, Map[Point, data]}, {}],
     If[pt == "c" || pt == "cjl" || pt == "cvl",
      {cic, Map[Disk[#, Offset[{cd / 2, cd / 2}]] &, data],
       AbsoluteThickness[cth], cc,
       Map[Circle[#, Offset[{cd / 2, cd / 2}]] &, data]}, {}]},
    Axes → True, FilterOptions[Graphics, opts]]]]
```

Depending on whether the value of the **plotType** option is **"p"**, **"c"**, **"jl"**, **"vl"**, **"pjl"**, **"pvl"**, **"cjl"**, or **"cvl"**, the program plots points, circles, joining lines, vertical lines, points and joining lines, points and vertical lines, circles and joining lines, or circles and vertical lines, respectively. The symbol || in the program stands for a logical "OR."

The program has several options to control the styles of lines, points, and circles. Line and circle thickness, line dashing, point diameter, and circle diameter are given in absolute units so that they can be used as arguments with **AbsoluteThickness**, **AbsoluteDashing**, **AbsolutePointSize**, and **Offset**. All of the options and their default values can be seen in the box above. The program also accepts all options of **Graphics**.

A key part of the program is the handling of the various options; this is considered in Section 14.3.4, p. 376.

■ An Example

In this example we consider the numbers of hare pelts sold to the Hudson Bay Trading Company in Canada from 1844 to 1934. These observations are from Burghes & Borrie (1981) (reproduced with the permission of the authors). First we read the numbers of hare pelts from a file. I have the data in a text file called **hare** in a folder called **MNData**:

```
haredata = Import["/Users/ruskeepaa/Documents/MNData/hare", "Table"]
```
```
{{1844, 30}, {1845, 25}, {1847, 25}, {1848, 15}, {1849, 30}, {1850, 55}, {1851, 80}, {1852, 80},
 {1853, 90}, {1854, 70}, {1855, 80}, {1856, 95}, {1857, 75}, {1858, 30}, {1859, 15},
 {1860, 20}, {1861, 40}, {1862, 5}, {1863, 155}, {1864, 140}, {1865, 105}, {1866, 45},
 {1867, 20}, {1868, 5}, {1869, 5}, {1870, 10}, {1871, 10}, {1872, 60}, {1873, 50}, {1874, 50},
 {1875, 105}, {1876, 85}, {1877, 60}, {1878, 15}, {1879, 10}, {1880, 15}, {1881, 10},
 {1882, 10}, {1883, 40}, {1884, 50}, {1885, 135}, {1886, 135}, {1887, 90}, {1888, 30},
 {1889, 20}, {1890, 50}, {1891, 55}, {1892, 60}, {1893, 55}, {1894, 80}, {1895, 95},
 {1896, 50}, {1897, 15}, {1898, 5}, {1899, 5}, {1900, 15}, {1901, 5}, {1902, 10}, {1903, 50},
 {1904, 70}, {1906, 20}, {1909, 25}, {1910, 50}, {1911, 55}, {1912, 75}, {1913, 70},
 {1914, 55}, {1915, 30}, {1916, 20}, {1917, 15}, {1918, 15}, {1919, 20}, {1920, 35},
 {1921, 60}, {1922, 80}, {1923, 85}, {1924, 60}, {1925, 30}, {1926, 20}, {1927, 10},
 {1928, 5}, {1929, 5}, {1930, 10}, {1931, 30}, {1932, 80}, {1933, 100}, {1934, 80}}
```

The file **hare** can be found on the CD-ROM that comes with this book. To **Import**, see Section 4.2.1, p. 90. The numbers are in thousands and are quoted to the nearest 5000. Observations regarding the number of hare pelts are lacking for the years 1846, 1905, 1907, and 1908. Now we plot the last 31 data points using all the possible methods offered by **dataPlot**:

```
{p1, p2, p3, p4, p5, p6, p7, p8} =
Map[dataPlot[Take[haredata, -31], plotType → #,
    pointDiameter → 2.5, circleDiameter → 2.2, Ticks → None,
    AxesOrigin → {1900, 0}, DisplayFunction → Identity] &,
  {"c", "p", "jl", "cjl", "pjl", "vl", "cvl", "pvl"}];
```

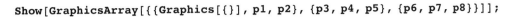

```
Show[GraphicsArray[{{Graphics[{}], p1, p2}, {p3, p4, p5}, {p6, p7, p8}}]];
```

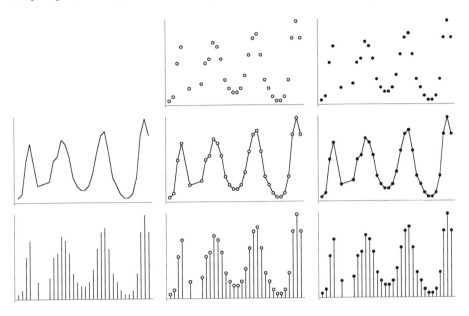

■ A Quality Plot

To get a plot of good quality for all of the hare pelt data, we define a smaller aspect ratio and our own ticks (now we do not use **dataPlot**; this example shows how well **Graphics** works with this type of plot):

```
p1 = Show[Graphics[
     {Line[haredata], AbsolutePointSize[2.3], Map[Point, haredata]},
     AspectRatio → 0.2, PlotRange → All, Axes → True, AxesOrigin → {1843, 0},
     Ticks → {Range[1850, 1930, 10], Range[20, 140, 20]}]];
```

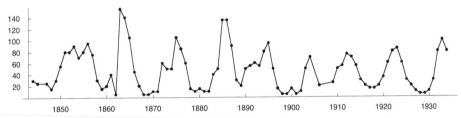

Generally a plot can be considered to consist of some *line segments* that go up or down. The slope of a segment tells the *orientation* of the segment. For example, if the slope is 1, we say that the line segment has an orientation of 45°. The orientations of the segments have an effect on how well the information contained in a plot can be perceived. Typically the judgments of a curve are optimized when the absolute values of the orientations of the line segments that make up the curve are about 45° (see Cleveland 1993, p. 89). The orientations can be adjusted by the aspect ratio of the plot. Choosing the aspect ratio to center the absolute orientations on 45° is *banking to 45°*.

Banking to 45° in the plot above would require an aspect ratio of about 0.05, but then the plot becomes very low. As a compromise, we have chosen the value 0.2.

9.1.3 Filled and Other Plots

■ Filled Plots

> *In the* `Graphics`FilledPlot`` *package:*
>
> `FilledListPlot[data]`   Fill the area between `data` and the *x* axis
> `FilledListPlot[data1, data2, … ]`   Fill the area between `data1, data2, …`

Data are given as they are for `ListPlot`. Options are the same as for `FilledPlot` (see Section 5.3.3, p. 126) except that the options relating to sampling of the function are lacking. A filled plot shows very clearly the shape of the data (we use the same data as we did in Section 9.1.2, p. 234):

```
<< Graphics`FilledPlot`
```

```
FilledListPlot[haredata, AspectRatio → 0.2, AxesOrigin → {1843, 0},
   Ticks → {Range[1850, 1930, 10], Range[20, 140, 20]},
   Epilog → {AbsolutePointSize[2.3], Map[Point, haredata]}];
```

A filled plot may be used to show how some factors develop when they are added:

```
data1 = Table[1 + 0.3 x + 0.2 Random[], {x, 0, 6, 0.2}];
data2 = Table[1 + 0.5 Sin[x] + 0.1 Random[], {x, 0, 6, 0.2}];
data3 = Table[1 + 0.5 Cos[x] + 0.1 Random[], {x, 0, 6, 0.2}];
d1 = Transpose[{Range[0, 6, 0.2], data1}];
d2 = Transpose[{Range[0, 6, 0.2], data1 + data2}];
d3 = Transpose[{Range[0, 6, 0.2], data1 + data2 + data3}];
```

```
FilledListPlot[d1, d2, d3, AspectRatio → 0.4,
   Fills → {{{Axis, 1}, RGBColor[0, 0, 1]},
      {{1, 2}, RGBColor[0, 1, 0]}, {{2, 3}, RGBColor[1, 0, 0]}},
   Epilog → {AbsolutePointSize[2], Map[Point, Union[d1, d2, d3]]}];
```

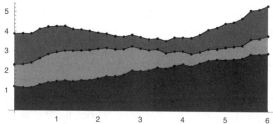

■ Error Plots

> *In the* **Graphics`Graphics`** *package:*
>
> **ErrorListPlot[data]**   Show data by points with error bars
>
> Data can be given in either of the following forms:
>
> **{{y₁ , yerr₁ }, {y₂ , yerr₂ }, …}**   Show the points **{i, y_i }** with error bars
> **{{x₁ , y₁ , yerr₁ }, {x₂ , y₂ , yerr₂ }, …}**   Show the points **{x_i , y_i }** with error bars

The error is shown by a vertical line centered at **{x_i , y_i }** and having a total length of two times **yerr_i** . For example:

```
data = Table[{x, 1.5 + Sin[x] + Random[Real, {-0.2, 0.2}],
     Random[Real, {0, 0.3}]}, {x, 0, 7, 0.2}];
```

```
<< Graphics`Graphics`
```

```
ErrorListPlot[data, AspectRatio → 0.4];
```

With **MultipleListPlot** we can form labeled plots for multiple data sets and more advanced error plots; see the manual of the packages.

■ Logarithmic Plots

> *In the* **Graphics`Graphics`** *package:*
>
> **LinearLogListPlot[data]**   A plot of **Log[yi]** as a function of **xi**
>   = **LogListPlot[data]**
> **LogLinearListPlot[data]**   A plot of **yi** as a function of **Log[xi]**
>   **LogLogListPlot[data]**   A plot of **Log[yi]** as a function of **Log[xi]**

The data are given in the same form as they are for **ListPlot**. The option **PlotJoined** can be used. In a logarithmic scale, exponential data are close to a line:

```
data = Table[{x, Exp[x] + Random[Real, {-0.5, 0.5}]}, {x, 0.1, 2, 0.1}];
```

```
LogListPlot[data, PlotStyle → AbsolutePointSize[1.5]];
```

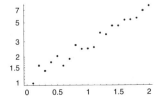

9.2 Plotting Several Time Series

9.2.1 Self-Made Plotting

■ Introduction to Plotting Several Data Sets

If we have several sets of data, we often want a single plot that shows them all so that we may compare the sets and deduce connections among them. However, **ListPlot** is not able to plot several data sets (remember that, with **Plot**, we can plot several curves at the same time). If we want to plot each data set with both points or circles *and* joining or vertical lines (as we often do), then we have two approaches:

- Use **Graphics** or **dataPlot** (see Section 9.1.2, p. 232) to plot each of the data sets and combine the plots with **Show**, or use **Graphics** to plot all of the data sets at the same time.
- Use **MultipleListPlot** from a package to plot all of the data sets at the same time.

These methods are considered in this section and Section 9.2.2.

Note that, if we plot the data sets by using either points or joining lines but not both, we can also use **ListPlot**; this involves plotting each data set separately and then combining the plots with **Show**. However, this method does not work if we want both points or circles and joining or vertical lines for each data set because of the way **Show** handles options: the values of the options in the *first* plot mentioned in **Show** are used, and all of the settings of the options in the other plots are ignored (see Section 5.1.2, p. 113). This means that, for example, the points plotted with **Epilog** in the first plot come into the combined plot while all other points mentioned in other **Epilog** options disappear. A way to get around this is to add, in **Show**, using **Epilog**, *all* of the primitives of the component plots. However, it seems simpler to use either of the two methods mentioned above.

■ Plotting Each Data Set Separately

We continue the hare pelt example of Section 9.1.2, p. 234, and now plot the numbers of lynx pelts sold to the Hudson Bay Trading Company. First we read the numbers of lynx pelts (the numbers are in thousands and are quoted to the nearest 1000):

```
lynxdata = Import["/Users/ruskeepaa/Documents/MNData/lynx", "Table"]
```
```
{{1844, 6}, {1845, 14}, {1846, 22}, {1847, 36}, {1848, 29}, {1849, 7}, {1850, 2}, {1851, 1},
{1852, 1}, {1853, 1}, {1854, 5}, {1855, 13}, {1856, 16}, {1857, 25}, {1858, 14},
{1859, 8}, {1860, 3}, {1861, 2}, {1862, 1}, {1863, 3}, {1864, 10}, {1865, 27}, {1866, 58},
{1867, 30}, {1868, 26}, {1869, 9}, {1870, 4}, {1871, 2}, {1872, 2}, {1873, 6}, {1874, 10},
{1875, 26}, {1876, 29}, {1877, 21}, {1878, 11}, {1879, 10}, {1880, 5}, {1881, 3},
{1882, 5}, {1883, 16}, {1884, 42}, {1885, 64}, {1886, 63}, {1887, 32}, {1888, 15},
{1889, 7}, {1890, 3}, {1891, 4}, {1897, 15}, {1898, 7}, {1899, 2}, {1900, 3}, {1901, 5},
{1902, 14}, {1903, 27}, {1904, 47}, {1905, 54}, {1906, 29}, {1907, 7}, {1908, 2}, {1909, 2},
{1910, 4}, {1911, 10}, {1912, 14}, {1913, 19}, {1915, 8}, {1916, 9}, {1917, 2}, {1918, 1},
{1919, 1}, {1920, 2}, {1921, 4}, {1922, 4}, {1923, 8}, {1924, 7}, {1925, 9}, {1926, 7},
{1927, 4}, {1928, 3}, {1929, 2}, {1930, 3}, {1931, 3}, {1932, 5}, {1933, 7}, {1934, 7}}
```

These data can also be found on the CD-ROM that accompanies this book. Observations on lynx pelts are lacking for the years 1892 to 1896 and 1914. Before continuing, we set the values of some options, because these values are used so often in this section:

```
SetOptions[Graphics, AspectRatio → 0.2,
   PlotRange → All, Axes → True, AxesOrigin → {1843, 0},
   Ticks → {Range[1850, 1930, 10], Range[20, 140, 20]}];
```

Often it is advantageous to use different styles for each data set so that the sets can be clearly seen from the combined plot. First we use dashed lines:

```
Show[Graphics[{AbsoluteDashing[{1.3}], Line[lynxdata],
   AbsolutePointSize[2.3], Map[Point, lynxdata]}]];
```

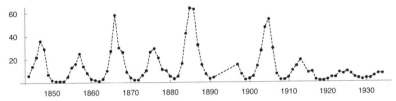

Using different colors is an efficient way to distinguish the data sets. A fine result can also be obtained with circles (see Section 9.1.2, p. 232). First we plot lines joining the points, then white disks to hide the lines inside the circles, and lastly black circles. The thickness, dashing, and color of the circles can be set with graphics directives. We choose circles having an absolute radius of 1 (in units of 1/72 inch):

```
p2 = Show[Graphics[{Line[lynxdata],
   {GrayLevel[1], Map[Disk[#, Offset[{1, 1}]] &, lynxdata]},
   Map[Circle[#, Offset[{1, 1}]] &, lynxdata]}]];
```

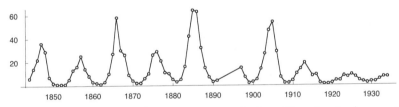

We now combine the plots of hare and lynx pelt data. The plot for hare pelts was given as follows in Section 9.1.2:

```
haredata = Import["/Users/ruskeepaa/Documents/MNData/hare", "Table"];
```

```
p1 = Show[Graphics[
   {Line[haredata], AbsolutePointSize[2.3], Map[Point, haredata]},
   AspectRatio → 0.2, PlotRange → All, Axes → True, AxesOrigin → {1843, 0},
   Ticks → {Range[1850, 1930, 10], Range[20, 140, 20]}],
   DisplayFunction → Identity];
```

For the combined plot, we add text primitives to identify the curves:

```
Show[p1, p2, DisplayFunction → $DisplayFunction, Epilog →
   {Text["hare", {1935, 80}, {-1, 0}], Text["lynx", {1935, 9}, {-1, 0}]}];
```

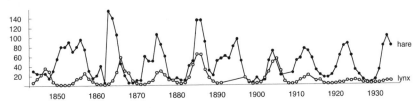

Both hare and lynx seem to have a cycle of about 10 years. More than 90% of the diet of the lynx is hare. When there are few hares available, lynx starve rather than eat other species.

■ Plotting Both Data Sets at the Same Time

We can also plot both data sets at the same time. For hare pelts, we can use disks instead of points (the radius of the black disks should be slightly larger than the radius of the circles so that they seem to be about the same size):

```
Show[Graphics[{Line[haredata], Line[lynxdata],
    Map[Disk[#, Offset[{1.2, 1.2}]] &, haredata],
    {GrayLevel[1], Map[Disk[#, Offset[{1, 1}]] &, lynxdata]},
    Map[Circle[#, Offset[{1, 1}]] &, lynxdata],
    Text["hare", {1935, 80}, {-1, 0}], Text["lynx", {1935, 9}, {-1, 0}]}]];
```

■ Adding a Legend

A simple and good way to identify the curves is to add some text next to each curve, as we did above. A true legend can be made as follows:

```
hareleg = {{1925, 142}, {1929, 142}};
lynxleg = {{1925, 125}, {1929, 125}};
haredl = Join[haredata, hareleg];
lynxdl = Join[lynxdata, lynxleg];
```

```
Show[Graphics[
    {Line[haredata], Line[hareleg], Line[lynxdata], Line[lynxleg],
    Map[Disk[#, Offset[{1.2, 1.2}]] &, haredl],
    {GrayLevel[1], Map[Disk[#, Offset[{1, 1}]] &, lynxdl]},
    Map[Circle[#, Offset[{1, 1}]] &, lynxdl],
    Text["hare", {1930, 142}, {-1, 0}],
    Text["lynx", {1930, 125}, {-1, 0}]}]];
```

■ A Phase Plot

Until now we have plotted two data sets—$(x_1, x_2, \ldots)$ and $(y_1, y_2, \ldots)$—as time series. Another kind of plot is obtained by plotting the pairs (x_i, y_i). In this way, we get a plot that is analogous to a phase plot of the solution of a pair of differential equations (see Section 23.3.2, p. 617). For the years 1909 to 1934, the numbers of hare and lynx pelts are as follows (the observation for the year 1914 is lacking):

```
harelynx = {{25, 2}, {50, 4}, {55, 10}, {75, 14},
    {70, 19}, {30, 8}, {20, 9}, {15, 2}, {15, 1}, {20, 1}, {35, 2},
    {60, 4}, {80, 4}, {85, 8}, {60, 7}, {30, 9}, {20, 7}, {10, 4},
    {5, 3}, {5, 2}, {10, 3}, {30, 3}, {80, 5}, {100, 7}, {80, 7}};
```

We construct a phase plot:

```
Show[
    Graphics[{Line[harelynx], AbsolutePointSize[3], Map[Point, harelynx],
        AbsolutePointSize[5], RGBColor[1, 0, 0], Point[First[harelynx]]},
        AspectRatio → 1 / GoldenRatio, Axes → True,
        AxesOrigin → {0, 0}, AxesLabel → {"hare", "lynx"},
        Ticks → {Range[10, 100, 10], Range[2, 18, 2]}]];
```

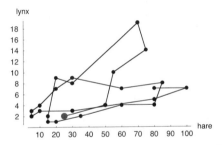

The starting point (25, 2) is red and larger than the other points. We can see a counter-clockwise cycle; this pattern is typical for a predator–prey system.

9.2.2 Built-in Plotting

In the **Graphics`MultipleListPlot`** *package:*

MultipleListPlot[data1, data2, …] Plot with symbols
MultipleListPlot[data1, data2, …, PlotJoined → True] Plot with symbols and
  joining lines

MultipleListPlot is a special plotting command used to plot several data sets at the same time. Each data set can be of the same form as is used for **ListPlot**. The sets of data are automatically marked with different symbols. There are five default symbols: diamond, star, box, triangle, and point. If points are connected with lines, then different line styles are also used. There are five default line styles: a continuous line and four different dashings. Symbols and line styles are used in a cyclic way if we have more than five data sets. Here are the default symbols and line styles:

```
data = Table[{j, 5 - i}, {i, 0, 4}, {j, 5}];
```

```
<< Graphics`MultipleListPlot`

MultipleListPlot[data[[1]], data[[2]],
   data[[3]], data[[4]], data[[5]], PlotJoined → True,
   AspectRatio → 0.4, PlotRange → {0.6, 5.4}, Axes → False];
```

Note that we can have, in the same plot, both joined and unjoined points. We could define, for example, **PlotJoined → {True, False}**.

■ Example 1

As an example, we plot the numbers of hare and lynx pelts; the data have been read in Sections 9.1.2, p. 234, and 9.2.1, p. 238. We first use the default symbols and line styles (we have to define a large enough plot range so that none of the symbols are cut):

```
MultipleListPlot[haredata, lynxdata, PlotJoined → True,
   AspectRatio → 0.2, PlotRange → {-5, 161}, AxesOrigin → {1843, 0},
   Ticks → {Range[1850, 1930, 10], Range[20, 140, 20]}];
```

To me, the shapes of the default symbols are not pleasing or professional (these shapes are not generally used in publications; the symbols are also too large in this plot). The best symbol—the point—is by default only used for the fifth data set, and circles are not used at all. The default dashing seems to be too sparse; the dashing needs not even be used if we use different symbols. However, the symbols and the line styles can be controlled with options.

■ Options

PlotStyle Styles of the lines; examples of values: **Automatic**, **{{Absolute⠐ Thickness[0.5]}, {AbsoluteThickness[0.5], AbsoluteDashing[{1.5}]}}**

SymbolShape Shapes of the symbols used in the data points; examples of values:
   Automatic, None, Stem (⊞5)**, Label, {Point, PlotSymbol[Box, 1]}, {PlotSymbol[⠐ Triangle, 1.7], MakeSymbol[RegularPolygon[4, 2]]}**

SymbolStyle Styles of the symbols; examples of values: **Automatic**, **{{AbsolutePoint⠐ Size[3]}, {RGBColor[1,0,0], AbsoluteThickness[0.25]}}**

SymbolLabel Labels for the data points; examples of values: **None, Automatic**, **{Range[20], CharacterRange["A", "T"]}**

Note that **MultipleListPlot** also accepts all options of **Graphics** (see Chapter 6) and all options of the **Graphics`Legend`** package (see Section 6.6.3, p. 173). Next we consider the options given in the box in some detail.

■ Style of Lines

The default value **Automatic** of **PlotStyle** means that the thickness of the lines is **Thickness[0.001]** and that the following dashings are used cyclically:

> **AbsoluteDashing[{}]**, **AbsoluteDashing[Dot]**, **AbsoluteDashing[Dot, Dash]**,
> **AbsoluteDashing[Dash]**, **AbsoluteDashing[Dot, Dash, Dot]**

In fact, with the package we can define dashings and absolute dashings by using the predefined objects **Dot**, **Dash**, and **LongDash** (instead of by using numerical values such as in **AbsoluteDashing[{1.5}]**).

■ Generating Symbols

> **PlotSymbol[type, radius]** Generates a graphic symbol of the given type having the given radius (the radius can be left out); valid types: **Diamond**, **Star**, **Box**, **Triangle**; with the option **Filled → False**, generates an outline of the symbol
> **MakeSymbol[{directives and primitives or a regular polygon}]** Generates a graphic symbol from the given directives and primitives or from the given regular polygon

With these commands we can define the symbols to be used in the plot. **PlotSymbol** only generates symbols of the built-in types **Diamond**, **Star**, **Box**, and **Triangle**, but with **MakeSymbol** we can form the symbols we want from graphic primitives or from a **Regular`Polygon[n, r]**. For example:

> **PlotSymbol[Box, 1]**
>
> **PlotSymbol[Box, 1, Filled → False]**

This defines a filled and an open square of approximate absolute radius 1 (in units of 1/72 inch). Another example:

> **s1 = MakeSymbol[{AbsolutePointSize[2.3], Point[{0, 0}]}];**
>
> **s2 = MakeSymbol[{{GrayLevel[1], Disk[{0, 0}, Offset[{1, 1}]]},**
> **AbsoluteThickness[0.5], Circle[{0, 0}, Offset[{1, 1}]]}];**
>
> **MakeSymbol[RegularPolygon[4, 1]];**
>
> **MakeSymbol[{Line[{{2, 2}, {-2, -2}}], Line[{{-2, 2}, {2, -2}}]}]**

This generates a point, a circle, a square of absolute diameter 2, and a cross (×).

■ Symbol Shapes

The default value **Automatic** of **SymbolShape** gives the following value:

> **{PlotSymbol[Diamond], PlotSymbol[Star],**
> **PlotSymbol[Box], PlotSymbol[Triangle], Point}**

This means that these symbols are used cyclically. The value **None** causes no symbol to be drawn, and the value **Label** causes labels defined with the option **SymbolLabel** to be used. With **PlotSymbol** and **MakeSymbol**, we can use other symbols.

The default value **Automatic** of **SymbolStyle** means that the symbols are black and nondashed and have **Thickness[0.0001]** and **PointSize[0.01]**.

The value **Automatic** of **SymbolLabel** generates labels like 1-1, 1-2, 1-3, …, 2-1, 2-2, 2-3, …. The value **Stem** (✻5) plots vertical lines with points at the top (**Stem[s]** defines the size of the points to be **s**):

```
MultipleListPlot[Table[Random[], {20}], SymbolShape → Stem];
```

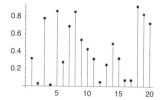

■ Example 2

In the following plot, we use points and circles (the shapes **s1** and **s2** defined above). We also add a legend:

```
MultipleListPlot[haredata, lynxdata, PlotJoined → True,
    PlotRange → All, AspectRatio → 0.2, AxesOrigin → {1843, 0},
    Ticks → {Range[1850, 1930, 10], Range[20, 140, 20]},
    PlotStyle → AbsoluteThickness[0.5], SymbolShape → {s1, s2},
    PlotLegend → {"hare", "lynx"}, LegendPosition → {0.67, 0.1},
    LegendSize → {0.5, 0.1}, LegendSpacing → -0.25, LegendShadow → None];
```

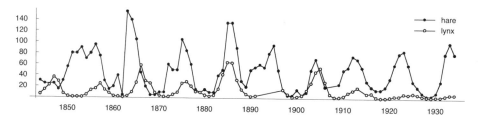

9.3 Bar and Pie Charts

9.3.1 Bars with Labels: One Data Set

The **Graphics`Graphics`** package contains several commands for bar charts. **BarChart**, **StackedBarChart**, and **PercentileBarChart** (see this section and Section 9.3.2, p. 248) are suitable for charts in which the bars are simply drawn side by side (without having specified positions) and in which there are labels (not coordinates) under the bars. **GeneralizedBarChart** (see Section 9.3.3, p. 250) is designed for charts in which the bars have specified positions and widths. **Histogram** (see Section 9.3.4, p. 251) is a special command to calculate frequencies and plot them as bar charts.

In the **Graphics`Graphics`** *package:*

BarChart[{y₁ , y₂ , … }]   Plot bars of heights **y₁ , y₂ ,** … and label the bars by **1, 2,** …

BarChart[{{y₁ , a₁ }, {y₂ , a₂ }, … }]   Label the bars by **a₁ , a₂ ,** …

BarChart[{y₁ , y₂ , … }, BarLabels → {a₁ , a₂ , … }]   Label the bars by **a₁ , a₂ ,** …

We have two methods to obtain the desired labels a_i under the bars: either form pairs of heights and labels or use the **BarLabels** option. The labels are sometimes numbers, but they can also be, for example, **"class A"**, **"class B"**, ….

■ Example 1

As an example, we plot the body and brain weights of some animals. The data are from a collection of data sets in http://lib.stat.cmu.edu/S/visualizing.data. All of the data are visualized in Cleveland (1993), and the data sets are also on the CD-ROM that comes with this book (the data sets are reproduced with the permission of the publishers, Hobart Press). On my computer, the animal data are in a text file **modAnimal** in a folder **visdata**. (Note that the original file **animal** contained spaces in the names of the animals, but we have now deleted the spaces.) We read the file and take only rows 38 through 48 (for reading data, see Section 4.2.1, p. 90):

```
braindata =
  Take[Import["/Users/ruskeepaa/Documents/MNData/visdata/modAnimal",
    "Table"], {38, 48}]

{{37, RoeDeer, 14830, 98.2},
 {38, Goat, 27660, 115.}, {39, Kangaroo, 35000, 56.},
 {40, GrayWolf, 36330, 119.5}, {41, Sheep, 55500, 175.},
 {42, GiantArmadillo, 60000, 81.}, {43, GraySeal, 85000, 325.},
 {44, Jaguar, 100000, 157.}, {45, BrazilianTapir, 160000, 169.},
 {46, Donkey, 187100, 419.}, {47, Pig, 192000, 180.}}
```

The numbers are the body and brain weights, respectively. Extract the columns of these measurements:

```
{no, name, body, brain} = Transpose[braindata]

{{37, 38, 39, 40, 41, 42, 43, 44, 45, 46, 47},
 {RoeDeer, Goat, Kangaroo, GrayWolf, Sheep, GiantArmadillo, GraySeal,
  Jaguar, BrazilianTapir, Donkey, Pig}, {14830, 27660, 35000,
  36330, 55500, 60000, 85000, 100000, 160000, 187100, 192000},
 {98.2, 115., 56., 119.5, 175., 81., 325., 157., 169., 419., 180.}}
```

Next we form pairs of brain weight and animal name and sort the pairs in ascending order according to brain weight:

```
data = Sort[Transpose[{brain, name}]]

{{56., Kangaroo}, {81., GiantArmadillo}, {98.2, RoeDeer}, {115., Goat},
 {119.5, GrayWolf}, {157., Jaguar}, {169., BrazilianTapir},
 {175., Sheep}, {180., Pig}, {325., GraySeal}, {419., Donkey}}
```

The bar chart is given as follows:

```
<< Graphics`Graphics`
```

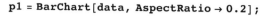

```
p1 = BarChart[data, AspectRatio → 0.2];
```

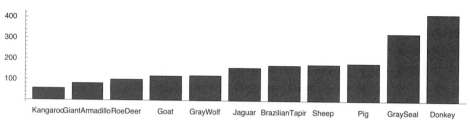

Here we have used a small aspect ratio and enlarged the plot with the mouse to minimize the overlapping of the labels. A good solution to overlapping labels is to use horizontal bars, as we do later on in this section. Another solution is to rotate the labels somewhat. Remember from Section 5.4.6, p. 142, that **Text** primitives can have any slope. We define the slope as {2, 1}; that is, the slope is 1/2. We also adjust the positioning of the labels: the old position in text coordinates is {0, 1} so that the labels are centered below the bars, but the new position is {0.8, 0.8} so that the labels are to the left of the bars:

```
p2 = FullGraphics[p1] /. Text[a_, b_, c_] → Text[a, b, {0.8, 0.8}, {2, 1}];

Show[p2, PlotRange → All];
```

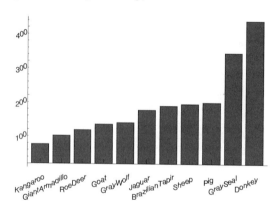

■ Options

BarChart accepts the following options and the options used with **Graphics**.

BarLabels Labels under the bars; examples of values: **Automatic**, {10, 20, 30}, {a, b, c}, {"class A", "class B", "class C"}

BarValues Whether to write the values on top of the bars; possible values: **False**, **True**

BarOrientation Orientation of the bars; possible values: **Vertical**, **Horizontal**

BarGroupSpacing The space between each group of bars as a fraction of the width of one bar; examples of values: **Automatic** (means **0.2**), **0**

BarSpacing The space between the bars within a group as a fraction of the width of one bar; examples of values: **Automatic** (means **0**), **0.1**, **-0.4**

BarStyle Style inside the edges of the bars; examples of values for one data set: **Automatic** (means **RGBColor[1,0,0]**), **GrayLevel[0.8]**, (**Hue[#/419] &**); an example for two data sets: {**GrayLevel[0.5]**, **GrayLevel[0.9]**}

BarEdgeStyle Style of the edges of the bars; examples of values for one data set:
GrayLevel[0], **AbsoluteThickness[1]**, **{{GrayLevel[0.5], Absolute﹀**
Thickness[1]}}; an example for two data sets: **{{AbsoluteThickness[1]}, {Gray﹀**
Level[0.5], AbsoluteThickness[1]}}
BarEdges Whether edges are drawn for the bars; possible values: **True**, **False**

BarGroupSpacing defines the space between each group of bars. Indeed, **BarChart** can generate bars for multiple data sets (see Section 9.3.2, p. 248), and then the bars are collected into groups; each group contains as many bars as there are data sets. Thus, if we have only one data set, then **BarGroupSpacing** simply defines the space between the bars. The default value **Automatic** means the value **0.2**. If you want no space between the bars, give the value **0**.

BarSpacing defines the space between bars within a group of bars. The default value **Automatic** means the value **0**: there is no space between the bars within a group. If you want a small space between the bars, give a small positive value for the option, and if you want the bars to overlap, give a small negative value (see Section 9.3.2).

The value of **BarStyle** can be a single color, but it can also be a pure function that defines the color of the bars as a function of the height of the bars.

■ Example 2

We use the same data as above and add some of the options (compare this with the dot plot of Section 9.4.1, p. 256):

```
BarChart[data, BarOrientation → Horizontal,
    BarGroupSpacing → 0, BarStyle → GrayLevel[0.7], PlotRange → All];
```

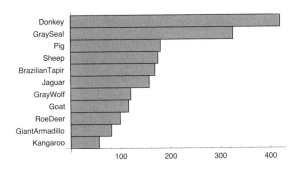

■ Example 3

We may want to define the y ticks for a vertical bar chart or the x ticks for a horizontal bar chart. This is not so straightforward as we would expect. In fact, we also have to take care of the labels. As an example we use custom x ticks so that we have to define the labels with y ticks:

```
sortedname = Transpose[data][[2]]
```

{Kangaroo, GiantArmadillo, RoeDeer, Goat, GrayWolf,
 Jaguar, BrazilianTapir, Sheep, Pig, GraySeal, Donkey}

```
yticks = Transpose[{Range[11], sortedname}]
```

```
{{1, Kangaroo}, {2, GiantArmadillo}, {3, RoeDeer},
 {4, Goat}, {5, GrayWolf}, {6, Jaguar}, {7, BrazilianTapir},
 {8, Sheep}, {9, Pig}, {10, GraySeal}, {11, Donkey}}
```

We use a black background:

```
BarChart[data, BarOrientation → Horizontal,
  BarGroupSpacing → 0, BarStyle → GrayLevel[1],
  PlotRange → All, PlotRegion → {{0.02, 0.95}, {0, 1}},
  Background → GrayLevel[0], TextStyle →
   {FontFamily → "Helvetica", FontWeight → "Bold", FontSize → 6},
  Axes → False, Frame → True, FrameStyle → AbsoluteThickness[1.3],
  FrameTicks → {Range[0, 400, 50], yticks, None, None},
  GridLines → {Table[{i, {GrayLevel[0.7]}}, {i, 0, 400, 50}], None},
  PlotLabel → StyleForm["Brain weights for some animals", FontSize → 9]];
```

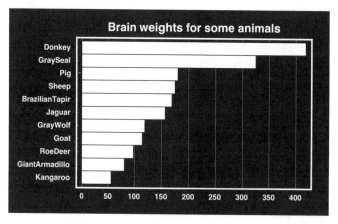

9.3.2 Bars with Labels: Several Data Sets

> *In the* `Graphics`Graphics`` *package:*
>
> `BarChart[data1, data2, …, BarLabels → labels]`
> `StackedBarChart[data1, data2, …, BarLabels → labels]`
> `PercentileBarChart[data1, data2, …, BarLabels → labels]`

BarChart can also be used for multiple data sets. The labels must now be given with the **BarLabels** option. As an example, we plot the running times of four algorithms before and after improvements (these are not real data; we illustrate the same data sets with a dot plot in Section 9.4.1, p. 256):

```
labels = {"Alg. 1", "Alg. 2", "Alg. 3", "Alg. 4"};
times1 = {8, 9, 11, 10};
times2 = {5, 7, 6, 8};
```

The default is that the bars are side by side, but we produce a plot in which the bars somewhat overlap:

```
<< Graphics`Graphics`
```

```
p = BarChart[times1, times2, BarLabels → labels, BarGroupSpacing → 0.6,
    BarSpacing → -0.4, BarStyle → {GrayLevel[0.5], GrayLevel[0.8]},
    PlotRange → {0, 12.3}, AspectRatio → 0.5,
    PlotLabel → StyleForm["Running times before (dark) and
     \n        after (light) improvements", FontSize → 8]];
```

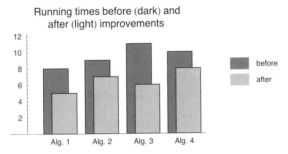

A legend can be added with **ShowLegend** (for more about legends, see Section 6.6.3, p. 173):

```
<< Graphics`Legend`
```

```
ShowLegend[p, {{{GrayLevel[0.5], "before"}, {GrayLevel[0.8], "after"}},
    LegendPosition → {0.85, -0.1}, LegendSize → {0.5, 0.25},
    LegendShadow → None, LegendSpacing → 0.2}];
```

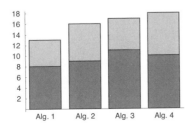

In a stacked bar chart, the bars are stacked instead of being placed side by side. In a percentile bar chart, the stacked bars are scaled so that each group has the height 100%. We use the same data as above but now interpret them so that we have made two improvements in the algorithms. The total savings in running times are as follows (we use custom ticks; see Section 9.3.1, p. 247):

```
StackedBarChart[times1, times2,
    BarStyle → {GrayLevel[0.5], GrayLevel[0.8]},
    Ticks → {Transpose[{Range[4], labels}], Range[2, 18, 2]}];
```

9.3.3 Bars with Positions

In the `Graphics`Graphics`` *package:*

`GeneralizedBarChart[{{x₁ , y₁ , w₁ }, {x₂ , y₂ , w₂ }, … }]`   Plot bars of heights y_1,
   y_2, … and widths w_1, w_2, … at the positions x_1, x_2, …

Recall that with `BarChart` you first gave the height of the bar and then its label. With `GeneralizedBarChart`, you first give the x position and then the height and width of the bar. The position must be a number that tells at which point on the x axis the bar is centered.

`GeneralizedBarChart` has all the same options as `BarChart` except for the options `BarLabels`, `BarGroupSpacing`, and `BarSpacing`. The options are in fact unnecessary, because the labels can be given with the `Ticks` option, and the spacing can be adjusted with the widths of the bars.

In this example, we toss a die 100 times and calculate the frequencies. Note that, for each frequency, we add the width 1 of the bar as the third component:

```
SeedRandom[5];
data = Table[Random[Integer, {1, 6}], {100}];

freq = Map[{#, Count[data, #], 1} &, Range[6]]

{{1, 15, 1}, {2, 16, 1}, {3, 14, 1}, {4, 16, 1}, {5, 20, 1}, {6, 19, 1}}
```

Then we plot the frequencies (note that now we do not have problems with ticks, as we did in Section 9.3.1, p. 247):

```
<< Graphics`Graphics`

GeneralizedBarChart[freq, BarStyle → {GrayLevel[0.6]},
    PlotRange → {{0.5, 6.5}, {0, Automatic}},
    AxesOrigin → {0.5, 0}, Ticks → {Range[6], Range[5, 20, 5]}];
```

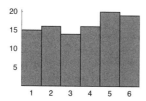

In this example, `Histogram` (see Section 9.3.4, p. 251) would be the right command, but we used `GeneralizedBarChart` just to illustrate this command.

`GeneralizedBarChart` can also be used for multiple data. Just give several data sets:
`GeneralizedBarChart[data1, data2, … ]`.

9.3.4 Bar Charts for Frequencies

In the `Graphics`Graphics`` *package:*

`Histogram[{x`$_1$`, x`$_2$`, … }]` Plot the frequencies of the given raw data
`Histogram[{f`$_1$`, …, f`$_n$`}, FrequencyData → True, HistogramCategories → value]`
  Plot the given frequencies

Options:

`HistogramCategories` How the data is categorized, that is, for which intervals the
  frequencies are calculated; possible values: `Automatic` (use an internal algorithm), a
  positive integer `n` (use exactly `n` categories of equal width, if `ApproximateIntervals`
  `→ False`, and about `n` categories, if `ApproximateIntervals → True`), or a list of cutoff
  values `{c`$_0$`, c`$_1$`, …, c`$_n$`}` (calculate the frequencies in the intervals $[c_0, c_1), [c_1, c_2), …$)
`ApproximateIntervals` Whether interval boundaries should be approximated by
  simple numbers; possible values: `Automatic` (usually means `True`), `True`, `False`
`HistogramScale` Whether to scale the heights of the bars; examples of values: `Auto`-
  `matic` (means `False` for categories with equal widths and `True` for categories with
  unequal widths), `False` (no scaling: plot frequencies as such), `True` (scale by dividing
  the heights by the widths of the bars to get a frequency density), `1` (scale to get the
  sum of the areas of the bars equal to 1 so that the histogram approximates the probability density function of the data; other constants can also be used)
`HistogramRange` Range of data to be included in the histogram; examples of values:
  `Automatic` (means that all data is included), `{0, 10}`
`BarOrientation`, `BarStyle`, `BarEdgeStyle`, `BarEdges` (see Section 9.3.1, p. 246)

`Histogram` also has the options of `Graphics`. With the `Ticks` option, we can use the
special values `IntervalBoundaries` and `IntervalCenters`.

With `Histogram` we can nicely plot frequencies as a bar chart; the data can be either
raw data or frequencies. In the former case, `Histogram` first calculates the frequencies. In
Section 27.2.1, p. 751, we consider the calculation of frequencies.

With `ParetoPlot` (✿5) from the `Statistics`StatisticsPlots`` package, we can plot
bars for the frequencies together with a line plot for the cumulative frequencies.

■ **Example 1**

We plot the frequencies of the same data that was used in Section 9.3.3, p. 250. First we
supply the raw data:

```
<< Graphics`Graphics`

SetOptions[Histogram, BarStyle → GrayLevel[0.5]];
```

```
Histogram[data, HistogramCategories → Range[0.5, 6.5, 1]];
```

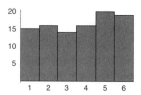

We know that all of the numbers in the data are from the set {1, 2, …, 6} so that if we define the categories to be [0.5, 1.5), [1.5, 2.5), …, [5.5, 6.5), we get the frequencies of 1, 2, …, 6. Next we plot frequencies we have calculated ourselves:

```
freq = Map[Count[data, #] &, Range[6]]
```

{15, 16, 14, 16, 20, 19}

```
Histogram[freq,
    HistogramCategories → Range[0.5, 6.5, 1], FrequencyData → True];
```

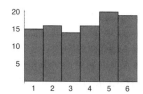

■ Example 2

To illustrate the **HistogramScale** option, we generate 2000 random numbers from (0, 10):

```
SeedRandom[2];
data2 = Table[Random[Real, {0, 10}], {2000}];
```

If we use categories of equal width 2, the frequencies are as follows:

```
<< Statistics`DataManipulation`
```

```
RangeCounts[data2, {2, 4, 6, 8}]
```

{383, 408, 409, 385, 415}

We plot the frequencies (with the raw data) by using three values of **Histogram**‹ **Scale**—the default value **Automatic** and the values **True** and **1**:

```
Block[{$DisplayFunction = Identity},
  p1 = Histogram[data2, HistogramCategories → 5];
  p2 = Histogram[data2, HistogramCategories → 5, HistogramScale → True];
  p3 = Histogram[data2, HistogramCategories → 5, HistogramScale → 1];]
```

```
Show[GraphicsArray[{p1, p2, p3}]];
```

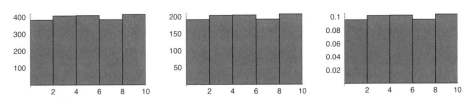

We see that the forms of the three histograms are the same but that the y ticks differ. With the default value **Automatic** of **HistogramScale** (the first plot), we get the frequencies as such: no scaling is done (the value **Automatic** here means **False**). With the value **True** (the second plot), we get a frequency density where the height of each bar is divided by the width of the bar (the area of each bar then equals the frequency). With the value **1** (the third plot), the area of the whole histogram equals 1, and the histogram approximates the probability density function of the data.

Then we use categories [0, 4), [4, 7), [7, 9), and [9, 10) of unequal widths. The frequencies and the histograms with the three values of **HistogramScale** are as follows:

```
Block[{$DisplayFunction = Identity},
  p4 = Histogram[data2, HistogramCategories → {0, 4, 7, 9, 10},
    HistogramScale → False, Ticks → IntervalBoundaries];
  p5 = Histogram[data2, HistogramCategories → {0, 4, 7, 9, 10},
    Ticks → IntervalBoundaries];
  p6 = Histogram[data2, HistogramCategories → {0, 4, 7, 9, 10},
    HistogramScale → 1, Ticks → IntervalBoundaries];]
```

```
Show[GraphicsArray[{p4, p5, p6}]];
```

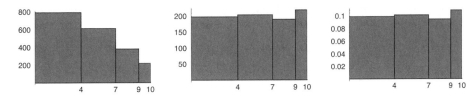

Now the form of the histogram varies depending of the value of **HistogramScale**. The value **False** (the first plot) gives the frequencies as such. With categories of unequal widths, the default is that the frequencies are scaled by dividing the heights by the widths to get a frequency density so that the value **Automatic** of **HistogramScale** here means **True** (the second plot). With the value **1** (the third plot), the heights are scaled so that the area of the histogram is 1.

9.3.5 Pie Charts

In the `Graphics`Graphics`` *package:*

`PieChart[{y₁ , y₂ , … }]` Plot a pie chart from the positive numbers y_1, y_2, …

Options:

`PieLabels` Labels in the wedges; examples of values: `Automatic`, `{2, 4, 3}`, `{a, b, c}`, `{"class A", "class B", "class C"}`

`PieStyle` Style(s) inside the borders of the wedges; examples of values: `Automatic`, `{GrayLevel[0.8]}`, `Table[GrayLevel[p], {p, 0.7, 1, 0.1}]`

`PieLineStyle` Style of the border of the wedges; examples of values: `Automatic`, `{AbsoluteThickness[1]}`, `{AbsoluteThickness[1], RGBColor[0,0,1]}`

`PieExploded` Whether some wedges are exploded; examples of values: `None`, `All`, `{4}`, `{4, 0.2}`, `{4, 5}`, `{{4, 0.2}, {5, 0.2}}`

`PieChart` also has the options of `Graphics`. An exploded wedge is set off from the pie. A value such as `{4, 0.2}` defines that the fourth wedge is set off by the amount 0.2.

A pie chart illustrates how a total amount is made up from certain components. The purpose is that the reader should obtain an impression of the relative magnitudes of the components. As an example, an algorithm was improved by four methods. Of the total savings in running time, the first method contributed 12%, the second 15%, the third 38%, and the fourth 35%. The corresponding pie chart is shown below:

```
<< Graphics`Graphics`

PieChart[{12, 15, 38, 35}, PieLabels →
    {"Method 1\n12%", "Method 2\n15%", "Method 3\n38%", "Method 4\n35%"},
  PieStyle → Table[GrayLevel[p], {p, 0.65, 0.95, 0.1}],
  PlotLabel → StyleForm["Savings by four methods", FontSize → 8]];
```

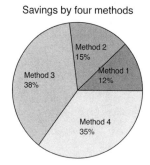
Savings by four methods

9.4 Dot and Box-and-Whisker Plots

9.4.1 Dot Plots

To illustrate dot plots, we use the same animal brain weight data we considered in Section 9.3.1, p. 245:

```
{no, name, body, brain} = Transpose[
  Take[Import["/Users/ruskeepaa/Documents/MNData/visdata/modAnimal",
    "Table"], {38, 48}]]
```

```
{{37, 38, 39, 40, 41, 42, 43, 44, 45, 46, 47},
 {RoeDeer, Goat, Kangaroo, GrayWolf, Sheep, GiantArmadillo, GraySeal,
  Jaguar, BrazilianTapir, Donkey, Pig}, {14830, 27660, 35000,
  36330, 55500, 60000, 85000, 100000, 160000, 187100, 192000},
 {98.2, 115., 56., 119.5, 175., 81., 325., 157., 169., 419., 180.}}
```

Sort the brain weights:

```
Sort[Transpose[{brain, name}]]
```

```
{{56., Kangaroo}, {81., GiantArmadillo}, {98.2, RoeDeer}, {115., Goat},
 {119.5, GrayWolf}, {157., Jaguar}, {169., BrazilianTapir},
 {175., Sheep}, {180., Pig}, {325., GraySeal}, {419., Donkey}}
```

```
{values, labels} = Transpose[%]
```

```
{{56., 81., 98.2, 115., 119.5, 157., 169., 175., 180., 325., 419.},
 {Kangaroo, GiantArmadillo, RoeDeer, Goat, GrayWolf,
  Jaguar, BrazilianTapir, Sheep, Pig, GraySeal, Donkey}}
```

The dot plot can be drawn with the following program:

```
dotPlot[values_, labels_, styles_,
  xmin_, xmax_, xticks_, vertgrid_, opts___] :=
Module[{n = Length[labels], points, yticks, hlines, ymin, ymax, vlines},
  points = Table[Map[{styles[[i, 1]], styles[[i, 2]], Point[#]} &,
    Transpose[{values[[i]], Range[n]}]], {i, Length[values]}];
  yticks = Transpose[{Range[n], labels}];
  hlines = Map[Line, Map[{{xmin, #}, {xmax, #}} &, Range[n]]];
  {ymin, ymax} = {0.3, n + 0.7};
  vlines = If[vertgrid != {}, Map[Line, Table[{{i, ymin}, {i, ymax}},
    {i, vertgrid[[1]], vertgrid[[2]], vertgrid[[3]]}]], {}];
  Show[Graphics[{{GrayLevel[0.7], vlines}, {AbsoluteThickness[0.2],
    AbsoluteDashing[{1, 1.5}], hlines}, points},
  PlotRange → {{xmin, xmax}, {ymin, ymax}}, Frame → True,
  FrameTicks → {xticks, yticks, None, None}, opts]]]
```

Here **values** is a list of one or more data sets, with each data set being a list of numbers; **labels** is a list containing the labels for the y axis. The variable **styles** is a list of styles for the points: the list has as many components as there are data sets, with each component being a list of two elements giving the point size and color of the points. The variables **xmin** and **xmax** define the x range, the variable **xticks** the ticks on the x axis, and the variable `vertgrid` a list of three numbers giving the position of the first vertical grid line, the increment of the grid lines, and the position of the last grid line (an empty list `{}` can also be given). For example:

```
dotPlot[{values}, labels, {{AbsolutePointSize[4], GrayLevel[0]}},
  0, 460, Automatic, {100, 400, 100},
  PlotLabel → StyleForm["Brain weights for some animals", FontSize → 9]];
```

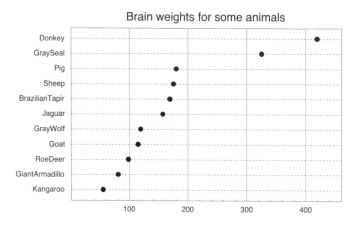

Quite a similar plot can be obtained with a horizontal bar chart (see Section 9.3.1, p. 247). The animal data set is also considered in Section 9.5.2, p. 263. With a dot plot we can also compare two or more data sets (another way is to use a bar chart, see Section 9.3.2, p. 248). As an example, we plot the running times of four algorithms before and after improvements (these are not real data):

```
labels = {"Algorithm 1", "Algorithm 2", "Algorithm 3", "Algorithm 4"};
times1 = {8, 9, 11, 10}; style1 = {AbsolutePointSize[5], GrayLevel[0]};
times2 = {5, 7, 6, 8}; style2 = {AbsolutePointSize[5], GrayLevel[0.5]};

dotPlot[{times1, times2}, labels, {style1, style2}, 0, 12,
  Range[11], {}, PlotLabel → StyleForm["Running times before (
    dark) and\n      after (light) improvements", FontSize → 8]];
```

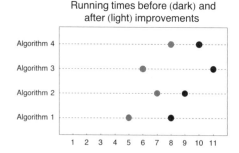

9.4.2 Multiway Dot Plots

An effective way to illustrate 3D data is by using a *multiway dot plot* (Cleveland 1993). As an example, we consider the data in the file **modBarley**. This file is from Cleveland (1993) and can be found on the CD-ROM accompanying this book. The file contains barley yields at 6 sites for 10 varieties in 1931. On my computer, the file is in a folder **visdata** in the folder **MNata** (for information about reading data, see Section 4.2.1, p. 90):

```
data =
  Import["/Users/ruskeepaa/Documents/MNData/visdata/modBarley", "Table"]
```
```
{{47.3, 40.5, 35., 35.1, 25.7, 29.7}, {48.9, 39.9, 34.4, 27., 29., 33.},
 {46.8, 44.1, 44.2, 24.7, 33.1, 19.7}, {55.2, 38.1, 35.1, 43.1, 29.7, 29.1},
 {50.2, 41.3, 38.8, 39.9, 26.3, 23.}, {48.6, 41.6, 43.2, 32.8, 32., 34.7},
 {63.8, 46.9, 46.6, 36.6, 33.9, 29.8}, {65.8, 48.6, 47., 36.6, 28.1, 24.9},
 {58.1, 45.7, 43.5, 43.3, 33.6, 32.2}, {58.8, 49.9, 47.2, 39.3, 31.6, 34.5}}
```

This file is a slightly modified version of the original file **barley**. In **modBarley**, we have somewhat rearranged the rows and columns of **barley**. The sites and varieties of the barley are as follows:

```
sites = {Waseca, Crookston, Morris, "Univ. Farm", Duluth, "Gr. Rapids"};
varieties = {Svansota, Manchuria, "No. 475", Glabron,
    Velvet, Peatland, Trebi, "No. 462", "No. 457", Wisconsin};
```

Calculate the total mean of all of the 60 yields:

```
tmean = Mean[Flatten[data]]
```
```
39.1183
```

We produce two multiway dot plots with the program `dotPlot` presented in Section 9.4.2. First, we make a multiway dot plot showing the yields of the 10 varieties at each of the 6 sites:

```
p1 = Map[dotPlot[{#[[1]]}, varieties,
    {{AbsolutePointSize[2.5], GrayLevel[0]}}, 0, 72, Automatic,
    {tmean, tmean, tmean}, DisplayFunction → Identity,
    TextStyle → {FontFamily -> "Helvetica", FontSize → 5},
    PlotLabel → StyleForm[#[[2]], FontSize → 7]] &,
  Transpose[{Transpose[data], sites}]];

p2 = GraphicsArray[Transpose[{p1}], GraphicsSpacing → 0];
```

Then we make a multiway dot plot showing the yields at the 6 sites of each of the 10 varieties:

```
p3 = Map[dotPlot[{#[[1]]}, Reverse[sites],
    {{AbsolutePointSize[2.5], GrayLevel[0]}}, 0, 72, Automatic,
    {tmean, tmean, tmean}, DisplayFunction → Identity,
    TextStyle → {FontFamily -> "Helvetica", FontSize → 5},
    AspectRatio → 0.5, PlotLabel → StyleForm[#[[2]], FontSize → 7]] &,
  Reverse[Transpose[{Map[Reverse, data], varieties}]]];

p4 = GraphicsArray[Transpose[{p3}], GraphicsSpacing → 0];
```

Now we show both multiway dot plots side by side:

```
Show[GraphicsArray[{p2, p4}]];
```

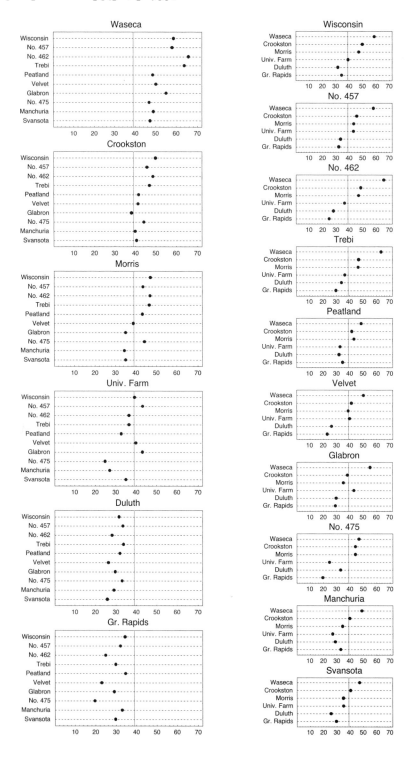

The multiway dot plot of the first column can be used to infer how the yields of the varieties vary within each site and how the yields vary among sites in general. The multiway dot plot of the second column can be used to infer how the yield of each variety varies among the sites. The grey line in both plots is the total mean. If we want to do detailed comparisons with the data, a multiway dot plot is among the best ways to show the data. The barley data is also considered in Sections 9.4.3, p. 259; 10.2.1, p. 274; and 10.2.2, p. 276.

9.4.3 Box-and-Whisker Plots

In the `Statistics`StatisticsPlots`` *package:*

`BoxWhiskerPlot[data]` (❆5) Show `data` as a box-and-whisker plot

Options:

`BoxQuantile` If set to *a*, a box shows data from $(0.5 - a)$-quantile to $(0.5 + a)$-quantile; examples of values: `0.25, 0.4`

`BoxLabels` Labels for the boxes; examples of values: `Automatic, {"Class A", "Class B", "Class C"}`

`BoxOrientation` Orientation of the graph; possible values: `Vertical, Horizontal`

`BoxOutliers` Whether to indicate outliers; possible values: `None, All, Automatic`

`BoxOutlierShapes` Shapes of the outliers; examples of values: `Automatic, Plot⌕ Symbol[Star]`

`BoxStyle` Colors of the boxes; examples of values: `Hue[0], {Hue[0], Hue[1/3], Hue[2/3]}`

`BoxLineStyle` Style of the lines in the graph; examples of values: `Automatic, Hue[2/3]`

`BoxMedianStyle` Additional styles for the median line; examples of values: `Auto⌕ matic, AbsoluteThickness[2]`

`BoxExtraSpacing` Extra space between the boxes; examples of values: `0, 0.1`

Before we explain the box-and-whisker plot, we present an example by considering the same barley data we investigated in Section 9.4.2, p. 257:

```
data = Import[
   "/Users/ruskeepaa/Documents/MNData/visdata/modBarley", "Table"];
```

First we plot the first column and show the yields of barley at Waseca:

```
d = Transpose[data][[1]] // Sort
```

```
{46.8, 47.3, 48.6, 48.9, 50.2, 55.2, 58.1, 58.8, 63.8, 65.8}
```

```
<< Statistics`StatisticsPlots`
```

```
BoxWhiskerPlot[d, BoxLabels → "Waseca"];
```

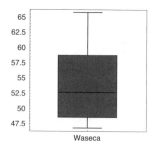

A box-and-whisker plot is simply a way to show the quartiles and the minimum and maximum of the data. In our example, these statistics are as follows:

```
{Min[d], Quantile[d, 0.25], Median[d], Quantile[d, 0.75], Max[d]}
```

`{46.8, 48.6, 52.7, 58.8, 65.8}`

The horizontal line inside the box is the median: the 0.5-quantile. Both below and above the median we have 50% of the data. The bottom and top of the box are at the 0.25- and 0.75-quantiles, so that inside the box we have 50% of the data. Both below and above the box, we have 0.25% of the data. The bottom and top horizontal lines of the "whiskers" are at the minimum and maximum of the data. This kind of plot gives a quick overview of the extent of a data set.

Next we plot all of the barley data:

```
BoxWhiskerPlot[data,
    BoxOrientation → Horizontal, AspectRatio → 0.4, BoxLabels →
    {Waseca, Crookston, Morris, "Univ. Farm", Duluth, "Gr. Rapids"}];
```

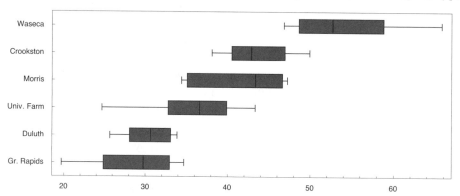

Now we plot the animal brain weight data we considered in Sections 9.3.1, p. 245, and 9.4.1, p. 255:

```
brain =
Transpose[Take[Import["/Users/ruskeepaa/Documents/MNData/visdata/
        modAnimal", "Table"], {38, 48}]][[4]] // Sort
```

`{56., 81., 98.2, 115., 119.5, 157., 169., 175., 180., 325., 419.}`

```
BoxWhiskerPlot[brain, BoxOrientation → Horizontal,
    AspectRatio → 0.2, BoxLabels → "brain", BoxOutliers → Automatic,
    BoxOutlierShapes → MakeSymbol[{AbsolutePointSize[4], Point[{0, 0}]}]];
```

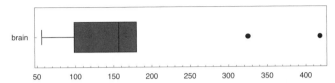

Here we indicated outliers with points. An outlier is a value beyond 1.5 times the interquantile range from the edge of the box. A far outlier is a value beyond 3 times the interquantile range. If the value of **BoxOutliers** is **All**, then all outliers are plotted in the same way, but if the value is **Automatic**, then near and far outliers are plotted differently as determined by the **BoxOutlierShapes** option. The shapes are defined in the same way as for **MultipleListPlot** (see Section 9.2.2, p. 242).

9.5 Comparing Dependent Variables

9.5.1 Pairwise Scatter Plots

Thus far we have presented plotting methods for showing each dependent variable separately—typically as time series but also as bar charts or dot plots. Here, in Section 9.5, we present some plotting methods that show two or even more dependent variables in the same plot. Such plots are useful when studying relationships among a number of dependent variables. We study pairwise scatter plots, labeled plots, and quantile-quantile plots.

In a scatter plot, we plot one variable against another variable. Such a plot may yield valuable information about the connections between the variables. A pairwise scatter plot, which is also called a scatter plot matrix or a correlation plot, is a collection of plots in which each plot shows one variable against another variable. A scatter plot matrix is among the best ways to illustrate multidimensional data.

As an example, we consider the data file **environmental**, which contains 111 observations of ozone, radiation, temperature, and wind in New York City from May to September of 1973. The data are from Cleveland (1993) (see Section 9.3.1, p. 245) and can be found on the CD-ROM of this book (we also consider this data set in Section 27.6.3, p. 782, when presenting local regression). First we read the data:

```
data = Rest[Import["/Users/ruskeepaa/
        Documents/MNData/visdata/environmental", "Table"]];
```

(**Rest** drops the first row containing the headings of the columns.) For example, the first row is as follows:

```
data[[1]]
```

```
{1, 41, 190, 67, 7.4}
```

Extract the columns of the data:

```
{no, ozone, radiation, temperature, wind} = Transpose[data];
```

We are interested in how ozone depends on the other variables. Form some pairs from the columns:

```
tempoz = Transpose[{temperature, ozone}];
radioz = Transpose[{radiation, ozone}];
windoz = Transpose[{wind, ozone}];
```

Plot those pairs:

```
Block[{$DisplayFunction = Identity}, p1 =
   ListPlot[radioz, Frame → True, FrameLabel → {"radiation", "ozone"}];
  p2 = ListPlot[tempoz, Frame → True,
     FrameLabel → {"temperature", "ozone"}, AxesOrigin → {56, 0}];
  p3 = ListPlot[windoz, Frame → True, FrameLabel → {"wind", "ozone"}];];

Show[GraphicsArray[{p1, p2, p3}]];
```

We can see that ozone values are high when radiation or temperature is high (a positive correlation) or wind is low (a negative correlation); however, a high value of radiation does not necessarily mean a high ozone value.

In the **Statistics`StatisticsPlots`** *package:*

PairwiseScatterPlot[data] (❀5) Plot multidimensional data as a pairwise scatter plot

Options:

DataLabels Labels for the variables; examples of values: **None**, **{"Class A", Class B", "Class C"}**

DataTicks Ticks for the variables; default value: **None**

DataSpacing Space between the subgraphs; examples of values: **0, 0.05**

DataRanges Ranges for the data; default value: **All**

PlotStyle Style of the points; examples of values: **Automatic, AbsolutePoint、Size[1]**

With **PairwiseScatterPlot**, we can plot the whole scatter plot matrix. First we form the data:

```
data2 = Transpose[{radiation, temperature, wind, ozone}];
```

Then we ask for the scatter plot matrix (of these 16 plots, we earlier plotted the first 3 in the top row):

```
<< Statistics`StatisticsPlots`
```

```
PairwiseScatterPlot[data2, DataTicks → Automatic,
    DataSpacing → 0.04, PlotStyle → AbsolutePointSize[1],
    DataLabels → {"radiation", "temperature", "wind", "ozone"}];
```

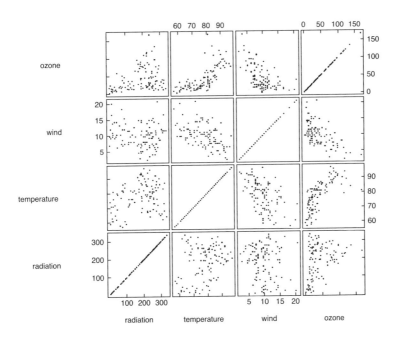

9.5.2 Labeled Plots

In the `Graphics`Graphics`` *package:*

`LabeledListPlot[data]` Show `data` as labels and points
`TextListPlot[data]` Show `data` as labels without points

Data can be given in one of the following forms:

$\{Y_1, Y_2, …\}$ Show the points $\{1, Y_1\}, \{2, Y_2\}, …$ as 1, 2, …
$\{\{x_1, Y_1\}, \{x_2, Y_2\}, …\}$ Show the given points as 1, 2, …
$\{\{x_1, Y_1, a_1\}, \{x_2, Y_2, a_2\}, …\}$ Show the given points as $a_1, a_1, …$

Labeled plots can be seen as special scatter plots in which each data point has a label. The automatic labels are consecutive integers, but we can specify our own labels (a_i). In **LabeledListPlot**, the labels are next to the points. In **TextListPlot**, the labels are centered at the data points.

As an example, we plot the body and brain weights of some animals. This data set has already been considered in Sections 9.3.1, p. 245, and 9.4.1, p. 255. We read the file and take only rows 38 through 48:

```
{no, name, body, brain} = Transpose[
  Take[Import["/Users/ruskeepaa/Documents/MNData/visdata/modAnimal",
    "Table"], {38, 48}]]
```

```
{{37, 38, 39, 40, 41, 42, 43, 44, 45, 46, 47},
 {RoeDeer, Goat, Kangaroo, GrayWolf, Sheep, GiantArmadillo, GraySeal,
  Jaguar, BrazilianTapir, Donkey, Pig}, {14830, 27660, 35000,
  36330, 55500, 60000, 85000, 100000, 160000, 187100, 192000},
 {98.2, 115., 56., 119.5, 175., 81., 325., 157., 169., 419., 180.}}
```

```
<< Graphics`Graphics`
```

```
LabeledListPlot[Transpose[{body, brain, name}],
  Frame → True, PlotRange → {{0, 225000}, {0, 460}}];
```

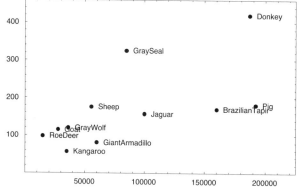

As can be seen, overlapping labels may be a problem with labeled list plots.

9.5.3 Quantile-Quantile Plots

In the **Statistics`StatisticsPlots`** *package:*

QuantilePlot[data1, data2] (⌘5)   Create a quantile-quantile plot for the two data
  sets

Options:

SymbolShape   Shape of the points; examples of values: **Automatic**, **PlotSymbol[Star]**
SymbolStyle   Style of the points; examples of values: **Automatic**, **Hue[0]**
PlotJoined   Whether the points are joined with lines; possible values: **False**, **True**
PlotStyle   Style of the joining line; examples of values: **Automatic**, **Hue[2/3]**
ReferenceLineStyle   Style of the reference line; examples of values: **Automatic**,
  GrayLevel[0.5]

A *quantile-quantile plot* or a *q-q plot* (see Cleveland 1993, p. 21) is a powerful method for comparing the distributions of two or more sets of univariate data. The plot simply shows the quantiles of one data set against the quantiles of another data set. If the resulting points are close to a line with a slope of 1, this supports the hypothesis that the distributions of the two data sets are the same.

QuantilePlot first determines the interpolated quantiles of the shorter of the two data sets at the equivalent positions in the longer data set. It then plots the two sets of quantiles against each other. **SymbolShape** is used in the same way as it was for **MultipleListPlot** in Section 9.2.2, p. 242.

As an example, generate data sets from a Student t-distribution with parameter 10 and from the standard normal distribution:

```
<< Statistics`NormalDistribution`

ran1 = RandomArray[StudentTDistribution[10], {2000}];
ran2 = RandomArray[NormalDistribution[0, 1], {2000}];
```

How close are the two distributions? Prepare a q-q plot:

```
QuantilePlot[ran1, ran2, AspectRatio → Automatic,
  Ticks → {Range[-6, 6], Range[-6, 6]},
  SymbolShape → MakeSymbol[{AbsolutePointSize[1], Point[{0, 0}]}]];
```

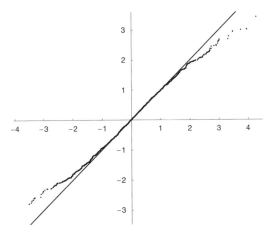

We see that the points are not sufficiently close to the reference line for supporting the hypothesis that the distributions are the same. Indeed, while we know that the t-distribution approaches the normal distribution as the parameter approaches infinity, the value 10 simply is not large enough. The tails of the t-distribution are fatter than the tails of the normal distribution.

10

3D Graphics for Data

Introduction

Cartesian geometry was a revolutionary discovery in mathematics. It seems that Descartes conceived the idea as he lay in bed watching a fly moving about in the air. In a flash, he realized that the fly's positions at any moment could be pinpointed by giving its distances from three mutually perpendicular planes in the bedroom in which he was resting.

In this chapter we consider data having two independent variables and one dependent variable. We try to create 3D graphics for such data.

ListPlot3D, **ListContourPlot**, and **ListDensityPlot** make up one collection of commands to be tried. An effective plot may be a 3D bar chart done with **BarChart3D**. Also, a scatter plot made with **ScatterPlot3D** may sometimes be effective, particularly if the plot can be viewed stereographically.

10.1 Surface, Contour, and Density Plots

10.1.1 Regular Data

■ Strongly Regular Data

> **ListPlot3D[data]** Plot **data** as a surface plot
> **ListContourPlot[data]** Plot **data** as a contour plot
> **ListDensityPlot[data]** Plot **data** as a density plot
>
> Data are given in the matrix form:
> $\{\{z_{11}, ..., z_{1n}\}, ..., \{z_{m1}, ..., z_{mn}\}\}$ (each row corresponds to a fixed value of y)

Note that only the heights or z values are given; values for x and y cannot be given. In fact, the three commands are designed to plot data in which the x *and* y *coordinates are equally spaced.* Taking this into account, we see immediately that the x and y values are actually unnecessary; all we need is to be able to tell the ranges of the x and y coordinates. This can be done with the **MeshRange** option, which is explained shortly.

For data of a rectangular form that also contain the x and y coordinates (which need not be equally spaced), we have the separate command **ListSurfacePlot3D** in a package, which is explained below. For data that are not of a rectangular form, we have **TriangularSurfacePlot** in a package (see Section 10.1.2, p. 272).

Here are some simple examples:

```
data1 = {{0, 0, 1, 1, 0}, {1, 0, 1, 1, 1}, {2, 1, 2, 2, 1}, {2, 1, 2, 2, 2}};
```

```
Block[{$DisplayFunction = Identity},
  p1 = ListPlot3D[data1, AxesLabel → {"x", "y", ""}];
  p2 = ListContourPlot[data1,
    FrameLabel → {"x", "y"}, RotateLabel → False];
  p3 = ListDensityPlot[data1, FrameLabel → {"x", "y"},
    RotateLabel → False];]
```

```
Show[GraphicsArray[{p1, p2, p3}]];
```

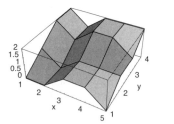 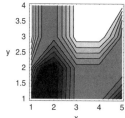 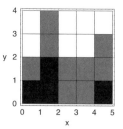

In the surface plot, the points are connected by surface pieces. The points themselves are not shown, but they are at the corners of the pieces. The first row {0, 0, 1, 1, 0} of the data is in front (parallel to the x axis), the second row {1, 0, 1, 1, 1} next, and so on.

In the contour plot, there are 10 contours that correspond to 10 equally spaced values between the minimum and maximum values. In the density plot, each data point has a rectangle that is colored by the value of the point; the points themselves can be thought of as being in the middle of the rectangles.

■ **Some Comments**

The three plotting commands are closely related to the corresponding commands **Plot3D**, **ContourPlot**, and **DensityPlot** for functions not only if we compare the resulting plots but also if we consider the methods they use. In fact, the commands for functions first sample the function on a 25×25 grid, whereas this stage is not needed in the commands for data. In all other ways the commands function in the same way. Indeed, we get the same result whether we plot a function or its values at 25×25 points.

The three commands also have almost the same options as the corresponding commands for functions: the options **Compiled** and **PlotPoints** are lacking, but the new option **MeshRange** is available (with the default value **Automatic**). Remember from Chapter 8 that the ticks of contour and density plots are set with **FrameTicks**. With **Contours**, we can define either the number of contours or the values for which contours of constant value are calculated. The form of the surface plot can be controlled with **BoxRatios** and the form of the contour and density plots with **AspectRatio**; for both options, the value **Automatic** specifies that one unit in each axis has the same length. Next we explain the use of **MeshRange** and **ColorFunction**.

■ Adjusting *x* and *y* Coordinates

> **ListPlot3D[data, MeshRange → {{xmin, xmax}, {ymin, ymax}}]** *x* and *y* values are
> evenly spaced in the given intervals (similarly for **ListContourPlot** and **List**
> **DensityPlot**)

As stated earlier, the data points contain only the *z* values. To plot ticks on the axes, *Mathematica* assumes that the *x* and *y* values are evenly spaced and are, in fact, the integers 1, 2, 3, …; you can see this from the plots above. If the true *x* and *y* values are not these integers, the option **MeshRange** should be used to input the true ranges within which the points lie. For example, suppose that *x* values are 0, 1, 2, 3, and 4 and *y* values 0, 2, 4, and 6; then the *x* range is {0, 4} and the *y* range is {0, 6}. We can get the correct *x* and *y* coordinates as follows:

```
Block[{$DisplayFunction = Identity},
  p1 = ListPlot3D[data1, MeshRange → {{0, 4}, {0, 6}}];
  p2 = ListContourPlot[data1, MeshRange → {{0, 4}, {0, 6}}];
  p3 = ListDensityPlot[data1, MeshRange → {{-0.5, 4.5}, {-1, 7}}]];

Show[GraphicsArray[{p1, p2, p3}]];
```

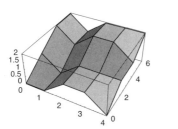 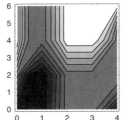 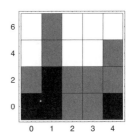

In the density plot, we extended the mesh range of *x* by 0.5 from {0, 4} to {–0.5, 4.5} and the mesh range of *y* by 1 from {0, 6} to {–1, 7} so that the ticks are at the centers of the sides of the rectangles (this is the same technique used for **DensityPlot** in Section 8.2.3, p. 227).

■ Coloring the Plots According to Height

We have already used the **ColorFunction** option in Sections 8.1.1, p. 206; 8.1.7, p. 217; 8.2.2, p. 223; and 8.2.3, p. 226. Now we use this option to color data plots according to the values of the data points.

```
data2 = Table[Random[Integer, {8 i + 5 j, 10 i + 7 j}], {i, 10}, {j, 15}];
```

```
Block[{$DisplayFunction = Identity},
  p1 = ListPlot3D[data2, BoxRatios → {15, 10, 10},
    ViewPoint → {-1.8, -2.5, 1.4}, ColorFunction → (Hue[(1 - #)] &)];
  p2 = ListContourPlot[data2, AspectRatio → Automatic,
    ColorFunction → (Hue[0.85 (1 - #)] &)];
  p3 = ListDensityPlot[data2, AspectRatio → Automatic, ColorFunction →
    (Hue[0.85 (1 - #)] &), MeshRange → {{0.5, 15.5}, {0.5, 10.5}}];]
```

```
Show[GraphicsArray[{p1, p2, p3}]];
```

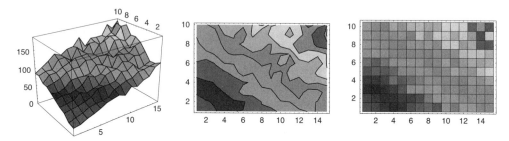

In general, a surface plot clearly shows the behavior of the data in a whole but may have the problem that the surface pieces are somewhat dominant: we may not recognize that the data are at the crossing points of the mesh lines. The contour lines for data are necessarily somewhat wild, because data normally do not behave as smoothly as a function. A density plot is a simple and good description of the data.

■ Weakly Regular Data

In the **Graphics`Graphics3D`** *package:*

ListSurfacePlot3D[data]   Plot **data** as a surface

Data are given in the rectangular form:
{ {{x₁₁ , y₁₁ , z₁₁ }, …, {x₁ₙ , y₁ₙ , z₁ₙ }},

    . . .

    {{xₘ₁ , yₘ₁ , zₘ₁ }, …, {xₘₙ , yₘₙ , zₘₙ }} }

This command has the same options as **Graphics3D** (note that **Axes** and **BoxRatios** have the default values **False** and **Automatic**). The data have to be in a rectangular form, but the x and y coordinates need not be evenly spaced. For example:

```
data3 = {{{0.1, 0.2, 0}, {1.2, 0.4, 0}, {2.2, 0.1, 1}, {2.8, 0.3, 1},
    {4.1, 0.0, 0}}, {{0.0, 2.3, 1}, {1.1, 1.7, 0}, {1.9, 2.5, 1},
    {3.1, 2.1, 1}, {3.8, 1.9, 1}}, {{0.2, 3.6, 2}, {0.8, 3.9, 1},
    {2.1, 4.3, 2}, {3.0, 4.0, 2}, {4.2, 3.8, 1}}, {{0.1, 6.1, 2},
    {1.2, 5.7, 1}, {2.0, 5.8, 2}, {2.9, 6.2, 2}, {4.1, 5.9, 2}}};
```

```
<< Graphics`Graphics3D`
```

```
ListSurfacePlot3D[data3, Axes → True];
```

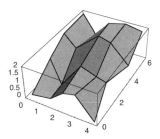

The command can also produce more complex surfaces, like the one produced with **ParametricPlot3D**.

Note that we can also form an interpolating surface with **ListInterpolation** or **Interpolation** (see Section 21.2.2, p. 558) (the data then has to be somewhat more regular than in the example above: in a column each point has the same *x* coordinate, and in a row each point has the same *y* coordinate).

10.1.2 Irregular Data

▪ Plotting Vertical Lines

A simple way to illustrate 3D data is to plot vertical lines. A stereographic pair of plots gives a clear view:

```
SeedRandom[5];
data4 = With[{p = N[Pi]}, Table[{x = Random[Real, {0, p}],
    y = Random[Real, {0, p}], Sin[x] + Sin[y]}, {30}]];

p1 = Show[Graphics3D[{Map[Line[{{#[[1]], #[[2]], 0}, #}] &, data4],
    Hue[0], AbsolutePointSize[2], Map[Point, data4]}], Axes → True,
    Ticks → None, BoxRatios → Automatic, DisplayFunction → Identity];

p2 = Show[p1, ViewPoint → {1.4, -2.3, 2.0}];

Show[GraphicsArray[{p1, p2}, GraphicsSpacing → 0]];
```

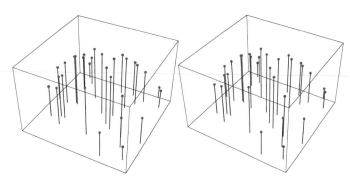

■ **Plotting a Surface**

In the `DiscreteMath`ComputationalGeometry`` *package:*

`TriangularSurfacePlot`[3D data]   Form the triangular surface plot
`PlanarGraphPlot`[2D data]   Plot the Delaunay triangulation

Data are given, correspondingly, in the following form:
`{{x₁ , y₁ , z₁}, {x₂ , y₂ , z₂}, ..., {xₙ , yₙ , zₙ}}`
`{{x₁ , y₁}, {x₂ , y₂}, ..., {xₙ , yₙ}}`

These commands can be used even if the data are not in the rectangular form discussed in Section 10.1.1. The former command has the same options as `Graphics3D`; the latter command has the same options as `Graphics` but has the additional option `LabelPoints`.

The first command works by first calculating the Delaunay triangulation for the (x, y) points. As an example, we consider `data4` and form a surface plot and the Delaunay triangulation:

```
<< DiscreteMath`ComputationalGeometry`

Block[{$DisplayFunction = Identity},
 p1 = TriangularSurfacePlot[data4];
 p2 = PlanarGraphPlot[Map[Drop[#, -1] &, data4], LabelPoints → False];]

Show[GraphicsArray[{p1, p2}]];
```

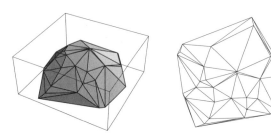

The `ExtendGraphics` collection of packages by Wickham-Jones (1994) (see Section 8.2.2, p. 225) contains packages for plotting irregular 3D data as surfaces and as contour plots as well as commands for irregular interpolation.

10.2 Bar Charts and Scatter Plots

10.2.1 Bar Charts

In the `Graphics`Graphics3D`` *package:*

`BarChart3D[data]`   Plot a 3D bar chart

Data are given in either of the following forms:
`{{z₁₁ , ..., z₁ₙ}, ..., {zₘ₁ , ..., zₘₙ}}` (each row corresponds to a fixed value of x)
`{{{z₁₁ ,style₁₁}, ..., {z₁ₙ ,style₁ₙ}}, ..., {{zₘ₁ ,styleₘ₁}, ..., {zₘₙ ,styleₘₙ}}}`

We use the same example as in Section 10.1.1, p. 267:

```
data1 = {{0, 0, 1, 1, 0}, {1, 0, 1, 1, 1}, {2, 1, 2, 2, 1}, {2, 1, 2, 2, 2}};
```

Some aspects of **BarChart3D** are illustrated by the following three plots:

```
xticks = Transpose[{Range[5], Range[0, 4]}]
```

```
{{1, 0}, {2, 1}, {3, 2}, {4, 3}, {5, 4}}
```

```
yticks = Transpose[{Range[4], Range[0, 6, 2]}]
```

```
{{1, 0}, {2, 2}, {3, 4}, {4, 6}}
```

```
<< Graphics`Graphics3D`
```

```
Block[{$DisplayFunction = Identity},
  p1 = BarChart3D[data1, AxesLabel -> {"x", "y", ""}];
  p2 = BarChart3D[Transpose[data1], AxesLabel -> {"x", "y", ""}];
  p3 = BarChart3D[Transpose[data1],
    BoxRatios -> {5, 6, 3}, ViewPoint -> {0.8, -3, 1.2},
    AxesLabel -> {"x", "y", ""}, Ticks -> {xticks, yticks, {0, 1, 2}}];]
```

```
Show[GraphicsArray[{p1, p2, p3}]];
```

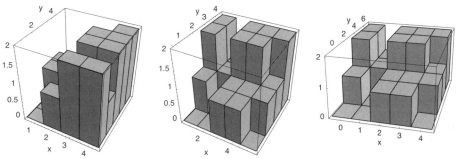

As can be seen, a bar chart is an efficient way of illustrating 3D data.

In the first plot, we see that the x and y axes are the usual ones for 3D graphics. However, the way the data are interpreted is different, for example, from **ListPlot3D**. With **ListPlot3D**, the rows of the data matrix correspond to fixed y values, whereas in **BarChart3D**, the rows correspond to fixed x values (e.g., the last row {2, 1, 2, 2, 2} is on the righthand side).

In the second plot, we have transposed the data, and then the result corresponds to the plots we obtained in Section 10.1.1. In the third plot, we have shown the bars from a different viewpoint, used different box ratios, and defined custom ticks. The default is that the x scale goes from 1 to m and the y scale from 1 to n, and if we want different ticks, we have to specify what symbols to use (e.g., here we decided that the number 0 is to be used for $x = 1$).

■ Coloring the Bars According to Height

In the second form of the data mentioned in the box above, we define the style of each bar separately. This requires the option setting **Lighting → False**. As an example, we use the same **data2** as in Section 10.1.1, p. 269, but we add a color for each datum:

```
data5 = Partition[Map[{#, Hue[1 - # / 140]} &, Flatten[data2]], 15];
```

Then we plot the bar chart using both the default illumination and our own colors:

```
Block[{$DisplayFunction = Identity},
  p1 = BarChart3D[Transpose[data2],
    BoxRatios → {15, 10, 10}, ViewPoint → {-1.8, -2.5, 1.4}];
  p2 = BarChart3D[Transpose[data5], BoxRatios → {15, 10, 10},
    ViewPoint → {-1.8, -2.5, 1.4}, Lighting → False];]

Show[GraphicsArray[{p1, p2}]];
```

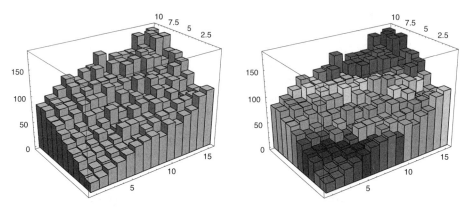

The first plot is attractive and stylish, but the second plot has the advantage that bars of about the same height can be easily identified because of the colors.

■ The Barley Example

In Sections 9.4.2, p. 257, and 9.4.3, p. 259, we considered barley yields data. Now we illustrate the same data with a bar chart (for information about reading data, see Section 4.2.1, p. 90):

```
data6 = Import[
  "/Users/ruskeepaa/Documents/MNData/visdata/modBarley", "Table"];

sites = {Waseca, Crookston, Morris, "Univ. Farm", Duluth, "Gr. Rapids"};
varieties = {Svansota, Manchuria, "No. 475", Glabron,
  Velvet, Peatland, Trebi, "No. 462", "No. 457", Wisconsin};

xticks = Transpose[{Range[6], sites}];
yticks = Transpose[{Range[10], varieties}];

xgrid = Range[1.5, 5.5, 1];
ygrid = Range[1.5, 9.5, 1]; zgrid = Range[10, 60, 10];
grids = {{{-1, 0, 0}, {ygrid, zgrid}},
  {{0, 1, 0}, {xgrid, zgrid}}, {{0, 0, -1}, {xgrid, ygrid}}};

<< Graphics`Graphics3D`
```

```
BarChart3D[Transpose[data6], BoxRatios → {6, 10, 7},
   ViewPoint → {1.8, -2.4, 1.6}, AxesEdge → {Automatic, {1, -1}, Automatic},
   Ticks → {xticks, yticks, Range[10, 60, 10]}, FaceGrids → grids];
```

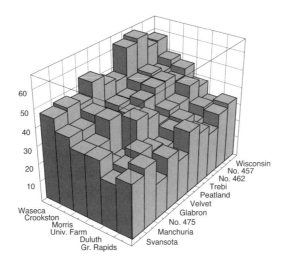

■ Options of BarChart3D

XSpacing Space between bars in x direction; examples of values: **0, 0.1**
YSpacing Space between bars in y direction; examples of values: **0, 0.1**
SolidBarEdges Whether to draw the edges of bars; possible values: **True, False**
SolidBarEdgeStyle Style of the edges of bars; examples of values: **GrayLevel[0]**,
  {AbsoluteThickness[0.75], RGBColor[0,0,1]}
SolidBarStyle Style of the faces of bars (requires **Ligthing → False**); examples of
  values: **GrayLevel[0.5], RGBColor[1,0,0]**

In addition to these options, **BarChart3D** has the options of **Graphics3D**. However, **BarChart3D** uses the default settings **Axes → Automatic**, **BoxRatios → {1,1,1}** (a cube), and **PlotRange → All**.

10.2.2 Scatter Plots

In the **Graphics`Graphics3D`** *package:*

ScatterPlot3D[data] Plot **data** as points
ScatterPlot3D[data, PlotJoined → True] Plot **data** as joining lines

Data are given in the following form:
{{x₁ , y₁ , z₁ }, {x₂ , y₂ , z₂ }, …, {xₙ , yₙ , zₙ }}

ScatterPlot3D has the same options as **Graphics3D** (**Axes** has now the default value **True**) plus the new options **PlotJoined** (with default value **False**) and **PlotStyle** (with default value **GrayLevel[0]**). **PlotStyle** controls the style of points if points are drawn and the style of lines if lines are drawn.

■ Example 1: Barley

The barley data was considered in Section 10.2.1, p. 274. For the scatter plot, we also need the x and y coordinates of the data:

```
Short[
 data7 = Flatten[Table[{j, i, data6[[i, j]]}, {i, 1, 10}, {j, 1, 6}], 1], 2]
{{1, 1, 47.3}, {2, 1, 40.5}, {3, 1, 35.},
 <<54>>, {4, 10, 39.3}, {5, 10, 31.6}, {6, 10, 34.5}}
```

```
<< Graphics`Graphics3D`
```

```
ScatterPlot3D[data7, BoxRatios → {6, 10, 7},
 PlotStyle → AbsolutePointSize[2]];
```

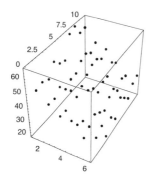

As can be seen, **ScatterPlot3D** is somewhat weak for the illustration of 3D data that are quite evenly spaced in the plotting area; this is because it is difficult to get a sufficiently clear idea of the locations of the points. **ScatterPlot3D** is better for illustrating data that have a more or less clear pattern or that are clustered to form data clouds; see the examples below.

■ Self-Made Scatter Plots

```
Show[Graphics3D[{AbsolutePointSize[d], Map[Point, data]}]]   Plot points
Show[Graphics3D[{AbsoluteThickness[d], Line[data]}]]   Plot lines
Show[Graphics3D[{ AbsoluteThickness[d1], Line[data], AbsolutePoint
 Size[d2], Map[Point, data]}]]   Plot points and lines
```

Scatter plots are very easy to produce with graphics primitives. In the place of points we can also use, for example, cuboids:

```
data8 = Table[{t Cos[t], t - t Sin[t], t}, {t, 5 Pi / 2, 5 Pi, Pi / 10.}];
```

```
Show[Graphics3D[{{RGBColor[1, 0, 0], Line[data8]},
    Map[Cuboid[# - 0.7 {1, 1, 1}, # + 0.7 {1, 1, 1}] &, data8]}]];
```

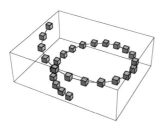

■ Example 2: Galaxy

We have in a text file **galaxy** various data about NGC 7531, a spiral galaxy in the Northern Hemisphere. The data are, again, from Cleveland (1993) (see Section 9.3.1, p. 245) and can be found on the CD-ROM accompanying this book.

```
data9 = Rest[
    Import["/Users/ruskeepaa/Documents/MNData/visdata/galaxy", "Table"]];
```

(**Rest** drops the first row, which contains the headings of the columns.) The file contains 323 rows. The first row is as follows:

```
data9[[1]]
```

```
{3, 8.46279, -38.1732, 102.5, 39.1, 1769}
```

The first item is the observation number (ranging from 3 to 417 but having missing observations), the second and third items the coordinates of a point of the galaxy, and the sixth item the velocity of the galaxy at the given point.

The next step is to extract the columns from the data:

```
{no, eastwest, southnorth, slitangle, radialposition, velocity} =
    Transpose[data9];
```

The velocity varies between the following numbers (given in kilometers per second):

```
{Min[velocity], Max[velocity]}
```

```
{1409, 1775}
```

Now we can plot the coordinates of the points where the velocity was measured. We see that velocities were measured along seven lines:

```
ListPlot[Transpose[{eastwest, southnorth}],
    PlotStyle → AbsolutePointSize[1], AspectRatio → Automatic];
```

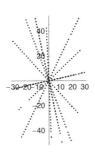

Then we form and plot the 3D data containing the coordinates and the velocities:

```
data11 = Transpose[{eastwest, southnorth, velocity}];

p = ScatterPlot3D[data11, BoxRatios → {6, 10, 10},
    ViewPoint → {-2.9, 1, 1.2}, AxesLabel → {"sn", "ew", ""}];
```

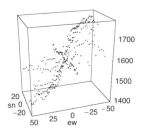

We can see that the velocity increases, going from the lower left front corner to the upper right back corner. A much better view can be obtained with a two-image stereogram (for information about this topic, see Section 7.3.2, p. 194):

```
p1 = Show[p, Ticks → None, AxesLabel → None, DisplayFunction → Identity];
p2 = Show[p1, ViewPoint → {-3.0, 0.9, 1.2}];

Show[GraphicsArray[{p1, p2}, GraphicsSpacing → 0.01]];
```

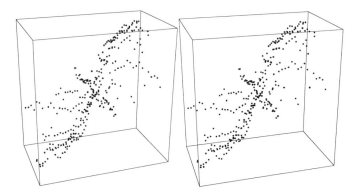

Another method for obtaining a better view is to use real-time or animated rotation (see Section 7.3.1, p. 193). Real-time rotation can be applied by using the following commands:

```
<< RealTime3D`
Show[p];
```

Then you can rotate the plot with the mouse. To go back to normal 3D plots, execute **<<Default3D`**. Animated rotation can be applied by entering the following:

```
<< Graphics`Animation`
SpinShow[p1, Frames → 50];
```

Then, double-click one of the plots with the mouse. We do not show the resulting plots here; they can be found on the CD-ROM of this book.

Numbers

Introduction

> *Here is a story about the theoretical physicists Hoyle and Bondi. It is usual in astrophysics to work with units that make all fundamental constants equal to unity. Then a conversion factor is applied to get the answer in standard units. Bondi was calculating some important number in physics and asked Hoyle for the conversion factor. "It's 10^{60}," said Hoyle. Bondi replied, "Do I multiply or divide?"*

Numbers are clear for us, but the representation and interaction of various kinds of numbers in the computer involve aspects worth careful study. Also, *Mathematica* has a richer assortment of numbers than is usually found in computer applications. We can use arbitrarily large integers, we can calculate with exact rational numbers, and we can ask to calculate with real numbers containing as many digits as we want.

We also study the precision and accuracy of real numbers. These relate to the relative and absolute error in the result. In Section 11.3, we study controlling the precision and accuracy of numerical routines in *Mathematica*. Note that we do not consider number theory in this book.

11.1 Introduction to Numbers

11.1.1 Types of Numbers

■ Four Types of Numbers

Mathematica has four types of numbers: *integers* like 38254, *rationals* like 41/7, *reals* like 58.723, and *complexes* like 9.45 + 3 *i*. The type of the number can be asked with **Head**:

> **{Head[38254], Head[41 / 7], Head[58.723], Head[9.45 + 3 I]}**
>
> {Integer, Rational, Real, Complex}

Although *Mathematica* has the four basic types of numbers, it recognizes more types with special tests. For example, *Mathematica* knows that $\sqrt{2}$ is an algebraic number:

> **Sqrt[2] ∈ Algebraics**   True

These tests are often used in simplifying expressions, and so we consider them in Section 12.2.1, p. 304.

■ Manipulating Integers

Prime[n] nth prime (when **Prime[1]** is 2)

FactorInteger[n] List of the prime factors of **n** together with their exponents

Divisors[n] List of the integers that divide **n**

Mod[n, m] **n** modulo **m** (remainder on division of **n** by **m**)

IntegerDigits[n] List of the decimal digits in **n**

Factor an integer and check that the result is correct:

> **FactorInteger[13689]**   {{3, 4}, {13, 2}}
>
> **3 ^ 4 * 13 ^ 2**   13689

For an application of **FactorInteger**, see Section 1.1.1, p. 2. If we allow complex factors, then even some primes can be factored. *Mathematica* uses the term Gaussian integer, which means a complex number with integer real and imaginary parts. We factor 13 and check the result:

> **FactorInteger[13, GaussianIntegers → True]**
>
> {{-i, 1}, {2 + 3 i, 1}, {3 + 2 i, 1}}
>
> **(-I) (2 + 3 I) (3 + 2 I)**   13

■ Manipulating Real Numbers

If an expression contains a real number, the whole result is a real number:

> **3. + (3 / 5) ^ 2 - Sin[π / 3] + Log[2]**   3.18712

To get a decimal result, we can also use **N** (see Section 11.1.2, p. 281). If all numbers are exact (i.e., integers, rationals or other exact quantities), the result is exact, too:

$$3 + (3/5)^2 - Sin[\pi/3] + Log[2] \qquad \frac{84}{25} - \frac{\sqrt{3}}{2} + Log[2]$$

Very large or very small reals can be entered by powers of 10:

$$325.7 / 10^{\wedge}8 \qquad 3.257`*^-6$$

We can also use the special form **325.7*^-8**.

Round[x] Integer closest to **x**

Floor[x] Greatest integer less than or equal to **x**

Ceiling[x] Smallest integer greater than or equal to **x**

IntegerPart[x] Integer part of **x**

FractionalPart[x] Fractional part of **x**

Chop[expr] Replace all real numbers in **expr** with magnitude less than 10^{-10} with 0

Chop[x, dx] Replace all real numbers in **expr** with magnitude less than **dx** with 0

Rationalize[x] Rational number close to **x**

ContinuedFraction[x, n] Continued fraction representation of **x** with **n** terms

RealDigits[x] List of the digits in **x** together with the number of digits to the left of the decimal point

---

Note that the result of **Round[x.5]** is the nearest even integer. Thus, **Round[2.5]** gives 2, but **Round[3.5]** gives 4.

■ Manipulating Complex Numbers

Re[z] Real part

Im[z] Imaginary part

Conjugate[z] Complex conjugate

Abs[z] Absolute value

Arg[z] Argument ϕ such that $z = |z| e^{i\phi}$

ComplexExpand[z] Expand **z** to real and imaginary parts

$$z = 3 - 2 I;$$

$$\{Re[z], Im[z], Conjugate[z], Abs[z], Arg[z]\}$$

$$\left\{3, -2, 3 + 2 i, \sqrt{13}, -ArcTan\left[\frac{2}{3}\right]\right\}$$

$$ComplexExpand[(-1)^{\wedge}(1/3)] \qquad \frac{1}{2} + \frac{i\sqrt{3}}{2}$$

$$ComplexExpand[Log[2 I]] \qquad \frac{i\pi}{2} + Log[2]$$

11.1.2 More about Real Numbers

Here we study real numbers in some detail, and we will continue to do so in Section 11.2, where we study especially the precision of real numbers and variable precision arithmetic.

■ Asking a Decimal Value

> **N[expr]** or **expr//N** Calculate decimal value of **expr**
> **N[expr, n]** Calculate decimal value of **expr** to **n**-digit precision

First we calculate a numerical value in the usual way:

   Sin[2] // N   0.909297

The result was calculated by using the normal floating-point numbers with 16 decimal digits of precision. Then we ask for a numerical value to 30-digit precision:

   N[Sin[2], 30]   0.909297426825681695396019865912

Note that **expr** in **N[expr, n]** must contain exact numbers or numbers of sufficiently high precision for the **n** to have an effect. Asking for, say, **N[2.3 Pi, 20]** is useless, because the expression **2.3 Pi** contains a low-precision number; the result the same as with **N[2.3 Pi]**. Indeed, normal 16-digit real numbers are used in the evaluation of an expression as soon as such a number is encountered in the expression. Ask for **N[23/10 Pi, 20]** or **N[2.3`20 Pi, 20]** instead (with **2.3`20** we give 2.3 to 20 digits of precision, as explained in Section 11.2.2, p. 289).

Form more about real numbers, see Section 11.2.

(Note that, in versions of *Mathematica* prior to 5, **N[expr, n]** is intended to be used for **n** values of at least 17. In fact, if **n** is at most 16, then *Mathematica* uses the standard decimal numbers and **n** has no effect.)

■ Adjusting the Number of Digits Shown

Mathematica normally shows 6 digits of decimal numbers; this is suitable in most cases. Sometimes we want to show more or less digits, and then we can use **NumberForm**.

> **NumberForm[expr, n]** Show **expr** using **n** digits
> **NumberForm[expr, {n, f}]** Show **expr** using **n** digits, of which **f** digits are to the right of the decimal point

   Exp[3.2]   24.5325

   NumberForm[Exp[3.2], 4]   24.53

   NumberForm[Exp[3.2], {4, 1}]   24.5

PaddedForm is a similar command but is used mainly when tabulating numbers (see Section 13.1.2, p. 328). The **PrintPrecision** option can also be used to set the number of digits shown. It can be set with the *Option Inspector:* go to **Formatting Options** ▷ **Expression Formatting** ▷ **Display Options**.

■ Extra Precision

If you are trying to find an *answer* satisfying a given requirement of precision, **N** can use extra precision in the *calculations*. However, the extra precision cannot exceed **$MaxExtra⸱ Precision**, which has a default value of 50. This means that if you ask for a value to 20-digit precision, *Mathematica* can use at most 70-digit precision during the computation. If a calculation does not succeed within this limit, we can increase the value of this constant (by typing, for example, **$MaxExtraPrecision = 100**) and then retry the calculation.

The limit of the extra precision can also appear somewhat unexpectedly. We solve an equation and then insert one solution into the equation to check that the solution is correct:

    eqn = 1 – 3 x + x^2 == 0;

    sol = Solve[eqn]

$$\left\{\left\{x \to \tfrac{1}{2}\,(3 - \sqrt{5})\right\},\ \left\{x \to \tfrac{1}{2}\,(3 + \sqrt{5})\right\}\right\}$$

    eqn /. sol[[1]]

    N::meprec : Internal precision limit $MaxExtraPrecision = 50.`

        reached while evaluating $1 - \tfrac{3}{2}\,(3 - \sqrt{5}) + \tfrac{1}{4}\,(3 - \sqrt{5})^2$. **More…**

$$1 - \tfrac{3}{2}\,(3 - \sqrt{5}) + \tfrac{1}{4}\,(3 - \sqrt{5})^2 == 0$$

Mathematica begins to investigate whether the equality is true or not but has to give up. In this example, increasing the extra precision would not help, but we can use **Simplify** to show that the solution is correct:

    % // Simplify   True

11.1.3 Constants, Units, and Elements

■ Mathematical Constants

Mathematica has the following mathematical constants:

    const = Select[Names["*"], MemberQ[Attributes[#], Constant] &]

    {Catalan, Degree, E, EulerGamma, Glaisher, GoldenRatio, Khinchin,
      MachinePrecision, Pi}

They have the following numerical values, respectively:

    const // ToExpression // N

    {0.915966, 0.0174533, 2.71828, 0.577216,
      1.28243, 1.61803, 2.68545, 15.9546, 3.14159}

Of these constants, **Degree**, **E**, **EulerGamma**, and **Pi** can also be entered as ⎡ESC⎤deg⎡ESC⎤, ⎡ESC⎤ee⎡ESC⎤, ⎡ESC⎤gg⎡ESC⎤, and ⎡ESC⎤p⎡ESC⎤, and the results are °, e, γ, and π. In traditional form, **Catalan** and **GoldenRatio** are written as C and ϕ:

    const // ToExpression // TraditionalForm

    {C, °, e, γ, Glaisher, ϕ, Khinchin, MachinePrecision, π}

Note that **Degree** is $\pi/180$ (the degrees-to-radians conversion factor), **EulerGamma** is $\lim_{m\to\infty}\left(\sum_{k=1}^{m}\frac{1}{k}-\log(m)\right)$, and **GoldenRatio** is $\frac{1}{2}\left(1+\sqrt{5}\right)$.

Other special symbols related to numbers include the following:

I $\sqrt{-1}$

Infinity ∞

ComplexInfinity An infinite quantity with an undetermined direction

Indeterminate An indeterminate numerical result

Of these, **I** and **Infinity** can be written as ESCiiESC and ESCinfESC, and the results are i and ∞. To demonstrate these special symbols, write the following:

    {Sqrt[-1], I^2, Exp[-π I]}   {i, -1, -1}

    {Limit[1 / x, x → 0, Direction → -1], Limit[1 / x, x → 0, Direction → 1], 1 / 0}

    Power::infy : Infinite expression $\frac{1}{0}$ encountered. **More…**

    {∞, -∞, ComplexInfinity}

    {0 / 0, 0^0, ∞ - ∞}

    Power::infy : Infinite expression $\frac{1}{0}$ encountered. **More…**

    ∞::indet :
     Indeterminate expression 0 ComplexInfinity encountered. **More…**

    Power::indet : Indeterminate expression 0^0 encountered. **More…**

    ∞::indet : Indeterminate expression $-\infty + \infty$ encountered. **More…**

    {Indeterminate, Indeterminate, Indeterminate}

The result of some calculations may also be an interval (for interval arithmetic, see Section 11.2.3, p. 291):

    Limit[Sin[1 / x], x → 0]   Interval[{-1, 1}]

■ Physical Constants

The **Miscellaneous`PhysicalConstants`** package contains the values of 51 physical constants. To get a list of the constants, type the following (the resulting list is not presented here):

    << Miscellaneous`PhysicalConstants`

    ? Miscellaneous`PhysicalConstants`*

Here are some examples:

    {IcePoint, SpeedOfLight, SpeedOfSound, AccelerationDueToGravity}

    $\left\{273.15\,\text{Kelvin}, \dfrac{299792458\,\text{Meter}}{\text{Second}}, \dfrac{340.292\,\text{Meter}}{\text{Second}}, \dfrac{9.80665\,\text{Meter}}{\text{Second}^2}\right\}$

■ Physical Units

In the `Miscellaneous`Units`` *package:*

`Convert[expr, newunits]`   Convert `expr` to `newunits`

`ConvertTemperature[temp, oldunits, newunits]`   Convert temperature `temp` to
  `newunits`

`SI[expr]`   Convert to SI units

`MKS[expr]`   Convert to MKS units (meter/kilogram/second)

`CGS[expr]`   Convert to CGS units (centimeter/gram/second)

With this package we can use more than 200 units related to electricity, length, informa-tion, time, mass, weight, force, inverse length, volume, viscosity, luminous energy and intensity, radiation, angles, power, area, amounts of substances, acceleration due to gravity, magnetism, pressure, energy frequency, and speed fineness for yam or thread. All names defined in the package can be seen by typing the following (the resulting list is not presented here):

    << Miscellaneous`Units`

    ? Miscellaneous`Units`*

Here are some examples of conversion:

`Convert[60 Mile / Hour, Kilo Meter / Hour]` $\dfrac{301752\ Kilo\ Meter}{3125\ Hour}$

`ConvertTemperature[100, Celsius, Fahrenheit]`   212

■ Chemical Elements

The package `Miscellaneous`ChemicalElements`` has information about chemical elements. The following information can be asked for the elements:

    << Miscellaneous`ChemicalElements`

    Drop[Complement[ToExpression[
      Names["Miscellaneous`ChemicalElements`*"]], Elements], -3]

    {Abbreviation, AtomicNumber, AtomicWeight, BoilingPoint, Density,
     EarthCrustAbundance, EarthOceanAbundance, ElectronConfiguration,
     ElectronConfigurationFormat, HeatOfFusion, HeatOfVaporization,
     IonizationPotential, MeltingPoint, SolarSystemAbundance,
     SpecificHeat, StableIsotopes, ThermalConductivity}

Giving the command `Elements` shows a list of the names of the elements, and `Element`
`Abbreviations` gives their abbreviations. For example:

`ThermalConductivity[Iron]` $\dfrac{80.2`\ Watt}{Kelvin\ Meter}$

11.2 Real Numbers

11.2.1 Precision and Accuracy

The following definitions are used for real numbers:

> *Precision:* The total number of significant decimal digits
> *Accuracy:* The number of significant decimal digits to the right of the decimal point

The precision and accuracy of a number can be asked with **Precision** and **Accuracy**. Before presenting examples, we define a function that gives the precision and accuracy of a number:

```
pa[x_] := {Precision[x], Accuracy[x]}
```

An example:

```
pa[11111.222223333344444]     {19.0458, 15.}
```

The precision of the number is about 20 and the accuracy is 15. Some other examples:

```
pa[0.00000111112222233333344444]     {19.0458, 25.}

pa[11111222233333.44444 10^10]     {19.0458, -5.}
```

Thus, the accuracy can be a negative number; in this example it tells us that there are five insignificant digits (zeros in this case) between the least significant digit and the decimal point. The precision and accuracy of exact numbers (e.g., integers, rational numbers, special constants) are infinity:

```
{pa[7], pa[3 / 4], pa[Pi], pa[Sin[2]]}

{{∞, ∞}, {∞, ∞}, {∞, ∞}, {∞, ∞}}
```

■ **Relative and Absolute Errors**

Precision and accuracy have interpretations in terms of relative and absolute errors.

> Relative error $\simeq 10^{-\text{precision}}$
> Absolute error $\simeq 10^{-\text{accuracy}}$

Thus, if the precision of a result is p, the relative error of the result is of the order 10^{-p}. Similarly, if the accuracy of a result is a, the absolute error of the result is of the order 10^{-a}.

Note that precision $\simeq -\log_{10}$(relative error) and accuracy $\simeq -\log_{10}$(absolute error). For example, the absolute error in 11111.22222 33333 44444 can be considered to be about 0.00000 00000 00000 5, and the precision and accuracy thus have the following approximate values:

```
-Log[10, 0.0000000000000005 / 11111.222223333344444]

19.3468
```

```
-Log[10, 0.000000000000005]
```

15.301

These are close to the values 19.0458 and 15. given by **Precision** and **Accuracy**. We can also see the precision of our number from the **InputForm**:

```
11111.222223333344444 // InputForm
```

11111.222223333344444`19.04576183352721

11.2.2 Two Types of Real Numbers

■ Two Types of Arithmetic

Mathematica has two types of floating-point arithmetic.

> *Fixed-precision arithmetic:* Implemented in the hardware
> *Variable-precision arithmetic:* Implemented in *Mathematica*

There are several ways to guide *Mathematica* to use the arithmetic we want, as we will see shortly. As an example, we calculate a decimal value by fixed-precision arithmetic and ask the precision of the result:

```
N[Sin[2]]    0.909297
```

```
Precision[%] // N    15.9546
```

Then we use variable-precision arithmetic:

```
N[Sin[2], 30]    0.909297426825681695396019865912
```

```
Precision[%]    30.
```

Usually fixed-precision arithmetic is used. The name of this system comes from the representation of real numbers in the computer hardware. Real numbers have a mantissa and an exponent, with the mantissa always containing a fixed number of bits; this usually means 16 decimal digits. There is no way to tell how precise such a number is, and *Mathematica* has adopted the convention that if you ask for the precision of such a number, the maximum precision (usually about 16, as in the example above) is given as the answer, independent of the true precision. Thus the name fixed-precision arithmetic: all numbers have the same precision, independent of what can be justified for them (i.e., whether or not all the digits in the result can be determined to be correct on the basis of the numbers in the input). So, the results you get with this arithmetic can contain insignificant digits, because *Mathematica* cannot say which digits are significant and which are insignificant.

Variable-precision arithmetic is what arithmetic should be: the precision of the result is what can be justified from the input and calculations. Only significant digits are then included in the result. The precision of such numbers varies—thus the name variable-precision arithmetic. This arithmetic is implemented in the software of *Mathematica*. Variable-precision arithmetic has two remarkable properties. First, all digits returned by *Mathematica* are correct if this arithmetic is used. Second, we can ask for the result to whatever precision we want.

■ Two Types of Real Numbers

There are two types of real numbers, which correspond with the two types of arithmetic:

Machine-precision numbers: Numbers produced by fixed-precision arithmetic
Arbitrary-precision numbers: Numbers produced by variable-precision arithmetic

Machine-precision numbers correspond to double-precision floating-point numbers in the underlying computer system. Arbitrary-precision numbers are handled with the software of *Mathematica*. Usually machine-precision numbers are used. Arbitrary-precision numbers can be formed in some special ways, which we consider below.

The precision of machine-precision numbers is indicated by the special symbol `MachinePrecision`.

`MachinePrecision` (❋5) Symbol used to indicate the precision of machine-precision numbers
`$MachinePrecision` Numerical value of `MachinePrecision`

When we calculate a numerical value with `N[expr]`, we actually ask the result as a machine-precision number. We could equally well write `N[expr, MachinePrecision]`:

```
{N[Pi], N[Pi, MachinePrecision]}
```

```
{3.14159, 3.14159}
```

The precision of these numbers is, indeed, `MachinePrecision`:

```
Map[Precision, %]
```

```
{MachinePrecision, MachinePrecision}
```

The numerical value of `MachinePrecision` is `$MachinePrecision`, which is about 16:

```
{MachinePrecision // N, $MachinePrecision}
```

```
{15.9546, 15.9546}
```

The fixed precision used in machine-precision numbers may vary between computer systems, but usually it is about 16. *Mathematica* knows the precision: it is the value of the constant `$MachinePrecision`.

■ Machine-Precision Numbers

Here are examples of machine-precision numbers:

```
{2.2, N[Pi], 1.2345678901234567, N[Sin[2.2], 20]}
```

```
{2.2, 3.14159, 1.23457, 0.808496}
```

The internal representations in *Mathematica* are as follows:

```
InputForm[%]
```

```
{2.2, 3.141592653589793, 1.2345678901234567,
  0.8084964038195901}
```

For example, `N[Pi]` is internally calculated with all of the standard 16 digits, but normally only 6 digits are shown. Note especially that, in `N[Sin[2.2], 20]`, the number 2.2 is a machine-precision number; this causes all calculations in this expression to be done with fixed-precision arithmetic. Thus, asking for the value of `Sin[2.2]` to 20-digit precision does not have the desired effect; the result is a 16-digit machine-precision number.

■ Arbitrary-Precision Numbers

Ways to form an arbitrary-precision number:

`2.2`9` Use ` to write a number with any precision (here with precision 9)

`N[22/10, 20]` Use `N` to form a number with any precision (here with precision 20)

`2.20000000000000000` Write at least 18 significant digits (in most computers)

`SetPrecision[2.2, 9]` Use `SetPrecision` to write a number with any precision

Note that *machine-precision numbers are used in a calculation as soon as a machine-precision number is encountered.* Thus, to use arbitrary-precision numbers in a computation, *all* numbers in the input have to be exact quantities or arbitrary-precision numbers.

Here are examples of arbitrary-precision numbers and their internal representations:

```
{2.2`9, N[Pi, 17], 1.234567890123456789, N[Sin[22 / 10], 20]}
```

```
{2.20000000, 3.1415926535897932,
 1.234567890123456789, 0.80849640381959018430}
```

```
InputForm[%]
```

```
{2.2`9.000000000000002, 3.1415926535897932384626433833`17.,
 1.234567890123456789`18.091514977212704,
 0.80849640381959018430403691041611906646`20.}
```

We see that the internal representation of arbitrary-precision numbers contains, after the mark `, the precision of the number. The input forms of the second and fourth number contain more digits than were requested. The accuracy of a number can be set with ``.

■ Printing

For machine-precision numbers, usually *six* digits are printed (with `InputForm` all 16 digits are shown).

For arbitrary-precision numbers, *significant* digits are printed.

Machine-precision numbers are usually printed with six digits (however, trailing zeros, even significant ones, are not printed). If you want to see all of the internal 16 digits, apply `InputForm` to the result. If you want to adjust the number of digits shown, use **Number⋅ Form** (see Section 11.1.2, p. 282).

An excellent aspect of arbitrary-precision numbers is that *Mathematica* shows for them only the digits for which it can be sure that they are significant. Thus, we can trust that all digits in such a result are correct.

■ Advantages and Disadvantages

The primary advantage of fixed-precision arithmetic is that it is fast; this is because the calculations are done by the hardware in the floating-point unit. Inversely, the primary disadvantage of variable-precision arithmetic is that it is slow; this is because the arithmetic is implemented in the software of *Mathematica*.

Advantages of variable-precision arithmetic include the following:

- we can use arbitrary precision in the calculations;
- no round-off errors are introduced by the arithmetic itself; and
- results contain only correct digits.

Inversely, disadvantages of fixed-precision arithmetic include the following:

- we cannot do high-precision calculations;
- round-off errors are introduced by the arithmetic (see Section 11.2.3, p. 290); and
- results may contain insignificant digits.

In addition, one disadvantage of variable-precision arithmetic has to be mentioned. It is good that the result contains only correct digits, but it is not so good that the rules used to determine the precision of the result may yield an overly pessimistic precision. This means that the result often has a better precision than the one given by *Mathematica*. The cause for this pessimism is the assumption that all errors are independent. For example, let a be a given arbitrary-precision number. Then $a - a$ should be exactly 0, but it is not:

```
a = N[Pi, 20];
```

```
a - a        0. × 10^-20
```

The errors in the two as are considered as independent instead of equal, and therefore $a - a$ cannot be assigned the value 0.

11.2.3 Round-off Errors and Interval Arithmetic

■ Round-off Errors

It is well known how round-off errors affect the results of fixed-precision arithmetic. Consider the following sum:

```
1.234567890123456 + 10.^-16 // InputForm
```

```
1.234567890123456
```

We see that the value of the sum is the same as the first summand; the second summand had no effect. The fixed-precision 16-digit system could not represent the result adequately. Such round-off errors are the primary sources of errors in this system. Round-off errors are also called representation errors. Instead, if we use variable-precision arithmetic, then no round-off error is introduced:

```
1.234567890123456`17 + 10.`17^-16
```

```
1.2345678901234561
```

As another example, define the following function:

```
f[x_] := (Log[1 - x] + x Exp[x / 2]) / x^3
```

Here is its limit at the origin:

```
Limit[f[x], x → 0.]        - 0.208333
```

Calculating values of the function near the origin results in huge errors:

```
Table[f[10.^-n], {n, 4, 8}]
```

$\{-0.208345, -0.162823, -28.9641, 52635.4, -5.02476 \times 10^7\}$

To get correct values, use arbitrary-precision numbers:

```
Table[f[10.`23^-n], {n, 4, 8}]
```

$\{-0.20835625197412, -0.20833562502, -0.208333563, -0.2083334, -0.20833\}$

Notice how we get less and less digits the nearer we get to the origin. This is because significant digits are lost more and more. If we try to plot the function near the origin, the result is far from good:

```
Plot[f[x], {x, 0, 0.001}, PlotPoints → 200];
```

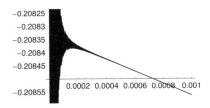

The plot shows clearly how wildly the machine-precision values of the function vary near the origin. If we use arbitrary-precision numbers, we get the correct plot:

```
Plot[f[SetPrecision[x, 25]], {x, 0, 0.001}, Compiled → False];
```

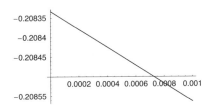

In Section 15.1.1, p. 382, we have a similar example.

Sometimes *Mathematica* may give numbers so near to 0 that we prefer to replace them with an exact zero; the nonzero digits are possibly only the result of round-off and other errors. We can use **Chop**, which is explained in Section 11.1.1, p. 281. **Chop[expr]** replaces all real numbers in **expr** with a magnitude of less than 10^{-10} with 0:

```
Exp[N[Pi / 2 I]]      6.12323 × 10^-17 + 1. i
```

```
Chop[%]      1. i
```

■ Interval Arithmetic

We can do interval arithmetic with *Mathematica*. **Interval[{min, max}]** represents the range of values between **min** and **max**. As an example, we consider linear algebra. First we form a square matrix that has intervals as elements, which reflects the uncertainty we have about these numbers:

```
a = {{2, -1}, {3, -5}};

aa = Map[Interval[{# - 0.01, # + 0.01}] &, a, {2}]
{{Interval[{1.99, 2.01}], Interval[{-1.01, -0.99}]},
 {Interval[{2.99, 3.01}], Interval[{-5.01, -4.99}]}}
```

Then we calculate the determinant and solve a system of linear equations:

```
Det[aa]     Interval[{-7.11, -6.89}]

Solve[aa.{x, y} == {4, 1}, {x, y}]
{{x → Interval[{2.67932, 2.75036}], y → Interval[{1.39944, 1.45864}]}}
```

The results belong to the intervals shown.

■ More about Computer Arithmetic

Not all machine-precision numbers can be distinguished because of the limited precision available. **$MachineEpsilon** is the smallest machine-precision number such that 1.0 + **$MachineEpsilon** is not equal to 1.0. **$MaxMachineNumber** is the largest positive machine-precision number and **$MinMachineNumber** the smallest positive machine-precision number:

```
{$MachineEpsilon, $MaxMachineNumber, $MinMachineNumber}
```
$\{2.22045 \times 10^{-16}, 1.79769 \times 10^{308}, 2.22507 \times 10^{-308}\}$

Note that, if the result of a calculation is a number outside the range specified by **$MinMachineNumber** and **$MaxMachineNumber**, the result is automatically converted to arbitrary-precision form.

With the **NumericalMath`Microscope`** package, you can investigate the machine arithmetic of your computer. The **NumericalMath`ComputerArithmetic`** package can be used to investigate floating-point systems with various rounding rules, bases, and precisions.

11.3 Options of Numerical Routines

11.3.1 Options for Precision

Although the main scope of *Mathematica* is symbolic calculation, sooner or later we meet a problem for which *Mathematica* cannot find a solution in a symbolic and exact form. Then we can resort to numerical routines such as **NIntegrate**, **FindRoot**, and **NDSolve** and obtain an approximate numerical solution.

The numerical routines have several options with which we can control and modify the routines. Most of the routines include one or more of the following three options. We consider them here for two reasons. First, they are closely related to round-off errors, precision, and accuracy, which we considered in Section 11.2. Second, the three options do not then need to be separately considered in detail for each routine.

> **WorkingPrecision → w** Calculations are done using numbers with **w**-digit precision
> **PrecisionGoal → p** Result should have **p**-digit precision
> **AccuracyGoal → a** Result should have **a**-digit accuracy

WorkingPrecision affects the precision of the _calculations_, while **PrecisionGoal** and **AccuracyGoal** affect the precision of the _result_. With **WorkingPrecision**, we can control the effect of _round-off error_. With **PrecisionGoal** and **AccuracyGoal**, we can control the effect of _truncation error_.

■ Precision and Accuracy Goals

Truncation errors are caused by the iterative method used, that is, by approximating the original problem with another, simpler one. Typically the iterative methods rely on calculating the value of a function at a finite set of points, whereas an infinite set would be required for the exact solution. For example, in calculating the value of an infinite sum by numerical methods, only a finite number of terms are summed, and the rest is estimated by various methods.

Precision and accuracy were considered in Section 11.2.1, p. 286. There we noted that, if the precision of a result is p, the relative error of the result is of the order 10^{-p}, and if the accuracy of a result is a, the absolute error of the result is of the order 10^{-a}. So we get the following interpretations of the two options:

> **PrecisionGoal → p** The _relative error_ of the result should be at most of the order 10^{-p}
> **AccuracyGoal → a** The _absolute error_ of the result should be at most of the order 10^{-a}

The goals are used in the stopping criteria for the iterative methods: _iterations are stopped as soon as either the accuracy goal or the precision goal is satisfied_. Note that the precision and accuracy goals are only goals; this is because the true relative and absolute errors are unknown and have to be estimated with the iterative methods. The true relative or absolute error may be much larger but also much smaller than the given goal.

Note that **p** and **a** can also be infinite. If **p** is infinity, the precision goal will never be satisfied, and so only absolute error is used as the criterion. Similarly, if **a** is infinity, the accuracy goal will never be satisfied, and so only relative error is used as the criterion.

■ Example

We calculate an integral:

```
f = Sin[4 x] Exp[-x];

Integrate[f, {x, 0, 50}]
```

$$\frac{1}{17}\left(4 - \frac{4\,\mathrm{Cos}[200] + \mathrm{Sin}[200]}{e^{50}}\right)$$

```
N[%, 20]
```

`0.23529411764705882353`

Numerical integration gives the following result:

```
NIntegrate[f, {x, 0, 50}] // InputForm
```

0.23529411763815106

The result has 10 correct decimals, so it is very good. However, we try to get a still better result by using 30 digits during the calculations (so arbitrary-precision numbers are used) and asking a result with an absolute error of at most 10^{-20}:

```
NIntegrate[f, {x, 0, 50}, WorkingPrecision → 30, AccuracyGoal → 20]
```

0.23529411764705882353

All 20 digits are, indeed, correct. If we are satisfied with a lower precision and accuracy, we can write the following:

```
NIntegrate[f, {x, 0, 10}, PrecisionGoal → 2, AccuracyGoal → 2] // InputForm
```

0.23529919027260296

■ Default Values of Options

The three options mentioned above usually have the following default values:

> **WorkingPrecision** → **MachinePrecision** (= 16 in most computers)
> **PrecisionGoal** → **Automatic** (= 8 in most computers)
> **AccuracyGoal** → **Automatic** (= 8 in most computers)

The default value **MachinePrecision** of **WorkingPrecision** means that normal machine-precision numbers are used in the calculations. As we saw in Section 11.2.2, p. 288, the numerical value of **MachinePrecision** is **$MachinePrecision**, which usually means a number close to 16.

The default values of the two goals mean that iterations are normally stopped when *either* the relative *or* the absolute error is less than 10^{-8} (10^{-6} in *Mathematica* 4). More detailed information about the default stopping criteria is as follows.

> **FindMinimum, FindMaximum**   The relative or absolute error of *the optimum point and of the value of the function at the optimum point* is less than 10^{-8}.
>
> **NMinimize, NMaximize**   The relative or absolute error of *the optimum point and of the value of a penalty function at the optimum point* is less than 10^{-8}.
>
> **FindFit**   The relative or absolute error of *the optimum point and of a norm function* (like squared residuals) is less than 10^{-8}.
>
> **FindRoot**   The relative or absolute error of *the root* is less than 10^{-8} and the absolute value of *the function at the root* is less than 10^{-8}.
>
> **NDSolve**   The relative or absolute error of *the solution of the differential equation at each chosen point* is less than 10^{-8}.
>
> **NIntegrate**   The relative error of *the integral* is less than 10^{-6}.
> **NSum**   The relative error of *the sum* is less than 10^{-6}.
> **NProduct**   The relative error of *the product* is less than 10^{-6}.

■ Adjusting the Options

With the default value **MachinePrecision** (which is about 16 in most computers) of **WorkingPrecision**, the usual fixed-precision numbers are used and the calculations are fast, but the precision may not suffice in a critical or ill-conditioned case (see Section 11.2.2, p. 289, for advantages and disadvantages of fixed-precision numbers).

By specifying a value for **WorkingPrecision** other than **MachinePrecision**, the calculations are done with variable-precision arithmetic. Using a high value like 20 or more generally gives more accurate results; the disadvantage is that the calculations take more time. Note that **WorkingPrecision** affects only the calculations; the result probably has a lower precision. To control the precision and accuracy of the result, use **PrecisionGoal** and **AccuracyGoal**.

The default value **Automatic** of **PrecisionGoal** and **AccuracyGoal** usually means 8 (6 in *Mathematica* 4) if **WorkingPrecision** has its default value. If **WorkingPrecision** has another value, then the default value of the precision and accuracy goals generally is **WorkingPrecision**/2. For example, if **WorkingPrecision** is 30, the default value of the precision and accuracy goals is usually 15. (For **NIntegrate**, **NSum**, and **NProduct**, the default value of **PrecisionGoal** is 6 or **WorkingPrecision** − 10.)

If you increase the precision or accuracy goal from the default value 8, you often also have to increase the value of **WorkingPrecision**; give it a value that is at least a few digits larger than the goal.

11.3.2 Other Common Options

■ StepMonitor and EvaluationMonitor

> **StepMonitor** (❀5) An option for various iterative numerical methods that gives a command to be executed after each step; examples of values: **None, ++n, Append To[iters, x]**
>
> **EvaluationMonitor** (❀5) An option for various numerical methods that gives a command to be executed after each evaluation of functions derived from the input; examples of values: **None, ++n, AppendTo[points, x]**

StepMonitor and **EvaluationMonitor** are useful when investigating how a numerical method proceeds (e.g., how many iterations are needed, what are all the points a numerical method generates). The following commands have both of these options: **FindFit**, **FindMaximum**, **FindMinimum**, **FindRoot**, **NMaximize**, **NMinimize**, and **NDSolve**. In addition, **NIntegrate**, **NProduct**, and **NSum** have the **EvaluationMonitor** option.

The values of these two options are set with a delayed setting by using **:>** instead of **->** (to avoid the immediate evaluation of the command given). Note that *Mathematica* automatically replaces **:>** with the special symbol **:→**. The following is a typical example:

```
f = Exp[-x] - x^2;
```

```
n = 0; FindRoot[f, {x, -1}, StepMonitor :> ++n]
```

{x → 0.703467}

n   6

We needed 6 iterations to find the root. Here is another example:

```
iters = {}; FindRoot[f, {x, -1}, StepMonitor :> AppendTo[iters, x]]
```

{x → 0.703467}

```
iters
```

{1.39221, 0.835088, 0.709834, 0.703483, 0.703467, 0.703467}

These are the 6 points generated by the iterative method used by **FindRoot**. We could also use **Sow** and **Reap** (🐾5):

```
Reap[FindRoot[f, {x, -1}, StepMonitor :> Sow[x]]]
```

{{x → 0.703467},
  {{1.39221, 0.835088, 0.709834, 0.703483, 0.703467, 0.703467}}}

Reap[expr] returns a list of two components: the value of **expr** (here the result of **Find-Root**) together with a list of values of the expression to which **Sow** has been applied during the calculation of **expr**.

■ Compiled

> **Compiled** An option for various numerical and plotting methods that tells whether to compile the expression with which they are working; possible values: **True**, **False**

The default value of **Compiled** is **True**; this means that the command *compiles* the expression to be manipulated. Compilation transforms the expression to a kind of pseudocode that contains simple instructions for evaluating the expression. This speeds up the computations.

Compiled expressions use only normal machine-precision numbers. If you want the command to use arbitrary-precision numbers when calculating the value of an expression, you have to turn the compiling off by giving the option **Compiled → False**. An example is in Section 11.2.3, p. 291, where we plot a badly behaving function using high-precision numbers.

Compiling is considered in more detail in Section 14.2.2, p. 365.

Expressions

Introduction

Algebra is generous: she often gives more than is asked for. — *J. d'Alembert*

Section 12.1 contains some very basic techniques for expressions: assigning values for variables, clearing values of variables, inserting a value of a variable into an expression, and picking parts of an expression. Read this section carefully.

Mathematica often writes the result in a nice form, but now and then we will want to perform some manipulations on an expression. *Mathematica* has many commands to this end, although a little experimenting is sometimes needed to get the desired result. This topic is considered in Sections 12.2 and 12.3.

You will probably find in *Mathematica* all of the usual and special functions you need; Section 12.4 gives some lists of these functions. *Mathematica* uses the functions effectively itself, too; many integrals, for example, can be written only in terms of certain special functions. Note that http://functions.wolfram.com/ is an excellent place to find information about functions.

Lists are very important expressions in *Mathematica*; they are considered in Chapter 13. *Patterns*, considered in Section 15.3, represent classes of expressions.

12.1 Basic Techniques

12.1.1 Assigning and Clearing Values

For referring to earlier results, you have at least three methods at your disposal: use %, use `Out[n]`, or give a name to the result to which you want to refer.

Using % (see Section 1.2.2, p. 9) is convenient provided that you refer to a very recent result and you do not need to execute the command containing % several times. If you have to execute several times until you obtain the desired result, then you need to use %%, %%%, and so on in the succeeding executions so that you refer to the right result.

A better possibility to refer to an earlier result may be the use of `Out[n]` (see Section 1.2.2, p. 9), because **n** remains the same even if you execute several times. (By the way, to start the numbering of results from 1 again, execute `$Line = 0`.)

Giving a name to the result and using it later on may be the best method. This is considered next.

■ Assigning Values to Symbols

> **x = a** Assign the value **a** to **x**
> **x = y = a** Assign the same value to several symbols
> **{x, y, …} = {a, b, …}** Assign **a** to **x**, **b** to **y**, …

Assigning values to symbols has a drawback. Suppose you want to consider the following expression:

   p = x + Sin[y];

You want to calculate the value of **p** when **x** is 3.7 and **y** is 1.2. One possibility is to assign the values to **x** and **y** and then ask the value of **p**:

   x = 3.7; y = 1.2; p   4.63204

This is a straightforward method but has a drawback: **x** and **y** have from here on the values 3.7 and 1.2, and **p** has from here on the value 4.63204, unless you assign a new value for **x** or **y** or clear the values of **x** and **y**. *The values of* **x** *and* **y** *are applied in all expressions where* **x** *or* **y** *appears*. This can cause trouble later on when you have perhaps forgotten that you assigned a value for **x** and **y**. Thus, remember to clear or remove a symbol when you no longer need it so that the value of the symbol does not cause trouble in later calculations. If the symbol is **x**, type **x =.**, `Clear[x]`, or `Remove[x]`; these commands are explained shortly.

A recommended rule is to assign values for variables sparingly. The preferred method for calculating values of expressions for particular values of variables is to use transformation rules (see Section 12.1.2, p. 300).

■ **Asking Information about Symbols**

> **x** Give the value of symbol **x**
>
> **?x** Give the context and definition of symbol **x**
>
> **?x*** Print all symbols beginning with **x**
>
> **?Global`*** Print all user-defined symbols
>
> **Names["Global`*"]** Give a list of all user-defined symbols
>
> **ToExpression[%]** Give a list of the values of the symbols given by **Names**

These commands are useful if you have forgotten what symbols you have used and what values they have been given. We ask the value and definition of **p**:

    p   4.63204

    ?p

    Global`p

    p = x + Sin[y]

Here are all the symbols we have used in this section:

    ?Global`*

Global`

p x y

Clicking the name of a symbol prints its definition. Next, we ask for our symbols and their values in a list form:

    Names["Global`*"]   {p, x, y}

    ToExpression[%]   {4.63204, 3.7, 1.2}

■ **Removing Symbols**

> **x =.** Clear the value of **x**
>
> **Remove[x]** Remove **x**
>
> **Remove[x, y ,…]** Remove **x, y, ...**
>
> **Remove["x*"]** Remove all symbols that start with **x**
>
> **Remove["Global`*"]** Remove all user-defined symbols

Clear is used in the same way as **Remove**. **Clear** as well as **=.** clears the value of a symbol. **Remove** not only clears the value of a symbol but removes the whole symbol. **Global`** is the context in which all user-defined symbols are stored (**System`** is the context containing all built-in symbols); contexts are considered in more detail in Section 14.3.1, p. 369. The command **Remove["Global`*"]** is handy for removing all user-defined symbols. A straightforward method to "clear the table" is also to quit the kernel from the **Kernel** menu and then start it anew from the same menu. Now we remove all of our symbols:

    Remove["Global`*"]

■ Possible Spelling Errors

If you use only slightly differing names, you may obtain a warning about a possible spelling error:

```
aaa = 1; aaaa = 2;
```

> General::spell1 : Possible spelling error: new symbol
>   name "aaaa" is similar to existing symbol "aaa". More…

This may be a useful warning: we may have written a name with a typing error. Note that this is only a warning; an error has not occurred. You can continue calculations if you have not made an error.

Sometimes you may use many slightly differing names; if so, you will get a lot of warnings about possible spelling errors. You may want to turn the warning messages off by entering the following:

```
Off[General::spell]
Off[General::spell1]
```

From now on you will not obtain warnings about possible spelling errors. You can turn the warnings back on by entering the following:

```
On[General::spell]
On[General::spell1]
```

If you work with the examples of this book, you will encounter messages about possible spelling errors, but please note that these messages have been removed from this text to save space.

12.1.2 Inserting Values

The preferred method for calculating the value of an expression **expr** for a specific value **a** of a variable **x** is to apply a *transformation rule* by writing **expr /. x → a** (or **Replace‑ All[expr, x → a]**).

Getting used to this transformation technique may take some time. However, it is worth spending the time on this topic, because it is very important in *Mathematica*. First, you can use the technique to insert values into expressions. Second, some important commands such as **Solve**, **NSolve**, **FindRoot**, **FindMinimum**, **Minimize**, **NMinimize**, **Find‑ Fit**, **DSolve**, and **NDSolve** give the result in the form of a transformation rule, so you have to know how to handle such results. Some examples were already worked in Sections 2.3.2, p. 41, and 2.3.4, p. 45. Transformation rules are also considered in Sections 15.3.1, p. 402, and 15.3.4, p. 411.

Note that the arrow (→) can be written as ->, because *Mathematica* automatically replaces these two marks with a genuine arrow. The arrow can also be written as ESC ->ESC.

■ Simple Rules

expr /. x → a Replace **x** with **a** in **expr**	
f = expr /. x → a Replace **x** with **a** and assign the result to **f**	

Here is an expression:

```
p = x + Sin[y];
```

Here is a value of the expression:

```
p /. x → 3.7      3.7 + Sin[y]
```

Note that, after the transformation **p /. x → 3.7**, the variable **p** still has its original value **x + Sin[y]** and **x** has no value, as can be seen by asking for the value of **{p, x}**:

```
{p, x}     {x + Sin[y], x}
```

In general, after a transformation like **expr /. x → a**:

- **x** has no value; and
- the value of **expr** has not changed.

Mathematica only calculates and shows the value of **expr** after the transformation is done; no assignment is made for **expr**. Write **f = expr /. x → a** if you want to store the result as a value of a variable.

■ **Several Rules**

> **expr /. {x → a, y → b, … }** Replace **x** with **a**, **y** with **b**, …
> **expr /. Thread[{x, y, … } → {a, b, … }]** Replace **x** with **a**, **y** with **b**, …
> **expr /. Thread[vars → vals]** Replace variables in the list **vars** with the corresponding values in the list **vals**
> **expr /. x → a /. y → b** Replace **x** with **a**; in the result, replace **y** with **b**

Try two rules:

```
p /. {x → 3.7, y → 1.2}      4.63204
```

Suppose we have some variables and some values:

```
vars = {x, y}; vals = {3.7, 1.2};
```

If we want to replace **vars** with **vals** in **expr**, we cannot write **expr /. vars → vals**. We have to apply **Thread** to obtain the list of rules:

```
Thread[vars → vals]      {x → 3.7, y → 1.2}
```

So, we can write the following:

```
p /. Thread[vars → vals]      4.63204
```

Note that, when calculating a transformation like **expr /. {x → a, y → b, … }**, *Mathematica* looks at each part of the expression, tries all the rules on it, and then goes on to the next part. The first rule that applies to a particular part is used; no other rules are tried on that part or on any of its subparts. For example:

```
{x, y} /. {x → y, y → x}     {y, x}
```

Here *Mathematica* tries the rules that may apply to **x**, finds that **x → y** is suitable, and applies it (**y → x** is no longer applied). Then it goes to **y**, finds that **y → x** is suitable, and applies it (**x → y** is no longer applied). On the other hand, consider the next example:

```
{x, y} /. x → y /. y → x     {x, x}
```

Mathematica first replaces **x** with **y** and then, in the resulting list **{y, y}**, replaces **y** with **x**.

■ **Forming a List of Values**

> **expr /. x → {a, b, … }** Replace **x** with **a, b,** …; the result is a list
> **expr /. {{x → a1, y → b1, … }, {x → a2, y → b2, … }, … }** Apply several sets of rules;
> the result is a list

We calculate the value of **p** with several values of **x**:

> **p /. x → {3.7, 3.8, 3.9}**
>
> {3.7 + Sin[y], 3.8 + Sin[y], 3.9 + Sin[y]}

The result is a list. We can apply several sets of rules:

> **p /. {{x → 3.7, y → 1.2}, {x → 3.8, y → 1.3}, {x → 3.9, y → 1.5}}**
>
> {4.63204, 4.76356, 4.89749}

Solve gives several sets of rules as the answer. As an example, we solve the following equation:

> **eqn = 6 – 5 x + x^2 == 0;**
>
> **sol = Solve[eqn, x]**   {{x → 2}, {x → 3}}

The result is a list of lists. Each sublist gives one solution of the equation. The form of the solution is useful for continuing the calculations. Because **sol** now has the value {{x → 2}, {x → 3}}, to get a list of values (instead of a list of transformation rules), we can write the following:

> **x /. sol**   {2, 3}

We can check that the solution is correct by inserting it into the equation:

> **eqn /. sol**   {True, True}

Mathematica could conclude that both solutions are correct, that is, that both solutions simplify the equation into the form 0 == 0, which is true. Similarly, we can calculate the value of any expressions with the solutions:

> **1 / (1 + x) /. sol** $\{\frac{1}{3}, \frac{1}{4}\}$

We could also ask directly for the values of **x**:

> **sol2 = x /. Solve[eqn, x]**   {2, 3}

If we now want to check the solution, we must directly write the necessary transformation rules:

> **eqn /. {{x → sol2[[1]]}, {x → sol2[[2]]}}**   {True, True}

Compare this with the simple command **eqn /. sol**. We see that the transformation rules given by **Solve** are handy when we continue the calculations.

■ **Using Rules Repeatedly**

> **expr //. {x → a, y → b, … }** Apply the rules repeatedly until the result no longer
> changes

Consider the following example:

> **p /. {x → y, y → a}**   y + Sin[a]

At each part of **p**, the rules are applied only once. Write then the following:

> **p /. x → y /. y → a**   a + Sin[a]

First **x** is replaced with **y**, and then **y** is replaced with **a**. *Mathematica* also has the command **ReplaceRepeated**, which is formed by **//.**. It applies the rules until the result no longer changes:

> **p //. {x → y, y → a}**   a + Sin[a]

12.1.3 Picking Parts

> **expr[[i]]**   The **i**th part of **expr**
>
> **expr[[i,j]]**   The **j**th part of the **i**th part of **expr**

Picking parts of expressions can be done with double brackets. Consider the following expression:

> **p = (y^2 + 2 y − 8) / (y + 4) ^2 + 2 / (x + 3)**
>
> $\dfrac{2}{3 + x} + \dfrac{-8 + 2\,y + y^2}{(4 + y)^2}$

We pick several parts:

> **{p[[1]], p[[2]], p[[1, 1]], p[[1, 2]], p[[1, 2, 1]], p[[1, 2, 2]]}**
>
> $\left\{ \dfrac{2}{3 + x},\ \dfrac{-8 + 2\,y + y^2}{(4 + y)^2},\ 2,\ \dfrac{1}{3 + x},\ 3 + x,\ -1 \right\}$

The result is not always the one you expect. For example, you might expect that **p[[1,2]]** is $3 + x$ instead of $1/(3 + x)$. Given that **p[[1,2]]** is $1/(3 + x)$, you might expect that **p[[1,2,1]]** is 1 instead of $3 + x$. The explanation is that the parts are picked from the representation *Mathematica* uses internally. You can look at this representation with **FullForm**:

> **FullForm[p[[1]]]**
>
> Times[2, Power[Plus[3, x], −1]]

From this we see that **p[[1,2]]** is in fact **Power[Plus[3, x], −1]**, that is, $1/(3 + x)$, and that **p[[1,2,1]]** is **Plus[3, x]**, that is, $3 + x$.

■ Example

Let us calculate the area enclosed by two curves. Here are the curves:

> **f1 = x^2 / 2 + x / 2 + 1;**
> **f2 = x^2 − 1 / 2;**

We want to calculate the filled area between the points of intersection:

> **<< Graphics`FilledPlot`**

```
FilledPlot[{f1, f2}, {x, -1.5, 2.5}];
```

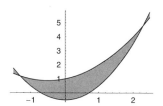

The points of intersection are as follows:

```
sol = x /. Solve[f1 == f2]
```
$\left\{ \frac{1}{2} (1 - \sqrt{13}), \frac{1}{2} (1 + \sqrt{13}) \right\}$

Then we integrate the difference of the curves between these points:

```
Integrate[f1 - f2, {x, sol[[1]], sol[[2]]}]
```
$\frac{13 \sqrt{13}}{12}$

12.2 Manipulating Expressions

12.2.1 Simplifying

Simplify[expr] Simplify by trying algebraic and trigonometric transformations
FullSimplify[expr] Try a much wider range of transformations

These commands can also be used in the form **expr//Simplify** as all commands with one argument. Often **Simplify** does a good job, but if we are not satisfied, we can use **FullSimplify**. It tries many more transformations, and it knows a lot of special functions; however, calculations can be long. As an example, **Simplify** does not do anything to the following expressions:

```
(n + 3) ! / n ! // Simplify
```
$\frac{(3 + n)\,!}{n\,!}$

```
Sqrt[5 + 2 Sqrt[6]] // Simplify
```
$\sqrt{5 + 2 \sqrt{6}}$

FullSimplify succeeds in the simplification:

```
(n + 3) ! / n ! // FullSimplify
```
$(1 + n)\,(2 + n)\,(3 + n)$

```
Sqrt[5 + 2 Sqrt[6]] // FullSimplify
```
$\sqrt{2} + \sqrt{3}$

Next we consider the use of assumptions with simplification. After that, we examine the options of the two commands.

■ Using Assumptions

Simplify[expr, assumptions]
FullSimplify[expr, assumptions]

Simplify and **FullSimplify** (and also **FunctionExpand**, see Section 12.2.2, p. 308) accept assumptions. The assumptions can be given by declaring domains of variables and by specifying equations and inequalities (note that a variable declared to satisfy an inequality is automatically assumed to be real). Various assumptions can be combined with logical operators like `&&` (and), `||` (or), and `!` (not) (see Section 12.3.5, p. 316). Domains can be declared as follows:

`x ∈ dom` Declare that `x` is an element of the domain **dom**

`{x, y, … } ∈ dom` or `(x|y| … ) ∈ dom` Declare that `x, y, …` are elements of the domain **dom**

Possible domains:

`Complexes, Reals, Algebraics, Rationals, Integers, Primes, Booleans`

Write $\in$ by pressing [ESC]elem[ESC]. In place if `x ∈ dom` and `{x, y, … } ∈ dom`, we can also write `Element[x, dom]` and `Element[{x, y, … }, dom]`. Next we consider several examples of using assumptions.

■ Square Roots

Mathematica does not automatically simplify $\sqrt{x^2}$ to $|x|$:

> `Sqrt[x^2] // Simplify` $\sqrt{x^2}$

The reason for this is that this simplification does not hold for complex numbers:

> `{Sqrt[(2 + 3 I)^2], Abs[2 + 3 I]}` $\{2 + 3\,i, \sqrt{13}\}$

However, if we assume that x is real, then the simplification is valid:

> `Simplify[Sqrt[x^2], x ∈ Reals]` $\mathrm{Abs}[x]$

If we assume that $x < 0$, then we get $-x$:

> `Simplify[Sqrt[x^2], x < 0]` $- x$

■ Powers of Powers

The expression $\left(a^b\right)^c$ cannot always be simplified as a^{bc}:

> `Simplify[(a^b)^c]` $\left(a^b\right)^c$

If c is an integer or if a and b are positive, then we get the simplified form:

> `Simplify[(a^b)^c, c ∈ Integers]` a^{bc}

> `Simplify[(a^b)^c, a > 0 && b > 0]` a^{bc}

■ Trigonometric Expressions

Trigonometric expressions often become simpler if some parameters are integers:

> `Integrate[Sin[m x] Cos[n x], {x, 0, Pi}]`
>
> $$\frac{m - m \cos[m\pi] \cos[n\pi] - n \sin[m\pi] \sin[n\pi]}{m^2 - n^2}$$

> `Simplify[%, {m, n} ∈ Integers]`
>
> $$-\frac{(-1 + (-1)^{m+n})\,m}{m^2 - n^2}$$

■ Inequalities

If the expression to be simplified is an inequality, then **Simplify**

- gives **True** if the inequality holds for all values of variables in the specified domain;
- gives **False** if the inequality does not hold at any point in the specified domain; and
- does not claim anything if the inequality holds for some points but does not hold for some other, or if the validity of the inequality cannot be decided.

Here are some examples:

```
Simplify[x ≤ x + a, a ≥ 0 && x ∈ Reals]       True

Simplify[x ≤ x + a, a < 0 && x ∈ Reals]       False

Simplify[x ≤ x + a, a ≤ 0 && x ∈ Reals]       a ≥ 0
```

Here is a less trivial example:

```
Simplify[Exp[x] > Log[x] + 2, x > 0]       True
```

■ Domains

We can ask whether a given number belongs to a given domain:

```
{Sqrt[5 + 2 Sqrt[7 ^ (3 / 8)]] ∈ Algebraics,
  E ∈ Algebraics, Pi ∈ Algebraics, E + Pi ∈ Algebraics}

{True, False, False, e + π ∈ Algebraics}
```

(Note that it is not yet known whether $e + \pi$ is an algebraic number.)

■ Specifying Default Assumptions

Assumptions can be used with some commands like **Simplify**, **FullSimplify**, **FunctionExpand**, **Limit**, and **Integrate**. Actually, we can distinguish two types of assumptions: default and specific. When we above wrote **Simplify[Sqrt[x^2], x < 0]**, we used a specific assumption. The default assumptions are represented by the variable **$Assumptions** (❀5). Its initial value is true:

```
$Assumptions       True
```

This has no effect, and so this means, in practice, that there are no default assumptions.

Default assumptions can be added by **Assuming** (❀5). As an example, we note that, in place of **Simplify[Sqrt[x^2], x < 0]**, we can also write the following:

```
Assuming[x < 0, Simplify[Sqrt[x^2]]]       - x
```

Here we added the assumption $x < 0$ into the current value of **$Assumptions**, giving the default assumption True && $(x < 0)$, that is, $x < 0$.

Assuming can be used to define local environments with given default assumptions in a way that is similar to the way **Module** and **Block** (see Section 14.1.5, p. 360) are used to define local environments with given variables or their values. Inside **Assuming**, we can write several commands: **Assuming[assum, expr1; expr2; …]**, and so we do not need to write the assumptions in each command. Each command can, however, have its specific assumptions; they are added to the default assumptions.

■ Options

> *Options of* `Simplify` *and* `FullSimplify`:
>
> `TimeConstraint` Time in seconds to try a particular transformation; default values:
>   300 (`Simplify`), ∞ (`FullSimplify`)
> `Trig` Whether to also use trigonometric transformations; possible values: `True`, `False`
> `TransformationFunctions` Functions tried in transforming the expression; examples
>   of values: `Automatic`, `{Automatic, f}`
> `ComplexityFunction` Function used in assessing the complexity of the transformed
>   expression; examples of values: `Automatic`, `(LeafCount[#]&)`
>
> *An additional option of* `FullSimplify`:
> `ExcludedForms` Forms of subexpressions not touched in simplification; examples of
>   values: `{}`, `Gamma[_]`

`Simplify` and `FullSimplify` do a sequence of transformations when searching for the simplest form of the expression. If `Simplify` does not complete a particular transformation in 300 seconds, it gives up, prints a warning, and goes to the next transformation. In such a case, you can consider trying the simplification anew with a larger value of `TimeConstraint`. `FullSimplify`, by default, does all transformations until completion.

Normally both algebraic and trigonometric transformations are tried. Set `Trig` to `False` if you want trigonometric transformations not to be tried.

`Simplify`—and in particular `FullSimplify`—have large collections of transformations to try. However, sometimes they are not able to simplify an expression. We can then help them by adding one or more transformation functions. For example, without the function `comp`, `FullSimplify` does not succeed in performing the following simplification:

```
comp[x_ ^ (2 n_)] := Expand[x ^ 2] ^ n

FullSimplify[(1 + Sqrt[2]) ^ (2 n) - (3 + 2 Sqrt[2]) ^ n,
  n ∈ Integers, TransformationFunctions → {Automatic, comp}]

0
```

Normally `Simplify` and `FullSimplify` assess the complexity of the transformed expressions primarily according to their `LeafCount` (the total number of indivisible subexpressions), with corrections to treat integers with more digits as more complex. For example, $4 \log(6)$ is considered as the most simple form, but if only `LeafCount` is used, then the most simple form is the following:

```
Simplify[4 Log[6], ComplexityFunction → (LeafCount[#] &)]

Log[1296]
```

`FullSimplify` allows the possibility to exclude some forms of subexpressions in the simplification. For example, if all subexpressions are taken into account in the simplification, we get the following:

```
FullSimplify[(n + 3) ! Gamma[n + 1] / (n ! Gamma[n + 2])]

(2 + n) (3 + n)
```

However, if gamma functions are not touched, we get the following:

> **FullSimplify[(n + 3)! Gamma[n + 1] / (n! Gamma[n + 2]),**
> **ExcludedForms → Gamma[_]]**
>
> $$\frac{(1 + n)\ (2 + n)\ (3 + n)\ \text{Gamma}[1 + n]}{\text{Gamma}[2 + n]}$$

■ Refine

> **Refine[expr, assumptions]** (❀5) Give the form of **expr** that would be obtained if
> symbols in it were replaced with explicit numerical expressions satisfying
> **assumptions**

Refine may sometimes be useful for simplifying or developing expressions under given assumptions. For example, in the following cases, **Simplify** does nothing, but **Refine** gives us a modified expression:

> **{Refine[Abs[x], x < 0], Refine[Log[x], x < 0],**
> **Refine[Sin[x + n π], n ∈ Integers], Refine[Sqrt[x y], x > 0]}**
>
> $\{-x,\ \mathrm{i}\,\pi + \text{Log}[-x],\ (-1)^n\ \text{Sin}[x],\ \sqrt{x}\ \sqrt{y}\}$

12.2.2 Expanding

> **Expand[expr]** Expand products and positive integer powers in the top level
> **ExpandAll[expr]** Expand products and positive integer powers in all levels
> **ExpandAll[expr, x]** Avoid expanding parts not containing **x**
> **FunctionExpand[expr]** Expand special functions
> **PowerExpand[expr]** Expand powers of products and powers of powers

Note that **FunctionExpand** accepts assumptions in a way that is similar to **Simplify** and **FullSimplify**. Note also that rational, polynomial, trigonometric, complex, and logical expressions as well as strings and characters are considered in Section 12.3.

FunctionExpand does not automatically write $\log(x\,y) = \log(x)\log(y)$:

> **Log[x y] // FunctionExpand**   Log[x y]

This happens because $\log(x\,y) = \log(x)\log(y)$ is not always correct. If x and y are positive, then the expansion can be made:

> **FunctionExpand[Log[x y], x > 0 && y > 0]**   Log[x] + Log[y]

FunctionExpand is sometimes able to convert expressions to forms in which the arguments are simpler:

> **Cos[ArcSin[x / 3]] // FunctionExpand** $\frac{1}{3}\ \sqrt{3 - x}\ \sqrt{3 + x}$

> **Log[ProductLog[3, z]] // FunctionExpand**
>
> $6\,\mathrm{i}\,\pi + \text{Log}[z] - \text{ProductLog}[3, z]$

PowerExpand does brutal expansions without taking into account the appropriate assumptions. For example:

> `{PowerExpand[Sqrt[x^2]], PowerExpand[(a^b)^c], PowerExpand[Log[x y]]}`
>
> $\{x, a^{bc}, Log[x] + Log[y]\}$

As we have already noted, these results are not always correct. In general, $(a^b)^c = a^{bc}$ and $(a\,b)^c = a^c\,b^c$ hold only if c is an integer or a and b are positive real numbers.

12.2.3 More about Expressions

■ Short Forms for Results

Sometimes the result of a computation is so long and complicated that we do not want to see it completely. On the other hand, we might sometimes be interested in the structure of the complicated expression. The following commands are useful in these cases:

Length[expr]	Give the number of topmost parts of **expr**
Short[expr, n]	Print **expr** in a shortened form that is about **n** lines long
Shallow[expr, depth]	Show all parts of **expr** below **depth** in skeleton form

If a second argument in **Short** and **Shallow** is not written, default values are used. Here is a long expression (we do not show the result):

> `q = Integrate[x^2 Sqrt[1 + x^2] / (1 - x + x^2), x] // Simplify;`

We ask for the length, a short form, and a skeleton form:

> `Length[q]`   2

> `Short[q, 1.8]`
>
> $\frac{1}{12}\left(12\sqrt{1 + x^2} + \ll 9 \gg + (-1)^{3/4}\sqrt{6\,(-i + \sqrt{3})}\,Log\left[(1 - x + x^2)\,(\ll 1 \gg)\right]\right)$

> `Shallow[q]`
>
> $\frac{1}{12}\,(Times[\ll 2 \gg] + Times[\ll 3 \gg] + Times[\ll 2 \gg] + Times[\ll 4 \gg] + $
>
> $Times[\ll 4 \gg] + Times[\ll 3 \gg] + Times[\ll 3 \gg] + Times[\ll 3 \gg] + Times[\ll 3 \gg])$

■ Everything Is an Expression

In earlier sections we have considered expressions in the ordinary mathematical sense. In *Mathematica*, however, an expression is a much wider concept. In fact, in *Mathematica*, *everything is an expression*. This fact is most clearly seen from the internal representation of the various objects of *Mathematica*. A typical expression is of the form **head[arguments]**. The internal form can be seen with **FullForm**, and the head can be seen with **Head**.

Mathematica has exactly 6 basic expressions, which are called *atoms*. Here are examples of all of them:

> `Map[Head, {2, 2/5, 3.7, 6 + 2 I, x, "message"}]`
>
> `{Integer, Rational, Real, Complex, Symbol, String}`

All expressions are made up of these basic elements. Here are some simple expressions:

```
Map[FullForm, {x + y, x y, x^y}]
```

```
{Plus[x, y], Times[x, y], Power[x, y]}
```

Some more complicated expressions:

```
Map[FullForm, {x - y, x / y, Sqrt[x]}]
```

```
{Plus[x, Times[-1, y]], Times[x, Power[y, -1]], Power[x, Rational[1, 2]]}
```

Some further examples:

```
Map[FullForm, {{x, y, z}, x == y, x < y, x → y}]
```

```
{List[x, y, z], Equal[x, y], Less[x, y], Rule[x, y]}
```

It is sometimes useful to know the internal form, particularly if we want to pick a part of an expression, as we saw in Section 12.1.3, p. 303. The parts of an expression are decided from the internal representation and not from the normal form.

■ Levels of Expressions

Here is a list:

```
p =.; t = {{0, 1}, {1, p}, {2, p^2}, {3, p^3}};
```

To illustrate the *levels* of an expression, we first define that the zeroth level of an expression is the expression itself. So, the zeroth level of **t** is **t**. At the first level, we have the four lists {0, 1}, ..., {3, p^3}. At the second level, we have the elements of the lists: 0, 1, 1, **p**, ..., **p^3**. At the third level, we have the components of the powers: **p**, 2, **p**, 3. Expressed in another way, the level of a part is the number of indices needed to show the position. For example:

```
{t[[4]], t[[4, 2]], t[[4, 2, 1]]}
```

$$\{\{3, p^3\}, p^3, p\}$$

The levels of the expressions {3,p^3}, p^3, and **p** (the **p** in **p^3**) in **t** are thus 1, 2, and 3. A list of the parts of an expression at a given level can be seen with **Level**:

```
Level[t, {3}]      {p, 2, p, 3}
```

In some commands (like **Position**, **Cases**, **Apply**, and **Map**, which are considered in Chapter 13) we can use a *level specification*. It defines the level or levels of the expression toward which the operation of the command is directed. Levels are specified as follows:

n	Levels 1 through *n*
∞	All levels 1, 2, ...
{n}	Level *n* only
{n,m}	Levels *n* through *m*

For example, the position of 2 in **t** at level 3 is as follows:

```
Position[t, 2, {3}]      {{3, 2, 2}}
```

12.3 Manipulating Special Expressions

12.3.1 Rational Expressions

`Factor[expr]` Factor polynomials in the top level of the expression

`Together[expr]` Put terms over a common denominator and cancel factors

`Cancel[expr]` Cancel out common factors in the numerator and denominator

`Apart[expr]` Give the partial fraction expansion

`Apart[expr, x]` Treat only **x** as a variable (treat other variables as constants)

`Horner[expr, x, y]` Put the numerator with respect to **x** and the denominator with respect to **y** in Horner form (in the **Algebra`Horner`** package)

`ExpandNumerator[expr]` Expand the numerator

`ExpandDenominator[expr]` Expand the denominator

`Numerator[expr]` Give the numerator

`Denominator[expr]` Give the denominator

`Variables[expr]` Give a list of all variables

Remember also the commands **Simplify** and **FullSimplify** (see Section 12.2.1, p. 304). We try some commands:

```
a = (x^2 + 2 x - 8) / (x + 4) ^2 + 2 / (x + 3)
```

$$\frac{2}{3+x} + \frac{-8+2x+x^2}{(4+x)^2}$$

```
{Simplify[a], Factor[a], Together[a], Cancel[a], Apart[a]}
```

$$\left\{ \frac{2+3x+x^2}{12+7x+x^2}, \frac{(1+x)(2+x)}{(3+x)(4+x)}, \right.$$

$$\left. \frac{2+3x+x^2}{(3+x)(4+x)}, \frac{2}{3+x} + \frac{-2+x}{4+x}, 1 + \frac{2}{3+x} - \frac{6}{4+x} \right\}$$

Factor has an option that allows Gaussian integers (complex numbers with integer real and imaginary parts):

```
Factor[1 + x^2, GaussianIntegers → True]
```
$(-i + x) (i + x)$

12.3.2 Polynomial Expressions

Below we list some commands for polynomials; see also Section 3.3 in Wolfram (2003).

`Factor[poly]` Factor **poly**

`FactorTerms[poly]` Pull out any overall numerical factor

`FactorTerms[poly, x]` Pull out any overall factor that does not depend on **x**

`Expand[poly]` Expand out products and powers

`Expand[poly, x]` Avoid expanding parts not containing **x**

`Collect[poly, x]` Collect together terms involving the same powers of **x**

`Collect[poly, x, h]` Apply function **h** to the coefficients of the powers of **x**

Decompose[poly, x]   Decompose **poly** into a composition of simpler polynomials
Horner[poly, x]   Put **poly** in Horner form (in the **Algebra`Horner`** package)

Coefficient[poly, expr]   Give the coefficient of **expr**
CoefficientList[poly, x]   Give a list of coefficients of powers of **x**
Exponent[poly, x]   Give the maximum power of **x**
Variables[poly]   Give a list of all variables

PolynomialQuotient[p, q, x]   Give the result of dividing **p** by **q**, with any remainder dropped
PolynomialRemainder[p, q, x]   Give the remainder from dividing **p** by **q**

Remember also the commands **Simplify** and **FullSimplify** (see Section 12.2.1, p. 304). Consider the following polynomial:

```
r = c^2 + 8 c x + 16 x^2 + 2 c x^2 + 8 x^3 + x^4;
```

We collect terms with respect to **c** and factor the coefficients:

```
Collect[r, c, Factor]
```

$$c^2 + 2 c x (4 + x) + x^2 (4 + x)^2$$

The Horner form is efficient and stable in numerical computations:

```
<< Algebra`Horner`
```

```
Horner[r /. c → 8]
```

$$64 + x (64 + x (32 + x (8 + x)))$$

12.3.3 Trigonometric and Hyperbolic Expressions

TrigExpand[expr]   Expand sums and multiple angles in arguments
TrigFactor[expr]   Factor **expr** in a product form
TrigReduce[expr]   Write **expr** into a sum with no products or powers
TrigToExp[expr]   Write trigonometric functions in terms of complex exponentials
ExpToTrig[expr]   Write complex exponentials in terms of trigonometric functions

Remember that **Simplify** and **FullSimplify** (see Section 12.2.1, p. 304) also simplify, by default, trigonometric expressions (write the option **Trig → False** if you do not want do trigonometric simplifications):

```
{Simplify[1 - Sin[x]^2], Simplify[(1 - Cos[x]) / Sin[x]]}
```

$$\left\{ Cos[x]^2, \ Tan\left[\frac{x}{2}\right] \right\}$$

Remember also that **Simplify** and **FullSimplify** accept assumptions:

```
Integrate[x^2 Sin[n x], {x, 0, Pi}]
```

$$\frac{-2 + (2 - n^2 \pi^2) Cos[n \pi] + 2 n \pi Sin[n \pi]}{n^3}$$

```
FullSimplify[%, n ∈ Integers]
```

$$\frac{-2 + (-1)^n (2 - n^2 \pi^2)}{n^3}$$

Here are some examples of the other commands:

> `{TrigReduce[2 Sin[x] ^2], TrigReduce[2 Sin[x] Cos[y]]}`
>
> $\{1 - \text{Cos}[2\,x]\,,\ \text{Sin}[x - y] + \text{Sin}[x + y]\}$

> `{TrigFactor[Sin[3 x]], TrigExpand[Sin[3 x]]}`
>
> $\{(1 + 2\,\text{Cos}[2\,x])\ \text{Sin}[x]\,,\ 3\,\text{Cos}[x]^2\,\text{Sin}[x] - \text{Sin}[x]^3\}$

> `{TrigToExp[2 Sin[x]], ExpToTrig[Exp[I x] + Exp[-I x]]}`
>
> $\{i\,e^{-i\,x} - i\,e^{i\,x}\,,\ 2\,\text{Cos}[x]\}$

With **TrigExpand**, we can calculate Chebyshev polynomials from a trigonometric expression:

> `Table[TrigExpand[Cos[n ArcCos[x]]], {n, 0, 4}]`
>
> $\{1\,,\ x\,,\ -1 + 2\,x^2\,,\ -3\,x + 4\,x^3\,,\ 1 - 8\,x^2 + 8\,x^4\}$

The same commands we considered above for trigonometric functions also work for hyperbolic functions. Some examples:

> `{Simplify[1 + Sinh[x] ^2], TrigReduce[2 Sinh[x] ^2], TrigFactor[Sinh[3 x]]}`
>
> $\{\text{Cosh}[x]^2\,,\ -1 + \text{Cosh}[2\,x]\,,\ (1 + 2\,\text{Cosh}[2\,x])\ \text{Sinh}[x]\}$

> `ExpToTrig[Exp[x / 2] + Exp[x] + Exp[2 I x]]`
>
> $\text{Cos}[2\,x] + \text{Cosh}\left[\dfrac{x}{2}\right] + \text{Cosh}[x] + i\,\text{Sin}[2\,x] + \text{Sinh}\left[\dfrac{x}{2}\right] + \text{Sinh}[x]$

12.3.4 Complex Expressions

> **ComplexExpand[expr]**   Expand to real and imaginary parts assuming all variables are real
>
> **ComplexExpand[expr, {x, y, … }]**   Expand assuming **x, y**, … are complex
>
> **ComplexExpand[expr, TargetFunctions → list]**   Try to expand in terms of functions in **list**

Possible target functions are **Re, Im, Abs, Arg, Conjugate**, and **Sign**; the default is typically to give results in terms of **Re** and **Im**. Some examples:

> `(-8) ^ (1 / 3)` $2\,(-1)^{1/3}$

> `ComplexExpand[%]` $1 + i\,\sqrt{3}$

> `ComplexExpand[Sin[x + I y]]`
>
> $\text{Cosh}[y]\,\text{Sin}[x] + i\,\text{Cos}[x]\,\text{Sinh}[y]$

We assume that **z** is complex:

> `ComplexExpand[z, z]` $i\,\text{Im}[z] + \text{Re}[z]$

The result is written, by default, in terms of **Re** and **Im**. Next we ask for the result in a polar form, that is, in terms of **Abs** and **Arg**:

```
ComplexExpand[z, z, TargetFunctions -> {Abs, Arg}]
```

Abs[z] Cos[Arg[z]] + i Abs[z] Sin[Arg[z]]

The **Algebra`ReIm`** package enhances the capabilities of *Mathematica* by providing a means for declaring a variable to be real and a function to be real for real arguments. Note also that in Section 5.3.4, p. 128, we considered plotting complex-valued functions.

For roots like $(-8)^{1/3}$, *Mathematica* gives one value, the *principal root* (the root with the last positive argument) (see Section 12.4.1, p. 318). For negative arguments, the principal value is complex. The package **Miscellaneous`RealOnly`** redefines the power function so that odd roots of negative numbers are negative instead of complex. More exactly, $b^{m/n}$ gives $\left(-(-b)^{1/n}\right)^m$ if b is negative and m and n are integers with n being odd. For example:

```
<< Miscellaneous`RealOnly`
```

```
(-8) ^ (1 / 3)      - 2
```

The package defines all unavoidable complex results as **Nonreal**:

```
3 + Sqrt[-2]
```

```
Nonreal::warning :  Nonreal number encountered.
Nonreal
```

12.3.5 Logical Expressions

▪ Logical Tests

For testing whether an expression has a given property, *Mathematica* has several built-in tests. Below is a collection of such tests. Tests are used, for example, with **Select** (see Section 13.2.5, p. 340), with **If** and **Which** (see Section 15.1.2, p. 387), and with patterns (see Section 15.3.5, p. 413).

== (Equal), != (Unequal), === (SameQ), =!= (UnsameQ)

< (Less), ≤ (LessEqual), > (Greater), ≥ (GreaterEqual)

Negative, NonPositive, NonNegative, Positive

NumericQ, NumberQ, IntegerQ, EvenQ, OddQ, PrimeQ

ListQ, VectorQ, MatrixQ, TensorQ

PolynomialQ, StringQ, OptionQ

FreeQ, MemberQ, MatchQ

A logical statement gives a result of **True** of **False** or, if *Mathematica* cannot decide the validity of the property, the test as such. Here are some examples:

```
{2 < 3, 4 < 3, x < 3}
```

{True, False, x < 3}

```
{Positive[-3], IntegerQ[3], EvenQ[3], PrimeQ[3]}
```

{False, True, False, True}

```
{ListQ[3], ListQ[{}], ListQ[{3, 5, 2}]}
```

{False, True, True}

        `IntegerQ[2] && EvenQ[2] && PrimeQ[2]`   `True`

NumericQ tests whether the expression has a numerical value, while **NumberQ** tests whether the expression is a number. Recall from Section 11.1.1, p. 280, that *Mathematica* has four kinds of numbers: integers, rationals, reals, and complexes. *Mathematica* considers exact expressions like **Pi**, **Sqrt[2]**, or **Sin[5]** not as numbers, but they do have numeric values:

        `{NumberQ[Pi], NumericQ[Pi]}`   `{False, True}`

VectorQ and **MatrixQ** accept a second argument defining a test to be satisfied by the elements (the test is written as a pure function; see Section 2.2.2, p. 36):

        `VectorQ[{2, a, Sqrt[5]}]`   `True`

        `VectorQ[{2, -3, Sqrt[5]}, NumericQ]`   `True`

        `VectorQ[{2, 3, Sqrt[5]}, NumericQ[#] && Positive[#] &]`   `True`

The following second-order polynomial does not contain **y**:

        `FreeQ[a + b x + c x^2, y]`   `True`

The attributes of **Pi** are as follows:

        `Attributes[Pi]`   `{Constant, Protected, ReadProtected}`

This means that **Constant** is a member of **Attributes[Pi]**:

        `MemberQ[Attributes[Pi], Constant]`   `True`

■ Testing Equality

Now we compare the tests **expr1 == expr2** (**Equal**) and **expr1 === expr2** (**SameQ**). In general, the latter test is more demanding. Both tests give **True** if the expressions are identical and **False** if they are not identical. If *Mathematica* cannot decide whether the expressions are identical, **==** gives the original test **expr1 == expr2** as such, but **===** gives **False**. The expression **expr1 == expr2** returned by **==** can be considered as an equation from which we can perhaps solve a variable. For example:

        `{2 == 2, 2 === 2, 2 - x == 0, 2 - x === 0}`

        `{True, True, 2 - x == 0, False}`

The tests differ somewhat in the way they treat numerical expressions. The test **==** gives **True** if the numerical values of the expressions differ in at most their eight binary digits, which correspond roughly to their last two decimal digits of the 16 standard digits. On the other hand, the test **===** gives **True** if the difference of the expressions is less than the uncertainty of either of them, which means in practice that the expressions must be equal to the last digit and that exact numbers are not considered equal to their decimal values. Here are some examples:

        `{2 == 2., 2 === 2.}`

        `{True, False}`

        `{2. + 10^-13 == 2., 2. + 10^-14 == 2.}`

        `{False, True}`

```
{2. + 10^-15 === 2., 2. + 10^-16 === 2.}
```

```
{False, True}
```

Note that **Equal**, **Unequal**, **SameQ**, and **UnsameQ** accept more than two expressions as arguments. For example:

```
UnsameQ[1, 2, 3, 4, 5, 2]    False
```

■ Logical Expressions

The logical tests can be combined with the following logical operations to form more complex logical expressions:

---

p && q True if both **p** and **q** are true (AND)
p || q True if one or both of **p** and **q** are true (OR)
!p True if **p** is false (NOT)
Xor[p, q] True if one and only one of **p** and **q** is true (exclusive OR)
Nand[p, q] Means **Not[And[p, q]]** (true if **p** or **q** is false, false if they are both true)
Nor[p, q] Means **Not[Or[p, q]]** (true if both **p** and **q** are false, false if either of them is true)

LogicalExpand[expr]   Expand a logical statement

```
LogicalExpand[Nand[p, q]]      !p || !q
```

```
LogicalExpand[(p || q) && (r || s)]
```

```
p && r || p && s || q && r || q && s
```

12.3.6 Strings and Characters

A string is an expression written inside quotation marks, such as `"Here is a message."` The quotation marks do not appear in the output:

```
s1 = "Here is a message."
```

```
Here is a message.
```

However, the quotation marks can be seen with **InputForm**:

```
s1 // InputForm
```

```
"Here is a message."
```

Mathematica has powerful string and character manipulation commands. Here are lists of names related to strings and characters:

```
Names["*String*"]
```

```
{DisplayString, ExportString, ImportString, InputString,
 InputStringPacket, InString, RepeatedString, ShowStringCharacters,
 String, StringBreak, StringByteCount, StringDrop, StringForm,
 StringInsert, StringJoin, StringLength, StringMatchQ, StringPosition,
 StringQ, StringReplace, StringReplacePart, StringReverse,
 StringSkeleton, StringTake, StringToStream, ToString, WriteString}
```

```
Complement[Names["*Character*"], Names["*Characteristic*"]]
```

{Character, CharacterEncoding, CharacterEncodingsPath,
 CharacterRange, Characters, ExternalDataCharacterEncoding,
 FromCharacterCode, MetaCharacters, ShowSpecialCharacters,
 ShowStringCharacters, SpanCharacterRounding, SpanningCharacters,
 ToCharacterCode, $CharacterEncoding, $SystemCharacterEncoding}

Let us mention only a few of these commands here. For a full exposition, see Section 2.8 in Wolfram (2003).

StringLength[s] Give the number of characters in a string

StringJoin[{s1, s2, … }] or **s1 <> s2 <> …** Join several strings together

Sort[{s1, s2, … }] Sort a list of strings

Characters[s] Convert a string to a list of characters

CharacterRange["c1", "c2"] Generate a list of all characters from **c1** to **c2**

In a string, a new line can be defined by **\n** and a tab with **\t**:

```
s1 = "\tHere is a message\n\tin two rows."
```

```
    Here is a message
    in two rows.
```

12.4 Mathematical Functions

12.4.1 Basic Functions

Some basic functions are **Sqrt**, **Exp**, and **Log**, together with the trigonometric and hyperbolic functions:

```
trig = {Sin, Cos, Tan, Cot, Sec, Csc};
invtrig = {ArcSin, ArcCos, ArcTan, ArcCot, ArcSec, ArcCsc};
hyp = {Sinh, Cosh, Tanh, Coth, Sech, Csch};
invhyp = {ArcSinh, ArcCosh, ArcTanh, ArcCoth, ArcSech, ArcCsch};
```

Note that **Sqrt[x]** can also be written as **x^(1/2)** or $\sqrt{x}$ and **Exp[x]** as **E^x** or e^x. Also note that **Log[x]** is the natural logarithm and **Log[b, x]** the logarithm to base **b**. The argument of the trigonometric functions is in radians.

For all mathematical functions of *Mathematica*, note the following:

- For integers, rational numbers, and special symbols, the functions give an exact result: the expression as such if *Mathematica* cannot simplify it or a simplified expression.
- An approximate decimal value is calculated if the argument contains a decimal number.
- Values can be calculated to any numerical precision by giving a high precision argument or with **N[expr, n]** (see Sections 11.1.2, p. 281, and 11.2.2, p. 287).
- Arguments can be complex numbers.
- Arguments can be lists.

For example:

```
{Exp[0.3`18], Exp[3.5 + I], Exp[{1., 2., 3.}]}
```

$\{1.349858807576003104, 17.8924 + 27.8657\,\mathrm{i}, \{2.71828, 7.38906, 20.0855\}\}$

Note that, to calculate exp(0.3) to high precision, the argument has to be written in a high precision form (see Section 11.2.2, p. 289). We could also have written **N[Exp[3/10], 19]**.

Some notes about roots and inverse trigonometric functions follow. If you use complex arguments, read about branch-cut discontinuities in Section 3.2.7 of Wolfram (2003).

■ Roots

Sqrt[x] is not a true inverse function of the function **x^2**. The true inverse is two-valued: the square root of 4 is a number x such that $x^2 = 4$, and there are two solutions for this equation: $x = 2$ and $x = -2$. However, **Sqrt[4]** gives only the positive value 2. If we want both the positive and the negative value for a square root, one possibility is to use **Solve**:

```
Solve[x^2 == 4]      {{x → -2}, {x → 2}}
```

The situation is similar for other roots. For example, the third root of 8 is a number x such that $x^3 = 8$. There are three solutions to this equation, but *Mathematica* gives only one, the *principal root* (the root with the last positive argument):

```
{8 ^ (1 / 3), (-8) ^ (1 / 3)}      {2, 2 (-1)^{1/3}}

% // ComplexExpand      {2, 1 + i √3 }
```

With **Solve**, we get all roots:

```
Solve[x^3 == 8]      {{x → 2}, {x → -1 - i √3 }, {x → -1 + i √3 }}

Solve[x^3 == -8]      {{x → -2}, {x → 1 - i √3 }, {x → 1 + i √3 }}
```

With the **Miscellaneous`RealOnly`** package, we get, for odd roots of negative numbers, a negative number as the result instead of a complex number (see Section 12.3.4, p. 314).

■ Inverse Trigonometric and Hyperbolic Functions

Inverse trigonometric functions and **ArcCosh** and **ArcSech** are multiple-valued, too. *Mathematica* gives the principal value for them. For easy reference, here are plots of all of the trigonometric and hyperbolic functions and their inverse functions:

```
trigPlot[functs_, interval_, ranges_, ticks_] := Map[
    Plot[#[x], interval, PlotLabel → #, Ticks → ticks, PlotRange → ranges,
        AspectRatio → Automatic, DisplayFunction → Identity] &, functs];

r = Range[-3, 3]; t = {-3.2, 3.2}; p1 = {-π / 2, "-π/2"}; p2 = {π / 2, "π/2"};

g1 = trigPlot[trig, {x, -π, π}, t, {{-π, p1, p2, π}, r}];
g2 =
    trigPlot[invtrig, {x, -3.2, 3.2}, {t, {-1.8, 3.4}}, {r, {p1, p2, π}}];
```

```
Show[GraphicsArray[{g1, g2}], GraphicsSpacing → {0, 0.1}];
```

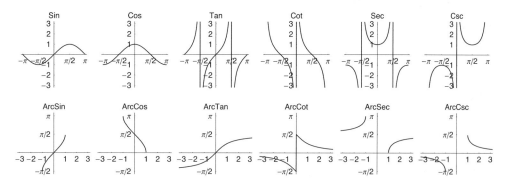

```
g3 = trigPlot[hyp, {x, -3.2, 3.2}, t, {r, r}];
g4 = trigPlot[invhyp, {x, -3, 3}, {t, t}, {r, r}];

Show[GraphicsArray[{g3, g4}], GraphicsSpacing → {0, 0.1}];
```

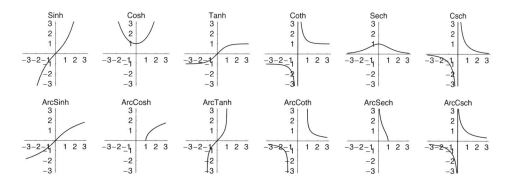

12.4.2 More Functions

■ Combinatorial Functions

n! Factorial $n(n-1)(n-2)\ldots 1$

n!! Double factorial $n(n-2)(n-4)\ldots$

Binomial[n, m] Binomial coefficient $\binom{n}{m} = \dfrac{n!}{m!\,(n-m)!}$

Multinomial[n1, …, nk] Multinomial coefficient $(N; n_1, \ldots, n_k) = \dfrac{N!}{n_1!\ldots n_k!}$, $N = \sum_{i=1}^{k} n_i$

Pochhammer[n, m] Pochhammer symbol $(n)_m = n(n+1)(n+2)\ldots(n+m-1)$

Fibonacci[n] Fibonacci number $F_n : F_n = F_{n-1} + F_{n-2}$, $F_1 = F_2 = 1$

HarmonicNumber[n] Harmonic number $H_n = \sum_{i=1}^{n} \dfrac{1}{i}$

Note that these functions can also be calculated for noninteger arguments. In addition we have such constants as **BernoulliB[n]**, **EulerE[n]**, **StirlingS1[n, m]**, and **StirlingS2[n, m]**. The **DiscreteMath`CombinatorialFunctions`** package defines

`CatalanNumber[n]`, `Hofstadter[n]`, and `Subfactorial[n]`. The `DiscreteMath`Combina` `torica`` package defines more than 230 functions used in combinatorics and graph theory.

■ Nonsmooth Functions

> `Max[x, y, … ]`, `Max[list]` The maximum of x, y, ... or of the elements in `list`
> `Min[x, y, … ]`, `Min[list]` The minimum of x, y, ... or of the elements in `list`
> `Abs[x]` $-x$ if $x < 0$, and x if $x \geq 0$
> `Sign[x]` -1, 0, or 1 if $x < 0$, $x = 0$, or $x > 0$, respectively
>
> `UnitStep[x]` 0 if $x < 0$, and 1 if $x \geq 0$
> `UnitStep[x, y, … ]` 0 if any of x, y, ... is negative, and 1 if x, y, ... are all nonnegative
> `DiracDelta[x]` 0 if $x \neq 0$
> `DiracDelta[x, y, … ]` 0 if any of x, y, ... is not 0.
> `DiscreteDelta[m, n, … ]` 1 if m, n, ... are all 0, and 0 otherwise
> `KroneckerDelta[m, n, … ]` 1 if m, n, ... are all equal, and 0 otherwise

The `DiscreteMath`DiscreteStep`` package also defines `DiscreteStep`. The `ProgrammingInMathematica`Abs`` package gives rules for the derivative and integral of `Abs` and `Sign`. Note that `Max` and `Min` also search maximum and minimum values from multidimensional lists. Here is an example of `Max`:

```
Plot[Abs[Max[Sin[x], Cos[x]]], {x, 0, 2 Pi}];
```

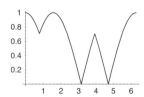

We plot a square wave with `UnitStep`:

```
Plot[UnitStep[Sin[2 x]], {x, 0, 10}, AxesOrigin → {-0.3, -0.1}];
```

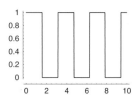

`DiracDelta` can be used, for example, with integrals, integral transformations, and differential equations:

```
Integrate[DiracDelta[x] , x]      UnitStep[x]

Integrate[DiracDelta''[x] f[x], {x, -∞, ∞}]      f''[0]
```

■ Orthogonal Polynomials

Below are some orthogonal polynomials (for even more, see Wolfram (2003), Section 3.2.9). If $P_n(x)$ is an orthogonal polynomial on (a, b) with respect to weight function $w(x)$, it satisfies the orthogonality condition $\int_a^b P_n(x) P_m(x) w(x)\, dx = 0$, $m \neq n$.

`LegendreP[n,x]`	Orthogonal in $(-1, 1)$ with respect to 1
`ChebyshevT[n,x]`	Orthogonal in $(-1, 1)$ with respect to $1/\sqrt{1-x^2}$
`ChebyshevU[n,x]`	Orthogonal in $(-1, 1)$ with respect to $\sqrt{1-x^2}$
`HermiteH[n,x]`	Orthogonal in $(-\infty, \infty)$ with respect to $\exp(-x^2)$
`LaguerreL[n,x]`	Orthogonal in $(0, \infty)$ with respect to $\exp(-x)$

```
t1 = Table[ChebyshevT[n, x], {n, 0, 4}]
```

$\{1,\ x,\ -1 + 2\,x^2,\ -3\,x + 4\,x^3,\ 1 - 8\,x^2 + 8\,x^4\}$

```
Plot[Evaluate[t1], {x, -1, 1}];
```

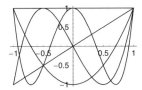

12.4.3 Special Functions

Below are some lists of special functions in *Mathematica*; for details and a complete list, see Sections 3.2.10, 3.2.11, and 3.2.12 in Wolfram (2003). We consider probability distribution functions in Section 26.1.

`Gamma[z]; Gamma[a,z]`	Gamma function $\Gamma(z)$; incomplete gamma function $\Gamma(a, z)$
`PolyGamma[z]; PolyGamma[n,z]`	Digamma function $\psi(z)$; the nth derivative of $\psi(z)$
`Beta[a,b]; Beta[z,a,b]`	Beta function $B(a, b)$; incomplete beta function $B_z(a, b)$
`Hypergeometric1F1[a,b,z]`	Confluent hypergeometric function $_1F_1(a; b; z)$
`Hypergeometric2F1[a,b,c,z]`	Hypergeometric function $_2F_1(a; b; c; z)$

$$\Gamma(z) = \int_0^\infty t^{z-1}\,e^{-t}\,dt, \quad \Gamma(a, z) = \int_z^\infty t^{a-1}\,e^{-t}\,dt, \quad \psi(z) = \frac{d}{dz}\log\Gamma(z) = \frac{\Gamma'(z)}{\Gamma(z)},$$

$$B(a, b) = \frac{\Gamma(a)\,\Gamma(b)}{\Gamma(a+b)} = \int_0^1 t^{a-1}\,(1-t)^{b-1}\,dt, \quad B_z(a, b) = \int_0^z t^{a-1}\,(1-t)^{b-1}\,dt,$$

$$_1F_1(a; b; z) = \frac{\int_0^1 e^{zt}\,t^{a-1}\,(1-t)^{b-a-1}\,dt}{B(a, b-a)}, \quad _2F_1(a; b; c; z) = \frac{\int_0^1 t^{b-1}\,(1-t)^{c-b-1}\,(1-tz)^{-a}\,dt}{B(b, c-b)}.$$

`Erf[z]; Erfc[z]`	Error function erf(z); complementary error function erfc(z)
`ExpIntegralE[n,z]; ExpIntegralEi[z]`	Exponential integrals $E_n(z)$ and Ei(z)
`LogIntegral[z]; PolyLog[n,z]`	Logarithmic integral function li(z); polylogarithm function $\mathrm{Li}_n(z)$
`SinIntegral[z]; CosIntegral[z]`	Sine and cosine integral functions Si(z) and Ci(z)
`Zeta[s]`	Riemann zeta function $\zeta(s)$
`ProductLog[z]`	Product log function $W(z)$ (solution of $z = w\,e^w$)

$$\mathrm{erf}\,z = \frac{2}{\sqrt{\pi}} \int_0^z e^{-t^2}\,dt, \quad \mathrm{erfc}(z) = 1 - \mathrm{erf}(z), \quad E_n(z) = \int_1^\infty \frac{e^{-zt}}{t^n}\,dt, \quad \mathrm{Ei}(z) = -\int_{-z}^\infty \frac{e^{-t}}{t}\,dt,$$

$$\mathrm{li}(z) = \int_0^z \frac{dt}{\log(t)}\,dt, \quad \mathrm{Li}_2(z) = \int_z^0 \frac{\log(1-t)}{t}\,dt, \quad \mathrm{Li}_n(z) = \sum_{k=1}^\infty \frac{z^k}{k^n},$$

$$\mathrm{Si}(z) = \int_0^z \frac{\sin(t)}{t}\,dt, \quad \mathrm{Ci}(z) = -\int_z^\infty \frac{\cos(t)}{t}\,dt, \quad \zeta(s) = \sum_{k=1}^\infty \frac{1}{k^s} \ (s > 1).$$

LegendreP[n,z]; LegendreQ[n,z]   Legendre function $P_n(z)$; Legendre function of the second kind $Q_n(z)$

LegendreP[n,m,z]; LegendreQ[n,m,z]   Associated Legendre function $P_n^m(z)$; associated Legendre function of the second kind $Q_n^m(z)$

BesselJ[n,z]; BesselY[n,z]   Bessel functions $J_n(z)$ and $Y_n(z)$

BesselI[n,z]; BesselK[n,z]   Modified Bessel functions $I_n(z)$ and $K_n(z)$

AiryAi[z]; AiryBi[z]   Airy functions Ai(z) and Bi(z)

$P_n(z)$, $Q_n(z)$: independent solutions of $(1 - z^2)\,y'' - 2\,z\,y' + n(n+1)\,y = 0$

$P_n^m(z)$, $Q_n^m(z)$: independent solutions of $(1 - z^2)\,y'' - 2\,z\,y' + [n(n+1) - m^2/(1-z^2)]\,y = 0$

$J_n(z)$, $Y_n(z)$: independent solutions of $z^2\,y'' + z\,y' + (z^2 - n^2)\,y = 0$

$I_n(z)$, $K_n(z)$: independent solutions of $z^2\,y'' + z\,y' - (z^2 + n^2)\,y = 0$

Ai(z), Bi(z): independent solutions of $y'' - z\,y = 0$

```
Plot[BesselJ[5, x], {x, 0, 50}];
```

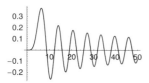

Remember that **FullSimplify** and **FunctionExpand** (see Sections 12.2.1, p. 304, and 12.2.2, p. 308) know a lot of special functions.

Lists

Introduction

> *Wiener once went to a doctor and told him that his memory was terrible and that he couldn't remember anything from one minute to the next. "How long has this been going on?" asked the doctor. "How long has what been going on?" said Wiener.*

A list is *Mathematica*'s way of storing information so that the pieces of information are well arranged and can, at any time, be easily "remembered" or retrieved. Lists are the bread and butter of *Mathematica*—you simply cannot live without them. Lists are used as the basic method of collecting numbers, symbols, and other objects. In addition, vectors and matrices are, in fact, lists. We would especially like to note that **Map**, which is considered in Section 13.3.1, is very useful for manipulating lists and is also central to functional program ming, which is considered in Section 15.2. Matrix calculus is discussed in Chapter 18.

13.1 Forming and Tabulating Lists

13.1.1 Forming Lists

■ Lists of Various Dimensions

A list is an ordered collection of zero or more elements. Here are some examples:

```
m1 = {32, 214, 5};
m2 = {{32.7, 8.39, -412.64}, {4.5, -56.2163, -7.606}};
m3 = {{{1, a}, {2, b}, {3, c}}, {{4, d}, {5, e}, {6, f}}};
```

The ordering means that, for example, **{3, 2, 5}** and **{2, 3, 5}** are not the same lists. An empty list is **{}**. A 1D list like **m1** can also be considered as a vector and a 2D list like **m2** as a matrix (each row is a list and so **m2** is a 2×3 matrix). Higher-dimensional lists like **m3** are tensors. Manipulating lists is considered in detail in Section 13.2, but here we present some important commands that you may need soon.

---

list[[i]], **list[[i,j]]**   ith and (i, j)th part of **list**
Transpose[list]   Transpose the first two levels of **list**

Picking of parts has already been considered in Section 12.1.3, p. 303.

```
m2[[1]]        {32.7, 8.39, -412.64}
```

```
m3[[1, 2, 2]]     b
```

Transposing a 2D list means converting columns to rows:

```
Transpose[m2]
```

```
{{32.7, 4.5}, {8.39, -56.2163}, {-412.64, -7.606}}
```

The following commands give information about the size and structure of a list.

Length[list]   Number of elements at the top level of **list**
Dimensions[list]   Dimensions of **list**
ArrayDepth[list] (❀5)   The depth to which **list** is a full array (use **TensorRank** in *Mathematica* 4)

```
{Length[m1], Length[m2], Length[m3]}
```

```
{3, 2, 2}
```

```
{Dimensions[m1], Dimensions[m2], Dimensions[m3]}
```

```
{{3}, {2, 3}, {2, 3, 2}}
```

```
{ArrayDepth[m1], ArrayDepth[m2], ArrayDepth[m3]}
```

```
{1, 2, 3}
```

■ Using Table and Range

Table is one of the most useful commands in *Mathematica*. If the elements of a list can be obtained from a formula, then **Table** is the right tool to use to form the list.

Table[expr, iterspec]   Form a list from **expr** according to the iteration specification **iterspec**

The following forms can be used for **iterspec**:

{b}   Make a list of **b** copies of **expr**

{i, b}   Make a list of the values of **expr** when **i** runs from 1 to **b**

{i, a, b}   **i** runs from **a** to **b**

{i, a, b, d}   **i** runs from **a** to **b** in steps of **d**

Iteration specifications extend to more indices. For example, the specification **{i, a, b}, {j, c, d}** means that **i** runs from **a** to **b** and, for each **i**, **j** runs from **c** to **d** (**c** and **d** can contain **i**). Here are some examples:

> **Table[1, {3}]**   {1, 1, 1}

> **Table[Random[], {3}]**   {0.727554, 0.0416672, 0.0570726}

> **Table[Cos[n Pi / 2], {n, 0, 6, 2}]**   {1, -1, 1, -1}

> **Table[Integrate[1 / x^i, x], {i, 0, 4}]**

$$\left\{x,\ \text{Log}[x],\ -\frac{1}{x},\ -\frac{1}{2\,x^2},\ -\frac{1}{3\,x^3}\right\}$$

> **Table[1 / (i + j - 1), {i, 3}, {j, 3}]**

$$\left\{\left\{1,\ \frac{1}{2},\ \frac{1}{3}\right\},\ \left\{\frac{1}{2},\ \frac{1}{3},\ \frac{1}{4}\right\},\ \left\{\frac{1}{3},\ \frac{1}{4},\ \frac{1}{5}\right\}\right\}$$

The index can also have decimal values, and the expression to be tabulated may be a list:

> **Table[Log[x], {x, 1, 2, 0.2}]**

> {0, 0.182322, 0.336472, 0.470004, 0.587787, 0.693147}

> **Table[{x, Log[x]}, {x, 1, 2, 0.2}]**

> {{1, 0}, {1.2, 0.182322}, {1.4, 0.336472},
> {1.6, 0.470004}, {1.8, 0.587787}, {2., 0.693147}}

Range[n] $\{1, 2, 3, \dots, n\}$

Range[m, n] $\{m, m+1, m+2, \dots, n\}$

Range[m, n, d] $\{m, m+d, m+2\,d, \dots, n\}$

With **Range**, it is easy to form lists of consecutive numbers:

> **{Range[4], Range[0, 4], Range[0, 4, 2]}**

> {{1, 2, 3, 4}, {0, 1, 2, 3, 4}, {0, 2, 4}}

Range even works with real numbers:

```
Range[2.6, 5.4, 0.5]

{2.6, 3.1, 3.6, 4.1, 4.6, 5.1}
```

Note that **Range** stopped at 5.1, because the next number, 5.6, would be larger than the upper bound given, 5.4.

13.1.2 Tabulating Lists

TableForm[list] Form a table from **list**

Options:

TableSpacing Space between rows and columns; examples of values: **Automatic** (means **{1, 3}** for 2D tables), **{2,4}**

TableAlignments Alignment of elements in horizontal and vertical directions; examples of values: **Automatic** (means **{Left, Center}** for 2D tables), **Right**, **{Right, Bottom}**

TableHeadings Labels for rows and columns; examples of values: **None**, **Automatic** (means consecutive integers), **{None, {"Col1", "Col2", "Col3\n"}}**

TableDepth Up to what level the tabular form is used; examples of values: ∞, **2**

TableDirections How to arrange the rows and columns; examples of values: **Column**, **{Row, Column}**

With **TableForm**, we can create attractive columns and tables from lists and nested lists:

```
m1 = {32, 214, 5};
m2 = {{32.7, 8.39, -412.64}, {4.5, -56.2163, -7.606}};
m3 = {{{1, a}, {2, b}, {3, c}}, {{4, d}, {5, e}, {6, f}}};

TableForm[m1]

32
214
5

m2 // TableForm

32.7        8.39          -412.64
4.5         -56.2163      -7.606

m3 // TableForm

1       2       3
a       b       c
4       5       6
d       e       f
```

Next we consider the options of **TableForm**.

■ TableSpacing

TableSpacing affects the space between rows and columns. The default value **Automatic** is **{1, 3}**, and this means that the space between rows is about the height of one character and the space between columns is about the width of 2×3 characters. An example:

```
TableForm[m2, TableSpacing → {0.2, 1.5}]
```

```
32.7    8.39        -412.64
4.5     -56.2163    -7.606
```

■ TableAlignments

TableAlignments specifies how each entry in the table should be aligned in the horizontal and vertical directions. Possible values for horizontal alignment are **Left**, **Center**, and **Right**. Possible values for vertical alignment are **Bottom**, **Center**, and **Top**. All combinations of horizontal and vertical alignments can be used. The default value **Automatic** means **{Left, Center}** or simply **Left** for 2D tables.

```
TableForm[m2, TableAlignments → Center]
```

```
32.7        8.39        -412.64
4.5       -56.2163      -7.606
```

```
TableForm[m2, TableAlignments → Right]
```

```
32.7         8.39       -412.64
4.5       -56.2163      -7.606
```

The value **Right** is suitable for tables of integers:

```
m4 = Table[{n!, (2 n)!, (3 n)!}, {n, 4}];
```

```
TableForm[m4, TableAlignments → Right]
```

```
1        2            6
2       24          720
6      720       362880
24   40320    479001600
```

For tables of real numbers, **PaddedForm** may be the most suitable command for horizontal alignment (see below).

■ TableHeadings

With **TableHeadings**, we can put labels on columns and rows. The default is **None**, which means that no headings are printed. The value **Automatic** gives successive integers as the headings:

```
TableForm[m2, TableHeadings → Automatic]
```

```
      1        2         3
1    32.7     8.39      -412.64
2    4.5     -56.2163   -7.606
```

Next we define our own headings; **\n** (the new line character) can be used to define an empty line below the headings of the columns:

```
TableForm[m2, TableHeadings → {None, {"Col. 1 ", "Col. 2 ", "Col. 3\n"}}]
```

```
Col. 1      Col. 2       Col. 3

32.7        8.39        -412.64
4.5        -56.2163     -7.606
```

```
TableForm[m2,
  TableHeadings → {Automatic, {"Col. 1 ", "Col. 2 ", "Col. 3 "}}]
```

	Col. 1	Col. 2	Col. 3
1	32.7	8.39	-412.64
2	4.5	-56.2163	-7.606

```
TableForm[m2,
  TableHeadings → {{"Row 1", "Row 2"}, {"Col. 1", "Col. 2", "Col. 3"}}]
```

	Col. 1	Col. 2	Col. 3
Row 1	32.7	8.39	-412.64
Row 2	4.5	-56.2163	-7.606

■ TableDepth

TableDepth specifies up to what level the tabular form is used; the default is ∞.

```
TableForm[m3, TableDepth → 2]
```

{1, a}	{2, b}	{3, c}
{4, d}	{5, e}	{6, f}

■ ColumnForm

> **ColumnForm[list]**   Form a column from **list**

For 1D lists, **ColumnForm** gives the same result as **TableForm**. For 2D lists, **ColumnForm** writes each row as such in its own row:

```
ColumnForm[m2]
```

```
{32.7, 8.39, -412.64}
{4.5, -56.2163, -7.606}
```

With **ColumnForm** we can easily add alignment specification **Left** (the default), **Center**, or **Right**:

```
ColumnForm[m1, Right]
```

```
 32
214
  5
```

■ PaddedForm

> **PaddedForm[TableForm[list], {n, f}]**   Align all numbers right; reserve space for **n** digits for all numbers, with **f** of them for decimals

Consider the matrix **m2**:

```
TableForm[m2, TableHeadings → {None, {"Col. 1", "Col. 2", "Col. 3"}}]
```

Col. 1	Col. 2	Col. 3
32.7	8.39	-412.64
4.5	-56.2163	-7.606

To align columns with the decimal point, use **PaddedForm** and a fixed space for the decimals:

```
PaddedForm[TableForm[m2, TableAlignments → Right,
  TableHeadings → {None, {"Col. 1", "Col. 2", "Col. 3"}}], {6, 3}]
```

Col. 1	Col. 2	Col. 3
32.700	8.390	-412.640
4.500	-56.216	-7.606

If necessary, the decimal digits are shortened and zeros are added to fill the fixed space of the decimals; the decimal point and the possible sign are not counted in the total space.

PaddedForm has the following options and default values:

```
Options[PaddedForm] // InputForm
```

```
{DigitBlock -> Infinity, ExponentFunction -> Automatic,
 ExponentStep -> 1, NumberFormat -> Automatic,
 NumberMultiplier -> "×", NumberPadding -> {" ", "0"},
 NumberPoint -> ".", NumberSeparator -> ",",
 NumberSigns -> {"-", ""}, SignPadding -> False}
```

If you do not want the filling zeros after short decimal parts, use the option **Number Padding → {" ", " "}**; if you want the minus sign in a fixed place, use **SignPadding → True** (then spaces are added after the minus sign, if necessary):

```
PaddedForm[TableForm[m2], {6, 3},
  NumberPadding → {" ", " "}, SignPadding → True]
```

32.7	8.39	-412.64
4.5	- 56.216	- 7.606

If the table contains both integers and real numbers or if the columns should use different padding specifications, then first transpose the matrix:

```
m4 = {{10, 8.39, -412.64}, {100, -56.2163, -7.606}};
m5 = Transpose[m4];
```

At this point, do nothing for integer columns, and use **Map** to pad the real columns with various specifications. Lastly, transpose back to the original form and align the integer columns right:

```
m6 = Transpose[{m5[[1]], Map[PaddedForm[#, {6, 2}] &, m5[[2]]],
    Map[PaddedForm[#, {6, 3}] &, m5[[3]]]}]
```

```
{{10,     8.39, -412.640}, {100,   -56.22,   -7.606}}
```

```
TableForm[m6, TableAlignments → Right,
  TableHeadings → {None, {"Col. 1", "Col. 2", "Col. 3"}}]
```

Col. 1	Col. 2	Col. 3
10	8.39	-412.640
100	-56.22	-7.606

Note that a simple way to align columns of a matrix is to use the traditional form:

```
m4 // TraditionalForm
```

$$\begin{pmatrix} 10 & 8.39 & -412.64 \\ 100 & -56.2163 & -7.606 \end{pmatrix}$$

13.1.3 Advanced Tabulating

■ Restrictions of TableForm

In Section 13.1.2, p. 326, we demonstrated how to make tables with **TableForm** (possibly with the aid of **PaddedForm**). As an example, consider again the data **m2**, which we now rename as **m**:

```
m = {{32.7, 8.39, -412.64}, {4.5, -56.2163, -7.606}};
```

We can form a table like the following:

```
mt = StyleForm[
    TableForm[m, TableSpacing → {1.5, 1.5}, TableAlignments → Right,
      TableHeadings → {None, Map[StyleForm[#, FontWeight → "Bold"] &,
        {"Col 1", "Col 2", "Col 3"}]}], FontFamily -> "Times", FontSize → 10]
```

Col 1	Col 2	Col 3
32.7	8.39	−412.64
4.5	−56.2163	−7.606

To get the columns aligned with the decimal point, use **PaddedForm**:

```
PaddedForm[mt, {6, 3}]
```

Col 1	Col 2	Col 3
32.700	8.390	−412.640
4.500	−56.216	−7.606

With **TableForm**, we cannot do more detailed adjustments. For example, we cannot use varying row or column spacings or add a line below the column headings.

Another way to form tables is with the use the menu command **Input ▷ Create Table/-Matrix/Palette....** We can ask for a table of a specified size that possibly contains lines between all rows and/or between all columns and that may have a frame. We get an empty table that contains placeholders for the entries. By filling in the placeholders, selecting the cell bracket of the resulting table, choosing **Cell ▷ Display As ▷ Traditional-Form**, selecting the numbers, and pressing ⌃b, we get the following:

Col. 1	Col. 2	Col. 3
32.500	8.390	−412.640
4.500	−56.216	−7.606

As is the case with the use of **TableForm**, this method also lacks ways to fine tune the table.

If we want to be able to do fine adjustments to a table, we have to form the table with **Grid** and use its options.

■ Grid

Grid[m]	Form a grid from a matrix **m**

With **Grid**, we get very much the same result as we did with **TableForm**:

```
TableForm[m]
```

```
32.7       8.39            -412.64
4.5        -56.2163        -7.606
```

```
Grid[m]
```

```
32.7     8.39     -412.64
4.5    -56.2163   -7.606
```

Note: **Grid** is an undocumented command. The introduction to the *Mathematica Reference Guide* of the manual by Wolfram (2003) says the following about undocumented features: "*You should not use any such features:* there is no certainty that features which are not documented will continue to be supported in future versions of *Mathematica*." However, I decided to use **Grid**, because the alternative available—namely **GridBox**—is not as nice to use. First, with **GridBox**, we need **DisplayForm** to actually show the grid box:

```
GridBox[m]
```

```
GridBox[{{32.7, 8.39, -412.64}, {4.5, -56.2163, -7.606}}]
```

```
% // DisplayForm
```

```
32.700000000000003  8.3900000000000006  -412.63999999999999
        4.5         -56.216299999999997  -7.6059999999999999
```

Second, as can be seen, most decimal numbers are not shown in the form in which we input them. Third, to get the original numbers, we have to transform the numbers into strings:

```
m2 = Map[ToString, m, {2}]
```

```
{{32.7, 8.39, -412.64}, {4.5, -56.2163, -7.606}}
```

With **InputForm**, we can see that the numbers in **m2** really are strings:

```
m2 // InputForm
```

```
{{"32.7", "8.39", "-412.64"}, {"4.5", "-56.2163", "-7.606"}}
```

Now **GridBox** gives the same result as **Grid**:

```
GridBox[m2] // DisplayForm
```

```
32.7     8.39     -412.64
4.5    -56.2163   -7.606
```

■ Options of Grid

Grid has a wealth of options for fine tuning a table (the options of **Grid** are the same as the options of **GridBox**).

RowAlignments How to align rows; possible single values (applied to all rows):
   Baseline, **Center**, **Top**, **Bottom**, **Axis**; an example of a list value: **{Bottom, Center}**
   (applied to the first row and to the rest of the rows, respectively)

ColumnAlignments How to align columns; possible single values: **Center**, **Left**,
   Right, **Decimal**, **"c"**; an example of a list value: **{Left, Decimal}**

RowSpacings Spacings between rows (in x heights); examples of values: **1**, **{2, 1}**

ColumnSpacings Spacings between columns (in ems); examples of values: **0.8**, **{2, 1}**

RowLines   Whether to draw lines between rows; examples of values: **False**, **True**, **{True, False}**, **3** (means to draw a line with a thickness that is 3 times the default thickness), **{3, 1}**

ColumnLines   Whether to draw lines between columns; examples of values: **False**, **True**, **{True, False}**, **3**, **{3, 1}**

RowMinHeight   Minimum total row height (in x heights); default value: **1**

ColumnWidths   Actual widths of columns (in ems); examples of values: **Automatic**, **{0.1, 0.15}**

RowsEqual   Whether all rows should have the same height; possible values: **False**, **True**

ColumnsEqual   Whether all columns should have the same width; possible values: **False**, **True**

GridFrame   Whether to draw a frame around the grid; examples of values: **False**, **True**, **3** (means to draw a frame with a thickness that is 3 times the default thickness), **{{3, 1}, {1, 3}}** (defines the thickness of the left, right, bottom, and top, respectively)

GridFrameMargins   Margins inside the grid frame (in em units); examples of values: **{{0.4, 0.4}, {0.5, 0.5}}** (defines the margins of the left, right, bottom, and top, respectively), **1**

GridBaseline   Vertical positioning of the grid box with respect to the baseline of the text surrounding it; default value: **Axes** (other values can be given as they are for **RowAlignments**)

The values of **RowAlignments**, **RowSpacings**, and **RowLines** and the corresponding options for columns can be of several forms:

val	use **val** in all cases
{val1, val2}	use **val1** in the first case and **val2** in all remaining cases
{val1, val2, val3, …}	use **vali** in the *i* th case

The general principle is that, if there are more cases in the table than there are given values for an option, the last value is used for all remaining cases. The second form **{val1, val2}** is often useful for row options: the first value is used for the heading row and the second for the rows of numbers. For example, **RowSpacings → {2, 1}** specifies space **2** below the column headings and space **1** below all other rows.

With **ColumnAlignments → Decimal**, we can align columns according to the decimal point. Columns can also be aligned according to the first occurrence of some other character **c**. The character can be an alignment marker ESC am ESC.

The x height is the height of an x character in the current font. An em is approximately the width of an M.

All of the options are explained in the *Help Browser* (go to **Front End** ▷ **Front End Options** ▷ **Formatting Options** ▷ **…Specific Box Options** ▷ **GridBoxOptions->{…}**).

■ **Using the Options of Grid**

Using **Grid** to form a table for a matrix **m** often includes the following two steps:

1. Add row and column headings (in the following example, **m** is a 3×3 matrix):

   ```
   cols = {{"Col 1", "Col 2", "Col 3"}}
   rows = {{"Row 1"}, {"Row 2"}, {"Row 2"}}
   << LinearAlgebra`MatrixManipulation`
   m2 = BlockMatrix[{{{{""}}, cols}, {rows, m}}]
   ```

2. Form the table:

   ```
   Grid[m2, opts]
   ```

With regard to step 1, note that **Grid** does not have an option to add column and row headings, (unlike **TableForm**, which has the **TableHeadings** option), so we have to do it ourselves with **BlockMatrix**, which creates larger matrices from given matrix blocks. For example, if A, B, C, and D are matrices (with suitable dimensions), the following command creates the matrix $\begin{pmatrix} A & B \\ C & D \end{pmatrix}$:

```
BlockMatrix[{{A, B}, {C, D}}]
```

The matrix `{{""}}` in the example of the box above, which contains only an empty string, comes to the upper left corner. With **StyleForm**, we can set font properties in the table:

```
StyleForm[Grid[m2, opts], FontFamily → "Times", FontSize → 10]
```

Next we consider some examples. For other examples, see Sections 20.2.1, p. 525; 26.1.5, p. 718; and 26.2.2, p. 731.

■ **Example 1**

We consider again the matrix **m**. First we add the column and row headings:

```
m = {{32.7, 8.39, -412.64}, {4.5, -56.2163, -7.606}};

cols = {{"Col 1", "Col 2", "Col 3"}};
rows = {{"Row 1"}, {"Row 2"}};

<< LinearAlgebra`MatrixManipulation`

m2 = BlockMatrix[{{{{""}}, cols}, {rows, m}}]
```

```
{{, Col 1, Col 2, Col 3},
 {Row 1, 32.7, 8.39, -412.64}, {Row 2, 4.5, -56.2163, -7.606}}
```

Then we can form the table:

```
Grid[m2,
   RowAlignments → {Bottom, Center}, ColumnAlignments → {Left, Decimal},
   RowSpacings → {2, 1}, ColumnSpacings → {2, 1},
   RowLines → {True, False}, ColumnLines → {True, False},
   GridFrame → 2, GridFrameMargins → 1] // TraditionalForm
```

	Col 1	Col 2	Col 3
Row 1	32.7	8.39	−412.64
Row 2	4.5	−56.2163	−7.606

■ Example 2

If we only want column headings, we can add them with **Join**:

```
cols =
 {Map[StyleForm[#, FontWeight → "Bold"] &, {"Col 1", "Col 2", "Col 3"}]}
{{Col 1, Col 2, Col 3}}
```

```
(m3 = Join[cols, m]) // TableForm
Col 1        Col 2         Col 3
32.7         8.39          -412.64
4.5          -56.2163      -7.606
```

Next we use **StyleForm** to set the family and size of the font in the table:

```
StyleForm[Grid[m3,
   RowAlignments → {Bottom, Center}, ColumnAlignments → Decimal,
   RowSpacings → {2, 1}, ColumnSpacings → 2,
   RowLines → {True, False}], FontFamily → "Times", FontSize → 10]
```

Col 1	Col 2	Col 3
32.7	8.39	−412.64
4.5	−56.2163	−7.606

13.1.4 Forming Indexed Variables

■ Defining and Using Indexed Variables

Forming a series of values is important in many mathematical calculations. Often it suffices to generate a list of values, for example, **xx**, and then to refer to its components with **xx[[i]]** (see Section 12.1.3, p. 303). We can also use indexed variables like **x[i]**:

```
xx = Table[x[i] = 2 + i 0.2, {i, 0, 5}]
{2, 2.2, 2.4, 2.6, 2.8, 3.}
```

Note that **xx[[1]]** and **x[0]** are the same. We can then refer both to the whole set of variables with **xx** and to separate variables with indices:

```
x[1]     2.2
```

On the other hand, if we only need the values of separate variables (i.e., if we do not need the indexed variables as a whole or as a list), we can use, for example, **Do**:

```
Do[y[i] = 2 + i 0.2, {i, 0, 5}]
```

If we want to see the values of all indexed variables with a certain name, for example, **x**, then type **?x**. To clear a single value, type **x[i] =.**, and to clear all values, type **Clear[x]**.

```
Clear[x, y]
```

■ General Indexed Variables

Sometimes we may want to consider general indexed variables without values. We can use **Table**:

```
vv = Table[v[i], {i, 0, 5}]
```
$\{v[0], v[1], v[2], v[3], v[4], v[5]\}$

```
ww = Table[w[i, j], {i, 2}, {j, 2}]
```
$\{\{w[1, 1], w[1, 2]\}, \{w[2, 1], w[2, 2]\}\}$

Another way of forming indexed variables is with the use of **Array**:

```
Array[f, n]     {f[1], …, f[n]}
Array[f, {m, n}]     {{f[1, 1], …, f[1, n]}, …, {f[m, 1], …, f[m, n]}}
```

A possible third argument gives the index origin (default is 1). For example:

```
rr = Array[r, 5]
```
$\{r[1], r[2], r[3], r[4], r[5]\}$

```
ss = Array[s, 6, 0]
```
$\{s[0], s[1], s[2], s[3], s[4], s[5]\}$

```
tt = Array[t, {2, 3}]
```
$\{\{t[1, 1], t[1, 2], t[1, 3]\}, \{t[2, 1], t[2, 2], t[2, 3]\}\}$

For the elements, we can define a function:

```
r[i_] := 2 + i 0.2
```

Now all elements of **rr** have a value:

```
rr     {2.2, 2.4, 2.6, 2.8, 3.}
```

■ Subscripts

True subscripts can be made with **Subscript** or by entering the subscripts in a 2D form. 2D subscripts can be written with the **BasicInput** palette (the first button in the last row) and with CTRL[_] or CTRL[-] (see Section 3.3.3, p. 70; to get out of the subscript position, press CTRL[.]). Here we use both **Subscript** and 2D input:

```
vv = Table[Subscript[v, i], {i, 0, 5}]
```
$\{v_0, v_1, v_2, v_3, v_4, v_5\}$

```
ww = Table[w_{i,j}, {i, 2}, {j, 3}]
```
$\{\{w_{1,1}, w_{1,2}, w_{1,3}\}, \{w_{2,1}, w_{2,2}, w_{2,3}\}\}$

■ **Symbolizing Subscripted Variables**

Subscripted variables can sometimes be used in the normal way:

>   Solve[x + a == b₀, x]   {{x → -a + b₀}}

However, sometimes such a variable is not recognized. For example, here we use a subscripted variable, and **DSolve** does not work:

>   DSolve[{y'[x] == y[x] / a + b, y[0] == y₀}, y[x], x]
>
>   DSolve::dvnoarg : The function y appears with no arguments. More…
>
>   DSolve[{y'[x] = b + $\frac{y[x]}{a}$, y[0] = y₀}, y[x], x]

One solution is to use another initial value like **c**. Another solution is to load a package:

>   << Utilities`Notation`

Now a palette appears on the screen. Click the **Symbolize** button, and type y_0 inside the brackets:

>   Symbolize[y₀]

Execute the resulting command. Now **DSolve** works:

>   DSolve[{y'[x] == y[x] / a + b, y[0] == y₀}, y[x], x]
>
>   {{y[x] → -a b + a b $e^{\frac{x}{a}}$ + $e^{\frac{x}{a}}$ y₀}}

13.2 Manipulating Lists

13.2.1 Selecting Parts

> list[[i]]   ith part of list
> list[[-i]]   ith part counted from the end
> list[[i, j]]   (i, j)th part
> list[[i, j, k]]   (i, j, k)th part
> list[[{i1, i2, …}]]   Parts i1, i2, …
>
> Extract[list, pos]   Part at position pos
> Extract[list, {pos1, pos2, … }]   Parts at positions pos1, pos2, …

We can also write **list[[i]]** as **Part[list, i]**. In **Extract**, the position is expressed as a list, as in **Extract[a, {3, 1}]**; this is equivalent to **a[[3, 1]]**. Selecting elements that satisfy a test is done with **Select**, and selecting elements that are of a given form is done with **Cases** (see Section 13.2.5, p. 340). For example:

>   m = {{11, 12, 13}, {21, 22, 23}, {31, 32, 33}};
>
>   m[[{1, 3}]]
>
>   {{11, 12, 13}, {31, 32, 33}}

```
Extract[m, {{1}, {3}}]
```

```
{{11, 12, 13}, {31, 32, 33}}
```

First[list], **Rest[list]** Take/drop the first element

Take[list, n], **Drop[list, n]** Take/drop the first **n** elements

Take[list, {m, n}], **Drop[list, {m, n}]** Take/drop elements **m** through **n**

Take[list, {m, n, s}], **Drop[list, {m, n, s}]** Take/drop elements **m** through **n** in steps of **s**

Take[list, -n], **Drop[list, -n]** Take/drop the last **n** elements

Last[list], **Most[list]** (❀5) Take/drop the last element

In place of **Most[list]**, *Mathematica* 4 users can write **Drop[list, -1]**. Note that the commands do not modify the original list. For example, if we want to replace **list** with a list where the last element is dropped, we have to write **list = Drop[list, -1]**. Manipulating the parts of matrices is considered in Section 18.2.3, p. 477.

13.2.2 Rearranging

Transpose[list] Transpose the first two levels of **list**

Sort[list] Sort the elements of **list** into a standard order

Sort[list, p] Sort the elements of **list** using the ordering function **p**

Union[list] Sort the elements and remove any duplicates

Union[list, SameTest → test] Use **test** to decide whether two elements are the same

Reverse[list] Reverse the order of elements

RotateLeft[list, n] Rotate the elements **n** positions to the left

RotateRight[list, n] Rotate the elements **n** positions to the right

Permutations[list] Give all orderings of the elements of **list**

KSubsets[list, n] Give all subsets of **list** containing **n** elements (in the **Discrete Math`Combinatorica`** package)

Here is an example of sorting into standard order:

```
a = {2.3, 3, 1, 2.99, 2, 3, 2, 1.6};
```

```
Sort[a]
```

```
{1, 1.6, 2, 2, 2.3, 2.99, 3, 3}
```

Next we sort into descending order (also **Sort[a]//Reverse**):

```
Sort[a, Greater[#1, #2] &]
```

```
{3, 3, 2.99, 2.3, 2, 2, 1.6, 1}
```

Select all distinct elements and sort them into standard order:

```
Union[a]
```

```
{1, 1.6, 2, 2.3, 2.99, 3}
```

Use a custom test to determine which elements are to be considered the same:

```
Union[a, SameTest → (Abs[#1 - #2] < 0.05 &)]
```

```
{1, 1.6, 2, 2.3, 2.99}
```

Ask for all possible permutations:

```
Permutations[{1, 2, 3}]
```

```
{{1, 2, 3}, {1, 3, 2}, {2, 1, 3}, {2, 3, 1}, {3, 1, 2}, {3, 2, 1}}
```

Ask for all possible unordered sets of two elements:

```
<< DiscreteMath`Combinatorica`
```

```
KSubsets[{1, 2, 3, 4}, 2]
```

```
{{1, 2}, {1, 3}, {1, 4}, {2, 3}, {2, 4}, {3, 4}}
```

13.2.3 Ungrouping and Grouping

> `Flatten[list]`   Flatten out all levels in `list`
> `Flatten[list, n]`   Flatten out the top `n` levels in `list`
>
> `Partition[list, n]`   Partition `list` into nonoverlapping sublists of `n` elements
> `Partition[list, n, d]`   Generate sublists with offset `d`
>
> `Split[list]`   Split `list` into pieces consisting of runs of identical elements
> `Split[list, test]`   Consider adjacent elements as identical if `test` gives `True`

(Note that `Partition` also has more advanced forms.)
We first make a table:

```
a = Table[{x, y, x y, x +y}, {x, 1, 2}, {y, 2, 4}]
```

```
{{{1, 2, 2, 3}, {1, 3, 3, 4}, {1, 4, 4, 5}},
 {{2, 2, 4, 4}, {2, 3, 6, 5}, {2, 4, 8, 6}}}
```

Flatten removes all inner curly braces { } and so ungroups the list:

```
at = Flatten[a]
```

```
{1, 2, 2, 3, 1, 3, 3, 4, 1, 4, 4, 5, 2, 2, 4, 4, 2, 3, 6, 5, 2, 4, 8, 6}
```

Next we flatten only the first level:

```
Flatten[a, 1]
```

```
{{1, 2, 2, 3}, {1, 3, 3, 4}, {1, 4, 4, 5},
 {2, 2, 4, 4}, {2, 3, 6, 5}, {2, 4, 8, 6}}
```

We will use this kind of flattening several times in the chapters to come. Now we partition the flattened list into groups of four elements:

```
Partition[at, 4]
```

```
{{1, 2, 2, 3}, {1, 3, 3, 4}, {1, 4, 4, 5},
 {2, 2, 4, 4}, {2, 3, 6, 5}, {2, 4, 8, 6}}
```

If we partition into groups of five elements, the remaining four elements (2, 4, 8, and 6) are dropped:

```
Partition[at, 5]
```

```
{{1, 2, 2, 3, 1}, {3, 3, 4, 1, 4}, {4, 5, 2, 2, 4}, {4, 2, 3, 6, 5}}
```

We then partition with offset 1:

```
Partition[at, 2, 1]
```

```
{{1, 2}, {2, 2}, {2, 3}, {3, 1}, {1, 3}, {3, 3}, {3, 4},
 {4, 1}, {1, 4}, {4, 4}, {4, 5}, {5, 2}, {2, 2}, {2, 4}, {4, 4},
 {4, 2}, {2, 3}, {3, 6}, {6, 5}, {5, 2}, {2, 4}, {4, 8}, {8, 6}}
```

Split finds runs of identical elements:

```
Sort[at]
```

```
{1, 1, 1, 2, 2, 2, 2, 2, 2, 3, 3, 3, 3, 4, 4, 4, 4, 4, 4, 5, 5, 6, 6, 8}
```

```
Split[%]
```

```
{{1, 1, 1}, {2, 2, 2, 2, 2, 2},
 {3, 3, 3, 3}, {4, 4, 4, 4, 4, 4}, {5, 5}, {6, 6}, {8}}
```

```
Map[{First[#], Length[#]} &, %]
```

```
{{1, 3}, {2, 6}, {3, 4}, {4, 6}, {5, 2}, {6, 2}, {8, 1}}
```

Next we find the points where the previous element is at most four and the next element is greater than four:

```
Split[at, ! (#1 ≤ 4 && #2 > 4) &]
```

```
{{1, 2, 2, 3, 1, 3, 3, 4, 1, 4, 4}, {5, 2, 2, 4, 4, 2, 3}, {6, 5, 2, 4}, {8, 6}}
```

```
a =.; at =.
```

13.2.4 Modifying Elements

Prepend[list, elem]	Add **elem** at the beginning of **list**
Append[list, elem]	Add **elem** at the end of **list**
PadLeft[list, n]	Pad **list** with zeros on the left to make a list of length **n**
PadRight[list, n]	Pad **list** with zeros on the right to make a list of length **n**
Insert[list, elem, i]	Insert **elem** at position **i** in **list**
Delete[list, i]	Delete the **i**th element of **list**
ReplacePart[list, elem, i]	Replace the **i**th element of **list** with **elem**

PadLeft and **PadRight** also have more general forms. The last argument in **Insert**, **Delete**, and **ReplacePart** can also be **-i**, meaning that the position is counted from the end of the list. The last argument can also be a more detailed definition of the position, such as **{i, j}**.

```
a = {{p, 1}, {q, 2}, {r, 3}};
```

```
Append[a, {s, 4}]        {{p, 1}, {q, 2}, {r, 3}, {s, 4}}
```

```
Insert[a, {u, 9}, 3]      {{p, 1}, {q, 2}, {u, 9}, {r, 3}}
```

```
ReplacePart[a, {u, 9}, 3]      {{p, 1}, {q, 2}, {u, 9}}
```

Note again that the commands above do not modify the original list. You could check that **a** still has the original three elements. If you want to replace the list with a modified list, you have to write, for example, **a = Append[a, {u, 6}]**. The following two commands do modify the list:

> **PrependTo[list, elem]**   Add **elem** at the beginning of **list** and reset **list** to the result
>
> **AppendTo[list, elem]**   Add **elem** at the end of **list** and reset **list** to the result

    AppendTo[a, {s, 4}]   {{p, 1}, {q, 2}, {r, 3}, {s, 4}}

The value of **a** is changed:

    a   {{p, 1}, {q, 2}, {r, 3}, {s, 4}}

One way to modify the elements of lists is to assign new values. We modify the last element of **a**:

    a[[4]] = {u, 9}   {u, 9}

Note that this method modifies the original list; **a** has become the following:

    a   {{p, 1}, {q, 2}, {r, 3}, {u, 9}}

    a = .

13.2.5 Searching Elements

■ Searching with a Test

> **Select[list, test]**   Select the elements of **list** for which **test** gives **True**

With **Select**, we can search for elements that satisfy a logical test. Logical tests were mentioned in Section 12.3.5. Tests useful with **Select** include the following:

    ==, !=, <, ≤, >, ≥, Negative, Nonnegative, Positive,
    NumericQ, IntegerQ, EvenQ, OddQ, FreeQ, MemberQ.

In general, a test used in **Select** is written as a pure function, such as **Select[c, EvenQ[#]&]** (for pure functions, see Sections 2.2.2, p. 36, and 14.1.5, p. 359). The argument of a pure function is written as **#**, and at the and we write **&**. However, simple built-in tests with one argument can be written without **#** and **&** so that we can write simply **Select[c, EvenQ]**. More complicated built-in tests that use two arguments have to be written as pure functions; for example: **Select[c, # < 2 &]**.

The built-in tests can be combined with logical operations like **&&** (and), **||** (or), and **!** (not) (see Section 12.3.5, p. 316). Such combined tests also have to be written as pure functions; for example: **Select[c, EvenQ[#] && Positive[#] &]**.

To demonstrate, we throw a die 20 times and select the results that satisfy various tests:

    a = Table[Random[Integer, {1, 6}], {20}]

    {2, 4, 3, 5, 4, 1, 4, 5, 6, 2, 4, 6, 4, 1, 3, 2, 4, 2, 1, 3}

    Select[a, EvenQ]   {2, 4, 4, 4, 6, 2, 4, 6, 4, 2, 4, 2}

```
Select[a, EvenQ[#] && # ≥ 4 &]     {4, 4, 4, 6, 4, 6, 4, 4}

Select[a, 2 ≤ # ≤ 4 &]     {2, 4, 3, 4, 4, 2, 4, 4, 3, 2, 4, 2, 3}

Select[a, # == 1 || # == 6 &]     {1, 6, 6, 1, 1}
```

In nested lists, a part specification may be needed:

```
b = Table[{i, Random[]}, {i, 5}]

{{1, 0.179938}, {2, 0.0702301},
 {3, 0.900999}, {4, 0.274275}, {5, 0.799332}}

Select[b, #[[2]] ≤ 0.5 &]

{{1, 0.179938}, {2, 0.0702301}, {4, 0.274275}}
```

■ Searching with a Pattern

Count[list, pattern]	Give the number of elements in **list** that match **pattern**
Cases[list, pattern]	Select the elements of **list** that match **pattern**
DeleteCases[list, pattern]	Remove the elements of **list** that match **pattern**

These commands are considered in Section 15.3.5, p. 414, so here we consider them only briefly. A true pattern is formed with the underscore (_), but simple expressions like **6**, **x**, or **{1, 2}** can also be considered as patterns: they are degenerate patterns. We count the number of sixes:

```
Count[a, 6]     2
```

13.2.6 Searching Positions

■ Searching Positions with a Pattern

Position[list, pattern]	Give the positions at which objects matching **pattern** occur in **list**
Position[list, pattern, levspec]	Give only the positions that are in levels specified by **levspec**

Note that here we presented **Position** in the context of lists, but the command works for other expressions as well. Further, **Position** accepts a fourth argument **n**, and then the command gives only the first **n** positions.

Level specifications are considered in Section 12.2.3, p. 310. Recall that a level specification **n** means all levels from 1 to n, **{n}** means only the level n, and **{n, m}** means levels **n** through **m**. If a level specification is not given in **Position**, the specification is assumed to be {0, ∞} which means all levels (0, 1, 2, …).

This command is considered in more detail in Section 15.3.5, p. 413. Here we give only one example. In the sequence of tosses presented in Section 13.2.5, sixes occurred at the following times:

```
Position[a, 6]     {{9}, {12}}
```

```
Clear[a, b]
```

■ **Searching the Positions of the Smallest through Largest Elements**

> **Ordering[list]**   Give the positions of all elements of **Sort[list]** in **list**
> **Ordering[list, n]**   Give the positions of the first **n** elements of **Sort[list]** in **list**
> **Ordering[list, -n]**   Give the positions of the last **n** elements in **Sort[list]** in **list**
> **Ordering[list, 1]**   Give the position of the smallest element in **list**
> **Ordering[list, -1]**   Give the position of the largest element in **list**

Consider the following list and its sorted version:

```
c = {13, 16, 14, 15, 11, 12};
```

```
sc = Sort[c]      {11, 12, 13, 14, 15, 16}
```

Ordering gives the following list:

```
Ordering[c]      {5, 6, 1, 3, 4, 2}
```

This means that the first element of **sc** (the smallest element of **c**), is the fifth element of **c**, the second smallest element of **c** is the sixth element of **c**, and so on, with the largest element of **c** being the second element of **c**. The sorted list can be obtained as follows:

```
c[[%]]      {11, 12, 13, 14, 15, 16}
```

The position of the smallest element and the smallest element itself are as follows:

```
{minpos = Ordering[c, 1][[1]], c[[minpos]]}      {5, 11}
```

Another way is to write the following:

```
Position[c, Min[c]][[1, 1]]      5
```

The position of the largest element and the largest element itself are as follows:

```
{maxpos = Ordering[c, -1][[1]], c[[maxpos]]}      {2, 16}
```

Another way is to write the following:

```
Position[c, Max[c]][[1, 1]]      2
```

```
c =.
```

13.2.7 Operations on Several Lists

> **Join[list1, list2, …]**   Lists concatenated together
> **Union[list1, list2, …]**   Sorted list of all distinct elements
> **Intersection[list1, list2, …]**   Sorted list of all distinct elements common to all lists
> **Complement[list0, list1, list2, …]**   Sorted list of all distinct elements of **list0** that are not in any of the other lists
> **Outer[List, list1, list2, …]**   Combine the elements of the lists in all ways
> **ListCorrelate[kernel,list]**   Form the correlation of **kernel** with **list**
> **ListConvolve[kernel,list]**   Form the convolution of **kernel** with **list**

```
e = {r, 3, 1, p, 2, r, 2, q};
f = {A, C, p, A, B, 3};

Join[e, f]     {r, 3, 1, p, 2, r, 2, q, A, C, p, A, B, 3}

Union[e, f]     {1, 2, 3, A, B, C, p, q, r}

Complement[e, f]     {1, 2, q, r}

Outer[List, {p, q}, {r, s, t}]

{{{p, r}, {p, s}, {p, t}}, {{q, r}, {q, s}, {q, t}}}
```

ListCorrelate and **ListConvolve** are considered in the context of the smoothing of data in Section 27.3.1, p. 757. Here, however, is a typical example:

```
ListCorrelate[{a, b, c}, {A, B, C, D, F, G}]

{a A + b B + c C, a B + b C + c D, a C + b D + c F, a D + b F + c G}

Clear[e, f]
```

13.3 Mapping Lists

13.3.1 Mapping the Elements of a List

■ Basic Mapping

> **Map[f, list]** or **f /@ list**   Apply **f** to each element on the first level of **list**

Map is one of the most useful commands of *Mathematica*. Here is a general example of how it can be used:

```
t = {a, b, c};

Map[f, t]     {f[a], f[b], f[c]}
```

Each element of the given list is mapped with the given function. The function **f** can be a built-in function like **Log**, and then no arguments are needed:

```
Map[Log, t]     {Log[a], Log[b], Log[c]}
```

Note, however, that these kinds of mappings are not necessary, because the built-in functions accept lists as arguments:

```
Log[{a, b, c}]     {Log[a], Log[b], Log[c]}
```

We can also build the function **f** by ourselves, and then the function is written as a pure function (for pure functions, see Section 14.1.5, p. 359):

```
Map[p #^2 + q &, t]     {a² p + q, b² p + q, c² p + q}
```

Even this example can be calculated directly with the list:

```
p t^2 + q     {a² p + q, b² p + q, c² p + q}
```

Map is very useful for more complicated list manipulation, and we will use it frequently in this book. For example, form pairs of the elements of **t** and their squares:

```
Map[{#, #^2} &, t]      {{a, a²}, {b, b²}, {c, c²}}
```

Pick the first column of a matrix:

```
m = {{1, 2, 3}, {a, b, c}, {A, B, C}};
```

```
Map[#[[1]] &, m]      {1, a, A}
```

An important application of **Map** is in the calculation of partial derivatives. Here we calculate the partial derivatives of the product $x\,y\,z$ with respect to x, y, and z:

```
Map[D[x y z, #] &, {x, y, z}]      {y z, x z, x y}
```

Another important application is in the calculation of frequencies. We toss a die 20 times:

```
u = Table[Random[Integer, {1, 6}], {20}]
```

```
{3, 4, 3, 1, 1, 4, 1, 2, 3, 2, 4, 4, 5, 4, 6, 1, 2, 4, 6, 4}
```

The frequencies are as follows:

```
Map[{#, Count[u, #]} &, {1, 2, 3, 4, 5, 6}]
```

```
{{1, 4}, {2, 3}, {3, 3}, {4, 7}, {5, 1}, {6, 2}}
```

■ Special Mappings

> **Map[f, list, levspec]** Apply **f** to each element on the specified levels of **list**
> **MapAt[f, list, parts]** Apply **f** to the specified parts of **list**
> **MapAll[f, list]** Apply **f** to all parts of **list**

These special mapping commands are not at all important. Level specifications were considered in Section 12.2.3, p. 310. The default level in **Map** is {1}, which means that each element at the first level is mapped. Here is an example of **MapAt**:

```
Map[ToString, m, {2}] // InputForm
```

```
{{"1", "2", "3"}, {"a", "b", "c"}, {"A", "B", "C"}}
```

```
MapAt[#^2 &, m, 3]
```

```
{{1, 2, 3}, {a, b, c}, {A², B², C²}}
```

> **MapIndexed[f, list]** Apply **f** to the elements of **list**, putting the part specification of each element as a second argument to **f**
> **Scan[f, list]** Apply **f** to each element in **list**, but do not form a list from the results

```
MapIndexed[f, {a, b, c}]
```

```
{f[a, {1}], f[b, {2}], f[c, {3}]}
```

```
MapIndexed[(1 / (1 + #1))^#2[[1]] &, {1, 2, 3}]
```

$$\left\{\frac{1}{2}, \frac{1}{9}, \frac{1}{64}\right\}$$

```
Scan[Print["The divisors of ", #, " are ", Divisors[#]] &, {15, 16, 17}]
```

The divisors of 15 are {1, 3, 5, 15}

The divisors of 16 are {1, 2, 4, 8, 16}

The divisors of 17 are {1, 17}

13.3.2 Mapping a List

■ Changing the Head

> **Apply[f, list]** or **f @@ list**   Replace the head **List** of **list** with the head **f**

The only thing **Apply** does is change the head (for heads, see Section 12.2.3, p. 309). As an example, consider the following list:

```
t = {a, b, c};
```

Its head is, of course, **List**, as can be seen in the **FullForm**:

```
FullForm[t]     List[a, b, c]
```

Other heads are, for example, **Plus**, **Times**, and **And**. We try these heads:

```
{Apply[Plus, t], Apply[Times, t], Apply[And, t]}
```

{a + b + c, a b c, a && b && c}

Now the heads are, indeed, **Plus**, **Times**, and **And**:

```
% // FullForm
```

List[Plus[a, b, c], Times[a, b, c], And[a, b, c]]

Multinomial coefficients are normally calculated as follows:

```
Multinomial[3, 2, 5, 4]     2522520
```

However, if you have the arguments in a list, use **Apply**:

```
t1 = {3, 2, 5, 4};
```

```
Apply[Multinomial, t1]     2522520
```

In general, if the new head is **f**, this is the result:

```
Apply[f, t]     f[a, b, c]
```

Thus, we can consider **Apply** as a tool to use to map the list as a whole with **f**. In the following table, we have collected useful applications of **Apply**.

> **Apply[Plus, list]**   Calculate the sum of the elements of **list**; in *Mathematica* 5, use
> **Total[list]** (see Section 18.1.1, p. 464)
> **Apply[Times, list]**   Calculate the product of the elements of **list**
> **Apply[And, list]**   Calculate the logical AND of the elements of **list**
> **Apply[List, sum]**   Form a list from the terms of **sum**

■ **Level Specifications**

> `Apply[f, list, levspec]`   Replace the head in the specified levels

Level specifications were considered in Section 12.2.3, p. 310. For example, level specification {0} means the whole expression, {1} means the parts of the expression at level 1, and {0, 1} means the parts at levels 0 and 1. The default level of **Apply** is {0}, which means that **Apply** replaces the head of the whole expression. As an example, we calculate column and row sums and the sum of all elements of a matrix:

```
m = {{1, 2, 3}, {a, b, c}, {A, B, C}};
```

```
Apply[Plus, m]      {1 + a + A, 2 + b + B, 3 + c + C}
```

```
Apply[Plus, m, {1}]      {6, a + b + c, A + B + C}
```

```
Apply[Plus, m, {0, 1}]      6 + a + A + b + B + c + C
```

■ **Using a Pure Function**

> `Apply[f[##]&, list]`   Replace the head **List** of **list** with the head **f**, that is, calculate the value of the multivariate function **f** for values of variables that are given in **list**

This more general form of **Apply** may be useful when working with more complex functions. Here **f[##]&** is a pure function (see Section 14.1.5, p. 359), and **##** represents *all* of the arguments of **f**. The **&** identifies **f[##]** as a pure function. We could apply this form to the examples above:

```
t = {a, b, c};
```

```
Apply[Plus[##] &, t]      a + b + c
```

13.3.3 Mapping Two Lists

`Thread[f[{a,b,c}, {A,B,C}]]`	`{f[a,A], f[b,B], f[c,C]}`
`MapThread[f, {{a,b,c}, {A,B,C}}]`	`{f[a,A], f[b,B], f[c,C]}`
`Inner[f, {a,b,c}, {A,B,C}]`	`f[a,A] + f[b,B] + f[c,C]`
`Inner[f, {a,b,c}, {A,B,C}, g]`	`g[f[a,A], f[b,B], f[c,C]]`
`Outer[f, {a,b,c}, {A,B,C}]`	`{{f[a,A], f[a,B], f[a,C]},`
	`{f[b,A], f[b,B], f[b,C]},`
	`{f[c,A], f[c,B], f[c,C]}}`

The results of **Thread** and **MapThread** are the same, but the way the two lists are inputted differs. A good way to apply **Thread** is to construct a list of substitutions:

```
Thread[{a, b, c} → {A, B, C}]
```

{a → A, b → B, c → C}

We could also write the following:

>   MapThread[Rule, {{a, b, c}, {A, B, C}}]

>   {a → A, b → B, c → C}

Next we calculate the pairwise maximums of two lists:

>   MapThread[Max, {{2, 3, 1, 4}, {3, 1, 2, 3}}]

>   {3, 3, 2, 4}

With **Inner**, we can form the inner product of two vectors:

>   Inner[Times, {a, b, c}, {A, B, C}]

>   a A + b B + c C

An easier way, though, is to use the dot:

>   {a, b, c}.{A, B, C}

>   a A + b B + c C

Inner can have a function as the fourth argument. The default of this function is **Plus**:

>   Inner[f, {a, b, c}, {A, B, C}, Plus]

>   f[a, A] + f[b, B] + f[c, C]

With **Outer**, we can calculate the outer product of two vectors:

>   Outer[Times, {a, b, c}, {A, B, C}]

>   {{a A, a B, a C}, {A b, b B, b C}, {A c, B c, c C}}

We then form all pairs of the elements of two lists:

>   Outer[List, {a, b, c}, {A, B, C}]

>   {{{a, A}, {a, B}, {a, C}}, {{b, A}, {b, B}, {b, C}}, {{c, A}, {c, B}, {c, C}}}

The Jacobian of a vector-valued function of several variables is nicely calculated with **Outer**:

>   Outer[D, {x y z, Sin[x] + Cos[y] + Tan[z]}, {x, y, z}]

>   {{y z, x z, x y}, {Cos[x], -Sin[y], Sec[z]^2}}

14

Functions

Introduction

> *Teacher: "If Tom gave you three apples and Bill gave you two apples, how many apples would you have then?" Mary: "Seven apples, teacher." Teacher: "Wrong, Mary, 3 + 2 = 5." Mary: "I know that, teacher, but I have two apples already."*

A function defined by `f[x_] := expr` is usually not needed to do calculations with expressions. Instead, give a name for the expression, like `f = expr`, and use this name. On the other hand, the syntax of a function is needed to write more complex functions, such as piecewise-defined or recursive functions. Such functions are considered in section 14.1. Functions are also used when writing programs (see Section 14.1.5 and Chapter 15).

Pure functions are important in *Mathematica*, however odd they may seem at first sight. With *tracing*, we can see how a calculation proceeds inside *Mathematica*; this may be useful to detect errors in functions and programs. By *compiling* a function, we can get a more efficient function and save computing time. We can also assign some *attributes* to a function and so give it some desired properties.

Functions are used in programming, and a collection of related functions can be gathered together into a *package*. To understand the structure of a package, we have to consider *contexts*. With contexts, we also get an explanation of the problems we encounter when we forget to load a package.

Sections 14.1.1 and 14.1.5 are the most important ones in this chapter.

14.1 User-Defined Functions

14.1.1 Defining a Function

Suppose we want to consider the function **x Cos[x]**. We want to calculate its value at some points and to differentiate, integrate, and plot it. To avoid having to write the expression again and again or use **%**, **%%**, **%%%**, and so on, we can either give a name to the expression or define a function:

1. Naming the expression:	2. Defining a function:
f = x Cos[x]	**f[x_] := x Cos[x]**
f /. x → 2	**f[2]**
Table[f, {x, 0, 1, 0.1}]	**Table[f[x], {x, 0, 1, 0.1}]**
D[f, x]	**f'[x]** (also **D[f[x], x]**)
Integrate[f, x]	**Integrate[f[x], x]**
Plot[f, {x, 0, Pi}]	**Plot[f[x], {x, 0, Pi}]**

The first method is often the most appropriate one. With it, we give a name to the expression, and later on we need only type this name. The second method is often not needed to perform calculations with an expression. However, function definitions are used to define more complicated functions and also programs.

■ Functions

> **f[x_] := expr**   A function of one variable
> **f[x_, y_] := expr**   A function of two variables
> **?f** Show the definition of **f**
> **Clear[f]**   Clear all definitions for **f**
> **Remove[f]**   Remove **f** completely

Functions are defined with **:=**, and each argument is followed by the underscore (_) (named "blank"). For example:

   f1[x_] := x Sin[x]

   f1[Pi / 3] $\dfrac{\pi}{2\sqrt{3}}$

   ? f1

   Global`f1

   f1[x_] := x Sin[x]

(The full name of **f1** is **Global`f1**, where **Global`** is the context of all user-defined symbols.) If you forget the underscore, the function does not work: *Mathematica* knows the value of the function only for the argument you have used in the definition and not for any other argument:

> **f2[x] := x Cos[x]**

> **f2[x]**   x Cos[x]

> **f2[Pi]**   f2[π]

Here is a function with two arguments:

> **charpoly[m_, x_] := Det[m - x IdentityMatrix[Length[m]]]**

> **charpoly[{{2, 5}, {6, 1}}, x]** $- 28 - 3 x + x^2$

One or more arguments can be used to index a function:

> **f3[n_, x_] := (n - 1) x^n**

> **Table[f3[n, y], {n, 4}]** $\{0, y^2, 2 y^3, 3 y^4\}$

■ Details

At this point, there are two things that require explanation: the underscore and the colon.

The underscore in **x_** makes the argument a *pattern*. The pattern **x_** is matched with *anything*: the value of the function is calculated using whatever you give as an argument. Indeed, this is what we want of a function. Patterns are considered in more detail in Section 15.3.2, p. 405. As we said above, if we define a function like **f2** without the underscore, the definition is nearly useless, because the value of the function is known only for the argument given in the definition and not for any other argument. Thus, *the crucial thing to remember when defining a function is to write the underscore.*

Recall that assigning values can be done as follows: **f = x Sin[x]**. This means that the value of **x Sin[x]** is *immediately* calculated and then assigned to **f** (if **x** happens to have a value when you define **f**, then **f** gets the corresponding special value of **x Sin[x]**). If you define a function **f[x_] := x Sin[x]**, the value of the righthand side is not evaluated *until* you ask for the value of the function with a specific argument. Indeed, the effect of the colon in **:=** is only to *delay* the evaluation of the righthand side. This may seem somewhat unimportant now, but later on you will encounter more complex functions in which the righthand side simply cannot be evaluated or its value is extremely complex and useless unless the argument has a numerical value. Hence, it makes sense to routinely use **:=** in function definitions (although **=** could work for simpler functions).

■ Example 1

We try to use **=** in the definition of a function:

> **g1[x_] = NIntegrate[Sin[t^2], {t, 0, x}]**

> NIntegrate::nlim : t = x is not a valid limit of integration. More…

> NIntegrate[Sin[t^2], {t, 0, x}]

Mathematica immediately tries to evaluate the righthand side of **g1**, but it cannot be evaluated: in **NIntegrate**, the limits of integration must be numerical (and not **x**, for example), and so we get the error message. We can calculate the value of **g1** at a point, though:

    g1[2]   0.804776

Then we use **:=** in the definition:

    g2[x_] := NIntegrate[Sin[t^2], {t, 0, x}]

Now the righthand side is not evaluated until a specific value of **x** is given:

    g2[2]   0.804776

■ Example 2

Sometimes we have to use **=** and not **:=**. For example, we define the following:

    g3[x_] = D[x Sin[x], x]   x Cos[x] + Sin[x]

The important point is that the derivative was calculated immediately, because we did not write the colon. Now **g3** contains the derivative, and we can calculate its value:

    g3[2.]   0.0770038

Let us then try **:=** in the definition:

    g4[x_] := D[x Sin[x], x]

    g4[2.]

    General::ivar : 2.` is not a valid variable. More…

    $\partial_{2.}$ 1.81859

We got an error message. The righthand side of **g4** was not evaluated when **g4** was defined, and when we then asked for the value at 2, we ended up with the impossible expression **D[1.81859, 2.]**. So, in this example, we should not use the colon in the definition of the function.

■ Example 3

Sometimes we may want to define a function from the result of a computation. For example, we may have calculated the following:

    D[x Sin[x] + Sinh[x], x]   x Cos[x] + Cosh[x] + Sin[x]

Now we want to define this result as the value of a function **g5** at **x**. This is another case in which we have to use **=** and not **:=**:

    g5[x_] = %   x Cos[x] + Cosh[x] + Sin[x]

    g5[1]   Cos[1] + Cosh[1] + Sin[1]

■ Differentiation

With regard to symbolic operations with piecewise-defined functions, symbolic differentiation can be done only if we have used **If** or **Which** in the definition:

    D[f2[x], x]

    If[x ≥ 0 && Sin[x] ≥ 0.8, Cos[x], If[x ≥ 0 && Sin[x] < 0.8, 0, 0]]

If we use **f1**, only numerical values can be asked for the derivative:

    D[f1[x], x] /. x → 2.   - 0.412743

However, the precision is far from good, at least in this example, because the exact derivative is -0.416147.

■ Indefinite Integration

For symbolic indefinite integration with piecewise-defined functions, one method is to calculate a definite integral up to a variable **x** and proceed step by step for each interval separately. Consider the following function for $x \geq 0$:

    f4[x_] := If[x < 1, x, x - 1]

Calculate its indefinite integral as follows:

    g4[x_] = If[x < 1, Evaluate[Integrate[t, {t, 0, x}]],
       Evaluate[Integrate[t, {t, 0, 1}] + Integrate[t - 1, {t, 1, x}]]]

$$\text{If}\left[x < 1, \frac{x^2}{2}, \frac{1}{2} + \frac{1}{2}(-1+x)^2\right]$$

Another method is to redefine the function **f4** with **UnitStep** (**UnitStep[x]** is 1 for $x \geq 0$ and 0 for $x < 0$):

    f5[x_] := x UnitStep[1 - x] + (x - 1) UnitStep[x - 1]

Now we can integrate in a straightforward way:

    g5[x_] = Integrate[f5[x], x]

$$\frac{x^2}{2} - (-1 + x) \text{UnitStep}[-1 + x]$$

The indefinite integrals **g4** and **g5** are equivalent.

■ Definite Integration

Symbolic definite integration may succeed with simple piecewise-defined functions:

    Integrate[f4[x], {x, 0, 2}]   1

A package may help with more complex integrals (see Section 17.1.3, p. 438):

    f6[x_] := Which[x ≤ 0, x^2, 0 ≤ x ≤ 1, 1 - x^3,
       x ≥ 1, (x - 1)^2]

    << Calculus`Integration`

    Integrate[f6[x], {x, -1, 2}] $\frac{17}{12}$

14.1.3 Periodic and Recursive Functions

■ Periodic Functions

We can define a periodic function very simply. If, for example, the function has a period of 2, simply define, with **If** or **Which**, the function in the basic interval [0, 2], and add the condition that the function is $f(x-2)$ for $x > 2$:

```
f7[x_] := If[0 ≤ x ≤ 2, x^2, f7[x - 2]]
```

The value of the function at a given point is calculated by using the recursion $f_7(x) = f_7(x-2)$ until the argument becomes onto the interval $0 \le x \le 2$. For example, $f_7(5) = f_7(3) = f_7(1) = 1^2 = 1$. Here is a plot:

```
Plot[f7[x], {x, 0, 10}];
```

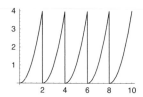

Note that indefinite integration with **f7** does not succeed, and definite integration only succeeds within the basic interval or does not succeed at all.

■ Recursive Functions: Dynamic Programming

A function can be recursive. In fact, the periodic function above contains a recursive part, which is that the value of $f(x)$ is $f(x-2)$ for $x > 2$. Often the recursion is with respect to an integer-valued variable. Here we define a function to calculate the factorial of an integer (this function is unnecessary, however, because we have the built-in function **n!** or **Factorial[n]**):

```
fac1[0] = 1;
fac1[n_] := n fac1[n - 1]

Table[fac1[n], {n, 0, 5}]
{1, 1, 2, 6, 24, 120}
```

Let us see what *Mathematica* knows about **fac1**:

```
? fac1

Global`fac1

fac1[0] = 1
fac1[n_] := n fac1[n - 1]
```

Here is another form for the function:

```
fac2[0] = 1;
fac2[n_] := fac2[n] = n fac2[n - 1]
```

Note the form **fac2[n_] := fac2[n] = ...** We calculate 3! and then ask for information about **fac2**:

```
fac2[3]     6
```

? fac2

```
Global`fac2
fac2[0] = 1
fac2[1] = 1
fac2[2] = 2
fac2[3] = 6
fac2[n_] := fac2[n] = n fac2[n - 1]
```

Thus, *Mathematica* has stored all the values that were needed to calculate **fac2[3]**. If we now want to calculate **fac2[4]**, *Mathematica* does not need to recalculate these values. This saves computing time. So, the form of **fac2** is the preferred one for recursive functions. This form is also known as *dynamic programming*. Its general form is as follows:

> **f[n_] := f[n] = expr**

Next we calculate Fibonacci numbers (this function is also unnecessary, because we have the built-in function **Fibonacci[n]**):

```
fib[1] = 1;
fib[2] = 1;
fib[n_] := fib[n] = fib[n - 1] + fib[n - 2]

fib[50]     12586269025
```

Here is the recursive definition of the Legendre orthogonal polynomials (again, this is an unnecessary function, because we have the built-in function **LegendreP[n,x]**):

```
leg[0, x_] := 1;
leg[1, x_] := x;
leg[n_, x_] :=
 leg[n, x] = Simplify[((2 n - 1) x leg[n - 1, x] - (n - 1) leg[n - 2, x]) / n]

leg[5, x]     1/8 x (15 - 70 x² + 63 x⁴)
```

$$\text{leg[5, x]} \quad \frac{1}{8}\, x\, (15 - 70\, x^2 + 63\, x^4)$$

■ **RecursionLimit**

To prevent infinite computations, *Mathematica* has a limit for recursive calculations:

```
$RecursionLimit     256
```

From this it follows that we cannot calculate **fac2[255]** if this is the first calculation with **fac2** (**fac2[254]** can be calculated). To show this, we remove **fac2**, redefine it, and try to calculate **fac2[255]**:

```
Remove[fac2];

fac2[0] = 1; fac2[n_] := fac2[n] = n fac2[n - 1]
```

```
fac2[255]
```

$RecursionLimit::reclim : Recursion depth of 256 exceeded. More...

335085068493297911765266512375481494202258406359174070257677988428620 8
7990357327710056261381267633142592808021185022824459265501355222518 56
7276925331930704128110833330325659322041700029792166250734253390513 754
4660457112403384627010340202629925813784231472766366436471553963053 52
5411055414394348401099150682854306750685916385819806041629403833565 86
7391982687821049246140766057935628652419821762074286209697768031494 67
4313868079724382476891586560000000000000000000000000000000000000000 000
00000000000000000000 Hold[1 fac2[1 - 1]]

The recursion limit 256 has been exceeded. The remaining steps are kept unevaluated with **Hold**. The calculations could, however, be completed with **ReleaseHold**:

```
ReleaseHold[%]
```

(We do not show the result here.) If you want to calculate large factorials with **fac2**, then increase the value of **$RecursionLimit**. You could, for example, write the following:

```
$RecursionLimit = 500
```

14.1.4 Implicit Functions

Implicit functions are defined via equations: the solution of an equation defines the value of the function. For example, the equation $x^3 - y^4 + x\,y^3 - y + 1 = 0$ implicitly defines a function y of x. Calculating derivatives of an implicit function is considered in Section 16.1.2, p. 419. Implicit functions can be plotted with **ImplicitPlot** or **ContourPlot** as is shown in Section 5.3.2, p. 123. Here are two additional examples.

■ Example 1

Consider the following polynomial expression:

```
expr = x^3 - y^4 + x y^3 - y + 1;
```

We plot the function **y** defined by the equation **expr == 0**:

```
<< Graphics`ImplicitPlot`
```

```
ImplicitPlot[expr == 0, {x, -2, 2}];
```

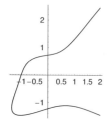

We can find values of **y** for a fixed **x** with **NSolve**:

```
y /. NSolve[expr == 0 /. x → 0, y]
```

{-1.22074, 0.248126 - 1.03398 i, 0.248126 + 1.03398 i, 0.724492}

■ **Example 2**

The implicit function can be quite complex, as in $\int_0^2 \sin(\sin(x\, y\, t))\, dt - 1 = 0$, and we can still succeed in plotting it. In using **ContourPlot** to plot an implicit function defined by **expr == 0**, we ask only for the contour in which **expr** takes on the value **0**:

```
ContourPlot[NIntegrate[Sin[Sin[x y t]], {t, 0, 2}, AccuracyGoal → 6] - 1,
   {x, 0, 4}, {y, 0, 3}, PlotPoints → 40, Frame → False, Axes → True,
   Contours → {0}, ContourShading → False, AspectRatio → Automatic];
```

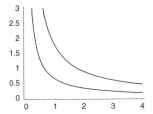

To calculate values of **y** for a fixed **x**, we can define a function:

```
f[x_, y0_, y1_] :=
   y /. FindRoot[NIntegrate[Sin[Sin[x y t]], {t, 0, 2}] == 1, {y, y0, y1}]
```

Here **y0** and **y1** are the starting values for the iterative method used by **FindRoot**. Giving different starting values may result in different values if the function has multiple values. We ask for the value of **y** when **x** is 1. First we start from 0.5 and 1, and then from 1 and 2 (we get a warning message but the result is correct):

```
{f[1, 0.5, 1], f[1, 1, 2]}
```

```
NIntegrate::inum :
   Integrand Sin[Sin[y t]] is not numerical at {t} = {1.}. More…
{0.622871, 1.73409}
```

```
Remove["Global`*"]
```

14.1.5 Pure Functions and Scoping Constructs

We already encountered *pure functions* in Sections 2.2.2, p. 36 (when considering **Map**), 13.2.5, p. 340 (the selecting criterion of **Select** was a pure function), and 13.3.1, p. 343 (when considering **Map** once again).

Mathematica has a number of *scoping constructs*, in which certain names are made local. The most important of these constructs is **Module**, but we also have **With** and **Block**. We have already used **Module** several times in this book, and **Block** has also been used to suppress the display of plots.

■ **Pure Functions**

We illustrate pure functions using **Select**. First we generate a table of 100 random numbers:

```
ran = Table[Random[], {100}];
```

Suppose we are interested in picking the numbers that are either less than 0.02 or greater than 0.98. **Select[list, test]** is the appropriate command. It selects from **ran** all the elements for which the test gives **True**. One possibility is to define a function to this end:

```
f[x_] := x < 0.02 || x > 0.98
```

The value of this function is **True** if $x < 0.02$ or $x > 0.98$ and **False** otherwise. For example:

```
{f[0.01], f[0.03]}
```

```
{True, False}
```

Then we can use this function in **Select**:

```
Select[ran, f]
```

```
{0.997796, 0.0179173, 0.987719, 0.0025816, 0.0096625, 0.987497}
```

This works fine. However, we used the function **f** only once, and it seems too ceremonious to define a function for such a temporary use. Pure functions are handy for temporary use. They need not be defined beforehand but instead only exactly where they are used. In our example, we can use a pure function as follows:

```
Select[ran, Function[x, x < 0.02 || x > 0.98]]
```

```
{0.997796, 0.0179173, 0.987719, 0.0025816, 0.0096625, 0.987497}
```

There is even a simpler construction in which **#** means the argument and **&** is the mark of a pure function:

```
Select[ran, # < 0.02 || # > 0.98 &]
```

```
{0.997796, 0.0179173, 0.987719, 0.0025816, 0.0096625, 0.987497}
```

In general, we have two ways of writing a pure function to perform a sequence **body** of commands for the arguments:

Function[x, body]   A pure function with the argument **x**

Function[{x, y, … }, body]   A pure function with the arguments **x, y, …**

body &   A pure function with the argument **#** or with the arguments **#1, #2, …** (**##** means all arguments supplied to a pure function)

For more examples of pure functions, see Sections 7.1.3, p. 184; 9.1.2, p. 233; 9.4.1, p. 255; 13.3.1, p. 343 (**Map**); 13.2.5, p. 340 (**Select**); and 13.3.2, p. 345 (**Apply**). Further examples will be found in Section 15.2 (**Nest, FixedPoint, Fold**) and throughout the coming chapters.

■ Module: Local Variables

Often the definition of a function consists of a sequence of operations. One possibility is to separate the steps with semicolons. Consider the following example in which we calculate the sum and product of integers 1, 2, ..., n:

```
f1[n_] := (s = 0; p = 1; Do[s = s + i; p = i p, {i, n}]; {s, p})
```

```
f1[7]    {28, 5040}
```

One drawback is that the temporary variables **s** and **p** retain their values outside of the function:

```
{s, p}    {28, 5040}
```

```
Remove[s, p]
```

This can cause confusion later on. It would be useful if the variables **s** and **d** were local to the function: outside the function they had no values. This can be achieved with a module.

```
f[x_] := Module[{local variables}, body]
```

Note the following:

- local variables are separated by commas;
- initial values can be given for the local variables;
- commands in the body are separated by semicolons; and
- the result of the last command of the module is printed automatically.

The preceding example is now as follows:

```
f2[n_] := Module[{s = 0, p = 1},
   Do[s = s + i; p = i p, {i, n}];
   {s, p}]
```

As we see, initial values of the local variables can be given at the same time as they are made local. Often each command in the module is written in its own row to make the module more readable. Before we use **f2**, we give some values for **s** and **p**:

```
s = -1; p = -2;
```

```
f2[7]    {28, 5040}
```

Now **s** and **p** have their old values:

```
{s, p}    {-1, -2}
```

This means that **s** and **p** in the module are not the same as **s** and **p** outside the module (indeed, inside the module *Mathematica* uses names of the form **s$nnn** and **p$nnn**, where **nnn** is increased incrementally by one every time we use the module).

The result of the last command of the module is automatically printed; in the example, the last command was **{s, p}**. If we want to print some intermediate results, we can use **Print** (see Section 15.1.2, p. 388).

Programming with *Mathematica* is essentially done by writing functions, often with **Module**. The programs can use all the commands of *Mathematica*. In Chapter 15, we consider some styles of programming and some special commands that are useful in these styles. You will encounter modules frequently in this book. We have already encountered them in Sections 7.1.3, p. 184; 9.1.2, p. 233; and 9.4.1, p. 255.

■ With: Local Constants

```
f[x_] := With[{x = x0, y = y0, ... }, body]
```

The **With** construction is formally similar to a **Module** but is actually more restricted: **With** is used to define local *constants*; these constants cannot be changed later on within the **With** construction. Note that a **Module** contains local *variables*, and their values can be changed in the module. **With** can be used inside a **Module** to define local constants. In the following example, we define **c** as having the numerical value of 2π:

```
f3[n_] := With[{c = N[2 π]}, Table[Random[Real, {0, c}], {n}]]

f3[1000000]; // Timing

{1.12 Second, Null}
```

The numerical value of 2π was calculated only once, and the 1,000,000 random numbers were generated quickly. Calculating without **With** or **Module** takes a longer time, because the numerical value of 2π is recalculated for each random number:

```
f4[n_] := Table[Random[Real, {0, N[2 π]}], {n}]

f4[1000000]; // Timing

{1.59 Second, Null}
```

■ Block: Local Values of Variables

```
Block[{x = x0, y = y0, ... }, body]
```

We have used blocks to suppress a display when plotting (see Section 5.2.1, p. 118). Indeed, this may be the most important use of blocks. In a **Block**, the variables mentioned within curly braces have only *local values*. For example, the variable **$DisplayFunction** normally has the following value:

```
$DisplayFunction      Display[$Display, #1] &
```

This causes the result of a plot to be displayed on the screen. We can use **Block** to give this variable a local value:

```
Block[{$DisplayFunction = Identity},
   g1 = Plot[Sin[x], {x, 0, Pi}];
   g2 = Plot[Cos[x], {x, 0, Pi}];
   g3 = Plot[Tan[x], {x, 0, Pi}]];
```

If the value **Identity** is given, the plot is not displayed. After executing the block, the variable still has the original value:

```
$DisplayFunction      Display[$Display, #1] &
```

Blocks are also important in that iteration commands like **Table**, **Sum**, and **Do** localize the values of the iterators with **Block**. For example:

```
i = 10; Sum[2^i, {i, 3}]; i      10
```

We see that **i** still has its original value 10.

However, there is still a possibility of confusion with iterator commands. Consider the following function:

```
i =.

f5[x_, n_] := Sum[x^i, {i, n}]

f5[x, 3]      x + x^2 + x^3
```

If we ask for **f5[i, 3]**, we do not get the desired result:

f5[i, 3]   32

We get the number $1^1 + 2^2 + 3^3$ instead of $i^1 + i^2 + i^3$. We can see this with **Trace** (see Section 14.2.1, p. 363):

Trace[f5[i, 3], TraceDepth → 1]

$$\left\{ \text{f5[i, 3]}, \sum_{i=1}^{3} i^i, 1 + 4 + 27, 32 \right\}$$

For this reason it is safe to use a module to localize the iterator **i**:

f6[x_, n_] := Module[{i}, Sum[x^i, {i, n}]]

f6[i, 3] $i + i^2 + i^3$

Remove["Global`*"]

14.2 Tracing, Compiling, and Attributes

14.2.1 Tracing

■ Tracing Expressions

If we are interested in seeing in detail how *Mathematica* calculates an expression, we can use tracing. When we write our own functions or programs, tracing may be useful in detecting coding errors and in making a function effective. We study tracing in some detail; for more, see Section 2.6.11 in Wolfram (2003).

Trace[expr]   Generate a list of all expressions used in the evaluation of **expr**   **Trace[expr, TraceDepth → n]**   Ignore steps that lead to lists nested more than **n** levels deep

For example:

Trace[5 (3 + 1) – 4]

{{{3 + 1, 4}, 5 4, 20}, 20 – 4, 16}

We see that *Mathematica* starts the calculation from 3 + 1, gets 4, forms the product 5 4, gets 20, forms the difference 20 – 4, and finally gets 16. As another example, consider the function **f3** we defined in Section 14.1.5, p. 362, to calculate random numbers:

f3[n_] := With[{c = N[2 π]}, Table[Random[Real, {0, c}], {n}]]

Now we investigate the function with **Trace** if three random numbers are generated:

f3[3] // Trace // ColumnForm

```
f3[3]
With[{c$ = N[2 π]}, Table[Random[Real, {0, c$}], {3}]]
{N[2 π], 6.28319}
Table[Random[Real, {0, 6.28319}], {3}]
{Random[Real, {0, 6.28319}], 2.20994}
{Random[Real, {0, 6.28319}], 0.456972}
{Random[Real, {0, 6.28319}], 1.80681}
{2.20994, 0.456972, 1.80681}
```

First a local variable called **c$** is formed, then **N[2π]** is calculated, and lastly the three random numbers are generated.

■ Tracing Assignments

> **Trace[expr, var = _]** Trace assignments to **var**
> **Trace[expr, _ = _]** Trace all assignments

Often it is useful to see all assignments to variables to check that they are done correctly. Remember that the underscore (_) means anything in *Mathematica* and so, for example, **var = _** means all assignments to **var** and **_ = _** means all assignments to all variables.

Here is a procedural program for calculating a factorial:

```
g[n_] := Module[{fac, i}, For[fac = 1; i = 1, i ≤ n, ++i, fac = fac i]; fac]
```

We trace all assignments for **i** or **fac** (with |, we can express alternative patterns):

```
Trace[g[3], (i = _) | (fac = _)]

{{{{{fac$15 = 1}, {i$15 = 1}}, {fac$15 = 1}, {{i$15 = 2}},
    {fac$15 = 2}, {{i$15 = 3}}, {fac$15 = 6}, {{i$15 = 4}}}}}}
```

■ Tracing Functions

> **Trace[expr, f]** Trace all calls to function **f**
> **Trace[expr, f[_]]** Give a list of intermediate expressions of the form **f[_]**
> **Trace[expr, f[x_] → x]** Give a list of all values of the argument of **f**
>
> **On[f]** Switch tracing on for the symbol **f**
> **Off[f]** Switch tracing off for the symbol **f**

Here is a function for calculating Fibonacci numbers (we do not use dynamic programming; see Section 14.1.3, p. 356):

```
fib[1] = 1;
fib[2] = 1;
fib[n_] := fib[n - 1] + fib[n - 2]
```

It is convenient to turn tracing on and off, particularly for recursive functions:

```
On[fib]
```

```
fib[3]
```

```
fib::trace :  fib[3]  --> fib[3 - 1] + fib[3 - 2] . More...

fib::trace :  fib[3 - 1]  --> fib[2] . More...

fib::trace :  fib[2]  --> 1 . More...

fib::trace :  fib[3 - 2]  --> fib[1] . More...

fib::trace :  fib[1]  --> 1 . More...

2
```

```
Off[fib]
```

Then we trace all calls to **fib**:

```
Trace[fib[3], fib]
```

```
{fib[3], fib[3 - 1] + fib[3 - 2], {fib[2], 1}, {fib[1], 1}}
```

```
Remove["Global`*"]
```

14.2.2 Compiling

■ Defining a Compiled Function

With compiling, we can speed up functions and programs. Normally *Mathematica* must handle many different kinds of arguments for functions (e.g., numbers, symbols, lists, more complex expressions), and this makes the evaluation process slower than in a situation in which *Mathematica* can assume that all arguments are numbers. In compiled functions, the arguments are assumed to be numbers, and this simplifies the evaluation process and speeds it up.

---

f = Compile[{x, y, … }, expr]   Create a compiled function **f** to evaluate **expr** for numerical values of the real variables **x**, **y**, …

Here is a simple compiled function:

```
f1 = Compile[{x}, x + 7.]
```

```
CompiledFunction[{x}, x + 7., –CompiledCode–]
```

The result of compiling is an object called `CompiledFunction`. The compiled code itself is not shown; `–CompiledCode–` represents it. We try the compiled function:

```
f1[10]      17.
```

With **InputForm**, we can see the compiled code:

```
f1 // InputForm
```

```
CompiledFunction[{_Real}, {{3, 0, 0}, {3, 0, 2}},
   {0, 0, 3, 0, 0}, {{1, 5}, {5, 7., 1}, {25, 0, 1, 2}, {2}},
   Function[{x}, x + 7.], Evaluate]
```

The code is not readable for us, but for *Mathematica*, it gives clear instructions about how to effectively calculate the value of the function. The instructions specify, for example, the types of arguments; the version of the compiler; the numbers of needed logical, integer, real, complex, and tensor registers; and what to load into these registers when the calculation proceeds. Thus, *Mathematica* compiles the function to form a kind of pseudocode that contains simple instructions for evaluating the compiled function. At the end of the code, we see the function as a pure function. This pure function is used if the compiled code for some reason cannot be used.

`Compile` handles numerical functions, list manipulation functions, matrix operations, procedural and functional programming constructs, and so on. Compiling is primarily designed for heavy numerical computations with machine-precision numbers. Note, however, that with compiled code we generally cannot get faster commands than the built-in commands of *Mathematica*. Also, a compiled code does not handle numerical precision in the same way as ordinary *Mathematica* code and may use somewhat restricted algorithms. Thus, use the built-in commands as much as possible, but remember compilation as a way to speed up your own heavy numerical programs.

Note that many built-in *Mathematica* commands compile their arguments automatically. These commands have the default setting `Compiled → True`.

■ Integer, Complex, and Logical Arguments

The default is that all arguments of compiled functions are real numbers. We can also specify integer, complex, and logical arguments.

> `f = Compile[{{x1, t1}, {x2, t2}, … }, expr]`   Arguments `x1`, `x2`, … are of types `t1`, `t2`, …; allowed types: `_Integer, _Real, _Complex, True|False`

The basic compiled function `Compile[{x, y, … }, expr]` considered above is the same as the following:

> `Compile[{{x, _Real}, {y, _Real}, … }, expr]`

Here is a compiled function with one real and one integer argument (the default real type need not be declared):

> `f2 = Compile[{x, {n, _Integer}}, x^n]`

> $\text{CompiledFunction}[\{x, n\}, x^n, -\text{CompiledCode}-]$

> `Table[f2[1.5, n], {n, 4}]`

> $\{1.5, 2.25, 3.375, 5.0625\}$

What if the value of **n** was 1751?

> `f2[1.5, 1751]`

> CompiledFunction::cfn : Numerical error encountered at
>   instruction 2; proceeding with uncompiled evaluation. More…
> $2.166679161861306 \times 10^{308}$

We got an error message, because the number exceeded the limit allowed for machine-precision numbers. The maximum machine-precision number is shown here:

> `$MaxMachineNumber` 1.79769×10^{308}

The value of the function was thus calculated with the usual uncompiled methods.

■ Tensor Arguments

> **f = Compile[{{x1, t1, r1}, {x2, t2, r2}, … }, expr]**   Arguments **x1**, **x2**, … are tensors, having elements of type **t1**, **t2**, … and ranks **r1**, **r2**, …

If an argument of a compiled function is a list, matrix, or, in general, a tensor, we define the type of its elements and also its rank. The rank of a tensor is the number of indices needed to specify each element. So the rank of a vector is one and that of a matrix is two.

We write a compiled function to calculate the product of a matrix and a vector:

```
prod = Compile[{{a, _Real, 2}, {b, _Real, 1}}, a.b]

CompiledFunction[{a, b}, a.b, -CompiledCode-]

prod[{{6, 8, 3}, {5, 1, 7}, {4, 2, 9}}, {2, 3, 1}]

{39., 20., 23.}
```

■ Types of Subexpressions

> **f = Compile[vars, expr, {{s1, t1}, {s2, t2}, … }]**   Subexpressions that match **s1**, **s2**, … are of type **t1**, **t2**, …

Mathematica knows the type of result from a calculation if the calculation only contains standard arithmetic operations. Thus, the result of $2 + 3$ is an integer, and the result of $2.0 + 3$ is a real number. If we use other functions, *Mathematica* may not know the type of an intermediate result and has to assume that the result is a real number. If we know that the result is not a real number, we can give this information at the end of the definition of the compiled function.

In the following example, the compiled code is not even used unless we declare that the result of **Eigenvalues[_]** is a complex tensor of rank 1 (i.e., a complex vector):

```
f3 = Compile[{{a, _Real, 2}},
    Eigenvalues[a.a], {{Eigenvalues[_], _Complex, 1}}];

f3[{{1., 2}, {-4, 1}}]

{-7. + 5.65685 i, -7. - 5.65685 i}
```

■ Compiling Expressions

Thus far we have considered forming a compiled function. We can also compile other expressions and then evaluate the resulting compiled code.

> **Experimental`CompileEvaluate[expr]**   Compile **expr** and then evaluate the compiled code

(**Experimental`** is a context of commands provided on an experimental basis. We could also first execute **<<Experimental`** and then **CompileEvaluate[expr]**.) Compiling an expression should speed up the evaluation, particularly for large numerical computations. For example:

```
Do[Random[], {1000000}]; // Timing
```
```
{2.3 Second, Null}
```

```
Experimental`CompileEvaluate[Do[Random[], {1000000}]]; // Timing
```
```
{0.62 Second, Null}
```

14.2.3 Attributes

Sometimes we may want a function to have a special property, such as being able to accept a list as an argument and calculate the value of the function for each element of that list. We have observed that the built-in functions have this property. For example:

```
Sin[{-2, -1, 0, 1}]
```
```
{-Sin[2], -Sin[1], 0, Sin[1]}
```

For this reason, the function **Sin** is said to have the **Listable** property or *attribute*. This can also be seen by asking for the attributes of **Sin**:

```
Attributes[Sin]
```
```
{Listable, NumericFunction, Protected}
```

If you form your own function, it may not necessarily be listable. The function **f** in the next example is not listable:

```
f[x_] := If[x < 0, Cos[x], Sin[x]]
```

```
f[{-2, -1, 0, 1}]
```
```
If[{-2, -1, 0, 1} < 0, Cos[{-2, -1, 0, 1}], Sin[{-2, -1, 0, 1}]]
```

One possibility for overcoming this problem is to use **Map**:

```
Map[f, {-2, -1, 0, 1}]
```
```
{Cos[2], Cos[1], 0, Sin[1]}
```

We can also assign the **Listable** attribute to **f**:

```
SetAttributes[f, Listable]
```

Now we can give a list as an argument:

```
f[{-2, -1, 0, 1}]
```
```
{Cos[2], Cos[1], 0, Sin[1]}
```

Attributes[f]   Give the attributes of **f**
SetAttributes[f, attr]   Add **attr** to the attributes of **f**
ClearAttributes[f, attr]   Remove **attr** from the attributes of **f**

The different attributes of the built-in commands are as follows:

```
Union[Flatten[Map[Attributes, Names["*"]]]]
```
```
{Constant, Flat, HoldAll, HoldAllComplete, HoldFirst, HoldRest,
  Listable, Locked, NHoldAll, NHoldFirst, NHoldRest, NumericFunction,
  OneIdentity, Orderless, Protected, ReadProtected, SequenceHold}
```

We do not explain all of the attributes; for more information, refer to Section 2.6.3 of Wolfram (2003). We just explain the **HoldAll** attribute. Many plotting commands have this attribute, and so do many others:

```
Select[Names["*Plot*"], MemberQ[Attributes[#], HoldAll] &]
```

```
{ContourPlot, DensityPlot,
  ParametricPlot, ParametricPlot3D, Plot, Plot3D}
```

This means that the expression to be plotted is not evaluated first but only at specific numerical sampling points as the plotting proceeds. On the other hand, we can force the expression to be evaluated first by enclosing the expression with **Evaluate** (see Section 5.1.2, p. 114). The **HoldAll** attribute is also a property of numerical commands like **Find- Minimum**, **FindRoot**, and **NIntegrate**, and so the expression is not evaluated first but only at the sampling points. In addition, iteration commands like **Product**, **NProduct**, **Sum**, **NSum**, **Table**, and **Do** have the **HoldAll** attribute. This again means that the expression to be processed is evaluated only at the specific values of the iteration variable.

14.3 Contexts and Packages

14.3.1 What is a Context?

When working with packages, we are also working with contexts. To put it simply, contexts are like directories. They enable us to arrange material hierarchically and to keep content from blending. Normally we do not need to concern ourselves with contexts; *Mathematica* takes care of them. However, in some cases contexts help us understand the way *Mathematica* works. Also, in writing packages, we need contexts.

We start a new session and ask what contexts we have right now:

```
$ContextPath     {Global`, System`}
```

The context **System`** contains *Mathematica*'s built-in definitions and the context **Global`** our own definitions. The **System`** context contains 1920 symbols:

```
Short[Names["System`*"], 2]
```

```
{Abort, AbortProtect, Above, Abs,
  <<1913>>, $UserName, $Version, $VersionNumber}
```

We have thus far defined no symbols, and so the **Global`** context is empty:

```
Names["Global`*"]     {}
```

We define some symbols:

```
a = 5; b = 8;
```

Now the **Global`** context contains these symbols:

```
Names["Global`*"]     {a, b}
```

We can ask for information about **a**:

```
? a
```

```
Global`a

a = 5
```

We see the context where **a** is defined and its value. In fact, we see the complete name of the symbol **a**, which is **Global`a**. In this way, all symbols have complete names that indicate the contexts in which they are defined. We can use the complete names if we want:

> **{Global`a, System`Exp[1]}**   {5, e}

However, this is normally not necessary.

Then we load a package:

> **<< Graphics`Animation`**

The contexts are now as follows:

> **$ContextPath**

> {Graphics`Animation`, Geometry`Rotations`,
> Utilities`FilterOptions`, Global`, System`}

Basically, packages have their own contexts: we have the new **Graphics`Animation`** context and, in addition, the contexts of two other packages. Indeed, a package may need some other packages, and so they are also loaded. Here are the names defined in the animation package:

> **Names["Graphics`Animation`*"]**

> {Animate, Animation, AnimationFunction, Frames, MovieContourPlot,
> MovieDensityPlot, MovieParametricPlot, MoviePlot,
> MoviePlot3D, RasterFunction, RotateLights, ShowAnimation,
> SpinDistance, SpinOrigin, SpinRange, SpinShow, SpinTilt}

$ContextPath	Give a list of all existing contexts
$Context	Give the current context
Context[name]	Give the context of **name**

Mathematica always has a current context. This is the context in which we are presently working and where all new names are stored. Normally it is **Global`**:

> **$Context**   Global`

Within a package, the current context changes, as we will see in Section 14.3.3, p. 372. We can ask for the context of a particular name:

> **{Context[Integrate], Context[a], Context[Animate]}**

> {System`, Global`, Graphics`Animation`}

■ Searching for a Name in the Contexts

Note the following important fact (we need it in Section 14.3.2, p. 371): when we have entered a short name (i.e., a name without the context), *Mathematica* searches the name from the contexts in the list **$ContextPath** *from left to right, but first from the current context* **$Context**. If the entered name is found somewhere, its value or definition is used. If the name is not found anywhere, the name is created in the current context. (If you enter a complete name, that is, a name with the context, then *Mathematica* searches for the name only in the given context.)

For example, write the following:

```
sin[π]
```

General::spell1 : Possible spelling error: new symbol
    name "sin" is similar to existing symbol "Sin". More...

```
sin[π]
```

Mathematica could not find the name **sin** anywhere and so creates it. From now on, the name **sin** stays in the **Global`** context:

Names["Global`*"]   {a, b, sin}

You did not intend to create a new name **sin**, but *Mathematica* had no other choice. If you want to get rid of **sin**, you can write **Remove[sin]**. If you want to remove all user-defined symbols, write the following:

Remove["Global`*"]

14.3.2 Forgetting to Load: Once Again

■ Problem

What to do when you forget to load a package was addressed in Section 4.1.1, p. 87, but let us now see what actually happens. We start a new session, and we want to use the command **CountRoots** but have forgotten to load the **Algebra`RootIsolation`** package:

CountRoots[2 x^4 - x^3 + 3 x^2 - 4, {x, -5, 5}]

CountRoots$[-4 + 3 x^2 - x^3 + 2 x^4, \{x, -5, 5\}]$

Mathematica searched the name **CountRoots** from all contexts, could not find such a name, and so created the name in the context **Global`**:

Names["Global`*"]   {CountRoots, x}

Next we load the package in the hope that we can use the command:

<< Algebra`RootIsolation`

CountRoots::shdw : Symbol CountRoots appears in
    multiple contexts {Algebra`RootIsolation`, Global`};
    definitions in context Algebra`RootIsolation` may
    shadow or be shadowed by other definitions. More...

We get the warning about multiple **CountRoots**: we have a **CountRoots** in the **Algebra`**﹂ **RootIsolation`** context and another in the **Global`** context. The multiple names may *shadow* each other, which means that *Mathematica* finds one of the names but not the other. Indeed, the name in the **Global`** context seems to shadow the name in the context of the package, because the command still does not work:

CountRoots[2 x^4 - x^3 + 3 x^2 - 4, {x, -5, 5}]

CountRoots$[-4 + 3 x^2 - x^3 + 2 x^4, \{x, -5, 5\}]$

■ Explanation

We now have two **CountRoots**:

Names["*`CountRoots"]

{Algebra`RootIsolation`CountRoots, CountRoots}

One **CountRoots** is in the context of the package and the other in the **Global`** context (*Mathematica* does not write the context for names in the current context). Here is the context path:

> **$ContextPath**
>
> {Algebra`RootIsolation`, Global`, System`}

When we want to use **CountRoots**, *Mathematica* searches for this name in the contexts in the context path from left to right, but *first* from the current context, **Global`**. *Mathematica* finds **CountRoots** in **Global`** and uses it rather than the **CountRoots** in **Algebra`Root:Isolation`**.

■ **Solution**

To solve the problem, we could use the complete name of **CountRoots**:

> **Algebra`RootIsolation`CountRoots[2 x^4 - x^3 + 3 x^2 - 4, {x, -5, 5}]**
>
> 2

Thus, if there is a name in multiple contexts, we have access to each of them by entering the complete name. In our situation, it is more convenient to remove the useless **Count:Roots** from **Global`**:

> **Remove[CountRoots]**

(We could also write **Remove[Global`CountRoots]**, but the context name is not needed for names in the current context.) Now we have only one **CountRoots**, and we can use the command without the context name:

> **CountRoots[2 x^4 - x^3 + 3 x^2 - 4, {x, -5, 5}]**
>
> 2

Here is a summary:

> If you forget to load a package before using a command that is in it, do as follows:
> • remove the name or names of the package you have used;
> • load the package; and
> • use the command of the package again.

Another solution is to quit the kernel from **Kernel ▷ Quit Kernel** and then restart the kernel from **Kernel ▷ Start Kernel**, but then you may need to do some calculations again. A way to avoid the problem is to automatically load packages of certain subjects as the packages are needed by using commands like **<<Algebra`**. Loading packages is considered in Section 4.1.

14.3.3 Writing a Package

■ **Converting a Notebook into a Package**

You may have developed some useful functions that you may need later on. Save them as a package, as follows. Collect useful functions into a notebook and then save the notebook with a name such as **programs.m** using **File ▷ Save As Special... ▷ Text**. See Section 4.1.3, p. 89, to find a *Mathematica* **Application** folder in which you can save the notebook.

To load the package and use the functions defined in it, write **<<programs.m**. In this way you can create notebooks that you can use like packages.

By the way, if you open the saved package with **File ▷ Open**, select the cell of the package, and choose **Cell ▷ Cell Properties**, you see that the cell has the property **Initialization Cell**. When a notebook is loaded, all initialization cells are automatically executed. In this way, the definitions of the programs in the package are automatically executed when the package is loaded.

A true package has a special structure that we now begin to study.

■ An Example of a Package

A package has a special structure in which contexts play a key role. To give an elementary example, we have written the following code in a new notebook:

```
BeginPackage["Own`newton`"]

newton::usage = "newton[f,x,x0,n] calculates a zero
    of f starting from x0 and using at most n iterations"

Begin["`Private`"]
(* one step *)
newtonStep[f_, df_, x_, xi_] := (x - f / df) /. x → N[xi]
(* iterate the step *)
newton[f_, x_, x0_, n_] :=
  With[{df = D[f, x]}, FixedPointList[newtonStep[f, df, x, #] &, N[x0], n]]
End[]

EndPackage[]
```

This is a single cell, and we have saved the notebook with the name **newton.m** into a new folder called **Own** in the **Applications** folder by selecting **File ▷ Save As Special... ▷ Text**. To try the package, we first load it:

```
<< Own`newton`
```

If we have already forgotten how **newton** is used, we can ask for information:

```
? newton
```

```
newton[f,x,x0,n] calculates a zero of
    f starting from x0 and using at most n iterations
```

We calculate a zero of an expression, starting from point 2:

```
newton[3 x^3 - E^x, x, 2, 20]
```

```
{2., 1.41942, 1.1019, 0.975117,
  0.953089, 0.952446, 0.952446, 0.952446, 0.952446}
```

■ The Structure of a Package

The definition of the package begins with **BeginPackage**, in which the package name is the argument. If the package needs other packages, these packages can also be mentioned as arguments of **BeginPackage** (an example is given below).

Also coming into play are the usage messages. These messages give information about the usage of the various functions. This information is displayed when the user writes **?name** where **name** is a name defined in the package. *The user of the package can use only the names for which a usage message exists.* In this way you can restrict the set of names available for use. In the example above, we did not define a usage message for the function **newton‹ Step**, and so this function cannot be used (it is used only within the package).

The program begins with the command **Begin["`Private`"]** and ends with the command **End[]**. Note that *Mathematica* commands can be annotated by inserting comments into the code. A comment starts with **(*** and ends with ***)**; a comment can be placed anywhere. Comments are useful especially in packages and in other longer codes to help with the reading of the code. The whole package ends with the command **EndPackage[]**.

The commands **BeginPackage**, **Begin**, **End**, and **EndPackage** affect the contexts. This topic will be discussed next.

■ Contexts in Packages

Packages are normally loaded or read in by **<<** in one step, and we do not see what actually happens during the loading. Let us now investigate what happens when we load a package. We will specifically investigate how the context changes during the reading. We start a new session and first load a package:

```
<< Miscellaneous`RealOnly`
```

Then we consider what happens when we load the following package:

```
BeginPackage["Own`newton`", "Algebra`RootIsolation`"]

newton::usage = "newton[f,x,x0,n] calculates a zero
   of f starting from x0 and using at most n iterations"

Begin["`Private`"]
newtonStep[f_, df_, x_, xi_] := (x - f / df) /. x → N[xi]
newton[f_, x_, x0_, n_] :=
 With[{df = D[f, x]}, FixedPointList[newtonStep[f, df, x, #] &, N[x0], n]]
End[]


EndPackage[]
```

This is almost the same example as before. Now, however, we only assumed that we also need the package **Algebra`RootIsolation`** (although we actually do not need it), so this package is mentioned in **BeginPackage**. We proceed step by step and observe how the context evolves. Here is a table of the evolution in terms of **$Context** (the current context) and **$ContextPath** (a list of contexts from which information is searched) after the commands **BeginPackage**, **Begin**, **End**, and **EndPackage**:

	$Context	$ContextPath
(initial state)	Global`	{Miscellaneous`RealOnly`, Global`, System`}
BeginPackage["Own`newton`", "Algebra`RootIsolation`"]	Own`newton`	{Own`newton`, Algebra`RootIsolation`, System`}
Begin["`Private`"]	Own`newton`Private`	Same as above
End[]	Own`newton`	Same as above
EndPackage[]	Global`	{Own`newton`, Algebra`RootIsolation`, Miscellaneous`RealOnly`, Global`, System`}

With regard to the current context, initially it is the usual **Global`**, where all user-defined names are stored. When the reading of the package begins, the context changes to **Own`newton`**. Thus, all names defined in the package are stored in this context. In particular, the names with a usage message are in this context. When the ordinary program begins, the current context gets a subcontext called **Private`**. Names defined in this context without usage messages are not available to the user. After the program ends, the current context changes back to **Own`newton`**; when the whole package is read in, we are again in the **Global`** context.

From the context paths, we see that several things happens when **BeginPackage** is read. First, the **Miscellaneous`RealOnly`** package that was loaded before **Own`newton`** disappears and thus is not available for the package. Consequently, a package can only use the packages declared in **BeginPackage**.

Second, the **Global`** context disappears from the context path; this means that the names in **Global`** are not available in the package. This is safe: we cannot accidentally use or change, within the package, the values of names in **Global`**. In fact, we can have the same names in **Global`** as we have in the package. After loading the package, the names in **Global`** still have their old values and not the values defined in the package.

Third, the **Own`newton`** and **Algebra`RootIsolation`** contexts are added to the context path so that new definitions made in the package are placed in the former context and the latter context makes available the commands in the corresponding package.

After **EndPackage**, we have access to the contexts that were available before reading the package and to the contexts **Own`newton`** and **Algebra`RootIsolation`**.

Maeder (1997) is an excellent source of information about programming and package development. The packages of Maeder's book are in **Mathematica** ▷ **AddOns** ▷ **ExtraPackages** ▷ **ProgrammingInMathematica**. If you open the **Skeleton.m** package in this folder, you can see the correct structure of a package. The notebook **Template.nb** in the same place can also be used to develop a package.

14.3.4 Handling Options and Messages

■ Handling Options

In this section, we give an example that shows you how to define options for your own functions. Another example is in the **dataPlot** program we presented in Section 9.1.2, p. 233. We again consider Newton's method and write the following package:

```
BeginPackage["Own`newton2`"]

newton2::usage = "newton2[f,x,x0,n,opts] calculates a zero
    of f starting from x0 and using at most n iterations."

dampingConstant::usage =
  "dampingConstant -> d is an option for newton2 that gives the
    damping factor. For example, if the zero is of multiplicity
    2, define dampingConstant -> 2 to accelerate the convergence."

stoppingCriterion::usage =
  "stoppingCriterion -> (pure function) is an option for newton2
    that gives the stopping criterion for the iteration. Examples
    of values: (Abs[#1 - #2] < 10^-8 &), (Abs[f /. x->#2] < 10^
    -6 &). Here #1 is the next to last point and #2 the last point."

Begin["`Private`"]

Options[newton2] = {dampingConstant → 1, stoppingCriterion → (#1 === #2 &)}

newtonStep2[f_, df_, x_, d_, xi_] := (x - d f / df) /. x → N[xi]

newton2[f_, x_, x0_?NumericQ, n_Integer, opts___?OptionQ] :=
  Module[{df = D[f, x], damp, stop},
    {damp, stop} =
      {dampingConstant, stoppingCriterion} /. {opts} /. Options[newton2];
    FixedPointList[newtonStep2[f, df, x, damp, #] &,
      N[x0], n, SameTest → stop]]

End[]

EndPackage[]
```

■ How It Works

The function **newton2** has two options: **dampingConstant** and **stoppingCriterion**. The package now contains usage messages for these options, too. After **Begin["`Private`"]** we define these options and give their default values as **1** and **(#1 === #2 &)**. The damping constant can be given a value higher than one if the zero to be found is a multiple zero. The default stopping criterion means that two successive points must be equal to 16-digit precision. Unless the user defines other values for the options, the default values are used.

The function **newtonStep2** contains the damping factor as the argument **d**. The function **newton2** contains **opts___?OptionQ** as the last argument. The three underscores (___)

after **opts** mean that there can be zero or more options. If the user writes no options, the default values of the options are used. We have also added some tests for the arguments: **x0** must be numeric, **n** must be an integer, and **opts** must contain options. If at least one of these conditions is not satisfied, **newton2** refuses to do anything.

In **newton2**, we first find out what the values of the options are. This is done as follows:

```
{damp, stop} =
  {dampingConstant, stoppingCriterion} /. {opts} /. Options[newton2]
```

The aim is that **damp** and **stop** should get the values of the options. To see how this works, let us first define the options here:

```
Options[newton2] = {dampingConstant → 1, stoppingCriterion → (#1 === #2 &)};
```

Suppose we write **newton2[Cos[x] - 1, x, 2, 20, dampingConstant → 2]**. Then {opts} = {dampingConstant → 2} so that, at the first step, we get as follows:

```
{dampingConstant, stoppingCriterion} /. {dampingConstant → 2}

{2, stoppingCriterion}
```

Now the custom values of options have been taken into account. The second step is then as follows:

```
% /. Options[newton2]

{2, #1 === #2 &}
```

Now the remaining options get the default values. In summary, **damp** gets the value **2** that we defined, and **stop** gets the default value **#1 === #2 &**. Now we can use **damp** and **stop** in **newton2** (**FixedPointList** has the option **SameTest**, and we give it the value **stop**).

■ Example

We have saved the package with the name **newton2.m** in the **Applications ▷ Own** folder. Then we start a new session and try the package:

```
<< Own`newton2`
```

We can ask for information about the package:

```
Names["Own`newton2`*"]

{dampingConstant, newton2, stoppingCriterion}
```

```
? dampingConstant

dampingConstant -> d is an option for newton2 that gives the
    damping factor. For example, if the zero is of multiplicity
    2, define dampingConstant -> 2 to accelerate the convergence.
```

We use a custom stopping criterion:

```
newton2[3 x^3 - E^x, x, 2, 20, stoppingCriterion → (Abs[#1 - #2] < 10^-6 &)]

{2., 1.41942, 1.1019, 0.975117, 0.953089, 0.952446, 0.952446}
```

If the arguments do not satisfy the conditions, then **newton2** does nothing:

```
newton2[3 x^3 - E^x, x, 2, 20.]

newton2[-e^x + 3 x^3, x, 2, 20.]
```

■ Filtering Options

If you build a program with several options for several commands, you may need the **Utilities`FilterOptions`** package. Suppose your program uses **FindRoot** and **Plot**. If **opts** contains all of the options given by the user of your program, you can pick up the options belonging to **FindRoot** and the options belonging to **Plot** by entering the following:

```
frOpts = FilterOptions[FindRoot, opts];
plOpts = FilterOptions[Plot, opts];
```

In your program you could then write, for example:

```
FindRoot[ …, frOpts];
Plot[ …, plOpts];
```

An example of filtering options is in the **dataPlot** program of Section 9.1.2, p. 233.

■ Handling Messages

A well-designed package prints messages if problems are encountered during the execution of the functions in the package. Messages can be printed with **Print** (see Section 15.1.2, p. 388), but there is a better method. Suppose an algorithm does not converge for the given values of **maxit** and **eps**. For this situation, the package contains, after the usage messages, the following message template:

```
myPackage::nonconv = "The algorithm did not converge with maxit = `1`
and eps = `2`"
```

The program also has, in the correct place, this command:

```
Message[myPackage::nonconv, maxit, eps]
```

The result is a message like this:

```
myPackage::nonconv :
  The algorithm did not converge with maxit = 50 and eps = 0.0001`
```

Programs

Introduction

> *There is a square room of side twenty feet with a pure mathematician in one corner and an applied mathematician in the opposite corner. In a third corner is a delicious apple. The mathematicians are allowed to approach the apple in bounces along the sides of the square, the first bounce a maximum of ten feet and every subsequent bounce a maximum of half the previous bounce. The pure mathematician, well versed in limits, quickly calculates that no matter how many bounces he takes he can never reach the apple, so he doesn't even begin to bounce. The applied mathematician sets off at once because he realizes that after five or six bounces he will be close enough to take the apple.*

Although *Mathematica* has a great many ready-to-use commands for almost all kinds of mathematical problems, we will, in this book, now and then present some short programs for doing calculations we find interesting, pedagogically worthwhile, or sometimes even practically useful. The same reasons may give you motivation to study programming with *Mathematica*. Although *Mathematica* is a kind of interactive calculator that does calculations step by step, a program, by combining the steps into one or a few logical blocks, may make the calculation vastly more effective and even simpler.

The material presented in this chapter introduces you to programming. Other sources of programming include Gray (1997), Maeder (1997) (the packages of this book come with *Mathematica*), and Wagner (1996).

Mathematica supports many styles of programming, such as procedural, functional, and rule-based. Even object-oriented programming can be implemented. Functional and rule-based programming make up the heart of programming in *Mathematica*. Let us briefly introduce the programming styles before we study them in some detail.

■ Procedural Programming

The procedural style is familiar from ordinary programming languages such as FOR-TRAN and C. You probably know such structures as **For**, **While**, and **If**; they all exist in *Mathematica*, too. However, as you study other styles of programming in *Mathematica*, you will find that they are often more effective than the procedural style. A large procedural program may be rewritten in a few lines of code of functional or rule-based programming. So, before you begin to code your problem, study in detail whether you can use functional or rule-based programming. Of course, these styles first require serious study, but the time you spend on them is very interesting and saves you time as you progress.

■ Functional Programming

With functional programming, we apply functions to arguments. The functions can be built-in functions or functions we have defined, and they can be applied in a nested way. The functions are applied to the arguments by special powerful commands such as **Map**, **Apply**, **Nest**, **FixedPoint**, and **Fold**. For example, Newton's method to calculate a zero of a function can be programmed with **FixedPoint** as follows (see Section 15.2.3, p. 399):

```
newton[f_, x_, x0_, max_] :=
  With[{df = D[f, x]}, FixedPoint[(x - f / df) /. x → # &, N[x0], max]]
```

We simply give the inputs, and **FixedPoint** does the iterations and stops when convergence has been achieved:

```
newton[3 x^3 - E^x, x, 2, 20]
```

```
0.952446
```

■ Rule-Based Programming

Rule-based programming uses *rules* and *patterns*. A function definition **f[x_] := expr** is an example of a *global rule*: whenever **f** is encountered with a specified argument, for example, **a**, this rule replaces **f[a]** with the value of **expr**, where **x** is replaced with **a**. In **expr /. x → a** we apply a *local rule*: whenever **x** is encountered in **expr**, replace it with **a**. The argument **x_** of a function is an example of a *pattern*. The pattern **x_** is very general and is, in fact, matched by anything; the name of the pattern is **x**. You can form more restrictive patterns such as the ones in the following example, where we give rules for a logarithm-like function:

```
loga[x_ y_] := loga[x] + loga[y]
loga[x_^y_] := y loga[x]
```

Now, whenever *Mathematica* encounters **loga**, it tries to use these rules:

```
loga[a^2 b^3 Sqrt[c] / d^4]
```

$$2 \, \text{loga}[a] + 3 \, \text{loga}[b] + \frac{\text{loga}[c]}{2} - 4 \, \text{loga}[d]$$

Generally in rule-based programming we give several rules for the same function. These rules cover several different situations or several patterns of argument.

■ Object-Oriented Programming

With object-oriented programming we are interested in certain data and their organization into hierarchies. In the following example, we consider objects **sin**, **cos**, and **tan** and associate two properties with each object, namely its derivative and integral:

```
der[sin] ^= cos; int[sin] ^= -cos;
der[cos] ^= -sin; int[cos] ^= sin;
der[tan] ^= sec^2; int[tan] ^= -ln[cos];
```

We can ask information about the **tan** object, for example:

```
? tan
```

```
Global`tan
```

$$der[tan] \ ^= sec^2$$
$$int[tan] \ ^= -ln[cos]$$

We do not consider object-oriented programming in this book (except the mention of upvalues in Section 15.3.1, p. 404); see Gray (1997) or Maeder (1994) instead.

15.1 Procedural Programming

15.1.1 Simple Programs

Before going to the ordinary commands of procedural programming, let us present some examples of simple programs where we do not use any special programming commands but rather familiar commands like **/.** (see Section 12.1.2, p. 300) and **Table** (see Section 13.1.1, p. 325). Our examples are from numerical analysis.

■ Approximating a Derivative

Because the derivative $f'(a)$ is defined to be $\lim_{h \to 0} [f(a+h) - f(a)]/h$, an approximation of the derivative is $[f(a+h) - f(a)]/h$ for a small h. We write a program for this approximation:

```
der[f_, x_, a_, h_] := (N[f /. x → a + h] - N[f /. x → a]) / h
```

Note that we have used **N** to calculate the decimal value of the approximation, because in numerical calculations we are not interested in "exact approximations." Furthermore, the results gotten when calculating with exact quantities grow easily to huge expressions, in iterative calculations especially, and the computation time may become very long. Thus, use decimal numbers from the start in a numerical program.

We try the program and also calculate the true derivative:

```
f = x Sin[x];
```

```
{der[f, x, 1, 10^-4], df = D[f, x] /. x → 1.}
```

```
{1.38179, 1.38177}
```

The approximation is quite good. Next, we use several values of h and investigate the error:

```
t1 = Table[{h = 10.^-n, d = der[f, x, 1, h], Abs[d - df]}, {n, 1, 15}];

TableForm[t1, TableHeadings → {None, {"h", "approx.", "error\n"}},
  TableSpacing → {0.3, 2}]
```

h	approx.	error
0.1	1.38857	0.00679782
0.01	1.38292	0.00114453
0.001	1.38189	0.000119056
0.0001	1.38179	0.0000119516
0.00001	1.38177	1.19562×10^{-6}
$1. \times 10^{-6}$	1.38177	1.1946×10^{-7}
$1. \times 10^{-7}$	1.38177	1.28788×10^{-8}
$1. \times 10^{-8}$	1.38177	5.995×10^{-9}
$1. \times 10^{-9}$	1.38177	7.17206×10^{-8}
$1. \times 10^{-10}$	1.38177	2.93765×10^{-7}
$1. \times 10^{-11}$	1.38177	8.16458×10^{-7}
$1. \times 10^{-12}$	1.38189	0.000121308
$1. \times 10^{-13}$	1.38001	0.00176607
$1. \times 10^{-14}$	1.38778	0.00600549
$1. \times 10^{-15}$	1.55431	0.172539

The error is smallest when $h = 10^{-8}$; after that the error grows due to increasing rounding errors. We plot the **der** as a function of h for small h:

```
Plot[der[f, x, 1, h], {h, 10^-15, 2 10^-7}, PlotPoints → 300];
```

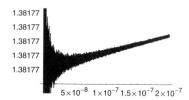

The plot shows the increasing difficulties seen when calculating the approximation as h becomes smaller; the usual fixed-precision decimal numbers simply cannot do better. If we use arbitrary-precision numbers, we have no problems (a similar example was in Section 11.2.3, p. 291):

```
der2[f_, x_, a_, h_] := (N[f /. x → a + h, 20] - N[f /. x → a, 20]) / h

Plot[der2[f, x, 1, SetPrecision[h, 20]],
  {h, 10^-15, 2 10^-7}, Compiled → False];
```

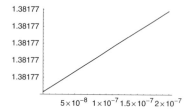

■ Approximating an Integral

The trapezoidal rule to approximate an integral is as follows:

$$\int_a^b f(x)\,dx \simeq \frac{h}{2}\left(f(a) + 2\sum_{i=1}^{n-1} f(a + ih) + f(b)\right)$$

Here $h = (b - a)/n$. We write a program for this rule:

```
trapez[f_, x_, a_, b_, n_] := With[{h = (b - a) / n},
  h / 2 (N[f /. x → a] + 2 Sum[N[f /. x → a + i h], {i, 1, n - 1}] + N[f /. x → b])]
```

Using the **With** scoping construct, we made h a local constant (see Section 14.1.5, p. 361). For example:

```
{trapez[f, x, 0, 2, 50], int = Integrate[f, {x, 0., 2}]}
```

```
{1.7416, 1.74159}
```

The following table shows how the error depends on the step size:

```
t2 = Table[{h = 3 10.^-n,
    InputForm[i = trapez[f, x, 0, 2, 10^n]], Abs[i - int]}, {n, 1, 4}];
```

```
TableForm[t2, TableHeadings → {None, {"h", "approx.", "error\n"}},
  TableSpacing → {0.3, 2}]
```

h	approx.	error
0.3	1.741851999412873	0.000260899
0.03	1.7415936671330097	2.56721×10^{-6}
0.003	1.7415911255879262	2.5668×10^{-8}
0.0003	1.741591001766418	2.56675×10^{-10}

```
h = .
```

■ Approximating a Zero

Newton's method for solving an equation $f(x) = 0$ uses the following recursion formula by starting from a given point x_0:

$$x_{i+1} = x_i - \frac{f(x_i)}{f'(x_i)}, \quad i = 0, 1, \dots$$

A program for it is as follows:

```
newton[f_, x_, x0_, n_] := Module[{newx = x - f / D[f, x], xi = x0},
  Table[xi = N[newx /. x → xi], {n}]]
```

The scoping construct **Module** makes **newx** and **xi** local variables (see Section 14.1.5, p. 360). The starting values of **newx** and **xi** are the righthand side of the recursion formula and the starting point **x0**. **Table** then does the iteration **n** times and stores the results in a list. At each iteration, the new value of **xi** is computed by inserting the old value of **xi** into **newx**. In this way, **Table** can be used in iterative methods, although we have more powerful functional iteration commands like **FixedPoint**. For example, define the following function:

```
g = 3 x^3 - E^x;
```

```
Plot[g, {x, -1, 2}, PlotRange → All];
```

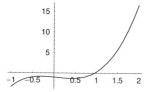

One zero seems to be near $x = 1$. We start from $x = 2$ and do 10 iterations:

```
zero = newton[g, x, 2, 10]
```

{1.41942, 1.1019, 0.975117, 0.953089, 0.952446,
 0.952446, 0.952446, 0.952446, 0.952446, 0.952446}

(Note that the starting point 2 is lacking from the list.) The function is zero very accurately at the last point:

```
g /. x -> Last[zero]     - 4.44089 × 10^-16
```

We investigate how the error in the zero and in the value of the function evolve as we do more and more iterations:

```
t3 = Transpose[{Range[10], zero, Abs[zero - Last[zero]], g /. x → zero}];
```

```
TableForm[t3, TableHeadings → {None, {"n", "zero", "error", "f(zero)\n"}},
  TableSpacing → {0.3, 2}]
```

n	zero	error	f(zero)
1	1.41942	0.466974	4.44462
2	1.1019	0.149457	1.00387
3	0.975117	0.0226711	0.1301
4	0.953089	0.000643299	0.00358769
5	0.952446	5.39696×10^{-7}	3.00737×10^{-6}
6	0.952446	3.80473×10^{-13}	2.11964×10^{-12}
7	0.952446	1.11022×10^{-16}	4.44089×10^{-16}
8	0.952446	0.	-4.44089×10^{-16}
9	0.952446	1.11022×10^{-16}	4.44089×10^{-16}
10	0.952446	0.	-4.44089×10^{-16}

Newton's method converges quickly.

We can also use a recursive function. First, define the function whose zero is to be sought:

```
h[x_] := 3 x^3 - E^x;
```

Then define the iteration points as a recursive function of the iteration index:

```
Clear[x]; x[0] = 2; x[n_] := x[n] = N[x[n - 1] - h[x[n - 1]] / h'[x[n - 1]]]
```

We have written **Clear[x]** at the beginning, because if you change the starting value, the earlier values of **x** have to be cleared so that new values are calculated. Note the method of dynamic programming (see Section 14.1.3, p. 356): **x[n_] := x[n] = ...** As a result, the iterations are stored to the variables x[n], and this speeds up the computations. For example:

```
Table[x[n], {n, 0, 7}]
{2, 1.41942, 1.1019, 0.975117, 0.953089, 0.952446, 0.952446, 0.952446}

Clear[x]
```

■ Approximating the Solution of a Differential Equation

Consider solving, with Euler's method, a differential equation $y'(x) = f[x, y(x)]$ with the initial value $y(x_0) = y_0$. The recursive formulas are as follows:

$$x_{n+1} = x_n + h$$
$$y_{n+1} = y_n + h\,f(x_n, y_n)$$

Here, h is a given step size. We again use **Table** to do the iterations:

```
euler[f_, x_, y_, x0_, y0_, x1_, n_] :=
  Module[{xi = x0, yi = y0, h = N[(x1 - x0) / n]},
    Prepend[Table[{xi, yi} = N[{x + h, y + h f} /. {x → xi, y → yi}], {n}],
    {x0, y0}]]
```

The solution is calculated from x0 to x1 using n steps. With **Prepend**, we add the starting point to the list. As an example, we solve the equation $y' = x - y^2$, $y(0) = 1$ in the interval [0, 1] using 10 steps:

```
eu = euler[x - y^2, x, y, 0, 1, 1, 10]

{{0, 1}, {0.1, 0.9}, {0.2, 0.829}, {0.3, 0.780276},
 {0.4, 0.749393}, {0.5, 0.733234}, {0.6, 0.729471},
 {0.7, 0.736258}, {0.8, 0.75205}, {0.9, 0.775492}, {1., 0.805354}}

ListPlot[eu, PlotJoined → True,
  Epilog → {AbsolutePointSize[2], Map[Point, eu]}];
```

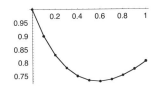

A very accurate value of the solution at $x = 1$ is as follows:

```
soln = y[1] /. NDSolve[{y'[x] == x - y[x]^2, y[0] == 1}, y, {x, 0, 1},
    WorkingPrecision → 25, PrecisionGoal → 20, AccuracyGoal → 20][[1]]

0.83338339146435446832753
```

We compare this value with various Euler approximations:

```
t4 = Table[{h = 10.^-n, eu = Last[euler[x - y^2, x, y, 0, 1, 1, 10^n]][[2]],
    Abs[eu - soln]}, {n, 1, 4}];
```

```
TableForm[t4, TableHeadings → {None, {"h", "approx.", "error\n"}},
  TableSpacing → {0.3, 2}]
```

h	approx.	error
0.1	0.805354	0.0280298
0.01	0.830746	0.0026377
0.001	0.833121	0.000262254
0.0001	0.833357	0.0000262104

The step size *h* has to be very small to get good precision with Euler's method. In fact, this method has almost only pedagogical value.

We can also use recursive functions:

```
v[x_, y_] := x - y^2

Clear[x, y]; x[0] = 0; y[0] = 1; h = 0.1;

x[n_] := x[n] = N[x[n - 1] + h]
y[n_] := y[n] = N[y[n - 1] + h v[x[n - 1], y[n - 1]]]

Table[{x[n], y[n]}, {n, 0, 10}]
```

```
{{0, 1}, {0.1, 0.9}, {0.2, 0.829}, {0.3, 0.780276},
 {0.4, 0.749393}, {0.5, 0.733234}, {0.6, 0.729471},
 {0.7, 0.736258}, {0.8, 0.75205}, {0.9, 0.775492}, {1., 0.805354}}
```

Before continuing, we remove all of our current definitions:

```
Remove["Global`*"]
```

15.1.2 Programming Commands

■ Doing

Do[body, {i, min, max}]   Do **body** while **i** goes from **min** to **max**
While[test, body]   Check **test**, then repeat **body** until **test** fails to give **True**
For[start, test, incr, body]   Do **start**; do **test**, **body**, and **incr** until **test** fails to give **True**

The iteration specification in **Do** can have all of the same forms as the one in **Table** (see Section 13.1.1, p. 325). Thus the specification can also be of the forms **{n}** (**body** is done **n** times); **{i, max}** (**i** goes from 1 to **max**); and **{i, min, max, step}**. In addition, multiple specifications like **{i, imax}, {j, jmax}** can be used. If **body, test, start**, or **incr** consists of a sequence of commands, they are separated by semicolons. An index **i** can be incremented with **++i**; it is equivalent to **i = i + 1**.

Note that **Do**, **While**, and **For** do not print anything as an answer; they only do what they are asked to do. Thus, after the calculation, remember to ask for the value of the variable in which you are interested. A prototype **Do**-calculation could be as follows:

```
x = init              (*give an initial value for x*)
Do[body, {i, max}]    (*do something for x*)
x                     (*ask the final value of x*)
```

As an example, we calculate the sum of the first 10 integers with all three commands:

```
s = 0; Do[s = s + i, {i, 10}]; s      55

s = 0; i = 1; While[i ≤ 10, s = s + i; i = i + 1]; s      55

For[s = 0; i = 1, i ≤ 10, i = i + 1, s = s + i]; s      55
```

Better ways, though, are as follows (**Total** is available in *Mathematica* 5):

```
Sum[i, {i, 10}]      55

Apply[Plus, Range[10]]      55

Total[Range[10]]      55
```

Do and **Table** are very similar commands; both do what we ask, but while **Do** does not print anything, **Table** gathers the results into a list and prints it. For example:

```
s = 0; Table[s = s + i, {i, 10}]

{1, 3, 6, 10, 15, 21, 28, 36, 45, 55}
```

■ Branching

If[test, then, else, otherwise] If **test** gives **True**, do **then**; else, if **test** gives **False**, do **else**; otherwise (if **test** is neither **True** nor **False**), do **otherwise**.

Which[test1, then1, test2, then2, … , True, otherwise] Evaluate each of the **testi** in turn and do the **theni** corresponding to the first test giving **True**. If all **testi** give **False**, do **otherwise**. If, in searching the first test giving **True**, a test is encountered giving neither **True** nor **False**, the searching is stopped and the **Which** command is returned with the remaining arguments unevaluated.

Switch[expr, form1, value1, form2, value2, … , _, otherwise] Compare **expr** with each of the **formi** in turn and return the **valuei** corresponding to the first match. If no matches are found, return **otherwise**.

The box contains the most complete forms of the three commands; shorter forms exist. Regarding **If**, it also accepts the following form:

```
If[test, then, else]
```

Then the **If** command is returned as such if **test** gives neither **True** nor **False**. Here is a still shorter form:

```
If[test, then]
```

This does nothing (meaning that the value of the command is **Null**) if **test** gives **False** and returns the **If** command if **test** gives neither **True** nor **False**. The **Which** command has the following shorter form:

```
Which[test1, then1, test2, then2, … ]
```

Now nothing is done (meaning that the value of the command is **Null**) if all tests give **False**. The **Switch** command has the following shorter form:

```
Switch[expr, form1, value1, form2, value2, … ]
```

Now the **Switch** command is returned if no matches are found. In the complete form, remember that _ means anything (see Section 14.1.1, p. 351) so that it matches all expressions.

The complete forms are safe to use, because then you have specified what to do in all possible cases (with the exception that in **Which** we cannot define what to do if a test is encountered that gives neither **True** nor **False**).

In the tests, we can apply operators like ==, !=, <, ≤, >, and ≥; use tests like **IntegerQ** and **OddQ**; and form more complex logical expressions with && (AND), || (OR), and ! (NOT) (see Section 12.3.5, p. 314).

Here are some examples of **If**:

```
{If[2 < 3, Yes], If[4 < 3, Yes], If[x < 3, Yes]}
```

```
{Yes, Null, If[x < 3, Yes]}
```

In the second example, the test $4 < 3$ gives **False**, so nothing is done. In the third example, we cannot say whether $x < 3$ or not, so the test gives $x < 3$ which is neither **True** nor **False** and, consequently, the **If** command is returned as such.

The following function gives the integral of x^n for each n:

```
f[n_] := If[n == -1, Log[x], x^(n + 1) / (n + 1)]
```

```
{f[-2], f[-1], f[0], f[1]}
```

$$\left\{-\frac{1}{x},\ \text{Log}[x],\ x,\ \frac{x^2}{2}\right\}$$

Next we use **Which**:

```
g[n_] := Which[n == -1, Log[x], True, x^(n + 1) / (n + 1)]
```

```
{g[-2], g[-1], g[0], g[1]}
```

$$\left\{-\frac{1}{x},\ \text{Log}[x],\ x,\ \frac{x^2}{2}\right\}$$

In this example, we could also write **n != -1** in place of **True**. Here is an example of **Switch**:

```
volume[name_] := Switch[name, cylinder, Pi r^2 h,
  sphere, 4 Pi r^3 / 3, ellipsoid, 4 Pi a b c / 3, _, unknown]
```

```
{volume[cylinder], volume[sphere], volume[cone]}
```

$$\left\{h\,\pi\,r^2,\ \frac{4\,\pi\,r^3}{3},\ \text{unknown}\right\}$$

■ **Communicating**

Print[expr1, expr2, …] Print the values of the expressions
x = Input["prompt"] Print the prompt, then read an expression as a value for **x**

As we noted in Section 14.1.5, p. 361, a module automatically prints the result of its last statement. If we want to print some intermediate results, we can use **Print**, although **Message** is the preferred command (see Section 14.3.4, p. 378). We can also build an interactive program that asks for some values with **Input**.

The function **div** asks for an integer and then prints its divisors. This is continued until the integer is negative:

```
div := Module[{n, cont = True},
    While[cont == True, n = Input["Give an integer"];
    If[n ≥ 0,
      Print["The divisors of ", n, " are ", Divisors[n]], cont = False]]]

div

The divisors of 1111 are {1, 11, 101, 1111}

The divisors of 11111 are {1, 41, 271, 11111}

The divisors of 111111 are {1, 3, 7, 11, 13, 21, 33, 37, 39,
  77, 91, 111, 143, 231, 259, 273, 407, 429, 481, 777, 1001, 1221,
  1443, 2849, 3003, 3367, 5291, 8547, 10101, 15873, 37037, 111111}
```

When **div** is entered, a new window appears in which we can enter the requested input. The result is then printed in the notebook.

Note that **Print** is useful for debugging a larger program. You can print with **Print** the values of some variables that play a central rule in a susceptible block of the program. This may help you to infer what the program actually does, and then you can more easily correct the code.

■ Controlling

> **Continue[]**   Go to the next step in the current loop of **Do**, **While**, or **For**
>
> **Break[]**   Exit the nearest enclosing loop of **Do**, **While**, or **For**
>
> **Return[expr]**   Return **expr**, exiting all procedures and loops in a function
>
> **Goto[name]**   Go to **Label[name]**

These commands are used to perform an exceptional operation. As an example, we program the tossing of an *n*-face die until the result is 1 or *n*. The following program is very clumsy:

```
die[n_] := Module[{r, i = 1},
  Label[start];
  r = Random[Integer, {1, n}];
  If[r == 1 || r == n,
    Print["We got ", r, " after ", i, " tosses"]; Goto[finish],
    ++i; Goto[start]];
  Label[finish];]
```

    die[6]   We got 6 after 2 tosses

We write a better program (here we assume that we get 1 or *n* in at most 100 tosses):

```
die2[n_] := Module[{r, i = 1},
  Do[r = Random[Integer, {1, n}];
    If[r == 1 || r == n, Return[{r, i}], ++i], {100}]]
```

    die2[6]   {6, 4}

Here the **Do** command allows for, at most, 100 tosses. If 1 or *n* is obtained sooner, **Return** stops the iterations and prints the result **{r, i}**. Here is a still better program:

```
die3[n_] := Module[{r = 0, i = 0},
  While[r != 1 && r != n, r = Random[Integer, {1, n}]; ++i];
  {r, i}]

die3[6]    {1, 5}
```

The command pair **Throw** and **Catch** and also the pair **Sow** (※5) and **Reap** (※5) may be used in programs (see the *Mathematica* manual).

15.1.3 Newton's Method

When we programmed Newton's method in Section 15.1.1, p. 383, we did a fixed, sufficiently large number of iterations to obtain the zero. Normally the convergence is controlled with the program: iterations are stopped once the present approximation to the solution is accurate enough. The following module checks the convergence:

```
newton2[f_, x_, x0_, eps_, max_] :=
 Module[{xi = x0, fi, df = D[f, x], dfi, iters = {x0}},
  Do[{fi, dfi} = N[{f, df} /. x → xi];
   If[Abs[fi] < eps, Break[]];
   xi = xi - fi / dfi;
   iters = {iters, xi}, {max}];
  Flatten[iters]]
```

The starting value is **x0**. If the value of the function becomes smaller than **eps**, iterations are stopped (note that, after **Break[]**, the program continues from **Flatten[iters]**). However, iterations are done at most **max** times (since we have written **Do[… , {max}]**). The iterations are gathered to **iters**. The successive values of **iters** are $\{x_0\}$, $\{\{x_0\}, x_1\}$, $\{\{\{x_0\}, x_1\}, x_2\}$, and so on, so that we have lastly used **Flatten** to get a simple list $\{x_0, x_1, x_2, … \}$ as the result (remember that a module automatically prints the result of the last command). For example:

```
f = 3 x^3 - E^x;
```

```
newton2[f, x, 2, 10^-14, 20]
```

```
{2, 1.41942, 1.1019, 0.975117, 0.953089, 0.952446, 0.952446, 0.952446}
```

```
f /. x → Last[%]    4.44089 × 10^-16
```

We see that the value of the function is, in fact, less than 10^{-14} at the last point. We can also find complex zeros if we start at a complex point:

```
newton2[f, x, -0.5 + 0.5 I, 10^-14, 20]
```

```
{-0.5 + 0.5 i, -0.400328 + 0.465629 i, -0.384087 + 0.473305 i,
 -0.38428 + 0.473739 i, -0.38428 + 0.473739 i, -0.38428 + 0.473739 i}
```

In Section 19.3.4, p. 511, we write a similar program for the secant method.

```
Remove["Global`*"]
```

■ Some Questions

1. Why have we written **iters = {iters, xi}** and not **AppendTo[iters, xi]**? The latter command directly gives a list of the simple form $\{x_0, x_1, x_2, \ldots\}$, so that at the end of the program we could simply write **iters** (flattening would not be needed).

The answer is that **AppendTo[iters, xi]** is slower than **iters = {iters, xi}**, even when we take into account the time required for the flattening:

Use **iters = {iters, xi}** *and lastly* **Flatten[iters]** *instead of* **AppendTo[iters, xi]**.

The time saved may not be noticeable in the small iteration of **newton2**, but for longer iterations, it may be considerable. If each iteration gives two numbers **{xi, yi}**, we can in the same way write **iters = {iters, {xi, yi}}** and lastly **Flatten[iters]**, but if the result must consist of the pairs **{xi, yi}**, the last step is **Partition[Flatten[iters], 2]**.

2. Instead of **iters = {iters, xi}**, could we also write **iters = {iters, %}**, as **xi** is calculated in the preceding command?

The answer is no:

We cannot use **%** *in programs.*

The symbol **%** is intended to be used only in interactive calculations.

3. Why have we introduced the variable **xi** in the program and given it the starting value **x0**? Why have we not directly used the variable **x0** containing the starting value?

An important point to note is the following:

The arguments of a program cannot be changed inside the program.

We might assume that we do not need the variable **xi** and instead write directly **x0 = x0 - fi / dfi**. This does not work, however, because when we use the program, **x0** has a specific numerical value like 2, and we cannot make an assignment like **2 = …** . We would obtain an error message such as the following:

```
Set::setraw : Cannot assign to raw object 2.
```

Thus, **x0** can be used only as the starting value, and the iterations have to be stored in another variable; we have used **xi**.

4. In defining the starting value of **iters**, we have written **iters = {x0}**. Could we write **iters = {xi}**? Remember that we have earlier defined that **xi = x0**.

Note the following:

In giving starting values for local variables in a module, earlier defined local variables cannot be used.

If we write **iters = {xi}**, we get the following result:

```
{xi, 1.41942, 1.1019, 0.975117, 0.953089, 0.952446, 0.952446}
```

We see that the starting value **xi** of **iters** was unknown to *Mathematica*.

15.2 Functional Programming

15.2.1 Introduction

Functional programming may be the most important programming style in *Mathematica*. It is suitable in many problems where we *manipulate lists* or *iterate functions*. In both types of tasks, we *apply functions to arguments*. The functions are often written as pure functions (see Section 14.1.5, p. 359).

In list manipulation, we typically modify the elements by applying a function to them so that, for example, we transform the list $\{a, b, c\}$ into $\{f(a), f(b), f(c)\}$ with **Map** or into $f(a, b, c)$ with **Apply**. In iterations, we start from a given value and then iterate it with a function, so that, for example, from the starting value x_0 we get new values with the iteration formula $x_{i+1} = f(x_i)$; this can be done with **Nest**. Next we consider list manipulation and function iteration in more detail.

■ List Manipulation

For list manipulation, the most important functional style programming commands are **Map** and **Apply**, but **Map** also has the variations **MapAt**, **MapAll**, **MapIndexed**, and **MapThread**; we also have **Thread**, **Inner**, and **Outer**. All these commands were considered in Section 13.3. We could have considered these commands here in the context of programming, but we felt it more suitable to consider them as one group of list manipulation commands, because they—or at least **Map** and **Apply**—are very useful in everyday calculations with *Mathematica*, not only in ordinary programs.

The key in functional list manipulation is that we do not explicitly treat each element of a list separately but rather *we treat the list as a whole*: we only indicate what we want do with the elements, and a functional command then does the operation to each element.

For example, to calculate the sum of the elements of a list, we do not use **Sum**:

```
t = {a, b, c};
Sum[t[[i]], {i, 1, 3}]     a + b + c
```

A still worse way is the use of **Do**:

```
s = 0; Do[s = s + t[[i]], {i, 1, 3}]; s     a + b + c
```

Instead, the command is simply the following:

```
Total[t]     a + b + c
```

Alternatively, in versions of *Mathematica* earlier than 5, we can use this:

```
Apply[Plus, t]     a + b + c
```

To square the elements, we do not use **Table**:

```
Table[t[[i]]^2, {i, 1, 3}]     {a², b², c²}
```

Instead, we enter the following simple command:

```
t^2     {a², b², c²}
```

To calculate the row sums of a matrix, we do not use **Table** and **Sum**:

```
m = {{1, 2, 3}, {a, b, c}, {A, B, C}};
Table[Sum[m[[i, j]], {j, 1, 3}], {i, 1, 3}]
```

    {6, a + b + c, A + B + C}

Instead, we write the following:

```
Map[Total, m]     {6, a + b + c, A + B + C}
```

Alternatively, in versions of *Mathematica* earlier than 5, we can use this:

```
Apply[Plus, m, 1]     {6, a + b + c, A + B + C}
```

The functional list manipulation commands have two advantages. First, once you have learned them, they are short to write and thus shorten the code needed to do a calculation. Second, they are fast. As an example, we calculate the row sums of a 1000×1000 matrix by three methods:

```
n = 1000; r = Table[Random[], {n}, {n}];

Table[s = 0; Do[s = s + r[[i, j]], {j, 1, n}]; s, {i, n}]; // Timing
```

    {14.35 Second, Null}

```
Table[Sum[r[[i, j]], {j, n}], {i, n}]; // Timing
```

    {1.27 Second, Null}

```
Map[Total, r]; // Timing
```

    {0.07 Second, Null}

We do not consider the functional list manipulation commands in more detail here (see Section 13.3). Examples of these commands appear frequently in the chapters to come.

■ Function Iteration

Many mathematical methods are iterative or recursive, having the typical form $x_{i+1} = f(x_i)$. Examples are several iterative numerical methods for nonlinear equations, nonlinear optimization, and differential, partial differential, and difference equations. *Mathematica* has special functional type commands like **Nest**, **FixedPoint**, and **Fold** for iterative calculations.

Nest is the basic iterating command, which does the iteration $x_{i+1} = f(x_i)$ a fixed number of times:

```
Nest[f, x0, 3]     f[f[f[x0]]]
```

FixedPoint does the iteration until the result does not change (the stopping criterion can be given). **Fold** also iterates a function, but now the function has two variables and, at each iteration, it takes one element from a given list as the second argument. We also have the commands **NestList**, **FixedPointList**, and **FoldList**, that print all intermediate steps, too.

For example, Newton's method can be written as follows:

```
newton[f_, x_, x0_, max_, opts___] := With[{df = D[f, x]},
    FixedPointList[(x - f / df) /. x → # &, N[x0], max, opts]]
```

Here is an application of it:

```
newton[3 x^3 - E^x, x, 2, 20, SameTest → (Abs[#1 - #2] < 10^-6 &)]
```

```
{2., 1.41942, 1.1019, 0.975117, 0.953089, 0.952446, 0.952446}
```

The first advantage of function iteration commands is that they shorten the code as compared with procedural programming. Compare the **newton** program above with the **newton2** program of Section 15.1.3, p. 390. Indeed, in **newton** we do not need to take care of setting suitable initial values to variables, implementing the stopping of the iteration according to the result of the stopping criterion, or adding iteration counters. We simply input the three or four items needed in the iteration: the function to be iterated, the starting point, the maximum number of steps, and (if the default stopping criterion is not suitable) a stopping criterion (as a pure function).

The second advantage is that the function iteration commands are fast. As an example, we compare two programs:

```
(x = 1.; Do[x = Cos[x] + Random[], {500000}]; x) // Timing
```

```
{6.75 Second, 1.45335}
```

```
Nest[Cos[#] + Random[] &, 1., 500000] // Timing
```

```
{0.69 Second, 0.826497}
```

The time needed by the procedural program is 10 times the time needed by the functional program.

Next we consider function iteration in more detail by studying **Nest**, **FixedPoint**, and **Fold** (and some other commands).

```
x =.
```

15.2.2 Iterating a Mapping

Nest[f, x0, n] Do n times the iteration $x_{i+1} = f(x_i)$, starting from x_0
NestList[f, x0, n] Give all iterations

Nest gives only the final result, while **NestList** also gives the starting point x_0 and all intermediate results (thus, a list of $n + 1$ elements). The function to be iterated is most naturally written as a pure function. With an unspecified function, we see a general result:

```
Nest[f[#] &, x0, 3]     f[f[f[x0]]]
```

```
NestList[f[#] &, x0, 3]     {x0, f[x0], f[f[x0]], f[f[f[x0]]]}
```

Note that **ComposeList** does not iterate the same function but possibly a different function each time:

```
ComposeList[{f, g, h}, x0]     {x0, f[x0], g[f[x0]], h[g[f[x0]]]}
```

■ Simple Examples

We do the iteration $x_{i+1} = \cos(x_i)$ four times by starting from $x_0 = 1$:

```
NestList[Cos[#] &, 1, 4]
```

```
{1, Cos[1], Cos[Cos[1]], Cos[Cos[Cos[1]]], Cos[Cos[Cos[Cos[1]]]]}
```

```
NestList[Cos[#] &, 1., 4]
```

$\{1., 0.540302, 0.857553, 0.65429, 0.79348\}$

Next we calculate continued fractions $x_{i+1} = \frac{1}{1+x_i}$, $x_0 = 3$:

```
NestList[1 / (1 + #) &, 3., 10]
```

$\{3., 0.25, 0.8, 0.555556, 0.642857, 0.608696,$
$0.621622, 0.616667, 0.618557, 0.617834, 0.61811\}$

With **HoldForm**, we can get the unsimplified results:

```
NestList[HoldForm[1 / (1 + #)] &, 3, 4]
```

$$\left\{3, \frac{1}{1+3}, \frac{1}{1+\frac{1}{1+3}}, \frac{1}{1+\frac{1}{1+\frac{1}{1+3}}}, \frac{1}{1+\frac{1}{1+\frac{1}{1+\frac{1}{1+3}}}}\right\}$$

The limit is as follows:

```
<< Calculus`Limit`
```

```
Limit[Nest[1 / (1 + #) &, x0, n], n → ∞]
```
$\frac{1}{2}(-1 + \sqrt{5})$

Then we calculate successive derivatives of 2^{x^2} by applying the iteration formula $f_{i+1}(x) = \partial_x f_i(x)$, $f_0(x) = 2^{x^2}$:

```
NestList[D[#, x] &, 2^(x^2), 3] // Simplify
```

$\left\{2^{x^2}, 2^{1+x^2} x \text{ Log}[2], 2^{1+x^2} \text{ Log}[2] (1 + x^2 \text{ Log}[4]), 2^{2+x^2} x \text{ Log}[2]^2 (3 + x^2 \text{ Log}[4])\right\}$

■ Newton's Method

```
newton3[f_, x_, x0_, n_] :=
  With[{df = D[f, x]}, NestList[(x - f / df) /. x → # &, N[x0], n]]
```

```
newton3[f[x], x, x0, 2]
```

$$\left\{x0, x0 - \frac{f[x0]}{f'[x0]}, x0 - \frac{f[x0]}{f'[x0]} - \frac{f[x0 - \frac{f[x0]}{f'[x0]}]}{f'[x0 - \frac{f[x0]}{f'[x0]}]}\right\}$$

```
newton3[3 x^3 - E^x, x, 2, 7]
```

$\{2., 1.41942, 1.1019, 0.975117, 0.953089, 0.952446, 0.952446, 0.952446\}$

We could also define a separate function for a single step of Newton's method.

```
newtonStep4[f_, df_, x_, xi_] := (x - f / df) /. x → N[xi]
newton4[f_, x_, x0_, n_] :=
  With[{df = D[f, x]}, NestList[newtonStep4[f, df, x, #] &, N[x0], n]]
```

Here **newtonStep4** gives the operation performed in a single step, and **newton4** nests this operation **n** times starting from **x0**. The function **newtonStep4** has several arguments, and so we have to express, in **newton4**, the argument with respect to which the iterations are to be done: this argument is **#** in the pure function. This method of defining separately a single step and the nesting of this step is used often in later chapters. Here are some examples:

```
newton4[f[x], x, x0, 2]
```

$$\left\{x0, \; x0 - \frac{f[x0]}{f'[x0]}, \; x0 - \frac{1.\,f[x0]}{f'[x0]} - \frac{f\left[x0 - \frac{1.\,f[x0]}{f'[x0]}\right]}{f'\left[x0 - \frac{1.\,f[x0]}{f'[x0]}\right]}\right\}$$

```
newton4[3 x^3 - E^x, x, 2, 7]
```

```
{2., 1.41942, 1.1019, 0.975117, 0.953089, 0.952446, 0.952446, 0.952446}
```

■ Euler's Method

```
eulerStep2[f_, x_, y_, {xi_, yi_}, h_] := {x + h, y + h f} /. {x → xi, y → yi}
euler2[f_, x_, y_, x0_, y0_, h_, n_] :=
NestList[eulerStep2[f, x, y, #, h] &, N[{x0, y0}], n]
```

Note that, because the pure function in **NestList** has to have only one argument (the #), we have to group the variables **xi** and **yi** into a list **{xi, yi}** in **eulerStep2**. We check **euler2** with the problem $y' = f(x, y)$, $y(x_0) = y_0$:

```
euler2[f[x, y], x, y, x0, y0, h, 2] // ColumnForm
```

```
{x0, y0}
{h + x0, y0 + h f[x0, y0]}
{2 h + x0, y0 + h f[x0, y0] + h f[h + x0, y0 + h f[x0, y0]]}
```

Here are some numerical values:

```
euler2[x - y^2, x, y, 0, 1, 0.1, 10]
```

```
{{0., 1.}, {0.1, 0.9}, {0.2, 0.829}, {0.3, 0.780276},
 {0.4, 0.749393}, {0.5, 0.733234}, {0.6, 0.729471},
 {0.7, 0.736258}, {0.8, 0.75205}, {0.9, 0.775492}, {1., 0.805354}}
```

■ Random Walks

Consider the following example:

```
NestList[# + (-1)^Random[Integer] &, 0, 10]
```

```
{0, 1, 0, -1, -2, -3, -2, -1, -2, -1, -2}
```

Note that **Random[Integer]** is 0 or 1, with each having the probability of $1/2$. We see that **(-1)^Random[Integer]** is then 1 or -1. The command thus calculates a list $\{0, 0 + r_1, (0 + r_1) + r_2, [(0 + r_1) + r_2] + r_3, ...\} = \{0, r_1, r_1 + r_2, r_1 + r_2 + r_3, ...\}$, where r_1, r_2, r_3, and so on are independent and each is 1 or -1, with equal probabilities. The command can thus be used to generate a random walk:

```
randomWalk[n_] := With[{t = NestList[# + (-1)^Random[Integer] &, 0, n]},
  ListPlot[Transpose[{Range[0, n], t}], PlotJoined → True,
    AspectRatio → 0.25, PlotStyle → AbsoluteThickness[0.2]]]
```

```
randomWalk[500];
```

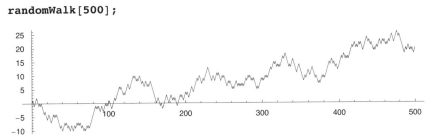

In a similar way, we can generate a random walk in higher dimensions. Here is a program for generating a random walk in the plane:

```
randomWalk2D[n_] :=
 With[{t = NestList[# + (-1) ^ Table[Random[Integer], {2}] &, {0, 0}, n]},
  ListPlot[t, PlotJoined → True, AspectRatio → Automatic,
   PlotStyle → AbsoluteThickness[0.2],
   Epilog → {RGBColor[1, 0, 0], AbsolutePointSize[4], Point[Last[t]]}]]
```

Here `(-1)^Table[Random[Integer], {2}]` gives a two-component list. For example, `(-1)^{0, 1}` gives the list `{1, −1}` (remember that *Mathematica* automatically does all computations element by element). Thus, we start at $(0, 0)$ and at each step add to the previous point one of the points $(1, 1)$, $(−1, 1)$, $(−1, −1)$, or $(1, −1)$. The `Epilog` shows the last point in red.

```
randomWalk2D[2000];
```

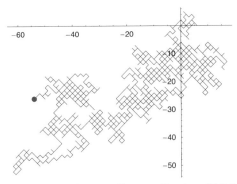

For more about random walks, see Section 26.2.1, p. 726.

15.2.3 Iterating until Convergence

`FixedPoint[f, x0]` Do the iteration $x_{i+1} = f(x_i)$, starting from x_0, until convergence
`FixedPointList[f, x0]` Give all iterations
`FixedPoint[f, x0, max]` Iterate until convergence but at most **max** times
`FixedPointList[f, x0, max]` Give all iterations

An option:

`SameTest` The test used as the stopping criterion; examples of values: **Automatic**
   (means `(#1 === #2 &)`), `(Abs[#1 − #2] < 10^−10 &)`, `(Abs[f /. x → #2] < 10^−5 &)`

FixedPoint is similar to **Nest**, but it applies the given function iteratively until the result no longer changes. In particular, **FixedPoint[f, x0, max]** is similar to **Nest[f, x0, n]**, although **FixedPoint** can stop before **max** iterations are done. If both **max** and a stopping criterion are given, then iterations are stopped as soon as the criterion gives **True**, but in all cases at most **max** iterations are done. The third argument **max** prevents infinite calculations. With an unspecified function, we see a general result:

```
FixedPoint[f[#] &, x0, 3]      f[f[f[x0]]]
```

```
FixedPointList[f[#] &, x0, 3]      {x0, f[x0], f[f[x0]], f[f[f[x0]]]}
```

The stopping criterion is a pure function of the last two iterations; the next to last iteration is denoted by **#1** and the last iteration by **#2**. The default stopping criterion is **SameQ[#1, #2] &**, that is, **(#1 === #2 &)**. This means that the last two iterations are the same to 16-digit precision (in most computers) (see Section 12.3.5, p. 315). This is a tight condition, but we can formulate milder criteria.

■ The Fixed-Point Method

If we have a nonlinear equation of the form $x = f(x)$, one possibility to solve the equation is to apply the *fixed-point method* by doing the iterations $x_{i+1} = f(x_i)$. **FixedPoint** does exactly the iterations of this method. The method converges at least when $|f'(x)| < 1$ near the solution. We try to solve the equation $x = \cos(x)$:

```
t = FixedPointList[Cos[#] &, 0.05];
```

The last point and the difference of the left- and righthand sides of the equation at this point are as follows:

```
{Last[t], x - Cos[x] /. x → Last[t]}
```

```
{0.739085, 0.}
```

The iterations proceed as follows:

```
ListPlot[Transpose[{Range[0, Length[t] - 1], t}],
    PlotJoined → True, PlotRange → {0, 1.05}, AspectRatio → 0.2];
```

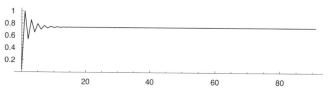

In about 15 iterations, we are already near the solution. However, to get it to the last decimal, we still need about 75 iterations. Next we use a custom stopping criterion:

```
FixedPointList[Cos[#] &, 0.05, SameTest → (Abs[#1 - #2] < 10^-2 &)]
```

```
{0.05, 0.99875, 0.541354, 0.857012, 0.654699, 0.793231, 0.701546,
  0.763845, 0.722182, 0.750365, 0.73144, 0.744213, 0.735621}
```

With the following program, we can illustrate the search of a fixed point (see also Sections 25.1.3, p. 678, and 25.2.1, p. 683):

```
cobwebPlot[f_, x_, x0_, n_, a_, b_, opts___] := Module[{p1, p2, xi = x0},
    p1 = Plot[{x, f}, {x, a, b}, DisplayFunction → Identity];
    p2 = Graphics[{Line[Prepend[Flatten[
        Table[{{xi, xi = f /. x → xi}, {xi, xi}}, {n}], 1], {x0, 0}]]}];
    Show[p1, p2, DisplayFunction → $DisplayFunction, opts]]
```

```
cobwebPlot[Cos[x], x, 0.05, 12, 0, 1.03, AspectRatio → Automatic];
```

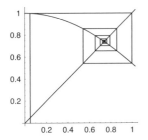

■ Newton's Method

```
newton5[f_, x_, x0_, max_, opts___] :=
    With[{df = D[f, x]}, FixedPointList[(x - f / df) /. x → # &, N[x0], max, opts]]
```

This formula is the same as for **newton3** in Section 15.2.2, p. 395, except that we now use **FixedPointList**, and we have also added the possibility of writing a stopping criterion as an option. The three underscores (___) after **opts** mean zero or more options (see Section 15.3.3, p. 408). Here is an application:

```
f = 3 x^3 - E^x;
```

```
newton5[f, x, 2, 20]
```

```
{2., 1.41942, 1.1019, 0.975117,
  0.953089, 0.952446, 0.952446, 0.952446, 0.952446}
```

The last two iterations are as follows:

```
{%[[-2]], %[[-1]]} // InputForm
```

```
{0.9524456794271664, 0.9524456794271663}
```

They are considered to be the same. Next we use a loose stopping criterion:

```
newton5[f, x, 2, 20, SameTest → (Abs[#1 - #2] < 10^-3 &)]
```

```
{2., 1.41942, 1.1019, 0.975117, 0.953089, 0.952446}
```

Now we require that the value of the function at the root should be sufficiently small at the last point calculated:

```
newton5[f, x, 2, 20, SameTest → (Abs[f /. x → #2] < 10^-6 &)]
```

```
{2., 1.41942, 1.1019, 0.975117, 0.953089, 0.952446, 0.952446}
```

A modification of **newton5** is also considered in Section 19.3.4, p. 509. There we also generalize the function to simultaneous nonlinear equations.

```
f =.
```

■ **Another Command**

> **NestWhile[f, x0, test]**   While **test** gives **True**, do the iteration $x_{i+1} = f(x_i)$, starting
> from x_0; give the first x_i, $i = 0, 1, ...$, for which **test** does not give **True**
> **NestWhile[f, x0, test, m]**   Use the most recent m results as arguments to **test**
> **NestWhile[f, x0, test, All]**   Use all results so far as arguments to **test**
> **NestWhile[f, x0, test, m, max]**   Do at most **max** iterations
> **NestWhile[f, x0, test, m, max, -n]**   Give the result found when **f** had been applied
> **n** fewer times

We also have **NestWhileList**, which gives all iterations. **NestWhile** is similar to **Fixed**‹
Point, but the tests used in these commands are the opposites of each other: iterations
stop in **FixedPoint** once the test (the default of **SameTest** is **SameQ**) gives **True**, but, in
NestWhile, they stop when the test no longer gives **True**. Indeed, the following commands are equivalent:

```
FixedPoint[f, x0, max]
```

```
NestWhile[f, x0, UnsameQ, 2, max]
```

Here is another example where we do the same calculation in two ways:

```
FixedPointList[Cos[#] &, 0.05, 10, SameTest → (Abs[#1 - #2] < 0.1 &)]
```
```
{0.05, 0.99875, 0.541354, 0.857012, 0.654699, 0.793231, 0.701546}
```

```
NestWhileList[Cos[#] &, 0.05, Abs[#1 - #2] ≥ 0.1 &, 2, 10]
```
```
{0.05, 0.99875, 0.541354, 0.857012, 0.654699, 0.793231, 0.701546}
```

Newton's method can be written as follows:

```
newton6[f_, x_, x0_, max_, test_] := With[{df = D[f, x]},
  NestWhileList[(x - f / df) /. x → # &, N[x0], test, 2, max]]
```

```
newton6[3 x^3 - E^x, x, 2, 20, Unequal]
```
```
{2., 1.41942, 1.1019, 0.975117,
  0.953089, 0.952446, 0.952446, 0.952446, 0.952446}
```

If we want to use more than the last two iterations in the test, then **NestWhile** is very
useful. As an example, we toss a die until we get a result the second time. Here we use all
iterations so far as arguments to the test (remember that **UnsameQ** accepts several
arguments):

```
NestWhileList[Random[Integer, {1, 6}] &, 0, UnsameQ, All] // Rest
```
```
{1, 4, 6, 2, 2}
```

15.2.4 Iterating with a Resource

> **Fold[f, x0, {y1, y2, ..., yn}]**   Do the iteration $x_{i+1} = f(x_i, y_{i+1})$, $i = 0, ..., n - 1$,
> starting from x_0, and give x_n
> **FoldList[f, x0, {y1, y2, ..., yn}]**   Give all iterations

Fold is similar to **Nest**, but while **Nest** uses a function of one variable, **Fold** uses a function of two variables. The second argument of **Fold** is the starting value. The third argument is a list from which one element at a time is fed in as the second argument of the function. The list can be considered as a resource from which a new element is drawn at each iteration. With an unspecified function, we see a general result:

```
Fold[f[#1, #2] &, x₀, {y₁, y₂, y₃}]
```

$f[f[f[x_0, y_1], y_2], y_3]$

```
FoldList[f[#1, #2] &, x₀, {y₁, y₂, y₃}]
```

$\{x_0, f[x_0, y_1], f[f[x_0, y_1], y_2], f[f[f[x_0, y_1], y_2], y_3]\}$

■ Cumulative Sums and Other Examples

Fold[Plus, 0, {y1, y2, …, yn}] //Rest   Compute the cumulative sums

---

FoldList is particularly useful in constructing cumulative sums of numbers $y_1, \ldots, y_n$ with the recursion $x_{i+1} = x_i + y_{i+1}$, $x_0 = 0$. With **Rest**, we drop the starting value 0:

```
FoldList[Plus, 0, {y₁, y₂, y₃}] // Rest
```

$\{y_1, y_1 + y_2, y_1 + y_2 + y_3\}$

Note that, instead of **Plus**, we could also have written **Plus[#1, #2]&**. In this example, **FoldList** works as follows:

```
{x₀ = 0, x₁ = Plus[0, y₁], x₂ = Plus[x₁, y₂], x₃ = Plus[x₂, y₃]}
```

$\{0, y_1, y_1 + y_2, y_1 + y_2 + y_3\}$

In place of **FoldList**, we can also use **CumulativeSums** from the **Statistics`DataManipu** **lation`** package. Cumulative products are calculated similarly as cumulative sums:

```
FoldList[Times, 1, {y₁, y₂, y₃}] // Rest
```

$\{y_1, y_1 y_2, y_1 y_2 y_3\}$

Next we find the elements of a given list that are records, that is, greater than all previous elements:

```
<< DiscreteMath`Combinatorica`
```

```
t = RandomPermutation[20]
```

$\{16, 12, 8, 14, 5, 10, 3, 2, 17, 7, 15, 18, 6, 1, 20, 13, 9, 11, 4, 19\}$

```
FoldList[Max, 0, t] // Rest
```

$\{16, 16, 16, 16, 16, 16, 16, 16, 17,$
$17, 17, 18, 18, 18, 20, 20, 20, 20, 20, 20\}$

```
Union[%]     {16, 17, 18, 20}
```

Next we write a third-degree polynomial $a + bx + cx^2 + dx^3$ in Horner form:

```
Fold[#1 x + #2 &, 0, {d, c, b, a}]     a + x (b + x (c + d x))
```

In this example, **Fold** forms the following expression:

```
(((0 x + d) x + c) x + b) x + a     a + x (b + x (c + d x))
```

■ Coin Tossing

Here we simulate coin tossing and investigate how the relative frequency of heads evolves as we toss the coin more and more times. It is expected that the relative frequency approaches the value 0.5 if the coin is unbiased.

```
coinTossing[n_] := Module[{tosses, cumSums, relFreqs},
   tosses = Table[Random[Integer], {n}];
   cumSums = FoldList[Plus, 0., tosses] // Rest;
   relFreqs = cumSums / Range[n];
   ListPlot[relFreqs, PlotJoined → True, AxesOrigin → {0, 0.5},
    PlotStyle → AbsoluteThickness[0.2], AspectRatio → 0.25]]
```

First we generate a list **tosses** of heads and tails; let us say that 1 means heads and 0 tails. Cumulative sums are then calculated, and relative frequencies of heads are obtained from the cumulative sums. Note that in **cumSums / Range[n]** we divide an n list by another n list. The division is automatically done element by element so that the result is a list of relative frequencies. We toss a coin 5000 times and so calculate a sequence of empirical estimates for the probability of a head:

```
coinTossing[5000];
```

As we see, the convergence to 0.5 may not be especially fast. It is better to just trust that the probability is 0.5.

15.3 Rule-Based Programming

15.3.1 Rules

We have thus far encountered operations as the following:

  =   assigning a value, e.g., **x = 3**

  :=   defining a function, e.g., **f[x_] := x Sin[x]**

  →   making a transformation, e.g., **a x + b /. x → 3**

All of these operations can be seen as *rules*. For example, **x = 3** is the rule that whenever **x** appears (in any expression during the rest of the current session), it will be replaced with 3. The rule **f[x_] := x Sin[x]** tells *Mathematica* that whenever **f[anything]** occurs during the rest of the current session, it should be replaced with **anything Sin[anything]**. The rule **a + b x /. x → 3** asks that wherever **x** appears in the expression **a + b x** (and only in this particular occurrence of the expression), it should be replaced with 3. *Rule-based programming* uses such rules.

In addition, an important concept in rule-based programming is a *pattern*. For example, in **f[x_]**, the argument is given in the form of a pattern: **x_** matches anything, and so this is a very general pattern. We can form a more restrictive pattern such as the one in **f[x_^y_] := ...**, in which the argument of **f** can be anything in the form **x^y**.

Here, in Section 15.3.1, we consider rules. The treatment is short because we have already studied most rules earlier. Patterns are considered in Sections 15.3.2 through 15.3.5.

■ Common Rules

	Global rules	Local rules
Evaluate the righthand side	= (**Set**)	→ (**Rule**)
Delay the evaluation of the righthand side	:= (**SetDelayed**)	:→ (**RuleDelayed**)

These are the four most common rules. Some rules evaluate the righthand side while others do not; the latter delay the evaluation until the rule is actually applied. Global rules are applied whenever the lefthand side is encountered. Local rules are applied only to the given expression.

We have already considered =, →, and := in Sections 12.1.1, p. 298; 12.1.2, p. 300; and 14.1.1, p. 350. These are the most important rules. The delayed local rule :> is quite seldom used. Some examples of rules follow.

■ Examples of Global Rules

With the rule **f = Integrate[1 + x, x]**, the righthand side is evaluated so that **f** gets the value $x + \frac{1}{2} x^2$. The rule is global: the value $x + \frac{1}{2} x^2$ is used for **f** whenever **f** is encountered in the current session.

With the rule **g[y_, x_] := Integrate[y, x]**, the righthand side is not evaluated at the time **g** is defined; it is evaluated only when we ask for the value of **g** for a specific **y** and **x**. This rule, too, is a global rule and is used whenever **g** is encountered.

■ Examples of Local Rules

Consider the following command:

```
a + b x /. x → Integrate[Sin[x], {x, 0, Pi}]
```

```
a + 2 b
```

Here the integral of the righthand side of the rule is evaluated and gives 2, and this value is then substituted for **x** in **a + b x**. The rule **x → ...** is local and is applied only to the present occurrence of the expression **a + b x**.

To see the effect of a delayed rule :> (which *Mathematica* transforms to the form :→), we consider the following two more complex examples:

```
1 + f[a + 2 b x] /. f[y_] → f[Integrate[y, x]]
```

```
1 + f[x (a + 2 b x)]
```

```
1 + f[a + 2 b x] /. f[y_] :→ f[Integrate[y, x]]
```

```
1 + f[a x + b x²]
```

In these examples, our aim is to replace the arguments of all functions `f[…]` with their integrals. The first example seems not to work. The reason is that the righthand side `f[Integrate[y, x]]` of the rule is evaluated and gives `f[x y]`, and when `f[anything]` is replaced with `f[x anything]` in `1 + f[a + 2 b x]`, we get the result shown.

On the other hand, the second example works, and this is because the righthand side of the rule is not evaluated until the rule is first applied. Thus we first get `1 + f[Integrate[a + 2 b x, x]]` and then the result shown.

■ Downvalues and Upvalues

	Downvalues	Upvalues
Evaluate the righthand side	= (**Set**)	^= (**UpSet**)
Delay the evaluation of the righthand side	:= (**SetDelayed**)	^:= (**UpSetDelayed**)

Setting a value like `x = 3` is also called defining a *downvalue* for `x`. Similarly, a function definition like `f[x_] := x Sin[x]` defines a downvalue for `f`. We also have *upvalues*. An example was in the introduction to this chapter: `der[sin] ^= cos` and `int[sin] ^= -cos`. These define two upvalues for `sin`. Down- and upvalues can also be defined with `/:` (**TagSet** or **TagSetDelayed**) (e.g., `cos /: der[cos] = -sin` and `cos /: int[cos] = sin`). For more information about down- and upvalues, see Section 2.5.10 in Wolfram (2003). Remember also the construction `f[x_] := expr /; cond` used to define the function for different situations (see Section 14.1.2, p. 354).

■ Dispatching

`dispatchrules = Dispatch[rules]` Generate an optimized dispatch table representation of a list of rules

`expr /. dispatchrules` Apply the dispatch rules

Dispatching allows `/.` to "dispatch" to potentially applicable rules immediately rather than testing all of the rules in turn. If the list of rules is long, this may save a significant amount of time. As an example, we form a list of rules and dispatch them:

```
rules = Table[f[i] → i^2, {i, 3000}];
```

```
dispatchrules = Dispatch[rules];
```

Then we do the same calculation with both sets of rules:

```
Do[f[i] /. rules, {i, 3000}] // Timing
{4.95 Second, Null}
```

```
Do[f[i] /. dispatchrules, {i, 3000}] // Timing
{0.03 Second, Null}
```

15.3.2 Patterns

A pattern represents *a class of expressions with a given structure*. Patterns are important in defining and restricting the arguments of functions, but patterns are useful also when using rules to transform expressions:

```
f[pattern] := expr
```

```
expr /. pattern → value
```

The pattern in **f** defines the class of arguments that the function is designed to accept. Often the pattern is simply, for example, **x_**, and then any expression is accepted as an argument, but we can also form more restrictive patterns like **x_Integer?Positive**.

The pattern in the transformation rule defines the class of subexpressions of **expr** that are replaced with **value**. Often the pattern is a degenerate pattern like **x**, and then **x** is replaced with **value** in all places **x** appears in **expr**; however, we can also form more restrictive patterns like **x_ /; x > 0.5**.

Patterns are also used in some commands like **Position**, **Cases**, **DeleteCases**, **Count**, and **Switch** (see Sections 13.2.5, p. 340; 13.2.6, p. 341; and 15.1.2, p. 387) and in some tests like **FreeQ**, **MemberQ**, and **MatchQ** (see Section 12.3.5, p. 314).

Here in Section 15.3.2 and also in Section 15.3.3, p. 408, we consider patterns mostly for the purpose of defining the arguments of functions. Patterns in transformation rules are considered in Section 15.3.4, p. 411, and patterns in **Position**, **Cases**, and so on in Section 15.3.5, p. 413.

■ A Pattern Matching Anything

_	A pattern representing any expression
x_	A pattern representing any expression; the expression is referred to with the name **x**

The most general pattern is **_**. It matches anything, and its internal name is **Blank**:

```
FullForm[_]     Blank[]
```

In Section 15.1.2, p. 388, we presented the following example of **Switch**:

```
volume[name_] := Switch[name, cylinder, Pi r^2 h,
    sphere, 4 Pi r^3 / 3, ellipsoid, 4 Pi a b c / 3, _, unknown]
```

The blank **_** matches anything, and so the result of volume is **unknown** if **name** is not cylinder, sphere, or ellipsoid.

The pattern **x_** also represents any expression, but now we give a name **x** for the expression so that we can refer to it later on. The most important use of this pattern is in defining the arguments of a function (see Section 14.1.1, p. 350). An example is **f[x_] := x Sin[x]**.

Next we consider ways to restrict a pattern. A restriction can be formed with heads, tests, and conditions. Note that, while a pattern represents a *structural* constraint, a restriction represents a *mathematical* constraint. A restricted pattern first accepts expressions that match the pattern but then does further mathematical tests.

■ Restricting Patterns with Heads

> **x_head**   Any expression having the head **head**

One type of restriction is to require that the expression must have a certain head. When we considered heads in Section 12.2.3, p. 309, we noted that every expression has a head, and we encountered such heads as **Integer**, **Rational**, **Real**, **Complex**, **Symbol**, **String**, **List**, **Plus**, **Times**, and **Power**.

As an example, we form a function that calculates the square of an integer:

```
f1[n_Integer] := n^2
```
```
{f1[3], f1[0], f1[-3], f1[2.7], f1[2 + 5 I], f1[a]}
```
```
{9, 0, 9, f1[2.7], f1[2 + 5 i], f1[a]}
```

As we can see, the function **f1** is applied only if the argument is an integer. The following function picks the first element of a list:

```
f2[x_List] := First[x]
```
```
{f2[{1, 2, 3}], f2[{{1, 2}, {3, 4}}], f2[8]}
```
```
{1, {1, 2}, f2[8]}
```

The expression 8 is not a list, and so **f2** is not applied.

■ Restricting Patterns with Tests

> **x_ ? test**   Any expression giving **True** from **test**
> **x_head ? test**   Any expression having the head **head** and giving **True** from **test**

(The internal name of **?** is **PatternTest**.) Tests were considered in Section 12.3.5, p. 314. We have tests like **==**, **!=**, **===**, **=!=**, **<**, **≤**, **>**, **≥**, **NonNegative**, **Positive**, **NumericQ**, **EvenQ**, **OddQ**, **VectorQ**, **MatrixQ**, and **OptionQ**. We can form more complex tests with operations like **&&** (AND), **||** (OR), and **!** (NOT).

As an example, we program the factorial function. We require that the argument **n** must be a positive integer:

```
fac[0] = 1;
fac[n_Integer?Positive] := n fac[n - 1]
```
```
{fac[5], fac[2.7], fac[-2]}
```
```
{120, fac[2.7], fac[-2]}
```

Next we transpose a matrix:

```
f3[x_?MatrixQ] := Transpose[x]
```
```
{f3[{{1, 2}, {3, 4}}], f3[{1, 2, 3}]}
```
```
{{{1, 3}, {2, 4}}, f3[{1, 2, 3}]}
```

The built-in tests can be used without an argument. If an argument is written, the test is written as a pure function in which the argument is **#**: **f3[x_? MatrixQ[#] &] := ... .**

User-defined tests are normally written as pure functions. As an example, we require that the argument is in the interval (0, 1):

```
f4[x_ ? (0 < # < 1 &)] := ArcSin[x]
```

```
{f4[0.3], f4[1.3]}
```

```
{0.304693, f4[1.3]}
```

However, it may be more natural to use conditions; these are considered next.

■ Restricting Patterns with Conditions

> **x_ /; cond** Any expression giving **True** from **cond**
> **x_head /; cond** Any expression having the head **head** and giving **True** from **cond**
>
> **x_ ? test /; cond** Any expression giving **True** from **test** and from **cond**
> **x_head ? test /; cond** Any expression having the head **head** and giving **True** from **test** and from **cond**

(The internal name of **/;** is **Condition**.) The following function accepts only real numbers as the argument, since the imaginary part of the argument is required to be 0:

```
f5[x_ /; Im[x] == 0] := x^2
```

```
{f5[-3], f5[Sqrt[2 - Sqrt[2]]], f5[Sqrt[1 - Sqrt[2]]], f5[2 + 5 I], f5[a]}
```

$$\left\{9, \ 2 - \sqrt{2}, \ f5\left[\dot{\imath} \ \sqrt{-1 + \sqrt{2}}\right], \ f5[2 + 5\,\dot{\imath}], \ f5[a]\right\}$$

The next function requires that the argument be positive:

```
f6[x_ /; Im[x] == 0 && x > 0] := x^2
```

```
{f6[3], f6[-7], f6[2 + I], f6[a]}
```

```
{9, f6[-7], f6[2 + i], f6[a]}
```

Now we require that **x** is a numerical matrix:

```
f7[x_ /; MatrixQ[x, NumericQ]] := Det[x]
```

```
{f7[{{3, 2}, {5, 1}}], f7[{{a, 2}, {5, 1}}]}
```

```
{-7, f7[{{a, 2}, {5, 1}}]}
```

A condition in the box above is written next to **x_**, but it can also be written in other places, such as at the end of the definition. The function **int** gives the integral of $\sin(m\,x)\cos(n\,x)$ from 0 to π:

```
int[m_Integer, n_Integer] := 0 /; EvenQ[m + n]
int[m_Integer, n_Integer] := 2 m / (m^2 - n^2) /; OddQ[m + n]
```

Here are some examples:

```
{int[0, 1], int[1, 0], int[1, 1], int[1, 2]}
```

$$\left\{0, \ 2, \ 0, \ -\frac{2}{3}\right\}$$

In the next example, we define a linear function. We restrict **a** and **b** to be free of **x**:

```
lin[a_. + b_. x_, x_] := a + b lin[x] /; (FreeQ[a, x] && FreeQ[b, x])
lin[a_, x_] := a /; FreeQ[a, x]
```

(The dot **.** is explained below.) For example:

```
{lin[2 + 3 x, x], lin[4, x], lin[5 x, x], lin[1 + Sin[x] , x]}
```

```
{2 + 3 lin[x], 4, 5 lin[x], lin[1 + Sin[x], x]}
```

The function **der** defines rules to be used to calculate derivatives:

```
der[p_ + q_, x_] := der[p, x] + der[q, x]
der[p_ q_, x_] := der[p, x] q + p der[q, x]
der[p_^q_, x_] := p^q (der[p, x] q / p + der[q, x] Log[p])
der[x_, x_] := 1
der[p_, x_] := 0 /; FreeQ[p, x]
```

Here are some examples:

```
{der[2 + 3 x - 7 x^2 + x^4, x], der[(1 + x^3) ^ (2 + x^2), x]}
```

$$\left\{3 - 14 x + 4 x^3, \ (1 + x^3)^{2+x^2} \left(\frac{3 x^2 (2 + x^2)}{1 + x^3} + 2 x \operatorname{Log}[1 + x^3]\right)\right\}$$

```
Remove["Global`*"]
```

15.3.3 More about Patterns

■ Variable Number of Arguments

x_	Any single expression
x__	Any sequence of *one* or more expressions
x___	Any sequence of *zero* or more expressions

Sometimes it is useful to form a function in which the number of arguments is unspecified. Such functions can be formed with a double underscore (**__**) (**BlankSequence**) or a triple underscore (**___**) (**BlankNullSequence**).

The functions **f1**, **f2**, and **f3** accept one, one or more, and zero or more arguments, respectively:

```
f1[x_] := Apply[Plus, {x}]
f2[x__] := Apply[Plus, {x}]
f3[x___] := Apply[Plus, {x}]
```

We try the functions for several numbers of arguments:

```
{f1[], f1[a], f1[a, b], f1[a, b, c]}     {f1[], a, f1[a, b], f1[a, b, c]}
```

```
{f2[], f2[a], f2[a, b], f2[a, b, c]}     {f2[], a, a + b, a + b + c}
```

```
{f3[], f3[a], f3[a, b], f3[a, b, c]}     {0, a, a + b, a + b + c}
```

We see that **f1** accepts only one argument, **f2** does not accept zero arguments, and **f3** accepts zero or more arguments.

The triple underscore is used especially for options. An example was shown in Section 15.2.3, p. 399:

```
newton5[f_, x_, x0_, max_, opts___] := With[{df = D[f, x]},
   FixedPointList[(x - f / df) /. x → # &, N[x0], max, opts]]
```

This function accepts zero or more options, so that we can write, for example, the following commands:

```
newton5[f, x, 2, 20]
```

```
newton5[f, x, 2, 20, SameTest → (Abs[#1 - #2] < 10^-3 &)]
```

■ Default Values

> **x_.**   Any expression which, if omitted, is taken to have a built-in default value
> **x_:v**   Any expression which, if omitted, is taken to have the default value **v**
> **x_head:v** Use the default value **v** for any omitted expression with the head **head**

(These constructs are internally formed with **Optional**.) Default values are useful, for example, in such expressions as **a^n_**. Consider the following three examples. In the first example, we do not have a default value for the exponent, and **a** or **a^1** is not transformed:

```
{a, a^2, a^3} /. a^n_ → b^(2 n)        {a, b⁴, b⁶}
```

In the following example, we use the built-in default value 1 for the exponent, and now **a**, too, is transformed:

```
{a, a^2, a^3} /. a^n_. → b^(2 n)        {b², b⁴, b⁶}
```

We can also define our own default value:

```
{a, a^2, a^3} /. a^(n_: 1) → b^(2 n)        {b², b⁴, b⁶}
```

In general, the default value of **y** in **x_ + y_.** is 0. In **x_ y_.** the default value of **y** is 1, and in **x_^y_.** it is also 1. As an example, the function **coeff** extracts the coefficients from a linear expression:

```
coeff[a_. + b_. x_, x_] := {a, b}
```

```
coeff[a_, x_] := {a, 0}
```

We try the function:

```
{coeff[u, z], coeff[u + z, z], coeff[v z, z], coeff[u + v z, z]}
```

```
{{u, 0}, {u, 1}, {0, v}, {u, v}}
```

■ Optional Arguments

A default value such as **v** in **x_:v** may be useful in function definitions. An example is the following function, which is also presented in Section 19.3.4, p. 509:

```
newtonSolve[f_, x_, x0_, d_: 1, n_: 20, opts___?OptionQ] :=
   With[{df = D[f, x]}, FixedPointList[(x - d f / df) /. x → # &, N[x0], n, opts]]
```

If we give only the first three arguments, as in **newtonSolve[f, x, 3]**, then the default value 1 is used for **d** and 20 for **n**. On the other hand, we can define **d** as in **newton‹ Solve[f, x, 3, 2]**, and then 2 is used as the damping factor. We can also define both **d** and **n** as in `newtonSolve[f, x, 3, 2, 25]`. In this way, we can define and use optional arguments.

An optional argument can sometimes be handy at the end of the parameter list. Another approach, which is useful in more complicated situations, is to define the function several times for several forms of arguments. Still another approach is to use options (see Section 14.3.4, p. 376). Options are good for situations in which the function has many adjustable features but where the default settings of the features work in most cases.

■ Alternative and Repeated Patterns

Thus far we have considered patterns of the form **x_** (added with tests and conditions). To form more complicated patterns, the following more general definition of a pattern is useful:

> **x : pattern** Represents any expression matching **pattern**; the expression is referred to with the name **x**

(The internal name of this construct is **Pattern**.) The pattern **x_** is a special case of this general definition. Indeed, **x_** is equivalent to **x:_**, and **x_head** is equivalent to **x:_head**. The general definition of a pattern is used in the following constructs:

> **x: (pattern1 | pattern2 | …)** Represents any expression matching any of the patterns
> **x: pattern ..** Represents any sequence of one or more expressions, each matching **pattern**
> **x: pattern ...** Represents any sequence of zero or more expressions, each matching **pattern**

Here is an example of **|**:

```
{a, b, c, d, e} /. a | b | c → x

{x, x, x, d, e}
```

Here **a|b|c** represents any of **a**, **b**, or **c**. The same effect is obtained with **Thread[{a, b, c} → x]**. In the next example, the argument of the function has to be a positive integer or a positive rational number:

```
f4[x : (_Integer | _Rational) ? (# > 0 &)] := x^2

{f4[3], f4[2 / 3], f4[-5]}
```

$$\left\{9, \frac{4}{9}, f4[-5]\right\}$$

The function **f5** accepts as an argument a nonempty list of two-element lists and then transposes the list:

```
f5[x : {{_, _} ..}] := Transpose[x]

f5[{{1, a}, {2, b}, {3, c}}]

{{1, 2, 3}, {a, b, c}}
```

15.3.4 Patterns in Transformation Rules

Thus far we have mostly used patterns in defining the arguments of functions. Now we will use patterns in transformation rules.

> **expr /. pattern → value** Apply the transformation once to each part of **expr**
>
> **expr //. pattern → value** Apply the transformation until the result no longer changes

(The internal names of /. and //. are **ReplaceAll** and **ReplaceRepeated**, respectively.) Some examples have already be presented in Section 15.3.3, p. 409 (see **Default Values** and **Alternative and Repeated Patterns**). Here are some other examples.

■ Example 1

Consider the following list of numbers:

```
t1 = Table[Random[], {7}]
```

$\{0.132529, 0.671609, 0.179236, 0.859771, 0.685538, 0.127227, 0.363438\}$

All of the pattern constructs can be used for replacement rules as well as for function definitions. For example, if we want to replace all numbers greater than 0.5 with 1, we can use a condition:

```
t1 /. (x_ /; x > 0.5) → 1
```

$\{0.132529, 1, 0.179236, 1, 1, 0.127227, 0.363438\}$

We could also use a test:

```
t1 /. (x_ ? (# > 0.5 &)) → 1
```

$\{0.132529, 1, 0.179236, 1, 1, 0.127227, 0.363438\}$

■ Example 2

Define a list:

```
vals = Table[{i, 2^i}, {i, 5}]
```

$\{\{1, 2\}, \{2, 4\}, \{3, 8\}, \{4, 16\}, \{5, 32\}\}$

To get square roots of the second elements, do as follows:

```
vals /. {x_, y_} → {x, Sqrt[y]}
```

$\{\{1, \sqrt{2}\}, \{2, 2\}, \{3, 2\sqrt{2}\}, \{4, 4\}, \{5, 4\sqrt{2}\}\}$

■ Example 3

Define an expression containing square roots:

```
t2 = Sqrt[x] + 1 / Sqrt[x];
```

If we want to transform the expression by replacing $\sqrt{x}$ with y, we do not get what we want:

```
t2 /. Sqrt[x] → y
```
$\dfrac{1}{\sqrt{x}} + y$

This is because, although the internal representation of $\sqrt{x}$ is $x^{1/2}$, that of $1/\sqrt{x}$ is not $1/x^{1/2}$ but rather $x^{-1/2}$ which does not match the pattern $x^{1/2}$:

> **FullForm[t2]**

> Plus[Power[x, Rational[-1, 2]], Power[x, Rational[1, 2]]]

We have to transform $\sqrt{x}$ and $1/\sqrt{x}$ separately:

> **t2 /. {Sqrt[x] → y, 1 / Sqrt[x] → 1 / y}** $\frac{1}{y} + y$

■ Example 4

Mathematica does not automatically expand logarithms like $\log(x^2)$ to $2\log(x)$:

> **t3 = Log[x^2] + Log[1 / Sqrt[y]];**

> **t3 // Expand** $\text{Log}[x^2] + \text{Log}\left[\dfrac{1}{\sqrt{y}}\right]$

This is because such an expansion is not always correct:

> **{Log[(-1)^2], 2 Log[-1]}** $\{0, 2\, \mathbb{i}\, \pi\}$

PowerExpand does the transformation, but we have to consider the conditions under which the transformation is correct:

> **t3 // PowerExpand** $2\, \text{Log}[x] - \dfrac{\text{Log}[y]}{2}$

The same effect can be obtained with a transformation rule:

> **t3 /. Log[a_^b_] → b Log[a]** $2\, \text{Log}[x] - \dfrac{\text{Log}[y]}{2}$

■ Example 5

Here is an ordinary replacement:

> **a + b + c /. a + b → a b** $a b + c$

Next we use a pattern:

> **a + b + c /. x_ + y_ → x y** $a\ (b + c)$

Mathematica finds the sum **a + (b + c)**, applies the transformation, and gets **a (b + c)**. If we apply the transformation repeatedly, we get the following:

> **a + b + c //. x_ + y_ → x y** $a b c$

Mathematica first gets **a (b + c)** and then **a (b c)**.

■ Replacing in All Possible Ways

> **ReplaceList[expr, pattern → value]** Apply the transformation once in all possible ways

We do a transformation in all possible ways and form a list from the results:

> **ReplaceList[a + b + c, x_ + y_ → x y]**

> $\{a\ (b + c),\ b\ (a + c),\ (a + b)\ c,\ (a + b)\ c,\ b\ (a + c),\ a\ (b + c)\}$

With **Union**, we get the results that are distinct:

```
% // Union    { (a + b) c, b (a + c), a (b + c) }
```

An application of **ReplaceList** can be found in Section 26.2.2, p. 732 (look at **Estimating the Transition Probabilities**).

15.3.5 Searching with Patterns

The commands **Position**, **Cases**, **DeleteCases**, and **Count** were introduced in Sections 13.2.5, p. 341, and 13.2.6, p. 341. Now that we know something about patterns, we consider the commands in more detail.

■ Searching Positions

Position[expr, pattern] Give the positions at which objects matching **pattern** occur in **expr**
Position[expr, pattern, levspec] Give only the positions that are in levels specified by **levspec**

Level specifications are considered in Section 12.2.3, p. 310. Recall that a level specification **n** means all levels 1, 2, ..., n; **{n}** means only the level n; and **{n, m}** means levels **n** through **m**. If a level specification is not given in **Position**, the specification is assumed to be {0, ∞}, which means all levels 0, 1, 2, and so on.

Consider the following list:

```
t1 = {3, 5.2, -6, 7.9, 4};
```

When applying **Position**, we can use all of the constructs we have learned that involve patterns. In the next examples, we use heads, tests, and conditions:

```
Position[t1, _Integer]    {{1}, {3}, {5}}

Position[t1, _Integer?Positive]    {{1}, {5}}

Position[t1, x_ /; x > 3]    {{2}, {4}, {5}}

Position[t1, x_Integer /; EvenQ[x] && x > 3]    {{5}}
```

Consider then the following list:

```
t2 = {{0, 1}, {1, p}, {2, p^2}, {3, p^3}};
```

The positions of **p** in **t2** are as follows:

```
Position[t2, p]    {{2, 2}, {3, 2, 1}, {4, 2, 1}}
```

Next we find all positions of p^{anything} :

```
Position[t2, p^_]    {{3, 2}, {4, 2}}
```

With **Extract**, we can see the corresponding objects:

```
Extract[t2, %]    {p^2, p^3}
```

Thus, the pattern **p^_** found the explicit powers **p^2** and **p^3**, but it failed to find the first power of **p**. To find the first power, we want a pattern of the form $p^{\text{anything or 1}}$; this can be expressed with **p^_:1**. Here **1** is a default value for the exponent: the value **1** is used if an explicit power is not found (see Section 15.3.3, p. 409). This pattern also matches the first power of **p**:

    Position[t2, p^_ : 1]

    {{2, 2}, {3, 2, 1}, {3, 2}, {4, 2, 1}, {4, 2}}

However, now we also obtained the plain **p** in **p^2** and **p^3** at positions {3, 2, 1}, and {4, 2, 1}. If we want only the power terms, we stop at the second level:

    Position[t2, p^_ : 1, 2]

    {{2, 2}, {3, 2}, {4, 2}}

■ Searching Elements

Select[list, test] Select the elements of **list** that satisfy **test**

Cases[list, pattern] Select the elements of **list** that match **pattern**

Cases[list, pattern → value] Select the elements of **list** that match **pattern**, and then apply the transformation to them

DeleteCases[list, pattern] Delete the elements of **list** that match **pattern**

Count[list, pattern] Give the number of times **pattern** appears as an element of **list**

Cases, **DeleteCases**, and **Count** accept a third argument that specifies the level from which matching objects are searched. Note that **Select** does not use patterns: it uses tests. We have, however, included it in the box above so that we can compare it with **Cases**. **Select** may be somewhat simpler to use in tasks like the following:

    Select[t1, # > 3 &]   {5.2, 7.9, 4}

    Cases[t1, x_ /; x > 3]   {5.2, 7.9, 4}

16

Differential Calculus

Introduction

> *Analysis takes back with one hand what it gives with the other. I recoil with fear and loathing from that deplorable evil, continuous functions with no derivatives. — Hermite*

Traditional differential calculus includes derivatives, Taylor series, and limits; the corresponding *Mathematica* commands are **D**, **Series**, and **Limit**. Later on, derivatives play a central role when we, for example, solve nonlinear equations by numerical methods (Chapter 19) and solve optimization problems (Chapter 20). Of course, derivatives are essential in differential and partial differential equations (Chapters 23 and 24).

As this chapter begins the mathematical part of the book, we would like to note that http://mathworld.wolfram.com/ is a place to find information about mathematical topics like calculus, algebra, applied mathematics, discrete mathematics, and probability and statistics.

16.1 Derivatives

16.1.1 Partial Derivatives

In the box below, we have some examples of calculating partial derivatives.

D[f, x]	$\partial f / \partial x$	**D[f, x, y]**	$\partial^2 f / (\partial x \, \partial y)$
D[f, x, x]	$\partial^2 f / \partial x^2$	**D[f, x, y, y, y]**	$\partial^4 f / (\partial x \, \partial y^3)$
D[f, x, x, x]	$\partial^3 f / \partial x^3$	**D[f, x, y, z]**	$\partial^3 f / (\partial x \, \partial y \, \partial z)$

You can write in **D** the whole expression to be differentiated, as in **D[a + b x, x]**, or you can first give a name to the expression, such as **h = a + b x**, and then use that name, such as in an expression like **D[h, x]**. Functions like **f[x]** can be differentiated simply by **f'[x]** (this is considered in more detail below).

Another way to calculate partial derivatives is to use the buttons $\partial_\square \blacksquare$ and $\partial_{\square,\square} \blacksquare$ of the **BasicInput** palette. Alternatively, you can do it yourself: write ∂ as ESC pd ESC, go to a subscript by pressing CTRL -, write the variable, go out of the subscript position with CTRL ⎵, and write the expression (see Section 3.3.3, p. 70):

   ∂_x **(a + b x)**   b

The command **D** also has another form in which the order of differentiation is expressed explicitly. For example, instead of **D[f, x]**, **D[f, x, x]**, and **D[f, x, y, y, y]**, we can also write **D[f, {x, 1}]**, **D[f, {x, 2}]**, and **D[f, {x, 1}, {y, 3}]**.

Note that the derivatives of **Abs** and **Sign** are not calculated. To get rules for the derivatives and integrals of these functions, load the **ProgrammingInMathematica`Abs`** package:

   << ProgrammingInMathematica`Abs`

   D[Abs[x], x]   Sign[x]

■ Example 1: Tangent

Consider the following function:

   f = (x + 2) (x^2 + 1) x (x - 1) (x - 2);

We calculate and plot a tangent for it at $x = 1.6$. First we calculate the corresponding values of the function and its derivative:

   x1 = 1.6;

   f1 = f /. x → x1   − 4.92134

   df1 = D[f, x] /. x → x1   − 4.76544

Then we form the tangent and plot it together with the function:

   tangent = f1 + df1 (x − x1)   − 4.92134 − 4.76544 (−1.6 + x)

   Plot[{f, tangent}, {x, 0, 2.1}, PlotStyle → {{}, RGBColor[0, 0, 1]},
   Epilog → {AbsolutePointSize[4], RGBColor[1, 0, 0], Point[{x1, f1}]}];

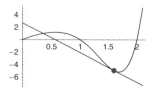

■ Example 2: Critical Points

Let us now examine the critical points of the function we considered in Example 1:

```
p = Plot[f, {x, -2.05, 2.1}];
```

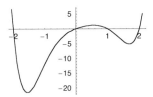

Critical points are characterized by the property that the first derivative is zero at these points. Critical points contain points of local maximum and minimum values and saddle points. Our function seems to have three critical points. To find them, we calculate the derivative and its zeros:

```
df = D[f, x] // Simplify
```

$4 - 8\,x + 9\,x^2 - 12\,x^3 - 5\,x^4 + 6\,x^5$

```
c = NSolve[df == 0, x]
```

$\{\{x \to -1.55135\}, \{x \to 0.065021 - 0.666791\,\mathrm{i}\},$
$\{x \to 0.065021 + 0.666791\,\mathrm{i}\}, \{x \to 0.567487\}, \{x \to 1.68715\}\}$

We are interested only in real critical points, and so we select such points:

```
crit = {c[[1]], c[[4]], c[[5]]}
```

$\{\{x \to -1.55135\}, \{x \to 0.567487\}, \{x \to 1.68715\}\}$

Then we form the corresponding points on the curve:

```
points = {x, f} /. crit
```

$\{\{-1.55135, -21.4839\}, \{0.567487, 1.19346\}, \{1.68715, -5.14393\}\}$

Lastly we show the points on the curve:

```
Show[p,
    Epilog → {AbsolutePointSize[4], RGBColor[1, 0, 0], Map[Point, points]}];
```

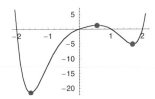

```
Remove["Global`*"]
```

■ Derivatives of Functions of One Variable

If you have defined a function **f[x]** of one variable, derivatives can be calculated with primes (this resembles the usual mathematical notation).

f'[x], f''[x], f'''[x], ...	The first, second, third, ... derivative of a function **f** at **x**

Consider, for example, the following function:

`f[x_] := a + b x + c x^2`

The first, second, and third derivatives are as follows:

`{f'[x], f''[x], f'''[x]}`   {b + 2 c x, 2 c, 0}

Note that we can at the same time also specify the point at which the derivative is calculated:

`f'[3]`   b + 6 c

However, if the function has several arguments, like **g** in this example:

`g[x_, y_] := Sin[x] Cos[y]`

then derivatives are again calculated with **D** (or **Derivative**, see below):

`D[g[x, y], y]`   - Sin[x] Sin[y]

■ Derivatives of Functions of Several Variables

Consider the function **g** defined above. If we want to calculate the value of its derivative at a point, we first have to calculate its derivative and then ask its value at the point:

`D[g[x, y], x, x, y] /. {x → 4, y → 5}`   Sin[4] Sin[5]

However, *Mathematica* has still one way to represent derivatives, which is **Derivative**, and with this command we can specify the orders and the point at the same time:

`Derivative[2, 1][g][4, 5]`   Sin[4] Sin[5]

Derivative[m, n, …][f][a, b, …] Differentiate function **f[x, y , …]** **m** times with respect to **x**, **n** times with respect to **y**, … and evaluate the derivative at the point **x** = **a, y** = **b,** …

When differentiating unspecified functions, *Mathematica* shows primes for functions of one variable:

`D[r[s[x]], x]`   r'[s[x]] s'[x]

Superscripts are used for functions of several variables:

`D[r[x, y], x, x, y]` $r^{(2,1)}[x, y]$

However, in internal representations, *Mathematica* uses **Derivative**:

`D[r[s[x]], x] // InputForm`

Derivative[1][r][s[x]]*Derivative[1][s][x]

`D[r[x, y], x, x, y] // InputForm`

Derivative[2, 1][r][x, y]

■ NonConstants

If a variable depends on another variable, it is simple to explicitly denote the dependency. In the next example, both **r** and **a** depend on **x**:

```
D[a[x] r[x], x]
```

$$r[x] \, a'[x] + a[x] \, r'[x]$$

Another way to express the dependency is to use the **NonConstants** option:

```
D[a r[x], x, NonConstants → a]
```

$$D[a, x, \text{NonConstants} \to \{a\}] \, r[x] + a \, r'[x]$$

D[f, x, NonConstants → {a, b, … }] Declare that **a, b**, … depend on **x**

16.1.2 Total Derivatives

Dt calculates total derivatives in which all variables in the expression are assumed to depend on all of the variables with respect to which the total derivative is calculated. The form of **Dt** is the same as the form of **D**. Here are some examples:

Dt[f, x], Dt[f, x, x], Dt[f, x, y]

We can also use forms like **Dt[f, {x, 2}]**. Here are two examples of total derivatives:

Dt[x y, x] $y + x \, \text{Dt}[y, x]$

Dt[x y, x, x] $2 \, \text{Dt}[y, x] + x \, \text{Dt}[y, \{x, 2\}]$

■ Example: Differentiating an Implicit Function

We define a function of two variables:

```
f = x^2 + 3 y^2 - x y - 1;
```

Then we define an equation where this function is equal to zero:

eqn = f == 0 $-1 + x^2 - x \, y + 3 \, y^2 == 0$

The equation **eqn** implicitly defines a function $y(x)$. We have already considered implicit functions in Section 14.1.4, p. 358. For example, we can plot the function with a package:

```
<< Graphics`ImplicitPlot`
```

```
p1 = ImplicitPlot[f == 0, {x, -1.1, 1.1}];
```

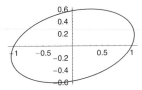

Now we calculate an equation for the derivative of the implicitly defined function:

deqn = Dt[eqn, x] $2 \, x - y - x \, \text{Dt}[y, x] + 6 \, y \, \text{Dt}[y, x] == 0$

Solving **Dt[y,x]** from here, we get the derivative $y'(x)$, and we give it the name **dy**:

Solve[deqn, Dt[y, x]] $\left\{\left\{\text{Dt}[y, x] \to \dfrac{2\,x - y}{x - 6\,y}\right\}\right\}$

$$\texttt{dy = Dt[y, x] /. \%[[1]]} \qquad \frac{2\,x - y}{x - 6\,y}$$

Another way to calculate the derivative is to use a result of analysis:

$$\texttt{-D[f, x] / D[f, y]} \qquad \frac{-2\,x + y}{-x + 6\,y}$$

Let us consider the derivative for $x = 0.5$. First, we solve the corresponding values of y:

```
x1 = 0.5;
```

```
y1 = y /. Solve[eqn /. x → x1, y]     {-0.423564, 0.59023}
```

Then we calculate the corresponding values of the derivative:

```
dy1 = dy /. {x → x1, y → y1}     {0.468065, -0.134731}
```

Next we form the tangents at these points:

```
tan1 = y1 + dy1 (x - x1)
```

$$\{-0.423564 + 0.468065\,(-0.5 + x),\ 0.59023 - 0.134731\,(-0.5 + x)\}$$

(Note how nicely we obtained both of the tangents with one command: *Mathematica* automatically does vector operations component by component.) Lastly we show the function, the tangents, and the points:

```
p2 = Plot[Evaluate[tan1], {x, -1.1, 1.1}, DisplayFunction → Identity];
```

```
Show[p1, p2, Epilog →
    {AbsolutePointSize[3], Point[{x1, y1[[1]]}], Point[{x1, y1[[2]]}]}];
```

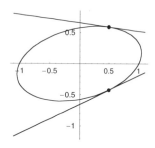

■ Constants

$\texttt{Dt[f, x, Constants} \to \texttt{\{a, b, ... \}]}$ Declare that $\texttt{a}$, $\texttt{b}$, ... do not depend on $\texttt{x}$

In this way we can tell that some symbols are treated as constants when we calculate a total derivative. For example:

```
Dt[a x^m, x, Constants → {a, m}]     a m x^{-1+m}
```

If we want to define some symbols permanently as constants, we can give them the attribute **Constant** (see Section 14.2.3, p. 368):

```
SetAttributes[{a, m}, Constant]
```

Now **a** and **m** are treated as constants:

```
Dt[a x^m, x]     a m x^{-1+m}
```

The attribute can be removed by writing **Remove[a, m]**.

We can define a certain derivative of a given symbol as having a certain value:

```
n /: Dt[n, x] = 0     0
```

This defines **n** as having the property that it does not depend on **x**. Now we get the following:

```
Dt[b x^n, x]     b n x^{-1+n} + x^n Dt[b, x]
```

The property `Dt[n, x] = 0` can be removed by writing `Remove[n]`.

■ Total Differentials

`Dt[f]`	Total differential of **f**

```
Dt[x^2 y^2]     2 x y^2 Dt[x] + 2 x^2 y Dt[y]

Remove["Global`*"]
```

16.1.3 Vector Analysis

For a scalar-valued function of several variables, we may want to calculate the following:

the *gradient* vector (vector of the first derivatives);
the *Hessian* matrix (matrix of the pure and mixed second derivatives); and
the *Laplacian* scalar (sum of the unmixed second derivatives).

For a vector-valued function of several variables, we may want to calculate the following:

the *Jacobian* matrix (the *i*th row is the gradient of the *i*th function); and
the *divergence* scalar (sum in which the *i*th term is the derivative of the *i*th function with respect to the *i*th variable).

Here are functions for calculating these.

```
gradient[f_, x_List] := Map[D[f, #] &, x]
hessian[f_, x_List] := Outer[D, gradient[f, x], x]
laplacian[f_, x_List] := Inner[D, gradient[f, x], x]
jacobian[f_List, x_List] := Outer[D, f, x]
divergence[f_List, x_List] := Inner[D, f, x]
```

For **Map**, **Outer**, and **Inner**, see Sections 13.3.1, p. 343, and 13.3.3, p. 346. Restrictions for the arguments, such as **x_List** (**x** must be a list), are considered in Section 15.3.2, p. 406. **Outer** and **Inner** are central in these functions. To understand them, examine the following generic examples of **Outer** and **Inner**:

```
Outer[f, {a, b, c}, {A, B, C}] // MatrixForm
```
$$\begin{pmatrix} f[a, A] & f[a, B] & f[a, C] \\ f[b, A] & f[b, B] & f[b, C] \\ f[c, A] & f[c, B] & f[c, C] \end{pmatrix}$$

```
Inner[f, {a, b, c}, {A, B, C}]
```
```
f[a, A] + f[b, B] + f[c, C]
```

You will then substitute the correct operator in place of **f** and the correct lists in place of **{a, b, c}** and **{A, B, C}**. In the following examples, we check that the functions really work:

> **gradient[f[x, y, z], {x, y, z}]**
>
> $\{f^{(1,0,0)}[x, y, z], f^{(0,1,0)}[x, y, z], f^{(0,0,1)}[x, y, z]\}$
>
> **hessian[f[x, y, z], {x, y, z}] // MatrixForm**
>
> $$\begin{pmatrix} f^{(2,0,0)}[x, y, z] & f^{(1,1,0)}[x, y, z] & f^{(1,0,1)}[x, y, z] \\ f^{(1,1,0)}[x, y, z] & f^{(0,2,0)}[x, y, z] & f^{(0,1,1)}[x, y, z] \\ f^{(1,0,1)}[x, y, z] & f^{(0,1,1)}[x, y, z] & f^{(0,0,2)}[x, y, z] \end{pmatrix}$$
>
> **laplacian[f[x, y, z], {x, y, z}]**
>
> $f^{(0,0,2)}[x, y, z] + f^{(0,2,0)}[x, y, z] + f^{(2,0,0)}[x, y, z]$
>
> **jacobian[{f[x, y, z], g[x, y, z], h[x, y, z]}, {x, y, z}] // MatrixForm**
>
> $$\begin{pmatrix} f^{(1,0,0)}[x, y, z] & f^{(0,1,0)}[x, y, z] & f^{(0,0,1)}[x, y, z] \\ g^{(1,0,0)}[x, y, z] & g^{(0,1,0)}[x, y, z] & g^{(0,0,1)}[x, y, z] \\ h^{(1,0,0)}[x, y, z] & h^{(0,1,0)}[x, y, z] & h^{(0,0,1)}[x, y, z] \end{pmatrix}$$
>
> **divergence[{f[x, y, z], g[x, y, z], h[x, y, z]}, {x, y, z}]**
>
> $h^{(0,0,1)}[x, y, z] + g^{(0,1,0)}[x, y, z] + f^{(1,0,0)}[x, y, z]$

Here are some more specific examples:

> **f = x^2 + x y^2 + x y z^2; g = Exp[x y z]; h = Sin[x y z];**
>
> **gradient[f, {x, y, z}]**
>
> $\{2 x + y^2 + y z^2, 2 x y + x z^2, 2 x y z\}$
>
> **jacobian[{f, g, h}, {x, y, z}] // MatrixForm**
>
> $$\begin{pmatrix} 2 x + y^2 + y z^2 & 2 x y + x z^2 & 2 x y z \\ e^{x y z} y z & e^{x y z} x z & e^{x y z} x y \\ y z \cos[x y z] & x z \cos[x y z] & x y \cos[x y z] \end{pmatrix}$$

■ A Package for Vector Analysis

In the **Calculus`VectorAnalysis`** package, there are many more commands for vector analysis. With this package, we can do calculations in various 3D coordinate systems. We will not give a thorough presentation of this package but rather a quick overview. First, load the package:

> **<< Calculus`VectorAnalysis`**

Then we can look at the names of this package with the following command:

> **? Calculus`VectorAnalysis`***

The resulting list of 47 names is not presented here. We can use 14 coordinate systems (e.g., **Cartesian**, **Cylindrical**, **Spherical**) and calculate **Grad**, **Laplacian**, **Biharmonic**, **Div**, **Curl**, **DotProduct**, **CrossProduct**, and **ScalarTripleProduct**, among others.

Let us use some commands. First, we ask for the current coordinate system and the default coordinates:

    `{CoordinateSystem, Coordinates[]}`   `{Cartesian, {Xx, Yy, Zz}}`

Now we can calculate, for example, the gradient of a function:

    `Grad[Xx + Sin[Yy Zz]]`   `{1, Zz Cos[Yy Zz], Yy Cos[Yy Zz]}`

We can set the coordinates:

    `SetCoordinates[Cartesian[x, y, z]]`   `Cartesian[x, y, z]`

Now we can use **x**, **y**, and **z**:

    `Div[{x y, x y z, Sin[x y z]}]`   `y + x z + x y Cos[x y z]`

Let us then move to spherical coordinates and ask for some information about this system:

    `SetCoordinates[Spherical[r, θ, ϕ]]`   `Spherical[r, θ, ϕ]`

    `CoordinateRanges[]` $\{0 \le r < \infty, \ 0 \le \theta \le \pi, \ -\pi < \phi \le \pi\}$

    `CoordinatesToCartesian[{r, θ, ϕ}]`

    `{r Cos[ϕ] Sin[θ], r Sin[θ] Sin[ϕ], r Cos[θ]}`

    `jdet = JacobianDeterminant[]` $r^2 \ \text{Sin}[\theta]$

As an application, we calculate the area and volume of a sphere of radius R:

    `Integrate[jdet, {θ, 0, Pi}, {ϕ, -Pi, Pi}] /. r → R` $4 \pi R^2$

    `Integrate[jdet, {θ, 0, Pi}, {ϕ, -Pi, Pi}, {r, 0, R}]` $\dfrac{4 \pi R^3}{3}$

16.1.4 Numerical Derivatives

■ Numerical Derivatives of Functions

Sometimes it may be too difficult or impossible to calculate a derivative symbolically. You can then use **ND** from a package (or the program we presented in Section 15.1.1, p. 381).

In the `NumericalMath`NLimit`` *package:*

`ND[f, x, a]` First derivative of **f** with respect to **x** at **a**
`ND[f, {x, n}, a]` nth derivative of **f** with respect to **x** at **a**

Options:

`WorkingPrecision` Precision used in internal computations; examples of values:
  `MachinePrecision` (❀5), **20**
`Method` Extrapolation method; possible values: `EulerSum`, `NIntegrate`
`Scale` Initial step size (`EulerSum`) or the radius of the circle of integration (`NInte`
  `grate`); default value: **1**
`Terms` Number of divided differences calculated (`EulerSum`); default value: **7**

ND has two methods. If `EulerSum` is used, then **ND** forms a sequence of divided differences with successively smaller step sizes and then extrapolates to the limit by calculating a numerical limit of the divided differences. The initial step size is `Scale`, and a total of `Terms` difference quotients is calculated by successively halving the previous step size. If

NIntegrate is used, then **ND** applies Cauchy's integral formula, and **Scale** is the radius of the circle of integration.

As an example, we calculate the first, second, and third derivatives of an expression numerically with **ND** using Cauchy's integral formula and compare the results with the numerical values of the exact derivatives given by **D**. The approximations are very good:

```
<< NumericalMath`NLimit`

f = Exp[-x^2];
a = Table[ND[f, {x, i}, 1, Method → NIntegrate], {i, 3}];
b = Table[D[f, {x, i}], {i, 3}] /. x → 1.;

a - b // Chop      {0, 0, 0}

Remove["Global`*"]
```

■ Numerical Derivatives of Data

If we have values of a function only at some given points, we can still calculate approximations to the derivative at most of the points. Here are some functions that calculate such approximations:

```
derData1a[data_, h_] := (Drop[data, 1] - Drop[data, -1]) / h
derData1b[data_, h_] := (Drop[data, 2] - Drop[data, -2]) / (2 h)
derData2[data_, h_] :=
  (Drop[data, 2] - 2 Take[data, {2, -2}] + Drop[data, -2]) / h^2
```

The functions **derData1a** and **derData1b** calculate first derivatives and **derData2** second derivatives. The list **data** contains the values of the function, and **h** is the step size. Remember that, for example, **Drop[data, 1]** drops the first element and **Drop[data, -1]** the last element. **Take[data, {2, -2}]** drops the first and last elements. Note that the functions rely on the fact that *Mathematica* automatically does all calculations with lists component by component. We try the approximation on a symbolic data set:

```
data = {a, b, c, d, e};
```

derData1a[data, h] $\left\{ \dfrac{-a+b}{h}, \dfrac{-b+c}{h}, \dfrac{-c+d}{h}, \dfrac{-d+e}{h} \right\}$

derData1b[data, h] $\left\{ \dfrac{-a+c}{2\,h}, \dfrac{-b+d}{2\,h}, \dfrac{-c+e}{2\,h} \right\}$

derData2[data, h] $\left\{ \dfrac{a-2\,b+c}{h^2}, \dfrac{b-2\,c+d}{h^2}, \dfrac{c-2\,d+e}{h^2} \right\}$

Note that **derData1a** does not calculate the approximation at the last point, whereas **derData1b** and **derData2** do not calculate the approximation at the first and last points.

16.2 Taylor Series

16.2.1 Taylor Series and Polynomials

> **Series[f, {x, a, n}]**   Taylor series of **f** with respect to **x** about the point **a** with terms up to the **n**th power of **x - a**

Series[Exp[c x], {x, 1, 2}]

$$e^c + c\, e^c\,(x-1) + \frac{1}{2}\,c^2\,e^c\,(x-1)^2 + O[x-1]^3$$

The remainder is in the form of a capital O. In normal mathematical notation, the remainder here would be written as $O((x-1)^3)$, indicating terms where the power of $x-1$ is at least 3. Here is another example:

t = Series[1 / Sqrt[1 + x], {x, 0, 4}]

$$1 - \frac{x}{2} + \frac{3\,x^2}{8} - \frac{5\,x^3}{16} + \frac{35\,x^4}{128} + O[x]^5$$

We can calculate with a Taylor series expansion:

t^2 $1 - x + x^2 - x^3 + x^4 + O[x]^5$

Here all terms of an order higher than four are gathered in the remainder. We can also calculate derivatives and integrals (note the change in the remainder):

D[t, x] $-\frac{1}{2} + \frac{3\,x}{4} - \frac{15\,x^2}{16} + \frac{35\,x^3}{32} + O[x]^4$

Integrate[t, x] $x - \frac{x^2}{4} + \frac{x^3}{8} - \frac{5\,x^4}{64} + \frac{7\,x^5}{128} + O[x]^6$

Additional functions are automatically expanded if they occur together with a series:

t + Sin[c x] $1 + \left(-\frac{1}{2} + c\right)x + \frac{3\,x^2}{8} + \left(-\frac{5}{16} - \frac{c^3}{6}\right)x^3 + \frac{35\,x^4}{128} + O[x]^5$

A series expansion is calculated automatically if we add a remainder:

Sin[c x] + O[x]^7 $c\,x - \frac{c^3\,x^3}{6} + \frac{c^5\,x^5}{120} + O[x]^7$

Unspecified functions are treated correctly:

Series[h[x], {x, 0, 4}]

$$h[0] + h'[0]\,x + \frac{1}{2}\,h''[0]\,x^2 + \frac{1}{6}\,h^{(3)}[0]\,x^3 + \frac{1}{24}\,h^{(4)}[0]\,x^4 + O[x]^5$$

■ Taylor Polynomial

> **Normal[series]**   Delete the remainder from the Taylor series

If we remove the remainder, we get the Taylor polynomial.

Normal[t] $1 - \frac{x}{2} + \frac{3\,x^2}{8} - \frac{5\,x^3}{16} + \frac{35\,x^4}{128}$

The result is an ordinary expression, and now all calculations are done as with usual expressions. The following figure shows Taylor polynomials of exp(x) at $x = 0$ up to degree 10:

```
t = Table[Series[Exp[x], {x, 0, i}] // Normal, {i, 0, 10}];

Plot[Evaluate[{Exp[x], t}], {x, -4, 4}, PlotRange → {-5.5, 20.5}];
```

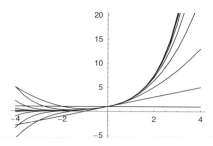

■ Special Power Series

Series can also calculate power series that contain negative and fractional powers:

```
Series[1 / Sin[x], {x, 0, 6}]
```
$$\frac{1}{x} + \frac{x}{6} + \frac{7\,x^3}{360} + \frac{31\,x^5}{15120} + O[x]^7$$

```
Series[Sin[Sqrt[x]], {x, 0, 4}]
```
$$\sqrt{x} - \frac{x^{3/2}}{6} + \frac{x^{5/2}}{120} - \frac{x^{7/2}}{5040} + O[x]^{9/2}$$

Sometimes an essential singularity is encountered, and the series cannot be calculated:

```
Series[Sin[1 / x], {x, 0, 3}]
```

Series::esss :

  Essential singularity encountered in $\mathrm{Sin}\left[\frac{1}{x} + O[x]^4\right]$. More…

$\mathrm{Sin}\left[\frac{1}{x}\right]$

However, this series can be calculated about infinity:

```
Series[Sin[1 / x], {x, Infinity, 8}]
```

$$\frac{1}{x} - \frac{1}{6}\left(\frac{1}{x}\right)^3 + \frac{1}{120}\left(\frac{1}{x}\right)^5 - \frac{\left(\frac{1}{x}\right)^7}{5040} + O\left[\frac{1}{x}\right]^9$$

If we want to develop a function of several variables into a series, we give the information about each variable in turn:

```
Series[f, {x, a, m}, {y, b, n}]
```

```
Series[Sin[x] Cos[y], {x, 0, 5}, {y, 0, 5}]
```

$$\left(1 - \frac{y^2}{2} + \frac{y^4}{24} + O[y]^6\right) x + \left(-\frac{1}{6} + \frac{y^2}{12} - \frac{y^4}{144} + O[y]^6\right) x^3 +$$
$$\left(\frac{1}{120} - \frac{y^2}{240} + \frac{y^4}{2880} + O[y]^6\right) x^5 + O[x]^6$$

> **ComposeSeries[series1, series2]** Replace the variable in **series1** with **series2**
>
> **InverseSeries[series, x]** Find the inverse series of **series**
>
> **Residue[f, {x, a}]** Residue of **f** when **x** equals **a**

With the **NumericalMath`NSeries`** package, we can calculate a numerical approximation to a power series expansion; with the **NumericalMath`NResidue`** package, we can calculate residues numerically.

16.2.2 Coefficients

> **SeriesCoefficient[series, n]** Give the coefficient of the nth-order term of **series**

    s = Series[Log[a + x], {x, 0, 4}]

$$\text{Log}[a] + \frac{x}{a} - \frac{x^2}{2\,a^2} + \frac{x^3}{3\,a^3} - \frac{x^4}{4\,a^4} + O[x]^5$$

    SeriesCoefficient[s, 4] $-\dfrac{1}{4\,a^4}$

If we want to see the coefficient of a *general* term of a power series expansion, we can use **SeriesTerm** from a package:

> *In the* **DiscreteMath`RSolve`** *package:*
>
> **SeriesTerm[f, {z, a, n}]** Give the coefficient of the **n**th term of the power series expansion of **f** expanded around **z** = **a**
>
> *Options:*
>
> **Assumptions** Assumptions about parameters; examples of values: **Automatic**, **{}**, **{n ≥ 0}**, **{n ≥ 0 && n ∈ Integers}**
>
> **IntegerFunctions** Whether to use **Mod**, **Even**, and **Odd** to express the result; possible values: **True**, **False**
>
> **SpecialFunctions** Whether to use special functions (such as Legendre polynomials) to express the result; possible values: **True**, **False**

Consider the following series:

    Series[1 / (1 + x^4), {x, 0, 12}]

    $1 - x^4 + x^8 - x^{12} + O[x]^{13}$

In this example, it is quite easy to write a general formula for the coefficients, but we try the package and ask for the *n*th coefficient:

    << DiscreteMath`RSolve`

    SeriesTerm[1 / (1 + x^4), {x, 0, n}]

    $\dfrac{1}{4}\,(-1)^{n/4}\,(1 + (-1)^n + \dot{\mathbb{i}}^n + \dot{\mathbb{i}}^{3n})$

Note that, unless otherwise stated with **Assumptions**, it is assumed that $n \geq 0$. In the next example, we have no assumptions:

```
SeriesTerm[1 / (1 + x^4), {x, 0, n}, Assumptions → {}]
```

$$\text{If}[n \ge 0, \text{If}[\text{Mod}[n, 4] == 0, (-1)^{n/4}, 0], 0]$$

This means that the nth coefficient is $(-1)^{n/4}$ for an n that is a multiple of four and zero otherwise. Lastly, we ask to not use **Mod**:

```
SeriesTerm[1 / (1 + x^4), {x, 0, n},
  Assumptions → {n ≥ 0 && n ∈ Integers}, IntegerFunctions → False]
```

$$\frac{1}{2} \left(\text{Cos}\left[\frac{n\pi}{4}\right] + \text{Cos}\left[\frac{3n\pi}{4}\right] \right)$$

16.2.3 Equations

Sometimes we want to determine conditions under which two power series are equivalent. For example, consider the following series:

```
s = Series[y[x], {x, 0, 3}] /. y[0] → 1
```

$$1 + y'[0] x + \frac{1}{2} y''[0] x^2 + \frac{1}{6} y^{(3)}[0] x^3 + O[x]^4$$

We want conditions for the derivatives of y at 0 under which the following equation is true (up to the remainder):

```
eqn = D[s, x] + s == Exp[x]
```

$$(1 + y'[0]) + (y'[0] + y''[0]) x + \left(\frac{y''[0]}{2} + \frac{1}{2} y^{(3)}[0] \right) x^2 + O[x]^3 == e^x$$

The expansion of e^x is as follows:

```
Series[Exp[x], {x, 0, 2}]       1 + x + x²/2 + O[x]³
```

We see that, for the equation **eqn** to be true, we must have $1 + y'(0) = 1$, and so on. With **LogicalExpand**, we can easily form such conditions.

LogicalExpand[series1 == series2]	Find conditions that make the equation true

We try this command with the preceding example:

```
LogicalExpand[eqn]
```

$$y'[0] == 0 \text{ \&\& } -1 + y'[0] + y''[0] == 0 \text{ \&\& } -\frac{1}{2} + \frac{y''[0]}{2} + \frac{1}{2} y^{(3)}[0] == 0$$

Now we can solve the equations:

```
sol = Solve[%]       {{y'[0] → 0, y''[0] → 1, y^(3)[0] → 0}}
```

We can also directly apply **Solve** to the equation **eqn** (without first using **LogicalExpand**):

```
Solve[eqn]       {{y'[0] → 0, y''[0] → 1, y^(3)[0] → 0}}
```

If we now insert these values into **s**, we get, in fact, a (crude) series solution for the differential equation $y'(x) + y(x) = e^x$ with the initial value $y(0) = 1$:

```
s /. sol[[1]] // Normal       1 + x²/2
```

For more about series solutions of differential equations, see Section 23.2.2, p. 605.

16.3 Limits

16.3.1 Symbolic Limits

> **Limit[f, x → a]** Limit of **f** as **x** approaches **a**
>
> *Options:*
> **Direction** Direction from which **a** is approached; possible values: **Automatic** (usually means **-1**), **-1** (from above), **1** (from below)
> **Assumptions** Assumptions for parameters; examples of values: **{}**, **a > 0**
> **Analytic** Whether unknown functions are treated as analytic; possible values: **False**, **True**

Write the arrow as **->** (*Mathematica* then replaces it with a true arrow →). The default value **Automatic** for **Direction** means **Direction → -1** (i.e., from above or from larger values) except for limits at infinity, where it means **Direction → 1**. For example:

$\qquad$ **Limit[(Cos[x] - 1) / x^2, x → 0]** $\qquad -\dfrac{1}{2}$

$\qquad$ **Limit[(1 + c / x) ^x, x → ∞]** $\qquad e^c$

$\qquad$ **Limit[x^y, x → ∞, Assumptions → y < 0]** $\qquad 0$

Derivatives are, by definition, obtained from limits:

$\qquad$ **Limit[(Sin[x + h] - Sin[x]) / h, h → 0]** $\qquad$ **Cos[x]**

The following limit is not unique, and we get a whole interval:

$\qquad$ **Limit[Sin[1 / x], x → 0]** $\qquad$ **Interval[{-1, 1}]**

■ **Direction**

Consider the following discontinuous function:

$\qquad$ **g = 1 / (2^ (1 / x) + 1);**

$\qquad$ **Plot[g, {x, -3, 3}];**

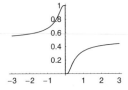

We see that the function has two different limits—1 and 0—at $x = 0$. *Mathematica* gives, by default, the limit from above:

$\qquad$ **Limit[g, x → 0]** $\qquad$ 0

If we suspect that the limit may be different depending on the direction, we can also calculate the limit from below:

$\qquad$ **Limit[g, x → 0, Direction → 1]** $\qquad$ 1

■ Unknown Functions

If we have an expression containing unknown functions, the default is that **Limit** does not assume that they are analytic and, consequently, the limit is not calculated:

$\texttt{Limit[(f[x + h] - f[x]) / h, h → 0]}$ $\quad \texttt{Limit}\left[\dfrac{-f[x] + f[h + x]}{h}, h → 0\right]$

Assuming that **f** is analytic, we get a result:

$\texttt{Limit[(f[x + h] - f[x]) / h, h → 0, Analytic → True]}$ $\quad \texttt{f'[x]}$

■ Expanding the Capabilities of Limit

> *In the* **Calculus`Limit`** *package:*
>
> Enhancements to **Limit** to calculate more limits, for example, limits of many special functions

The **Calculus`Limit`** package expands the capabilities of **Limit** to handle, in particular, a wide variety of special functions. Thus, if **Limit** cannot handle your limit, try loading this package and then using **Limit** again. An example is in Section 15.2.2, p. 395.

16.3.2 Numerical Limits

> *In the* **NumericalMath`NLimit`** *package:*
>
> **NLimit[f, x → a]** Numerical limit of **f** as **x** approaches **a**
>
> *Options:*
> **Direction** Direction from which **a** is approached; possible values: **Automatic** (usually means **-1**), **-1** (from above), **1** (from below)
> **WorkingPrecision** Precision used in internal computations; examples of values: **MachinePrecision** (✿5), **20**
> **Scale** Initial step size; default value: **1**
> **Terms** Number of values calculated; default value: **7**
> **Method** Extrapolation method; possible values: **EulerSum**, **SequenceLimit**
> **WynnDegree** Degree to use in **SequenceLimit**; default value: **1**

NLimit works by calculating a sequence of values for the function with successively smaller step sizes and then extrapolating to the limit. The initial step size is **Scale** (default is 1), and a total of **Terms** (default is 7) values is calculated by successively dividing the previous step size by 10. The sequence of values is then extrapolated by applying either a generalized Euler transformation (**Method → EulerSum**; this is the default) or Wynn's ϵ-algorithm (**Method → SequenceLimit**).

It turns out that the default method does not calculate the limit of the first example in Section 16.3.1, but the other method works:

```
<< NumericalMath`NLimit`
```

```
NLimit[(Cos[x] - 1) / x^2, x → 0, Method → SequenceLimit]     - 0.5
```

Integral Calculus

Introduction

> *But just as easy as it is to find the differential of a given quantity, so it is difficult to find the integral of a given differential. Moreover, sometimes we cannot say with certainty whether the integral of a given quantity can be found or not. — Johann Bernoulli*

Bernoulli was right. Even with the extensive capability of *Mathematica*'s **Integrate**, your integral may remain unevaluated. However, the command has such power and knowledge that you will certainly be more than satisfied with its performance.

With definite integration, we have a better situation than with indefinite integration: if we do not get an exact answer for a definite integral, we can resort to numerical methods (numerical quadrature) and get an approximate answer with **NIntegrate**. This command uses an advanced adaptive Gauss–Kronrod method.

In this chapter, we will also consider sums and products.

Many transforms of functions are based on integrals or sums. The best known transform is the Laplace transform. Solving differential and partial differential equations with this transform is considered in Sections 23.2.1, p. 604, and 24.1.2, p. 643. In Section 25.1.3, p. 677, we show you how to use the Z transform to solve difference equations. In Section 27.3.2, p. 760, we use the discrete Fourier transform to smooth data.

We now relate a story about Riemann. Riemann consulted a doctor about his diet. He was told to reduce the amount of food he ate at each meal but to increase the number of meals. He proceeded to do so and ultimately ate infinitesimal amounts infinitely often, and he found that his weight did not change. Shortly after this, he gave a precise definition of a definite integral.

17.1 Integration

17.1.1 Indefinite Integration

> `Integrate[f, x]` Indefinite integral of `f` with respect to `x`

Note that the constant of integration is not shown:

> `Integrate[x / (a + b x), x]` $\dfrac{x}{b} - \dfrac{a \, \mathrm{Log}[a + b \, x]}{b^2}$

If the derivative of the integral is the same as the integrand, this supports the correctness of the integral:

> `D[%, x] // Simplify` $\dfrac{x}{a + b \, x}$

Remember that integrals can also be entered with the aid of the **BasicInput** palette (see Section 1.4.1, p. 17). Still another way to enter integrals is to write 2D input directly with the keyboard (see Section 3.3.3, p. 70): the integral sign $\int$ can be written as ESCintESC, and the d appearing before the variable of the integration is entered as ESCddESC:

> $\int (a + x \, \mathbf{Exp[x]}) \; d\mathbf{x}$ $e^x \, (-1 + x) + a \, x$

Derivative can also be used to calculate integrals: `Derivative[-n][f][x]` gives the nth antiderivative (or indefinite integral) of `f` at `x` (recall from Section 16.1.1, p. 418, that `Derivative[n][f][x]` gives the nth derivative of `f` at `x`). For example:

> `f[x_] := x^2`

> `Table[Derivative[-n][f][x], {n, 0, 3}]`

> $\left\{ x^2, \; \dfrac{x^3}{3}, \; \dfrac{x^4}{12}, \; \dfrac{x^5}{60} \right\}$

■ Special Values of Parameters

An important point is the following: *Mathematica* assumes that all parameters in the integrand have *generic* values. For example, the first integral above is correct only if b is not 0. If $b = 0$, the integral is, of course, $x^2 / (2 a)$. *Mathematica* does not tell you for what values the result holds; you have to check special cases directly. Here is another example:

> **Integrate[Log[x]^n / x, x]** $\quad \dfrac{\text{Log}[x]^{1+n}}{1+n}$

This holds for a generic n, that is, for an n not equal to -1. If $n = -1$, the result is different:

> **Integrate[Log[x]^(-1)/x, x]** $\quad$ Log[Log[x]]

However, for *definite* integrals, *Mathematica* can give conditions under which the integral converges (see Section 17.1.3, p. 437).

■ Special Functions

Many functions do not have integrals in terms of elementary functions. The integral may, however, be representable in terms of some special functions:

> **Integrate[Exp[-x^2], x]** $\quad \dfrac{1}{2}\sqrt{\pi}\ \text{Erf}[x]$

The result of the following integral contains an elliptic integral:

> **Integrate[1 / Sqrt[1 + x^3], x]**

$$\frac{1}{3^{1/4}\sqrt{1+x^3}}\left(2\ (-1)^{1/6}\sqrt{(-1)^{5/6}\ (-1+(-1)^{1/3}\ x)}\right.$$

$$\left.\sqrt{1+(-1)^{1/3}\ x+(-1)^{2/3}\ x^2}\ \text{EllipticF}\left[\text{ArcSin}\left[\frac{\sqrt{-(-1)^{5/6}\ (1+x)}}{3^{1/4}}\right],\ (-1)^{1/3}\right]\right)$$

The integrals of **Abs** and **Sign** are not calculated. Load the **ProgrammingIn`** **Mathematica`Abs`** package to get rules for the integral and derivative of these functions:

> **<< ProgrammingInMathematica`Abs`**

> **{Integrate[Abs[x], x], Integrate[Sign[x], x]}**

$$\left\{\frac{1}{2}\ x\ \text{Abs}[x],\ \text{Abs}[x]\right\}$$

■ No Results

If *Mathematica* does not do the integration, there are two possibilities: the integral really cannot be calculated in terms of any of the built-in elementary and special functions of *Mathematica*, or the integral can be calculated but *Mathematica* did not succeed at doing so. If you think the integral could be calculated, try helping *Mathematica*. For example, try special values for the parameters.

If *Mathematica* cannot calculate the integral, it simply writes the given command as such:

> **Integrate[Log[x^x]^n, x]** $\quad \displaystyle\int \text{Log}[x^x]^n\ dx$

Mathematica can calculate this integral for any given positive integer value of n. For example:

> **Integrate[Log[x^x]^2, x]**

$$\frac{1}{54}\ x$$

$$\left(4\ x^2 + 18\ x^2\ \text{Log}[x]^2 + 3\ x\ \text{Log}[x]\ (5\ x - 18\ \text{Log}[x^x]) - 27\ x\ \text{Log}[x^x] + 54\ \text{Log}[x^x]^2\right)$$

You can also try to write the integrand in another form (e.g., with **Apart** you get partial fraction expansions), or try integration by parts or change of variable (see examples in Section 17.1.2). If you have a definite integral, you can also try numerical quadrature (see Section 17.2).

If the integrand contains unknown functions, *Mathematica* may not be able to calculate the integral:

> **Integrate[p[x] p''[x], x]** $\int p[x]\, p''[x]\, dx$

However, sometimes it succeeds:

> **Integrate[2 p[x] p'[x], x]** $p[x]^2$

■ Wrong Results

All systems, including *Mathematica*, can sometimes give wrong results due to inadequate or erroneous code. For an example of a wrong integral, see **Generalizing the Integrand** in Section 17.1.2, p. 435. It is always good to check the result given by *Mathematica*. For an indefinite integral, we can differentiate the result (although this test is not sufficient; see the mentioned example). For a definite integral, we can also use numerical quadrature (for specific values of the parameters) and compare the results.

17.1.2 Special Techniques

■ Integration by Parts

Mathematica can do the following integral:

> **Integrate[Log[x] x^n, x]** $\dfrac{x^{1+n}\,(-1 + (1+n)\,\text{Log}[x])}{(1+n)^2}$

However, let us try integration by parts (our example is from Spiegel 1963, p. 91). The rule can be written as follows:

> **Integrate[u dv, x] = u v - Integrate[v du, x]**

Here, **du** and **dv** are derivatives of **u** and **v**. Define **u** and **dv**:

> **u = Log[x]; dv = x^n;**

Then calculate the derivative of **u** and the integral of **dv**:

> **{du = D[u, x], v = Integrate[dv, x]}** $\left\{ \dfrac{1}{x},\ \dfrac{x^{1+n}}{1+n} \right\}$

The original integral is then as follows:

> **u v - Integrate[v du, x]** $-\dfrac{x^{1+n}}{(1+n)^2} + \dfrac{x^{1+n}\,\text{Log}[x]}{1+n}$

■ Change of Variable

Let us integrate, by change of variable, the following function:

> **f = a^Sqrt[b + c x];**

(This example is a generalization of an example in Spiegel 1963, p. 91.) Denote $y = \sqrt{b + c\,x}$ and express this as an equation:

```
eqn = y == Sqrt[b + c x]      y == √b + c x
```

Solve this for **x**:

```
xx = x /. Solve[eqn, x][[1]]
```
$$\frac{-b + y^2}{c}$$

Insert this into the integrand and into the differential:

```
g = PowerExpand[(f /. x → xx) D[xx, y]]
```
$$\frac{2 \, a^y \, y}{c}$$

(Here we used **PowerExpand** to simplify $\sqrt{y^2}$ to y, which is true if $y \geq 0$.) Let us now try integration:

```
iy = Integrate[g, y] // Simplify
```
$$\frac{2 \, a^y \, (-1 + y \, \text{Log}[a])}{c \, \text{Log}[a]^2}$$

Lastly we go back to the variable x:

```
ix = iy /. ToRules[eqn] // Simplify
```
$$\frac{2 \, a^{\sqrt{b+c\,x}} \, (-1 + \sqrt{b + c\,x} \, \text{Log}[a])}{c \, \text{Log}[a]^2}$$

(Here **ToRules[eqn]** writes the equation **eqn** as the rule **{y → Sqrt[b + c x]}**.) By the way, *Mathematica* does this integral, too:

```
Integrate[f, x] // Simplify
```
$$\frac{2 \, a^{\sqrt{b+c\,x}} \, (-1 + \sqrt{b + c\,x} \, \text{Log}[a])}{c \, \text{Log}[a]^2}$$

■ Generalizing the Integrand

Consider the following function:

```
f = Sqrt[1 + Sin[x]];
```

```
Plot[f, {x, -π / 2, 7 π / 2}];
```

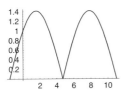

Mathematica gives the following integral:

```
int = Integrate[f, x] // Simplify
```
$$\left(-2 + \frac{4 \, \text{Sin}[\frac{x}{2}]}{\text{Cos}[\frac{x}{2}] + \text{Sin}[\frac{x}{2}]}\right) \sqrt{1 + \text{Sin}[x]}$$

The integral passes the derivative test:

```
D[int, x] // FullSimplify        √1 + Sin[x]
```

However, the integral is wrong. This may be seen from the plot of the integral:

```
Plot[int, {x, -π / 2, 7 π / 2}];
```

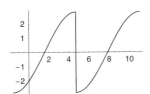

The integral is not continuous, as it should be, and not nondecreasing, as it also should be (as the integrand is nonnegative). To get the correct result, ask for the integral for a slightly *more general* integrand containing a parameter *a* (if *a* = 1, we get the original integrand):

```
int1 = Integrate[Sqrt[a + Sin[x]], x] // Simplify
```

$$-\frac{2 \, \text{EllipticE}[\frac{1}{4} \, (\pi - 2\,x), \frac{2}{1+a}] \, \sqrt{a + \text{Sin}[x]}}{\sqrt{\frac{a + \text{Sin}[x]}{1+a}}}$$

Give *a* the value 1:

```
int2 = int1 /. a → 1
```
$$-2\sqrt{2} \, \text{EllipticE}\left[\frac{1}{4} \, (\pi - 2\,x), 1\right]$$

This is correct, as can be seen from the derivative test and from the plot:

```
Simplify[D[int2, x]]
```
$$\sqrt{1 + \text{Sin}[x]}$$

```
Plot[int2, {x, -π / 2, 7 π / 2}];
```

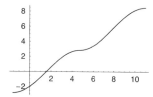

The slightly more general integrand apparently forces **Integrate** to use more advanced methods. In this example, these methods succeeded in giving the correct result.

```
Remove["Global`*"]
```

17.1.3 Definite Integration

> **Integrate[f, {x, a, b}]** Definite integral of **f** when **x** varies from **a** to **b**

```
Integrate[Exp[-x^2], {x, -∞, ∞}]     √π
```

(Recall that ∞ can be written as [ESC]inf[ESC].) Definite integrals can also be calculated by the **BasicInput** palette, or you can write 2D input directly with the keyboard (see Section 3.3.3, p. 70). For example, consider the following integral:

$$\int_0^\pi (\text{Sin}[x] + \text{Log}[x]) \, dx$$

To write this expression, type [ESC]int[ESC][CTRL]-[-]0[CTRL]%[ESC]p[ESC][CTRL].(Sin[x]+Log[x])[ESC]dd[ESC]x.

If a definite integral is not calculated, we can ask for a numerical value (see Section 17.2.1, p. 442):

```
Integrate[Sin[Sin[x]], {x, 0, π / 3}]     
```
$$\int_0^{\frac{\pi}{3}} \text{Sin}[\text{Sin}[x]] \, dx$$

```
% // N     0.466185
```

■ Simplifying the Result

Let us try to derive formula 18.26 in Spiegel (1999, p. 107). This formula says that the integral of $\sin(m\,x)\sin(n\,x)$ (m and n are integers) over $(0, \pi)$ is zero unless $m = n$, in which case the integral is $\pi/2$. First we compute the integral with the general m and n (❀4!):

```
int = Integrate[Sin[m x] Sin[n x], {x, 0, π}]
```

$$\frac{n\,\text{Cos}[n\,\pi]\,\text{Sin}[m\,\pi] - m\,\text{Cos}[m\,\pi]\,\text{Sin}[n\,\pi]}{m^2 - n^2}$$

Then we assume that m and n are integers:

```
Simplify[int, {m, n} ∈ Integers]     0
```

However, the result obtained is a generic result that is valid only for most values of m and n. Specifically, it does not hold for the case $m = n$. We calculate and simplify this integral separately:

```
int2 = Integrate[Sin[m x]^2, {x, 0, π}]
```
$$\frac{\pi}{2} - \frac{\text{Sin}[2\,m\,\pi]}{4\,m}$$

```
Simplify[int2, m ∈ Integers]
```
$$\frac{\pi}{2}$$

■ Conditions of Convergence

For definite integrals, *Mathematica* can give conditions under which the integral converges. The following integral converges only if the real part of **a** is positive (❀4!):

```
Integrate[Exp[-x / a], {x, 0, ∞}]
```

$$\text{If}\left[\text{Re}[a] > 0,\ a,\ \text{Integrate}\left[e^{-\frac{x}{a}},\ \{x, 0, \infty\},\ \text{Assumptions} \to \text{Re}[a] \le 0\right]\right]$$

The result is, however, a *generic* result valid only if **a** is not zero (*Mathematica* does not tell you such conditions). Here is another example (the definition of the beta function) (❀4!):

```
Integrate[t ^ (a - 1) (1 - t) ^ (b - 1), {t, 0, 1}]
```

$$\text{If}\left[\text{Re}[a] > 0\ \&\&\ \text{Re}[b] > 0,\ \frac{\text{Gamma}[a]\,\text{Gamma}[b]}{\text{Gamma}[a + b]},\right.$$
$$\left.\text{Integrate}\left[(1 - t)^{-1+b}\,t^{-1+a},\ \{t, 0, 1\},\ \text{Assumptions} \to \text{Re}[a] \le 0\ ||\ \text{Re}[b] \le 0\right]\right]$$

The integral converges only if the real parts of **a** and **b** are positive. Here is a third example (❀4!):

```
Integrate[1 / x^2, {x, a, 1}, Assumptions → a ∈ Reals]
```

$$\text{If}\left[a \ge 0,\ -1 + \frac{1}{a},\ \text{Integrate}\left[\frac{1}{x^2},\ \{x, a, 1\},\ \text{Assumptions} \to a \in \text{Reals}\ \&\&\ a < 0\right]\right]$$

The output tells us that the integral is $-1 + \frac{1}{a}$ if $a \ge 0$. However, for $a = 0$, the integral does not converge:

```
Integrate[1 / x^2, {x, 0, 1}]
```

Integrate::idiv : Integral of $\frac{1}{x^2}$ does not converge on $\{0, 1\}$. **More…**

$$\int_0^1 \frac{1}{x^2}\,dx$$

This example shows that the conditions of convergence given by *Mathematica* should be checked.

■ Options

> *Options for* **Integrate** *in definite integration:*
>
> **Assumptions** Assumptions about parameters; examples of values: **{ }, n > 0, n ∈ Integers, n > 0 && n ∈ Integers**
>
> **GenerateConditions** Whether to generate answers containing conditions for the parameters; possible values: **Automatic** (usually means **True**), **True**, **False**
>
> **PrincipalValue** Whether to find the Cauchy principal value; possible values: **False**, **True**

The default value for **Assumptions** is an empty list (**{ }**) and means that no assumptions are made. The assumptions can be equations, inequalities, and domain specifications (see Section 12.2.1, p. 305, for various domains) and their logical combinations. An assumption could be, for example, **x ∈ Reals** (write ∈ as ⎡ESC⎤elem⎡ESC⎤). The default value **Automatic** for **GenerateConditions** essentially means **True**.

We try some assumptions for the integrals we calculated above in **Conditions of Convergence**:

    Integrate[Exp[-x / a], {x, 0, ∞}, Assumptions → a > 0]   a

    Integrate[t ^ (a - 1) (1 - t) ^ (b - 1), {t, 0, 1}, Assumptions → a > 0 && b > 0]
$$\frac{\text{Gamma}[a]\ \text{Gamma}[b]}{\text{Gamma}[a + b]}$$

We can ask to not print conditions of convergence:

    Integrate[Exp[-x / a], {x, 0, ∞}, GenerateConditions → False]   a

Some integrals that do not have a finite value in the usual (Riemannian) sense over intervals containing a point of singularity may have a finite Cauchy principal value. This value for the integral is obtained by deleting a small interval centered at the singular point and then taking the limit of the integral as the length of the interval goes to zero. With the **NumericalMath`CauchyPrincipalValue`** package, this calculation can also be done numerically.

■ Advanced Integrals

The integrand may contain unknown functions:

    Integrate[f[x, t], {x, a[t], b[t]}] $\int_{a[t]}^{b[t]} f[x, t]\ dx$

We can calculate the derivative of this with respect to *t*:

    D[%, t] $\int_{a[t]}^{b[t]} f^{(0,1)}[x, t]\ dx - f[a[t], t]\ a'[t] + f[b[t], t]\ b'[t]$

Here, $f^{(0,1)}[x, t]$ means derivative with respect to *t*.

The integrand may also contain such functions as **Abs**, **Sign**, **UnitStep**, **Min**, and **Max**:

    Integrate[Abs[x] + Sign[x] + Min[1 / 2, Sin[x]], {x, -1, 1}]
$$\frac{3}{2} - \frac{\sqrt{3}}{2} - \frac{\pi}{12} + \text{Cos}[1]$$

However, if the integrand is a piecewise-defined function written with **If** or **Which**, then **Integrate** does not work. To integrate such functions, load the package **Calculus`** **Integration`** (see Section 17.1.4, p. 440):

> `<< Calculus`Integration``

> `Integrate[If[0 ≤ x < 1, 1, Exp[x]], {x, 0, 2}]` $1 + (-1 + e) \, e$

> `Integrate[Which[x^2 > x^3 - x, x^2, True, x^3 - 1], {x, 0, 3}]`

> $\frac{1}{3} \, (2 + \sqrt{5}\,) + \frac{1}{8} \, (135 + \sqrt{5}\,)$

■ Change of Variable

Let us calculate the integral of the following function over $(0, \pi)$:

> `f = x Sin[x] / (1 + Cos[x]^2);`

According to Spiegel (1963, p. 92), the integral has the value $\pi^2/4$. For this integral, *Mathematica* gives a long expression of about one page containing **ArcSin**, **ArcCos**, **ArcTan**, **Log**, and **PolyLog** (we do not show the result here):

> `int = Integrate[f, {x, 0, π}]; // Timing`

> `{471.81 Second, Null}`

The simplified result is

> `(int2 = int // ComplexExpand // FullSimplify) // Timing`

> $\{24.73 \text{ Second},$
>
> $\frac{1}{16} \left(4\,\pi^2 - \text{Log}[2]^2 + \text{Log}[3 - 2\,\sqrt{2}\,]^2 + (\text{Log}[16] - 4\,\text{Log}[2 + \sqrt{2}\,]) \, \text{Log}[2 + \sqrt{2}\,] \right) \}$

(*Mathematica* 4.2 gives the simple result $\pi^2/4$.) *Mathematica* was not able to simplify the result to $\pi^2/4$, but the result is correct:

> `int2 - π^2/4 // N` 4.44089×10^{-16}

Let us try the same technique as Spiegel used to directly obtain the simple result $\pi^2/4$. Do a change of variable $y = \pi - x$:

> `g = (f /. x → π - y) D[π - y, y]` $- \dfrac{(\pi - y)\,\text{Sin}[y]}{1 + \text{Cos}[y]^2}$

Expand this expression:

> `g = Expand[g]` $- \dfrac{\pi\,\text{Sin}[y]}{1 + \text{Cos}[y]^2} + \dfrac{y\,\text{Sin}[y]}{1 + \text{Cos}[y]^2}$

The last term is in the same form as the original function. If the original integral is **int**, then the integral of the last term is **-int**, because the integration with respect to y goes from π to 0. Integrate the first term:

> `i1 = Integrate[First[g], {y, π, 0}]` $\dfrac{\pi^2}{2}$

We now have the equation **int == i1 - int**. Solving the equation for **int**, we get $\pi^2/4$.

17.1.4 Multiple Integrals

■ Basic Multiple Integrals

Consider the following multiple integral:

$$\int_a^b \left(\int_x^{x+1} 12\, x\, y\, dy \right) dx$$

$$-3\, a^2 - 4\, a^3 + b^2\, (3 + 4\, b)$$

This can be calculated with nested **Integrate** commands:

```
Integrate[Integrate[12 x y, {y, x, x + 1}], {x, a, b}]
```

$$-3\, a^2 - 4\, a^3 + b^2\, (3 + 4\, b)$$

However, **Integrate** also has special forms for multiple integrals:

```
Integrate[f, {x, a, b}, {y, c, d}]
Integrate[f, {x, a, b}, {y, c, d}, {z, e, f}]
```

In the first command, the integral is first calculated with respect to **y** and then with respect to **x**. The endpoints **c** and **d** may be functions of **x**. The first command thus means the following:

$$\int_a^b \left(\int_c^d f\, dy \right) dx$$

For example:

```
Integrate[12 x y, {x, a, b}, {y, x, x + 1}]
```

$$-3\, a^2 - 4\, a^3 + b^2\, (3 + 4\, b)$$

■ Advanced Multiple Integrals

> *In the* **Calculus`Integration`** *package:*
>
> Enhancements to **Integrate** and **NIntegrate** to calculate definite integrals 1) of piece-wise functions and 2) over regions described by logical combinations of algebraic inequalities
>
> **Boole[ineqs]**   The characteristic function of the set defined by inequalities **ineqs** (i.e., **Boole[True]** is 1 and **Boole[False]** is 0)

The **Calculus`Integration`** package extends the capabilities of **Integrate** and **NIntegrate**. In both univariate and multivariate cases we can integrate piecewise-defined functions containing **If** and **Which** (see Section 17.1.3, p. 439). In multivariate cases, the package also allows for the easy definition of regions of integration by **Boole** and easy multiple integration of functions over such regions. The package uses **Cylindrical`** **AlgebraicDecomposition** to handle algebraic inequalities.

To try the package, first define two inequalities that determine the area of integration:

```
ineqs = x^2 + y^2 < 1 && (x - 1 / 2) ^2 + (y - 1 / 2) ^2 > 1 / 30
```

$$x^2 + y^2 < 1 \&\& \left(-\frac{1}{2} + x\right)^2 + \left(-\frac{1}{2} + y\right)^2 > \frac{1}{30}$$

With a package, we can nicely plot the corresponding area (see Section 5.3.3, p. 127):

```
<< Graphics`InequalityGraphics`

InequalityPlot[ineqs, {x, 0, 1}, {y, 0, 1}];
```

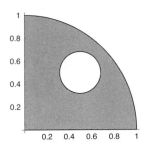

Next we define the function to be integrated. It is $x^3 + y^3$ in the area where the inequalities hold; elsewhere the function is 0:

```
f := If[ineqs, x^3 + y^3, 0]

Plot3D[f, {x, 0, 1}, {y, 0, 1}, Ticks → None];
```

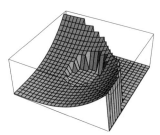

Now we load the integration package and calculate the integral:

```
<< Calculus`Integration`

Integrate[f, {x, 0, 1}, {y, 0, 1}] // FullSimplify // Timing
```

$$\left\{73.45 \text{ Second}, \frac{4}{15} - \frac{11\,\pi}{1200}\right\}$$

Another way to define the function is the use of **Boole**:

```
g := (x^3 + y^3) Boole[ineqs]

Integrate[g, {x, 0, 1}, {y, 0, 1}] // FullSimplify
```

$$\frac{4}{15} - \frac{11\,\pi}{1200}$$

Here the region of integration is the intersection of the region defined by **Boole** and the region defined by the integration bounds.

As another example, we calculate the volume of the unit sphere:

Integrate[Boole[x^2+y^2+z^2 ≤ 1], {x, -1, 1}, {y, -1, 1}, {z, -1, 1}]

$$\frac{4\,\pi}{3}$$

For more about the integration package, look at the **Calculus`Integration`** package in the *Help Browser*.

17.2 Numerical Quadrature

17.2.1 Basics

If **Integrate** does not calculate your definite integral, apply numerical integration. You can first try exact integration, and if a result is not obtained, ask for a numerical value.

Integrate[f, {x, a, b}] Try exact integration
% // N If exact integration did not succeed, use numerical methods

We try the following integral:

i = Integrate[Exp[x^x], {x, 0, 1}] $\int_0^1 e^{x^x}\, dx$

Exact integration did not succeed, and so we ask for a numerical value:

N[i]   2.19754

We can also directly resort to numerical methods.

NIntegrate[f, {x, a, b}] Calculate the integral using numerical methods

NIntegrate[Exp[x^x], {x, 0, 1}]   2.19754

If we use the method **% // N**, eventually **NIntegrate** will be applied.

The integration interval can be infinite, and the endpoints can be singular (for singular midpoints, see Section 17.2.3, p. 448):

NIntegrate[Exp[-x^2] Log[x], {x, 0, ∞}]   −0.870058

Multiple integrals are calculated in the familiar way.

NIntegrate[f, {x, a, b}, {y, c, d}] Integrate first with respect to **y** and then with respect to **x**

NIntegrate[Sin[x y], {x, 0, π}, {y, 0, x}]   1.45034

For advanced multiple numerical integration, load the package **Calculus`Integra‹ tion`** and use **NIntegrate** (see Section 17.1.4, p. 440).

17.2.2 Options

With **NIntegrate**, we can use many quadrature methods. The options of **NIntegrate** can be divided into two groups: options that are common for all methods and options that are available for special methods. We divide the latter options further into two groups: options for methods not of the Monte Carlo type and options for methods of the Monte Carlo type. In addition, we have an option to specify the method.

Options for **NIntegrate** *common for all methods:*

WorkingPrecision Precision used in internal computations; examples of values:
  MachinePrecision (❁5), **20**
PrecisionGoal If the value of the option is **p**, the relative error of the integral should
  be of the order 10^{-p}; examples of values: **Automatic** (usually means **6**; is **2** for Monte
  Carlo methods), **10**
AccuracyGoal If the value of the option is **a**, the absolute error of the integral should
  be of the order 10^{-a}; examples of values: ∞, **10**
EvaluationMonitor (❁5) Command to be executed after each evaluation of the
  function to be integrated; examples of values: **None**, **++n**, **AppendTo[points, x]**
Compiled Whether the integrand should be compiled or not; possible values: **True**,
  False

An option for **NIntegrate** *specifying the method:*

Method Method to use; possible values: **Automatic** (means **GaussKronrod** or **Multi**-
  Dimensional), **GaussKronrod**, **Oscillatory**, **MultiDimensional**, **DoubleExponen**-
  tial, **Trapezoidal**, **MonteCarlo**, **QuasiMonteCarlo**

Additional options for **NIntegrate**, *if the method is* **GaussKronrod**, **Oscillatory**, **Multi**-
  Dimensional, **DoubleExponential**, *or* **Trapezoidal**:

MinRecursion Minimum number of recursive subdivisions; examples of values: **0**, **3**
MaxRecursion Maximum number of recursive subdivisions; examples of values: **6**, **10**
SingularityDepth Number of recursive subdivisions before changing variables;
  examples of values: **4**, **6**; this option is available for **GaussKronrod**, **Oscillatory**, and
  MultiDimensional
GaussPoints Initial number of sample points; examples of values: **Automatic** (usually
  means **5**), **7**; this option is available for **GaussKronrod** and **Oscillatory**

An additional option for **NIntegrate**, *if the method is* **MonteCarlo** *or* **QuasiMonteCarlo**:

MaxPoints Maximum total number of sample points; examples of values: **Automatic**
  (usually means **50000**), **70000**

■ Options Related to Precision

The default value of **WorkingPrecision** is **MachinePrecision** (16 in *Mathematica* 4) so that **NIntegrate** uses, by default, the usual fixed-precision arithmetic with 16-digit precision. Give another value for this option if you want the calculations to be done with variable-precision arithmetic (see Section 11.3.1, p. 292).

The method used by **NIntegrate** stops improving the result as soon as either **PrecisionGoal** or **AccuracyGoal** is met.

The default value **Automatic** of **PrecisionGoal** means 6 so that **NIntegrate** tries to give you an answer having a *relative error* of the order 10^{-6} (this is only a goal; the actual relative error can be smaller or larger). If you have defined your own value for **WorkingPrecision**, then **PrecisionGoal** has the default value **WorkingPrecision - 10**.

The default value of **AccuracyGoal** is ∞. Such a goal will be never satisfied, so the method actually stops as soon as **PrecisionGoal** is met. In practice, this means that the *absolute error* is, by default, not used as a goal. If you define a finite value of, for example, 6 for **AccuracyGoal**, then the method stops if it gets a result having an absolute error at most of the order 10^{-6}.

These three options are explained in detail in Section 11.3.1, p. 292. If you have skipped this section, now is a good time to visit it.

■ Options Related to the Method

The default value of the option **Method** is **Automatic**, and this means that for one-dimensional integrals the method is **GaussKronrod** and for multidimensional integrals the method is **MultiDimensional**. These two methods (and only these) are adaptive, which means that they are able to put special effort into difficult subintervals in which the function varies greatly. It is generally wise to let **Method** have the default value **Automatic**.

In all but the two Monte Carlo methods, **NIntegrate** first samples the integrand at a sequence of points (in the Gauss-Kronrod method, the number of points can be set with the option **GaussPoints**). If the precision or accuracy goal is not satisfied, more points are sampled by recursively subdividing the integration region. **MinRecursion** is the minimum number of subdivisions (default value is 0), and **MaxRecursion** is the maximum number of subdivisions (default value is 6). With some methods, **SingularityDepth** (default value is 4) can be used to determine how many subdivisions **NIntegrate** should try before it concludes that the integrand is "blowing up" at one of the endpoints and makes a change of variables (for singular midpoints, see Section 17.2.3, p. 448).

In the Monte Carlo methods, the function is sampled at a large number of points. The maximum number of points is determined by **MaxPoints**; its default value is **Automatic** and means 50,000. If you give a value for **MaxPoints**, then automatically **QuasiMonteCarlo** is used (unless you do not specify **MonteCarlo** as the method). If we specify the method to be **MonteCarlo[n]** with a given number **n**, the random number generator is used with the seed **n**. The two Monte Carlo methods have 2 as the default value of **PrecisionGoal**, which means that the result will not be very precise.

Next we consider each method in turn.

■ Methods

GaussKronrod is an adaptive Gaussian quadrature with error estimation based on evaluation at Kronrod points. The option **GaussPoints** gives the initial number of sample points. Its default value **Automatic** means **Floor[WorkingPrecision/3]**. For example, if the calculations are done with the usual 16-digit precision, then **GaussPoints** has the value 5. If these points are not sufficient to obtain the desired precision, the interval of integration is bisected, and the integrand is sampled anew on each subinterval. If the integration does not succeed with sufficient precision in a particular subinterval, this subinterval is bisected again. This procedure of bisection is continued until sufficient precision is obtained in the whole integration interval.

    Oscillatory uses a transformation to handle integrals containing trigonometric and Bessel functions. This method is designed for integrands of the form $f(x)\,w(x)$, where w is **Sin**, **Cos**, **BesselJ**, or **BesselY** (the last two with positive parameters) and the range of integration is in the form $(-\infty, a]$, $[a, \infty)$, or $(-\infty, \infty)$. The function f can be any nonconstant analytic function. Oscillatory integrands are considered in Section 17.2.4, p. 449.

    MultiDimensional is an adaptive Genz–Malik algorithm. Note that the other methods can also be used for multidimensional integrals.

    DoubleExponential is a nonadaptive double-exponential quadrature. This method may be used if the integral does not decay very fast: the method does a transformation that makes the integral decay rapidly.

    Trapezoidal is the elementary trapezoidal method; it can be useful for integrating analytic periodic functions over one period.

    MonteCarlo is a nonadaptive Monte Carlo method and uses pseudorandom points in the integration domain. **QuasiMonteCarlo** is a nonadaptive Halton–Hammersley–Wozniakowski algorithm and uses quasi-random number sequences (nonrandom sequences having certain advantageous properties for quadrature). These two methods may be useful in high-dimensional integrals.

■ Example 1: Showing the Sample Points

To show the points where **NIntegrate** evaluates the function, use the **Evaluation⌐ Monitor** (✾5) option:

```
f[x_] := Sin[1 / x]

points = {}; NIntegrate[f[x], {x, π / 10, π},
  EvaluationMonitor :→ AppendTo[points, {x, f[x]}]]
```

1.64952

```
Length[points]     77
```

The 77 points where the function was evaluated are as follows:

```
Show[Graphics[
   {AbsoluteThickness[0.2], Map[Line[{{#[[1]], 0}, #}] &, points]}],
   Axes → True, AspectRatio → 0.2];
```

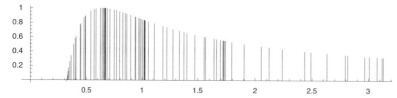

This plot demonstrates the adaptivity property of the Gauss–Kronrod method: more points are sampled where the function changes rapidly. In the following way we see in what order the points were sampled:

```
xpoints = {}; NIntegrate[f[x], {x, π / 10, π},
   EvaluationMonitor :> AppendTo[xpoints, x]];
```

```
ListPlot[xpoints, PlotStyle → AbsolutePointSize[2]];
```

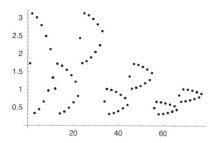

■ Example 2: Comparing Exact and Numerical Results

Define the following:

```
f = Sin[1 / x]; a = π / 90; b = π / 4;
```

Calculate both the exact and the numerical integral:

```
i1 = Integrate[f, {x, a, b}];
i2 = NIntegrate[f, {x, a, b}];
```

Here are the absolute and relative errors of the numerical integral:

```
{Abs[i1 - i2], Abs[i1 - i2] / i1}
```

$\{9.45159 \times 10^{-13}, 3.02804 \times 10^{-12}\}$

The actual relative error is much better than its goal of 10^{-6} (the default value of **Preci** **sionGoal** is 6). Use then a tighter precision goal (❖4!):

```
i3 = NIntegrate[f, {x, a, b}, PrecisionGoal → 8];
```

```
{Abs[i1 - i3], Abs[i1 - i3] / i1}
```

$\{1.96024 \times 10^{-16}, 6.28009 \times 10^{-16}\}$

Now both errors are practically zero. Note that, if we increase **PrecisionGoal**, we often also have to increase **WorkingPrecision**; otherwise convergence may not be reached.

■ Example 3: A Zero Integral

If the true value of the integral is zero, **NIntegrate** tells us that it cannot reach convergence (❀4!):

```
g = Sin[x] - 2 / π;
```

```
NIntegrate[g, {x, 0, π}]
```

NIntegrate::ploss :
 Numerical integration stopping due to loss of precision. Achieved
    neither the requested PrecisionGoal nor AccuracyGoal; suspect
    one of the following: highly oscillatory integrand or the true
    value of the integral is 0. If your integrand is oscillatory
    try using the option Method->Oscillatory in NIntegrate. More…

0.

If we want to get rid of the warning, we can set the value of **AccuracyGoal** to be smaller than ∞ (❀4!):

```
NIntegrate[g, {x, 0, π}, AccuracyGoal → 15]      0
```

■ Example 4: Failing Convergence

```
NIntegrate[Sin[1 / x], {x, π / 90, π / 4}, PrecisionGoal → 14]
```

NIntegrate::ncvb :
 NIntegrate failed to converge to prescribed accuracy after 7
    recursive bisections in x near x = 0.08474391641519344`. More…

0.312135

Seven recursive bisections do not suffice for this integral (the default value of **MaxRecur**‑ **sion** is 6, but apparently 7 is allowed). Increasing the value of **MaxRecursion** from 6 to 7 solves the problem:

```
NIntegrate[Sin[1 / x], {x, π / 90, π / 4},
 PrecisionGoal → 14, MaxRecursion → 7]
```

0.312135

■ Example 5: Monte Carlo Methods

Consider the following integral:

```
Integrate[Sin[x y], {x, 0, Pi}, {y, 0, x}]
```

$$\frac{1}{2} \ (\text{EulerGamma} - \text{CosIntegral}[\pi^2]) + \text{Log}[\pi]$$

```
% // N      1.45034
```

The **QuasiMonteCarlo** method gives a good result, and the result remains the same in repeated calculations:

```
NIntegrate[Sin[x y], {x, 0, π}, {y, 0, x},
 Method → QuasiMonteCarlo, MaxPoints → 70000]
```

1.45025

We then apply the **MonteCarlo** method with seed numbers 1, ..., 8:

```
Table[NIntegrate[Sin[x y], {x, 0, π}, {y, 0, x},
   Method → MonteCarlo[n], MaxPoints → 70000], {n, 1, 8}]
```

{1.45327, 1.46116, 1.42874, 1.47518, 1.45674, 1.43831, 1.46563, 1.45627}

The result here varies between about 1.43 and 1.47.

17.2.3 Singularities and Long Intervals

NIntegrate[f, {x, a, s1, s2, …, sn, b}]   Integrate in several pieces

We can ask **NIntegrate** to do the integration in several pieces. In the example of the box, **NIntegrate** integrates separately on each of the intervals (a, s_1), (s_1, s_2), ..., (s_n, b). One application of this technique is to take into account the possible singularity of the points s_1, s_2, ..., s_n in addition to the possible singularity of the points a and b. Another application is to divide very long intervals into smaller pieces to get a more precise result. The technique can also be used to specify a piecewise linear integration contour in the complex plane.

Note also that, with the package **NumericalMath`CauchyPrincipalValue`**, we can calculate the Cauchy principal value of an integral that has a nonintegrable singularity.

■ Singularities

If the integration interval contains points of discontinuity or singularity of the integrand, **NIntegrate** may not give the correct result or any result at all. In the following example, we get no result (❀4!):

```
NIntegrate[1 / (2 ^ (1 / x) + 1), {x, -1, 1}]
```

NIntegrate::inum :

Integrand $\frac{1}{2^{1/x} + 1}$ is not numerical at {x} = {0.}. **More…**

NIntegrate$\left[\frac{1}{2^{1/x} + 1}, \{x, -1, 1\}\right]$

The integration succeeds if we use 0 as an intermediate point:

```
NIntegrate[1 / (2 ^ (1 / x) + 1), {x, -1, 0, 1}]     1.
```

■ Long Intervals

If the interval of integration is long and there is a rapid change somewhere, **NIntegrate** may not detect the critical small interval, and the result may be inaccurate. We first calculate the numeric value of an exact integral:

```
f = Exp[-x^2] Cos[x];
```

```
i1 = Integrate[f, {x, -100, 100}] // N     1.38039 + 0. i
```

NIntegrate prints a warning message (although the result is very good):

```
NIntegrate[f, {x, -100, 100}]
```

```
NIntegrate::ncvb :
  NIntegrate failed to converge to prescribed accuracy
    after 7 recursive bisections in x near x = 0.78125`. More…
  1.38039
```

Because the rapid change is near the point 0, we next integrate from –100 to 0 and then from 0 to 100. In this way, we get the correct result:

```
i2 = NIntegrate[f, {x, -100, 0, 100}]     1.38039
```

```
i1 - i2     - 3.03757 × 10⁻¹³ + 0. i
```

$$i1 - i2 \quad -3.03757 \times 10^{-13} + 0.\, i$$

Another possibility for long intervals and rapid change is to increase the default value 0 of **MinRecursion** and the default value 6 of **MaxRecursion**. In the example, the following settings work:

```
i3 = NIntegrate[f, {x, -400, 400}, MinRecursion → 2, MaxRecursion → 9]
```

```
1.38039
```

$$i1 - i3 \quad 1.06026 \times 10^{-12} + 0.\, i$$

17.2.4 Oscillatory Integrands

Recall from Section 17.2.2, p. 445, that if we specify **Method → Oscillatory**, then **NInte¿ grate** uses a transformation to handle integrals containing trigonometric and Bessel functions. The method is designed for integrands of the form $f(x)\, w(x)$, where w is **Sin**, **Cos**, **BesselJ**, or **BesselY** (the last two with positive parameters) and the range of integration is in the form $(-\infty, a]$, $[a, \infty)$, or $(-\infty, \infty)$. The function f can be any nonconstant analytic function.

■ Example 1

A highly oscillatory integrand can cause trouble in the convergence of numerical integration. Consider the following function:

```
f = x Sin[x] / (x^2 + 1);
```

```
Plot[f, {x, 0, 50}];
```

This is a highly oscillatory function. Its integral from zero to infinity is as follows:

```
i1 = Integrate[f, {x, 0, ∞}] // N     0.577864
```

Try then **NIntegrate**:

```
NIntegrate[f, {x, 0, ∞}]
```

```
NIntegrate::slwcon :
 Numerical integration converging too slowly; suspect one
    of the following: singularity, value of the integration
    being 0, oscillatory integrand, or insufficient
    WorkingPrecision. If your integrand is oscillatory try
    using the option Method->Oscillatory in NIntegrate. More…
NIntegrate::ncvb :
 NIntegrate failed to converge to prescribed accuracy after 7
    recursive bisections in x near x = 2.288332793335697`*^56. More…
222.71
```

NIntegrate fails. The method designed for oscillatory functions gives the correct result:

```
i2 = NIntegrate[f, {x, 0, ∞}, Method → Oscillatory]       0.577864
```

```
i1 - i2       - 3.33067 × 10⁻¹⁶
```

Keiper (1993) presents the following method for oscillatory functions. Calculate the integral between successive zeros of the integrand, and sum these integrals (the warning we get is harmless):

```
i3 = NSum[NIntegrate[f, {x, n π, (n + 1) π}], {n, 0, ∞}]
```

```
NIntegrate::nlim :
 x = 3.14159 n is not a valid limit of integration. More…
0.577864
```

```
i1 - i3       - 2.77556 × 10⁻¹⁵
```

■ Example 2

Consider a Bessel function:

```
f = BesselJ[2, x];
```

This, too, is a highly oscillatory function (try a plot). Its integral from zero to infinity is one:

```
i1 = Integrate[f, {x, 0, ∞}]       1
```

NIntegrate fails again. The method **Oscillatory** cannot now be used, because the integrand is not in a suitable form; we use the summing method presented in **Example 1**. Calculate the first 30 zeros of **BesselJ[2,x]** (actually 27 zeros would suffice):

```
<< NumericalMath`BesselZeros`
```

```
bz = Prepend[BesselJZeros[2, 30], 0];
```

Here we added the first zero (0) to the list of zeros. **NSum** is now able to produce the correct result (we get some harmless warnings not shown here) (※4!):

```
i2 = NSum[NIntegrate[f, {x, bz[[n]], bz[[n + 1]]}], {n, ∞}, Method → Fit]
```

```
1.
```

```
i1 - i2       - 2.33147 × 10⁻¹⁴
```

```
Remove["Global`*"]
```

17.2.5 Newton–Cotes and Gaussian Quadrature

■ Newton–Cotes Quadrature

Newton–Cotes formulas are of the following form:

$$\int_a^b f(x)\,dx \simeq \sum_{i=1}^n w_i\,f(x_i)$$

With a package, we can derive various Newton–Cotes formulas (both closed, which is the default, and open) and their error formulas.

In the **NumericalMath`NewtonCotes`** *package:*

NewtonCotesWeights[n, a, b]   Give a list of points and weights $\{\{x_1, w_1\}, \ldots, \{x_n, w_n\}\}$

NewtonCotesError[n, f, a, b]   Give the error formula

An option:

QuadratureType   The type of the quadrature; possible values: **Closed**, **Open**

For example, let us ask for information about a rule having n equal to two:

```
<< NumericalMath`NewtonCotes`
```

NewtonCotesWeights[2, a, a + h] $\left\{\left\{a, \dfrac{h}{2}\right\}, \left\{a + h, \dfrac{h}{2}\right\}\right\}$

NewtonCotesError[2, f, a, a + h] $\dfrac{h^3\,f''}{12}$

From the error formula, we see that the rule gives correct results for all linear functions. If a composite form of this rule is made, we see that the formula has an overall coefficient $h/2$ and that the values of the function have coefficients 1, 2, 2, 2, ..., 2, 2, 1. So we get the trapezoidal rule we programmed in Section 15.1.1:

$$\int_a^b f(x)\,dx \simeq \frac{h}{2}\left(f(a) + 2\sum_{i=1}^{n-1} f(a + i\,h) + f(b)\right).$$

An illustration of the trapezoidal rule is as follows:

```
a = Range[0.5, 5.5]; f = Map[Sin[#] + 1.1 &, a];
joiningLines = Line[Transpose[{a, f}]];
verticalLines = Map[Line[{{#, 0}, {#, Sin[#] + 1.1}}] &, a];
xticks = Table[{a[[i]], x_{i-1}}, {i, Length[a]}]; a =.; f =.;

Plot[Sin[x] + 1.1, {x, 0.5, 5.5},
  Ticks → {xticks, None}, Epilog → {joiningLines, verticalLines}];
```

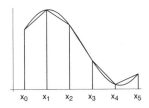

■ Gaussian Quadrature

In n-point Gaussian quadrature on the interval $(-1, 1)$, the points x_i are the zeros of the nth degree Legendre polynomial $P_n(x)$ and the weights w_i are $2/[P_n'(x_i)^2 (1 - x_i^2)]$. Programs for the points and weights are as follows:

```
gaussianPoints[n_] := x /. NSolve[LegendreP[n, x] == 0, x];
gaussianWeights[n_] := With[{points = gaussianPoints[n]},
  2 / ((D[LegendreP[n, x], x] /. x → points) ^2 (1 - points^2))]
```

(Note here that we have made use of the fact that *Mathematica* automatically calculates elementwise with vectors, see Section 2.3.2, p. 40.) For example:

```
gaussianPoints[3]        {-0.774597, 0., 0.774597}
```

```
gaussianWeights[3]       {0.555556, 0.888889, 0.555556}
```

Gaussian quadrature on an interval (a, b) is as follows:

$$\int_a^b f(x)\,dx \simeq d \sum_{i=1}^n w_i\, f(c + d\,x_i)$$

Here, $c = (a + b)/2$ and $d = (b - a)/2$. We then write a program:

```
gaussianQuadrature[f_, x_, a_, b_, n_] := With[{points = gaussianPoints[n],
  weights = gaussianWeights[n], c = (a + b) / 2, d = (b - a) / 2},
  d weights . (f /. x → c + d points)]
```

(Here we calculated the sum with ., which is an inner product.) For example:

```
Integrate[Sin[x], {x, 0, 2}] - gaussianQuadrature[Sin[x], x, 0, 2, 3]
```

```
-0.0000518162
```

■ A Package for Gaussian Quadrature

In the `NumericalMath`GaussianQuadrature`` *package:*

`GaussianQuadratureWeights[n, a, b]` Give a list of points and weights $\{\{x_1, w_1\}, ..., \{x_n, w_n\}\}$

`GaussianQuadratureError[n, f, a, b]` Give the error formula

(These commands also accept an additional argument that gives the wanted precision of the results.) We derive three-point Gaussian quadrature points and weights and the error in the interval $(-1, 1)$:

```
<< NumericalMath`GaussianQuadrature`
```

```
g = GaussianQuadratureWeights[3, -1, 1]
```

```
{{-0.774597, 0.555556}, {0., 0.888889}, {0.774597, 0.555556}}
```

```
GaussianQuadratureError[3, f, -1, 1]
```

```
-0.0000634921 f^{(6)}
```

Thus, the integral of f over $(-1, 1)$ is approximated by the following:

```
Apply[Plus, Map[#[[2]] f[#[[1]]] &, g]]
```

$0.555556\ f[-0.774597] + 0.888889\ f[0.] + 0.555556\ f[0.774597]$

The formula gives correct results for all fifth-order polynomials. The following program calculates an integral using n-point Gaussian quadrature.

```
gaussianQuadrature2[f_, x_, a_, b_, n_] :=
 With[{g = GaussianQuadratureWeights[n, a, b]},
  Apply[Plus, Map[#[[2]] (f /. x → #[[1]]) &, g]]]
```

We try the same example as above with **gaussianQuadrature**:

```
Integrate[Sin[x], {x, 0, 2}] - gaussianQuadrature2[Sin[x], x, 0, 2, 3]
```

-0.0000518162

17.2.6 Quadrature for Data

If we do not know the whole function to be integrated but only its values at a collection of points, we have some possibilities. If the data contain observational errors, then a good method is first to fit a function to the data using **Fit** or **FindFit** (see Section 22.1) and then integrate this function (using **Integrate** or **NIntegrate**). Otherwise, we can form a piecewise-interpolating polynomial using **Interpolation** or **ListInterpolation** (see Section 21.2) and then integrate this function with **Integrate**.

If we want to integrate a *function* of an interpolating function (instead of the interpolating function itself), we could use **NIntegrate**, but this method is slow (and may have convergence problems). Indeed, an interpolating function, while continuous, is only piecewise smooth, and **NIntegrate** requires a lot of time to integrate nonsmooth functions. A more efficient method can be found in a package:

In the **NumericalMath`NIntegrateInterpolatingFunct`** *package:*

NIntegrateInterpolatingFunction[expr, {x, a, b}] Integrate **expr** containing interpolating functions

---

This method is speedier, because the command breaks up the interval of integration into sections in which the integrand is smooth and integrates (with **NIntegrate**) separately in each section. Multidimensional integrals can also be calculated with this command.

17.3 Sums and Products

17.3.1 Exact Sums

Sum[expr, {i, a, b}] Sum of the values of **expr** when **i** varies from **a** to **b**

The iteration specification can also be of the form **{b}** if we sum **b** copies of **expr**, **{i, b}** if the starting value of **i** is 1, and **{i, a, b, d}** if **i** goes from **a** to **b** in steps of **d**.

Remember that sums can also be entered with the aid of the **BasicInput** palette (see Section 1.4.1, p. 17). The keyboard can also be used (see Section 3.3.3, p. 70). Consider the following sum:

$$\sum_{i=1}^{10} 2^i$$

To write this expression, type [ESC]sum[ESC][CTRL]+]i=1[CTRL]%]10[CTRL]_]2[CTRL]^]i[CTRL]_]. Here are some examples of infinite sums:

Sum[1 / i^2, {i, ∞}] $\dfrac{\pi^2}{6}$

Sum[(1 + i) / (2 + i)^3, {i, 0, ∞}] $\dfrac{1}{6}(\pi^2 - 6\,\text{Zeta}[3])$

Sum[(-1)^(n - 1) x^n / n, {n, ∞}] $\text{Log}[1 + x]$

In the last example, we succeeded in obtaining the function that corresponds to a given power series. Next we calculate symbolic sums, which are sums in which the upper bound is a symbol:

Sum[i^2 + 2^i, {i, n}] $2(-1 + 2^n) + \dfrac{1}{6} n(1 + n)(1 + 2n)$

Sum[i^2 Binomial[m, i], {i, m}] $2^{-2+m} m(1 + m)$

Sum[1 / i, {i, n}] $\text{HarmonicNumber}[n]$

■ Problematic Sums

If you use a named expression in a sum in which the upper bound is infinity or a symbol, the result is incorrect. Here is a named expression and a sum:

expr = t^i t^i

Sum[expr, {i, 0, n}] $(1 + n)\,t^i$

The answer is incorrect. To get the correct result, use **Evaluate**, or write the expression directly into **Sum**:

Sum[Evaluate[expr], {i, 0, n}] $\dfrac{-1 + t^{1+n}}{-1 + t}$

Sum[t^i, {i, 0, n}] $\dfrac{-1 + t^{1+n}}{-1 + t}$

Here is an example in which the upper bound is infinity (❀4!):

Sum[expr, {i, 0, ∞}] $t^i \infty$

The answer is incorrect. To get the correct result, again use **Evaluate**, or write the expression directly into **Sum**:

Sum[Evaluate[expr], {i, 0, ∞}] $\dfrac{1}{1 - t}$

As always, it is best to check the results. For example, the following sum is wrong:

```
Sum[i Binomial[2 n - i, n] / 2^(2 n - i), {i, 0, n}]
```

$((-1)^n \, 2^{-2n} \, \pi \, \text{Csc}[n \, \pi] \, \text{HypergeometricPFQRegularized}[\{2, 1 - n\}, \{1 - 2n\}, 2]) /$
$(\text{Gamma}[n] \, \text{Gamma}[1 + n])$

The true value is $\binom{2n}{n} \dfrac{2 n+1}{2^{2n}} - 1$.

■ Multiple Sums

```
Sum[expr, {i, a, b}, {j, c, d}]
```

Multiple sums are calculated in much the same way as multiple integrals. Above, **c** and **d** may depend on **i**, that is, the ranges of the indices are given in the familiar mathematical notation: the outer index is given first.

```
Sum[x^i y^j, {i, 1, 3}, {j, 1, i}]
```

$x \, y + x^2 \, y + x^3 \, y + x^2 \, y^2 + x^3 \, y^2 + x^3 \, y^3$

17.3.2 Numerical Sums

```
NSum[expr, {i, a, b}]   Calculate the sum with numerical methods
```

The idea of **NSum** is to sum a certain number of the first terms and then estimate accurately the sum of the terms neglected. Numerical summation is useful for such infinite sums that cannot be calculated with **Sum**. Also, sums with a great many terms can be calculated nicely using numerical methods. One application of numerical summation is in numerical integration of highly oscillatory functions (see Section 17.2.4, p. 450).

We calculate several partial sums of the harmonic series with both **Sum** and **NSum**:

```
(s1 = Table[Sum[N[1 / i], {i, 10^n}], {n, 1, 6}]) // Timing
```

$\{19.26 \text{ Second}, \{2.92897, 5.18738, 7.48547, 9.78761, 12.0901, 14.3927\}\}$

```
(s2 = Table[NSum[1 / i, {i, 10^n}], {n, 1, 6}]) // Timing
```

$\{0.07 \text{ Second}, \{2.92897, 5.18738, 7.48547, 9.78761, 12.0901, 14.3927\}\}$

The difference in computing times is considerable, but the differences in the results are very small:

```
s1 - s2
```

$\{0., 2.35504 \times 10^{-10}, 3.05667 \times 10^{-11},$
$2.39993 \times 10^{-10}, 2.36627 \times 10^{-10}, 2.26909 \times 10^{-9}\}$

Here is a sum for which **Sum** does not give a result but **NSum** works:

```
NSum[Log[i^2] / (2^i i!), {i, ∞}]        0.227205
```

We can also apply **% // N** to the result of **Sum** if **Sum** does not succeed. The sum is then actually calculated by **NSum**.

■ Options

Options for **NSum**:

WorkingPrecision Precision used in internal computations; examples of values:
  MachinePrecision (⌘5), **20**

NSumTerms Number of terms summed explicitly; examples of values: **15, 20**

VerifyConvergence Whether to test explicitly for convergence of infinite sums;
  possible values: **True, False**

EvaluationMonitor (⌘5) Command to be executed after each evaluation of the
  expression to be summed; examples of values: **None, ++n, AppendTo[points, i]**

Compiled Whether the summand should be compiled or not; possible values: **True,
  False**

Method Method to use; possible values: **Automatic** (means **Integrate** or **Fit**), **Inte**
  grate, Fit

PrecisionGoal If the value of the option is **p**, the relative error of the sum should be
  of the order 10^{-p}; examples of values: **Automatic** (usually means **6**), **10**; this option is
  applicable if **Method** is **Integrate**

AccuracyGoal If the value of the option is **a**, the relative error of the integral should
  be of the order 10^{-a}; examples of values: ∞, **10**; this option is applicable if **Method** is
  Integrate

NSumExtraTerms Number of additional terms used in extrapolation; examples of
  values: **12, 20**; this option is applicable if **Method** is **Fit**

WynnDegree Degree to use in the Wynn method; examples of values: **1, 2**; this option is
  applicable if **Method** is **Fit**

NSum first calculates **NSumTerms** (default is 15) terms of the sum and then estimates the rest by one of two methods: **Integrate** or **Fit**. The default value of **Method** is **Automatic**, which means that some heuristics are used to decide between the two methods.

The method used by **Integrate** is Euler–Maclaurin summation. The value of the sum of the neglected terms is estimated by integration. The integration is done by **Integrate**, and, if it fails, by **NIntegrate**. With this method, we can set our own **PrecisionGoal** and **AccuracyGoal**.

The second method is **Fit** (which actually means **SequenceLimit**; **Fit** does not do anything with the fitting command **Fit**). This method forms a sequence of partial sums from **NSumExtraTerms** (default is 12) terms. **SequenceLimit** is then used to calculate the limit from these partial sums. **SequenceLimit** is an implementation of Wynn's epsilon algorithm. The degree of Wynn's method can be set with the option **WynnDegree**. This method is especially good for alternating series.

We calculate two sums and compare them with the exact result:

```
Sum[1 / i^2, {i, ∞}] - NSum[1 / i^2, {i, ∞}]      8.82157 × 10⁻¹¹

Sum[(-1)^i / i^2, {i, ∞}] - NSum[(-1)^i / i^2, {i, ∞}]      0.
```

Both numerical results are very good. (In the former sum, the method was **Integrate** and in the latter sum **Fit**.)

The command `EulerSum[f, {i, a, ∞}]` in the `NumericalMath`NLimit`` package can be used as an alternative to `NSum`, especially for alternating series.

17.3.3 Products

> `Product[expr, {i, a, b}]`
> `NProduct[expr, {i, a, b}]`

Products can also be entered with the aid of the **BasicInput** palette or with the keyboard ($\prod$ is entered as ESCprodESC). Here are some examples:

> `Product[(a + i), {i, 4}]` $(1 + a)\ (2 + a)\ (3 + a)\ (4 + a)$

> `Product[1 - 1 / (2 i^2), {i, ∞}]` $\dfrac{\sqrt{2}\ \text{Sin}\left[\frac{\pi}{\sqrt{2}}\right]}{\pi}$

`NProduct` has the same options as `NSum` except that `NSumTerms` is replaced with `NProductFactors` and `NSumExtraTerms` with `NProductExtraFactors`. With the `Inte`grate method, the product is first written as a sum with a logarithmic transformation, and then `NSum` is used. With the `Fit` method, a sequence of partial products is calculated, and these values are passed to `SequenceLimit`.

17.4 Transforms

17.4.1 Laplace Transform

> `LaplaceTransform[F, t, s]`   Laplace transform of **F** (a function of **t**); the transform will be a function of **s**
> `InverseLaplaceTransform[f, s, t]`   Inverse Laplace transform of **f** (a function of **s**); the original function will be a function of **t**

The Laplace transform of a function $F(t)$ is $f(s) = \int_0^\infty F(t)\, e^{-st}\, dt$. For example:

> `p = LaplaceTransform[Sin[3 t], t, s]` $\dfrac{3}{9 + s^2}$

> `InverseLaplaceTransform[p, s, t]` $\text{Sin}[3\,t]$

Here is another example (✤4!):

> `p = LaplaceTransform[Exp[a t] Cosh[b t], t, s]` $\dfrac{-a + s}{a^2 - b^2 - 2\,a\,s + s^2}$

> `q = InverseLaplaceTransform[p, s, t]` $\dfrac{1}{2}\ e^{(a-b)\,t}\ (1 + e^{2\,b\,t})$

> `FullSimplify[q]` $e^{a\,t}\ \text{Cosh}[b\,t]$

Another example (✤4!):

> `p = LaplaceTransform[UnitStep[t - a], t, s]`
> $\dfrac{\text{UnitStep}[-a] + e^{-a\,s}\ \text{UnitStep}[a]}{s}$

```
InverseLaplaceTransform[p, s, t]
```

UnitStep[-a] + UnitStep[a] UnitStep[-a + t]

And another example:

```
p = LaplaceTransform[DiracDelta[t - 3], t, s]      e^{-3 s}
```

```
InverseLaplaceTransform[p, s, t]
```

DiracDelta[-3 + t] UnitStep[-3 + t]

We can also calculate transforms of some expressions that contain unspecified functions. Here is an example:

```
p = LaplaceTransform[F'[t], t, s]
```

-F[0] + s LaplaceTransform[F[t], t, s]

```
InverseLaplaceTransform[p, s, t]      F'[t]
```

And another example (✿4!):

```
p = LaplaceTransform[Integrate[F[u], {u, 0, t}], t, s]
```

$$\frac{LaplaceTransform[F[t], t, s]}{s}$$

```
InverseLaplaceTransform[p, s, t]
```

Integrate[F[s], {s, 0, t}, Assumptions → True]

If the original function is an infinite sum (see Spiegel [1999, p. 187]), *Mathematica* cannot calculate the inverse transform:

```
InverseLaplaceTransform[Sinh[s x] / (s Sinh[s a]), s, t]
```

$$InverseLaplaceTransform\left[\frac{Csch[a s] Sinh[s x]}{s}, s, t\right]$$

With **LaplaceTransform** and **InverseLaplaceTransform**, we can use the same options as we did with **Integrate** (see Section 17.1.3, p. 438). Multidimensional Laplace transforms and their inverse transforms can also be calculated.

For application of the Laplace transform to the solution of ordinary differential equations, integral equations, and partial differential equations, see Sections 23.2.1, p. 604; 23.2.4, p. 609; and 24.1.2, p. 643.

■ Numerical Inversion

For the numerical inversion of Laplace transforms, see Cheng, Sidauruk, and Abousleiman (1994). Here is one such method: the Stehfest method.

```
c[n_, i_] := (-1) ^ (i + n / 2) *
  Sum[N[k ^ (n / 2) (2 k) !] / N[(n / 2 - k) ! k ! (k - 1) ! (i - k) ! (2 k - i) !],
    {k, Floor[(i + 1) / 2], Min[i, n / 2]}]

stehfestILT[f_, s_, t_, n_ ? EvenQ] :=
  Log[2.] / t Sum[c[n, i] f /. s → i Log[2.] / t, {i, 1, n}]
```

Note that **n** must be even (**n** is the number of terms in the sum of **stehfestILT**; a value between 6 and 20 is recommended). We calculate the value of the inverse transform of $1/(1 + s^2)$ when **t** is 1 using 20 terms and compare the result with the true inverse sin(1):

```
stehfestILT[1 / (1 + s^2), s, 1, 20]     0.841474
```

```
Sin[1.]     0.841471
```

The numerical inverse is quite good. We can plot the inverse transform:

```
Plot[Evaluate[stehfestILT[1 / (1 + s^2), s, t, 20]], {t, 0, 2 π}];
```

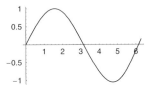

17.4.2 Fourier Transforms and Series

■ Fourier Transforms

FourierTransform[f, t, w] Fourier transform of **f** (a function of **t**); the transform will be a function of **w**

InverseFourierTransform[F, w, t] Inverse Fourier transform of **F** (a function of **w**); the original function will be a function of **t**

An option:

FourierParameters Parameters of the Fourier transform; examples of values: **{0, 1}**, **{1, -1}**, **{-1, 1}**, **{0, -2π}**

The Fourier transform of $f(t)$ in *Mathematica* is $F(w) = \frac{1}{\sqrt{2\pi}} \int_{-\infty}^{\infty} f(t)\, e^{iwt}\, dt$ if the option **FourierTransform** has the default value **{0, 1}**. Various other Fourier transforms are applied in different disciplines. They can be used by giving **FourierParameters** a suitable value. If the value is **{a, b}**, the transform is $\left(\frac{|b|}{(2\pi)^{1-a}}\right)^{1/2} \int_{-\infty}^{\infty} f(t)\, e^{ibwt}\, dt$. For example:

```
p = Simplify[FourierTransform[1 / (b^2 + t^2), t, w], b > 0]
```

$$\frac{e^{-b\, \text{Abs}[w]}\, \sqrt{\frac{\pi}{2}}}{b}$$

```
InverseFourierTransform[p, w, t]          1/(b² + t²)
```

FourierSinTransform[f, t, w], **InverseFourierSinTransform[f, t, w]**
FourierCosTransform[f, t, w], **InverseFourierCosTransform[f, t, w]**

The default Fourier sine transform is $\sqrt{\frac{2}{\pi}} \int_0^{\infty} f(t) \sin(w\, t)\, dt$. Other Fourier sine transforms can be obtained from the most general form $2\left(\frac{|b|}{(2\pi)^{1-a}}\right)^{1/2} \int_0^{\infty} f(t) \sin(b\, w\, t)\, dt$ by giving the parameters a and b suitable values with **FourierParameters**. Similarly, we get Fourier cosine transforms. For example:

```
p = FourierSinTransform[Exp[b t],
    t, w, FourierParameters → {1, -1}] // Simplify
```

$$-\frac{2 w}{b^2 + w^2}$$

```
q = Simplify[
    InverseFourierSinTransform[p, w, t, FourierParameters → {1, -1}], b > 0]
```

$e^{-b t}$

The package `Calculus`FourierTransform`` also has the commands `DTFourier`. `Transform` and `InverseDTFourierTransform` and numerical versions of the eight commands mentioned above (e.g., `NFourierTransform`).

■ Fourier Series

In the `Calculus`FourierTransform`` *package:*

`FourierTrigSeries[f, t, k]` Fourier trigonometric series expansion to order **k** of **f**;
   by default **f** is treated as a periodic function of **t** with one period on $(-\frac{1}{2}, \frac{1}{2})$
`FourierSeries[f, t, n]` Fourier exponential series expansion to order **k** of **f**; by
   default **f** is treated as a periodic function of **t** with one period on $(-\frac{1}{2}, \frac{1}{2})$

An option:

`FourierParameters` Parameters of the series expansion; default value: `{0, 1}`

If the option `FourierParameters` has the default value `{0, 1}`, the Fourier trigonometric series is $c_0 + \sum_{n=1}^{k} [c_n \cos(2\pi n t) + d_n \sin(2\pi n t)]$ for a periodic function with one period on $(-\frac{1}{2}, \frac{1}{2})$. If `FourierParameters` is given the value `{0, b}`, then the series is (if b is positive) $\sqrt{b} \{c_0 + \sum_{n=1}^{k} [c_n \cos(2\pi b n t) + d_n \sin(2\pi b n t)]\}$ for a periodic function with one period on $(-\frac{1}{2b}, \frac{1}{2b})$. The coefficients c_0, c_n, and d_n are as follows:

$$c_0 = \sqrt{b} \int_{-\frac{1}{2b}}^{\frac{1}{2b}} f(t)\, dt, \quad c_n = 2\sqrt{b} \int_{-\frac{1}{2b}}^{\frac{1}{2b}} f(t)\cos(2\pi b n t)\, dt, \quad d_n = 2\sqrt{b} \int_{-\frac{1}{2b}}^{\frac{1}{2b}} f(t)\sin(2\pi b n t)\, dt$$

As an example, we calculate the third-order Fourier trigonometric series of $t + 2$ when this function is treated as periodic with one period on $(-2, 2)$. Because we want $\frac{1}{2b}$ to be 2, we choose b to be $1/4$:

```
<< Calculus`FourierTransform`
```

```
ser = FourierTrigSeries[t + 2, t, 3, FourierParameters → {0, 1/4}]
```

$$\frac{1}{2} \left(4 + \frac{8 \sin[\frac{\pi t}{2}]}{\pi} - \frac{4 \sin[\pi t]}{\pi} + \frac{8 \sin[\frac{3\pi t}{2}]}{3\pi} \right)$$

To compare this function with the original periodic function, we define the periodic function (see Section 14.1.3, p. 356):

```
f[t_] := If[-2 ≤ t < 2, t + 2, f[t - 4]]
```

Then we plot the original function and the series:

```
Plot[{f[t], ser}, {t, -2, 10}, AspectRatio → Automatic];
```

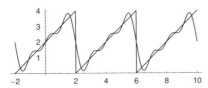

We can also calculate coefficients using direct integration. Here are the coefficients c_0, c_n, and d_n (we use the function $t + 2$, because the integration of the recursive function `f[t]` does not succeed):

```
1 / 2 Integrate[t + 2, {t, -2, 2}]      4
```

```
Simplify[2 * 1 / 2 Integrate[(t + 2) Cos[π n t / 2], {t, -2, 2}], n ∈ Integers]
```

0

```
Simplify[2 * 1 / 2 Integrate[(t + 2) Sin[π n t / 2], {t, -2, 2}], n ∈ Integers]
```

$$-\frac{8 \, (-1)^n}{n \, \pi}$$

So we see that, in the interval $(-2, 2)$, we have $t + 2 = \frac{1}{2}\left(4 - \sum_{n=1}^{\infty} \frac{8}{n \pi}(-1)^n \sin(\frac{n \pi t}{2})\right)$.

The package also has `FourierSinCoefficient[f, t, n]`, `FourierCosCoefficient[f, t, n]`, `FourierCoefficient[f, t, n]`, and `InverseFourierCoefficient[f, n, t]`; it also has numeric versions of the five commands mentioned above (e.g., `NFourierTrig`. `Series`).

17.4.3 Discrete Fourier Transform

`Fourier[data]`	Fourier transform
`InverseFourier[data]`	Inverse Fourier transform

An option:

`FourierParameters` Parameters of the transform; examples of values: `{0, 1}`, `{-1, 1}`, `{1, -1}`

Data are often analyzed by calculating the discrete Fourier transform or the spectrum. For data $\{u_1, ..., u_n\}$, the transform is $\frac{1}{\sqrt{n}} \sum_{r=1}^{n} u_r \, e^{2\pi i (r-1)(s-1)/n}$ if the option `Fourier`. `Parameters` has the default value `{0, 1}`. If `FourierParameters` is given the value `{a, b}`, then the transform is $n^{-(1-a)/2} \sum_{r=1}^{n} u_r \, e^{2\pi i b (r-1)(s-1)/n}$. `Fourier` can also find the transformation for higher-dimensional data.

As a simple example, we calculate the Fourier transform of a list having 3 elements; `Chop` can be used to replace near-zero real or imaginary parts with an exact zero:

```
d = {1, 0, 2};
```

```
fd = Fourier[d]      {1.73205 + 0. i, 0. - 1. i, 0. + 1. i}
```

```
fd = Chop[fd]      {1.73205, -1. i, 1. i}
```

Then we calculate the inverse transform:

> **ifd = InverseFourier[fd]** $\{1., 2.56395 \times 10^{-16}, 2.\}$

> **Chop[ifd]** $\{1., 0, 2.\}$

In Section 27.3.2, p. 760, we present an example that shows how the discrete Fourier transform can be used to smooth or filter data.

The **LinearAlgebra`FourierTrig`** package defines discrete cosine and sine transformations.

17.4.4 Z Transform

> **ZTransform[f, n, z]**   Z transform of **f** (a function of **n**); the transform will be a function of **z**
>
> **InverseZTransform[g, z, n]**   Inverse Fourier transform of **g** (a function of **z**); the original function will be a function of **n**

The Z transform is defined by $g(z) = \sum_{n=0}^{\infty} f(n) z^{-n}$. For example:

> **ZTransform[a^n n^2, n, z]** $-\dfrac{a\,z\,(a+z)}{(a-z)^3}$

> **InverseZTransform[%, z, n]** $a^n\,n^2$

The Z transform can be used to solve difference equations (see Section 25.1.3, p. 677).

18

Matrices

Introduction

> *In some colleges of music, part of the doctoral requirement is to compose an original full length symphony. Because modern music sounds so weird, a good ploy is to take a well-known classical symphony, write it backwards and submit it as an original work. One student took the daring step of taking his professor's doctoral symphony and reversing it. He failed to receive his degree, the examiners remarking that he had reproduced Sibelius' Fourth Symphony with not a single note changed.*

The treatment of vectors and matrices is somewhat short here, but remember that systems of linear equations are considered in Section 19.1, and linear programming is addressed in Section 20.2.1. In addition, remember that vectors and matrices are lists, so the material of Chapter 13, in which we considered lists, is useful for vectors and matrices, too.

18.1 Vectors

18.1.1 Calculating with Vectors

A vector is a one-dimensional list. Here are two vectors:

```
r = {2, 5}; s = {x, y};
```

Calculating with vectors is easy, because *Mathematica* automatically does all operations element by element:

```
{5 s, 1 + s, s^2, 1 / s, Log[s]}
```

$$\left\{\{5\,x,\ 5\,y\},\ \{1+x,\ 1+y\},\ \{x^2,\ y^2\},\ \left\{\frac{1}{x},\ \frac{1}{y}\right\},\ \{\text{Log}[x],\ \text{Log}[y]\}\right\}$$

```
{r + s, s r, s / r, s^r}
```

$$\{\{2 + x, 5 + y\}, \{2\,x, 5\,y\}, \{\tfrac{x}{2}, \tfrac{y}{5}\}, \{x^2, y^5\}\}$$

Table and **Array** can be used to construct vectors from formulas.

Table[expr, {i,n}] Create an **n** vector from **expr**
Array[f, {n}] Create an **n** vector with elements **f[i]**

   Table[Prime[n], {n, 10}]   {2, 3, 5, 7, 11, 13, 17, 19, 23, 29}

   Array[f, {4}]   {f[1], f[2], f[3], f[4]}

■ Sum of Elements

Total[list] (❀5) The sum of the elements of **list**
Apply[Plus, list] The sum of the elements of **list**

Users of *Mathematica* versions prior to version 5 have to use **Apply**. Calculating the sum of the elements of a list is a common task in many mathematical calculations, and so **Total** is a useful command. Note, however, that in this book we avoid the use of **Total** to make the code accessible for those of us not using *Mathematica* 5. For example:

   Total[{a, b, c}]   a + b + c

   Apply[Plus, {a, b, c}]   a + b + c

■ Products

---
u v Element-by-element product (the result is a vector)
u.v Inner product (the result is a scalar)
u×v Cross product (the result is a vector)
Outer[Times, u, v] Outer product (the result is a matrix)

The inner and cross products can also be written as **Dot[u, v]** and **Cross[u, v]**. The cross (×) can be written as [ESC]cross[ESC]. Consider the following vectors:

   u = {a, b, c}; v = {P, Q, R};

Here are all four products:

   u v   {a P, b Q, c R}

   u.v   a P + b Q + c R

   u × v   {−c Q + b R, c P − a R, −b P + a Q}

   Outer[Times, u, v]   {{a P, a Q, a R}, {b P, b Q, b R}, {c P, c Q, c R}}

Note especially that the normal product (written as a space) cannot be used to calculate the inner product; we have to use the dot.

Note also that a vector like *{a, b, c}* looks like a *row* vector. The truth is, however, that *Mathematica* does not distinguish between row and column vectors; all vectors are written in the same way. *Mathematica* can work in this way, because it is almost always clear to *Mathematica* what should be done in a computation that contains vectors and matrices. To illustrate this further, we go somewhat ahead and introduce a matrix:

```
T = {{3, 1}, {4, 6}};
```

To multiply **T** and **s**, just write the following:

```
T.s      {3 x + y, 4 x + 6 y}
```

To multiply **s** and **T**, just write this:

```
s.T      {3 x + 4 y, x + 6 y}
```

However, if we want the product of **T**, **r** (considered as a column vector), **r** (considered as a row vector), and **T**, we have to be careful and use **Outer**:

```
T.Outer[Times, r, r].T      {{286, 352}, {988, 1216}}
```

■ **Norms**

> **Norm[v, p]** (✻5)   The *p*-norm of vector **v**, $1 \leq p \leq \infty$

The 2-norm can also be written simply as **Norm[v]**. **Norm** accepts all real values of *p* in the interval $[1, \infty]$; the general *p*-norm of a vector is $(\sum |v_i|^p)^{1/p}$ for $1 \leq p < \infty$, and the maximum of the absolute values of the elements if *p* is ∞.

```
v = {a, b, c};

{Norm[v, 1], Norm[v], Norm[v, ∞]}
```

$\{\mathtt{Abs[a] + Abs[b] + Abs[c]},$

$\sqrt{\mathtt{Abs[a]}^2 + \mathtt{Abs[b]}^2 + \mathtt{Abs[c]}^2}, \mathtt{Max[Abs[a], Abs[b], Abs[c]]}\}$

Users of *Mathematica* 4 have first to load the **LinearAlgebra`MatrixManipulation** package and then use **VectorNorm[v, p]** (note that this command requires that **v** is numerical and has at least one decimal point).

The 1-, 2-, and ∞-norms are easy to program:

```
vectorNorm[v_, p_] := Which[
  p === 1, Apply[Plus, Abs[v]],
  p === 2, Sqrt[Abs[v].Abs[v]],
  p === ∞, Max[Abs[v]]]
```

■ **Orthogonalization**

*In the **LinearAlgebra`Orthogonalization`** package:*

GramSchmidt[{v1, v2, … }]   Generate an orthonormal set from the given vectors

Householder[{v1, v2, … }]   Generate an orthonormal set from the given numeric vectors

Normalize[v]   Normalize the vector

Projection[v1, v2]   Orthogonal projection of **v1** onto **v2**

Here is an example of orthonormalization:

```
<< LinearAlgebra`Orthogonalization`
```

```
{v1, v2} = GramSchmidt[{{3, 4}, {1, 1}}]
```

$\left\{\left\{\dfrac{3}{5}, \dfrac{4}{5}\right\}, \left\{\dfrac{4}{5}, -\dfrac{3}{5}\right\}\right\}$

(If you want to orthogonalize, use the option **Normalized → False**.) The inner product of these vectors is zero, and their norm is one:

```
{v1.v2, Norm[v1], Norm[v2]}    {0, 1, 1}
```

The usual inner product is **v1.v2**, and this is used unless otherwise specified. We can, however, use other inner products with the **InnerProduct** option. The inner product is defined as a pure function with two arguments **#1** and **#2**. For example, the default usual inner product could be defined as **InnerProduct → (#1.#2 &)**. We use another inner product and orthogonalize a set of monomials to demonstrate:

```
GramSchmidt[{1, x, x^2, x^3, x^4}, Normalized → False,
    InnerProduct → (Integrate[#1 #2 / Sqrt[1 - x^2], {x, -1, 1}] &)]
```

$\left\{1, \; x, \; -\dfrac{1}{2} + x^2, \; -\dfrac{3x}{4} + x^3, \; \dfrac{1}{8} - x^2 + x^4\right\}$

Compare these with the Chebyshev polynomials:

```
Table[ChebyshevT[n, x], {n, 0, 4}]
```

$\{1, \; x, \; -1 + 2x^2, \; -3x + 4x^3, \; 1 - 8x^2 + 8x^4\}$

We see that by orthogonalization, we got the Chebyshev polynomials except for a constant multiplier.

■ Rotation

With the **Geometry`Rotations`** package we can rotate vectors.

18.2. Matrices

18.2.1 Constructing Matrices

A matrix is represented as a list; each row of the matrix is a separate sublist. Here is a matrix with four rows and five columns:

```
p = {{0, 0, 1, 1, 0}, {1, 0, 1, 2, 1}, {2, 1, 2, 1, 1}, {2, 1, 1, 3, 2}}
```

$\{\{0, 0, 1, 1, 0\}, \{1, 0, 1, 2, 1\}, \{2, 1, 2, 1, 1\}, \{2, 1, 1, 3, 2\}\}$

MatrixForm[m]	Write matrix **m** in a two-dimensional form
Length[m]	The number of rows of **m**
Dimensions[m]	The dimensions (numbers of rows and columns) of **m**

```
p // MatrixForm
```

$$\begin{pmatrix} 0 & 0 & 1 & 1 & 0 \\ 1 & 0 & 1 & 2 & 1 \\ 2 & 1 & 2 & 1 & 1 \\ 2 & 1 & 1 & 3 & 2 \end{pmatrix}$$

```
Dimensions[p]      {4, 5}
```

■ MatrixForm

A warning about **MatrixForm**: *with a matrix in a matrix form, we cannot do any calculations.* For example, write the following:

```
q = {{3, 1}, {2, 5}} // MatrixForm
```
$\begin{pmatrix} 3 & 1 \\ 2 & 5 \end{pmatrix}$

We defined **q** to be not the matrix itself but its matrix form. *Mathematica* will not do any calculations with such a form:

```
{2 + q, 3 q, Transpose[q], Inverse[q]}
```

$\left\{ 2 + \begin{pmatrix} 3 & 1 \\ 2 & 5 \end{pmatrix}, \ 3 \begin{pmatrix} 3 & 1 \\ 2 & 5 \end{pmatrix}, \ \text{Transpose}\left[\begin{pmatrix} 3 & 1 \\ 2 & 5 \end{pmatrix}\right], \ \text{Inverse}\left[\begin{pmatrix} 3 & 1 \\ 2 & 5 \end{pmatrix}\right] \right\}$

So, when defining **q**, we have to be careful so that the value of **q** will be the matrix itself and not its matrix form. This can be done by writing in either of the following two ways:

```
MatrixForm[q = {{3, 1}, {2, 5}}]
```
$\begin{pmatrix} 3 & 1 \\ 2 & 5 \end{pmatrix}$

```
(q = {{3, 1}, {2, 5}}) // MatrixForm
```
$\begin{pmatrix} 3 & 1 \\ 2 & 5 \end{pmatrix}$

We can also first define the matrix and then ask for its matrix form, as we did with **p** above.

A nice way to show all matrices in the matrix form—without problems in calculations—is to define **$PrePrint = If[MatrixQ[#], MatrixForm[#], #]&**. Now all matrices are shown in the matrix form. By defining **$PrePrint =.**, we can cancel the definition.

■ Constructing Matrices

```
Table[expr, {i,m}, {j,n}]  Create an (m×n) matrix
Array[f, {m, n}]  Create an (m×n) matrix with elements f[i, j]
Array[f, {m, n}, {m0, n0}]  Index origins are m0 and n0
IdentityMatrix[n]  An (n×n) identity matrix
DiagonalMatrix[list]  A diagonal matrix with diagonal elements from list
```

In the **LinearAlgebra`MatrixManipulation`** *package:*

```
UpperDiagonalMatrix[f, n]  Elements on or above the diagonal are f[i, j]
LowerDiagonalMatrix[f, n]  Elements on or below the diagonal are f[i, j]
ZeroMatrix[n]  An (n×n) zero matrix
HilbertMatrix[n]  Elements are 1/(i + j − 1)
HankelMatrix[n]  First row and first column are 1, ..., n
```

See Section 13.1.1, p. 325, for more about **Table**. Here is a Hilbert matrix:

```
Table[1 / (i + j - 1), {i, 3}, {j, 3}]
```

$$\{\{1, \frac{1}{2}, \frac{1}{3}\}, \{\frac{1}{2}, \frac{1}{3}, \frac{1}{4}\}, \{\frac{1}{3}, \frac{1}{4}, \frac{1}{5}\}\}$$

```
Array[1 / (#1 + #2 - 1) &, {3, 3}]
```

$$\{\{1, \frac{1}{2}, \frac{1}{3}\}, \{\frac{1}{2}, \frac{1}{3}, \frac{1}{4}\}, \{\frac{1}{3}, \frac{1}{4}, \frac{1}{5}\}\}$$

Here are some general elements:

```
Array[f, {2, 2}]
```

```
{{f[1, 1], f[1, 2]}, {f[2, 1], f[2, 2]}}
```

Next we use **If**, **Which**, and **Switch** (see Section 15.1.2, p. 387). Here are examples of upper-triangular matrices:

```
Table[If[i ≤ j, i + j, 0], {i, 3}, {j, 3}] // MatrixForm
```

$$\begin{pmatrix} 2 & 3 & 4 \\ 0 & 4 & 5 \\ 0 & 0 & 6 \end{pmatrix}$$

```
Table[Which[i > j, 0, i == j, 1, i == j - 1, 2, i ≤ j - 2, 3], {i, 3}, {j, 3}] //
 MatrixForm
```

$$\begin{pmatrix} 1 & 2 & 3 \\ 0 & 1 & 2 \\ 0 & 0 & 1 \end{pmatrix}$$

```
<< LinearAlgebra`MatrixManipulation`
```

```
UpperDiagonalMatrix[#1 + #2 - 1 &, 3] // MatrixForm
```

$$\begin{pmatrix} 1 & 2 & 3 \\ 0 & 3 & 4 \\ 0 & 0 & 5 \end{pmatrix}$$

Next we show a tridiagonal matrix, in which the element is a, b, c, or 0 depending on whether $i - j$ is $-1, 0, 1$, or something else:

```
Table[Switch[i - j, -1, a, 0, b, 1, c, _, 0], {i, 3}, {j, 3}] // MatrixForm
```

$$\begin{pmatrix} b & a & 0 \\ c & b & a \\ 0 & c & b \end{pmatrix}$$

A Vandermonde matrix is displayed as follows:

```
Outer[Power, {x, y, z}, Range[0, 2]] // MatrixForm
```

$$\begin{pmatrix} 1 & x & x^2 \\ 1 & y & y^2 \\ 1 & z & z^2 \end{pmatrix}$$

■ Sparse Arrays

In sparse vectors or matrices, we typically have few nonzero elements as compared with the number of zero elements. Using **SparseArray**, we can nicely specify only the nonzero elements.

> **SparseArray[rules, dims, default]** (❋5)   Create a sparse array from **rules** that has dimensions **dims** and that takes unspecified elements to be **default**

The **rules** can be of either of the following forms:

   {pos_1 → val_1, pos_2 → val_2, …}

   {pos_1, pos_2, …} → {val_1, val_2, …}

Here we specify the positions and values of some elements; elements in other positions are set to **default**. The position specifications can contain patterns (see Section 15.3.2, p. 405). Dimensions **dims** is, for example, of the form {d_1} for vectors and of the form {d_1, d_2} for matrices.

   SparseArray also accepts the shorter form **SparseArray[rules, dims]**, and then unspecified elements are taken to be 0. A still shorter form is **SparseArray[rules]**, and then the dimensions are exactly large enough to include elements with positions that have been explicitly specified.

   As an example, we generate a sparse array that is a 3×3 diagonal matrix with 5 as the diagonal element in three different ways:

   s1 = SparseArray[{{1, 1} → 5, {2, 2} → 5, {3, 3} → 5}]

   SparseArray[<3>, {3, 3}]

   s2 = SparseArray[{{1, 1}, {2, 2}, {3, 3}} → {5, 5, 5}]

   SparseArray[<3>, {3, 3}]

   s3 = SparseArray[{{i_, i_} → 5}, {3, 3}]

   SparseArray[<3>, {3, 3}]

The result is a **SparseArray** object. We only see the number of rules (three here) and the dimensions of the array ({3, 3} here). With **Normal**, we can see the array in the usual matrix form:

   s1 // Normal   {{5, 0, 0}, {0, 5, 0}, {0, 0, 5}}

   As another example, we generate a tridiagonal matrix:

   s4 = SparseArray[
     {{i_, i_} → b, {i_, j_} /; i - j == -1 → a, {i_, j_} /; i - j == 1 → c}, {3, 3}]

   SparseArray[<7>, {3, 3}]

   s4 // Normal // MatrixForm

   $$\begin{pmatrix} b & a & 0 \\ c & b & a \\ 0 & c & b \end{pmatrix}$$

Sparse arrays can be used in calculations as normal arrays. For example:

s1.s4   SparseArray[< 7 >, {3, 3}]

Normal[%]

{{5 b, 5 a, 0}, {5 c, 5 b, 5 a}, {0, 5 c, 5 b}}

In general, matrix calculus like **Eigenvalues**, **LinearSolve**, and **LinearProgramming** work with sparse arrays as they do with normal matrices.

Normal[sparseArray] (❀5) Create a list version of **sparseArray**

SparseArray[list] (❀5) Create a sparse array version of **list**

ArrayRules[sparseArray] (❀5) Give the rules of **sparseArray**

As we saw above, **Normal** converts a sparse array into a usual list. The inverse is done with **SparseArray**, which means that this command is also able to convert usual lists into sparse arrays. With **ArrayRules**, we get the list of rules that specify the elements of a sparse array. For example:

SparseArray[{{b, a, 0}, {c, b, a}, {0, c, b}}]

SparseArray[<7>, {3, 3}]

ArrayRules[%]

{{1, 1} → b, {1, 2} → a, {2, 1} → c, {2, 2} → b,
  {2, 3} → a, {3, 2} → c, {3, 3} → b, {_, _} → 0}

■ Plotting Matrices

In the **LinearAlgebra`MatrixManipulation`** *package:*

MatrixPlot[m] (❀5) Plot the matrix by showing zero elements in white and nonzero elements in black

An option:

MaxMatrixSize Matrices with at most this many rows or columns are shown elementwise; larger matrices are downsampled to this size or less (in this case, a dark cell indicates that at least one of the covered elements is nonzero); value ∞ prevents downsampling; default value: **512**

MatrixPlot[m, w] can be used to show elements with value **w** as white and other elements as black. In addition to **MaxMatrixSize**, the command also has the same options as **ListDensityPlot** (see Section 8.2.3, p. 226), but the default value of some options is different. As an example, a plot of the matrix **p**, which was defined earlier, is as follows:

p // MatrixForm

$$\begin{pmatrix} 0 & 0 & 1 & 1 & 0 \\ 1 & 0 & 1 & 2 & 1 \\ 2 & 1 & 2 & 1 & 1 \\ 2 & 1 & 1 & 3 & 2 \end{pmatrix}$$

<< LinearAlgebra`MatrixManipulation`

```
MatrixPlot[p, Mesh → True, MeshStyle → {GrayLevel[0.5]}];
```

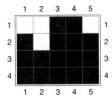

Consider then a matrix of size 100×200:

```
<< Statistics`MultiDiscreteDistributions`
```

```
dist = MultinomialDistribution[10,
  Table[1 / i, {i, 11, 210}] / Sum[1 / i, {i, 11, 210}]];
```

```
ran = RandomArray[dist, {100}];
```

```
MatrixPlot[ran];
```

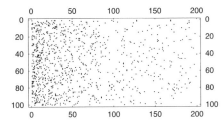

Now we dowsample the matrix to 50×50:

```
MatrixPlot[ran, MaxMatrixSize → 50];
```

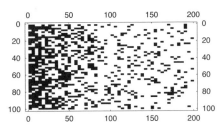

If a matrix is dowsampled, we can use a color function to show the varying density of nonzero elements:

```
MatrixPlot[ran, MaxMatrixSize → 10,
  ColorFunction → (GrayLevel[1 - #] &), ColorFunctionScaling → True];
```

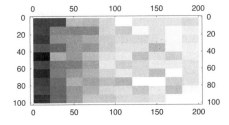

■ Tensors

A tensor of rank k is a k-dimensional table of values. Tensors of rank zero, one, and two are scalars, vectors, and matrices, respectively. With `ArrayDepth[t]` (※5) (with `Tensor⟍ Rank[t]` in *Mathematica* 4) we can obtain the depth of a tensor. For details of tensors, see Section 3.7.11 in Wolfram (2003).

18.2.2 Matrix Calculus

Basic calculations with matrices are easy:

```
a = {{3, 2}, {4, 1}};
b = {{2, 3}, {1, 2}};

3 a       {{9, 6}, {12, 3}}

a + 1     {{4, 3}, {5, 2}}

a + b     {{5, 5}, {5, 3}}
```

However, be careful with usual products and powers: they function element by element, which is probably not what you want. For example, you probably do not want these results:

```
{a b, a^2, a^-1}
```

$$\left\{\{\{6, 6\}, \{4, 2\}\}, \{\{9, 4\}, \{16, 1\}\}, \left\{\left\{\frac{1}{3}, \frac{1}{2}\right\}, \left\{\frac{1}{4}, 1\right\}\right\}\right\}$$

You need the dot (.), `MatrixPower`, and `Inverse` if you want true products, powers, and inverses of matrices.

`Transpose[m]`   Transpose
`Tr[m]`   Trace (the sum of diagonal elements)
`Tr[m, List]`   List of diagonal elements

`Det[m]`   Determinant of a square matrix
`Minors[m]`   Minors of a square matrix
`Minors[m, k]`   **k**th minors

`m1.m2`   Product of matrices
`MatrixPower[m, n]`   The **n**th power of a square matrix
`MatrixExp[m]`   Matrix exponential of a square matrix

`Inverse[m]`   Inverse of a square matrix
`PseudoInverse[m]`   Pseudoinverse (of a possibly rectangular matrix)
`SingularValueList[m]` (※5)   Singular values (of a possibly rectangular matrix)

`Eigenvalues[m]`   Eigenvalues of a square matrix
`Eigenvectors[m]`   Eigenvectors of a square matrix
`Eigensystem[m]`   Eigenvalues and eigenvectors of a square matrix
`CharacteristicPolynomial[m, x]`   Characteristic polynomial of a square matrix

`RowReduce[m]`   Do Gaussian elimination to produce a reduced row echelon form
`NullSpace[m]`   Basis vectors of the null space

> **MatrixRank[m]** (⌘5) Rank
>
> **Norm[m, p]** (⌘5) The p-norm (p is a number in $[1, \infty]$ or **Frobenius**)

All of these commands—except **SingularValueList**—work for both numerical matrices and matrices that contain symbols. Note that singular values and eigenvalues are given in the order of decreasing absolute value (⌘5). **RowReduce** and **PseudoInverse** are also considered in Section 19.1.2, p. 488. Next we consider some of the commands in the box in more detail.

■ Product and Power

Note that, if we write **a b**, the product is formed element by element, and this is probably not a product you want. The normal matrix product is formed with the use of the dot:

>   **a.b**   {{8, 13}, {9, 14}}

Also, powers of matrices like **a^3** are formed element by element, and this is again something you most likely do not want. If you want to calculate the true matrix power, write **a.a.a** or **MatrixPower[a, 3]**:

>   **a.a.a**   {{83, 42}, {84, 41}}

Next we calculate the matrix exponential $e^{\mathbf{a}} = \sum_{i=0}^{\infty} \frac{1}{i!} \mathbf{a}^i$ (⌘4!):

>   **MatrixExp[a]** $\left\{\left\{\dfrac{1 + 2\,e^6}{3\,e}, \dfrac{-1 + e^6}{3\,e}\right\}, \left\{\dfrac{2\,(-1 + e^6)}{3\,e}, \dfrac{2 + e^6}{3\,e}\right\}\right\}$

■ Inverse

Note that **a^-1** only inverts each element. Use **Inverse** to calculate the true inverse matrix:

>   **Inverse[a]** $\left\{\left\{-\dfrac{1}{5}, \dfrac{2}{5}\right\}, \left\{\dfrac{4}{5}, -\dfrac{3}{5}\right\}\right\}$

Matrix inversion is numerically an exacting and risky task. Avoid it whenever possible. Often we do not really need an inverse, because the problem can often be restated into a form where the problem is to solve a linear system of equations. Hilbert matrices are examples of matrices that behave badly, but no problems arise if we calculate with exact numbers:

>   **h[n_] := Table[1 / (i + j - 1), {i, n}, {j, n}]**
>
>   **(h[6].Inverse[h[6]])[[2]]**
>
>   {0, 1, 0, 0, 0, 0}

The result is as it should be. With decimal numbers, however, the result is not so good:

>   **(N[h[6]].Inverse[N[h[6]]])[[2]]**
>
> $\{6.40196 \times 10^{-11}, 1., -9.93259 \times 10^{-11},$
> $1.67937 \times 10^{-11}, -2.35703 \times 10^{-11}, -3.25646 \times 10^{-11}\}$

We can increase the precision of the calculations; here we use 22-digit precision, which gives a good result again:

>   **(N[h[6], 22].Inverse[N[h[6], 22]])[[2]]**
>
> $\{0. \times 10^{-19}, 1.0000000000000000, 0. \times 10^{-16}, -0. \times 10^{-16}, 0. \times 10^{-16}, 0. \times 10^{-16}\}$

■ **Pseudoinverse**

Even if the normal inverse does not exist, we can calculate a pseudoinverse:

```
c = {{3, 2}, {6, 4}};
```

```
Inverse[c]
```

Inverse::sing : Matrix {{3, 2}, {6, 4}} is singular. More…

Inverse[{{3, 2}, {6, 4}}]

```
p = PseudoInverse[c]
```
$\{\{\frac{3}{65}, \frac{6}{65}\}, \{\frac{2}{65}, \frac{4}{65}\}\}$

The pseudoinverse is calculated with singular value decomposition and satisfies the following four identities:

```
{c.p.c == c, p.c.p == p, Transpose[c.p] == c.p, Transpose[p.c] == p.c}
```

{True, True, True, True}

■ **Singular Values**

Singular values are listed from largest to smallest (note that the matrix has to contain at least one decimal point):

```
SingularValueList[a // N]
```
{5.39835, 0.92621}

Singular values can also be calculated as follows:

```
Sqrt[Eigenvalues[a.Conjugate[Transpose[a]]]]
```

$\{\sqrt{5 (3 + 2 \sqrt{2})}, \sqrt{5 (3 - 2 \sqrt{2})}\}$

```
% // N
```
{5.39835, 0.92621}

Users of *Mathematica* with versions prior to version 5 have to use **Singular**-**Values[m][[2]]** in place of **SingularValueList[m]**.

■ **Eigenvalues**

Eigenvalues and eigenvectors of **a** are as follows:

```
Eigenvalues[a]
```
{5, -1}

```
{u, v} = Eigenvectors[a]
```
{{1, 1}, {-1, 2}}

Eigenvalues are given in the order of decreasing absolute value (✿5). Eigenvalues and eigenvectors can also be calculated at the same time:

```
{{λ, μ}, {u, v}} = Eigensystem[a]
```
{{5, -1}, {{1, 1}, {-1, 2}}}

Then we verify:

```
{a.u == λ u, a.v == μ v}
```
{True, True}

With **Eigenvalues[a, k]** (✿5), we can ask for the *k* largest eigenvalues:

```
Eigenvalues[a, 1]
```
{5}

Roots of cubic and quadratic equations are often not written explicitly, because they are often long expressions (✿4!):

```
Eigenvalues[{{3, 2, 4}, {2, 2, 4}, {6, 4, 3}}]
```

$\{\text{Root}[10 - 23\,\#1 - 8\,\#1^2 + \#1^3\ \&,\ 3],$
$\quad \text{Root}[10 - 23\,\#1 - 8\,\#1^2 + \#1^3\ \&,\ 1],\ \text{Root}[10 - 23\,\#1 - 8\,\#1^2 + \#1^3\ \&,\ 2]\}$

Give the option **Cubics → True** (❀5) or **Quartics → True** (❀5) if you want explicit eigenvalues. **Eigenvalues[{a, b}]** (❀5) gives the generalized eigenvalues of **a** with respect to matrix **b**.

Eigenvalues can be easily calculated in the following ways:

```
Solve[Det[a - x IdentityMatrix[2]] == 0, x]     {{x → -1}, {x → 5}}
```

```
Solve[CharacteristicPolynomial[a, x] == 0, x]       {{x → -1}, {x → 5}}
```

If you do not want exact eigenvalues and eigenvectors, it is better to calculate with decimal numbers from the start; this can be done with **Eigenvalues[N[a]]**. If you want to start the calculations with k-digit precision, use **Eigenvalues[N[a, k]]**.

It may happen that a matrix has fewer eigenvectors than the number of rows; in this case, null vectors are added to get the same number of vectors as there are rows.

■ Null Space

The null space of matrix **a** consists of vectors **x** satisfying **a.x == 0**. **NullSpace[a]** gives the basis vectors of this space. For nonsingular matrices, the null space is empty. Define the following matrix:

```
d = {{0, 1, 0, 0, 0}, {0, 0, 0, 1, 1},
     {0, 1, 0, 1, 1}, {0, 0, 0, 0, 0}, {1, 1, 0, 0, 1}};
```

This matrix has a null space spanned by the following vectors:

```
n = NullSpace[d]     {{-1, 0, 0, -1, 1}, {0, 0, 1, 0, 0}}
```

Any linear combination of these vectors multiplied by **d** is a zero vector:

```
d.(α n[[1]] + β n[[2]])     {0, 0, 0, 0, 0}
```

The rank (the number of linearly independent rows or columns) of the matrix is as follows:

```
MatrixRank[d]     3
```

Users of *Mathematica* with versions prior to version 5 have to use **Dimensions[m][[2]] - Length[NullSpace[m]]** in place of **MatrixRank[m]**.

```
Remove["Global`*"]
```

■ Sums

Total[m] (❀5)	The column sums of matrix **m**
Map[Total, m] (❀5)	The row sums of matrix **m**
Total[m, 2] (❀5)	The sum of all elements of matrix **m**
Total[list, n] (❀5)	The sum of the elements of **list** down to level **n**

Users of *Mathematica* versions prior to version 5 have to use **Apply**:

Apply[Plus, m]   The column sums of matrix **m**

Apply[Plus, m, 1]   The row sums of matrix **m**

Apply[Plus, Flatten[m]]   The sum of all elements of matrix **m**

Total[m] is equivalent to **Apply[Plus, m]**. Consider the following matrix:

m = {{1, 2, 3}, {a, b, c}, {A, B, C}};

The column sums can be calculated as follows:

Total[m]   {1 + a + A, 2 + b + B, 3 + c + C}

Apply[Plus, m]   {1 + a + A, 2 + b + B, 3 + c + C}

The row sums can be calculated as follows:

Map[Total, m]   {6, a + b + c, A + B + C}

Apply[Plus, m, 1]   {6, a + b + c, A + B + C}

Calculate the sum of all elements:

Total[m, 2]   6 + a + A + b + B + c + C

Apply[Plus, Flatten[m]]   6 + a + A + b + B + c + C

For sums in which numerical errors cause problems, use **Total** with the option **Method** → **"CompensatedSummation"** to reduce the error.

■ Norms

Norm[m, p] (❀5) The p-norm (p is a number in $[1, \infty]$ or **Frobenius**)

In the **LinearAlgebra`MatrixManipulation`** *package:*

InverseMatrixNorm[m, p]   The p-norm of the inverse of **m**

InverseMatrixNorm[lu, p]   The p-norm of the inverse of **m** when **m** has LU decomposition **lu**

MatrixConditionNumber[m, p]   The p-norm condition number of **m**, that is,
$\| m \|_p \, \| m^{-1} \|_p$

In **Norm**, p can be any number in $[1, \infty]$ or **Frobenius**, but in the commands of the package, p can be 1, 2, or ∞. **Norm[m, 2]** can be written simply as **Norm[m]**. For example:

a = {{3, 2}, {4, 1}};

{Norm[a, 1], Norm[a // N], Norm[a, Frobenius], Norm[a, ∞]}

{7, 5.39835, $\sqrt{30}$, 5}

In place of **Norm**, users of *Mathematica* 4 have to use **MatrixNorm** from the mentioned package (the matrix has to be numerical and contain at least one decimal point).

The basic matrix norms are easy to program (here we have not used **Total**):

```
matrixNorm[m_, p_] := Which[
  p === 1, Max[Apply[Plus, Abs[m]]],
  p === 2, Max[SingularValueList[N[m]]],
  p === "Frobenius", Sqrt[Apply[Plus, Flatten[Abs[m]^2]]],
  p === ∞, Max[Apply[Plus, Abs[m], 1]]]
```

The 1-norm is the largest of the absolute column sums, and the ∞-norm is the largest of the absolute row sums. The 2-norm is the largest of the singular values. The Frobenius norm is the square root of the sum of squared absolute values. For example:

```
Map[matrixNorm[a, #] &, {1, 2, "Frobenius", ∞}]
```

$\{7, 5.39835, \sqrt{30}, 5\}$

18.2.3 Matrix Manipulations

■ Taking Columns and Rows

```
m[[i]]   Take row i
m[[{i1, i2, … }]]   Take rows i1, i2, …
Take[m, spec]   Take the rows specified by spec
Drop[m, spec]   Drop the rows specified by spec
```

```
m[[All, j]]   Take column j (also Transpose[m][[j]])
m[[All, {j1, j2, … }]]   Take columns j1, j2, …
Map[Take[#, spec]&, m]   Take the columns specified by spec
Map[Drop[#, spec]&, m]   Drop the columns specified by spec

In the LinearAlgebra`MatrixManipulation` package:
TakeColumns[m, spec]   Take the columns specified by spec
```

The **spec** in **Take**, **Drop**, and **TakeColumns** can be, for example, of the forms **n**, **-n**, or **{n1, n2}** (see Section 13.2.1, p. 337). Consider the following matrix:

```
MatrixForm[a = {{1, 2, 3}, {p, q, r}, {P, Q, R}}]
```

$$\begin{pmatrix} 1 & 2 & 3 \\ p & q & r \\ P & Q & R \end{pmatrix}$$

We take rows one and three and also columns one and three:

```
a[[{1, 3}]]      {{1, 2, 3}, {P, Q, R}}

a[[All, {1, 3}]]      {{1, 3}, {p, r}, {P, R}}
```

■ Taking Submatrices

> `m[[{i1, i2, … }, {j1, j2, … }]]`   Take the submatrix that has elements with the given row and column indices
>
> `Take[m, {imin, imax}, {jmin, jmax}]`   Take the submatrix that has elements with row and column indices in the given ranges
>
> `Drop[m, {imin, imax}, {jmin, jmax}]`   Drop rows `imin` to `imax` and columns `jmin` to `jmax`
>
> `Partition[list, {n, m}]`   Partition a matrix into blocks of size n×m

    `a[[{2, 3}, {1, 3}]]`   {{p, r}, {P, R}}

    `Take[a, {2, 3}, {1, 3}]`   {{p, q, r}, {P, Q, R}}

    `Partition[a, {2, 1}] // MatrixForm` $\left(\begin{pmatrix} 1 \\ p \end{pmatrix} \begin{pmatrix} 2 \\ q \end{pmatrix} \begin{pmatrix} 3 \\ r \end{pmatrix} \right)$

Note that, in the last command, any leftover elements are dropped.

■ Combining Matrices

> *In the* `LinearAlgebra`MatrixManipulation`` *package:*
>
> `AppendRows[m1, m2, … ]`   Put matrices side by side
>
> `AppendColumns[m1, m2, … ]`   Put matrices one below the other
>
> `BlockMatrix[blocks]`   Form a matrix from the blocks

Define another matrix:

    `b = {{4, 5, 6}, {s, t, u}, {S, T, U}};`

Then join the matrices **a** and **b** in various ways:

    `<< LinearAlgebra`MatrixManipulation` `

    `AppendRows[a, b] // MatrixForm`

$$\begin{pmatrix} 1 & 2 & 3 & 4 & 5 & 6 \\ p & q & r & s & t & u \\ P & Q & R & S & T & U \end{pmatrix}$$

    `AppendColumns[a, b] // MatrixForm`

$$\begin{pmatrix} 1 & 2 & 3 \\ p & q & r \\ P & Q & R \\ 4 & 5 & 6 \\ s & t & u \\ S & T & U \end{pmatrix}$$

```
BlockMatrix[{{a, IdentityMatrix[3]}, {2 IdentityMatrix[3], b}}] //
  MatrixForm
```

$$\begin{pmatrix} 1 & 2 & 3 & 1 & 0 & 0 \\ p & q & r & 0 & 1 & 0 \\ P & Q & R & 0 & 0 & 1 \\ 2 & 0 & 0 & 4 & 5 & 6 \\ 0 & 2 & 0 & s & t & u \\ 0 & 0 & 2 & S & T & U \end{pmatrix}$$

Note that the effect of **AppendRows** and **AppendColumns** is also obtained with **Map** **Thread[Join, {m1, m2, … }]** and **Join[m1, m2, …]**. The effect of **BlockMatrix** can then be obtained by combining these commands.

18.2.4 Matrix Decompositions

Command to decompose M	Form	Conditions for M
LUDecomposition	**L U**	Square
CholeskyDecomposition (❀5)	**L´L**	Square, symmetric, positive definite
QRDecomposition	**Q´R**	–
PolarDecomposition	**U S**	Numeric, at least one decimal point
SingularValueDecomposition (❀5)	**U´W V**	Numeric, at least one decimal point
SchurDecomposition	**Q T Q´**	Square, numeric, at least one decimal point
JordanDecomposition	**S J S⁻¹**	Square

PolarDecomposition is in the **LinearAlgebra`MatrixManipulation`** package. As to Cholesky and singular value decompositions, users of *Mathematica* 4 have to use **Experimental`Cholesky** and **SingularValues**. Some of the decompositions require that the matrix is numeric with at least one decimal point. Such decompositions can thus not be calculated for matrices that have symbols or that are of infinite precision (remember that integers, rational numbers, and special constants have infinite precision).

Before studying the decompositions, we define a matrix:

```
M = {{0.53, 0.88, 0.18}, {0.70, 0.44, 0.17}, {0.17, 0.56, 0.36}};
```

■ LU Decomposition

LU decomposition is used in *Mathematica* to solve linear systems. L and U are lower- and upper-triangular matrices with L having one as the diagonal elements. If M is the matrix to be decomposed, then $M = L U$ (if rows are not permuted). Once the decomposition has been done, the linear system is easily solved by back substitution; this is done with **LUBack** **Substitution**. The result of the decomposition is not the two matrices L and U but rather one matrix containing both L and U. The result also contains a list of permutations representing the permutations of rows (rows are permuted to increase precision). A third component of the result is an estimate of the ∞ norm condition number of the decomposed matrix.

We decompose **M** and solve the system **M.x == c**, where **c** is {5, 2, 4}:

Map[MatrixForm, {LU, perm, cond} = LUDecomposition[M]]

$$\left\{ \begin{pmatrix} 0.7 & 0.757143 & 0.242857 \\ 0.44 & 0.546857 & 0.828631 \\ 0.17 & 0.0512857 & 0.276217 \end{pmatrix}, \begin{pmatrix} 2 \\ 1 \\ 3 \end{pmatrix}, 12.7397 \right\}$$

x = LUBackSubstitution[{LU, perm, cond}, {5, 2, 4}]

{-1.56616, 6.16157, 2.26602}

This is really the solution of the system:

M.x - {5, 2, 4} // Chop   {0, 0, 0}

If we have several different righthand sides **c**, we need to find the LU decomposition only once; just apply the back substitution for each righthand side. With a package, we can get explicit *L* and *U* matrices:

<< LinearAlgebra`MatrixManipulation`

Map[MatrixForm, {L, U} = LUMatrices[LU]]

$$\left\{ \begin{pmatrix} 1. & 0. & 0. \\ 0.757143 & 1. & 0. \\ 0.242857 & 0.828631 & 1. \end{pmatrix}, \begin{pmatrix} 0.7 & 0.44 & 0.17 \\ 0 & 0.546857 & 0.0512857 \\ 0 & 0 & 0.276217 \end{pmatrix} \right\}$$

Now we can check that the decomposition is correct:

M[[perm]] - L.U // Chop

{{0, 0, 0}, {0, 0, 0}, {0, 0, 0}}

LUFactor, which is defined in the **LinearAlgebra`GaussianElimination`** package, works in the same way as **LUDecomposition** (the condition number is not given). The back substitution is done with **LUSolve**.

■ Cholesky Decomposition

The Cholesky decomposition is $M = L'L$, where L is an upper triangular matrix. Define the following symmetric matrix:

M2 = {{0.64, 0.61, 0.66}, {0.61, 0.76, 0.69}, {0.66, 0.69, 0.80}};

It is positive definite:

Eigenvalues[M2]

{2.04494, 0.0999083, 0.0551525}

The *L*-matrix of the Cholesky decomposition is as follows:

(L = CholeskyDecomposition[M2]) // MatrixForm

$$\begin{pmatrix} 0.8 & 0.7625 & 0.825 \\ 0. & 0.422604 & 0.144195 \\ 0. & 0. & 0.313979 \end{pmatrix}$$

M2 - Transpose[L].L // Chop

{{0, 0, 0}, {0, 0, 0}, {0, 0, 0}}

■ QR Decomposition

The QR decomposition is $M = Q' R$, where Q is an orthogonal matrix and R is an upper-triangular matrix:

```
Map[MatrixForm, {Q, R} = QRDecomposition[M]]
```

$$\left\{ \begin{pmatrix} -0.592632 & -0.782722 & -0.19009 \\ 0.523927 & -0.553844 & 0.647115 \\ -0.611791 & 0.283908 & 0.738315 \end{pmatrix}, \begin{pmatrix} -0.894315 & -0.972364 & -0.308169 \\ 0. & 0.579749 & 0.233115 \\ 0. & 0. & 0.203935 \end{pmatrix} \right\}$$

```
M - Conjugate[Transpose[Q]].R // Chop
```

{{0, 0, 0}, {0, 0, 0}, {0, 0, 0}}

Q is orthogonal:

```
Inverse[Q] - Transpose[Q] // Chop
```

{{0, 0, 0}, {0, 0, 0}, {0, 0, 0}}

■ Polar Decomposition

The polar decomposition is $M = U S$, where $U U' = I$ and S is a positive definite matrix:

```
<< LinearAlgebra`MatrixManipulation`
```

```
Map[MatrixForm, {U, S} = PolarDecomposition[M]]
```

$$\left\{ \begin{pmatrix} 0.184446 & 0.912983 & -0.363925 \\ 0.965021 & -0.0980354 & 0.243154 \\ -0.186318 & 0.396044 & 0.89913 \end{pmatrix}, \begin{pmatrix} 0.741597 & 0.482584 & 0.130179 \\ 0.482584 & 0.982074 & 0.290247 \\ 0.130179 & 0.290247 & 0.299517 \end{pmatrix} \right\}$$

```
M - U.S // Chop
```

{{0, 0, 0}, {0, 0, 0}, {0, 0, 0}}

```
U.Conjugate[Transpose[U]] // Chop
```

{{1., 0, 0}, {0, 1., 0}, {0, 0, 1.}}

```
Eigenvalues[S]
```

{1.44299, 0.394413, 0.185784}

■ Singular Value Decomposition

In the singular value decomposition $M = U W V'$, U and V are row orthonormal matrices, and W is a diagonal matrix, with the diagonal elements being the singular values. **Pseudo∢ Inverse** uses singular value decomposition and returns $V W^{-1} U'$.

```
Map[MatrixForm, {U, W, V} = SingularValueDecomposition[M]]
```

$$\left\{ \begin{pmatrix} -0.716157 & 0.156269 & 0.68022 \\ -0.54731 & -0.730522 & -0.408399 \\ -0.433096 & 0.664769 & -0.608695 \end{pmatrix}, \right.$$

$$\left. \begin{pmatrix} 1.44299 & 0. & 0. \\ 0. & 0.394413 & 0. \\ 0. & 0. & 0.185784 \end{pmatrix}, \begin{pmatrix} -0.579564 & -0.800004 & -0.155239 \\ -0.771708 & 0.477566 & 0.419997 \\ -0.261863 & 0.363215 & -0.894149 \end{pmatrix} \right\}$$

```
M - U.W.Conjugate[Transpose[V]] // Chop
```

{{0, 0, 0}, {0, 0, 0}, {0, 0, 0}}

The ratio of the largest to smallest singular value is the 2-norm condition number of M:

```
Max[Tr[W, List]] / Min[Tr[W, List]]        7.76704
```

■ Schur Decomposition

In the Schur decomposition $M = Q T Q'$, Q is an orthogonal matrix, and T is a block upper-triangular matrix.

```
Map[MatrixForm, {Q, T} = SchurDecomposition[M]]
```

$$\left\{ \begin{pmatrix} -0.687844 & -0.486333 & -0.538843 \\ -0.584 & 0.811651 & 0.0129322 \\ -0.431063 & -0.32358 & 0.842307 \end{pmatrix}, \begin{pmatrix} 1.38995 & -0.361635 & 0.123309 \\ 0. & -0.30741 & -0.196508 \\ 0. & 0. & 0.24746 \end{pmatrix} \right\}$$

```
M - Q.T.Conjugate[Transpose[Q]] // Chop
```

{{0, 0, 0}, {0, 0, 0}, {0, 0, 0}}

With **Developer`HessenbergDecomposition[M]** we get, for a square, numerical matrix **M** with at least one decimal point, a decomposition of the form **{P, H}**, satisfying $M = P H P'$, where P is a unitary matrix.

■ Jordan Decomposition

In the result $M = S J S^{-1}$ of the Jordan decomposition, S is a similarity matrix, and J is the Jordan canonical form of M (J is usually a diagonal matrix of eigenvalues, but generally J can also have ones directly above the diagonal).

```
Map[MatrixForm, {S, J} = JordanDecomposition[M]]
```

$$\left\{ \begin{pmatrix} -0.687844 & -0.61899 & -0.198684 \\ -0.584 & 0.672139 & -0.134772 \\ -0.431063 & -0.406301 & 0.970753 \end{pmatrix}, \begin{pmatrix} 1.38995 & 0 & 0 \\ 0 & -0.30741 & 0 \\ 0 & 0 & 0.24746 \end{pmatrix} \right\}$$

```
M - S.J.Inverse[S] // Chop
```

{{0, 0, 0}, {0, 0, 0}, {0, 0, 0}}

Equations

Introduction

> *Someone told me that each equation I included in the book would halve the sales.* — *Stephen Hawking*

Equations can be classified as *linear, polynomial, radical,* or *transcendental*. Polynomial equations consist of sums of integer powers of variables, while radical equations may contain rational powers. Transcendental equations contain transcendental functions like $\sin(x)$ and $\log(x + y)$. The main command for linear, polynomial, and radical equations is **Solve**, and for transcendental equations it is **FindRoot**. However, other commands can also be used; these are explained next.

If we have linear equations in the form of a coefficient matrix and righthand-side vector, then **LinearSolve** is easy to use. On the other hand, for sparse linear systems, **Solve** is recommended.

For polynomial equations, **Solve** gives an answer for *generic* values of the possible parameters of the equations. If an exhaustive analysis of the solutions is wanted for *all* possible values of the parameters in polynomial equations, then **Reduce** can be used. If **Solve** cannot obtain exact solutions for polynomial equations, then **NSolve** can be used to calculate the solutions numerically.

For transcendental equations, we can sometimes apply **Solve** or **Reduce** (see Section 19.3.1, p. 501), but usually we have to resort to the iterative methods provided by **FindRoot** (Newton's, Brent's, or the secant method).

Here is a summary of the commands for equations:

- *linear equations:* **Solve**, **LinearSolve**;
- *polynomial and radical equations:* **Solve**, **Reduce**, **NSolve**; and
- *transcendental equations:* **Solve**, **Reduce**, **FindRoot**.

Exact symbolic solutions can be obtained with **Solve**, **LinearSolve**, and **Reduce**. Numeric methods are used with **NSolve** and **FindRoot**.

Note that we have considered the fixed-point method in Section 15.2.3, p. 398, and Newton's method in Sections 14.3.3, p. 373; 14.3.4, p. 376; 15.1.1, p. 383; 15.1.3, p. 390; 15.2.1, p. 393; 15.2.2, p. 395; and 15.2.3, p. 399. In Section 5.3.2, p. 125, we used **FindRoot** to plot a function of an implicit function. Note also that *inequalities* are considered in Section 19.2.4, p. 499.

19.1 Linear Equations

19.1.1 Two Representations

Linear systems can be represented by either writing down the equations explicitly with variables or by giving the lefthand-side coefficient matrix and the righthand-side vector. Appropriate commands for these representations are **Solve** and **LinearSolve**, respectively.

■ Giving the Equations

> **Solve[eqns]**   Solve the equations **eqns** for the symbols therein
> **Solve[eqns, vars]**   Solve the equations **eqns** for the variables **vars**

Here **eqns** is a list of equations, and **vars** is a list of variables. As an example, we write a system of two linear equations and then solve the system (remember that equations are defined with two equal signs (==), but *Mathematica* replaces them with the special symbol ==):

    eqns = {x + 2 y == 7, 2 x - y == 4};

    sol = Solve[eqns]   {{x → 3, y → 2}}

The result is a list of transformation rules. A system of equations may have several solutions, and, accordingly, the general form of the result of **Solve** is {sol_1, sol_2, ... }, where each sol_i is a list of rules for the variables. In our example, the equations have only one solution, and so the result is {sol_1}.

For unique solutions, it may be convenient to get rid of the outermost curly braces. We can do this by picking the first and only element of **sol**:

    sol2 = sol[[1]]   {x → 3, y → 2}

We can check the solution:

    eqns /. sol2   {True, True}

We can also insert the solution into other expressions:

    x y /. sol2   6

If a list of values is wanted as the solution, we can write the following:

    {x, y} /. sol2   {3, 2}

And if we want to assign the solution into **{x, y}**, we can write the following:

    {x, y} = {x, y} /. sol2   {3, 2}

Now $x = 3$ and $y = 2$:

    {x, y}   {3, 2}

You may want to reread Section 12.1.2, p. 300, where we considered transformation rules. We now clear all of our assignments:

    Clear[a, b, x, y]

If symbols appear as coefficients, then the variables have to be given:

    Solve[{x + y == 2 a, x - y == 2 b}, {x, y}]   {{x → a + b, y → a - b}}

Here we have a system with no solutions:

    Solve[{3 x + y == 9, 6 x + 2 y == 4}]   {}

The result is an empty list, which indicates that no solutions exist. Here we have infinitely many solutions:

    Solve[{3 x + y == 9, 6 x + 2 y == 18}]

    Solve::svars :
    Equations may not give solutions for all "solve" variables. More…

$$\left\{\left\{x \to 3 - \frac{y}{3}\right\}\right\}$$

This means that y can be arbitrary and $x = 3 - y/3$.

    For *sparse* systems, **Solve** uses special methods that are efficient for such systems (a linear system is sparse if the coefficient matrix contains many zeros). The special methods are available if the coefficients are real or complex machine numbers. (**LinearSolve** does not have special methods for sparse systems.)

■ Giving the Coefficients

LinearSolve[a, b] Solve the system **a.x == b** of linear equations

---

In the simplest case, **x** is the vector of unknowns, **a** is a square matrix, and **b** is a vector (generalizations are considered in Section 19.1.2, p. 486). For example, let **a** and **b** be as follows:

    a = {{1, 2}, {2, -1}};
    b = {7, 4};

The equations are then $x + 2y = 7$ and $2x - y = 4$. Solve the equations:

    sol = LinearSolve[a, b]   {3, 2}

Thus, $x = 3$ and $y = 2$. We could also use **Solve**:

> **Solve[a.{x, y} == b]**   {{x → 3, y → 2}}

To check the solution, write this:

> **a.sol – b**   {0, 0}

If the coefficient matrix is singular (i.e., has a zero determinant), the system usually has no solutions:

> **LinearSolve[{{3, 1}, {6, 2}}, {9, 4}]**
>
> LinearSolve::nosol :
>   Linear equation encountered which has no solution. More...
> LinearSolve[{{3, 1}, {6, 2}}, {9, 4}]

Sometimes an infinite number of solutions exists:

> **LinearSolve[{{3, 1}, {6, 2}}, {9, 18}]**   {3, 0}

In this example, all solutions are of the form {3 – y/3, y}; **LinearSolve** gives one of the possible solutions (**Solve**, as we saw above, gives all solutions).

Inverting the coefficient matrix is one possibility for solving linear equations, but this method is not recommended (inversion of a matrix is more demanding than solving linear equations):

> **Inverse[a].b**   {3, 2}

With the **LinearAlgebra`MatrixManipulation`** package, we can find the coefficient matrix and righthand-side vector of explicitly written equations:

> **<< LinearAlgebra`MatrixManipulation`**
>
> **eqns = {x + 2 y == 7, 2 x – y == 4};**
>
> **{a, b} = LinearEquationsToMatrices[eqns, {x, y}]**
>
> {{{1, 2}, {2, –1}}, {7, 4}}

On the other hand, if we have the matrices, we get explicit equations by writing the following:

> **Thread[a.{x, y} == b]**
>
> {x + 2 y == 7, 2 x – y == 4}

19.1.2 Special Topics

■ Eliminating Variables

> **Solve[eqns, vars, elims]**   Attempt to solve the equations **eqns** for the variables **vars**, eliminating the variables **elims**

Define two equations:

> **eqns = {x – y == c, x + y == 2 c};**

We can solve for x and y:

> **Solve[eqns, {x, y}]** $\{\{x \to \dfrac{3\,c}{2}, y \to \dfrac{c}{2}\}\}$

We can also ask to eliminate c:

```
Solve[eqns, {x, y}, {c}]
```

```
Solve::svars :
 Equations may not give solutions for all "solve" variables. More…
```
$\{\{x \to 3\, y\}\}$

■ Several Systems

> $\mathbf{f = LinearSolve[a]}$ (❀5) Give a function $\mathbf{f}$ for which $\mathbf{f[b]}$ solves the equation
> $\mathbf{a.x == b}$
> $\mathbf{f[b1]}, \mathbf{f[b2]}, \dots$ Solve the systems $\mathbf{a.x == b1}$, $\mathbf{a.x == b2}$, …

This is a handy way to solve several systems that have the same lefthand-side matrix but different righthand-side vectors. In the following example, we have two righthand sides: $\{1, 2, 3\}$ and $\{-4, 5, 6\}$:

```
a = {{2, 1, 1}, {1, 1, 1}, {1, 0, 1}};
b1 = {1, 2, 3};
b2 = {-4, 5, 6};
```
First, ask for the solution as a function:

```
f = LinearSolve[a]    LinearSolveFunction[{3, 3}, <>]
```
We are then able to solve systems with various righthand-side vectors:

```
{f[b1], f[b2]}     {{-1, -1, 4}, {-9, -1, 15}}
```
So, the corresponding solutions are $\{-1, -1, 4\}$ and $\{-9, -1, 15\}$.

Another way to solve several systems with the same lefthand-side matrix is to use **LUDecomposition**, which is considered next.

■ LU Decomposition

In Section 18.2.4, p. 479, we considered matrix decompositions. One of them was LU decomposition, which is used to solve linear systems.

> $\mathbf{lu = LUDecomposition[a]}$ Find the LU decomposition of the square matrix $\mathbf{a}$
> $\mathbf{LUBackSubstitution[lu, b]}$ Solve $\mathbf{a.x = b}$

Section 18.2.4 contained an example, but here is one more. We use $\mathbf{a}$, $\mathbf{b1}$, and $\mathbf{b2}$ as defined above:

```
lu = LUDecomposition[a]
```
$\{\{\{1, 1, 1\}, \{2, -1, -1\}, \{1, 1, 1\}\}, \{2, 1, 3\}, 1\}$

```
LUBackSubstitution[lu, b1]     {-1, -1, 4}
```

```
LUBackSubstitution[lu, b2]     {-9, -1, 15}
```
We see that we can nicely solve several systems with the same lefthand-side matrix without having to calculate the decomposition several times.

The logic of the LU decomposition for the system $a\,x = b$ is that when L and U have been obtained, the original system is replaced with two very easy triangular systems: first we solve $L y = b$ for y and then $U x = y$ for x; then $a x = (L U) x = L y = b$. This means that x is the solution. Triangular systems are solved very easily: y is solved by forward substitution, and then x is solved by backward substitution.

The package **LinearAlgebra`GaussianElimination`** defines **LUFactor** and **LUSolve** that work like **LUDecomposition** and **LUBackSubstitution**.

■ Gaussian Elimination

> **RowReduce[a]**   Do Gaussian elimination to **a** to produce a row-reduced echelon form

RowReduce can be used to solve linear systems. We solve the system **a.x == b1** we defined earlier. First we append the elements of the righthand-side vector to the rows of the lefthand-side matrix:

```
m = Transpose[Append[Transpose[a], b1]]
```
```
{{2, 1, 1, 1}, {1, 1, 1, 2}, {1, 0, 1, 3}}
```

Then we do a Gaussian elimination:

```
r = RowReduce[m]
```
```
{{1, 0, 0, -1}, {0, 1, 0, -1}, {0, 0, 1, 4}}
```

The solution is the last column of this matrix:

```
Map[Last, r]     {-1, -1, 4}
```

■ Tridiagonal Systems

> *In the* **LinearAlgebra`Tridiagonal`** *package:*
>
> **TridiagonalSolve[sub, main, super, rhs]**   Solve a tridiagonal system

Here **sub**, **main**, and **super** are the sub-, main, and superdiagonals of the coefficient matrix (all other elements of the matrix are zero), and **rhs** is the righthand side. Such tridiagonal systems can be solved very efficiently with Gaussian elimination and back substitution. As an example, we solve a tridiagonal system where the coefficient matrix is as follows:

```
{{4, 2, 0}, {1, 4, 2}, {0, 1, 4}} // MatrixForm
```
$$\begin{pmatrix} 4 & 2 & 0 \\ 1 & 4 & 2 \\ 0 & 1 & 4 \end{pmatrix}$$

The righthand-side vector is $\{6, 4, 7\}$:

```
sub = {1, 1}; main = {4, 4, 4}; super = {2, 2}; rhs = {6, 4, 7};
```

```
<< LinearAlgebra`Tridiagonal`
```

```
TridiagonalSolve[sub, main, super, rhs]
```
$$\left\{ \frac{5}{3}, -\frac{1}{3}, \frac{11}{6} \right\}$$

■ Overdetermined Systems

In addition to square linear systems, *Mathematica* can handle rectangular linear systems. If there are more equations than variables, the system is *overdetermined,* and solutions usually do not exist. **Solve** and **LinearSolve** can be tried: they give a solution if it exists and otherwise give an empty list (**Solve**) or tell us that no solutions exist (**LinearSolve**). A fairly good approximating solution can, however, always be obtained with a pseudo-inverse (or a generalized inverse or Moore–Penrose inverse).

> **PseudoInverse[a].b** Give an approximating solution of an overdetermined system
>   **a.x == b**

This can be compared with **Inverse[a].b** (see Section 19.1.1, p. 486). Consider the following overdetermined system:

    a = {{3, 1}, {2, 5}, {8, 1}};
    b = {2, 7, 5};

An approximating solution is as follows:

    apprsol = PseudoInverse[a].b // N   {0.450549, 1.20513}

The lefthand sides of the equations with these values for the variables are close to the righthand sides $\{2, 7, 5\}$:

    a.apprsol   {2.55678, 6.92674, 4.80952}

Indeed, the solution x has the property that if $r = ax - b$ is the residual vector, then the sum of the squares of the residuals, $r.r$, is minimized by this x.

■ Underdetermined Systems

If there are more variables than equations, the system is *underdetermined,* and an infinite number of solutions usually exists. For example:

    a = {{4, 3, 1}, {3, 2, 5}};
    b = {9, 4};

Solve gives all solutions:

    Solve[a.{x, y, z} == b]

    Solve::svars :
     Equations may not give solutions for all "solve" variables. More…
     {{x → -6 - 13 z, y → 11 + 17 z}}

LinearSolve gives one solution:

    s = LinearSolve[a, b]   {-6, 11, 0}

PseudoInverse also gives one solution (exact this time):

    PseudoInverse[a].b // N   {1.50545, 1.18519, -0.577342}

19.1.3 Iterative Methods

There are a number of iterative solution methods for linear systems. Here we implement Jacobi's method and the Gauss–Seidel method. Consider the following diagonally dominant system:

```
a = {{4, 1, -2}, {-1, 4, 3}, {2, 1, -3}}; b = {5, 2, 2};
```

```
n = 3; y = Array[x, n]      {x[1], x[2], x[3]}
```

```
eqns = Thread[a.y == b]
```

$\{4\,x[1] + x[2] - 2\,x[3] == 5,\ -x[1] + 4\,x[2] + 3\,x[3] == 2,\ 2\,x[1] + x[2] - 3\,x[3] == 2\}$

The solution of this system is as follows:

```
sol = y /. Solve[eqns][[1]] // N
```

```
{1.30769, 0.538462, 0.384615}
```

■ **Jacobi's Method**

First we solve x_i from the ith equation:

```
newy = y /. Table[Solve[eqns[[i]], x[i]][[1, 1]], {i, n}]
```

$\{\frac{1}{4}\,(5 - x[2] + 2\,x[3]),\ \frac{1}{4}\,(2 + x[1] - 3\,x[3]),\ \frac{1}{3}\,(-2 + 2\,x[1] + x[2])\}$

With Jacobi's method, we start from a point, for example, $(0, 0, ..., 0)$, and then iteratively calculate new values for $x_1, ..., x_n$ from **newy**. We do 30 iterations:

```
yi = {0, 0, 0};
Do[yi = newy /. Thread[y → yi] // N, {30}];
yi
```

```
{1.30769, 0.538462, 0.384615}
```

This is quite a good approximation to the solution:

```
sol - yi
```

$\{2.96988 \times 10^{-8},\ 1.02619 \times 10^{-9},\ 2.91744 \times 10^{-8}\}$

In summary, here is a simple implementation of Jacobi's method with fixed number **iters** of iterations:

```
Do[yi = newy /. Thread[y → yi] // N, {iters}]
```

■ **Gauss–Seidel Method**

To accelerate convergence of Jacobi's method, the already calculated new values for $x_1, ...,$ x_i can be used in the calculation of the new values for x_{i+1}, $i = 1, ..., n - 1$. This is the Gauss–Seidel method. We again start from $(0, 0, 0)$ and do 30 iterations:

```
yi = {0, 0, 0};
Do[yi = Table[yi[[j]] = newy[[j]] /. Thread[y → yi], {j, n}] // N, {30}];
yi
```

```
{1.30769, 0.538462, 0.384615}
```

The point here is to do the update equation by equation, with **Table**, and to assign the new value **newy[[j]]** directly to **yi[[j]]** so that this value is used in the remaining updates when the substitution **y → yi** is made. The solution is much better than the solution offered by Jacobi's method:

    sol - yi

    $\{8.92619 \times 10^{-14}, -5.50671 \times 10^{-14}, 4.12448 \times 10^{-14}\}$

In summary, here is a simple implementation of the Gauss–Seidel method with fixed number **iters** of iterations:

```
Do[yi = Table[yi[[j]] = newy[[j]] /. Thread[y → yi], {j, n}] // N, {iters}]
```

We can form a better program by adding a stopping test. We use the functional style (for **FixedPoint**, see Section 15.2.3, p. 397).

```
gaussSeidelStep[newy_, y_, yi_, n_] :=
  Module[{zi = yi}, Table[zi[[j]] = newy[[j]] /. Thread[y → zi], {j, n}] // N]

gaussSeidel[eqns_, y_, y0_, eps_: 10^-10, maxit_: 100] :=
  Module[{n = Length[y], newy},
    newy = y /. Table[Solve[eqns[[i]], y[[i]]][[1, 1]], {i, n}];
    FixedPoint[gaussSeidelStep[newy, y, #, n] &, y0, maxit,
      SameTest → (Sqrt[(#1 - #2).(#1 - #2)] < eps &)]]
```

Here, **gaussSeidelStep** does one step of the method. We have substituted the present iteration **yi** into a new variable **zi**, because the arguments of a function cannot be changed inside the function (note that **yi** is one of the arguments of **gaussSeidelStep**). The result of this function is the new approximation.

The function **gaussSeidel** organizes the whole calculation. The iterations start from **y0** and are continued until the two last approximations differ by less than **eps**; however, at most **maxit** iterations are done. The default values of **eps** and **maxit** are 10^{-10} and 100, respectively (for default values, see Section 15.3.3, p. 409). If these values suit you, you need not specify the values of **eps** and **maxit** at all (note that if you want to modify the value of **maxit**, you also have to give a value for **eps**). For example:

    gs = gaussSeidel[eqns, y, {0, 0, 0}]

    $\{1.30769, 0.538462, 0.384615\}$

    sol - gs

    $\{2.69744 \times 10^{-11}, -1.58236 \times 10^{-11}, 1.27084 \times 10^{-11}\}$

    Remove["Global`*"]

19.2 Polynomial and Radical Equations

19.2.1 Polynomial Equations

■ Exact Solution

> **Solve[eqn]**   Solve the equation **eqn** for the symbol in the equation
> **Solve[eqn, x]**   Solve the equation **eqn** for the variable **x**

If the equation contains only one symbol, it need not be mentioned in the command. Here is the familiar second-order equation:

> **eqn = a x^2 + b x + c == 0;**

> **sol = Solve[eqn, x]**

$$\left\{\left\{x \to \frac{-b - \sqrt{b^2 - 4\,a\,c}}{2\,a}\right\}, \left\{x \to \frac{-b + \sqrt{b^2 - 4\,a\,c}}{2\,a}\right\}\right\}$$

The solutions are given as transformation rules (see Section 12.1.2, p. 300). We can verify the solutions:

> **eqn /. sol // Simplify**

> {True, True}

(For both solutions, the equation simplifies to 0 == 0, which is **True**).

If you want a list of values (rather than a list of rules), enter the following:

> **x /. sol**

$$\left\{\frac{-b - \sqrt{b^2 - 4\,a\,c}}{2\,a}, \frac{-b + \sqrt{b^2 - 4\,a\,c}}{2\,a}\right\}$$

You can also type this:

> **x /. Solve[eqn, x]**

$$\left\{\frac{-b - \sqrt{b^2 - 4\,a\,c}}{2\,a}, \frac{-b + \sqrt{b^2 - 4\,a\,c}}{2\,a}\right\}$$

Note, however, that the transformation rules are handy in that they can easily be used to calculate the value of any expression:

> **b + 2 a x /. sol**

$$\left\{-\sqrt{b^2 - 4\,a\,c}, \sqrt{b^2 - 4\,a\,c}\right\}$$

■ Special Questions

Solve is able to solve all polynomials up to order four. Polynomials of order five and greater can also sometimes be solved:

> **Solve[-6 + 23 x - 34 x^2 + 24 x^3 - 8 x^4 + x^5 == 0]**

> {{x → 1}, {x → 1}, {x → 1}, {x → 2}, {x → 3}}

(Here 1 is a zero of multiplicity 3.) However, it often happens that such high-order equations cannot be solved exactly:

```
sol = Solve[x^5 - x + 1 == 0]
```

$$\{\{x \rightarrow \text{Root}[1 - \#1 + \#1^5 \,\&,\, 1]\},$$
$$\{x \rightarrow \text{Root}[1 - \#1 + \#1^5 \,\&,\, 2]\},\, \{x \rightarrow \text{Root}[1 - \#1 + \#1^5 \,\&,\, 3]\},$$
$$\{x \rightarrow \text{Root}[1 - \#1 + \#1^5 \,\&,\, 4]\},\, \{x \rightarrow \text{Root}[1 - \#1 + \#1^5 \,\&,\, 5]\}\}$$

Solve gives a symbolic representation for the solution with the **Root** object. **Root[f, k]** represents the **k**th root of the equation **f == 0** (for **Root**, see Section 19.2.2, p. 496). We can, however, ask for numerical values for the roots:

```
sol // N
```

$$\{\{x \rightarrow -1.1673\},\, \{x \rightarrow -0.181232 - 1.08395\, i\},\, \{x \rightarrow -0.181232 + 1.08395\, i\},$$
$$\{x \rightarrow 0.764884 - 0.352472\, i\},\, \{x \rightarrow 0.764884 + 0.352472\, i\}\}$$

If the solution contains powers of –1, we can transform them with **ComplexExpand** to expressions that contain trigonometric functions, which in turn sometimes automatically reduce to radicals (i.e., to arithmetic combinations of various roots):

```
sol = x /. Solve[x^3 + 1 == 0]
```

$$\{-1,\, (-1)^{1/3},\, -(-1)^{2/3}\}$$

```
sol // ComplexExpand
```

$$\left\{-1,\, \frac{1}{2} + \frac{i\sqrt{3}}{2},\, \frac{1}{2} - \frac{i\sqrt{3}}{2}\right\}$$

Solve uses explicit formulas up to degree four. For higher-order polynomials, **Solve** attempts to reduce polynomials using **Factor** and **Decompose**, and **Solve** recognizes cyclotomic and other special polynomials.

With the **Algebra`RootIsolation`** package, we can count roots and find isolating intervals for them.

■ Numerical Solution

Solve[eqn] Try exact solution

% // N If this did not succeed, use numerical methods (**NSolve** is eventually used)

NSolve[eqn] Solve **eqn** numerically for the symbol therein

NSolve[eqn, x] Solve **eqn** numerically for **x**

NSolve[eqn, x, n] Solve **eqn** numerically for **x**, using **n**-digit precision in the calculations

NSolve is based on the Jenkins–Traub algorithm. For example:

```
p = x^5 - x + 1;
```

```
sol = NSolve[p == 0]
```

$$\{\{x \rightarrow -1.1673\},\, \{x \rightarrow -0.181232 - 1.08395\, i\},\, \{x \rightarrow -0.181232 + 1.08395\, i\},$$
$$\{x \rightarrow 0.764884 - 0.352472\, i\},\, \{x \rightarrow 0.764884 + 0.352472\, i\}\}$$

The solution does not pass the test that we should have **p == 0** at the roots:

```
p == 0 /. sol
```

```
{False, False, False, False, False}
```

This is a normal situation with numerical solutions. If we instead ask for the value of the polynomial at the roots, we observe that the roots are very good (✻4!):

```
p /. sol
```

$\{-2.22045 \times 10^{-15}, 0. - 2.22045 \times 10^{-16}\, i, 0. + 2.22045 \times 10^{-16}\, i,$
$3.60822 \times 10^{-16} - 4.44089 \times 10^{-16}\, i, 3.60822 \times 10^{-16} + 4.44089 \times 10^{-16}\, i\}$

Here is a plot of the solutions in the complex plane:

```
ListPlot[{Re[x], Im[x]} /. sol, PlotStyle → AbsolutePointSize[3]];
```

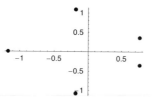

■ Several Polynomial Equations

`Solve[eqns, vars]`	Solve the equations for **vars**
`NSolve[eqns, vars]`	Solve numerically for the given variables

For simultaneous equations we write a list of equations and a list of variables (the latter is not needed if the solution is wanted for all symbols in the equations). When solving systems of polynomial equations, **Solve** actually constructs a Gröbner basis with **GroebnerBasis**.

In the following example, we get two solutions:

```
Solve[{x^2 + y^2 == 5, x + y == 1}]
```

$\{\{x \to -1, y \to 2\}, \{x \to 2, y \to -1\}\}$

The following system has six solutions:

```
NSolve[{x^2 + y^3 - x y == 0, x + y + x^2 - 1 == 0}]
```

$\{\{x \to 1.13665, y \to -1.42864\},$
$\{x \to -1.43152 + 0.695043\, i, y \to 0.865346 + 1.2949\, i\},$
$\{x \to -1.43152 - 0.695043\, i, y \to 0.865346 - 1.2949\, i\},$
$\{x \to -2.06867, y \to -1.21074\},$
$\{x \to 0.397534 - 0.0995355\, i, y \to 0.454339 + 0.178673\, i\},$
$\{x \to 0.397534 + 0.0995355\, i, y \to 0.454339 - 0.178673\, i\}\}$

19.2.2 Special Topics

■ Eliminating Variables

`Solve[eqns, vars, elims]`	Solve **eqns** for **vars**, eliminating **elims**
`Eliminate[eqns, elims]`	Eliminate **elims** from **eqns**

In the following example, we ask to eliminate **y** and solve for **x**:

```
eqns = {x^2 + y^2 == a, x + y == b};
```

```
Solve[eqns, x, y]
```

$$\left\{\left\{x \rightarrow \frac{1}{2}\left(b-\sqrt{2\,a-b^2}\,\right)\right\},\left\{x \rightarrow \frac{1}{2}\left(b+\sqrt{2\,a-b^2}\,\right)\right\}\right\}$$

Next we ask for an elimination only. The result is an equation or several equations:

```
Eliminate[eqns, y]    a - 2 x² == b² - 2 b x
```

Here is another example:

```
Eliminate[{x^2 + y^3 == x y, x + y + x^2 == 1}, x]
```

$$-3\,y + y^2 + 4\,y^3 - y^4 + y^6 == -1$$

■ Making Equations Valid for All Values of Some Variables

> **SolveAlways[eqns, vars]** Give conditions for the parameters appearing in **eqns** that make the equations valid for all values of the variables **vars**

SolveAlways may be useful, for example, in the method of undetermined coefficients. In this method, we want to find conditions under which a trial expression is a solution of an equation. Consider the following differential equation:

```
eqn = y'[t] + a y[t] + b == 0    b + a y[t] + y'[t] == 0
```

Define the following function:

```
z[t_] := c + d Exp[e t]
```

We examine whether this function could be a solution of the equation for some values of the parameters c, d, and e. First we insert the function into the equation:

```
test = eqn[[1]] /. y → z    b + d e e^(e t) + a (c + d e^(e t))
```

Because **SolveAlways** does not handle transcendental equations, we form a series expansion:

```
test2 = Series[test, {t, 0, 6}]
```

$$(b + a\,c + a\,d + d\,e) + (a\,d\,e + d\,e^2)\,t + \left(\frac{1}{2}\,a\,d\,e^2 + \frac{d\,e^3}{2}\right)t^2 + \left(\frac{1}{6}\,a\,d\,e^3 + \frac{d\,e^4}{6}\right)t^3 +$$
$$\left(\frac{1}{24}\,a\,d\,e^4 + \frac{d\,e^5}{24}\right)t^4 + \left(\frac{1}{120}\,a\,d\,e^5 + \frac{d\,e^6}{120}\right)t^5 + \left(\frac{1}{720}\,a\,d\,e^6 + \frac{d\,e^7}{720}\right)t^6 + O[t]^7$$

Then we find conditions under which this expression is identically zero for all t:

```
cond = SolveAlways[test2 == 0, t]
```

$$\{\{b \rightarrow -a\,c,\ d \rightarrow 0\},\ \{b \rightarrow -a\,c,\ e \rightarrow -a\},\ \{b \rightarrow -a\,c - a\,d,\ e \rightarrow 0\},$$
$$\{b \rightarrow 0,\ e \rightarrow 0,\ a \rightarrow 0\}\}$$

The second solution gives the solution $-\frac{b}{a} + d\,e^{-a\,t}$ for the differential equation. This solution is the same as that obtained with **DSolve**:

```
DSolve[eqn, y[t], t]
```

$$\left\{\left\{y[t] \rightarrow -\frac{b}{a} + e^{-a\,t}\,C[1]\right\}\right\}$$

Another example is in Section 23.2.4, p. 610, where we solve a Fredholm integral equation.

■ **Unstable Systems**

> *An option for* **Solve**:
>
> **VerifySolutions** Whether solutions are verified: extraneous solutions rejected or
> inaccurate numerical solutions improved; possible values: **Automatic**, **True**, **False**

When solving a *single* equation with **Solve**, the solution is automatically verified. The verification is important when solving radical equations, where extraneous solutions easily emerge (see Section 19.2.3, p. 499). For *systems* of equations, the verification is not done automatically. If we want to verify solutions for systems, we can set **Verify⌐ Solutions → True**. For numerically unstable systems, this setting may improve the solution. Consider the following system:

```
p1 = -49.8333 + 703.3295 x^2 + 1022.7811 x y + 895.0554 y^2;
p2 = -54.8990 + 791.4604 x^2 + 1150.9409 x y + 959.1353 y^2;
```

We use **Solve** and **NSolve** and calculate the values of the polynomials **p1** and **p2** at the solutions (we do not show the four solutions, and we show the values of the polynomials at only the first solution) (❀4!):

```
{p1, p2} /. Solve[{p1 == 0, p2 == 0}][[1]]
```

$\{-0.00125683, -0.00141431\}$

```
{p1, p2} /. NSolve[{p1 == 0, p2 == 0}][[1]]
```

$\{1.15734 \times 10^{-9}, 1.3422 \times 10^{-9}\}$

NSolve gave a better solution. To get more accurate solutions with **Solve**, we use the **VerifySolutions** option (note that **NSolve** does not have this option):

```
{p1, p2} /. Solve[{p1 == 0, p2 == 0}, VerifySolutions → True][[1]]
```

$\{1.42109 \times 10^{-14}, 1.06581 \times 10^{-14}\}$

■ **Root Objects**

> **Root[f, k]** The **k**th root of the polynomial equation **f[x] == 0** (**f** must be a pure
>   function)
> **RootReduce[expr]** Attempts to reduce **expr** to a single **Root** object
> **RootSum[f, form]** The sum of **form[x]** for all **x** that satisfy the polynomial equation
>   **f[x] == 0**

If **Solve** cannot solve an equation, it represents the solution by the **Root** object. Consider the following equation and its solution:

```
sol = x /. Solve[a - x + x^5 == 0, x]
```

$\{\text{Root}[a - \#1 + \#1^5 \&, 1], \text{Root}[a - \#1 + \#1^5 \&, 2],$
$\text{Root}[a - \#1 + \#1^5 \&, 3], \text{Root}[a - \#1 + \#1^5 \&, 4], \text{Root}[a - \#1 + \#1^5 \&, 5]\}$

Root objects are exact though implicit representations for the roots. We can plot a root as a function of **a**:

```
Plot[Evaluate[sol[[1]]], {a, -2, 1}];
```

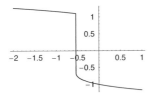

RootSum can be used to calculate the sum of the values of a function at the solutions of a polynomial equation. We calculate the sum of roots and their inverses:

```
RootSum[a - # + #^5 &, # &]     0
```

$$\text{RootSum[a - \# + \#^5 \&, 1 / \# \&]} \qquad \frac{1}{a}$$

Some integrals are expressed in terms of **RootSum**:

```
Integrate[1 / (1 + x + x^3), {x, 0, ∞}]
```

$$-\text{RootSum}\left[1 + \#1 + \#1^3 \&, \frac{\text{Log}[-\#1]}{1 + 3\,\#1^2} \&\right]$$

```
% // N     0.921763 + 0. i
```

■ Detailed Solution

When solving equations, **Solve** produces solution candidates and then verifies which ones are correct. Note that **Solve** rejects only solution candidates that are incorrect for *all* values of parameters; candidates that are valid for at least some values of the parameters are accepted. Also note that the candidates that **Solve** accepts are *generic* solutions, which means that they are solutions that are valid for *general* values of the parameters; for special values of parameters, the solution may be different. For example, the solution for the general quadratic equation $ax^2 + bx + c = 0$ is valid only if a is not 0. If a is 0, the solution is $x = -c/b$. If b is also 0, then c must be zero for the equation to be satisfied. Such a detailed solution can be obtained with **Reduce**.

Reduce[eqns, vars]   Give a detailed analysis of the solutions of given equations

Options:
Cubics (❀5) Whether cubic equations are solved explicitly; possible values: **False**,
   True
Quartics (❀5) Whether quartic equations are solved explicitly; possible values:
   False, True
Backsubstitution (❀5) Whether values of later variables in the result are allowed to
   depend on earlier variables (**False**) or are given explicitly (**True**); possible values:
   False, True

Actually, **Reduce** can solve much more general conditions (❀5) containing logical combinations of equations (==), inequations (!=), inequalities (<, ≤, etc.), domain specifications (x ∈ **Reals**, etc., see Section 12.2.1, p. 305), and universal (∀ or **ForAll** [❀5]) and existential (∃ or **Exists** [❀5]) quantifiers. For example (❀4!):

Reduce[a x^2 + b x + c == 0, x]

$$a \neq 0 \;\&\&\; \left(x == \frac{-b - \sqrt{b^2 - 4\,a\,c}}{2\,a} \;||\; x == \frac{-b + \sqrt{b^2 - 4\,a\,c}}{2\,a} \right) \;||$$

$$a == 0 \;\&\&\; b \neq 0 \;\&\&\; x == -\frac{c}{b} \;||\; c == 0 \;\&\&\; b == 0 \;\&\&\; a == 0$$

Remember that **&&** means logical AND and **||** means logical OR. Later variables may be expressed in terms of earlier variables (✳4!):

Reduce[{x + y == 1, x^2 - y == 2}, {x, y}]

$$\left(x == \frac{1}{2}\,(-1 - \sqrt{13}\,) \;||\; x == \frac{1}{2}\,(-1 + \sqrt{13}\,) \right) \;\&\&\; y == 1 - x$$

With the **Backsubstitution** option, however, we get explicit values:

Reduce[{x + y == 1, x^2 - y == 2}, Backsubstitution → True]

$$y == \frac{1}{2}\,(3 - \sqrt{13}\,) \;\&\&\; x == \frac{1}{2}\,(-1 + \sqrt{13}\,) \;||\; y == \frac{1}{2}\,(3 + \sqrt{13}\,) \;\&\&\; x == \frac{1}{2}\,(-1 - \sqrt{13}\,)$$

In Sections 19.2.3, 19.2.4, and 19.3.1, we use **Reduce** to solve radical equations, inequalities, and transcendental equations.

19.2.3 Radical Equations

Solve, **NSolve**, **Eliminate**, and **Reduce** can also be used to solve radical equations, which are equations that contain rational powers. For example (✳4!):

eqn = Sqrt[x + 1] + x^(1 / 3) == 2;

Solve[eqn]

$$\left\{ \left\{ x \rightarrow \frac{1}{3}\left(-2 - 161\left(\frac{2}{1703 + 459\,\sqrt{93}} \right)^{1/3} + \left(\frac{1}{2}\,(1703 + 459\,\sqrt{93}\,) \right)^{1/3} \right) \right\} \right\}$$

Reduce does not write an explicit solution for cubic or quartic equations (✳4!):

Reduce[eqn]

$$x == \text{Root}\left[-27 + 55\,\#1 + 2\,\#1^2 + \#1^3\, \&,\, 1 \right]$$

With the **Cubics** (✳5) and **Quartics** (✳5) options, however, we get an explicit expression:

Reduce[eqn, Cubics → True]

$$x == \frac{1}{6}\left(-4 - 322\left(\frac{2}{1703 + 459\,\sqrt{93}} \right)^{1/3} + 2^{2/3}\,(1703 + 459\,\sqrt{93}\,)^{1/3} \right)$$

Here is another example:

f = Sqrt[3 x + 2] + Sqrt[2 x - 1] - 3 Sqrt[x - 1];

Solve[f == 0] $\left\{ \left\{ x \rightarrow \frac{3}{4}\,(-7 - \sqrt{73}\,) \right\} \right\}$

■ **Extraneous Solutions**

When solving radical equations, extraneous solutions easily emerge (as the result of, for example, squaring). **Solve** automatically verifies solution candidates and rejects extraneous solutions. All solution candidates can be seen if the candidates are not verified. We continue with the preceding example:

sol = Solve[f == 0, VerifySolutions → False]

$$\left\{\left\{x \to \frac{3}{4}\ (-7 - \sqrt{73}\ )\right\}, \left\{x \to \frac{3}{4}\ (-7 + \sqrt{73}\ )\right\}\right\}$$

sol // N $\{\{x \to -11.658\}, \{x \to 1.158\}\}$

In the preceding example, **Solve** rejected the second solution candidate. To graphically show the real root, we could plot f. However, because f is complex valued for $x < 1$, we plot $|f|$: f has a zero whenever $|f|$ has. The figure shows that the first candidate really is a solution but the second candidate is not:

Plot[Abs[f], {x, -20, 20}, AspectRatio → 0.2];

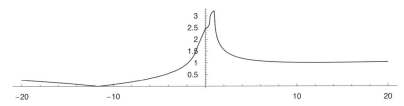

19.2.4 Inequalities

■ **Solving Inequalities**

> **Reduce[expr, vars]** (❀5)   Give a detailed analysis of the solutions of given equations and inequalities
>
> **Reduce[expr, vars, dom]** (❀5)   Assume **dom** to be the domain of variables, parameters, and function values

In Section 19.2.2, p. 497, we used **Reduce** to get a detailed solution for equations. **Reduce** is also useful when solving inequalities and sets of equations and inequalities. Users of *Mathematica* 4 can solve inequalities with **InequalitySolve[expr, vars]** from the package **Algebra`InequalitySolve`**. In our first examples, we do not have any parameters, and so the variable need not be declared:

Reduce[x^2 - 3 x - 2 > 0]

$$x < \frac{1}{2}\ (3 - \sqrt{17}\ )\ ||\ x > \frac{1}{2}\ (3 + \sqrt{17}\ )$$

Remember that **&&** means logical AND and **||** means logical OR. All numbers should be exact. For problems with inexact numbers, **Reduce** solves a corresponding exact problem and numericizes the result:

```
Reduce[x^2 - 3 x - 2. > 0]
```

Reduce::ratnz :
  Reduce was unable to solve the system with inexact coefficients.
    The answer was obtained by solving a corresponding
    exact system and numericizing the result. More…
x < -0.561553 || x > 3.56155

Next we solve a system of inequalities:

```
Reduce[{x^2 - 3 x - 2 > 0, Abs[x] < 1}]
```

$-1 < x < \frac{1}{2} (3 - \sqrt{17})$

■ Specifying a Domain

Reduce assumes that all quantities that appear algebraically in inequalities are real and that all other quantities are complex. With a third argument given to **Reduce**, we can restrict all variables, parameters, and function values to a given domain such as **Reals**, **Integers**, or **Complexes**. The domain **Reals** may be suitable in many problems with inequalities. In the following example, we get the condition that **a** and **b** should be real:

```
Reduce[{x + y == a, y < a b}, {x, y}]
```

$(a | b) \in \text{Reals} \, \&\& \, \text{Re}[x] > a - a b \, \&\& \, \text{Im}[x] == 0 \, \&\& \, y == a - x$

It is useful to declare that all variables are real:

```
Reduce[{x + y == a, y < a b}, {x, y}, Reals]
```

$x > a - a b \, \&\& \, y == a - x$

In Section 25.3.1, p. 695, we solve the following system of inequalities:

```
ineqs = {Abs[2 - a] < 1, Abs[b (a - 1) / a] < 1, b ≥ 0};
```

```
Reduce[ineqs, {a, b}, Reals]
```

$1 < a < 3 \, \&\& \, 0 \le b < \dfrac{a}{-1 + a}$

Inequalities can be plotted with **InequalityPlot** (see Section 5.3.3, p. 127):

```
<< Graphics`InequalityGraphics`
```

```
InequalityPlot[ineqs, {a, 0, 4},
  {b, -10, 10}, AspectRatio → 1 / GoldenRatio];
```

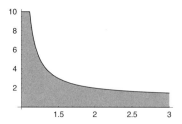

Next we restrict variables to be integers:

> **Reduce[{0 ≤ x ≤ 2, 0 ≤ y ≤ x}, {x, y}, Integers]**

x == 0 && y == 0 || x == 1 && y == 0 || x == 1 && y == 1 ||
x == 2 && y == 0 || x == 2 && y == 1 || x == 2 && y == 2

We ask under what conditions is a quadratic expression positive for all real x:

> **Reduce[ForAll[x, {x, a, b, c} ∈ Reals, a x^2 + b x + c > 0]]**

$$c > 0 \;\&\&\; \left(b < 0 \;\&\&\; a > \frac{b^2}{4\,c} \;||\; b == 0 \;\&\&\; a \geq 0 \;||\; b > 0 \;\&\&\; a > \frac{b^2}{4\,c} \right)$$

In Section 19.3.1, p. 501, we use **Reduce** to solve transcendental equations.

■ Finding Instances

> **FindInstance[expr, vars]** (❀5) Find an instance of **vars** that makes **expr** be true
> **FindInstance[expr, vars, dom]** Assume **dom** to be the domain of variables, parameters, and function values
> **FindInstance[expr, vars, dom, n]** Find **n** instances

A typical use of this command is to find a point that satisfies a set of inequalities (or, in terms of optimization, to find a feasible point). Inequalities in the following examples were considered above with **Reduce**:

> **FindInstance[{x^2 - 3 x - 2 > 0, Abs[x] < 1}, {x}, Reals]**

$$\left\{\left\{x \to -\frac{3}{4}\right\}\right\}$$

> **FindInstance[{x^2 - 3 x - 2 > 0, Abs[x] < 1}, {x}, Reals, 4]**

$$\left\{\left\{x \to -\frac{878}{915}\right\}, \left\{x \to -\frac{862}{915}\right\}, \left\{x \to -\frac{824}{915}\right\}, \left\{x \to -\frac{212}{305}\right\}\right\}$$

> **FindInstance[ineqs, {a, b}, Reals]**

$$\left\{\left\{a \to \frac{3}{2}, b \to \frac{1}{2}\right\}\right\}$$

19.3 Transcendental Equations

19.3.1 Exact Solutions

If a transcendental equation or inequality is simple enough, **Solve** or **Reduce** may be able to get a solution; such equations are considered here in Section 19.3.1. In most cases, we have to resort to iterative methods provided by **FindRoot**. These equations are considered in Sections 19.3.2 to 19.3.4.

> **Solve[eqns, vars]** Try to give some solutions of the given transcendental equations
> **Reduce[expr, vars, dom]** (❀5) Try to give a complete solution of the given transcendental equations and inequalities

■ An Example

A transcendental equation can be solved symbolically if the equation can be transformed into an equation in which a single transcendental function can be taken to be the variable and if the transformed equation can be solved for this transcendental function. The solution of the original equation is then obtained with the inverse function. As an example, we first try **Solve**:

```
Solve[Sin[x] == Cos[x], x]
```

```
Solve::ifun :
  Inverse functions are being used by Solve, so some solutions may not
     be found; use Reduce for complete solution information. More…
```

$$\left\{\left\{x \to -\frac{3\pi}{4}\right\}, \left\{x \to \frac{\pi}{4}\right\}\right\}$$

We obtained two solutions, with the warning that other solutions may exist. Indeed, the equation is valid whenever $x = \frac{\pi}{4} + n\pi$ and n is an integer. This is typical for solutions of transcendental equations given by **Solve**: we get some solutions, but other solutions may (and often do) exist. Then we try **Reduce**:

```
Reduce[Sin[x] == Cos[x], x]
```

$$C[1] \in \text{Integers \&\&}$$
$$(x == -2 \, \text{ArcTan}[1 + \sqrt{2}] + 2\pi C[1] \mid\mid x == -2 \, \text{ArcTan}[1 - \sqrt{2}] + 2\pi C[1])$$

```
% // FullSimplify
```

$$(\pi + 8\pi C[1] == 4 x \mid\mid \pi (-3 + 8 C[1]) == 4 x) \&\& C[1] \in \text{Integers}$$

Reduce was able to give all of the solutions: for an integer $C[1]$, solutions are $x = \frac{\pi}{4} + 2 C[1]\pi$ and $x = -\frac{3\pi}{4} + 2 C[1]\pi$, which is the same as $x = \frac{\pi}{4} + n\pi$.

By the way, **Solve** has the **InverseFunctions** option. The default value **Automatic** of this option means that inverse functions are used, and a warning message about possibly missing roots is printed. If we give the option the value **True**, then inverse functions are also used, but the warning is not printed. With the value **False**, inverse functions are not used.

To save space, we turn warnings about inverse functions off:

```
Off[Solve::"ifun"]
Off[InverseFunction::"ifun"]
```

■ More Examples

In the first example, we have hyperbolic functions:

```
Solve[Sinh[x] == Cosh[x] - 2, x] // FullSimplify
```

$$\{\{x \to -\text{Log}[2]\}\}$$

```
Reduce[Sinh[x] == Cosh[x] - 2, x]
```

$$C[1] \in \text{Integers \&\& } x == 2 \, \dot{\imath} \, \pi C[1] - \text{Log}[2]$$

With **Reduce**, we can restrict variables and functions to be real:

```
Reduce[Sinh[x] == Cosh[x] - 2, x, Reals]
```

$x == -\text{Log}[2]$

Sometimes the solution contains a product log function. The next example gives the definition of product log:

```
Solve[z == w Exp[w], w]
```

$\{\{w \to \text{ProductLog}[z]\}\}$

This shows that product log at z is the (principal) solution for w of $z = w\,e^w$. Here is another example:

```
f = a - x Exp[x^2];
```

```
Solve[f == 0, x]
```

$$\left\{\left\{x \to -\frac{\sqrt{\text{ProductLog}[2\,a^2]}}{\sqrt{2}}\right\}, \left\{x \to \frac{\sqrt{\text{ProductLog}[2\,a^2]}}{\sqrt{2}}\right\}\right\}$$

From the result of **Reduce**, we see that the first solution is correct for $a \leq 0$ and the second solution for $a \geq 0$.

```
Reduce[f == 0, x, Reals]
```

$$a > 0\ \&\&\ x == \frac{\sqrt{\text{ProductLog}[2\,a^2]}}{\sqrt{2}}\ ||$$

$$a == 0\ \&\&\ x == 0\ ||\ a \neq 0\ \&\&\ a < 0\ \&\&\ x == -\frac{\sqrt{\text{ProductLog}[2\,a^2]}}{\sqrt{2}}$$

Now we turn the messages on:

```
On[Solve::"ifun"]
On[InverseFunction::"ifun"]
```

■ Towards Numerical Methods

The next equation is not solvable with **Solve** (or **Reduce**):

```
Solve[Sin[x] == x, x]
```

```
Solve::tdep :
  The equations appear to involve the variables to be solved
    for in an essentially non-algebraic way. More…
Solve[Sin[x] == x, x]
```

However, this equation has the simple solution $x = 0$. If **Solve** or **Reduce** does not succeed, we can use numerical methods especially developed for transcendental equations. These methods include Newton's method, the secant method, and the bisection method, among others. These methods are considered next.

19.3.2 Numerical Solutions

FindRoot is used to solve transcendental equations with iterative methods. The methods can be divided into two groups:

- a method that requires derivatives (Newton's method) (this method needs one starting point); and
- two methods not requiring derivatives (the secant method and Brent's method) (these methods need two starting points).

FindRoot decides the type of method to use from the number of starting points. Note that, even when **FindRoot** starts with, for example, Newton's method, it may later on move to other methods.

■ Newton's Method

The best-known method for solving a transcendental equation $f(x) = 0$ is Newton's method:

$$x_{i+1} = x_i - \frac{f(x_i)}{f'(x_i)}$$

It can be used in the following way:

FindRoot[eqn, {x, x0}]   Find a solution for the equation starting from the point **x0**

If we write, in place of **{x, x0}**, a list **{x, x0, xmin, xmax}**, then iterations are stopped if the solution goes outside the interval (**xmin, xmax**). In place of an equation **expr1 == expr2**, we can also write a single expression **expr**, and then it is understood that the equation is **expr == 0**.

As an example, we solve the equation $e^{-x} - x^2 = 0$ by first defining and plotting the lefthand-side function:

```
f = Exp[-x] - x^2;

Plot[f, {x, -1, 2}];
```

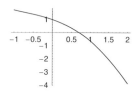

A root seems to be about 0.7. Apply Newton's method starting from 0:

```
sol = FindRoot[f, {x, 0}]      {x → 0.703467}
```

We could also have written an explicit equation:

```
sol = FindRoot[f == 0, {x, 0}]      {x → 0.703467}
```

The value of the function at the solution is, in fact, zero very accurately (❄4!):

```
f /. sol      - 1.4988 × 10^-15
```

If you only want the value of the zero (instead of the rule), then write the following:

```
x /. sol      0.703467
```

You can also write this:

```
sol = x /. FindRoot[f, {x, 0}]      0.703467
```

Complex zeros, too, can be searched by giving complex starting points:

    FindRoot[f, {x, -1 + I}]   {x → -1.58805 + 1.54022 i}

■ The Secant Method and Brent's Method

Newton's method requires the derivative of the function, but we can also use methods that do not need the derivative. The secant method and Brent's method are some of the best-known methods of this kind (Brent's method uses inverse quadratic interpolation and bisection). They can be used in the following way:

FindRoot[eqn, {x, x0, x1}]   Find a solution starting from the points x0 and x1

If we write, in place of {x, x0, x1}, a list {x, x0, x1, xmin, xmax}, then iterations are stopped if the solution goes outside the interval (xmin, xmax). *Mathematica* uses Brent's method if the values of the function at the points x0 and x1 are real and of opposite sign; otherwise, *Mathematica* uses the secant method. The secant method applies the following recursion formula:

$$x_{n+1} = x_n - \frac{f(x_n)(x_n - x_{n-1})}{f(x_n) - f(x_{n-1})}$$

For example:

    FindRoot[f, {x, 0, 1}]   {x → 0.703467}

■ Several Transcendental Equations

FindRoot[{eqn1, eqn2, … }, {x, x0}, {y, y0}, …]   Use Newton's method
FindRoot[{eqn1, eqn2, … }, {x, x0, x1}, {y, y0, y1}, …]   Use the secant method

As an example, we find simultaneous zeros for the following two functions (or, equivalently, solve the corresponding pair of equations):

    f1 = x^2 + y^2 - 1;
    f2 = Sin[x] - y;

The situation can be visualized with contour plots. For each function, we demand only the single contour for which the function is zero:

    p = Map[ContourPlot[#, {x, -2, 2}, {y, -2, 2}, DisplayFunction → Identity,
        Contours → {0}, ContourShading → False, PlotPoints → 40] &, {f1, f2}];

    Show[p, AspectRatio → Automatic, Frame → False,
        Axes → True, DisplayFunction → $DisplayFunction];

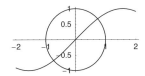

The system seems to have two solutions:

```
FindRoot[{f1, f2}, {x, 1}, {y, 1}]
```
$\{x \rightarrow 0.739085, y \rightarrow 0.673612\}$

```
FindRoot[{f1, f2}, {x, -1}, {y, -1}]
```
$\{x \rightarrow -0.739085, y \rightarrow -0.673612\}$

■ Formulation by Matrices

Simultaneous equations can also be formulated with matrices (✻5). For example:

```
a = {{2, 1}, {1, 2}}; b = {{1, -1}, {2, 1}}; c = {{3, 2}, {1, 4}};
```

```
eqns = x.a.x + b.x == c
```
$\{\{1, -1\}, \{2, 1\}\}.x + x.\{\{2, 1\}, \{1, 2\}\}.x == \{\{3, 2\}, \{1, 4\}\}$

```
FindRoot[eqns, {x, {{1, 1}, {1, 1}}}]
```
$\{x \rightarrow \{\{0.990833, 0.358932\}, \{-0.437387, 1.13018\}\}\}$

FindRoot takes the dimensions of **x** from the starting point.

■ Options

Options for **FindRoot**:

WorkingPrecision   Precision used in internal computations; examples of values:
   MachinePrecision (✻5), **20**

PrecisionGoal (✻5)   If the value of the option is **p**, the relative error of the root
   should be of the order 10^{-p}; examples of values: **Automatic** (usually means **8**), **10**

AccuracyGoal   If the value of the option is **a**, the absolute error or the root and the
   absolute value of the function at the root should be of the order 10^{-a}; examples of
   values: **Automatic** (usually means **8**), **10**

Method   Method used; possible values: **Automatic**, **Newton** (✻5), **Secant** (✻5), **Brent**
   (✻5)

MaxIterations   Maximum number of iterations used; default value: **100**

DampingFactor   Damping factor; examples of values: **1**, **2**

Jacobian   Jacobian of the system in Newton's method; examples of values: **Automatic**,
   Symbolic (✻5), **FiniteDifference** (✻5)

Compiled   Whether the function should be compiled; possible values: **True**, **False**

StepMonitor (✻5)   Command to be executed after each step of the iterative method;
   examples of values: **None**, **++n**, **AppendTo[iters, x]**

EvaluationMonitor (✻5)   Command to be executed after each evaluation of the
   equation; examples of values: **None**, **++n**, **AppendTo[points, x]**

As usual, the default values of options are mentioned first. The default is that iterations
are stopped when the relative or absolute error of *the root* is less than 10^{-8} and the absolute
value of *the function at the root* is less than 10^{-8} (in *Mathematica* 4, iterations are stopped
when the absolute value of the function is less than 10^{-6}). The general default value of
PrecisionGoal and **AccuracyGoal** is **WorkingPrecision**/2. For more about these
options, see Section 11.3.1, p. 292.

A slow convergence may be an indication of a multiple zero. If the zero is of multiplicity d, we get a quadratic convergence again if we give **DampingFactor** the value d. The iteration formula for Newton's method is now $x_{i+1} = x_i - d\, f(x_i)/f'(x_i)$. For equations in which the calculation of the Jacobian may cause problems, we can give **Jacobian** the value **FiniteDifference**, and then the Jacobian is approximated by numeric methods.

19.3.3 Special Topics

■ Looking at the Iterations

Consider the following function:

```
f = Exp[-x] - x^2;
```

To see the iterations, write the following (❀5):

```
iters = {1};
FindRoot[f, {x, 1}, StepMonitor :> AppendTo[iters, x]]
```

```
{x → 0.703467}
```

```
iters    {1, 0.733044, 0.703808, 0.703467, 0.703467}
```

In the following way we get a table of the iterations and the values of the function (❀5):

```
n = 0; iters = {{0, x, f}} /. x → 1.;
FindRoot[f, {x, 1}, StepMonitor :> AppendTo[iters, {++n, x, f}]];

TableForm[iters, TableSpacing → {0.4, 1.5},
   TableHeadings → {None, {" n", " x", " f"}}] // PaddedForm
```

n	x	f
0	1.	−0.632121
1	0.733044	−0.0569084
2	0.703808	−0.000647392
3	0.703467	-8.7166×10^{-8}
4	0.703467	-1.4988×10^{-15}

■ Zeros of Bessel Functions

With a package, we can get a list of the first n positive zeros of various Bessel functions and their derivatives and also of some special expressions containing Bessel functions and their derivatives.

In the **NumericalMath`BesselZeros`** *package:*

BesselJZeros[v, n], **BesselJPrimeZeros[v, n]**   Zeros of $J_v(x)$ and $J_v'(x)$

BesselYZeros[v, n], **BesselYPrimeZeros[v, n]**   Zeros of $Y_v(x)$ and $Y_v'(x)$

BesselJYJYZeros[v, λ, n]   Zeros of $J_v(x)\,Y_v(\lambda x) - J_v(\lambda x)\,Y_v(x)$

BesselJPrimeYJYPrimeZeros[v, λ, n]   Zeros of $J_v'(x)\,Y_v(\lambda x) - J_v(\lambda x)\,Y_v'(x)$

BesselJPrimeYPrimeJPrimeYPrimeZeros[v, λ, n]   Zeros of
$J_v'(x)\,Y_v'(\lambda x) - J_v'(\lambda x)\,Y_v'(x)$

The package eventually uses **FindRoot**, but the key point is in finding two starting values close enough to the zero requested. The package has a special method for this. The commands have the options **WorkingPrecision** and **AccuracyGoal**.

We can also ask for a list of zeros with, for example, `BesselJZeros[v, {m, n}]`; this gives the *m*th through *n*th zeros. Zeros in a given interval (*a*, *b*) can be asked by commands like `BesselJZerosInterval[v, {a, b}]` (⌘5).

Zeros of Bessel functions are sometimes needed in the solutions of partial differential equations (see Section 24.2.4, p. 654). The zeros can also be used to numerically calculate integrals of Bessel functions (see Section 17.2.4, p. 450).

Consider, for example, `BesselJ[2,x]`:

```
Plot[BesselJ[2, x], {x, 0, 20}];
```

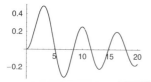

```
<< NumericalMath`BesselZeros`
```

```
BesselJZeros[2, 5]
```

　　$\{5.13562, 8.41724, 11.6198, 14.796, 17.9598\}$

In addition, 0 is a zero.

■ Inverse Cubic Interpolation

In the `NumericalMath`InterpolateRoot`` *package:*

`InterpolateRoot[f, {x, a, b}]`   Find a zero for **f** near the starting points **a** and **b**

This package uses inverse cubic interpolation. The idea of this method is to take the four last points—for example *a*, *b*, *c*, and *d*—to calculate a cubic interpolation polynomial through the points ($f(a)$, *a*), ($f(b)$, *b*), ($f(c)$, *c*), and ($f(d)$, *d*) and to calculate the value of this polynomial at 0. This package is designed only for simple roots of a single function or an equation that is well behaved. The method can be advantageous in comparison to `Find` `Root` in cases where evaluating the function is extremely laborious, particularly for very high precision.

```
<< NumericalMath`InterpolateRoot`
```

```
f = Exp[-x] - x^2;
```

```
r = InterpolateRoot[f, {x, 0, -2.5}]
```

　　$\{x \to 0.703467422498391652049818602\}$

```
f /. r       - 0. × 10^-27
```

The default value of `WorkingPrecision` is 40.

■ Interval Arithmetic

The **NumericalMath`IntervalRoots`** package implements the bisection, the secant, and Newton's methods using interval arithmetic. The methods start with a given interval and find all roots of a function on this interval. The roots are specified not by points but by small intervals. The methods are continued until the length of all of the intervals is less than a given small number.

19.3.4 Own Programs

■ Own Newton

In Chapters 14 and 15, we have presented several implementations for Newton's method; see the references in the introduction to this chapter. Here we present a version of the function **newton5** that was presented in Section 15.2.3, p. 399. We add a damping factor, which is used for multiple zeros and explained in Section 19.3.2, p. 507.

```
newtonSolve[f_, x_, x0_, d_: 1, n_: 20, opts___?OptionQ] :=
  With[{df = D[f, x]}, FixedPointList[(x - d f / df) /. x → # &, N[x0], n, opts]]
```

You may want to read Section 15.2.3 for explanations of this program. The iterations are stopped if two successive points are the same to 16-digit precision. The damping factor **d** has the default value 1. At most **n** iterations are done; **n** has the default value 20. We can write zero or more options (note the three underscores after **opts**). The option we can use is **SameTest** (see Section 15.2.3). (Note that, if you want to use a different **n**, you must also enter a value for **d**, and if you want to write an option, also write values for **d** and **n**.) We try **newtonSolve** for the equation $f = 0$ that we considered earlier:

```
f = Exp[-x] - x^2;
```

```
newtonSolve[f, x, 1]
```
```
{1., 0.733044, 0.703808, 0.703467, 0.703467, 0.703467, 0.703467}
```

Complex starting values can be given:

```
newtonSolve[f, x, -1.5 + 1.5 I]
```
```
{-1.5 + 1.5 i, -1.59551 + 1.54133 i, -1.58809 + 1.54021 i,
 -1.58805 + 1.54022 i, -1.58805 + 1.54022 i, -1.58805 + 1.54022 i}
```

Consider the following function:

```
f2 = x^3 / 3 - x + 2 / 3;
```

```
Plot[f2, {x, -3, 3}];
```

It seems that a point at about 1 is a zero of multiplicity two, so we use a damping factor of two. We also use a custom stopping criterion. We stop the iterations after two successive approximations to the zero (represented by **#1** and **#2** in the **SameTest** option) differ by at most 10^{-6}:

```
newtonSolve[f2, x, 4, 2, 20, SameTest → (Abs[#1 - #2] < 10^-6 &)]
```

```
{4., 1.6, 1.04615, 1.00035, 1., 1.}
```

Then we stop after the value of the function at the last point is less than 10^{-14}:

```
newtonSolve[f2, x, 4, 2, 20, SameTest → (Abs[f2 /. x → #2] < 10^-14 &)]
```

```
{4., 1.6, 1.04615, 1.00035, 1.}
```

■ Illustrating Newton's Method

Newton's method for an equation $f(x) = 0$ is interpreted as follows. A tangent to $f(x)$ is drawn at the present point x_i. The next point x_{i+1} is where the tangent intersects the x axis. A new tangent is drawn at this point, and so we continue. To graphically show this process, we first form a set of points that consist of the iteration points at the x axis and at the function:

```
it = newtonSolve[f, x, -2.5];
```

```
points = Flatten[Map[{{#, 0}, {#, f /. x → #}} &, it], 1];
```

Then we show how Newton's method proceeds (the starting point is at the left: $x_0 = -2.5$):

```
Plot[f, {x, -2.7, 1.7},
  PlotStyle → AbsoluteThickness[1], Epilog → Line[points]];
```

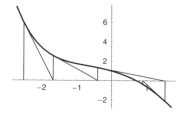

■ Own Newton for Several Equations

```
newtonSolveSystem[f_List, x_List, x0_List, eps_: 10^-6, n_: 20] :=
 With[{jac = N[Outer[D, f, x]]}, FixedPointList[
   (# + LinearSolve[jac /. Thread[x → N[#]], -f /. Thread[x → N[#]]]) &,
   N[x0], n, SameTest → (Sqrt[(#1 - #2).(#1 - #2)] < eps &)]]
```

This program is a generalization of **newtonSolve** for several equations. The stopping criterion is now the 2-norm (or the square root of the sum of the squares) of the difference between the last two iterations; we can give the **eps** for this criterion (users of *Mathematica* 5 can use **(Norm[#1 - #2] < eps &)** in the stopping test). The new point could be calculated from $x_{n+1} = x_n - J_n^{-1} f_n$, where J_n is the Jacobian of the system (**jac** in the program). However, we can avoid calculating the inverse of J_n by solving the linear system $J_n \, \delta_x = -f_n$ and then calculating $x_{n+1} = x_n + \delta_x$. This method is faster. For example:

```
newtonSolveSystem[{x^2 + y^2 - 1, Sin[x] - y}, {x, y}, {-1, -1}]
```

```
{{-1., -1.}, {-0.778309, -0.721691}, {-0.740277, -0.674994},
 {-0.739086, -0.673613}, {-0.739085, -0.673612}, {-0.739085, -0.673612}}
```

■ Own Secant in Procedural Style

```
secantSolve[f_, x_, x0_, x1_, eps_: 10^-10, d_: 1, n_: 20] :=
 Module[{y0 = x0, y1 = x1, f0 = N[f /. x → x0], f1, newy, iters = {x0, x1}},
  Do[f1 = N[f /. x → y1];
   If[Abs[f1] < eps, Break[]];
   newy = y1 - d f1 (y1 - y0) / (f1 - f0);
   iters = {iters, newy};
   y0 = y1; y1 = newy; f0 = f1, {n}];
  Flatten[iters]]
```

This program is in the procedural style. If the value of the function becomes smaller than **eps** (default value is 10^{-10}), then iterations are stopped. Iterations are done at most **n** times (the default value is 20). The damping factor **d** has the default value 1.

The iterations are gathered to **iters**. In general, a module prints the result of the last command so that we have placed the simple command **Flatten[iters]** as the last command to get a list of all of the iterations. **Flatten** is needed because **iters = {iters, newy}** creates a nested list.

Note that we have used **y0** and **y1** to store the two successive iterations. We cannot use **x0** and **x1**, because **x0** and **x1** are arguments of **secantSolve**, and the arguments of a function cannot be changed inside the function (see Section 15.1.3, p. 391). For example:

```
f = Exp[-x] - x^2;
```

```
secantSolve[f, x, 0.2, 0.5]
```

```
{0.2, 0.5, 0.753338, 0.699275, 0.703386, 0.703468, 0.703467}
```

■ Illustrating the Secant Method

The next point is where the secant through the last two points intersects the *x* axis.

```
it = secantSolve[f, x, -2.5, -2.0]; points = {};
```

```
Do[AppendTo[points, {it[[i]], 0}];
 AppendTo[points, {it[[i]], f /. x → it[[i]]}];
 AppendTo[points, {it[[i + 1]], f /. x → it[[i + 1]]}];
 AppendTo[points, {it[[i + 2]], 0}], {i, Length[it] - 2}]
```

```
Plot[f, {x, -2.7, 1.7},
 PlotStyle → AbsoluteThickness[1], Epilog → Line[points]];
```

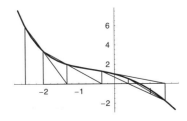

■ **Own Secant in Functional Style**

```
secantStep2[f_, x_, d_, {x0_, x1_, f0_, f1_}] :=
 With[{newx = x1 - d f1 (x1 - x0) / (f1 - f0)}, {x1, newx, f1, N[f /. x → newx]}]

secantSolve2[f_, x_, x0_, x1_, eps_: 10^-10, d_: 1, n_: 20] :=
 Transpose[FixedPointList[secantStep2[f, x, d, #] &,
    {x0, x1, N[f /. x → x0], N[f /. x → x1]}, n,
    SameTest → (Abs[#2[[4]]] < eps &)]][[2]]
```

This program is in the functional style. Now we have four values to be iterated, and we enclose them in a list `{x0, x1, f0, f1}` in `secantStep2`, because `FixedPointList` requires one iteration variable (denoted by `#`). Note how nicely one step can be done: just calculate the new value (denoted by **newx**) and return the new values `{x1, newx, f1, N[f /. x → newx]}`.

The function `secantSolve2` does the needed iterations. The starting point is `{x0, x1, N[f /. x → x0], N[f /. x → x1]}`. The stopping criterion is that the value of the function at the latest approximation is less than **eps**. The variable `#2` represents the latest iteration that consists of the four values `{x0, x1, N[f /. x → x0], N[f /. x → x1]}` so that `#2[[4]]` is the value of the function at the latest approximation.

The result of `FixedPointList` is an $(m \times 4)$-matrix if m iterations were done. By taking the second row of its transpose, the values of the approximations to the root are obtained (the initial guess **x0** is, however, lacking). For example:

```
secantSolve2[f, x, 0.2, 0.5]
```

```
{0.5, 0.753338, 0.699275, 0.703386, 0.703468, 0.703467}
```

Optimization

Introduction

> *Because the shape of the whole universe is most perfect and, in fact, designed by the wisest creator, nothing in the world will occur in which no maximum or minimum rule is somehow shining forth.* — Leonhard Euler

Mathematica has four main commands for minimization (for most of them, we also have the corresponding commands for maximization):

FindMinimum	unconstrained *nonlinear* problems (local optimum)
LinearProgramming	constrained *linear* problems (global optimum)
Minimize	constrained *polynomial* problems (global optimum)
NMinimize	constrained *nonlinear* and *integer* problems (global optimum)

FindMinimum, which is considered in Section 20.1, has several iterative methods (some requiring derivatives, others not) to approximate a local minimum of a nonlinear function without constraints. Constrained optimization is considered in Section 20.2. The simplex algorithm used by **LinearProgramming** finds the global minimum of constrained linear problems. **Minimize** (which was new in *Mathematica* 4.0) uses cylindrical algebraic decomposition to calculate the global minimum of constrained polynomial (and also linear) problems. **NMinimize** (which was new in *Mathematica* 4.2) can be used for approximate global solution of constrained nonlinear and integer problems (e.g., with a genetic algorithm).

If the problem is exact (with no decimal numbers), **LinearProgramming** and **Minimize** also give an exact answer. **FindMinimum** and **NMinimize** use iterative methods and thus give an approximate decimal answer. All of the commands require that the problem is numeric, that is, no symbolic parameters are allowed.

Note that, if your problem is unconstrained, it is not necessary to use **FindMinimum**. Indeed, for an unconstrained problem, you may want to use **Minimize** or **NMinimize**, because the former is able to give an *exact* result (unlike **FindMinimum**), the latter is also able to solve *nonsmooth* and *integer* problems (unlike **FindMinimum**), and both are able to find a *global* minimum (also unlike **FindMinimum**).

In Section 20.3, we consider classical optimization in which necessary conditions are used to find the optimum. The conditions are usually equations that contain derivatives of the object function or of a modified function if constraints are present. By solving the (nonlinear) equations, we get solution candidates for the problem. Often we can use some sufficiency conditions to check whether the solution candidates are maximum or minimum points. With **D**, **Solve**, and **Eigenvalues**, we can use the results of classical optimization. The problem can also contain symbolic parameters (unlike **Minimize**).

Note that, in the **DiscreteMath`Combinatorica`** package, there are several optimization functions for graph-theoretical problems. Those worth mentioning include functions like **Dijkstra**, **MinimumSpanningTree**, **ShortestPath**, and **TravelingSalesman**. For dynamic programming with *Mathematica*, see Wagner (1995). For genetic programming, see Nachbar (1995). A multiplier method for constrained nonlinear problems can be found at http://library.wolfram.com/database/MathSource/795/. For optimization with *Mathematica*, see Bhatti (2000). For integer programming, see Bulmer and Carter (1996).

Note also that the following commercial products are available: *Global Optimization* (global optimization for constrained and unconstrained nonlinear functions), *Industrial Optimization* (local optimization for linear, nonlinear, and queuing problems), *MathOptimizer* (global and local numerical solution of a very general class of continuous optimization problems), and *Operations Research* (constrained optimization with applications from operations research); look at http://www.wolfram.com/products/index.html#apps.

20.1 Unconstrained Optimization

20.1.1 Functions of One Variable

> **FindMinimum[f, {x, x0}]**   Find a local minimum of **f** by using its derivative; start from **x0**
>
> **FindMinimum[f, {x, x0, x1}]**   Find a local minimum of **f** by not using its derivative; start from **x0** and **x1**

We also have **FindMaximum** (✽5), which is used in the same way as **FindMinimum**. In place of {x, x0}, we can also write {x, x0, xmin, xmax}, and then iterations are stopped if the solution is going outside the interval (xmin, xmax). Likewise, in {x, x0, x1}, we can also add **xmin** and **xmax**.

FindMinimum uses iterative methods to *approximate* a *local* minimum point (note that **Minimize**, introduced in Section 20.2.2, finds an *exact*, *global* minimum). If we give only one starting point **x0**, a method (quasi-Newton method) is used that utilizes the derivative of the function **f** to be minimized. If we give two starting points **x0** and **x1**, a method is used that does not need the derivative. The starting point(s) should be near the minimum point we are searching, or the method may diverge or converge to another minimum. For example:

```
f = x Cos[x];

Plot[f, {x, 0, 8}];
```

One of the local minimums seems to be near 3.5:

```
mi = FindMinimum[f, {x, 3}]
```

$\{-3.28837, \{x \rightarrow 3.42562\}\}$

The corresponding minimum value is –3.28837. We can check whether the derivative is zero (a necessary condition) and the second derivative positive (a sufficient condition) at the point found (❀4!):

```
{D[f, x], D[f, x, x]} /. mi[[2]]
```

$\{5.80336 \times 10^{-12}, 3.84882\}$

Next we find a local point of maximum:

```
FindMaximum[f, {x, 6}]
```

$\{6.361, \{x \rightarrow 6.4373\}\}$

■ Options

The options enable the choice of the method, among other things.

Options for **FindMinimum** *and* **FindMaximum**:

WorkingPrecision Precision used in internal computations; examples of values:
   MachinePrecision (❀5), **20**

PrecisionGoal If the value of the option is **p**, the relative error of the optimum point and of the value of the function at the optimum point should be of the order 10^{-p}; examples of values: **Automatic** (usually means **8**), **10**

AccuracyGoal If the value of the option is **a**, the absolute error of the optimum point and of the value of the function at the optimum point should be of the order 10^{-a}; examples of values: **Automatic** (usually means **8**), **10**

Method Method used; possible values: **Automatic**, **Gradient**, **ConjugateGradient** (❀5), **Newton**, **QuasiNewton**, **LevenbergMarquardt**

MaxIterations Maximum number of iterations used; examples of values: **100**, **200**

Gradient Gradient of the function; examples of values: **Automatic**, **Symbolic** (❄5),
    FiniteDifference (❄5)
Compiled Whether the function should be compiled; possible values: **True**, **False**
StepMonitor (❄5) Command to be executed after each step of the iterative method;
    examples of values: **None**, **++n**, **AppendTo[iters, x]**
EvaluationMonitor (❄5) Command to be executed after each evaluation of the
    function to be minimized; examples of values: **None**, **++n**, **AppendTo[points, x]**

The default value of **PrecisionGoal** and **AccuracyGoal** is usually 8 (6 in *Mathematica* 4); their general default value is **WorkingPrecision**/2. They both refer to both the minimum point and the minimum value of the function (in *Mathematica* 4, they only refer to the minimum value). So, iteration is, by default, stopped when the estimated relative or absolute error of the optimum point and of the optimum value is less than 10^{-8}. (For more about these three options, see Section 11.3.1, p. 292.)

The default value **Automatic** of **Method** means that **FindMinimum** uses quasi-Newton methods if one starting point is given and Brent's principal axis method if two starting points are given. However, if the function to be minimized is a sum of squares, then, by default, the **LevenbergMarquardt** method is used (this method is the basis of **FindFit** and **NonlinearRegress**; see Sections 22.1.4, p. 580, and 27.6.2, p. 779).

■ **Detailed Information**

We can get detailed information about the iterations in the following way (❄5):

```
f = x Cos[x]; df = D[f, x];

n = 0; iters = {{0, x, f, df}} /. x → 3.;
FindMinimum[f, {x, 4}, StepMonitor :→ AppendTo[iters, {++n, x, f, df}]]
{-3.28837, {x → 3.42562}}

TableForm[iters, TableSpacing → {0.3, 1.5},
    TableHeadings → {None, {" n", " x", " f", " f'"}}] // PaddedForm
```

n	x	f	f'
0	3.	-2.96998	-1.41335
1	3.46189	-3.28582	0.140828
2	3.42795	-3.28836	0.00897361
3	3.42564	-3.28837	0.0000779862
4	3.42562	-3.28837	4.52149×10^{-8}
5	3.42562	-3.28837	2.27929×10^{-13}

20.1.2 Functions of Several Variables

FindMinimum[f, {{x, x0}, {y, y0}, … }] Use the gradient
FindMinimum[f, {{x, x0, x1}, {y, y0, y1}, … }] Do not use the gradient

■ **Example 1**

Consider the following function:

```
f = x^4 + 3 x^2 y + 5 y^2 + x + y;
```

A local minimum point is as follows:

```
sol = FindMinimum[f, {{x, 1}, {y, -2}}]
```

$\{-0.832579, \{x \to -0.886324, y \to -0.335671\}\}$

A necessary condition for a minimum or maximum is that the gradient is zero. A sufficient condition for a minimum [maximum] is that the Hessian is positive [negative] definite (a symmetric matrix is positive [negative] definite if and only if all eigenvalues are positive [negative]). Calculate the gradient (✿4!) and the eigenvalues of the Hessian:

```
{D[f, x], D[f, y]} /. Last[sol]
```

$\{-2.96061 \times 10^{-11}, 1.79594 \times 10^{-10}\}$

```
{{D[f, x, x], D[f, x, y]}, {D[f, y, x], D[f, y, y]}} /. Last[sol]
```

$\{\{7.41282, -5.31795\}, \{-5.31795, 10\}\}$

```
Eigenvalues[%]     {14.1794, 3.23339}
```

We could have also used the following functions (see Section 16.1.3, p. 421):

```
gradient[f_, vars_List] := Map[D[f, #] &, vars]
hessian[f_, vars_List] := Outer[D, gradient[f, vars], vars]
```

```
gradient[f, {x, y}] /. Last[sol]
```

$\{-2.96061 \times 10^{-11}, 1.79594 \times 10^{-10}\}$

```
hessian[f, {x, y}] /. Last[sol]
```

$\{\{7.41282, -5.31795\}, \{-5.31795, 10\}\}$

■ Detailed Information

We can see how the method proceeds in the following way (✿5):

```
iters = {{1, -2}}; FindMinimum[f, {{x, 1}, {y, -2}},
  EvaluationMonitor :→ AppendTo[iters, {x, y}]];

ContourPlot[f, {x, -1.2, 1.6}, {y, -2, 0.6},
  PlotPoints → 50, Contours → 30, ContourShading → False,
  PlotRange → All, AspectRatio → Automatic, Epilog → Line[iters]];
```

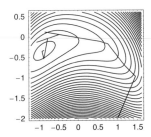

Here the quasi-Newton method was applied. Next we use two starting points, and then Brent's principal axis method is applied (✿5):

```
iters = {{1, -2}}; FindMinimum[f, {{x, 1, 1.1}, {y, -2, -2.1}},
  EvaluationMonitor :→ AppendTo[iters, {x, y}]];
```

```
ContourPlot[f, {x, -2.9, 2}, {y, -2.2, 0.6},
   PlotPoints → 50, Contours → 30, ContourShading → False,
   PlotRange → All, AspectRatio → Automatic, Epilog → Line[iters]];
```

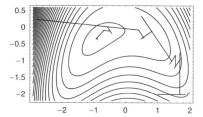

■ Example 2

The following function has a local minimum, a local maximum, and two saddle points, as can be seen from the contours:

```
f = x^3 + y^3 + 2 x^2 + 4 y^2 + 6;
```

```
ContourPlot[f, {x, -2.5, 1}, {y, -3.5, 1},
   AspectRatio → Automatic, Contours → 34, PlotPoints → 60,
   FrameTicks → {Range[-2, 1], Range[-3, 1], None, None}];
```

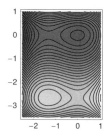

We calculate the minimum and the maximum points and check the nature of the points by calculating the Hessian; see **hessian** in Example 1 above (❀4!):

```
FindMinimum[f, {{x, 1}, {y, 1}}]
```

$\{6., \{x \to 2.07017 \times 10^{-9}, y \to 6.2673 \times 10^{-9}\}\}$

```
Eigenvalues[hessian[f, {x, y}] /. Last[%]]      {8., 4.}
```

```
FindMaximum[f, {{x, -1}, {y, -2}}]
```

$\{16.6667, \{x \to -1.33333, y \to -2.66667\}\}$

```
Eigenvalues[hessian[f, {x, y}] /. Last[%]]      {-8., -4.}
```

One of the saddle points is $(0, -8/3)$. If we start from this point, we get a warning. *Mathematica* finds that this point may be a saddle point:

```
FindMinimum[f, {{x, 0}, {y, -8/3}}]
```

```
FindMinimum::fmgz :
  Encountered a gradient which is effectively zero. The result returned
    may not be a minimum; it may be a maximum or a saddle point. More…
  {15.4815, {x → 0., y → -2.66667}}
```

```
Eigenvalues[hessian[f, {x, y}] /. Last[%]]    {-8., 4.}
```

From the eigenvalues we see that the Hessian is neither positive definite nor negative definite. The other saddle point is $(-4/3, 0)$. Even if the starting point is not a saddle point, it may happen that the search is stopped at a saddle point without a warning (calculating the eigenvalues of the Hessian reveals the nature of the point, however) (❄4!):

```
FindMinimum[f, {{x, -4 / 3}, {y, -2}}]
```

$$\{7.18519, \{x \to -1.33333, y \to 8.79606 \times 10^{-14}\}\}$$

```
Eigenvalues[hessian[f, {x, y}] /. Last[%]]    {8., -4.}
```

A slight modification of the starting point helps to avoid the saddle point (❄4!):

```
FindMinimum[f, {{x, -1.33333}, {y, -2}}]
```

$$\{6., \{x \to 2.00118 \times 10^{-9}, y \to -6.59593 \times 10^{-10}\}\}$$

In general, it is recommended that no "special" points (like $-4/3$) be chosen as starting points. A good starting point is near the minimum point but is otherwise rather random.

■ Example 3

The function to be minimized can be formulated with matrices and vectors (❄5). The dimension of the variable is then taken from the dimension of the initial value. For example:

```
A = {{4, 1, 1}, {1, 1, 1}, {1, 1, 3}}; b = {3, 1, 2}; c = 3;
```

```
FindMinimum[x.A.x + b.x + c, {x, {1, 1, 1}}]
```

$$\{2.29167, \{x \to \{-0.333333, 0.0833332, -0.25\}\}\}$$

20.1.3 Own Programs

■ Newton's Method

In Section 19.3.4, p. 509, we presented the following program to solve nonlinear equations $f(x) = 0$ using Newton's method.

```
newtonSolve[f_, x_, x0_, d_: 1, n_: 20, opts___?OptionQ] :=
 With[{df = D[f, x]}, FixedPointList[(x - d f / df) /. x -> # &, N[x0], n, opts]]
```

Now we are interested in maximum and minimum points, and so we want to solve equations of the form $f'(x) = 0$. Thus, **newtonSolve** can also be used for optimization. The program stops if two successive points are the same to 16-digit precision. For example:

```
f = x Cos[x]; df = D[f, x];
```

```
Plot[f, {x, 0, 8}];
```

Here is a point of minimum:

```
newtonSolve[df, x, 4]
```

```
{4., 3.42503, 3.42562, 3.42562, 3.42562, 3.42562}
```

Here is a point of maximum:

```
newtonSolve[df, x, 1]
```

```
{1., 0.864536, 0.860339, 0.860334, 0.860334, 0.860334}
```

■ Dawidon–Fletcher–Powell Method

Here we write a program for the Dawidon–Fletcher–Powell (DFP) method to minimize a
function of several variables. First, we define functions for a norm, for a substitution and
for Newton's method (the optimal step size is calculated by Newton's method; we use a
simplified version of the program **newtonSolve** we presented above).

```
norm[x_List] := Sqrt[x.x]
subst[f_, x_List, x0_List] := N[f /. Thread[x → x0]]
newton[f_, x_, x0_] :=
 With[{df = D[f, x]}, FixedPoint[(x - f / df) /. x → # &, x0, 15]]
```

Note that *Mathematica* 5 has **Norm**, but we have written our own norm function to make
the code accessible for users of *Mathematica* 4.

Then we write a program **dfpStep** for one step of the DFP method. Let $f(x)$ be the
function to be minimized and $g(x)$ its gradient. Let x_i be the present point, $g_i = g(x_i)$, and
H_i an approximation to the inverse of the Hessian at x_i. We calculate the next values x_{i+1},
g_{i+1}, and H_{i+1} as follows.

The direction of search is $d_i = -H_i g_i$ (normalized to have length 1). The optimal step
size λ is calculated in the program **oneDim** by minimizing $h(\lambda) = f(x_i + \lambda d_i)$ with respect to
λ. The minimization is done by first sampling $h(\lambda)$ at a set of values for λ. The best value
λ_0 is chosen as the starting point for Newton's method to calculate the optimal value λ_i.

The updates are then calculated in the program **update**. The new step is $dx_i = \lambda_i d_i$, the
new point is $x_{i+1} = x_i + dx_i$, and the new gradient is $g_{i+1} = g(x_{i+1})$. If we denote $dg_i =
g_{i+1} - g_i$, the new approximation to the inverse of the Hessian is as follows:

$$H_{i+1} = H_i + \frac{dx_i \, dx_i{}^t}{dx_i{}^t \, dg_i} - \frac{H_i \, dg_i \, dg_i{}^t \, H_i}{dg_i{}^t \, H_i \, dg_i}.$$

(Here, for example, $dx_i{}^t$ is the transpose of dx_i.)

```
dfpStep[f_, g_, x_, λsample_, {xi_, gi_, Hi_}] := Module[{di, ndi, λi},
  di = -Hi.gi;        (*direction of search*)
  ndi = norm[di];     (*norm of the direction*)
  If[ndi == 0., Return[{xi, gi, Hi}], di = di / ndi]; (*normalize the dir.*)
  λi = oneDim[f, x, xi, di, λsample];      (*λi is the optimal λ*)
  If[λi == 0., Return[{xi, gi, Hi}]];      (*test for stopping*)
  update[g, x, λi, di, xi, gi, Hi]]        (*return updated values*)
```

```
oneDim[f_, x_, xi_, di_, λsample_] := Module[{h, hsample, λ0},
  h = subst[f, x, xi + λ di];      (*step size λ is chosen optimally*)
  hsample = N[h /. λ → λsample];      (*sample h at points λsample*)
  λ0 = λsample[[ Ordering[hsample, 1][[1]] ]];      (*best of λsample*)
  newton[D[h, λ], λ, λ0]]      (*the optimal λ*)
```

```
update[g_, x_, λi_, di_, xi_, gi_, Hi_] :=
 Module[{dx, xnew, gnew, dg, Hnew},
  dx = λi di;      (*the optimal step*)
  xnew = xi + dx;      (*update x*)
  gnew = subst[g, x, xnew];      (*update g*)
  dg = gnew - gi;      (*needed next*)
  Hnew = Hi + 1 / (dx.dg) Outer[Times, dx, dx] -
    1 / (dg.Hi.dg) Outer[Times, Hi.dg, dg.Hi];      (*update H*)
  {xnew, gnew, Hnew}]      (*return updated values*)
```

In **dfpStep**, **f** is the function to be minimized, **g** its gradient, **λsample** a set of points (used to get a good starting value of λ for **newton**; see **dfpMinimize** below), and **{xi, gi, Hi}** the present state of the algorithm. The result of **dfpStep** is **{xnew, gnew, Hnew}**. Note that **Ordering[hsample, 1][1]**, which is used in **oneDim**, gives the position of the smallest element in **hsample**.

We then write the main program **dfpMinimize**. We apply the functional programming style in which **dfpStep** is iterated with **FixedPointList**:

```
dfpMinimize[f_, x_List, startx_List, eps1_: 10^-6, eps2_: 10^-6] :=
 Module[{g, λsample, x0, g0, H0, result},
  g = Map[D[f, #] &, x];      (*gradient of f*)
  λsample = Range[0., 10., 0.2];      (*dfpStep uses this to sample h*)
  x0 = N[startx];      (*initial x*)
  g0 = subst[g, x, x0];      (*initial g*)
  H0 = IdentityMatrix[Length[x]];      (*initial H*)

  result = FixedPointList[dfpStep[f, g, x, λsample, #] &, {x0, g0, H0}, 100,
    SameTest → (norm[#2[[2]]] < eps1 && norm[#1[[1]]] - #2[[1]]] < eps2 &)];
      (*solve the problem*)

  iters = Transpose[result][[1]];      (*values of x*)
  {Length[iters] - 1, subst[f, x, Last[iters]], Last[iters]}]
      (*number of iterations, optimal f, optimal x*)
```

The stopping criterion requires that the norm of the gradient is at most **eps1** (default value is 10^{-6}) and the norm of the difference between the last two points is less than **eps2** (default value is 10^{-6}). Remember that **#1** is the next to last and **#2** the last iteration, and that, for example, **#2[[2]]** is the second component of the last iteration, that is, the gradient **gnew**. We have not defined **iters** (the whole list of points generated by **dfpMinimize**) as a local variable, so it is available outside the module. We can then, for example, plot the points.

■ Example

Let us minimize the same function as we minimized with **FindMinimum** in Example 1 of Section 20.1.2, p. 516:

```
f = x^4 + 3 x^2 y + 5 y^2 + x + y;

dfpMinimize[f, {x, y}, {1, -2}]

{9, -0.832579, {-0.886324, -0.335671}}
```

We needed 9 iterations, the minimum value of the function is -0.832579, and the minimum point is $(-0.886324, -0.335671)$. The iterations are in the variable **iters**. The following program can be used to show the iterations:

```
showIterations[f_, x_, y_, iters_List, opts___] :=
  Module[{X, Y, e = 0.2}, {X, Y} = Transpose[iters];
   ContourPlot[f, {x, Min[X] - e, Max[X] + e}, {y, Min[Y] - e, Max[Y] + e},
    ContourShading → False, PlotRange → All, opts,
    Epilog → {AbsolutePointSize[2], Map[Point, iters], Line[iters]}]]
```

```
showIterations[f, x, y, iters, Contours → 30,
  PlotPoints → 30, AspectRatio → Automatic];
```

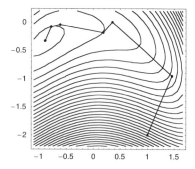

20.2 Constrained Optimization

20.2.1 Linear Problems

We can formulate a linear programming problem either by writing explicit expressions with variables or by giving only coefficient matrices and vectors. The corresponding commands are **Minimize** and **LinearProgramming**, respectively.

■ Formulation with Variables

Minimize[{f, cons}, vars] (✳5) Give the global minimum of **f** with respect to variables **vars** subject to constraints **cons**

Maximize (✳5) is used in the same way. (Here, these commands are presented in the context of linear optimization, but, more generally, these commands can solve *polynomial* optimization problems, see Section 20.2.2, p. 529.) The problem cannot contain any symbolic parameters. For example:

```
Maximize[{x + y, 2 x + y ≤ 2, x - y ≥ -1 / 2, x + 2 y ≤ 2}, {x, y}]
```

$\{\frac{4}{3}, \{x \to \frac{2}{3}, y \to \frac{2}{3}\}\}$

Note that `ConstrainedMin[obj, cons, vars]` and `ConstrainedMax`, used through *Mathematica* version 4.2, are now obsolete. For example:

```
ConstrainedMax[x + y, {2 x + y ≤ 2, x - y ≥ -1 / 2, x + 2 y ≤ 2}, {x, y}]
```

```
ConstrainedMax::deprec :
  ConstrainedMax is deprecated and will not be supported in future
    versions of Mathematica.  Use NMaximize or Maximize instead. More…
```

$\{\frac{4}{3}, \{x \to \frac{2}{3}, y \to \frac{2}{3}\}\}$

■ Visualization

Linear problems with two variables can easily be visualized. We consider the same problem we have solved above. First we define the inequalities and plot the corresponding feasible region (plotting inequalities is considered in Section 5.3.3, p. 127):

```
ineqs = {2 x + y ≤ 2, x - y ≥ -1 / 2, x + 2 y ≤ 2};
```

```
<< Graphics`InequalityGraphics`
```

```
constr =
   InequalityPlot[ineqs, {x, 0, 1}, {y, 0, 0.9}, Fills → GrayLevel[0.8]];
```

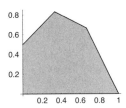

Then we plot lines where the object function takes on some constant values. We choose the values $1/3$, $2/3$, 1, $4/3$, and $5/3$:

```
obj = ContourPlot[x + y, {x, 0, 1}, {y, 0, 0.9},
   ContourShading → False, Contours → Range[1 / 3, 5 / 3, 1 / 3],
   ContourStyle → {RGBColor[0, 0, 1], AbsoluteDashing[{2}]}];
```

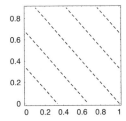

Lastly we combine the plots and add a point where we have the optimal solution:

```
Show[constr, obj,
  Epilog → {RGBColor[1, 0, 0], AbsolutePointSize[4], Point[{2 / 3, 2 / 3}]}];
```

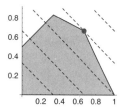

The contour where the object function has the value 4/3 is the highest that still has a point in the feasible region; this point, shown in red, is the solution of the problem.

■ Example: A Transportation Problem

Consider the following transportation problem. Plants 1, 2, and 3 have supply of 47, 36, and 52 units, respectively, and cities 1, 2, 3, and 4 have demand of 38, 34, 29, and 34 units, respectively. The problem is to decide the amounts to be transported from the plants to the cities so as to minimize transportation costs. We can use a program presented in Section 5.4.5, p. 140, to illustrate the situation:

```
points = {{0, 1}, {0, 3}, {0, 5}, {7, 0}, {7, 2}, {7, 4}, {7, 6}};
plants = {"Plant 1", "Plant 2", "Plant 3"};
cities = {"City 1", "City 2", "City 3", "City 4"};
labels = Join[plants, cities];
lines = Flatten[Outer[List, {1, 2, 3}, {4, 5, 6, 7}], 1];

<< MathematicaNavigator2`MN1`

networkPlot[points, labels, lines, 11, PlotRange → {{-1, 8}, {-1, 7}}];
```

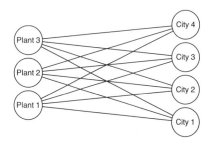

The transportation costs are as follows:

```
supply = {47, 36, 52}; demand = {38, 34, 29, 34};
costs = {{5, 7, 6, 10}, {9, 4, 6, 7}, {5, 8, 6, 6}};
```

To write a table for the problem, proceed as follows (for **Grid**, see Section 13.1.3, p. 330):

```
<< LinearAlgebra`MatrixManipulation`
```

```
data = BlockMatrix[{
    { {{""}}, {cities}, {{"Supply"}} },
    { Transpose[{plants}], costs, Transpose[{supply}] },
    { {{"Demand"}}, {demand}, {{Apply[Plus, supply]}} } }]
```

$\{\{, \text{City 1, City 2, City 3, City 4, Supply}\},$
$\{\text{Plant 1, 5, 7, 6, 10, 47}\}, \{\text{Plant 2, 9, 4, 6, 7, 36}\},$
$\{\text{Plant 3, 5, 8, 6, 6, 52}\}, \{\text{Demand, 38, 34, 29, 34, 135}\}\}$

```
Grid[data, ColumnAlignments → {Left, Right},
    RowLines → {True, False, False, True},
    ColumnLines → {True, False, False, False, True}] // TraditionalForm
```

	City 1	City 2	City 3	City 4	Supply
Plant 1	5	7	6	10	47
Plant 2	9	4	6	7	36
Plant 3	5	8	6	6	52
Demand	38	34	29	34	135

For example, transportation of one unit from Plant 1 to City 1 costs \$5. Let $x_{i,j}$ be the amount transported from Plant i to City j:

```
vars = Table[x_{i,j}, {i, 3}, {j, 4}]
```

$\{\{x_{1,1}, x_{1,2}, x_{1,3}, x_{1,4}\}, \{x_{2,1}, x_{2,2}, x_{2,3}, x_{2,4}\}, \{x_{3,1}, x_{3,2}, x_{3,3}, x_{3,4}\}\}$

The total cost is as follows:

```
obj = {Apply[Plus, Flatten[costs vars]]}
```

$\{5 x_{1,1} + 7 x_{1,2} + 6 x_{1,3} + 10 x_{1,4} + 9 x_{2,1} +$
$4 x_{2,2} + 6 x_{2,3} + 7 x_{2,4} + 5 x_{3,1} + 8 x_{3,2} + 6 x_{3,3} + 6 x_{3,4}\}$

(Note that here we did not use the dot in matrix multiplication but rather the space: the space does an element-by-element multiplication.) Supply, demand, and nonnegativity constraints are as follows:

```
supplyConstr = Thread[Apply[Plus, vars, 1] ≤ supply]
```

$\{x_{1,1} + x_{1,2} + x_{1,3} + x_{1,4} \le 47, x_{2,1} + x_{2,2} + x_{2,3} + x_{2,4} \le 36, x_{3,1} + x_{3,2} + x_{3,3} + x_{3,4} \le 52\}$

```
demandConstr = Thread[Apply[Plus, vars] ≥ demand]
```

$\{x_{1,1} + x_{2,1} + x_{3,1} \ge 38, x_{1,2} + x_{2,2} + x_{3,2} \ge 34,$
$x_{1,3} + x_{2,3} + x_{3,3} \ge 29, x_{1,4} + x_{2,4} + x_{3,4} \ge 34\}$

```
nonneg = Thread[Flatten[vars] ≥ 0]
```

$\{x_{1,1} \ge 0, x_{1,2} \ge 0, x_{1,3} \ge 0, x_{1,4} \ge 0, x_{2,1} \ge 0,$
$x_{2,2} \ge 0, x_{2,3} \ge 0, x_{2,4} \ge 0, x_{3,1} \ge 0, x_{3,2} \ge 0, x_{3,3} \ge 0, x_{3,4} \ge 0\}$

The problem to be solved is now as follows:

```
problem = Join[obj, supplyConstr, demandConstr, nonneg]
```

$\{5\,x_{1,1} + 7\,x_{1,2} + 6\,x_{1,3} + 10\,x_{1,4} + 9\,x_{2,1} + 4\,x_{2,2} + 6\,x_{2,3} + 7\,x_{2,4} + 5\,x_{3,1} + 8\,x_{3,2} +$
$6\,x_{3,3} + 6\,x_{3,4},\ x_{1,1} + x_{1,2} + x_{1,3} + x_{1,4} \le 47,\ x_{2,1} + x_{2,2} + x_{2,3} + x_{2,4} \le 36,$
$x_{3,1} + x_{3,2} + x_{3,3} + x_{3,4} \le 52,\ x_{1,1} + x_{2,1} + x_{3,1} \ge 38,\ x_{1,2} + x_{2,2} + x_{3,2} \ge 34,$
$x_{1,3} + x_{2,3} + x_{3,3} \ge 29,\ x_{1,4} + x_{2,4} + x_{3,4} \ge 34,\ x_{1,1} \ge 0,\ x_{1,2} \ge 0,\ x_{1,3} \ge 0,\ x_{1,4} \ge 0,$
$x_{2,1} \ge 0,\ x_{2,2} \ge 0,\ x_{2,3} \ge 0,\ x_{2,4} \ge 0,\ x_{3,1} \ge 0,\ x_{3,2} \ge 0,\ x_{3,3} \ge 0,\ x_{3,4} \ge 0\}$

We solve the problem:

```
sol = Minimize[problem, Flatten[vars]]
```

$\{704,\ \{x_{1,1} \to 20,\ x_{1,2} \to 0,\ x_{1,3} \to 27,\ x_{1,4} \to 0,\ x_{2,1} \to 0,$
$x_{2,2} \to 34,\ x_{2,3} \to 2,\ x_{2,4} \to 0,\ x_{3,1} \to 18,\ x_{3,2} \to 0,\ x_{3,3} \to 0,\ x_{3,4} \to 34\}\}$

■ Integer Problems

Some problems with integer variables can be solved with **Minimize** and **Maximize**. For example:

```
Maximize[{x + y, 2 x + y ≤ 2, x - y ≥ -1 / 2,
   x + 2 y ≤ 2, x ≥ 0, y ≥ 0, {x, y} ∈ Integers}, {x, y}]
```

$\{1,\ \{x \to 1,\ y \to 0\}\}$

However, integer problems are generally solved with **NMinimize** or **NMaximize**. These commands, which are applicable to general nonlinear problems, are considered in some detail in Section 20.2.3. Next we give a brief example of how to use them to solve integer problems.

NMinimize[{f, cons}, vars] (❀5) Give the global minimum of **f** with respect to variables **vars** subject to constraints **cons**

■ Example: A Knapsack Problem

Let us solve a knapsack problem where the benefits of four items are 14, 10, 15, 8, and 9 and the corresponding weights are 6, 8, 5, 6, and 4. What is the optimal collection of items when we want to maximize the total benefit subject to the constraint that the total weight has to be at most 18? Let x_i be 1 if the ith item is included and 0 otherwise. 0–1 variables can be formulated in *Mathematica* by constraining the variables in the interval [0, 1] and requiring that the variables are integers. Write the following:

```
NMaximize[{14 x1 + 10 x2 + 15 x3 + 8 x4 + 9 x5, 6 x1 + 8 x2 + 5 x3 + 6 x4 + 4 x5 ≤ 18,
   0 ≤ x1 ≤ 1, 0 ≤ x2 ≤ 1, 0 ≤ x3 ≤ 1, 0 ≤ x4 ≤ 1, 0 ≤ x5 ≤ 1,
   {x1, x2, x3, x4, x5} ∈ Integers}, {x1, x2, x3, x4, x5}]
```

$\{38.,\ \{x1 \to 1,\ x2 \to 0,\ x3 \to 1,\ x4 \to 0,\ x5 \to 1\}\}$

It is optimal to take the first, third, and fifth item; the total benefit is then 38.

With larger problems, it is useful to use vectors. As an example, we solve the same problem anew, now using vectors:

```
benefits = {14, 10, 15, 8, 9};
weights = {6, 8, 5, 6, 4}; vars = Table[x_i, {i, 5}];
obj = {benefits.vars}; constr = {weights.vars ≤ 18};
bounds = Thread[0 ≤ vars ≤ 1]; dom = {vars ∈ Integers};
```

```
problem = Join[obj, constr, bounds, dom]
```

$\{14\, x_1 + 10\, x_2 + 15\, x_3 + 8\, x_4 + 9\, x_5,\ 6\, x_1 + 8\, x_2 + 5\, x_3 + 6\, x_4 + 4\, x_5 \le 18,\ 0 \le x_1 \le 1,$
$0 \le x_2 \le 1,\ 0 \le x_3 \le 1,\ 0 \le x_4 \le 1,\ 0 \le x_5 \le 1,\ (x_1 \mid x_2 \mid x_3 \mid x_4 \mid x_5) \in \text{Integers}\}$

```
sol = NMaximize[problem, vars]
```

$\{38.,\ \{x_1 \to 1,\ x_2 \to 0,\ x_3 \to 1,\ x_4 \to 0,\ x_5 \to 1\}\}$

The total weight is as follows:

```
constr[[1, 1]] /. sol[[2]]        15
```

This means that 3 weight units are left unused.

Next we introduce **LinearProgramming**, where the problem to be solved is formulated using vectors and matrices without explicit variables.

■ Formulation with Matrices

> **LinearProgramming[c, m, b]** Minimize **c.x** subject to **m.x** ≥ **b** and **x** ≥ 0
> **LinearProgramming[c, m, {{b₁, s₁}, {b₂, s₂}, … }]** The ith constraint is
> **m**$_i$ **.x** ≤ **b**$_i$, **m**$_i$ **.x** == **b**$_i$, or **m**$_i$ **.x** ≥ **b**$_i$ according to whether **s**$_i$ is < 0, = 0, or > 0
> **LinearProgramming[c, m, b, {l₁, l₂, … }]** Add the constraints **x**$_i$ ≥ **l**$_i$ (a single number can also be supplied for the lower bound; it is then applied for each variable)
> **LinearProgramming[c, m, b, {{l₁, u₁}, {l₂, u₂}, … }]** Add the constraints
> **l**$_i$ ≤ **x**$_i$ ≤ **u**$_i$

The **Method → "InteriorPoint"** (✿5) option is available.

If we have to maximize **c.x**, then we can minimize **-c.x**, instead. As an example, we solve the same problem we solved with **Maximize**: minimize $-x - y$ subject to $-2x - y \ge -2$, $x - y \ge -1/2$, and $-x - 2y \ge -2$ together with $x \ge 0$ and $y \ge 0$:

```
c = {-1, -1}; m = {{-2, -1}, {1, -1}, {-1, -2}};
b = {-2, -1/2, -2};
```

```
sol = LinearProgramming[c, m, b]        {2/3, 2/3}
```

Thus $x = y = 2/3$. The value of the original objective function $x + y$ is as follows:

```
-c.sol        4/3
```

If we want to write the constraints as $2x + y \le 2$, $x - y \ge -1/2$, and $x + 2y \le 2$, we have to use the s_i numbers. We denote a ≤ inequality with $s_i = -1$ and a ≥ inequality with $s_i = 1$:

```
c = {-1, -1}; m = {{2, 1}, {1, -1}, {1, 2}};
b = {2, -1/2, 2}; s = {-1, 1, -1};
bs = Transpose[{b, s}]
```

$\left\{\{2, -1\}, \left\{-\dfrac{1}{2}, 1\right\}, \{2, -1\}\right\}$

```
LinearProgramming[c, m, bs]      {2/3, 2/3}
```

In the following example, we add the bounds $\frac{1}{4} \le x \le \frac{1}{2}$ and $\frac{1}{5} \le y \le \frac{2}{5}$:

```
l = {1 / 4, 1 / 5}; u = {1 / 2, 2 / 5};
lu = Transpose[{l, u}]
```

$$\{\{\frac{1}{4}, \frac{1}{2}\}, \{\frac{1}{5}, \frac{2}{5}\}\}$$

```
LinearProgramming[c, m, bs, lu]      {1/2, 2/5}
```

■ Example: A Transportation Problem

Let us solve the transportation problem anew, now using vectors and matrices. The data were

```
supply = {47, 36, 52}; demand = {38, 34, 29, 34};
costs = {{5, 7, 6, 10}, {9, 4, 6, 7}, {5, 8, 6, 6}};
```

Although variables will not be used in the formulation, they are as follows:

```
vars = Flatten[Table[x_{i,j}, {i, 3}, {j, 4}]]
```

$$\{x_{1,1}, x_{1,2}, x_{1,3}, x_{1,4}, x_{2,1}, x_{2,2}, x_{2,3}, x_{2,4}, x_{3,1}, x_{3,2}, x_{3,3}, x_{3,4}\}$$

The cost vector, the lefthand-side matrix, and the righthand-side matrix are shown here:

```
c = Flatten[costs]
```

$\{5, 7, 6, 10, 9, 4, 6, 7, 5, 8, 6, 6\}$

```
m = {{1, 1, 1, 1, 0, 0, 0, 0, 0, 0, 0, 0},
     {0, 0, 0, 0, 1, 1, 1, 1, 0, 0, 0, 0},
     {0, 0, 0, 0, 0, 0, 0, 0, 1, 1, 1, 1},
     {1, 0, 0, 0, 1, 0, 0, 0, 1, 0, 0, 0},
     {0, 1, 0, 0, 0, 1, 0, 0, 0, 1, 0, 0},
     {0, 0, 1, 0, 0, 0, 1, 0, 0, 0, 1, 0},
     {0, 0, 0, 1, 0, 0, 0, 1, 0, 0, 0, 1}};
```

```
bs = Transpose[{Join[supply, demand], {-1, -1, -1, 1, 1, 1, 1}}]
```

$\{\{47, -1\}, \{36, -1\}, \{52, -1\}, \{38, 1\}, \{34, 1\}, \{29, 1\}, \{34, 1\}\}$

The solution is as follows:

```
sol = LinearProgramming[c, m, bs]
```

$\{20, 0, 27, 0, 0, 34, 2, 0, 18, 0, 0, 34\}$

We pick the variables that have a positive value:

```
Select[Thread[vars → sol], #[[2]] ≠ 0 &]
```

$\{x_{1,1} \to 20, x_{1,3} \to 27, x_{2,2} \to 34, x_{2,3} \to 2, x_{3,1} \to 18, x_{3,4} \to 34\}$

The minimum cost is as follows:

```
Apply[Plus, Flatten[costs Partition[sol, 4]]]      704
```

20.2.2 Polynomial Problems

> `Minimize[f, vars]` (⌘5) Give the global minimum of `f` with respect to variables `vars`
> `Minimize[{f, cons}, vars]` (⌘5) Give the global minimum of `f` with respect to
>   variables `vars` subject to constraints `cons`

We also have `Maximize` (⌘5), which is used in the same way. Note that users of *Mathematica* version 4 must use the command in the form `Experimental`Minimize` or first execute `<<Experimental`` and then use `Minimize` (`Maximize` does not exist in version 4). However, the commands are not so powerful in *Mathematica* 4 than in *Mathematica* 5; for example, integer problems cannot be solved in *Mathematica* 4. The use of `Minimize` to solve linear problems is considered in Section 20.2.1, p. 522.

The function to be minimized and the equality and inequality constraints can be, for example, polynomials (of possibly several variables). They can also be rational expressions. In general, they can be radical expressions, which are expressions that contain rational powers. In addition, the expressions are not allowed to contain symbolic constants like *a*. The constraints are written as lists or as logical expressions containing, for example, the logical AND (`&&`). Variables can be constrained to be integer valued by writing a constraint `{x, y, … } ∈ Integers` (all variables have to be either real or integer); for linear integer problems, see Section 20.2.1, p. 526.

`Minimize` gives the *global* minimum of `expr` in the region in which the constraints `cons` hold (recall that `FindMinimum` only searches a local minimum). If the global minimum is not unique, `Minimize` picks one of them. If we are interested in local minimums, we have to add suitable constraints so that, in the resulting feasible region, the local minimum we are searching for is also the global minimum.

Note also that `Minimize` finds an *exact* solution (i.e., the solution is not a decimal number) if the problem is exact. Indeed, unlike `FindMinimum`, `Minimize` does not use an iterative numerical method but rather an exact algebraic method.

■ Example 1: Local Optima

Consider the following function:

```
f = 5 + 40 x^3 - 45 x^4 + 12 x^5;
```

```
Plot[f, {x, -0.7, 2.4}];
```

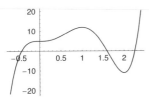

When using `Minimize`, we should remember that it calculates the *global* minimum. Thus, without any restrictions, the global minimum value of our function is $-\infty$:

```
Minimize[f, x]
```

```
Minimize::natt : The minimum is not attained
    at any point satisfying the given constraints. More...
{-∞, {x → -∞}}
```

If we want to calculate the local minimum at about $x = 2$, we have to write suitable constraints to force the search to a region in which that local minimum is also the global minimum:

```
Minimize[{f, x > 0}, x]       {-11, {x → 2}}
```

So, at $x = 2$, we have a local minimum where the value of the function is -11. Next we find a local maximum:

```
Maximize[{f, x < 3 / 2}, x]       {12, {x → 1}}
```

If we have a strict inequality constraint, the result may be that an optimum does not exist. In such a case, **Minimize** gives the point on the boundary:

```
Minimize[{f, x > 2}, x]
```

```
Minimize::wksol :
  Warning: There is no minimum in the region described by
    the contraints; returning a result on the boundary. More...
  {-11, {x → 2}}
```

The following function was also considered in Example 2 of Section 20.1.2. It has a local minimum at (0, 0):

```
f = x^3 + y^3 + 2 x^2 + 4 y^2 + 6;
```

```
Minimize[{f, -1 < x < 1, -1 < y < 1}, {x, y}]
```

```
{6, {x → 0, y → 0}}
```

■ Example 2: Global Constrained Optima

The following problem has both equality and inequality constraints:

```
Minimize[{(x - y)^2 + 5 z, x + y + z == -5, y - 3 z == 1, x ≥ 0}, {x, y, z}]
```

$$\left\{ \frac{19}{4}, \left\{ x \to 0, y \to -\frac{7}{2}, z \to -\frac{3}{2} \right\} \right\}$$

Here is a nonlinear integer problem (❀5):

```
Minimize[
  {x^2 + x y + z, x + x y ≥ 5, x ≥ 0, y ≥ 0, z ≥ 0, {x, y, z} ∈ Integers}, {x, y, z}]
{5, {x → 1, y → 4, z → 0}}
```

The next problem is somewhat difficult (*Mathematica* 4 does not solve this problem at all):

```
Minimize[{x^2 / 2 + (y^2 + z^2 + v^2) / 6, x ≥ 2,
  x + y ≥ 5, x + z ≥ 2, x + v ≥ 1, v ≥ 0}, {x, y, z, v}] // Timing
```

$$\left\{ 10.58 \text{ Second}, \left\{ \frac{7}{2}, \{x \to 2, v \to 0, y \to 3, z \to 0\} \right\} \right\}$$

Consider the following problem (prepare to wait for the solution for a long time in *Mathematica* 4):

```
sol = Minimize[{1 / (5 x y z) + 4 / x + 3 / z, 2 x z + x y ≤ 10, x ≥ 0, y ≥ 0, z ≥ 0},
   {x, y, z}] // FullSimplify
```

$$\left\{ \frac{4}{5} \sqrt{\frac{1}{5} \, (76 + \sqrt{151})} \, , \right.$$

$$\left. \left\{ y \to \frac{1}{\sqrt{10}} \, , \ z \to \frac{\sqrt{\frac{151}{10}}}{2} \, , \ x \to \text{Root}[2000 - 608 \, \#1^2 + 45 \, \#1^4 \, \&, \, 3] \right\} \right\}$$

The value of x is expressed as a **Root** object (see Section 19.2.2, p. 496). Decimal values are as follows:

```
sol // N      {3.36168, {y → 0.316228, z → 1.94294, x → 2.37976}}
```

The first inequality constraint is active:

```
2 x z + x y /. sol[[2]] // FullSimplify      10
```

If no points can be found to satisfy the constraints, the result is as follows:

```
Minimize[{x^2, x < -1, x > 1}, x]
```

```
Minimize::natt : The minimum is not attained
   at any point satisfying the given constraints. More…
```

$\{\infty, \{x \to \text{Indeterminate}\}\}$

In Section 20.3.2, p. 539, and in Example 4 of Section 20.3.3, p. 546, we minimize, by classical optimization, the surface area $\pi r \sqrt{r^2 + h^2}$ of a cone given that its volume $\frac{1}{3} \pi r^2 h$ has a given value v. This problem cannot be solved by **Minimize**, because it contains the symbolic constant v.

20.2.3 Numerical Optimization

If **Minimize** is not able to solve an optimization problem, we can resort to **NMinimize**. Like **Minimize**, **NMinimize** also finds *global* optimum points. However, unlike **Minimize**, **NMinimize** only gives an *approximate* decimal answer.

NMinimize[f, vars] (❀5) Give the global minimum of **f** with respect to variables **vars**
NMinimize[{f, cons}, vars] (❀5) Give the global minimum of **f** with respect to variables **vars** subject to constraints **cons**

We also have **NMaximize** (❀5), which is used in the same way. Note that users of *Mathematica* 4.2 must first load the package **NumericalMath`NMinimize`** before using **NMinimize**.

NMinimize is very flexible and powerful for the finding of minima. Its methods are derivative-free, and it solves unconstrained and constrained problems and problems with continuous and integer variables. When saying that **NMinimize** finds global optimum points, we have to add that this is not guaranteed but only happens with high probability.

In Sections 20.2.3 and 20.2.4, we give an overview of **NMinimize**. For advanced information and examples of these commands, look at **NMinimize** in the *Help Browser* and read the document http://library.wolfram.com/database/Conferences/4033/.

■ Examples

Here is an example without constraints:

```
f = x^4 + 3 x^2 y + 5 y^2 + x + y;
```

```
NMinimize[f, {x, y}]
```

$\{-0.832579, \{x \to -0.886324, y \to -0.335671\}\}$

The result is essentially the same as what we obtained with **FindMinimum** in Example 1 of Section 20.1.2, p. 516. Here is an example with constraints:

```
NMinimize[{(x - y)^2 + 5 z, x + y + z + 5 == 0, y - 3 z - 1 == 0, x ≥ 0}, {x, y, z}]
```

$\{4.75, \{x \to 5.2184 \times 10^{-19}, y \to -3.5, z \to -1.5\}\}$

The result is essentially the same one we found with **Minimize** in Example 2 of Section 20.2.2, p. 530. If no points can be found to satisfy the constraints, the result is as follows:

```
NMinimize[{x^2, x < -1, x > 1}, x]
```

```
NMinimize::nsol : There are no points
    that satisfy the constraints {1 - x ≤ 0, 1 + x ≤ 0}. More…
```
$\{\infty, \{x \to \text{Indeterminate}\}\}$

■ Initial Intervals

NMinimize is an iterative method and needs, for each variable, an initial interval in which to start. If an interval is not given for a variable, the interval $[-1, 1]$ is used. An interval, for example, $[1, 3]$, can be given as $\{x, 1, 3\}$ in the list of variables or as $1 \le x \le 3$ in the constraints. In the list of variables, an interval is used to start, but in the constraints, an interval is used both to start and to constraint the optimum. Thus, in the following example, the intervals $[1, 3]$ and $[0, 1]$ are used only to start:

```
NMinimize[f, {{x, 1, 3}, {y, 0, 1}}]
```

$\{-0.832579, \{x \to -0.886324, y \to -0.335671\}\}$

However, in the next example, the intervals are used both to start and as constraints:

```
NMinimize[{f, 1 ≤ x ≤ 3, 0 ≤ y ≤ 1}, {x, y}]
```

$\{2., \{x \to 1., y \to -7.93016 \times 10^{-18}\}\}$

If we give intervals in both the variables and in the constraints, the intervals in the variables are used to start and the intervals in the constraints are treated as constraints:

```
NMinimize[{f, 0 ≤ x ≤ 4, 0 ≤ y ≤ 5}, {{x, 1, 3}, {y, 0, 1}}]
```

$\{0., \{x \to 0., y \to 0.\}\}$

An interval can be specified in different ways for different variables.

■ Constraints and Integer Variables

NMinimize handles constraints with penalty functions: a penalty is given to points that violate the constraints. Constraints can be written either as lists or as logical expressions (a logical expression can contain, e.g., **&&** (and), **||** (or), and **!** (not)).

Constraints can contain (possibly nonlinear) equalities and inequalities and also domain specifications. Strong inequalities with < or > are converted to weak inequalities with ≤ and ≥. A domain specification can be **x ∈ Reals** (the default) or **x ∈ Integers**. Variables with only zero or one as the value can be formed with the constraints **x ∈ Integers** and **0 ≤ x ≤ 1**. Here is an example in which the constraints are written as a logical expression:

> **NMinimize[{ (x - y) ^2 + 5 z, x + y + z + 5 == 0 && y - 3 z - 1 == 0 && x ≥ 0}, {x, y, z}]**

> $\{4.75, \{x \to 5.2184 \times 10^{-19}, y \to -3.5, z \to -1.5\}\}$

Here is a linear example with 0-1 variables:

> **NMaximize[{3 x + 2 y - 5 z - 2 v + 3 w, x + y + z + 2 v + w ≤ 4,**
> **7 x + 3 z - 4 v + 3 w ≤ 8, 11 x - 6 y + 3 v - 3 w ≥ 3, 0 ≤ x ≤ 1, 0 ≤ y ≤ 1, 0 ≤ z ≤ 1,**
> **0 ≤ v ≤ 1, 0 ≤ w ≤ 1, {x, y, z, v, w} ∈ Integers}, {x, y, z, v, w}]**

> $\{5., \{v \to 0, w \to 0, x \to 1, y \to 1, z \to 0\}\}$

For linear integer problems, see also Section 20.2.1, p. 526.

20.2.4 Options

NMinimize an **NMaximize** can resort to several methods. With regard to the options of the commands, some options are common to all methods, while others are specific to each method. First we consider the common options.

■ Common Options

Options for **NMinimize** *and* **NMaximize**:

WorkingPrecision Precision used in internal computations; examples of values:
  MachinePrecision (❀5), **20**
PrecisionGoal If the value of the option is **p**, the relative error of the optimum point
  and of the value of the function (or a penalty function) at the optimum point should
  be of the order 10^{-p}; examples of values: **Automatic** (usually means **8**), **10**
AccuracyGoal If the value of the option is **a**, the absolute error of the optimum point
  and of the value of the function (or a penalty function) at the optimum point should
  be of the order 10^{-a}; examples of values: **Automatic** (usually means **8**), **10**
Method Method used; possible values: **Automatic**, **"DifferentialEvolution"**,
  "NelderMead", **"RandomSearch"**, **"SimulatedAnnealing"**
MaxIterations Maximum number of iterations used; examples of values: **Automatic**
  (usually means **100**), **200**
StepMonitor (❀5) Command to be executed after each step of the iterative method;
  examples of values: **None**, **++n**, **AppendTo[iters, x]**
EvaluationMonitor (❀5) Command to be executed after each evaluation of the
  function to be minimized; examples of values: **None**, **++n**, **AppendTo[points, x]**

Iterations are stopped when either the precision goal or the accuracy goal is satisfied. It may be advantageous to try some values of **MaxIterations** that are larger than the default value 100 and also some values of **Method** to see if the solution gets still better. For example, in the next problem simulated annealing works somewhat poorly:

```
problem = {1 / (5 x y z) + 4 / x + 3 / z, 2 x z + x y ≤ 10, x ≥ 0, y ≥ 0, z ≥ 0};

NMinimize[problem, {x, y, z}, Method → "NelderMead"]
{3.36168, {x → 2.37976, y → 0.316228, z → 1.94294}}

NMinimize[problem, {x, y, z}, Method → "SimulatedAnnealing"]
{3.39923, {x → 2.2338, y → 0.353932, z → 2.02229}}
```

However, using more iterations we obtain a good solution:

```
NMinimize[problem, {x, y, z},
  Method → "SimulatedAnnealing", MaxIterations → 400]
{3.36168, {x → 2.37976, y → 0.316228, z → 1.94294}}
```

▪ Methods

We can use four methods:

- **DifferentialEvolution** is a genetic method developed by K. Price and R. Storn. It may be the most robust method of the four, but it is also computationally demanding and often slower than other methods. This method is suggested, for example, for problems with integer variables.
- **NelderMead** is the simplex method by J. A. Nelder and R. Mead. It is generally the fastest of the four methods, and it suits problems with continuous variables.
- **RandomSearch** first generates a large number of points in the initial region and then uses each point as the starting point for **FindMinimum**. This method requires that the objective function is locally continuous. The method is not well suited for discrete problems.
- **SimulatedAnnealing** starts from many points, and in each iteration it moves from each current point to a random direction. If, for a present point, the move results in a better point, it is accepted; otherwise the point is accepted with a certain probability. The method is often faster than **DifferentialEvolution**. The method can also be used for discrete problems.

The default setting **Method → Automatic** tries to choose a good method, as follows:

- If the problem is linear (and does not have integer variables), use **LinearProgram⹂ming**.
- If any of the variables are integer-valued, use **DifferentialEvolution**.
- Otherwise, that is, if the problem is nonlinear and continuous, use **NelderMead**, and if it does poorly, switch to **DifferentialEvolution**.

Each method has special options that we study next.

▪ Method-Specific Options

The method-specific options are used inside the value of the **Method** option, as follows:

```
NMinimize[{f, cons}, vars, opts, Method → {"method", methodSpecOpts}]
```

Among the method-specific options, six options are not so specific but are shared with all four methods (with one exception). These options are considered first.

■ Common Method-Specific Options

"**RandomSeed**" Seed for random number generator; default value: **0**
"**SearchPoints**" Number of initial points (not for **NelderMead**); default value:
  Automatic, which means, for d variables:
    min(10 d, 50) for **DifferentialEvolution**
    min(10 d, 100) for **RandomSearch**
    min(2 d, 50) for **SimulatedAnnealing**
  (**NelderMead** uses $d + 1$ points)
"**InitialPoints**" Set of initial points; examples of values: **Automatic**, **{{x1,y1},**
  {x2,y2}, … }
"**PenaltyFunction**" Function applied to constraints to penalize invalid points; default
  value: **Automatic**
"**Tolerance**" Tolerance for accepting constraint violations. Default value: **0.001**
"**PostProcess**" Whether to postprocess using local search methods; possible values:
  Automatic (means **True**), **True**, **False**

All of the methods use random numbers: either in choosing initial points and/or in each iteration. The default value of "**RandomSeed**" is zero (note that *Mathematica* also has **SeedRandom**, see Section 26.1.1, p. 707). Thus, we get the same result if we solve a problem several times. However, with different seeds, we may get different results. In fact, it is advisable to solve a problem with several seeds and pick the best result. For example, here we use six different seeds:

```
problem = {1 / (5 x y z) + 4 / x + 3 / z, 2 x z + x y ≤ 10, x ≥ 0, y ≥ 0, z ≥ 0};

Table[{i, NMinimize[problem, {x, y, z},
    Method → {"NelderMead", "RandomSeed" → i}]}, {i, 0, 5}]

{{0, {3.36168, {x → 2.37976, y → 0.316228, z → 1.94294}}},
 {1, {3.42212, {x → 2.0552, y → 0.392257, z → 2.20085}}},
 {2, {3.36168, {x → 2.37976, y → 0.316228, z → 1.94294}}},
 {3, {3.36969, {x → 2.23809, y → 0.300825, z → 2.08351}}},
 {4, {3.36187, {x → 2.35628, y → 0.32217, z → 1.96089}}},
 {5, {3.36168, {x → 2.37977, y → 0.316228, z → 1.94293}}}}
```

In this example, the best result is obtained with the seeds 0, 2, and 5.

All of the four methods start from a set of points. We can either give the number of initial points as the value of **SearchPoints** and let the algorithms choose the points at random from the initial intervals or give the points themselves as the value of **Initial**⸱**Points**. **NelderMead** and **RandomSearch** proceed deterministically after the initial points are chosen, but **DifferentialEvolution** and **SimulatedAnnealing** use random numbers in each iteration. In the next example, we use different amounts of initial points:

```
Table[{i, NMinimize[problem, {x, y, z}, Method →
    {"DifferentialEvolution", "SearchPoints" → i}]}, {i, 10, 50, 10}]
```

$\{\{10, \{3.44951, \{x \to 2.03582, y \to 0.657435, z \to 2.12125\}\}\},$
$\{20, \{3.37194, \{x \to 2.28808, y \to 0.266014, z \to 2.04995\}\}\},$
$\{30, \{3.36168, \{x \to 2.37977, y \to 0.316228, z \to 1.94293\}\}\},$
$\{40, \{3.36337, \{x \to 2.32638, y \to 0.31005, z \to 1.99352\}\}\},$
$\{50, \{3.36168, \{x \to 2.37977, y \to 0.316225, z \to 1.94293\}\}\}\}$

The best result is obtained with 30 or 50 starting points.

■ Special Method-Specific Options

Some method-specific options are special for each method (however, random search does not have special method specific options). For more about these options, see **NMinimize** in the *Help Browser*.

> *Special options for* **DifferentialEvolution**:
>
> **"CrossProbability"** Probability that a gene is taken from the parent; default value: **0.5**
>
> **"ScalingFactor"** Scale applied to the difference vector when creating a mate; default value: **0.6**

In integer problems, a larger value of **ScalingFactor** like one may be tried in an effort to get better mobility with respect to the integer variables.

> *Special options for* **NelderMead**:
>
> **"ContractRatio"** Ratio used for contraction; default value: **0.5**
> **"ExpandRatio"** Ratio used for expansion; default value: **2.0**
> **"ReflectRatio"** Ratio used for reflection; default value: **1.0**
> **"ShrinkRatio"** Ratio used for shrinking; default value: **0.5**

> *Special options for* **SimulatedAnnealing**:
>
> **"BoltzmannExponent"** Exponent for the probability function; default value: **Automatic**
>
> **"LevelIterations"** Maximum number of iterations to stay at a given point; default value: **50**
>
> **"PerturbationScale"** Scale for the random jump; default value: **1.0**

20.3 Classical Optimization

20.3.1 No Constraints

■ Example 1: One Variable

We have already considered, in Example 2 of Section 16.1.1, p. 417, optimization of a function of one variable. Let us now examine the critical points of the following function:

```
f = 5 + 40 x^3 - 45 x^4 + 12 x^5;
```

(See also Example 1 in Section 20.2.2, p. 529.) The critical points are points where the derivative is zero. Among these points are points of minimum and maximum. The critical points are as follows:

```
c = Solve[D[f, x] == 0]
```

$\{\{x \to 0\}, \{x \to 0\}, \{x \to 1\}, \{x \to 2\}\}$

If your function is a polynomial of high order, **Solve** may not be able to solve the equation, and you may want to use **NSolve** instead (or, if the function to be optimized is transcendental, you may need **FindRoot**). In general, the critical points may also be complex. However, we pick real and distinct points:

```
crit = {c[[2]], c[[3]], c[[4]]}
```

$\{\{x \to 0\}, \{x \to 1\}, \{x \to 2\}\}$

We check the second derivative:

```
D[f, x, x] /. crit        {0, -60, 240}
```

Thus, $x = 1$ is a maximum point and $x = 2$ a minimum point. At the point $x = 0$, the third derivative is nonzero:

```
D[f, x, x, x] /. crit[[1]]        240
```

So, this point is an inflection point. The critical points and their corresponding function values are as follows:

```
points = {x, f} /. crit
```

$\{\{0, 5\}, \{1, 12\}, \{2, -11\}\}$

Lastly we plot the function and the critical points:

```
Plot[f, {x, -0.7, 2.4},
    Epilog → {AbsolutePointSize[3], RGBColor[1, 0, 0], Map[Point, points]}];
```

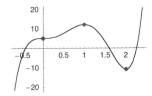

■ **Example 2: Several Variables**

Here is a function of two variables (see also Example 1 in Section 20.2.2, p. 530):

```
f = x^3 + y^3 + 2 x^2 + 4 y^2 + 6;
```

Calculate the gradient:

```
grad = {D[f, x], D[f, y]}        {4 x + 3 x², 8 y + 3 y²}
```

Then calculate the critical points:

```
c = Solve[grad == 0]
```

$\left\{\left\{x \to -\frac{4}{3}, y \to -\frac{8}{3}\right\}, \left\{x \to -\frac{4}{3}, y \to 0\right\}, \left\{x \to 0, y \to -\frac{8}{3}\right\}, \{x \to 0, y \to 0\}\right\}$

These are all real and distinct, so we accept them all:

```
crit = c;
```

We plot the function and the points:

```
ContourPlot[f, {x, -2.5, 1}, {y, -3.5, 1},
   AspectRatio → Automatic, Contours → 34, PlotPoints → 60,
   FrameTicks → {Range[-2, 1], Range[-3, 1], None, None},
   Epilog → {PointSize[0.03], Map[Point, {x, y} /. crit]}];
```

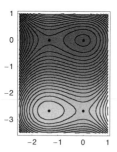

From this plot, we see that $(0, 0)$ is a minimum point, $(-\frac{4}{3}, -\frac{8}{3})$ is a maximum point, and $(-\frac{4}{3}, 0)$ and $(0, -\frac{8}{3})$ are saddle points. Values of the function at the critical points are as follows:

```
f /. crit // N        {16.6667, 7.18519, 15.4815, 6.}
```

A sufficient condition for a minimum [maximum] is that the Hessian is positive [negative] definite. Recall that a symmetric matrix is positive [negative] definite if and only if all eigenvalues are positive [negative]). Calculate the Hessian:

```
hess = {{D[f, x, x], D[f, x, y]}, {D[f, y, x], D[f, y, y]}}

{{4 + 6 x, 0}, {0, 8 + 6 y}}
```

Next we write a table that contains the eigenvalues of the Hessian at each critical point:

```
t = {x, y, f, Eigenvalues[hess]} /. N[crit];
```

```
PaddedForm[
  TableForm[t, TableSpacing → {0, 1}, TableDepth → 2, TableHeadings →
    {None, {"   x", "   y", "   f", "eigenvalues of Hessian\n"}}], {7, 5}]
```

x	y	f	eigenvalues of Hessian
-1.33333	-2.66667	16.66667	{ -4.00000, -8.00000}
-1.33333	0.00000	7.18519	{ -4.00000, 8.00000}
0.00000	-2.66667	15.48148	{ 4.00000, -8.00000}
0.00000	0.00000	6.00000	{ 4.00000, 8.00000}

From this, we can infer that the first point is a maximum, the last point is a minimum, and the rest are saddle points.

Note that we can also calculate the gradient and the Hessian with the following functions (see Section 16.1.3, p. 421):

```
gradient[f_, vars_List] := Map[D[f, #] & , vars]
hessian[f_, vars_List] := Outer[D, gradient[f, vars], vars]
```

```
gradient[f, {x, y}]        {4 x + 3 x², 8 y + 3 y²}

hessian[f, {x, y}]         {{4 + 6 x, 0}, {0, 8 + 6 y}}
```

20.3.2 Equality Constraints

Consider the following problem: Given that the volume of a cone should be v, what should the height h and radius r of the cone be if we want to minimize the surface area of the cone? The surface area A and the volume V are as follows:

A = Pi r Sqrt[h^2 + r^2] $\pi r \sqrt{h^2 + r^2}$

V = Pi h r^2 / 3 $\frac{1}{3} h \pi r^2$

We want to minimize A with respect to h and r given that $V = v$. We solve the problem with three methods:

- graphically (for a numerical value of v);
- by the method of substitution; and
- by Lagrange's method.

■ A Graphical Solution

With the graphical method, we assume that $v = 150$. First we plot contours of constant value of the surface area: we plot, on the (h, r)-surface, the contours where the surface area has the values 18, 38, 58, ..., 258:

```
p1 = ContourPlot[A, {h, 0, 10}, {r, 0, 7}, ContourShading → False,
    Contours → Range[18, 258, 20], AspectRatio → Automatic,
    FrameLabel → {"h", "r"}, RotateLabel → False];
```

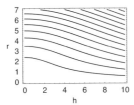

Then we plot the constraint $V - 150 = 0$ by plotting the contour of $V - 150$ where it has the value 0:

```
p2 = ContourPlot[V - 150, {h, 0, 10}, {r, 0, 7}, ContourShading → False,
    Contours → {0}, AspectRatio → Automatic, FrameLabel → {"h", "r"},
    RotateLabel → False, ContourStyle → RGBColor[1, 0, 0]];
```

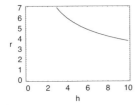

All points on this curve satisfy the constraint. Combine the plots:

```
Show[p1, p2, FrameTicks → {Range[10], Range[7], None, None},
  Epilog → {AbsolutePointSize[3], Point[{6.6, 4.7}]}];
```

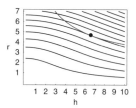

In the combined plot, we have added (by trial and error) a point that is approximately the solution of the problem. The point is about $(h, r) = (6.6, 4.7)$. Why is this point the solution? The point of solution is such that the constraint curve and one of the contours of constant value of the surface area have the same tangent. Such a contour seems to be the one that has the value 118; the smallest surface area is thus about 118.

The plot also shows how sensitive the solution is: how much the surface area increases if we go away from the optimum point but stay on the constraint curve. We see that the surface area does not increase much if h is in an interval of about $(5.5, 7.5)$ and r is adjusted accordingly such that the volume has the value 150 (r changes from about 5.1 to about 4.4).

■ The Method of Substitution

Now we solve one variable from the constraint and substitute it into the object function, thereby reducing a two-dimensional problem to a one-dimensional one. Indeed, we solve h from the constraint and substitute it into the surface area:

```
h1 = Solve[V == v, h][[1, 1]]
```
$\quad h \to \dfrac{3\,v}{\pi\,r^2}$

```
A1 = A /. h1
```
$\quad \pi\,r\,\sqrt{r^2 + \dfrac{9\,v^2}{\pi^2\,r^4}}$

We find the optimum value of r:

```
r1 = Solve[D[A1, r] == 0, r]
```

$\left\{\left\{r \to \dfrac{\left(-\frac{3}{\pi}\right)^{1/3}\,v^{1/3}}{2^{1/6}}\right\},\ \left\{r \to -\dfrac{\left(\frac{3}{\pi}\right)^{1/3}\,v^{1/3}}{2^{1/6}}\right\},\ \left\{r \to \dfrac{\left(\frac{3}{\pi}\right)^{1/3}\,v^{1/3}}{2^{1/6}}\right\},\right.$

$\left.\left\{r \to \dfrac{(-1)^{2/3}\,\left(\frac{3}{\pi}\right)^{1/3}\,v^{1/3}}{2^{1/6}}\right\},\ \left\{r \to -\dfrac{(-3)^{1/3}\,v^{1/3}}{2^{1/6}\,\pi^{1/3}}\right\},\ \left\{r \to -\dfrac{(-1)^{2/3}\,3^{1/3}\,v^{1/3}}{2^{1/6}\,\pi^{1/3}}\right\}\right\}$

Of these values, only the third is real and positive. We choose this value as the optimum r and calculate the corresponding values of h and A:

```
{ropt = r1[[3, 1]], hopt = h1 /. ropt, Aopt = A1 /. ropt /. hopt} // Simplify
```

$\left\{r \to \dfrac{\left(\frac{3}{\pi}\right)^{1/3}\,v^{1/3}}{2^{1/6}},\ h \to \left(\dfrac{6}{\pi}\right)^{1/3}\,v^{1/3},\ \dfrac{3\,3^{1/6}\,\pi^{1/3}\,v^{2/3}}{2^{1/3}}\right\}$

■ Lagrange's Method

Lastly we form Lagrange's function, in which we have the object function and the lefthand-side of the constraint $V - v = 0$ multiplied by a constant λ (Lagrange's multiplier):

$$\textbf{L = A} + \lambda \ (\textbf{V} - \textbf{v}) \qquad \pi \, r \, \sqrt{h^2 + r^2} + \left(\frac{1}{3} \, h \, \pi \, r^2 - v\right) \lambda$$

A necessary condition for the optimum solution is that partial derivatives of Lagrange's function with respect to h and r are zero and that the equality constraint is satisfied:

```
eqns = {D[L, h] == 0, D[L, r] == 0, V - v == 0}
```

$$\left\{ \frac{h \, \pi \, r}{\sqrt{h^2 + r^2}} + \frac{1}{3} \, \pi \, r^2 \, \lambda == 0, \ \frac{\pi \, r^2}{\sqrt{h^2 + r^2}} + \pi \, \sqrt{h^2 + r^2} + \frac{2}{3} \, h \, \pi \, r \, \lambda == 0, \ \frac{1}{3} \, h \, \pi \, r^2 - v == 0 \right\}$$

Solve these equations:

```
sol = Solve[eqns, {λ, h, r}] // Simplify
```

$$\left\{ \left\{ \lambda \to -\frac{2^{2/3} \, 3^{1/6} \, \pi^{1/3}}{v^{1/3}}, \ h \to \left(\frac{6}{\pi}\right)^{1/3} v^{1/3}, \ r \to \frac{\left(\frac{3}{\pi}\right)^{1/3} v^{1/3}}{2^{1/6}} \right\}, \right.$$

$$\left\{ \lambda \to \frac{2^{2/3} \, 3^{1/6} \, \pi^{1/3}}{v^{1/3}}, \ h \to \left(\frac{6}{\pi}\right)^{1/3} v^{1/3}, \ r \to -\frac{\left(\frac{3}{\pi}\right)^{1/3} v^{1/3}}{2^{1/6}} \right\},$$

$$\left\{ \lambda \to \frac{(-2)^{2/3} \, 3^{1/6} \, \pi^{1/3} \, \sqrt{-(-1)^{1/3} \, v^{2/3}}}{v^{2/3}}, \right.$$

$$\left. h \to (-1)^{2/3} \left(\frac{6}{\pi}\right)^{1/3} v^{1/3}, \ r \to -\frac{(-1)^{2/3} \, 3^{1/3} \, v^{1/3}}{2^{1/6} \, \pi^{1/3}} \right\},$$

$$\left\{ \lambda \to -\frac{2^{2/3} \, 3^{1/6} \, \pi^{1/3}}{\sqrt{-(-1)^{1/3} \, v^{2/3}}}, \ h \to (-1)^{2/3} \left(\frac{6}{\pi}\right)^{1/3} v^{1/3}, \ r \to \frac{(-1)^{2/3} \left(\frac{3}{\pi}\right)^{1/3} v^{1/3}}{2^{1/6}} \right\},$$

$$\left\{ \lambda \to -\frac{2 \, 3^{1/6}}{\sqrt{\left(-\frac{2}{\pi}\right)^{2/3} v^{2/3}}}, \ h \to -\left(-\frac{6}{\pi}\right)^{1/3} v^{1/3}, \ r \to -\frac{(-3)^{1/3} \, v^{1/3}}{2^{1/6} \, \pi^{1/3}} \right\},$$

$$\left\{ \lambda \to \frac{2 \, 3^{1/6}}{\sqrt{\left(-\frac{2}{\pi}\right)^{2/3} v^{2/3}}}, \ h \to -\left(-\frac{6}{\pi}\right)^{1/3} v^{1/3}, \ r \to \frac{\left(-\frac{3}{\pi}\right)^{1/3} v^{1/3}}{2^{1/6}} \right\} \right\}$$

Some solutions seem to be negative and some complex. To see which ones are real and positive, we calculate the numerical values:

```
{h, r} /. sol // N
```

$$\{ \{ 1.2407 \, v^{1/3}, \ 0.877308 \, v^{1/3} \}, \ \{ 1.2407 \, v^{1/3}, \ -0.877308 \, v^{1/3} \},$$
$$\{ (-0.62035 + 1.07448 \, i) \, v^{1/3}, \ (0.438654 - 0.759771 \, i) \, v^{1/3} \},$$
$$\{ (-0.62035 + 1.07448 \, i) \, v^{1/3}, \ (-0.438654 + 0.759771 \, i) \, v^{1/3} \},$$
$$\{ (-0.62035 - 1.07448 \, i) \, v^{1/3}, \ (-0.438654 - 0.759771 \, i) \, v^{1/3} \},$$
$$\{ (-0.62035 - 1.07448 \, i) \, v^{1/3}, \ (0.438654 + 0.759771 \, i) \, v^{1/3} \} \}$$

Only the first solution has real and positive values for h and r, and so we choose this as the optimal solution:

```
λhropt = sol[[1]]
```

$$\left\{ \lambda \to -\frac{2^{2/3} \, 3^{1/6} \, \pi^{1/3}}{v^{1/3}}, \ h \to \left(\frac{6}{\pi}\right)^{1/3} v^{1/3}, \ r \to \frac{\left(\frac{3}{\pi}\right)^{1/3} v^{1/3}}{2^{1/6}} \right\}$$

Another way to say this is $h_{\text{opt}} = \sqrt[3]{6\,v/\pi}$ and $r_{\text{opt}} = \sqrt[3]{3\,v/(\sqrt{2}\,\pi)}$. The smallest surface area is as follows:

```
Aopt = A /. λhropt // Simplify
```

$$\frac{3\;3^{1/6}\,\pi^{1/3}\,v^{2/3}}{2^{1/3}}$$

Another way to say this is $A_{\text{opt}} = 3\sqrt[3]{\sqrt{3}\,\pi v^2/2}$. We may plot the optimal r and h as functions of the volume v:

```
Plot[Evaluate[{r, h} /. λhropt], {v, 0, 200}, PlotRange -> All,
    Epilog -> {Text[ropt, {220, 5.2}], Text[hopt, {220, 7.3}]}];
```

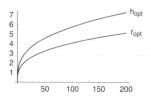

We also plot the optimal surface area as a function of the volume v:

```
Plot[Aopt, {v, 0, 200}];
```

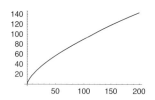

If the volume of the cone is, for example, 150, the optimal solution is as follows:

```
{A0, λhr0} = {Aopt, λhropt} /. v → 150.
```

$\{118.234,\ \{\lambda \to -0.525484,\ h \to 6.59221,\ r \to 4.66139\}\}$

This optimal cone is displayed like this:

```
<< Graphics`Shapes`
```

```
Show[Graphics3D[Cone[r, h / 2, 30] /. λhr0],
    Axes → True, ViewPoint → {1.6, -2.8, 1.0},
    Ticks → {{-4, 0, 4}, {-4, 0, 4}, {-3, 0, 3}}];
```

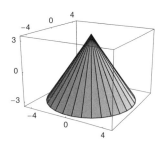

20.3.3 Equality and Inequality Constraints

■ Example 1

We minimize $(x - y)^2 + 5z$ subject to $x + y + z + 5 = 0$, $y - 3z - 1 = 0$, and $-x \leq 0$ (see also Example 2 in Section 20.2.2, p. 530). First we define the corresponding functions:

```
f = (x - y)^2 + 5 z;
g1 = x + y + z + 5;
g2 = y - 3 z - 1;
h1 = -x;
```

Then we form Lagrange's function:

```
L = f + λ1 g1 + λ2 g2 + μ1 h1
```

$$(x - y)^2 + 5 z + (5 + x + y + z) \, \lambda1 + (-1 + y - 3 z) \, \lambda2 - x \, \mu1$$

A necessary condition for the optimum is that the derivatives of L with respect to x, y, and z are 0. Another necessary condition is that the equality constraints hold. A third necessary condition is the *complementary slackness condition*: $\mu_1 h_1 = 0$, which means that a given multiplier is 0 if the corresponding inequality constraint is satisfied with strict inequality ($h_1(x) < 0$), and a multiplier can be positive only if the corresponding inequality constraint is satisfied as an equality. We form these three types of equality conditions:

```
eqns = {D[L, x] == 0, D[L, y] == 0, D[L, z] == 0, g1 == 0, g2 == 0, μ1 h1 == 0}
```

$$\{2 \, (x - y) + \lambda1 - \mu1 == 0, \ -2 \, (x - y) + \lambda1 + \lambda2 == 0,$$
$$5 + \lambda1 - 3 \lambda2 == 0, \ 5 + x + y + z == 0, \ -1 + y - 3 z == 0, \ -x \, \mu1 == 0\}$$

Further necessary conditions are that the original inequality conditions hold and that the Lagrange multipliers are nonnegative. We form these two types of inequality conditions:

```
ineqs = {h1 ≤ 0, μ1 ≥ 0}      {-x ≤ 0, μ1 ≥ 0}
```

The equality and inequality conditions together form the *Karush–Kuhn–Tucker conditions*. First we solve the equality conditions:

```
sol = Solve[eqns]
```

$$\left\{ \left\{ x \to 0, \ y \to -\frac{7}{2}, \ z \to -\frac{3}{2}, \ \lambda1 \to 4, \ \lambda2 \to 3, \ \mu1 \to 11 \right\},\right.$$
$$\left.\left\{ y \to -\frac{211}{98}, \ z \to -\frac{103}{98}, \ x \to -\frac{88}{49}, \ \lambda1 \to -\frac{5}{7}, \ \lambda2 \to \frac{10}{7}, \ \mu1 \to 0 \right\} \right\}$$

We then check which of the solutions also satisfy the inequality conditions:

```
ineqs /. sol      {{True, True}, {False, True}}
```

Only the first solution satisfies all inequality conditions. This can also be seen as follows (for **Apply**, see Section 13.3.2, p. 345):

```
ineqsCheck = Map[Apply[And, #] &, %]      {True, False}
```

```
Position[ineqsCheck, True]      {{1}}
```

The candidate for a point of minimum and the corresponding minimum value are as follows:

```
cand = Extract[sol, %]
```

$$\left\{\left\{x \to 0, \ y \to -\frac{7}{2}, \ z \to -\frac{3}{2}, \ \lambda 1 \to 4, \ \lambda 2 \to 3, \ \mu 1 \to 11\right\}\right\}$$

```
f /. %[[1]]
```

$$\frac{19}{4}$$

■ A General Program

Consider the general problem of optimizing $f(x)$ (x is a vector) with respect to the equality constraints $g(x) = 0$ ($g(x)$ is a vector) and the inequality constraints $h(x) \le 0$ ($h(x)$ is a vector). The Lagrangian is $L = f(x) + \lambda^t g(x) + \mu^t h(x)$. The Karush–Kuhn–Tucker necessary conditions for a local optimum are as follows: a first set of conditions is that $\partial L / \partial x = 0$, $g(x) = 0$, and $\mu_i h_i(x) = 0$ for all i (the last condition is the complementary slackness condition); a second set of conditions is that $h(x) \le 0$ and $\mu \ge 0$.

The following program finds candidates for an optimum point, that is, points satisfying the necessary conditions. In the program, we proceed as we did in Example 1.

```
kktOptimize[f_, g_List, h_List, x_List, ε_: 10.^-12] :=
 Module[{xλμ, L, λλ = Array[λ, Length[g]], μμ = Array[μ, Length[h]], gradL,
    eqns, sol, realsol, nrealsol, ineqs, ineqsCheck, pos, cand},
  xλμ = Join[x, λλ, μμ];
  L = f + λλ.g + μμ.h;
  gradL = Map[D[L, #] &, x];
  eqns = Thread[Join[gradL, g, μμ h] == 0];
  sol = Union[Solve[eqns, xλμ]];
  realsol = Select[sol, FreeQ[Chop[N[#], ε], Complex] &];
  If[realsol == {}, Return["No real solutions; try a larger ε"]];
  nrealsol = Map[Thread[xλμ → #] &, Chop[N[xλμ /. realsol], ε]];
  ineqs = Thread[Join[h, -μμ] ≤ ε];
  ineqsCheck = Map[Apply[And, #] &, ineqs /. nrealsol];
  pos = Position[ineqsCheck, True];
  If[pos == {}, Return["Inequalities not satisfied; try a larger ε"]];
  cand = Union[Map[Sort, Extract[realsol, pos]]];
  Transpose[{f /. cand, cand}]]
```

Here we first form the set of variables **xλμ** and the Lagrangian **L**, and then we calculate the gradient **gradL** of the Lagrangian, form the necessary equality conditions **eqns**, find the solution **sol** of the equations, select real solutions into **realsol**, and form a numerical version **nrealsol** of the set of real solutions (to be used to check the inequality conditions).

Then we form the set of necessary inequality conditions **ineqs** and check which solutions in **nrealsol** satisfy all the inequalities; the result is a list **ineqsCheck** with the ith component **True** or **False** according to whether the ith solution in **nrealsol** satisfies all of the inequalities. The solutions of the equations that give **True** for all inequalities form the set **cand** of candidates for an optimum point. Lastly we attach to each candidate the corresponding value of the object function.

Note that we have used a small number ϵ for three points. When selecting the real solutions into **realsol**, we use ϵ to ignore (with **Chop**) small imaginary parts that may be the result of numerical inaccuracy (and not "true" imaginary parts). When forming the numerical solutions **nrealsol**, we use ϵ for the same reason. When forming the inequality conditions **ineqs**, we use ϵ to make the conditions slightly less tight: we replace the original conditions $h(x) \le 0$ and $\mu \ge 0$ with $h(x) \le \epsilon$ and $\mu \ge -\epsilon$; otherwise numerical inaccuracy may cause the rejection of a feasible point when the inequalities are tested (see Example 7 below). (On the other hand, the less-tight conditions probably do not cause the acceptance of solutions that do not satisfy the original tight inequality conditions.)

The default value 10^{-12} of ϵ has shown itself to be suitable in many problems. Although in most problems ϵ could be exactly 0, in some problems this value is too small: due to numerical inaccuracies, the inequality conditions do not hold exactly. When using **kktOptimize**, you can give a new value for ϵ by writing a fifth argument for the command.

If your problem does not contain some types of constraints, replace **g** and/or **h** with an empty list (**{}**).

■ Some Notes

The result of **kktOptimize** for problems that have only exact numbers may be a very long expression (even several pages); this is because **Solve** writes explicit expressions for the solutions of third- and fourth-order polynomials. Such exact results may be useless, so it is better to solve such problems with decimal numbers (simply insert a decimal point into the problem). A decimal solution is also speedier to calculate.

In the program, we do not simplify the candidates. If appropriate, you may apply **Simplify** or **FullSimplify** to the solution (note, though, that **Simplify** and especially **FullSimplify** may take a very long time for a long expression).

The multiplier method package mentioned in the introduction to this chapter contains a notebook of test problems for nonlinear constrained optimization. **kktOptimize** solved all of the test problems (the longest computation time was about 2 seconds). Some of these test problems are solved in the examples that follow. **Minimize** (see Section 20.2.2, p. 529) also solves all of the test problems.

The strong advantage of **Minimize** is that it gives a definite result: the global minimum. **kktOptimize** only gives candidates for optimum points (some of them may be minimums, some maximums, and some saddle points). A negative point of **Minimize** is that it accepts only numerical coefficients; **kktOptimize** also solves problems with symbolic coefficients.

■ Example 2

First we solve the problem of Example 1:

> **kktOptimize[(x - y) ^2 + 5 z, {x + y + z + 5, y - 3 z - 1}, {-x}, {x, y, z}]**

$$\left\{\left\{\frac{19}{4}, \left\{x \to 0, y \to -\frac{7}{2}, z \to -\frac{3}{2}, \lambda[1] \to 4, \lambda[2] \to 3, \mu[1] \to 11\right\}\right\}\right\}$$

Minimize gives the same result:

> **Minimize[{ (x - y) ^2 + 5 z, x + y + z == -5, y - 3 z == 1, x ≥ 0}, {x, y, z}]**

$$\left\{\frac{19}{4}, \left\{x \to 0, y \to -\frac{7}{2}, z \to -\frac{3}{2}\right\}\right\}$$

■ Example 3

The following problem has only inequality constraints:

```
f = (x - 2)^2 + (y - 3)^2;
h1 = x + y - 4;
h2 = x - y - 2;

kktOptimize[f, {}, {h1, h2}, {x, y}]
```

$$\left\{\left\{\frac{1}{2}, \left\{x \to \frac{3}{2}, y \to \frac{5}{2}, \mu[1] \to 1, \mu[2] \to 0\right\}\right\}\right\}$$

A figure confirms that $(\frac{3}{2}, \frac{5}{2})$ really is the solution of the constrained problem:

```
<< Graphics`InequalityGraphics`

p1 = InequalityPlot[{h1 ≤ 0, h2 ≤ 0}, {x, 0, 4.1}, {y, 0, 4.1},
    Fills → GrayLevel[0.9], DisplayFunction → Identity];

p2 = ContourPlot[f, {x, 0, 4.1},
    {y, 0, 4.1}, Contours → 20, ContourShading → False,
    ContourStyle → {AbsoluteDashing[{2}]}, DisplayFunction → Identity];

Show[p1, p2, PlotRange → All,
    Epilog → {AbsolutePointSize[3], Point[{3 / 2, 5 / 2}]},
    DisplayFunction → $DisplayFunction];
```

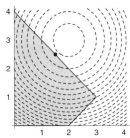

■ Example 4

Next we solve the problem of Section 20.3.2, p. 539, which has an equality constraint:

```
kktOptimize[Pi r Sqrt[h^2 + r^2], {Pi h r^2 / 3 - v}, {}, {h, r}] // Simplify
```

$$\left\{\left\{-\frac{3 \cdot 3^{1/6} \pi^{1/3} v^{2/3}}{2^{1/3}}, \left\{h \to \left(\frac{6}{\pi}\right)^{1/3} v^{1/3}, r \to -\frac{\left(\frac{3}{\pi}\right)^{1/3} v^{1/3}}{2^{1/6}}, \lambda[1] \to \frac{2^{2/3} \cdot 3^{1/6} \pi^{1/3}}{v^{1/3}}\right\}\right\},\right.$$

$$\left.\left\{\frac{3 \cdot 3^{1/6} \pi^{1/3} v^{2/3}}{2^{1/3}}, \left\{h \to \left(\frac{6}{\pi}\right)^{1/3} v^{1/3}, r \to \frac{\left(\frac{3}{\pi}\right)^{1/3} v^{1/3}}{2^{1/6}}, \lambda[1] \to -\frac{2^{2/3} \cdot 3^{1/6} \pi^{1/3}}{v^{1/3}}\right\}\right\}\right\}$$

We obtained two candidates, but only the second one has a positive value for r, and so we choose this candidate as the solution.

■ Example 5

We previously looked at the following unconstrained problem in Sections 20.1.2, p. 518; 20.2.2, p. 530; and 20.3.1, p. 537:

```
kktOptimize[x^3 + y^3 + 2 x^2 + 4 y^2 + 6, {}, {}, {x, y}]
```

$$\left\{ \left\{ \frac{50}{3}, \left\{ x \to -\frac{4}{3}, y \to -\frac{8}{3} \right\} \right\}, \left\{ \frac{194}{27}, \left\{ x \to -\frac{4}{3}, y \to 0 \right\} \right\}, \right.$$
$$\left. \left\{ \frac{418}{27}, \left\{ x \to 0, y \to -\frac{8}{3} \right\} \right\}, \{6, \{x \to 0, y \to 0\}\} \right\}$$

The program gives us four candidates (these can be investigated as in Section 20.3.1; the candidates contain local minimum and maximum points and saddle points).

■ Example 6

The following two examples were first encountered in Section 20.2.2, p. 530:

```
kktOptimize[x^2 / 2 + (y^2 + z^2 + v^2) / 6, {},
   {2 - x, 5 - x - y, 2 - x - z, 1 - x - v, -v}, {x, y, z, v}] // Timing
```

$$\left\{ 0.12 \text{ Second}, \left\{ \left\{ \frac{7}{2}, \{v \to 0, x \to 2, y \to 3, \right. \right. \right.$$
$$\left. \left. \left. z \to 0, \mu[1] \to 1, \mu[2] \to 1, \mu[3] \to 0, \mu[4] \to 0, \mu[5] \to 0\} \right\} \right\} \right\}$$

```
kktOptimize[1 / (5 x y z) + 4 / x + 3 / z, {},
   {2 x z + x y - 10, -x, -y, -z}, {x, y, z}] // FullSimplify
```

$$\left\{ \left\{ \frac{4}{5} \sqrt{\frac{1}{5} (76 + \sqrt{151})}, \left\{ x \to \text{Root}[2000 - 608 \#1^2 + 45 \#1^4 \&, 3], y \to \frac{1}{\sqrt{10}}, \right. \right. \right.$$
$$\left. \left. z \to \frac{\sqrt{\frac{151}{10}}}{2}, \mu[1] \to \frac{\sqrt{\frac{2}{5}} (151 + 76 \sqrt{151})}{3775}, \mu[2] \to 0, \mu[3] \to 0, \mu[4] \to 0 \right\} \right\} \right\}$$

■ Example 7

In the following problem, the value 10^{-15} for ϵ is too small:

```
f = 100 (x^2 - y) ^2 + (x - 1) ^2;
```

```
kktOptimize[f, {}, {3 / 2 - y}, {x, y}, 10^-15]
```

```
Inequalities not satisfied; try a larger ∈
```

Due to numerical inaccuracies, the candidates do not satisfy the inequality conditions. If we use a larger ϵ, we get a candidate:

```
kktOptimize[f, {}, {3 / 2 - y}, {x, y}, 10^-14] // N // Chop
```

```
{{226.003, {x → -0.00334451, y → 1.5, μ[1.] → 299.998}}}
```

Giving still a larger ϵ, we get several candidates:

```
kktOptimize[f, {}, {3 / 2 - y}, {x, y}, 10^-12] // N // Chop
```

```
{{0.0504262, {x → 1.22437, y → 1.5, μ[1.] → 0.183254}},
 {226.003, {x → -0.00334451, y → 1.5, μ[1.] → 299.998}},
 {4.94123, {x → -1.22103, y → 1.5, μ[1.] → 1.81898}}}
```

By the way, **Minimize** gives the global minimum:

```
Minimize[{f, 3 / 2 - y ≤ 0}, {x, y}] // N
```

```
{0.0504262, {x → 1.22437, y → 1.5}}
```

20.3.4 Calculus of Variations

In a package we have commands that relate to the calculus of variations. We present here only one command from this package: **EulerEquations**. This command forms Euler's equation (or equations) for a problem in which we want to find the extremum for the integral $\int_a^b f(x, u(x), u'(x))\, dx$. The command also accepts several independent variables x, y, ... and several dependent variables $u(x, y, ...), v(x, y, ...)$. Here are some examples of the command:

In the `Calculus`VariationalMethods`` *package:*

```
EulerEquations[f, u[x], x]
EulerEquations[f, u[x,y], {x,y}]
EulerEquations[f, {u[x,y],v[x,y]}, {x,y}]
```

Let us find the curve $y(x)$ of minimum length between $(0, 0)$ and (a, b) (the solution is, of course, the straight line that connects the points). The length of the curve is $\int_0^a \sqrt{1 + y'(x)^2}\, dx$. Euler's equation is as follows:

```
<< Calculus`VariationalMethods`

eqn = EulerEquations[Sqrt[1 + y'[x]^2], y[x], x]
```

$$-\frac{y''[x]}{(1 + y'[x]^2)^{3/2}} == 0$$

The solution with the given boundary conditions is as follows:

```
DSolve[{eqn, y[0] == 0, y[a] == b}, y[x], x]
```

$$\left\{\left\{y[x] \rightarrow \frac{b\,x}{a}\right\}\right\}$$

The package also has the commands **VariationalD**, **FirstIntegrals**, **Variational Bound**, **NVariationalBound**.

21

Interpolation

Introduction

> *Life is the art of drawing sufficient conclusions from insufficient premises.* — *Samuel Butler*

With interpolation, we represent a set of points with a curve that passes exactly through all of the points. Interpolation can also be applied to functions by first sampling the function at some points. In this way, we can obtain for the data or function a suitable representation that may be sufficiently precise for practical purposes. However, if the observations contain errors (as they often do), then you may not want a function that passes exactly through the points but rather a simple function that represents the overall behavior of the observations. Approximation is then the right technique to use (see Section 22.1).

■ Interpolation of Data

With *Mathematica*, we can do three kinds of interpolation of data. First, the usual *interpolating polynomial*, which is calculated with **InterpolatingPolynomial**, gives, for $n + 1$ data points, the unique polynomial of at most degree n that passes exactly through all of the points.

The object produced by **ListInterpolation** and **Interpolation** is called an *interpolating function*. It is a set of piecewise-calculated interpolating polynomials between successive points (the result is a continuous curve, having, however, a discontinuous derivative).

We can choose the degree of the piecewise polynomials. The commands also work for multidimensional data, producing, for example, an interpolating surface.

SplineFit calculates various *splines*, such as cubic splines that pass through all points and have a continuous first and second derivative.

■ Interpolation of Functions

For the interpolation of functions we have two commands. **RationalInterpolation** calculates, for a given function, a rational interpolating function (i.e., a quotient of two polynomials). We can give the interpolation points ourselves or let the command choose them; in the latter case we obtain, in fact, a rational Chebyshev approximation (see Section 22.2.1, p. 586).

FunctionInterpolation forms, for mathematical expressions, interpolating functions that consist of a set of piecewise-calculated interpolating polynomials. For a complex expression, such an interpolating function may be a useful and efficient representation. Indeed, if an approximation for a function is required, then piecewise interpolation by **FunctionInterpolation** is a strong candidate, despite the fact that this command falls into the category of interpolation.

21.1 Usual Interpolation

21.1.1 Interpolating Polynomial

> **InterpolatingPolynomial[data, x]**   The interpolating polynomial through the points in **data**; the result is a function of **x**
>
> The data points can be given in either of the following forms:
> {f_1, f_2, ... }   Interpolate through the points {1, f_1}, {2, f_2}, ...
> {{x_1, f_1}, {x_2, f_2}, ... }   Interpolate through the points {x_1, f_1}, {x_2, f_2}, ...

The command gives the interpolating polynomial in the Newton form, using divided differences.

■ Simple Examples

We calculate the line that goes through the points (1, *f*) and (2, *g*):

```
InterpolatingPolynomial[{f, g}, x]
```

$$f + (-f + g)\,(-1 + x)$$

The next polynomial goes through the points (*a*, *f*) and (*b*, *g*):

```
InterpolatingPolynomial[{{a, f}, {b, g}}, x]
```

$$f + \frac{(-f + g)\,(-a + x)}{-a + b}$$

Here is the quadratic polynomial that goes through the points (*a*, *f*), (*b*, *g*), and (*c*, *h*):

```
InterpolatingPolynomial[{{a, f}, {b, g}, {c, h}}, x]
```

$$f + (-a + x)\left(\frac{-f + g}{-a + b} + \frac{\left(-\frac{-f+g}{-a+b} + \frac{-g+h}{-b+c}\right)\,(-b + x)}{-a + c}\right)$$

We check that this really goes through the given points:

```
% /. x → {a, b, c} // Simplify
```

```
{f, g, h}
```

■ A Numerical Example

Next we consider numerical data:

```
data = Transpose[{Range[14], {1, 2, 0, 2, 2, 2, 0, 0, 2, 3, 5, 4, 3, 1} // N}]
```

```
{{1, 1.}, {2, 2.}, {3, 0.}, {4, 2.}, {5, 2.}, {6, 2.}, {7, 0.},
 {8, 0.}, {9, 2.}, {10, 3.}, {11, 5.}, {12, 4.}, {13, 3.}, {14, 1.}}
```

```
p = ListPlot[data, PlotStyle → AbsolutePointSize[2]];
```

The interpolating polynomial is as follows:

```
int = InterpolatingPolynomial[data, x]
```

$$1. + (1. + (-1.5 + (1.16667 +$$
$$(-0.541667 + (0.175 + (-0.0458333 + (0.0109127 + (-0.00240575 +$$
$$(0.000468474 + (-0.0000766093 + (9.99579 \times 10^{-6} +$$
$$(-9.22753 \times 10^{-7} + 2.50521 \times 10^{-8} (-13 +$$
$$x)) (-12 + x)) (-11 + x)) (-10 + x))$$
$$(-9 + x)) (-8 + x)) (-7 + x)) (-6 + x))$$
$$(-5 + x)) (-4 + x)) (-3 + x)) (-2 + x)) (-1 + x)$$

Note that the interpolating polynomial should not be simplified or expanded, because the unsimplified, nested form is the best for numerical computations. However, to show that the result in our example of 14 points is a polynomial of degree 13, we expand it:

```
int // Expand
```

$$-1674. + 4948.28 \, x - 6052.37 \, x^2 + 4139.98 \, x^3 - 1790.19 \, x^4 +$$
$$521.379 \, x^5 - 105.873 \, x^6 + 15.2428 \, x^7 - 1.56035 \, x^8 + 0.112324 \, x^9 -$$
$$0.00552547 \, x^{10} + 0.00017544 \, x^{11} - 3.20249 \times 10^{-6} \, x^{12} + 2.50521 \times 10^{-8} \, x^{13}$$

Here are some values of the polynomial:

```
int /. x → {1, 1.5, 2, 2.5, 3}
```

```
{1., 17.9203, 2., -2.27272, 0.}
```

We see that at the points 1, 2, 3, … the polynomial really has the values 1, 2, 0, …, but, between these points, the value may be far from the neighboring values. We plot the polynomial and also show the points:

```
Plot[int, {x, 0.97, 14.03},
  Epilog → {AbsolutePointSize[2], Map[Point, data]}];
```

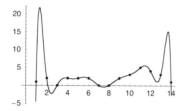

As can be seen, the polynomial goes through all of the points and is quite a good representation of the data in an interval of, for example, (3, 12). Outside of this interval, that is,
near the endpoints, the polynomial behaves badly. Indeed, high-order interpolating
polynomials should be used with caution. It may be better to use piecewise-interpolating
polynomials (see Section 21.2), splines (see Section 21.3), or least-squares fits (see Section
22.1).

■ Using Values of Derivatives

`InterpolatingPolynomial` also accepts values of derivatives. For example, the data
could be in the following forms:

$$\{\{x_1, \{f_1, df_1\}\}, \{x_2, \{f_2, df_2\}\}, \dots \}$$
$$\{\{x_1, \{f_1, df_1, ddf_1\}\}, \{x_2, \{f_2, df_2, ddf_2\}\}, \dots \}$$

Here df_1 is the value of the derivative at point x_1, and ddf_1 is the value of the second
derivative. Higher-order derivatives can also be given.

As an example, we calculate an interpolating polynomial for given values of the function and its first derivative at two points:

```
int = InterpolatingPolynomial[{{a, {f, df}}, {b, {g, dg}}}, x]
```

$$f + (-a + x)\left(df + (-a + x)\left(\frac{-df + \frac{-f+g}{-a+b}}{-a+b} + \frac{\left(\frac{dg - \frac{-f+g}{-a+b}}{-a+b} - \frac{-df + \frac{-f+g}{-a+b}}{-a+b}\right)(-b+x)}{-a+b}\right)\right)$$

The result is now a third-degree polynomial.

As another example, we try to improve the interpolating polynomial of the numerical
example by defining that the derivative of the polynomial at points 1, 2, 13, and 14 is 2, –3,
–1, and –3, respectively (the values were found by trial and error):

```
data2 = {{1, {1, 2}}, {2, {2, -3}}, {3, 0}, {4, 2}, {5, 2}, {6, 2}, {7, 0},
    {8, 0}, {9, 2}, {10, 3}, {11, 5}, {12, 4}, {13, {3, -1}}, {14, {1, -3}}};
```

The result is now much better:

```
int2 = InterpolatingPolynomial[data2, x];
```

```
Plot[int2, {x, 0.97, 14.03},
  Epilog → {AbsolutePointSize[2], Map[Point, data]}];
```

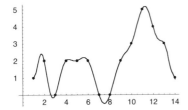

21.1.2 Own Programs

■ Lagrange Form

```
lagrangeInterpolation[xx_List, yy_List, x_] :=
  Sum[yy[[i]] Apply[Times, (x - Drop[xx, {i}]) / (xx[[i]] - Drop[xx, {i}])],
    {i, Length[xx]}]
```

Define some x values and their corresponding y values:

```
xx = {a, b, c}; yy = {f, g, h};
```

The interpolation polynomial in the Lagrange form is as follows:

```
lagrangeInterpolation[xx, yy, x]
```

$$\frac{h\,(-a+x)\,(-b+x)}{(-a+c)\,(-b+c)} + \frac{g\,(-a+x)\,(-c+x)}{(-a+b)\,(b-c)} + \frac{f\,(-b+x)\,(-c+x)}{(a-b)\,(a-c)}$$

To understand the program, note first that `Drop[xx, {i}]` deletes the ith element of `xx`. Note then that calculations with lists are done automatically element by element:

```
x - Drop[xx, {3}]     {-a + x, -b + x}
```

```
xx[[3]] - Drop[xx, {3}]     {-a + c, -b + c}
```

Here is the quotient of these terms:

```
(x - Drop[xx, {3}]) / (xx[[3]] - Drop[xx, {3}])
```

$$\left\{ \frac{-a+x}{-a+c}, \frac{-b+x}{-b+c} \right\}$$

Note again that the division of the two lists was done automatically element by element. Then we multiply the elements of the last list:

```
Apply[Times, (x - Drop[xx, {3}]) / (xx[[3]] - Drop[xx, {3}])]
```

$$\frac{(-a+x)\,(-b+x)}{(-a+c)\,(-b+c)}$$

If we multiply this term by `yy[[3]]`, we get one term of the Lagrange interpolating polynomial. Summing all such terms gives the whole polynomial.

■ **Newton Form**

```
div[z_List] :=
 div[z] = (div[Drop[z, 1]] - div[Drop[z, -1]]) / (Last[z] - First[z])

newtonInterpolation[xx_List, yy_List, x_] := With[{n = Length[xx]},
  Do[div[{xx[[i]]}] = yy[[i]], {i, n}];
  Sum[div[Take[xx, i]] Product[x - xx[[j]], {j, i - 1}], {i, n}]]
```

Here is an example (we use the same data as above):

newtonInterpolation[xx, yy, x]

$$f + \frac{(-f + g)(-a + x)}{-a + b} + \frac{(-\frac{-f+g}{-a+b} + \frac{-g+h}{-b+c})(-a + x)(-b + x)}{-a + c}$$

This is of the form $\mathrm{div}(a) + \mathrm{div}(a, b)(x - a) + \mathrm{div}(a, b, c)(x - a)(x - b)$, where div denotes divided differences (note that **InterpolatingPolynomial** writes the polynomial in a nested form, see Section 21.1.1, p. 550).

The function **div** calculates the divided differences. We demonstrate how **div** works. First note that we have used dynamic programming: **div[z_List] := div[z] = …** to speed up the computations (see Section 14.1.3, p. 356). The starting values for **div** are calculated as follows:

Do[div[{xx[[i]]}] = yy[[i]], {i, 3}]

Now **div** is defined for **{a}**, **{b}**, and **{c}**:

{div[{a}], div[{b}], div[{c}]}   {f, g, h}

Then we can calculate the first- and second-order divided differences:

{div[{a, b}], div[{b, c}]} $\left\{ \frac{-f + g}{-a + b}, \frac{-g + h}{-b + c} \right\}$

div[{a, b, c}] $\dfrac{-\frac{-f+g}{-a+b} + \frac{-g+h}{-b+c}}{-a + c}$

We can show the divided differences in the form of a table:

```
TableForm[
 Transpose[{{a, "", b, "", c}, {div[{a}], "", div[{b}], "", div[{c}]}, {"",
   div[{a, b}], "", div[{b, c}], ""}, {"", "", div[{a, b, c}], "", ""}}]]
```

a	f		
		$\frac{-f+g}{-a+b}$	
b	g		$\frac{-\frac{-f+g}{-a+b} + \frac{-g+h}{-b+c}}{-a+c}$
		$\frac{-g+h}{-b+c}$	
c	h		

21.2 Piecewise Interpolation

21.2.1 Two-Dimensional Data

As shown in the numerical example of Section 21.1.1, p. 551, if we have many points and thus an interpolating polynomial of a high order, the result may be bad, that is, the polynomial behaves badly, particularly near the endpoints. Often it is better to proceed piecewise: calculate low-order polynomials between successive points. This can be done with `ListInterpolation` or `Interpolation`.

These two commands differ in the way we specify the data. With 2D data, we have values f_i at some points x_i. With `ListInterpolation`, we specify separately the f_i values and the corresponding x_i values, whereas with `Interpolation` we form a list that consists of pairs $\{x_i, f_i\}$. Here is a summary of how to calculate an interpolating function that consists of polynomials between successive points (the default is that the polynomials are of third degree).

`ListInterpolation[{f₁ , …, fₘ }]` x coordinates are assumed to be $\{1, …, m\}$
`ListInterpolation[{f₁ , …, fₘ }, {{xₘᵢₙ , xₘₐₓ }}]` x coordinates are assumed to be
  evenly spaced between $[x_{\min}, x_{\max}]$
`ListInterpolation[{f₁ , …, fₘ }, {{x₁ , …, xₘ }}]` x coordinates are $\{x_1, …, x_m\}$

`Interpolation[{{x₁ , f₁ }, …, {xₘ , fₘ }}]`

The values `fᵢ` can be real or complex (or even symbolic), whereas the values of `xᵢ` must be real (and numeric).

■ Example 1: Piecewise-Cubic Interpolation

As an example, we consider the same data we used in Section 21.1.1, p. 551:

    data = {1, 2, 0, 2, 2, 2, 0, 0, 2, 3, 5, 4, 3, 1};

We calculate a piecewise-cubic interpolating function, assuming that the x coordinates are 1, …, 14:

    int = ListInterpolation[data]
    InterpolatingFunction[{{1, 14}}, <>]

The result of `ListInterpolation` (and `Interpolation`) is an object called `Interpo-latingFunction`; it contains all of the information needed to handle the piecewise polynomial. Only the interval of definition of the piecewise polynomial is shown; all other information is hidden inside the marks `<>`.

We can calculate with the piecewise-interpolating function in all possible ways. Just give a name to the function, like `int` above, and remember that its value, for example, at **a** is `int[a]` (similarly as the value of `Sin` at **a** is `Sin[a]`).

> *A typical use of an interpolating function:*
>
> `int = ListInterpolation[data]`   Calculate an interpolating function
> `int[a]`   Calculate the value of the interpolating function at `a`

We calculate the value of the function at some points:

`{int[1], int[1.5], int[2]}`

`{1, 2.3125, 2}`

The function goes exactly through all data points and interpolates between them. If we ask the value at a point outside of the interval of definition, extrapolation is used, and we get a warning:

`int[0.5]`

```
InterpolatingFunction::dmval :
  Input value {0.5} lies outside the range of data in the
    interpolating function. Extrapolation will be used. More...
-2.8125
```

The interpolating function is displayed as follows:

`xf = Transpose[{Range[14], data}];`

`Plot[int[x], {x, 1, 14}, Epilog → {AbsolutePointSize[2], Map[Point, xf]}];`

A cubic polynomial is calculated for each pair of points by using these points and their nearest neighbors. The resulting curve is smooth enough to make it an effective and useful way of summarizing and using large data sets.

Here is the derivative function:

`Show[Table[Plot[int'[x], {x, i, i + 1}, DisplayFunction → Identity],`
`    {i, 1, 13}], DisplayFunction → $DisplayFunction];`

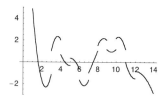

We see that the derivative is discontinuous at 3, 4, 5, …, 12. Next we integrate the function (integration of interpolating functions is considered in Section 17.2.6, p. 453):

`Integrate[int[x], {x, 1, 14}] // N`   `26.75`

Above we assumed that the x coordinates are 1, ..., 14. If the x coordinates are, for example, 0, 1, ..., 13, then we have to write one of the following three commands:

```
int = ListInterpolation[data, {{0, 13}}]
```

```
int = ListInterpolation[data, {Range[0, 13]}]
```

```
int = Interpolation[Transpose[{Range[0, 13], data}]]
```

If we solve a differential equation numerically with **NDSolve**, the result is simply an **InterpolatingFunction** object (see Section 23.3.1, p. 611),

■ Options

> *Options for* **ListInterpolation** *and* **Interpolation**:
>
> **InterpolationOrder** Degree of the piecewise-interpolating polynomials; examples of values: **3, 1**
>
> **PeriodicInterpolation** Whether a periodic interpolating function is formed; possible values: **False, True**

The default value of **InterpolationOrder** is 3, so that if you intend to use third-order polynomials, the option need not be written. Give the option the value 1 if you want a piecewise-linear interpolation (see the next example). If we give **PeriodicInterpolation** the value **True**, then the interpolating function is considered as a periodic function, with one period being the same as the range of the data (see Example 3).

■ Example 2: Piecewise-Linear Interpolation

We calculate the piecewise-linear interpolating function for the data given in Example 1:

```
int = ListInterpolation[data, InterpolationOrder → 1]
```

InterpolatingFunction[{{1, 14}}, <>]

```
Plot[int[x], {x, 1, 14}, Epilog → {AbsolutePointSize[2], Map[Point, xf]}];
```

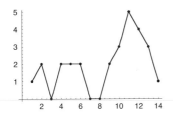

■ Example 3: Periodic Interpolation

Now we ask for a periodic interpolating function:

```
int = ListInterpolation[data, PeriodicInterpolation → True]
```

InterpolatingFunction[{{..., 1, 14, ...}}, <>]

```
Plot[int[x], {x, 1, 56}, AspectRatio → 0.2];
```

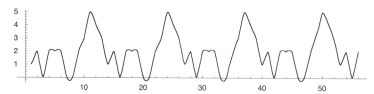

Note that, for a periodic interpolating function, the data at the endpoints of the fundamental period must match: if the first and last data points are (x_1, f_1) and (x_n, f_n), we must have $f_1 = f_n$.

■ Using Values of Derivatives

We can also input the values of first and higher derivatives in a way that is similar to what we did for **InterpolatingPolynomial**. For example, if we input the values of the first derivative, the commands are in the following forms:

```
ListInterpolation[{{f₁ , df₁ }, …, {fₘ , dfₘ }}]
Interpolation[{{x₁ , {f₁ , df₁ }}, …, {xₘ , {fₘ , dfₘ }}}]
```

Here df_i is the value of the first derivative at x_i. We try to improve the interpolating function of **data** by specifying that the derivative of the function at 1, 5, 6, and 7 is 1.3, 0, 0, and 0, respectively:

```
data2 = {{1, 1.3}, 2, 0, 2, {2, 0}, {2, 0}, {0, 0}, 0, 2, 3, 5, 4, 3, 1};

int2 = ListInterpolation[data2];

Plot[int2[x], {x, 1, 14},
  Epilog → {AbsolutePointSize[2], Map[Point, xf]}];
```

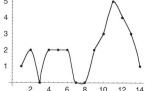

21.2.2 Higher-Dimensional Data

■ Regular 3D Data

ListInterpolation and **Interpolation** can also be used for higher-dimensional data to calculate piecewise-interpolating surfaces. The choice of command depends on the form of the data.

For example, with 3D data, we have values f_i at some points (x_i, y_i). With **ListInterpolation**, we define separately the f_i values in a matrix form and the values of x_i and y_i in one of several easy ways, while **Interpolation** requires the points in the form $\{x_i, y_i, f_i\}$. Here is a summary of the ways of calculating a piecewise-interpolating surface for 3D data (the summary extends readily to higher-dimensional data):

ListInterpolation[data] x and y coordinates are $\{1, ..., m\}$ and $\{1, ..., n\}$

ListInterpolation[data, {{x$_{min}$, x$_{max}$ }, {y$_{min}$, y$_{max}$ }}] x and y coordinates are
   evenly spaced between $[x_{min}, x_{max}]$ and $[y_{min}, y_{max}]$

ListInterpolation[data, {{x$_1$, ..., x$_m$ }, {y$_1$, ..., y$_n$ }}] x and y coordinates are
   $\{x_1, ..., x_m\}$ and $\{y_1, ..., y_n\}$

data is of the matrix form:
   {{f$_{11}$, ..., f$_{1n}$ }, ..., {f$_{m1}$, ..., f$_{mn}$ }}   (each row corresponds to a fixed value of x)

Interpolation[data]

data is of the list form:
   {{x$_1$, y$_1$, f$_1$ }, ..., {x$_k$, y$_k$, f$_k$ }}

With both commands, the x and y coordinates must eventually form a *regular rectangular grid* on the (x, y) plane, as is seen in the following figure:

```
xy = Outer[List, {0, 1, 4, 6, 7}, {0, 2, 5, 6}]
```

```
{{{0, 0}, {0, 2}, {0, 5}, {0, 6}},
 {{1, 0}, {1, 2}, {1, 5}, {1, 6}}, {{4, 0}, {4, 2}, {4, 5}, {4, 6}},
 {{6, 0}, {6, 2}, {6, 5}, {6, 6}}, {{7, 0}, {7, 2}, {7, 5}, {7, 6}}}
```

```
ListPlot[Flatten[xy, 1], PlotStyle → AbsolutePointSize[3],
   Axes → None, AspectRatio → Automatic, PlotRange → All,
   Epilog → {Map[Line, xy], Map[Line, Transpose[xy]]}];
```

Note that, unlike **ListInterpolation**, **Interpolation** does not accept the points in a *matrix* form: the points must be in a *flattened* list form in which the rows are not separated by curly braces; if your data is in a matrix form, use **Flatten[data, 1]** to remove the curly braces of the rows.

The options **InterpolationOrder** and **PeriodicInterpolation** can be used as they are for 2D data (see Section 21.2.1, p. 557). The default order is 3. The order can also be set separately for each independent variable (e.g., **InterpolationOrder → {2, 1}**). If there are not enough data for a requested order, the order is lowered automatically (with a warning). The periodicity can also be defined separately for each independent variable (e.g., **PeriodicInterpolation → {True, False}**).

■ An Example

Consider the following data:

```
data = {{5, 6, 5, 7}, {4, 6, 6, 4}, {6, 4, 6, 3}, {2, 3, 3, 5}};
```

Here each row of f_i values corresponds to a fixed value of x. Before calculating an interpolating surface for these data, we plot the data (see Section 10.1.1, p. 267). Data with only the f_i values can be plotted with **ListPlot3D**, **ListContourPlot**, or **ListDensityPlot** (data with x, y, and f coordinates can be plotted with **ListSurfacePlot3D** from the **Graphics`Graphics3D`** package). If x and y are both in the interval $(0, 3)$, we can plot the surface as follows:

```
ListPlot3D[Transpose[data],
    MeshRange → {{0, 3}, {0, 3}}, AxesLabel → {"x", "y", None}];
```

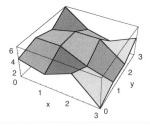

Note that we have to transpose the data, because the plotting commands interpret the data in such a way that each row corresponds to a fixed value of y (whereas, with **ListInter-polate**, each row corresponds to a fixed value of x). Then we calculate a piecewise third-order interpolating surface and plot it:

```
int = ListInterpolation[data, {{0, 3}, {0, 3}}]
```

```
InterpolatingFunction[{{0, 3}, {0, 3}}, <>]
```

```
Plot3D[int[x, y], {x, 0, 3}, {y, 0, 3}, AxesLabel → {"x", "y", ""}];
```

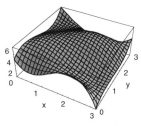

The surface goes exactly through all of the given points and interpolates between them. We calculate a value of the surface and integrate it in a region:

```
int[1.7, 2.1]      6.24158
```

```
Integrate[int[x, y], {x, 1, 2}, {y, 0, 1}] // N      4.76563
```

■ Giving Derivatives

```
ListInterpolation[ …, {f_ij , {dfx_ij , dfy_ij }}, … ]
Interpolation[ …, {x_i , y_i , {f_i , {dfx_i , dfy_i }}}, … ]
```

Derivatives are specified in **ListInterpolation** by replacing f_{ij} with a list {f_{ij}, {dfx_{ij}, dfy_{ij}}} in which dfx_{ij} and dfy_{ij} are the derivatives with respect to x and y, respectively (if a derivative is lacking at a point, then give the value **Automatic** for the derivative).

■ Irregular 3D Data

We noted that, for **ListInterpolation** and **Interpolation**, the 3D points must form a regular rectangular grid on the (x, y) plane (however, neither the x points nor the y points need be evenly spaced). If the points are spaced irregularly, there is no built-in command to calculate a piecewise-interpolating surface as a mathematical function. In the **Discrete\`Math\`ComputationalGeometry\`** package, however, there is the command **Triangular\`SurfacePlot** to *show* such a surface (see Section 10.1.2, p. 272).

The **ExtendGraphics** collection of packages by Wickham-Jones (1994) (see Section 8.2.2, p. 225) contains advanced techniques for interpolating surfaces through irregular 3D points. With the packages, we can both show such surfaces and define them as mathematical functions.

21.3 Splines

21.3.1 Cubic Splines

■ Introduction to Splines

We observed, in Section 21.1.1, p. 551, that if one polynomial is required to pass through many points, the resulting polynomial may fluctuate in an undesirable manner. Piecewise interpolation is often better; such interpolating functions were considered in Section 21.2.1, p. 555. The resulting function does not have unnecessary fluctuations, but its derivative is not continuous, and so the function lacks this smoothness condition. *Splines* are piecewise-interpolating functions that are smooth.

In the **NumericalMath\`SplineFit\`** *package:*

SplineFit[data, type]   A spline of type **type** through the points in **data**

The points are given in the form **{{x₁ , f₁ }, {x₂ , f₂ }, … }**, and the type of spline can be **Cubic**, **Bezier**, or **CompositeBezier**. A *cubic spline* is made up of a set of cubic polynomials in such a way that the resulting function passes through each point, the first and second derivatives of the function are continuous, and the second derivative is zero at the endpoints. *Bezier splines* interpolate only the endpoints; other points "control" the curve. A *composite Bezier* spline interpolates the first, third, fifth, … points, while the other points control the curve.

■ Example 1

Consider again the numeric data that we have used earlier in this chapter:

```
data = Transpose[{Range[14], {1, 2, 0, 2, 2, 2, 0, 0, 2, 3, 5, 4, 3, 1}}]
```

```
{{1, 1}, {2, 2}, {3, 0}, {4, 2}, {5, 2}, {6, 2}, {7, 0},
 {8, 0}, {9, 2}, {10, 3}, {11, 5}, {12, 4}, {13, 3}, {14, 1}}
```

Calculate the cubic spline:

```
<< NumericalMath`SplineFit`
```

```
cub = SplineFit[data, Cubic]
```

SplineFunction[Cubic, {0., 13.}, <>]

The result is a **SplineFunction** object; it contains all of the information about the spline. Only the interval in which the spline is defined is shown. You may wonder about the interval {0., 13.}, because our points were in the interval (1, 14). Generally, if you have n points, the interval of the spline is {0., (n – 1).}. Thus the interval shown emerges simply by giving each observation an ordinal number, starting with 0. This may seem odd, but there are reasons for it, which will be seen when we consider multiple-valued splines in Example 2.

If we want to calculate the value of the spline at a particular point, we have to reparameterize the point so that it complies with the numbering system used by **SplineFit**. For example, the point 3.5 is halfway between the third and fourth points, and so the appropriate argument is 2.5:

```
cub[2.5]     {3.5, 0.797698}
```

The result shows, besides the value of the spline (0.797698), also the coordinate of the point in the normal x axis (3.5). So, **cub** is a parametric function. Accordingly, the spline can be plotted with **ParametricPlot** (the option **Compiled → False** is needed for splines):

```
ParametricPlot[cub[t], {t, 0, 13}, Compiled → False,
   Epilog → {AbsolutePointSize[2], Map[Point, data]}];
```

If we want to integrate the spline, we have to take the second component **cub[t][[2]]** that contains the value of the spline. We integrate the spline numerically when x is in (1, 14), which also means that t is in (0, 13) (the warning we get is harmless):

```
NIntegrate[cub[t][[2]], {t, 0, 13}]
```

Part::partw : Part 2 of
   (SplineFunction[Cubic, {0., 13.}, <>])[t] does not exist. More...
26.3721

Derivatives can also be calculated numerically. We calculate the derivative when x is 2.5, that is, when t is 1.5:

```
<< NumericalMath`NLimit`
```

```
ND[cub[t], t, 1.5][[2]]     – 2.66597
```

■ Example 2

A spline can be drawn through any set of points in the (x, y) plane. Accordingly, the resulting curve may well be multiple-valued. Here is an example:

```
data2 = {{0, 1}, {1, 1}, {2, 1}, {2, 2}, {1, 2},
    {1, 1}, {1, 0}, {1, -1}, {1, -2}, {2, -2}, {2, -1}, {1, -1},
    {0, -1}, {-1, -1}, {-2, -1}, {-2, -2}, {-1, -2}, {-1, -1},
    {-1, 0}, {-1, 1}, {-1, 2}, {-2, 2}, {-2, 1}, {-1, 1}, {0, 1}};
```

```
cub2 = SplineFit[data2, Cubic]
```

SplineFunction[Cubic, {0., 24.}, <>]

```
ParametricPlot[cub2[t], {t, 0, 24},
    Compiled → False, AspectRatio → Automatic,
    Epilog → {AbsolutePointSize[2], Map[Point, data2]}];
```

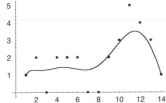

With multiple-valued splines, the reparameterization of the argument becomes clear. If, for example, we want the value of the spline at a point in the highest part of the top right loop, we must inform *Mathematica* that we want a value between the third and fourth points (when the counting begins from zero):

```
cub2[3.5]     {1.47835, 2.17933}
```

21.3.2 Bezier Splines

■ Ordinary Bezier Splines

Consider again the same data introduced earlier, and calculate a Bezier spline:

```
<< NumericalMath`SplineFit`
```

```
b = SplineFit[data, Bezier]
```

SplineFunction[Bezier, {0., 13.}, <>]

```
ParametricPlot[b[t], {t, 0, 13}, Compiled → False, PlotRange → {-0.3, 5.3},
    Epilog → {AbsolutePointSize[2], Map[Point, data]}];
```

As can be seen, a Bezier spline interpolates only the endpoints; other points control the curve.

■ Composite Bezier Splines

Now we calculate a composite Bezier spline:

```
cb = SplineFit[data, CompositeBezier]

SplineFunction[CompositeBezier, {0., 13.}, <>]

ParametricPlot[cb[t], {t, 0, 13},
  Compiled → False, PlotRange → {{-0.3, 14.3}, All},
  Epilog → {AbsolutePointSize[2], Map[Point, data]}];
```

As can be seen, a composite Bezier spline interpolates the first, third, fifth, ... points, while the other points control the curve.

21.4 Interpolation of Functions

21.4.1 Usual Interpolation

Thus far we have considered the interpolation of given data. Another situation is the case where we want to build an interpolating polynomial for a given function. One possibility is that the interpolation points are given (i.e., we cannot determine them ourselves), and another possibility is that we can choose the interpolation points. In the latter case, we can define the points so that the error of interpolation (between interpolation points) becomes smaller; the result is a Chebyshev approximation. With a package, we can calculate both polynomial and rational interpolating functions.

In the **NumericalMath`Approximations`** *package:*

RationalInterpolation[f, {x, m, k}, {x₁, x₂, ..., x_{m+k+1}}] Rational interpolating function of degree (**m**, **k**) for **f** through the given points

RationalInterpolation[f, {x, m, k}, {x, a, b}] Rational interpolating function of degree (**m**, **k**) for **f** in the interval (**a**, **b**) (i.e., rational Chebyshev approximation)

Here, **m** and **k** are the desired degrees of the numerator and the denominator. Giving **k** the value zero, we can calculate polynomial interpolating functions.

In Chebyshev approximation, the interpolation points are chosen in a special way: they are the zeros of the $(m + k + 1)$th degree Chebyshev polynomial. It has turned out that, by choosing the point in this way, we get a good approximation to the function: the error is small throughout the interval (a, b). This means that the result is near the best approximation, which is the minimax approximation. Approximation of functions is considered in more detail in Section 22.2.

■ Example 1: Interpolation

Suppose we have to interpolate the following function (the cumulative distribution function of the standard normal distribution):

```
f = (1 + Erf[x / Sqrt[2]]) / 2;
```

The interpolation points are given as 0, 1/3, 2/3, ..., 2. We form the sixth-degree interpolating polynomial:

```
<< NumericalMath`Approximations`

int = RationalInterpolation[f, {x, 6, 0}, Range[0, 2, 1 / 3]]
```

$0.5 + 0.398489\,x + 0.00309722\,x^2 - 0.0739335\,x^3 +$
$0.00750692\,x^4 + 0.00801905\,x^5 - 0.0018339\,x^6$

Here is the absolute error:

```
Plot[f - int, {x, 0, 2}];
```

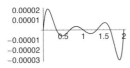

The error has relatively large values near the endpoints of the interval.

■ Example 2: Chebyshev Approximation

Now we calculate the sixth-degree Chebyshev approximation:

```
cheb = RationalInterpolation[f, {x, 6, 0}, {x, 0, 2}]
```

$0.500007 + 0.398604\,x + 0.00255252\,x^2 -$
$0.0732252\,x^3 + 0.00721336\,x^4 + 0.00800419\,x^5 - 0.00181145\,x^6$

```
Plot[f - cheb, {x, 0, 2}];
```

The points where the error is zero are the zeros of the Chebyshev polynomial. Notice how small the error is over the whole interval. We could use the option **Bias** to fine-tune the points where the error is zero to produce an even more uniform error (see Section 22.2.2, p. 589).

21.4.2 Piecewise Interpolation

FunctionInterpolation forms a piecewise-interpolating function for a mathematical expression. The expression can contain built-in mathematical functions and possibly also interpolating functions. For example, if an expression is so complicated that working with it takes some time, we may consider forming an interpolating function for it, because working with these latter functions is fast. Another example is an expression containing several interpolating functions. We may again consider forming a single interpolating function for the expression, thus making computations faster.

Of course, we could form an interpolating function manually by sampling the mathematical expression at some points and then using, for example, **ListInterpolation**. However, **FunctionInterpolation** does the job automatically, is adaptive, and offers some options for controlling the precision.

> **FunctionInterpolation[expr, {x, a, b}]** Form an interpolating function for **expr** by sampling **expr** at sufficiently many points in (**a**, **b**)

The command generalizes for multivariate expressions. For example, if **expr** has two independent variables **x** and **y**, the command is of the following form:

FunctionInterpolation[expr, {x, a, b}, {y, c, d}]

The command has some options:

Options[FunctionInterpolation]

{InterpolationOrder → 3, InterpolationPrecision → Automatic,
 AccuracyGoal → Automatic, PrecisionGoal → Automatic,
 InterpolationPoints → 11, MaxRecursion → 6}

Of these, only **InterpolationOrder** and **InterpolationPrecision** are documented. The former is the usual order of the polynomial pieces, whereas the latter is the precision of the values to be returned by the interpolating function generated. **InterpolationPoints** is probably the initial number of evenly spaced points (in each dimension) at which the expression is evaluated and **MaxRecursion** the maximum number of times a subinterval can be bisected (to achieve the desired precision). **PrecisionGoal** and **AccuracyGoal** are the standard options for controlling the precision and accuracy of the result.

■ Example 1: A Definite Integral

Suppose we want to treat the definite integral of $\sin(\sin(t^2))$ over $(0, x)$ as a function of x. First we define this function:

g[x_?NumberQ] := NIntegrate[Sin[Sin[t^2]], {t, 0, x}]

We then form an interpolating function for this function:

int = FunctionInterpolation[g[x], {x, 0, π}]

InterpolatingFunction[{{0., 3.14159}}, <>]

Now we can, for example, plot the function:

```
Plot[int[x], {x, 0, π}];
```

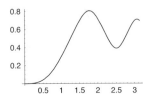

(The time taken to make this plot is a fraction of the time needed to make the plot from the original definite integral as **Plot[g[x], {x, 0, π}]**.) The error is small:

```
Plot[g[x] - int[x], {x, 0, π}, PlotRange → All];
```

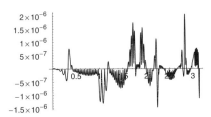

Below, we show the points at which the expression has been sampled. We see that the sampling is adaptive: more points are taken where the expression changes more rapidly.

```
p = Transpose[{InputForm[int][[1, 3, 1]], InputForm[int][[1, 4, 2]]}];
```

```
Show[
  Graphics[{AbsoluteThickness[0.25], Map[Line[{{#[[1]], 0}, #}] &, p]}],
  Axes → True, AspectRatio → 0.2];
```

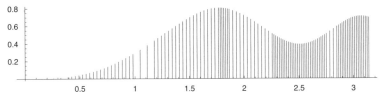

In Section 27.7.3, p. 792, we use **FunctionInterpolation** in Bayesian statistics in the same way as in the example here.

▪ Example 2: The Solution of a Nonlinear Equation

In Section 5.3.2, p. 125, we considered the following equation:

```
eqn = y Exp[r (1 - y)] == 2 - y;
```

The equation defined a function $y(r)$. We define that function:

```
yr[r_?NumberQ] := y /. FindRoot[eqn, {y, 0.1}]
```

Then we form an interpolating function for this function (we do not show here a series of warning messages) and plot it:

```
yy = FunctionInterpolation[yr[r], {r, 2, 2.8}]
```

```
InterpolatingFunction[{{2., 2.8}}, <>]
```

```
Plot[yy[r], {r, 2, 2.8}];
```

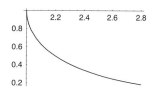

■ Example 3: A Function of an Interpolating Function

As a continuation of Example 2, we note that, in Section 5.3.2 we also plotted the following expression as a function of r:

```
delta = Abs[(1 - r yy[r]) (1 - 2 r + r yy[r])];
```

We can first form an interpolating function for **delta** and then plot this function (we choose a large number of interpolation points to get a sharp enough corner in the figure):

```
delta2 =
  FunctionInterpolation[delta, {r, 2, 2.8}, InterpolationPoints → 60]
InterpolatingFunction[{{2., 2.8}}, <>]

Plot[delta2[r], {r, 2, 2.8}, Epilog → Line[{{0, 1}, {2.8, 1}}]];
```

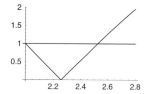

■ Example 4: A Combination of Two Interpolating Functions

In Section 23.4.3, p. 632, we present a module **linearBVP** that solves linear boundary value problems. The solution can be written as a linear combination of two interpolating functions (they are the results of solving two initial value problems with **NDSolve**). **Function** **Interpolation** can then be used to form a single interpolating function.

Approximation

Introduction

If the facts do not fit the theory, change the facts. — Albert Einstein

We will let Einstein rest (relatively speaking) with his strange advice, and we will do just the opposite. Indeed, we have facts in the form of data or functions, and we have theory in the form of approximating functions. We will try to find the approximations that fit the facts as well as possible. There are two basic areas in approximation: approximation of *data* by a function and approximation of a *function* by another function. Approximation of data is often done to get a good summary of the overall behavior of the data, whereas the reason to approximate a function may be to get an expression that it takes less time to evaluate.

■ Approximation of Data

Approximation of data is useful if the data contain errors and we want to find a simple representation of the data by a function. We choose the *form* of the function by looking at the overall behavior of the data. The chosen function contains *parameters*, and for these parameters we try to find the best values according to a chosen criterion. The most widely used method is the *least-squares method*. There are two types of least-squares problems:

- *Linear* least squares: The parameters a, b, … appear linearly in the function, which means that the function is of the form $a\,f(x) + b\,g(x) + …$; for example, $a + b\,x + c\,x^2$ or $a\exp(-x) + b\sin(x)$.
- *Nonlinear* least squares: The parameters appear nonlinearly in the function, which means that the function is of the form $f(x, a, b, …)$; for example, $\exp(a + b\,x)$ or $a\,x^b\exp(c\,x)$.

For linear least squares, we have **Fit**. Nonlinear least squares are done with **FindFit** (✻5). If we want statistical analysis of the fits, we can use **Regress** and **Nonlinear** **Regress** from some packages; these commands are considered in Chapter 27. There we also consider *local regression* and smoothing of data.

If the data do not contain observational errors, then interpolation or piecewise interpolation may be the appropriate technique for summarizing and using the data (see Chapter 21).

In Sections 23.4.4, p. 634, and 25.3.2, p. 699, we estimate differential and difference equation models from data.

■ Approximation of Functions

Approximation of a function is useful if we have a complicated function that is difficult or time-consuming to evaluate and handle and we want to find a simpler function that is close enough to the original function for practical purposes. We can distinguish two types of situations: approximation near a point and approximation in an interval.

For approximation near a point, we have, for example, Taylor polynomials. Another method is Padé approximation; in that case, the approximating function is a rational function.

For approximation in an interval, we can use interpolation and approximation techniques. In Chapter 21, we noted that piecewise interpolation by **FunctionInterpolation** is a strong candidate for an approximation of a complex function in an interval. There we also noted that **RationalInterpolation** gives Chebyshev approximations that are close to minimax approximations. The minimax approximation is calculated by **MinimaxApprox** **imation**; now the maximum error of the approximation over the whole interval is made as small as possible. The approximating function is a rational function, and the starting point for the iterative method is a Chebyshev approximation.

22.1 Approximation of Data

22.1.1 Linear Least Squares

Fit[data, basis, var] Fit **data** with a linear combination of functions of **var** in **basis**

data can be given in either of the following forms:
{f$_1$, f$_2$, … } Fit using the points **{1, f$_1$ }, {2, f$_2$ },** …
{{x$_1$, f$_1$ }, {x$_2$, f$_2$ }, … } Fit using the given points

Examples of **basis**:
{1, x} The fitting function is of the form $a + b x$
{1, x, x^2} The fitting function is of the form $a + b x + c x^2$
{Exp[x], Cos[x]} The fitting function is of the form $a \exp(x) + b \cos(x)$

Fit is designed for linear least-squares fitting. Note that we also have **FindFit**, which suits both linear and nonlinear fitting. We will consider **FindFit** mainly in Section 22.1.4 in the context of nonlinear fitting.

> **FindFit[data, funct, params, var]** (❀5) When **funct** is an expression of the
> variable **var** and contains the parameters **params**, find values for the parameters such
> that the function fits **data** in the best way (in the sense of least squares, by default)

Before we consider some examples, we turn off messages about possible spelling errors. We will use only slightly differing names, and such names generate warnings about possible spelling errors. We do not want to see these warnings:

```
Off[General::spell]; Off[General::spell1]
```

■ A First Fit

In this example, we will use simulated data. First we load a statistics package, then generate the data from a quadratic expression, and lastly show a few of the points:

```
<< Statistics`NormalDistribution`

SeedRandom[2]; data = Table[
    {x, 2 + x - 0.004 x^2 + 2 Random[NormalDistribution[0, 1]]}, {x, 0, 50}];

Short[data, 2]
```

$$\{\{0, 3.89452\}, \{1, 0.843111\}, \ll 48 \gg, \{50, 42.9281\}\}$$

We could plot the data simply as **ListPlot[data]**, but a better look at the data is obtained as follows:

```
pdata = ListPlot[data, PlotJoined → True, AspectRatio → 0.4,
    Epilog → {AbsolutePointSize[2], Map[Point, data]}];
```

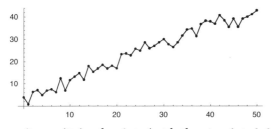

We try a linear fit for the data (with the simulated data in mind, we know that a quadratic fit would be better):

```
fit = Fit[data, {1, x}, x]
```

$$3.69809 + 0.801487 x$$

We could also use **FindFit**:

```
FindFit[data, a + b x, {a, b}, x]
```

$$\{a \to 3.69809, b \to 0.801487\}$$

```
fit = a + b x /. %
```

$$3.69809 + 0.801487 x$$

We plot the fit:

```
pfit = Plot[fit, {x, 0, 50}, DisplayFunction → Identity];
```

Lastly, we show the fit together with the data:

Show[pdata, pfit];

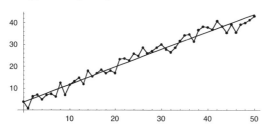

The fit seems quite good. Next we do a simple graphical residual analysis to get information about the quality of the fit. But first we write, for later use, a program for fitting and showing the data and the fit:

```
showFit[data_, basis_, var_, opts___] :=
  Module[{xx, ff, pdata, fit, pfit},
    {xx, ff} = Transpose[data];
    pdata = ListPlot[data, PlotJoined → True, DisplayFunction → Identity,
      Epilog → {AbsolutePointSize[2], Map[Point, data]}];
    fit = Fit[data, basis, var];
    pfit = Plot[fit, {var, Min[xx], Max[xx]}, DisplayFunction → Identity];
    Show[pdata, pfit, DisplayFunction → $DisplayFunction, opts];
    fit]
```

■ **Graphical Residual Analysis**

First we extract the x and f values:

{xx, ff} = Transpose[data];

Now **xx** contains the x values and **ff** the f values. We then calculate the predicted values, that is, the values of the fit at the data points:

pred = fit /. x → xx;

We calculate the residuals:

resf = ff - pred;

The sum of the squared residuals is as follows:

resf.resf   239.337

(The parameters of **fit** were chosen by **Fit** such that the sum of the squared residuals is as small as possible. The minimum value is thus 239.337.) To plot the residuals, add the x values:

res = Transpose[{xx, resf}];

Here is a plot of the residuals:

```
pres = ListPlot[res, PlotJoined → True,
   Epilog → {AbsolutePointSize[2], Map[Point, res]}];
```

The residuals are quite random but not wholly random: a roughly quadratic pattern can be seen. We investigate the situation by fitting a cubic polynomial to the residuals:

```
resfit = Fit[res, {1, x, x^2, x^3}, x]
```

$-0.9185 - 0.00452759\,x + 0.00701618\,x^2 - 0.000152897\,x^3$

```
presfit = Plot[resfit, {x, 0, 50}, DisplayFunction → Identity];

Show[pres, presfit, PlotRange → All];
```

This plot confirms that the linear fit to the data is not adequate; the residuals contain some information.

For later use, we write a program for this kind of graphical residual analysis:

```
showResiduals[data_, fit_, var_, opts___] :=
 Module[{xx, ff, resf, res, pres, resfit, presfit},
  {xx, ff} = Transpose[data];
  resf = ff - (fit /. var → xx);
  res = Transpose[{xx, resf}];
  pres = ListPlot[res, PlotJoined → True, DisplayFunction → Identity,
    Epilog → {AbsolutePointSize[2], Map[Point, res]}];
  resfit = Fit[res, {1, var, var^2, var^3}, var];
  presfit =
   Plot[resfit, {var, Min[xx], Max[xx]}, DisplayFunction → Identity];
  Show[pres, presfit, DisplayFunction → $DisplayFunction,
   PlotRange → All, opts];
  Print["Sum of squared residuals is ", resf.resf]]
```

■ A Second Fit

Since the linear fit was not adequate, we next try a quadratic fit:

```
fit = showFit[data, {1, x, x^2}, x, AspectRatio → 0.4]
```

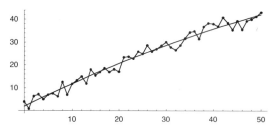

$$1.88055 + 1.02404\,x - 0.00445111\,x^2$$

The fit seems very good. We also show the residuals:

```
showResiduals[data, fit, x]
```

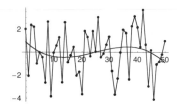

```
Sum of squared residuals is 201.433
```

It seems that the residuals do not contain significant information any more.

■ Remarks

If you want statistical information about the fit, use **Regress** from the **Statistics`** **LinearRegression`** package (see Section 27.6.1).

For polynomial fits, we also have the special command **PolynomialFit** in the **Numeri`** **calMath`PolynomialFit`** package. High-order polynomial fits may suffer from numerical stability problems when calculated in the usual manner with **Fit**, but **PolynomialFit** avoids these problems by using a special algorithm. The value of the fitting polynomial is also calculated in a stable way.

Outliers can cause problems in a least-squares fit. An outlier is an observation that has a value that differs markedly from the general trend of the data. Because the least-squares fit is calculated by minimizing the squared residuals, an outlier can have a considerable unwanted effect on the fit. An excellent illustration of outliers is given in Shaw and Tigg (1994, pp. 315–319).

22.1.2 Trigonometric Fits

If data shows periodic patterns, then a good fit may possibly be obtained by using trigonometric functions in the basis. We can use **Fit**, as in Example 1 below, but we can also use **TrigFit** from a package, as is done in Example 2.

■ Example 1

First we generate data:

```
<< Statistics`NormalDistribution`
```

```
SeedRandom[1];
data =
    Table[{x, 0.5 + Sin[x] - Cos[2 x] + 0.4 Random[NormalDistribution[0.1]]},
    {x, 0, 10, 0.1}];

pdata = ListPlot[data, PlotJoined → True, AspectRatio → 0.2,
    Epilog → {AbsolutePointSize[2], Map[Point, data]}];
```

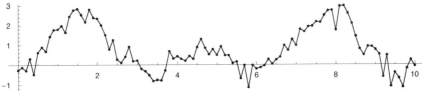

Now we try to look at the data as if we did not know how they emerged. The data seem to contain some cyclical components, and so we try a trigonometric fit. We guess that a good fit could contain, for example, functions 1, $\sin(x)$, $\sin(2x)$, $\cos(x)$, and $\cos(2x)$:

```
fit = Fit[data, {1, Sin[x], Sin[2 x], Cos[x], Cos[2 x]}, x]
```

$0.683544 + 0.0469772 \, \text{Cos}[x] -$
$0.993764 \, \text{Cos}[2 \, x] + 0.876562 \, \text{Sin}[x] + 0.026848 \, \text{Sin}[2 \, x]$

Because the coefficients of $\cos(x)$ and $\sin(2x)$ are small, we try a fit without these functions (here we use the program **showFit** that we presented in Section 22.1.1, p. 572). The result seems to be good:

```
fit = showFit[data, {1, Sin[x], Cos[2 x]}, x, AspectRatio → 0.2]
```

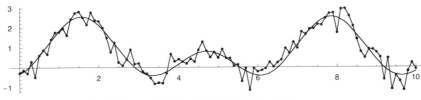

$0.681853 - 0.995907 \, \text{Cos}[2 \, x] + 0.877773 \, \text{Sin}[x]$

The problem in this approach is the choice of the trigonometric functions. Which one should we use in the basis? We have a package to calculate trigonometric fits.

■ A Package

In the **NumericalMath`TrigFit`** *package:*

TrigFit[data, n, {x, x$_0$, x$_1$}]   Find a least-squares approximation of the form
$a_0 + \sum_{i=1}^{n} \left[a_i \cos\left(2\pi i \frac{x - x_0}{x_1 - x_0}\right) + b_i \sin\left(2\pi i \frac{x - x_0}{x_1 - x_0}\right) \right]$ with fundamental period $x_1 - x_0$

We have only to specify the order n; we need not give a list of all of the trigonometric functions in the sum.

Because **TrigFit** uses a special method (**Fourier**), the observations have some restrictions. First, **data** contains only the f values; it is assumed that x values are equally spaced. Second, **data** has to cover exactly one period of a periodic function. To explain this, assume that one period is the interval $[x_0, x_1)$, with length $L = x_1 - x_0$. If **data** contains N

elements, then the spacing between x values is $\frac{L}{N-1}$. The first datum then corresponds with x_0 and the last with $x_1 - \frac{L}{N-1}$.

■ Example 2

We continue with the preceding example. By looking at the data, we can perhaps agree that one period is about the interval $(1.5, 8)$. We take observations 17 through 79:

```
data2 = Take[data, {17, 79}];
```

```
{data2 // First, data2 // Last}
```

```
{{1.6, 2.52082}, {7.8, 2.71651}}
```

```
pdata2 = ListPlot[data2, PlotJoined → True, AspectRatio → 0.2,
    Epilog → {AbsolutePointSize[2], Map[Point, data2]},
    PlotRange → {{0, 10}, All}];
```

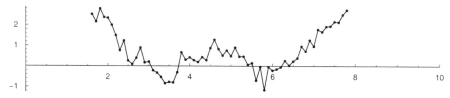

This looks like one period. So, we decide that one period is about the interval $[1.6, 7.9]$. Thus, $x_0 = 1.6$, $x_1 = 7.9$, and $L = 6.3$. Before applying **TrigFit**, we extract x and f components from **data2**:

```
{xx2, ff2} = Transpose[data2];
```

We try several values for the order n:

```
<< NumericalMath`TrigFit`
```

```
fit0 = TrigFit[ff2, 0, {x, 1.6, 7.9}]
```

```
0.706502
```

```
fit1 = TrigFit[ff2, 1, {x, 1.6, 7.9}]
```

```
0.706502 + 0.884036 Cos[0.997331 (-1.6 + x)] -
  0.130893 Sin[0.997331 (-1.6 + x)]
```

```
fit2 = TrigFit[ff2, 2, {x, 1.6, 7.9}]
```

```
0.706502 + 0.884036 Cos[0.997331 (-1.6 + x)] +
  0.921571 Cos[1.99466 (-1.6 + x)] -
  0.130893 Sin[0.997331 (-1.6 + x)] - 0.157805 Sin[1.99466 (-1.6 + x)]
```

```
fit3 = TrigFit[ff2, 3, {x, 1.6, 7.9}]
```

```
0.706502 + 0.884036 Cos[0.997331 (-1.6 + x)] +
  0.921571 Cos[1.99466 (-1.6 + x)] - 0.0558122 Cos[2.99199 (-1.6 + x)] -
  0.130893 Sin[0.997331 (-1.6 + x)] -
  0.157805 Sin[1.99466 (-1.6 + x)] + 0.0589622 Sin[2.99199 (-1.6 + x)]
```

Each time we increase the order by one, a new sin and a new cos term appear; earlier terms remain the same. In **fit3**, the coefficients of the new terms are small, and if we would continue by using orders 4, 5, and so on, we would observe that the coefficients of the new terms remain small. We can accept `fit2`.

However, if we expand `fit2`, we see that even here some coefficients are small:

`fit2 // TrigExpand`

```
0.706502 + 0.108813 Cos[0.997331 x] - 0.928292 Cos[1.99466 x] +
   0.887024 Sin[0.997331 x] + 0.111672 Sin[1.99466 x]
```

We continue by picking up only the terms we consider to be significant:

`basis = {%[[1]], %[[3, 2]], %[[4, 2]]}`

```
{0.706502, Cos[1.99466 x], Sin[0.997331 x]}
```

We can then use `Fit` with this basis and with all of the data to get a simplified fit:

`fit2a = showFit[data, basis, x, AspectRatio → 0.2]`

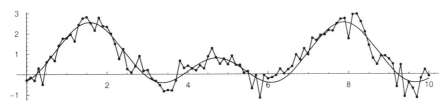

```
0.679925 - 0.995785 Cos[1.99466 x] + 0.874345 Sin[0.997331 x]
```

The fit is quite near to the expression $0.5 + \sin(x) - \cos(2x)$ from which we generated the data.

22.1.3 Special Topics

■ Multidimensional Data

With `Fit`, the data can be multidimensional. If we have, for example, two independent variables x and y and a response variable f, we can find a fitting surface for the data. Below is a summary of 3D fitting, but the summary generalizes readily for higher-dimensional data.

`Fit[data, basis, vars]`   Fit `data` with a linear combination of functions of `vars` in `basis`

`data` is given in the following form:

`{{x₁ , y₁ , f₁ }, {x₂ , y₂ , f₂ }, … }`

Examples of `basis`:

`{1, x, y}`   The fitting function is of the form $a + bx + cy$

`{1, x, y, x y, x^2, y^2}`   The fitting function is of the form $a + bx + cy + dxy + ex^2 + fy^2$

Consider the following 3D data:

```
data =
   {{6, 4, 7.92}, {6, 5, 9.31}, {6, 6, 9.74}, {7, 4, 11.24}, {7, 5, 12.09},
    {7, 6, 12.62}, {8, 4, 14.31}, {8, 5, 14.58}, {8, 6, 16.16}};
```

(In our example, the x and y arguments form a regular grid, but generally the points may be irregular.) We then fit a plane:

```
fit = Fit[data, {1, x, y}, {x, y}]
```

$-13.305 + 3.01333\, x + 0.841667\, y$

```
Plot3D[fit, {x, 6, 8}, {y, 4, 6},
   AxesLabel → {"x", "y", ""}, Ticks → {{6, 7, 8}, {4, 5, 6}, {8, 16}}];
```

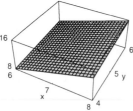

The sum of the squared residuals is as follows:

```
{xx, yy, ff} = Transpose[data];
pred = fit /. {x → xx, y → yy};
(ff - pred).(ff - pred)
```

```
0.526717
```

■ Own Least Squares

The following program is based on the equation $X'\, X\, a = X'\, f$, where vector a contains the parameters to be estimated, vector f contains the f_i values, and the ith row of matrix X contains the values of the basis functions at x_i. If we solve the linear equations for a, we get the least-squares parameters.

```
dataLSQ[xx_List, ff_List, basis_List, t_] := Module[{X, Xt},
  X = Map[basis /. t → # &, xx];
  Xt = Transpose[X];
  LinearSolve[Xt.X, Xt.ff].basis]
```

Here, **xx** and **ff** contain the x and f values of the data, **basis** contains the basis functions, and **t** is the variable of the basis functions. As an example, we find a quadratic fitting curve for the same data we considered in Section 22.1.1, p. 571:

```
<< Statistics`NormalDistribution`
xx = Range[0, 50]; SeedRandom[2];
ff =
  Table[2 + x - 0.004 x^2 + 2 Random[NormalDistribution[0, 1]], {x, 0, 50}];
```

```
dataLSQ[xx, ff, {1, x, x^2}, x]
```

$1.88055 + 1.02404\, x - 0.00445111\, x^2$

■ Logarithmic Transform

If the data show an exponential growth, then one candidate for the fitting function is $f(x) = \exp(a + bx)$, but this is nonlinear in the parameters a and b, so **Fit** cannot be applied in the standard way. However, we can take logarithms of the values of the function: $\log(f(x)) = a + bx$. Thus, the logarithms of the data have a simple linear form to which we can apply **Fit**. After the fit $\hat{a} + \hat{b}x$ is found for $(x_i, \log(f_i))$, we do the inverse transform to find the fit $\exp(\hat{a} + \hat{b}x)$ for the original data.

Taking logarithms and then using linear least squares is a widely used procedure, but note that the resulting fit is not the best possible one. To find a true least-squares fit to the exponential model, **FindFit** should be used (see Section 22.1.4, p. 581).

First we generate points that show an exponential growth:

```
<< Statistics`NormalDistribution`

SeedRandom[2];
data = Table[{x, Exp[0.3 + 0.2 x] + 0.5 Random[NormalDistribution[0, 1]]},
    {x, 0, 10, 0.2}];
```

Then we take logarithms of the f values:

```
{xx, ff} = Transpose[data];
logff = Log[ff];
logdata = Transpose[{xx, logff}];
```

Now we fit a linear function to this data and make the inverse transform:

```
logfit = Fit[logdata, {1, x}, x]
```

$0.265563 + 0.204719 x$

```
fit = Exp[logfit]
```

$e^{0.265563 + 0.204719 x}$

The coefficients 0.266 and 0.205 are quite near the values 0.3 and 0.2 that we used in the simulation. We plot the fit and the data:

```
Plot[fit, {x, 0, 10},
    Epilog → {AbsolutePointSize[2], Map[Point, data], Line[data]}];
```

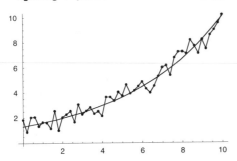

The fit seems to be good. We show the residuals (using the program **showResiduals** we presented in Section 22.1.1, p. 573):

```
showResiduals[data, fit, x]
```

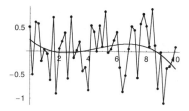

```
Sum of squared residuals is 12.9083
```

The residuals seem to be quite random. We will apply `FindFit` in Section 22.1.4, and then we get a slightly better fit.

22.1.4 Nonlinear Fitting

> `FindFit[data, funct, params, var]` (⌘5) When `funct` is an expression of the variable `var` and contains the parameters `params`, find values for the parameters such that the function fits `data` in the best way (in the sense of least squares, by default)
>
> `data` can be given in either of the following forms:
> `{f₁ , f₂ , … }` Fit using the points `{1, f₁ }, {2, f₂ }, …`
> `{{x₁ , f₁ }, {x₂ , f₂ }, … }` Fit using the given points
>
> Examples of `funct`: `Exp[a + bx], a/(1 + b Exp[-c x])`
>
> The parameter specification `params` is of the form `{a, b, … }` or of the form `{{a, a₀ }, {b, b₀ }, … }`. In the former case, starting values for the parameters are chosen by a default method, and, in the latter case, a_0, b_0, and so on are used as the starting values (all parameters need not have the same form of specification).

`FindFit` is new in *Mathematica* 5. Users of *Mathematica* versions prior to version 5 have to load a package and then use `NonlinearFit`:

```
<< Statistics`NonlinearFit`
```

```
NonlinearFit[data, funct, var, params]
```

(Note here that the variable and the parameters are written in the opposite order as compared with `FindFit`.) If you want statistics of the model, use `NonlinearRegress` (see Section 27.6.2, p. 779).

`FindFit` uses, by default, the Levenberg–Marquardt method to find the best values for the parameters of a function that is nonlinear in the parameters. The default criterion is to minimize the square root of the sum of the squares of the residuals; the result is a least squares fit. The method is an iterative procedure, and initial guesses for the parameters can be provided. In general, `FindFit` only finds a locally optimal fit.

Multidimensional data can be entered in the same way as it is for `Fit`. For example, if we have two independent variables, the command is of the following form:

```
FindFit[{{x₁ , y₁ , f₁ }, {x₂ , y₂ , f₂ }, … }, funct, {a, b, … }, {x, y}]
```

■ Example 1: Exponential Growth

We consider the same data that was used when we introduced the logarithmic transform in Section 22.1.3, p. 579, and we fit the same model $\exp(a + b\,x)$, which we used there. We try the default starting value of zero for both parameters:

```
f = Exp[a + b x];

ab = FindFit[data, f, {a, b}, x]
{a → 0.325637, b → 0.197218}

fit = f /. ab
```
$e^{0.325637 + 0.197218\,x}$

In Section 22.1.3 we obtained, by the logarithmic transform, the fit $e^{0.265563\,x + 0.204719\,x}$. As you can see, the two fits differ somewhat. We plot the fit and show the residuals (using the program **showResiduals** we presented in Section 22.1.1, p. 573):

```
Plot[fit, {x, 0, 10},
    Epilog → {AbsolutePointSize[2], Map[Point, data], Line[data]}];
```

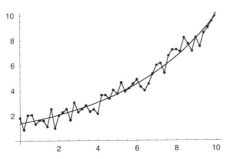

```
showResiduals[data, fit, x]
```

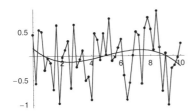

```
Sum of squared residuals is 12.583
```

The sum of the squared residuals, 12.583, is somewhat smaller than the value 12.908 obtained by the logarithmic transform.

■ Example 2: Logistic Growth

In an experiment, the growth of a yeast culture was measured at time instances 0, 1, 2, ..., 18 (hours). The measurements were as follows (see Pearl 1927):

```
yeast = {9.6, 18.3, 29.0, 47.2, 71.1, 119.1, 174.6, 257.3, 350.7, 441.0,
    513.3, 559.7, 594.8, 629.4, 640.8, 651.1, 655.9, 659.6, 661.8};
```

```
tt = Range[0, 18];
data2 = Transpose[{tt, yeast}];

p1 = ListPlot[data2, PlotStyle → AbsolutePointSize[3],
    Ticks → {Range[18], Automatic}, Epilog → Line[{{0, 663}, {18, 663}}]];
```

The growth seems to follow the logistic pattern $a/(1 + b\,e^{-cx})$. We try this model:

```
f = a / (1 + b Exp[-c t]);

abc = FindFit[data2, f, {a, b, c}, t]

{a → 663.022, b → 71.5763, c → 0.546995}

fit = f /. abc
```

$$\frac{663.022}{1 + 71.5763\, e^{-0.546995\, t}}$$

```
p2 = Plot[fit, {t, 0, 18}, DisplayFunction → Identity];

Show[p1, p2];
```

Another parametrization of the logistic model is $M/\left[1 + \left(\frac{M}{y_0} - 1\right)\exp(-r\,M\,t)\right]$. This is the form of the solution of the logistic differential equation model $y'(t) = r\,y(t)\,(M - y(t))$ with $y(0) = y_0$. The parameter M is the limiting value of $y(t)$ as time approaches infinity. Now we give our own starting values, but we get the same fit:

```
f2 = M / (1 + (M / y0 - 1) Exp[-r M t]);

FindFit[data2, f2, {{M, 1}, {y0, 1}, {r, 0.1}}, t]

{M → 663.022, y0 → 9.13552, r → 0.000825002}

f2 /. %
```

$$\frac{663.022}{1 + 71.5763\, e^{-0.546995\, t}}$$

■ Remarks

If the starting values are not good enough, the result may be wrong:

```
FindFit[data2, f2, {M, y0, r}, t, MaxIterations → 150]
```

$\{M \to 393.039, y0 \to 9.6, r \to 0.819405\}$

Because of this, it is wise to try several starting values. We can also plot the criterion function by fixing all but two parameters. We fix y_0 to be 9.6 and form and plot the criterion (the sum of the squares of the residuals, which is called the χ^2 merit function) as a function of M and r:

```
res = yeast - (f2 /. {y0 → 9.6, t → tt});
khi2 = res.res;

ContourPlot[khi2, {M, 200, 1000}, {r, 0, 0.003},
    ContourShading → False, Contours → 30, PlotPoints → 50];
```

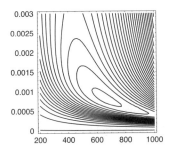

We see that the optimum values of M and r are near 700 and 0.001.

Note that we can also use **NMinimize** (see Section 20.2.3, p. 531). It often finds a global minimum:

```
res = yeast - (f2 /. t → tt);
khi2 = res.res;

NMinimize[khi2, {{M, 0, 1000}, {y0, 0, 100}, {r, 0, 0.05}}]
```

$\{194.325, \{M \to 663.022, r \to 0.000825002, y0 \to 9.13552\}\}$

■ Options

With the options, we can choose the method and norm function, among other things.

Options for **FindFit**:

WorkingPrecision Precision used in internal computations; examples of values: **MachinePrecision**, **20**

PrecisionGoal If the value of the option is **p**, the relative error of the optimum point and of the value of the norm of the residuals should be of the order 10^{-p}; examples of values: **Automatic** (usually means 8), **10**

AccuracyGoal If the value of the option is **a**, the absolute error of the optimum point and of the value of the norm of the residuals should be of the order 10^{-a}; examples of values: **Automatic** (usually means 8), **10**

Method　Method used; possible values: **Automatic** (usually means **Levenberg⸱ Marquardt**) **LevenbergMarquardt**, **Gradient**, **ConjugateGradient**, **Newton**, **QuasiNewton**

MaxIterations　Maximum number of iterations; examples of values: **100, 200**

NormFunction　Norm of the residuals to be minimized; examples of values: **Norm** (means **(Norm[#, 2] &)**), **(Norm[#, 1] &)**

Gradient　How the gradient is calculated; examples of values: **Automatic**, **Symbolic**, **FiniteDifference**

StepMonitor　Command to be executed after each step of the iterative method; examples of values: **None, n++, AppendTo[iters, {a, b}]**

EvaluationMonitor　Command to be executed after each evaluation of the expression to be fitted; examples of values: **None, n++, AppendTo[points, {a, b}]**

The default is that iterations are stopped if the estimated relative or absolute error in the point obtained and in the value of the norm of the residuals is less than 10^{-8}.

The norm of the residuals is minimized with **Method**. The default value **Automatic** of this option means **LevenbergMarquardt** if the 2-norm is used (which is the default). The Levenberg–Marquardt method initially uses the steepest descent method, but it shifts gradually to quadratic minimization.

The default norm is **Norm**, which is the 2-norm (the square root of the sum of the squares of the residuals); we could also write this norm as **(Norm[#, 2] &)**. With the **NormFunction** option, we can define other norms, like the 1-norm or **(Norm[#, 1] &)** (the sum of the absolute values of the residuals) or the ∞-norm or **(Norm[#, ∞] &)** (the maximum of the absolute values of the residuals; the result is a minimax approximation).

By using the **StepMonitor** option, we can see all of the steps of the iterative method. At the same time, we can also calculate the successive values of the norm of the errors. As an example, we again consider the exponential growth of Example 1. Now we use 1-norm as the criterion and the gradient method to minimize the value of the criterion:

```
f = Exp[a + b x]; {xx, ff} = Transpose[data]; fxx = f /. x → xx;

i = 0; iters = {}; fit = FindFit[data, f, {a, b},
   x, Method → Gradient, NormFunction → (Norm[#, 1] &),
   StepMonitor :→ {AppendTo[iters, {++i, a, b, Norm[ff - fxx, 1]}]}]]
{a → 0.321575, b → 0.198355}

TableForm[iters, TableSpacing → {0.6, 2},
  TableHeadings → {None, {"i", "a", "b", "norm of the residuals"}}]
```

i	a	b	norm of the residuals
1	0.904015	0.126822	45.4382
2	0.40929	0.181192	24.087
3	0.351889	0.194987	20.7631
4	0.322147	0.198986	20.7321
5	0.321575	0.198355	20.6528
6	0.321575	0.198355	20.6528
7	0.321575	0.198355	20.6528

22.2 Approximation of Functions

22.2.1 Simple Methods

When we next explain several methods for the approximation of functions, we will use the cumulative distribution function of the standard normal distribution as an example:

```
f = (1 + Erf[x / Sqrt[2]]) / 2
```

$$\frac{1}{2}\left(1 + \text{Erf}\left[\frac{x}{\sqrt{2}}\right]\right)$$

```
Plot[f, {x, 0, 4}, AxesOrigin → {0, 0.5}];
```

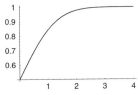

We apply different approximation methods either at point 1 or in the interval (0, 2). If we want an approximation over a wider interval, for example, (0, 4), it is probably better to find at least two approximations, one for (0, 2) and the other for (2, 4).

We will use the following module to show the absolute and relative errors of the approximation **appr** for **f** in the interval from **a** to **b**.

```
showError[f_, appr_, x_, a_, b_, opts___?OptionQ] := Show[
  GraphicsArray[Map[Plot[#, {x, a, b}, opts, DisplayFunction → Identity] &,
  {f - appr, 1 - appr / f}]]]
```

First we consider some simple methods of approximation: Taylor polynomials and least squares and Chebyshev approximation. In Section 22.2.2, p. 587, we explain the minimax approximation. This method minimizes the maximum error in the interval considered to get an error that is evenly spread over the whole interval.

Note that in Section 21.4.2, p. 566, we discussed **FunctionInterpolation**. This command, although it uses interpolation, is very good for approximating functions, because it is adaptive (i.e., gives special care to regions where the function changes rapidly), its precision can be controlled, and it is easy to calculate and use.

■ Taylor Polynomials and Padé Approximation

A Taylor polynomial (see Section 16.2.1, p. 425) gives an approximation at a point. For example:

```
taylor6 = Normal[Series[f, {x, 1., 6}]] // Expand
```

$0.500569 + 0.395219\, x + 0.0100821\, x^2 -$
$0.0806569\, x^3 + 0.0100821\, x^4 + 0.00806569\, x^5 - 0.00201642\, x^6$

The absolute and relative errors show that the approximation is good in the interval
(0.5, 1.5):

```
showError[f, taylor6, x, 0, 2,
  PlotRange → All, Ticks → {{1, 2}, {-0.0005, 0.0005}}];
```

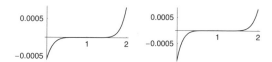

Padé approximation can be used to approximate a function around a point by using a
rational expression; use **Pade** from the **Calculus`Pade`** package.

■ Least-Squares Approximation

With the least-squares method, we minimize the integral of the square of the difference
between the function and the approximation on the interval considered. If the approximating function is a linear combination of certain functions, the coefficients are obtained from
a system of linear equations. Here is a module that calculates a least-squares approximation for **f** in the interval from **a** to **b**; the approximating function is a linear combination of
the functions in the list **basis**.

```
functionLSQ[f_, x_, a_, b_, basis_] :=
  LinearSolve[NIntegrate[Evaluate[Outer[Times, basis, basis]], {x, a, b}],
    NIntegrate[Evaluate[f basis], {x, a, b}]].basis
```

We calculate a sixth degree least-squares approximation:

```
lsq6 = functionLSQ[f, x, 0, 2, Table[x^i, {i, 0, 6}]]
```

$0.500019 + 0.398426\,x + 0.00319491\,x^2 - $
$0.0741471\,x^3 + 0.00780805\,x^4 + 0.00784075\,x^5 - 0.00179815\,x^6$

```
showError[f, lsq6, x, 0, 2, Ticks → {{1, 2}, {-0.00002, 0.00002}}];
```

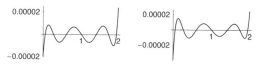

The approximation is good, but at the endpoints the errors are considerably larger than at
other points.

■ Chebyshev Approximation

In the **NumericalMath`Approximations`** *package:*

RationalInterpolation[f, {x, m, k}, {x, a, b}] Rational interpolating function of
degree (**m, k**) for **f** in the interval (**a, b**) (i.e., rational Chebyshev approximation)

In Section 21.4.1, p. 564, we already considered rational Chebyshev approximation. We apply this method to our function by using degrees six (the degree of the numerator) and zero (the degree of the denominator), that is, the approximating function is a sixth-degree polynomial:

```
<< NumericalMath`Approximations`

cheb60 = RationalInterpolation[f, {x, 6, 0}, {x, 0, 2}]
```

$0.500007 + 0.398604\,x + 0.00255252\,x^2 -$
$0.0732252\,x^3 + 0.00721336\,x^4 + 0.00800419\,x^5 - 0.00181145\,x^6$

```
showError[f, cheb60, x, 0, 2, Ticks → {{1, 2}, {-10.^-5, 10.^-5}}];
```

The approximation is very accurate, although the errors are not spread perfectly evenly over the interval.

With the option **Bias**, we can fine-tune the approximation. The value 0 means that the interpolation points are chosen symmetrically in the interval. A positive [negative] value causes the points to be shifted towards the right [left]. The value has to be between −1 and 1.

We try to make the relative error of the approximation **cheb60** more even. Using trial and error, we find that a bias of −0.016 is appropriate:

```
cheb60b = RationalInterpolation[f, {x, 6, 0}, {x, 0, 2}, Bias → -0.016]
```

$0.500006 + 0.398638\,x + 0.00234536\,x^2 -$
$0.072771\,x^3 + 0.00676662\,x^4 + 0.00820574\,x^5 - 0.00184544\,x^6$

```
showError[f, cheb60b, x, 0, 2, Ticks → {{1, 2}, {-10.^-5, 10.^-5}}];
```

With **GeneralRationalInterpolation**, we can calculate Chebyshev approximations for parametrically defined functions.

22.2.2 Minimax Approximation

The goal of minimax approximation is to minimize the maximum error (absolute or relative) in an interval. This is clearly a very desirable goal: the error is then evenly low over the whole interval, which is in contrast with an error that is low over a subinterval but large elsewhere.

In the **NumericalMath`Approximations`** *package:*

MiniMaxApproximation[f, {x, {a, b}, m, k}]   Rational minimax approximation of
   degree (**m, k**) for **f** in the interval (**a, b**)

Here, **m** and **k** are the degrees of the numerator and the denominator. Giving **k** the
value zero, we can calculate polynomial interpolating functions.

The procedure starts with a rational interpolating function $r(x)$ using **RationalInterpo‌**
lation. This function is then iteratively modified according to Remes' algorithm: the
interpolation points are adjusted to make the maximum relative error as small as possible.

The procedure uses the relative error $|1 - r(x)/f(x)|$ as the criterion. This means that
$f(x)$ cannot have a zero in the interval. However, we can overcome this problem by
dividing the zero out of the function (see the manual of the packages). Singularities must
also first be eliminated. In addition, it is better to calculate several approximations for
small intervals rather than one approximation for a long interval.

■ Example

Here is the (6, 0) degree rational minimax approximation (i.e., the sixth-degree polynomial
minimax approximation) for our familiar function:

```
f = (1 + Erf[x / Sqrt[2]]) / 2;

<< NumericalMath`Approximations`

{maxPoints, {miniMax60, maxError}} =
 MiniMaxApproximation[f, {x, {0, 2}, 6, 0}]
```

$\{\{0., 0.0946758, 0.364016, 0.761322, 1.20893, 1.61548, 1.89853, 2.\},$
$\quad\{0.500006 + 0.398643\,x + 0.00232708\,x^2 - 0.0727457\,x^3 +$
$\qquad 0.00675247\,x^4 + 0.00820846\,x^5 - 0.00184541\,x^6, -0.0000118083\}\}$

The result is of the following form: {{points where the maximum relative error occurs},
{the minimax approximation, the maximum relative error}}. The relative error is perfectly
even in the interval (0, 2):

```
showError[f, miniMax60, x, 0, 2, Ticks → {{1, 2}, {-0.00001, 0.00001}}];
```

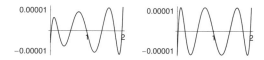

The Chebyshev approximations **cheb60** and **cheb60b** that we calculated in Section 22.2.1,
p. 586, are close to the minimax approximation.

■ Options

Options for **MiniMaxApproximation**:

WorkingPrecision Precision used in internal computations; examples of values: **MachinePrecision** (❀5), **20**

Bias Bias in the symmetry of the initial interpolation points; examples of values: **0**, **-0.1, 0.26**

Brake Defines the braking properties of the algorithm; default value: **{5, 5}**

MaxIterations Maximum number of iterations after braking has ended; default value: **20**

Derivatives Specifies a function to use for the derivatives; default value: **Automatic**

PlotFlag Whether to plot the relative error at each step; possible values: **False, True**

PrintFlag Whether to print information about the relative error at each step; possible values: **False, True**

The value of **Bias** is a number between –1 and 1. The default value 0 means that the initial interpolation points are chosen symmetrically in the interval. A positive [negative] value of **Bias** causes the points to shift toward the right [left].

Brake controls the braking of the iterations. If the change from one iteration to the next is too large, the procedure may go astray. Braking can prevent this. The default value of the option is {5, 5}. The first value in the list tells us how many iterations are to be affected by the braking, and the second value tells us how much braking is to be applied to the first iteration. The braking automatically decreases with the iterations.

The manual of the packages contains much more information about minimax approximation than is presented here. The package also defines **GeneralMinimaxApproximation**, which can be used to approximate parametrically defined functions.

Differential Equations

Introduction

> *God is not so cruel as to create situations described by*
> *nonlinear differential equations. — Edward Sexton*

Solving ordinary differential equations with *Mathematica* is straightforward: we have **DSolve** for symbolic solution and **NDSolve** for numerical solution. Both commands accept one or more equations, first- or higher-order equations, and linear and nonlinear equations, and they solve both initial and boundary value problems (although **NDSolve** only solves linear boundary value problems).

In addition, we consider solving differential equations with the Laplace transform, finding series solutions, and solving integral equations. We implement the Runge–Kutta method and also some methods for boundary value problems. Some well-known nonlinear systems are considered, such as a predator–prey model, a competing species model, and the Lorenz model. We also consider estimating parameters of nonlinear differential equations.

For more about differential equations with *Mathematica*, see Abell and Braselton (1997).

23.1 Symbolic Solutions

23.1.1 First-Order Equations

Here are some common commands for first-order differential equations:

> `sol = y[t] /. DSolve[eqn, y[t], t]` Give the general solution
> `sol = y[t] /. DSolve[{eqn, y[a] == α}, y[t], t]` Solve an initial value problem
> `Plot[sol, {t, a, b}]` Plot the solution of an initial value problem

An example of a differential equation is `y'[t] == a y[t] + b t + c`. The dependent variable, which here is `y`, must contain the independent variable, here `t`, as the argument; this means that we cannot write the equation as `y' == a y + b t + c`. Note, too, that the equation must contain `==` (not `=`) and that the initial condition must also contain `==` (not `=`) (remember that *Mathematica* replaces `==` with `=`).

■ Example 1: General Solution

Consider the following logistic equation:

> `eqn = y'[t] == r y[t] (M - y[t])`
>
> $y'[t] == r (M - y[t]) y[t]$

The name of the equation is **eqn**. The solution is as follows:

> `DSolve[eqn, y[t], t]`
>
> $\left\{ \left\{ y[t] \to \dfrac{e^{Mrt+MC[1]} M}{-1 + e^{Mrt+MC[1]}} \right\} \right\}$

The solution is in the form of a transformation rule (for more about rules, see Section 12.1.2, p. 300). The arbitrary constant is `C[1]` (we can give it whatever value we want).

In general, the solution given by **DSolve** consists of a list of solutions:

> `{{solution 1}, {solution 2}, ...}`

Indeed, a given equation can have several solutions. Each solution is again a list that consists of as many elements as there are dependent variables. In our example, we have only one dependent variable, **y**, and we obtained only one solution. Thus, the solution is of the form `{{solution 1 for y}}`.

Often it is convenient to ask for the value of `y[t]`:

> `y[t] /. DSolve[eqn, y[t], t]`
>
> $\left\{ \dfrac{e^{Mrt+MC[1]} M}{-1 + e^{Mrt+MC[1]}} \right\}$

If there is only one solution, we may also want to get rid of the curly braces by asking for the first component of the solution:

```
y[t] /. DSolve[eqn, y[t], t][[1]]
```

$$\frac{e^{M r t + M C[1]} M}{-1 + e^{M r t + M C[1]}}$$

We will use this method from now on.

Note that the solution **sol** is a *generic* solution, which is a solution that is valid for general values of the parameters **r** and **M**. For some particular values, the solution may be of a different form. For example, when **M** is zero, the solution is as follows:

```
y[t] /. DSolve[eqn /. M → 0, y[t], t][[1]]
```

$$\frac{1}{r t - C[1]}$$

■ Example 2: Initial Value Problem

Next we solve an initial value problem:

```
y[t] /. DSolve[{eqn, y[0] == α}, y[t], t][[1]]
```

```
Solve::ifun :
  Inverse functions are being used by Solve, so some solutions may not
    be found; use Reduce for complete solution information. More…
```

$$\frac{e^{M r t} M \alpha}{M - \alpha + e^{M r t} \alpha}$$

For all transcendental equations, **Solve** gives the warning of using inverse functions. We turn the message off:

```
Off[Solve::ifun]
```

Next we give specific values for all constants:

```
sol = y[t] /. DSolve[{eqn /. {r → 1 / 10, M → 10}, y[0] == 1 / 4}, y[t], t][[1]]
```

$$\frac{10 \, e^t}{39 + e^t}$$

This solution can be plotted, because it does not contain any parameters:

```
p1 = Plot[{10, sol}, {t, 0, 10}];
```

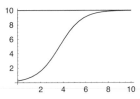

This is a logistic curve. We also plotted the asymptote 10 of this curve. Here is a value of the solution and a table of values:

```
sol /. t → 0.0        0.25
```

```
Table[{t, sol}, {t, 0., 4, 1}]
```

```
{{0., 0.25}, {1., 0.65158}, {2., 1.59284}, {3., 3.3994}, {4., 5.83325}}
```

■ Example 3: Direction Field

We can learn the behavior of the solution of a differential equation by plotting a set of arrows that are tangent to the solution. The plot is called a direction field. It can be plotted with **PlotVectorField** from a package (see Section 7.2.3, p. 189). This command plots vectors **{exprx, expry}** for some values of **x** and **y**. In our example, we can choose **exprx** to be **1** and **expry** to be **y'[t]**:

```
<< Graphics`PlotField`
```

```
p2 = PlotVectorField[{1, 1 / 10 y (10 - y)}, {t, 0, 10}, {y, 0, 10},
    PlotPoints → 11, Axes → True, DisplayFunction → Identity];
```

We show both the direction field and one solution:

```
Show[p1, p2];
```

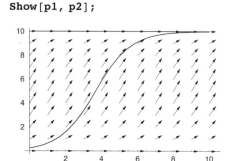

■ Example 4: A Set of Trajectories

Next we plot a set of trajectories, which is a set of solutions to the equation with different starting points. To begin, we compute the solution with general values **a** and α for the starting time and starting value:

```
sol = y[t] /. DSolve[{eqn /. {r → 1 / 10, M → 10}, y[a] == α}, y[t], t][[1]]
```

$$\frac{10\,e^t\,\alpha}{10\,e^a - e^a\,\alpha + e^t\,\alpha}$$

First we plot some trajectories starting from $t = 0$. We fix **a** to be 0 and give α various values:

```
solset = Table[sol /. a → 0, {α, 0.1, 15.1, 0.5}];
```

```
Plot[Evaluate[solset], {t, 0, 8},
    PlotRange → All, Ticks → {Range[8], {5, 10, 15}}];
```

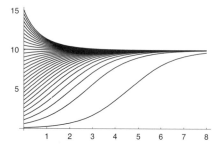

If we want an area filled with curves, it is best to fix α and give **a** various values. First we give α the value 0.01 and let **a** vary from –13 to 7.5:

```
p1 = Table[Plot[Evaluate[sol /. α → 0.01],
    {t, a, 8}, DisplayFunction → Identity], {a, -13, 7.5, 0.5}];

p11 = Show[p1, DisplayFunction → $DisplayFunction];
```

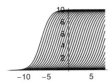

Then we give α the value 15 and let **a** vary from –6 to 7.5:

```
p2 = Table[Plot[Evaluate[sol /. α → 15],
    {t, a, 8}, DisplayFunction → Identity], {a, -6, 7.5, 0.5}];

p22 = Show[p2, DisplayFunction → $DisplayFunction];
```

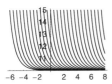

Lastly we give α the value –0.01 and let **a** vary from –6 to 7.5:

```
p3 = Table[Plot[Evaluate[sol /. α → -0.01],
    {t, a, a + 6}, DisplayFunction → Identity], {a, -6, 7.5, 0.5}];

p33 = Show[p3, DisplayFunction → $DisplayFunction, PlotRange → {-5, 2}];
```

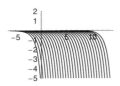

Lastly we combine the plots:

```
Show[p11, p22, p33, PlotRange → {{0, 8}, {-5.1, 15.1}}];
```

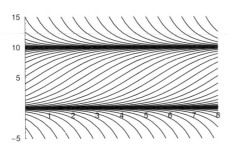

■ Example 5: Equilibrium Points

Recall from a course of differential equations that, if the differential equation is $y' = f(y)$, then the equilibrium points y^* are the solutions of the equation $f(y^*) = 0$ and an equilibrium point is asymptotically stable if $f'(y^*) < 0$. In our model, the f function is as follows:

```
f = eqn[[2]]      r (M - y[t]) y[t]
```

The equilibrium points are as follows:

```
equi = Solve[f == 0, y[t]]      {{y[t] → 0}, {y[t] → M}}
```

We check the stability:

```
D[f, y[t]] /. equi      {M r, -M r}
```

So, if $Mr > 0$, then 0 is an unstable and M an asymptotically stable equilibrium point. These properties can also be seen in the last figure.

■ Example 6: Several Solutions and Nonlinear Equations

Sometimes the problem has several solutions:

```
DSolve[y'[t] == 1 / y[t], y[t], t]
```

$$\{\{y[t] \to -\sqrt{2} \ \sqrt{t + C[1]}\ \}, \{y[t] \to \sqrt{2} \ \sqrt{t + C[1]}\ \}\}$$

The solution of a nonlinear equation may be quite complicated:

```
DSolve[y'[t] == y[t]^2 + t, y[t], t] // FullSimplify
```

$$\left\{\left\{y[t] \to \frac{\sqrt{t}\ \left(-\text{BesselJ}\left[-\frac{2}{3}, \frac{2\,t^{3/2}}{3}\right] + \text{BesselJ}\left[\frac{2}{3}, \frac{2\,t^{3/2}}{3}\right]C[1]\right)}{\text{BesselJ}\left[\frac{1}{3}, \frac{2\,t^{3/2}}{3}\right] + \text{BesselJ}\left[-\frac{1}{3}, \frac{2\,t^{3/2}}{3}\right]C[1]}\right\}\right\}$$

■ Example 7: Equations Are Defined by ==

A common problem encountered when solving differential equations is the following:

```
eqn = {y'[t] == y[t] + t + 1, y[0] = 2};
```

```
DSolve[eqn, y[t], t]
```

```
DSolve::deqn :
  Equation or list of equations expected instead of 2 in the
     first argument {y'[t] == 1 + t + y[t], 2}. More…
DSolve[{y'[t] == 1 + t + y[t], 2}, y[t], t]
```

DSolve tells you that it found, from the first argument, the element 2 and that this is not an equation. Indeed, we observe that the initial condition **y[0] = 2** is not a correct equation. It must be written as **y[0] == 2**. When we wrote **y[0] = 2**, we actually assigned the value **2** for **y[0]**, and this causes the error message. Before we solve the initial value problem, we must clear the value of **y[0]** and correct the initial condition:

```
y[0] =.
```

```
eqn = {y'[t] == y[t] + t + 1, y[0] == 2};
```

```
DSolve[eqn, y[t], t]      {{y[t] → -2 + 4 e^t - t}}
```

23.1.2 Second- and Higher-Order Equations

> `sol = y[t] /. DSolve[eqn, y[t], t]` General solution
> `sol = y[t] /. DSolve[{eqn, y[a] == α, y'[a] == β}, y[t], t]` Initial value problem
> `sol = y[t] /. DSolve[{eqn, y[a] == α, y[b] == β}, y[t], t]` Boundary value problem
> `Plot[sol, {t, a, b}]` Plot the solution of an initial or boundary value problem

These commands apply to second-order equations. They generalize directly to higher-order equations. The initial and boundary conditions mentioned are the simplest ones; the conditions can be more complex equations.

■ Example 1: Basic Techniques

We ask for a general solution of a second-order equation:

```
eqn = y''[t] + y[t] == 1;
```

```
y[t] /. DSolve[eqn, y[t], t][[1]]
```

$1 + C[1] \, Cos[t] + C[2] \, Sin[t]$

The arbitrary constants are `C[1]` and `C[2]` (by the way, with the option **GeneratedParame·** **ters** [**DSolveConstants** in *Mathematica* versions prior to 5], you can give the constants another name). Next we give two initial conditions:

```
sol = y[t] /. DSolve[{eqn, y[0] == 0, y'[0] == 1}, y[t], t][[1]]
```

$1 - Cos[t] + Sin[t]$

```
Plot[sol, {t, 0, 4 π}];
```

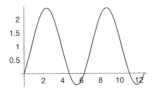

Now we give two boundary conditions, one at $t = 0$ and the other at $t = 5$:

```
y[t] /. DSolve[{eqn, y[0] == 0, y[5] == 7}, y[t], t][[1]]
```

$1 - Cos[t] + Cot[5] \, Sin[t] + 6 \, Csc[5] \, Sin[t]$

The boundary conditions can be even more complex:

```
y[t] /. DSolve[{eqn, y[0] == 2, y[5] + y'[5] == 1}, y[t], t][[1]] //
  FullSimplify
```

$$1 + Cos[t] + \frac{(-Cos[5] + Sin[5]) \, Sin[t]}{Cos[5] + Sin[5]}$$

■ Example 2: Constant Coefficients

All linear second-order equations with constant coefficients can be solved. For example:

```
eqn = y''[t] == a y'[t] + b y[t] + c;
```

```
y[t] /. DSolve[eqn, y[t], t][[1]]
```

$$-\frac{c}{b} + e^{\frac{1}{2}\left(a-\sqrt{a^2+4b}\right)t} C[1] + e^{\frac{1}{2}\left(a+\sqrt{a^2+4b}\right)t} C[2]$$

Note that this is a generic solution. For special values of the parameters, the solution can be of another form. The following solution is of the preceding form:

```
y[t] /.
  DSolve[{eqn /. {a → 2, b → 3, c → 2}, y[0] == 2, y'[0] == 0}, y[t], t][[1]]
```

$$\frac{2}{3} e^{-t} \left(3 - e^t + e^{4t}\right)$$

However, this solution is not:

```
y[t] /.
  DSolve[{eqn /. {a → 2, b → -1, c → 1}, y[0] == 2, y'[0] == 0}, y[t], t][[1]]
```

$$1 + e^t - e^t t$$

Neither is this:

```
sol = y[t] /.
  DSolve[{eqn /. {a → -1, b → -1, c → 1}, y[0] == 2, y'[0] == 0}, y[t], t][[1]]
```

$$\frac{1}{3} e^{-t/2} \left(3 e^{t/2} + 3 \cos\left[\frac{\sqrt{3}\,t}{2}\right] + \sqrt{3} \sin\left[\frac{\sqrt{3}\,t}{2}\right]\right)$$

■ Example 3: A Set of Trajectories

We continue the last example by varying `y[0]`:

```
sol =
  y[t] /. DSolve[{eqn /. {a → -1, b → -1, c → 1}, y[0] == α, y'[0] == 0}, y[t],
      t][[1]] // Simplify
```

$$\frac{1}{3} e^{-t/2} \left(3 e^{t/2} + 3 (-1 + α) \cos\left[\frac{\sqrt{3}\,t}{2}\right] + \sqrt{3} (-1 + α) \sin\left[\frac{\sqrt{3}\,t}{2}\right]\right)$$

```
solset = Table[sol, {α, 0, 5}];
```

```
Plot[Evaluate[solset], {t, 0, 10}, PlotRange → All];
```

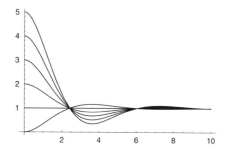

23.1.3 Simultaneous Equations

In simultaneous equations, we have several dependent variables. Here are some typical commands for two equations. They generalize easily to more equations and to different initial and boundary conditions.

> `vars = {x[t], y[t]}` Define the dependent variables
>
> `eqns = {eqn1, eqn2}` Define the differential equations
>
> `inits = {x[a] == α, y[a] == β}` Define the initial (or boundary) conditions
>
> `sol = vars /. DSolve[eqns, vars, t]` Give the general solution
>
> `sol = vars /. DSolve[Join[eqns, inits], vars, t]` Solve an initial value problem
>
> `Plot[Evaluate[sol], {t, a, b}]` Plot $x[t]$ and $y[t]$
>
> `ParametricPlot[sol, {t, a, b}]` Plot a phase trajectory

Note that we can also solve differential-algebraic equations (❀5); an example:

```
DSolve[{x'[t] == 2 x[t] + y[t], x[t] + y[t] == 1, x[0] == 1}, {x[t], y[t]}, t]
```

$$\{\{x[t] \to -1 + 2\, e^t,\ y[t] \to -2\,(-1 + e^t)\}\}$$

■ Example 1: Basic Techniques

First we ask for a general solution:

```
vars = {x[t], y[t]};

eqns = {x'[t] == y[t], y'[t] == x[t]};

DSolve[eqns, vars, t] // Simplify
```

$$\left\{\left\{x[t] \to \frac{1}{2}\, e^{-t}\, ((1 + e^{2t})\, C[1] + (-1 + e^{2t})\, C[2]),\right.\right.$$
$$\left.\left. y[t] \to \frac{1}{2}\, e^{-t}\, ((-1 + e^{2t})\, C[1] + (1 + e^{2t})\, C[2])\right\}\right\}$$

We may want to directly ask for the values of `x[t]` and `y[t]`:

```
vars /. DSolve[eqns, vars, t][[1]] // Simplify
```

$$\left\{\frac{1}{2}\, e^{-t}\, ((1 + e^{2t})\, C[1] + (-1 + e^{2t})\, C[2]),\right.$$
$$\left.\frac{1}{2}\, e^{-t}\, ((-1 + e^{2t})\, C[1] + (1 + e^{2t})\, C[2])\right\}$$

The solution of this constant coefficient system can also be obtained using the matrix exponential:

```
MatrixExp[{{0, 1}, {1, 0}} t].{C[1], C[2]} // Simplify
```

$$\left\{\frac{1}{2}\, e^{-t}\, ((1 + e^{2t})\, C[1] + (-1 + e^{2t})\, C[2]),\right.$$
$$\left.\frac{1}{2}\, e^{-t}\, ((-1 + e^{2t})\, C[1] + (1 + e^{2t})\, C[2])\right\}$$

Next we solve and plot an initial value problem:

```
inits = {x[0] == 1, y[0] == 0};

sol = vars /. DSolve[Join[eqns, inits], vars, t][[1]]
```

$$\left\{\frac{1}{2}\, e^{-t}\,(1 + e^{2t}),\ \frac{1}{2}\, e^{-t}\,(-1 + e^{2t})\right\}$$

```
Plot[Evaluate[sol], {t, 0, 2}, PlotStyle → {{}, AbsoluteDashing[{2}]}];
```

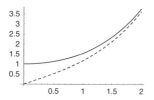

■ Example 2: Phase Trajectories

Consider the following linear system:

```
vars = {x[t], y[t]};

eqns = {x'[t] == x[t] - y[t], y'[t] == 3 x[t] - 2 y[t]};

inits = {x[0] == 10, y[0] == 0};

sol = vars /. DSolve[Join[eqns, inits], vars, t][[1]]
```

$$\left\{10\, e^{-t/2}\left(\text{Cos}\left[\frac{\sqrt{3}\ t}{2}\right] + \sqrt{3}\ \text{Sin}\left[\frac{\sqrt{3}\ t}{2}\right]\right),\ 20\,\sqrt{3}\ e^{-t/2}\ \text{Sin}\left[\frac{\sqrt{3}\ t}{2}\right]\right\}$$

(In versions of *Mathematica* prior to 5, this solution can be obtained by simplifying the result with **ComplexExpand** and **Simplify**). Plot the solution:

```
Plot[Evaluate[sol], {t, 0, 10}, PlotStyle → {{}, AbsoluteDashing[{2}]}];
```

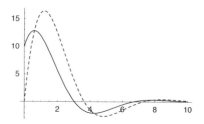

With **ParametricPlot**, we can plot phase trajectories for equations with two dependent variables. A plot of this kind describes how the point $(x(t), y(t))$ moves on the (x, y) plane:

```
ParametricPlot[sol, {t, 0, 30}, PlotRange → All, AxesLabel → {x, y}];
```

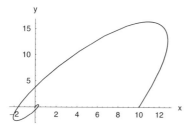

The curve approaches the point (0, 0) like a spiral (origin is a stable focus). Plotting individual points shows the speed of the point as it moves on a curve:

```
p = Table[sol, {t, 0, 20, 0.2}];
```

```
ListPlot[p, PlotRange → All, AxesLabel → {x, y}];
```

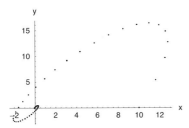

For a linear matrix system $y' = A\,y$, the origin is an equilibrium point. The nature of this point can be seen from the eigenvalues of the coefficient matrix A:

```
Eigenvalues[{{1, -1}, {3, -2}}] // ComplexExpand
```

$$\left\{-\frac{1}{2} + \frac{i\sqrt{3}}{2}, \, -\frac{1}{2} - \frac{i\sqrt{3}}{2}\right\}$$

Because the eigenvalues are complex with a negative real part, origin is a stable focus.

■ Example 3: A Set of Trajectories

Plotting a direction field gives an impression about how the trajectories behave:

```
<< Graphics`PlotField`
```

```
PlotVectorField[{x - y, 3 x - 2 y}, {x, -20, 20}, {y, -30, 30},
    PlotPoints → 11, Axes → True, AspectRatio → 1 / GoldenRatio];
```

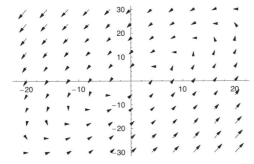

However, plotting several trajectories from different starting points gives a better description of the behavior of the system. We consider the system of Example 2 and give general initial values **x0** and **y0**:

```
inits = {x[0] == x0, y[0] == y0};
```

```
sol = vars /. DSolve[Join[eqns, inits], vars, t][[1]] // Simplify
```

$$\left\{\frac{1}{3}\,e^{-t/2}\left(3\,x0\,\text{Cos}\left[\frac{\sqrt{3}\,t}{2}\right] + \sqrt{3}\,(3\,x0 - 2\,y0)\,\text{Sin}\left[\frac{\sqrt{3}\,t}{2}\right]\right),\right.$$

$$\left. e^{-t/2}\left(y0\,\text{Cos}\left[\frac{\sqrt{3}\,t}{2}\right] + \sqrt{3}\,(2\,x0 - y0)\,\text{Sin}\left[\frac{\sqrt{3}\,t}{2}\right]\right)\right\}$$

(In *Mathematica* 4, use **//ComplexExpand //Simplify**.) When **y0** is 30, we let **x0** vary from –6 to 14 in steps of 2; when **y0** is –30, we let **x0** vary from –14 to 6 in steps of 2:

```
solset1 = Table[sol, {x0, -6, 14, 2}];

p1 = ParametricPlot[Evaluate[solset1 /. y0 → 30],
    {t, 0, 15}, PlotStyle → AbsoluteThickness[0.25]];
```

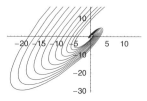

```
solset2 = Table[sol, {x0, -14, 6, 2}];

p2 = ParametricPlot[Evaluate[solset2 /. y0 → -30],
    {t, 0, 15}, PlotStyle → AbsoluteThickness[0.25]];
```

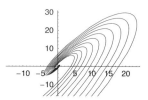

The combined plot is displayed as follows:

```
Show[p1, p2, PlotRange → All];
```

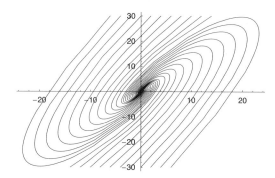

■ Example 4: Three Equations

Let x, y, and z stand for the amount of lead in blood, tissues, and bones. Define $v = (x, y, z)^t$. Borrelli and Coleman (1998, p. 339) present the model $v' = A v + b$, in which the coefficients are as follows:

```
A = {{-0.0361, 0.0124, 0.000035},
    {0.0111, -0.0286, 0}, {0.0039, 0, -0.000035}};

b = {49.3, 0, 0};
```

The equilibrium point is the v that satisfies $A v + b = 0$, that is, $A v = -b$:

```
LinearSolve[A, -b]
```

```
{1800.1, 698.639, 200582.}
```

To solve the system of differential equations, define the variables and the equations:

```
vars = {x[t], y[t], z[t]};
```

```
eqns = Thread[D[vars, t] == A.vars + b]
```

$\{x'[t] == 49.3 - 0.0361 x[t] + 0.0124 y[t] + 0.000035 z[t],$
$y'[t] == 0.0111 x[t] - 0.0286 y[t], z'[t] == 0.0039 x[t] - 0.000035 z[t]\}$

```
inits = {x[0] == 0, y[0] == 0, z[0] == 0};
```

Time is measured in days. Then solve the system:

```
sol = vars /. DSolve[Join[eqns, inits], vars, t] // Expand // N // Chop
```

$\{\{1800.1 - 719.885\ 2.71828^{-0.0446688\ t} -$
$855.314\ 2.71828^{-0.0200356\ t} - 224.898\ 2.71828^{-0.0000306322\ t},$
$698.639 + 497.283\ 2.71828^{-0.0446688\ t} - 1108.54\ 2.71828^{-0.0200356\ t} -$
$87.3791\ 2.71828^{-0.0000306322\ t}, 200582. + 62.902\ 2.71828^{-0.0446688\ t} +$
$166.781\ 2.71828^{-0.0200356\ t} - 200812.\ 2.71828^{-0.0000306322\ t}\}\}$

(Here we applied **Chop** to get rid of some negligible terms that originated from rounding errors.) The solution can be written as follows:

$$\begin{pmatrix} x(t) \\ y(t) \\ z(t) \end{pmatrix} = \begin{pmatrix} 1800.1 \\ 698.639 \\ 200582. \end{pmatrix} + \begin{pmatrix} -719.885 & -855.314 & -224.898 \\ 497.283 & -1108.54 & -87.3791 \\ 62.902 & 166.781 & -200812. \end{pmatrix} \begin{pmatrix} e^{-0.0446688\ t} \\ e^{-0.0200356\ t} \\ e^{-0.0000306322\ t} \end{pmatrix}$$

The solution is displayed as follows:

```
Plot[Evaluate[sol], {t, 0, 400},
    PlotStyle → {{}, AbsoluteDashing[{2}], GrayLevel[0.5]}];
```

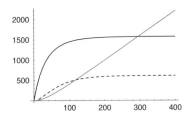

The amounts of lead in blood and tissues reach the equilibrium quite rapidly (in about a year), but it takes a very long time (several hundred years) before the amount of lead in bones would reach the equilibrium:

```
Plot[Evaluate[sol[[1, 3]]], {t, 0, 200000}, PlotStyle → GrayLevel[0.5],
    PlotRange → All, Ticks → {{100000, 200000}, Automatic}];
```

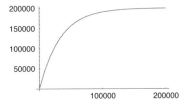

23.2 More about Symbolic Solutions

23.2.1 Using the Laplace Transform

■ One Equation

We try to solve, using the Laplace transform (see Section 17.4.1), the following problem:

```
eqn = y''[t] + y[t] == t;
```

```
inits = {y[0] → α, y'[0] → β};
```

First, take the Laplace transform of the equation:

```
LaplaceTransform[eqn, t, s]
```

```
LaplaceTransform[y[t], t, s] +
```
$$s^2 \text{ LaplaceTransform}[y[t], t, s] - s\, y[0] - y'[0] == \frac{1}{s^2}$$

Insert the initial values:

```
lapeqn = % /. inits
```
$$-s\,\alpha - \beta + \text{LaplaceTransform}[y[t], t, s] +$$
$$s^2 \text{ LaplaceTransform}[y[t], t, s] == \frac{1}{s^2}$$

From this equation, solve the transform:

```
lap = Solve[lapeqn, LaplaceTransform[y[t], t, s]]
```
$$\left\{\left\{\text{LaplaceTransform}[y[t], t, s] \to \frac{1 + s^3\,\alpha + s^2\,\beta}{s^2\,(1 + s^2)}\right\}\right\}$$

Lastly, take the inverse transform, which is then the solution to the initial value problem:

```
sol = InverseLaplaceTransform[lap, s, t]
```
$$\{\{y[t] \to t + \alpha \cos[t] + (-1 + \beta) \sin[t]\}\}$$

The same solution is obtained with **DSolve**:

```
DSolve[{eqn, y[0] == α, y'[0] == β}, y[t], t] // Simplify
```
$$\{\{y[t] \to t + \alpha \cos[t] + (-1 + \beta) \sin[t]\}\}$$

■ Simultaneous Equations

Now we solve the same initial value problem as was seen in Example 2 of Section 23.1.3, p. 600:

```
eqns = {y'[t] == y[t] - z[t], z'[t] == 3 y[t] - 2 z[t]};
```

```
inits = {y[0] → 10, z[0] → 0};
```

```
lapeqns = LaplaceTransform[eqns, t, s] /. inits
```
```
{-10 + s LaplaceTransform[y[t], t, s] ==
   LaplaceTransform[y[t], t, s] - LaplaceTransform[z[t], t, s],
 s LaplaceTransform[z[t], t, s] ==
   3 LaplaceTransform[y[t], t, s] - 2 LaplaceTransform[z[t], t, s]}
```

```
lap = Solve[lapeqns,
  {LaplaceTransform[y[t], t, s], LaplaceTransform[z[t], t, s]}]
```

$$\left\{\left\{\text{LaplaceTransform}[y[t], t, s] \rightarrow \frac{10\,(2 + s)}{1 + s + s^2},\right.\right.$$

$$\left.\left.\text{LaplaceTransform}[z[t], t, s] \rightarrow \frac{30}{1 + s + s^2}\right\}\right\}$$

```
sol = InverseLaplaceTransform[lap, s, t]
```

$$\left\{\left\{y[t] \rightarrow 10\,e^{-t/2}\left(\text{Cos}\left[\frac{\sqrt{3}\,t}{2}\right] + \sqrt{3}\,\text{Sin}\left[\frac{\sqrt{3}\,t}{2}\right]\right),\right.\right.$$

$$\left.\left.z[t] \rightarrow 20\,\sqrt{3}\,e^{-t/2}\,\text{Sin}\left[\frac{\sqrt{3}\,t}{2}\right]\right\}\right\}$$

The solution is the same as the one obtained in Section 23.1.3.

■ The Step Function

Consider the following equation:

```
eqn = y'[t] == r y[t] - h UnitStep[30 - t];
```

It describes an exponential growth in which a harvesting of h units per time unit occurs for $0 \le t \le 30$. Apply the Laplace transform:

```
lapeqn = LaplaceTransform[eqn, t, s] /. y[0] → α
```

$$-\alpha + s\,\text{LaplaceTransform}[y[t], t, s] ==$$
$$-\frac{(1 - e^{-30\,s})\,h}{s} + r\,\text{LaplaceTransform}[y[t], t, s]$$

```
lap = Solve[lapeqn, LaplaceTransform[y[t], t, s]]
```

$$\left\{\left\{\text{LaplaceTransform}[y[t], t, s] \rightarrow \frac{e^{-30\,s}\,(h - e^{30\,s}\,h + e^{30\,s}\,s\,\alpha)}{s\,(-r + s)}\right\}\right\}$$

```
lapsol = y[t] /. InverseLaplaceTransform[lap, s, t][[1]]
```

$$\frac{h - e^{r\,t}\,h + e^{r\,t}\,r\,\alpha + (-1 + e^{r\,(-30+t)})\,h\,\text{UnitStep}[-30 + t]}{r}$$

```
Plot[lapsol /. {α → 100, r → 0.01, h → 2}, {t, 0, 100}];
```

23.2.2 Series Solutions

■ Infinite Number of Terms

We have already considered series solutions in Section 16.2.3, p. 428. Now we consider the equation $y''(t) - 2\,t\,y'(t) - 2\,y(t) = 0$. Let us try to find the solution in the form $y(t) = \sum_{i=0}^{\infty} a_i\,t^i$ by inserting this into the equation. First note the following:

$$t\,y'(t) = \sum_{i=1}^{\infty} i\,a_i\,t^i = \sum_{i=0}^{\infty} i\,a_i\,t^i, \quad y''(t) = \sum_{i=2}^{\infty} i(i-1)\,a_i\,t^{i-2} = \sum_{i=0}^{\infty} (i+2)(i+1)\,a_{i+2}\,t^i.$$

The equation then becomes $\sum_{i=0}^{\infty} [(i+2)(i+1)\,a_{i+2} - 2\,i\,a_i - 2\,a_i]\,t^i = 0$. So, the coefficients a_i satisfy the difference equation $(i+2)(i+1)\,a_{i+2} - 2(i+1)\,a_i = 0$, that is, $a_i = \frac{2}{i}\,a_{i-2}$. In addition, $y(0) = a_0$ and $y'(0) = a_1$.

Assume first that $y(0) = 1$ and $y'(0) = 0$ so that $a_0 = 1$ and $a_1 = 0$. It is easy to see that the solution of $a_i = \frac{2}{i}\,a_{i-2}$ is $a_i = 1/(\frac{i}{2})!$ for i even and $a_i = 0$ for i odd (we could also try **RSolve** when solving the difference equation [see Section 25.1.1], but in this example **RSolve** of *Mathematica* 5 was not able to give a useful answer). So, we get the series solution $y(t) = \sum_{i\,even} t^i\,/(i/2)!$. The value of the sum is as follows:

```
sol = Sum[t^i / (i / 2) !, {i, 0, ∞, 2}]
```
e^{t^2}

Thus, $y(t) = e^{t^2}$ is the solution of the problem.

Assume then that $y(0) = 0$ and $y'(0) = 1$ so that $a_0 = 0$ and $a_1 = 1$. From $a_i = \frac{2}{i}\,a_{i-2}$, it is easy to see that $a_{2i+1} = 2^i\,/(2i+1)!!$, $i = 0, 1, 2, \ldots$, and other a_i values are zero. So, we know the series expansion of $y(t)$, but *Mathematica* is not able to calculate the infinite sum:

```
sol = Sum[2^i / (2 i + 1) !! t^ (2 i + 1), {i, 0, ∞}]
```
$\displaystyle\sum_{i=0}^{\infty} \frac{2^i\,t^{1+2\,i}}{(1+2\,i)\,!!}$

However, if we are satisfied with a finite series, we can get an approximate solution, as is seen in the next two examples.

■ A Finite Number of Terms 1

We continue with the preceding example and calculate an approximate solution as a finite sum by using the formula $a_{2i+1} = 2^i\,/(2i+1)!!$:

```
apprsol = Sum[2^i / (2 i + 1) !! t^ (2 i + 1), {i, 0, 5}]
```
$t + \dfrac{2\,t^3}{3} + \dfrac{4\,t^5}{15} + \dfrac{8\,t^7}{105} + \dfrac{16\,t^9}{945} + \dfrac{32\,t^{11}}{10395}$

We can also directly use the difference equation $a_i = \frac{2}{i}\,a_{i-2}$ of the coefficients by first defining the equation:

```
a[0] = 0; a[1] = 1;

a[i_] := a[i] = 2 a[i - 2] / i
```

Then we calculate a finite sum:

```
apprsol = Sum[a[i] t^i, {i, 0, 12}]
```
$t + \dfrac{2\,t^3}{3} + \dfrac{4\,t^5}{15} + \dfrac{8\,t^7}{105} + \dfrac{16\,t^9}{945} + \dfrac{32\,t^{11}}{10395}$

Before we continue, we have to remove the definition of the coefficients:

```
Remove[a]
```

■ A Finite Number of Terms 2

If we do not know the difference equation of the coefficients, we can proceed as follows. First, we define the differential equation and the initial conditions (it is now advantageous to use **&&** rather than a list, because later on we will use **LogicalExpand**):

> **eqn = y''[t] – 2 t y'[t] – 2 y[t] == 0 && y[0] == 0 && y'[0] == 1;**

Before we find an approximate series solution, we check whether **DSolve** is able to solve the equation:

> **sol = y[t] /. DSolve[eqn, y[t], t][[1]]**

$$\frac{1}{2} e^{t^2} \sqrt{\pi} \, \text{Erf}[t]$$

Then we try to find an approximate solution in the form of a twelfth-degree power series:

> **y[t_] = Sum[a[i] t^i, {i, 0, 12}] + O[t]^13**

$a[0] + a[1] \, t + a[2] \, t^2 + a[3] \, t^3 + a[4] \, t^4 + a[5] \, t^5 + a[6] \, t^6 +$
$a[7] \, t^7 + a[8] \, t^8 + a[9] \, t^9 + a[10] \, t^{10} + a[11] \, t^{11} + a[12] \, t^{12} + O[t]^{13}$

The equation is now as follows:

> **eqn**

$(-2\,a[0] + 2\,a[2]) + (-4\,a[1] + 6\,a[3])\,t +$
$\quad (-6\,a[2] + 12\,a[4])\,t^2 + (-8\,a[3] + 20\,a[5])\,t^3 + (-10\,a[4] + 30\,a[6])\,t^4 +$
$\quad (-12\,a[5] + 42\,a[7])\,t^5 + (-14\,a[6] + 56\,a[8])\,t^6 + (-16\,a[7] + 72\,a[9])\,t^7 +$
$\quad (-18\,a[8] + 90\,a[10])\,t^8 + (-20\,a[9] + 110\,a[11])\,t^9 +$
$\quad (-22\,a[10] + 132\,a[12])\,t^{10} + O[t]^{11} == 0 \text{ && } a[0] == 0 \text{ && } a[1] == 1$

We could find the conditions under which the equation is true:

> **cond = LogicalExpand[eqn]**

$a[0] == 0 \text{ && } a[1] == 1 \text{ && } -2\,a[0] + 2\,a[2] == 0 \text{ && } -4\,a[1] + 6\,a[3] == 0 \text{ && }$
$\quad -6\,a[2] + 12\,a[4] == 0 \text{ && } -8\,a[3] + 20\,a[5] == 0 \text{ && } -10\,a[4] + 30\,a[6] == 0 \text{ && }$
$\quad -12\,a[5] + 42\,a[7] == 0 \text{ && } -14\,a[6] + 56\,a[8] == 0 \text{ && } -16\,a[7] + 72\,a[9] == 0 \text{ && }$
$\quad -18\,a[8] + 90\,a[10] == 0 \text{ && } -20\,a[9] + 110\,a[11] == 0 \text{ && } -22\,a[10] + 132\,a[12] == 0$

Then we could solve the conditions with **Solve[cond]**, but actually **Solve** can be applied directly to **eqn**:

> **aa = Solve[eqn]**

$\{\{a[0] \to 0,\ a[1] \to 1,\ a[2] \to 0,\ a[3] \to \frac{2}{3},\ a[4] \to 0,\ a[5] \to \frac{4}{15},\ a[6] \to 0,$
$\quad a[7] \to \frac{8}{105},\ a[8] \to 0,\ a[9] \to \frac{16}{945},\ a[10] \to 0,\ a[11] \to \frac{32}{10395},\ a[12] \to 0\}\}$

The corresponding approximate series solution of the original equation is as follows:

> **apprsol = y[t] /. aa[[1]]**

$t + \frac{2\,t^3}{3} + \frac{4\,t^5}{15} + \frac{8\,t^7}{105} + \frac{16\,t^9}{945} + \frac{32\,t^{11}}{10395} + O[t]^{13}$

This is exactly the same as the series expansion of the exact solution:

```
Series[sol, {t, 0, 12}]
```

$$t + \frac{2\,t^3}{3} + \frac{4\,t^5}{15} + \frac{8\,t^7}{105} + \frac{16\,t^9}{945} + \frac{32\,t^{11}}{10395} + O[t]^{13}$$

From a figure, we see that the approximate solution is quite good near the origin:

```
Plot[Evaluate[{apprsol // Normal, sol}],
  {t, 0, 2}, PlotStyle → {GrayLevel[0.5], {}}];
```

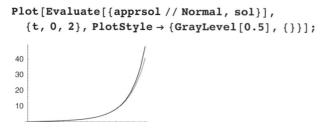

```
Remove["Global`*"]
```

23.2.3 Solution as a Pure Function

> `sol = y /. DSolve[eqn, y, t]`   Give the solution in the form of a pure function
> `sol[a]`   Calculate the value of the solution at **a**
> `eqn /. y → sol`   Check the solution
> `Plot[sol[t], {t, a, b}]`   Plot the solution

Thus far we have used **DSolve** in the form **DSolve[eqn, y[t], t]**. **DSolve** also has another form, where the dependent variable is not declared as **y[t]** but as **y**. So, we can also write **DSolve[eqn, y, t]**. The solution is then expressed as a function—more specifically, a pure function (see Section 14.1.5, p. 359). As an example, we define an equation:

```
eqn = {y'[t] == y[t] + t + 1, y[0] == 1};
```

We calculate the solution in both ways:

```
DSolve[eqn, y[t], t]     {{y[t] → -2 + 3 e^t - t}}
```

```
DSolve[eqn, y, t]     {{y → Function[{t}, -2 + 3 e^t - t]}}
```

It may seem as if the latter form of solution would not be as nice as the former. In fact, the latter form has a clear advantage in that we can use the solution similarly as we can other functions. Next we compare the methods in more detail.

First we declare the dependent variable to be **y[t]**:

```
sol = y[t] /. DSolve[eqn, y[t], t][[1]]     -2 + 3 e^t - t
```

A value calculation, differentiation, and checking of the correctness of the solution are done as follows:

```
sol /. t → 2     -4 + 3 e^2
```

```
D[sol, t]     -1 + 3 e^t
```

```
eqn /. {y[t] → sol, y'[t] → D[sol, t], y[0] → sol /. t → 0}
```

```
{True, True}
```

When checking correctness, we inserted the solution into the equation and into the initial condition. The result of the checking shows that both the equation and the initial condition are satisfied.

Then we declare the dependent variable to be **y**:

> **sol = y /. DSolve[eqn, y, t][[1]]**   Function[{t}, -2 + 3 e^t - t]

Now we can use **sol** in the same way as any other function of *Mathematica*, like **Sin** or **Exp**. We calculate a value and derivative and check the correctness of the solution as follows:

> **sol[2]** $-4 + 3 e^2$

> **sol'[t]** $-1 + 3 e^t$

> **eqn /. y → sol**   {True, True}

The checking in particular is very easy with this latter method.

23.2.4 Integral and Integro-Differential Equations

▪ Volterra Integral Equation

The following equation is a Volterra integral equation of the second kind:

> **eqn = y[t] == t^2 + Integrate[Sin[t - u] y[u], {u, 0, t}]**

$$y[t] == t^2 + \int_0^t \text{Sin}[t - u] \, y[u] \, du$$

This can be solved using Laplace transform (❀4!):

> **Solve[LaplaceTransform[eqn, t, s],LaplaceTransform[y[t],t,s]]**

$$\left\{\left\{\text{LaplaceTransform}[y[t], t, s] \to \frac{2 (1 + s^2)}{s^5}\right\}\right\}$$

> **sol = y[t] /. InverseLaplaceTransform[%, s, t][[1]] // Simplify**

$$t^2 + \frac{t^4}{12}$$

We perform a check:

> **eqn /. {y[t] → sol, y[u] → (sol /. t → u)}**   True

> **sol =.**

▪ Fredholm Integral Equation

The next equation is a Fredholm equation of the second kind:

> **eqn = y[t] == 2 + 3 t + 4 Integrate[(1 + t u^2 + t^2 u + t^3) y[u], {u, 0, 1}]**

$$y[t] == 2 + 3 t + 4 \int_0^1 (1 + t^3 + t^2 u + t u^2) \, y[u] \, du$$

We guess (from the form of the equation) that the solution is of the following form:

> **sol[t_] := 2 + 3 t + 4 (a + b t + c t^2 + d t^3)**

We then insert the guess into the equation:

```
eqn /. y → sol
```

$$2 + 3\,t + 4\,(a + b\,t + c\,t^2 + d\,t^3) ==$$
$$2 + 3\,t + 4\,\left(\frac{7}{2} + \frac{4\,c}{3} + d + \frac{17\,t}{12} + \frac{4\,c\,t}{5} + \frac{2\,d\,t}{3} + 2\,t^2 + c\,t^2 + \frac{4\,d\,t^2}{5} + \frac{7\,t^3}{2} + \right.$$
$$\left. \frac{4\,c\,t^3}{3} + d\,t^3 + b\,\left(2 + t + \frac{4\,t^2}{3} + 2\,t^3\right) + a\,\left(4 + \frac{4\,t}{3} + 2\,t^2 + 4\,t^3\right)\right)$$

We solve the coefficients that satisfy the equation identically (for **SolveAlways**, see Section 19.2.2, p. 495):

```
coeff = SolveAlways[%, t]
```

$$\left\{\left\{a \to -\frac{335}{636},\ b \to -\frac{167}{424},\ c \to -\frac{385}{848},\ d \to -\frac{335}{636}\right\}\right\}$$

So, the solution is as follows:

```
sol2 = sol[t] /. coeff[[1]]
```

$$2 + 3\,t + 4\,\left(-\frac{335}{636} - \frac{167\,t}{424} - \frac{385\,t^2}{848} - \frac{335\,t^3}{636}\right)$$

We check that the solution is correct:

```
eqn /. {y[t] → sol2, y[u] → (sol2 /. t → u)} // Simplify
True
```

■ Integro-Differential Equation

Define an integro-differential equation:

```
eqn = y[t] + y'[t] == 1 - Integrate[Exp[t - u] y[u], {u, a, t}]
```

$$y[t] + y'[t] == 1 - \int_a^t e^{t-u}\,y[u]\,du$$

It has the initial condition $y(a) = c$. Differentiate the equation:

```
eqn2 = D[eqn, t]
```

$$y'[t] + y''[t] == -\int_a^t e^{t-u}\,y[u]\,du - y[t]$$

Eliminate the integral from the two equations (for **Eliminate**, see Section 19.2.2, p. 494):

```
eqn3 = Eliminate[{eqn, eqn2}, Integrate[Exp[t - u] y[u], {u, a, t}]]
```

$$y''[t] == -1$$

We arrived at a differential equation. To get a second initial condition, substitute a for t in the original equation:

```
cond = eqn /. t → a          y[a] + y'[a] == 1
```

The solution of the initial value problem is as follows:

```
sol = y /. DSolve[{eqn3, y[a] == c, cond}, y, t][[1]]
```

$$\text{Function}\left[\{t\},\ \frac{1}{2}\,(-2\,a - a^2 + 2\,c + 2\,a\,c + 2\,t + 2\,a\,t - 2\,c\,t - t^2)\right]$$

```
Collect[sol[t], t, Factor]
```

$$\frac{1}{2}\,(-2\,a - a^2 + 2\,c + 2\,a\,c) + (1 + a - c)\,t - \frac{t^2}{2}$$

This solves the original equation:

```
eqn /. y → sol // Simplify     True
```

23.3 Numerical Solutions

23.3.1 One Equation

Here are typical commands that are used when we numerically solve one differential equation with initial or boundary conditions.

> ```
> sol = y[t] /. NDSolve[{eqn, conds}, y[t], {t, a, b}]   Solve the problem
> Plot[sol, {t, a, b}]   Plot the solution
> ```

The arguments of **NDSolve** are otherwise the same as the ones of **DSolve** except that, in place of **t**, we have to define the interval **{t, a, b}**, where we ask for the numerical solution.

The conditions in **{eqn, conds}** are often initial conditions, but they also can be boundary conditions. The conditions need not be given at the endpoints of the solution interval. For example, if you ask for the solution in $(0, 3)$, you can state all conditions at $t = 1$ or some conditions at $t = 1$ and other conditions at $t = 2$. Generally, if the highest order derivative of the equation is n, a total of n conditions must be given so that the solution can be computed. In the simplest case, the conditions give values of $y(a)$, $y'(a)$, ..., $y^{(n-1)}(a)$.

In boundary value problems, the equation has to be linear in the dependent variable. For example, a linear second-order equation is of the form $p(t)\,y''(t) + q(t)\,y'(t) + r(t)\,y(t) = s(t)$. Nonlinear boundary value problems (containing, for example, $y(t)^2$) are not solved.

NDSolve normally uses an adaptive Adams predictor-corrector method; the methods are considered in more detail in Section 23.4.1, p. 627. The basic principle of the solution method is that the solution is computed at a finite set of t-points, and piecewise interpolation is used to calculate the values of the solution at other points. Note that options of **NDSolve** are considered in Section 23.4.1.

■ Example 1: A First-Order Initial Value Problem

We solve a first-order nonlinear equation:

```
eqn = 1000 y[t] y '[t] == 160 - 3 y[t]^3;

sol = NDSolve[{eqn, y[0] == 0.00001}, y[t], {t, 0, 150}]

{{y[t] → InterpolatingFunction[{{0., 150.}}, <>][t]}}
```

The solution is an **InterpolatingFunction** object (see Section 21.2.1, p. 555). The object represents the solution as a collection of cubic interpolating polynomials. Only the interval of the solution is shown (**{0., 150.}** here); all other information is hidden behind the marks <>.

It may be preferable to ask directly for the value of `y[t]`:

```
sol = y[t] /. NDSolve[{eqn, y[0] == 0.00001}, y[t], {t, 0, 150}]
```

```
{InterpolatingFunction[{{0., 150.}}, <>][t]}
```

If there is only one solution, we may want to get rid of the curly braces by asking for the first component of the solution:

```
sol = y[t] /. NDSolve[{eqn, y[0] == 0.00001}, y[t], {t, 0, 150}][[1]]
```

```
InterpolatingFunction[{{0., 150.}}, <>][t]
```

The solution looks like this:

```
Plot[sol, {t, 0, 150}];
```

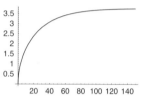

Next we can ask for the value of the solution at a point and tabulate the solution at a set of points:

```
sol /. t → 100        3.6752
```

```
Table[{t, sol}, {t, 1, 4}]
```

```
{{1, 0.565302}, {2, 0.798468}, {3, 0.976357}, {4, 1.12527}}
```

We can plot the solution for several initial values:

```
solset = Table[y[t] /. NDSolve[{eqn, y[0] == y0}, y[t], {t, 0, 150}],
    {y0, 0.00001, 6.0001, 0.5}];
```

```
Plot[Evaluate[solset], {t, 0, 150}, PlotRange → All];
```

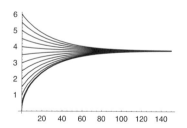

■ Example 2: A Second-Order Initial Value Problem

Consider the following second-order equation (the first Painlevé transcendent):

```
eqn = y''[t] == y[t]^2 - t;
```

Solve the equation when $y(0) = 0$ and $y'(0)$ varies from –3.4 to 1 in steps of 0.25:

```
solset =
    Table[y[t] /. NDSolve[{eqn, y[0] == 0, y'[0] == d0}, y[t], {t, 0, 12}][[1]],
    {d0, -3.4, 1, 0.25}];
```

```
Plot[Evaluate[solset], {t, 0, 12}];
```

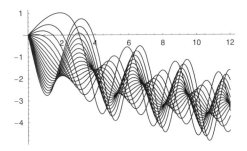

■ Example 3: A Second-Order Boundary Value Problem

Now we give boundary conditions:

```
eqn = y''[t] == y'[t] - t y[t]
```

$$y''[t] == -t\,y[t] + y'[t]$$

```
sol = y[t] /. NDSolve[{eqn, y[0] == 2, y[4] == 2}, y[t], {t, 0, 4}];
```

```
Plot[sol, {t, 0, 4}];
```

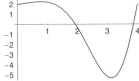

Boundary value problems are also considered in Section 23.4.3, p. 632.

■ Example 4: Several Solutions

More than one solution can be obtained:

```
sol = y[t] /. NDSolve[{y'[t]^2 == t, y[0] == 0}, y[t], {t, 0, π}]
```

```
{InterpolatingFunction[{{0., 3.14159}}, <>][t],
 InterpolatingFunction[{{0., 3.14159}}, <>][t]}
```

```
Plot[Evaluate[sol], {t, 0, π}];
```

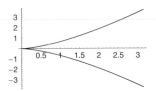

■ Example 5: Piecewise-Defined Equations

Consider the following equation:

```
eqn = y'[t] == 0.01 y[t] - 2 UnitStep[30 - t];
```

It describes an exponential growth in which a harvesting of h units per time unit occurs for $0 \le t \le 30$. In place of **2 UnitStep[30 - t]**, we could also write **If[t ≤ 30, 2, 0]**. We could solve the equation with the Laplace transform (see Section 23.2.1, p. 605), but now we solve the equation numerically:

```
sol = y[t] /. NDSolve[{eqn, y[0] == 100}, y[t], {t, 0, 100}];

Plot[sol, {t, 0, 100}];
```

■ Solution as a Pure Function

> **sol = y /. NDSolve[{eqn, conds}, y, {t, a, b}]**   Give the solution in the form of a
>   pure function
> **sol[a]**   Calculate the value of the solution at **a**
> **Plot[sol[t], {t, a, b}]**   Plot the solution

Thus far, we have used **NDSolve** in the form **DSolve[{eqn, conds}, y[t], {t, a, b}]**. **NDSolve**, like **DSolve** (see Section 23.2.3, p. 608), also has another form in which the dependent variable is not declared as **y[t]** but as **y**. The solution is then expressed as a function, more exactly as a pure function (see Section 14.1.5, p. 359). As an example, define an equation:

```
eqn = y'[t] == -t y[t] + Exp[-t];
```

Ask for the solution as a pure function:

```
sol = y /. NDSolve[{eqn, y[0] == 1}, y, {t, 0, 5}][[1]]

InterpolatingFunction[{{0., 5.}}, <>]
```

Now **sol** can be used like other functions. We calculate a value and plot the solution:

```
sol[1]      1.04613

Plot[sol[t], {t, 0, 5}];
```

■ Delay-Differential Equations

For delay-differential equations, look at http://library.wolfram.com/database/Math-Source/725/. There, you can download a package defining **NDelayDSolve**.

23.3.2 Two Equations

`vars = {x[t], y[t]}` Define the dependent variables

`eqns = {eqn1, eqn2}` Define the differential equations

`inits = {x[a] == α, y[a] == β}` Define the initial (or boundary) conditions

`sol = vars /. NDSolve[Join[eqns, inits], vars, {t, a, b}]` Solve the problem

`Plot[Evaluate[sol], {t, a, b}]` Plot `x[t]` and `y[t]`

`ParametricPlot[sol, {t, a, b}]` Plot a phase trajectory

We present two examples of systems of two nonlinear differential equations. For more about the examples, see Borrelli & Coleman (1998, p. 276–298). For more about studying systems of nonlinear equations with *Mathematica*, see Murrell (1994).

■ A Predator–Prey Model

Define the predator–prey model by Lotka and Volterra:

```
vars = {x[t], y[t]};
```

```
f = x[t] (p - q y[t]);
g = y[t] (-P + Q x[t]);
```

```
eqns = {x'[t] == f, y'[t] == g}
```

$\{x'[t] == x[t] (p - q y[t]), y'[t] == (-P + Q x[t]) y[t]\}$

The variables $x(t)$ and $y(t)$ are magnitudes of the prey and predator, respectively, at time t.

First we calculate the equilibrium points. At these points the derivatives are simultaneously zero:

```
equi = Solve[{f == 0, g == 0}, vars]
```

$\left\{\{x[t] \to 0, y[t] \to 0\}, \left\{x[t] \to \frac{P}{Q}, y[t] \to \frac{p}{q}\right\}\right\}$

To check the nature of these points, calculate the Jacobian of **f** and **g** (see Section 16.1.3, p. 421):

```
jac = Outer[D, {f, g}, vars]
```

$\{\{p - q y[t], -q x[t]\}, \{Q y[t], -P + Q x[t]\}\}$

The eigenvalues of the Jacobian at the equilibrium points are as follows:

```
eig = Map[Eigenvalues[jac /. #] &, equi]
```

$\{\{p, -P\}, \{-i \sqrt{p} \sqrt{P}, i \sqrt{p} \sqrt{P}\}\}$

So, if p and P are positive, the equilibrium point $(0, 0)$ is a saddle point and the point $\left(\frac{P}{Q}, \frac{p}{q}\right)$ a center or a focus.

Consider the following numerical values for the constants:

```
p = 3.0; q = 2;
P = 2.5; Q = 1;
```

The equilibrium points are now as follows:

```
equi
```

$$\{\{x[t] \to 0, y[t] \to 0\}, \{x[t] \to 2.5, y[t] \to 1.5\}\}$$

Calculate a solution of the system:

```
inits = {x[0] == 2.5, y[0] == 0.5};
```

```
sol = vars /. NDSolve[Join[eqns, inits], vars, {t, 0, 10}][[1]]
```

$$\{\text{InterpolatingFunction}[\{\{0., 10.\}\}, <>][t],$$
$$\text{InterpolatingFunction}[\{\{0., 10.\}\}, <>][t]\}$$

We plot this solution in three ways. First we plot $x(t)$ and $y(t)$ as functions of t:

```
Plot[Evaluate[sol], {t, 0, 10},
  PlotStyle → {{}, AbsoluteDashing[{1.5}]}, AspectRatio → 0.2];
```

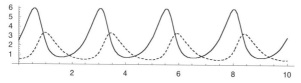

The sizes of the populations vary cyclically. Then we plot a phase trajectory in (x, y) plane:

```
ParametricPlot[sol, {t, 0, 2.5}, AxesLabel → {"prey", "predator"}];
```

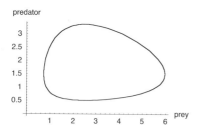

The trajectories seem to be closed curves. Lastly we plot the phase trajectory as a collection of points:

```
ListPlot[Table[sol, {t, 0, 2.5, 0.1}], AxesLabel → {"prey", "predator"}];
```

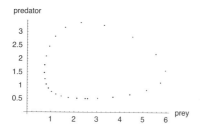

This figure gives an impression of the speed of motion in the plane (the speed can also be seen from the first figure, but not from the second figure).

A direction field gives an impression of the behavior of the model in an area:

```
<< Graphics`PlotField`
```

```
PlotVectorField[{f, g}, {x[t], 0, 12},
  {y[t], 0, 6}, PlotPoints → 11, Axes → True];
```

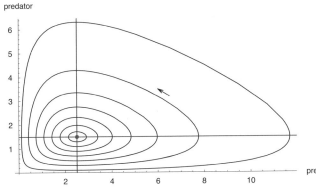

A collection of phase trajectories is even more illustrative. We calculate solutions when $x(0) = 2.5$ and $y(0)$ gets a set of values:

```
solset = Table[
    vars /. NDSolve[Join[eqns, {x[0] == 2.5, y[0] == y0}], vars, {t, 0, 3.1}],
    {y0, 0.1, 1.5, 0.2}];
```

```
<< Graphics`Arrow`
```

```
ParametricPlot[Evaluate[solset], {t, 0, 3.1},
  Compiled → False, PlotRange → All, AspectRatio → Automatic,
  AxesLabel → {"prey", "predator"}, Epilog ->
    {Arrow[{6.5, 3.2}, {6.0, 3.5}, HeadLength → 0.02, HeadCenter → 0.3],
    RGBColor[0, 0, 1], Line[{{0, p / q}, {12, p / q}}],
    Line[{{P / Q, 0}, {P / Q, 6.5}}], AbsolutePointSize[3],
    RGBColor[1, 0, 0], Point[{2.5, 1.5}]}];
```

The red point is the center and the blue lines are the nullclines, which are the curves where $x'(t) = 0$ or $y'(t) = 0$. The direction of motion is counterclockwise.

```
Remove["Global`*"]
```

■ Competing Species

Define the following model that describes competing species:

```
vars = {x[t], y[t]};
```

```
f = x[t] (p - q x[t] - r y[t]);
g = y[t] (P - Q x[t] - R y[t]);
```

```
eqns = {x'[t] == f, y'[t] == g};
```

The variables $x(t)$ and $y(t)$ are the population magnitudes at time t.

First, we calculate the nullclines where the derivatives $x'(t)$ and $y'(t)$ are zero:

```
h1 = y[t] /. Solve[f == 0, y[t]][[1]] /. x[t] → x // Simplify
```

$$\frac{p - q x}{r}$$

```
h2 = y[t] /. Solve[g == 0, y[t]][[2]] /. x[t] → x
```

$$\frac{P - Q x}{R}$$

Then we calculate the equilibrium points:

```
equi = Solve[{f == 0, g == 0}, vars] // Simplify
```

$$\left\{ \{x[t] \to 0, \ y[t] \to 0\}, \ \left\{x[t] \to \frac{p}{q}, \ y[t] \to 0\right\}, \right.$$
$$\left. \left\{x[t] \to \frac{P r - p R}{Q r - q R}, \ y[t] \to \frac{P q - p Q}{-Q r + q R}\right\}, \ \left\{y[t] \to \frac{P}{R}, \ x[t] \to 0\right\}\right\}$$

Calculate the Jacobian of **f** and **g**:

```
jac = Outer[D, {f, g}, vars]
```

$$\{\{p - 2 q x[t] - r y[t], -r x[t]\}, \{-Q y[t], P - Q x[t] - 2 R y[t]\}\}$$

The eigenvalues of the Jacobian at the equilibrium points are as follows:

```
eig = Map[Eigenvalues[jac /. #] &, equi] // Simplify
```

$$\left\{ \{p, P\}, \ \left\{-p, P - \frac{p Q}{q}\right\}, \ \left\{- \frac{1}{2 Q r - 2 q R} \left(P q r - p q R - P q R + p Q R + \right. \right. \right.$$
$$\left. \sqrt{4 (P q - p Q) (P r - p R) (-Q r + q R) + (P q (r - R) + p (-q + Q) R)^2} \right),$$
$$\frac{1}{2 Q r - 2 q R} \left(-P q r + p q R + P q R - p Q R + \right.$$
$$\left. \left. \sqrt{4 (P q - p Q) (P r - p R) (-Q r + q R) + (P q (r - R) + p (-q + Q) R)^2} \right) \right\},$$
$$\left\{-P, \ p - \frac{P r}{R}\right\}\right\}$$

Define the following numerical values for the constants:

```
p = 2; q = 2 / 3; r = 2;
P = 2; Q = 4 / 3; R = 1;
```

The equilibrium points are now as follows:

```
equi
```

$$\left\{ \{x[t] \to 0, \ y[t] \to 0\}, \ \{x[t] \to 3, \ y[t] \to 0\}, \right.$$
$$\left. \left\{x[t] \to 1, \ y[t] \to \frac{2}{3}\right\}, \ \{y[t] \to 2, \ x[t] \to 0\}\right\}$$

Here are the eigenvalues:

```
eig
```

$$\left\{ \{2, 2\}, \ \{-2, -2\}, \ \left\{-2, \frac{2}{3}\right\}, \ \{-2, -2\}\right\}$$

Thus, the equilibrium points are an unstable node, a stable node, a saddle point, and a stable node, respectively.

Calculate a solution of the system:

```
inits = {x[0] == 3.5, y[0] == 2};
```

```
sol = vars /. NDSolve[Join[eqns, inits], vars, {t, 0, 7}][[1]]
```

```
{InterpolatingFunction[{{0., 7.}}, <>][t],
 InterpolatingFunction[{{0., 7.}}, <>][t]}
```

We plot this solution in three ways, as we did in Example 1. First, we plot $x(t)$ and $y(t)$ as functions of t:

```
Plot[Evaluate[sol], {t, 0, 7}, PlotStyle → {{}, AbsoluteDashing[{1.5}]}];
```

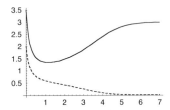

We see that the x population wins the competition and the y population vanishes; the limiting values are 3 and 0, respectively. Then we plot a phase trajectory in (x, y) plane:

```
ParametricPlot[sol, {t, 0, 7}, PlotRange → All, AxesLabel → {"x", "y"}];
```

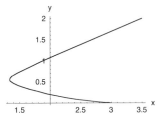

Lastly we plot the phase trajectory as a collection of points to get an impression of the speed of motion in the plane:

```
ListPlot[Table[sol, {t, 0, 7, 0.1}],
    PlotRange → All, AxesLabel → {"x", "y"}];
```

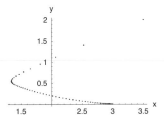

Next we plot a direction field:

```
<< Graphics`PlotField`
```

```
PlotVectorField[{f, g}, {x[t], 0, 3.5},
    {y[t], 0, 2.5}, PlotPoints → 11, Axes → True];
```

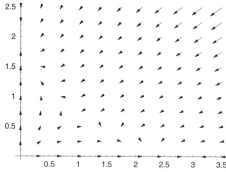

To plot a collection of phase trajectories, we first calculate solutions that go through the line $y = 1 - x$ at time $t = 4$ for some values of $x(4)$:

```
solset1 = Table[
    vars /. NDSolve[Join[eqns, {x[4] == x0, y[4] == 1 - x0}], vars, {t, 0, 20}],
    {x0, 0.05, 0.95, 0.1}];

p1 = ParametricPlot[Evaluate[solset1], {t, 0, 20}, Compiled → False];
```

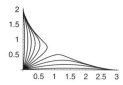

The time instance $t = 4$ was chosen by trial and error such that, when the solution starts at $t = 0$, the starting points are close enough to the origin. Then we start from the line $y = 7 - x$ for some values of $x(0)$:

```
solset2 = Table[
    vars /. NDSolve[Join[eqns, {x[0] == x0, y[0] == 7 - x0}], vars, {t, 0, 10}],
    {x0, 0.1, 6.9, 0.25}];

p2 = ParametricPlot[Evaluate[solset2], {t, 0, 10}, Compiled → False];
```

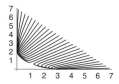

We also plot the nullclines:

```
p3 = Plot[{h1, h2}, {x, 0, 3.5},
    PlotStyle → RGBColor[0, 0, 1], DisplayFunction → Identity];
```

Then we show all of the plots in one figure:

```
<< Graphics`Arrow`
```

```
Show[p1, p2, p3, PlotRange → {{-0.04, 3.6}, {-0.04, 2.6}}, Epilog →
    {Arrow[{2, 1.3}, {1.85, 1.2}, HeadLength → 0.02, HeadCenter → 0.3],
     Arrow[{0.5, 0.33}, {0.65, 0.43}, HeadLength → 0.02, HeadCenter → 0.3],
     AbsolutePointSize[4], RGBColor[1, 0, 0], Point[{3, 0}],
     Point[{0, 2}], RGBColor[0, 1, 0], Point[{0, 0}], Point[{1, 2 / 3}]}}];
```

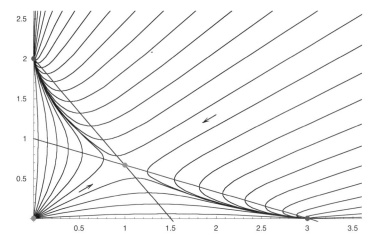

The red points are the stable equilibrium points, the green points are unstable equilibrium points, and the blue lines are the nullclines. The direction of motion is toward the stable equilibrium points $(3, 0)$ and $(0, 2)$. Note that, for some other values of the parameters p, q, r, P, Q, and R, the behavior of the model may be wholly different.

■ A Matrix Equation

NDSolve accepts equations formulated with matrices (✿5). For example:

```
a = {{0.1, -0.5}, {0.3, -0.2}}; x0 = {15, 15};
```

```
sol = x[t] /. NDSolve[{x'[t] == a.x[t], x[0] == x0}, x[t], {t, 0, 80}][[1]]
```

```
InterpolatingFunction[{{0., 80.}}, <>][t]
```

```
ParametricPlot[sol, {t, 0, 80}, Compiled → False, PlotRange → All];
```

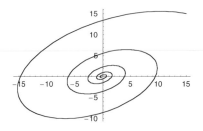

In this example, **x[t]** is a two-component vector. The dependent variable can also be a matrix.

■ A Differential-Algebraic System

NDSolve can solve differential-algebraic systems (※5). Here is a simple example:

```
sol = {x[t], y[t]} /.
  NDSolve[{x'[t] == x[t]^2 + y[t] + 0.5, x[t] + y[t] == 1, x[0] == 0},
    {x[t], y[t]}, {t, 0, 1.5}][[1]]
{InterpolatingFunction[{{0., 1.5}}, <>][t],
 InterpolatingFunction[{{0., 1.5}}, <>][t]}

Plot[Evaluate[sol], {t, 0, 1.5}];
```

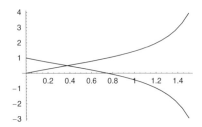

```
Remove["Global`*"]
```

23.3.3 Three Equations

```
vars = {x[t], y[t], z[t]}   Define the dependent variables
eqns = {eqn1, eqn2, eqn3}   Define the differential equations
inits = {x[a] == α, y[a] == β, z[a] == γ}   Define the initial (or boundary) conditions

sol = {sx, sy, sz} = vars /. DSolve[Join[eqns, inits], vars, {t, a, b}][[1]]
  Solve the problem

Plot[Evaluate[sol], {t, a, b}]   Plot x, y, and z in one figure
Map[Plot[#, {t, a, b}]&, sol]   Plot x, y, and z in three figures

Map[ParametricPlot[#, {t, a, b}]&, {{sx, sy}, {sx, sz}, {sy, sz}}]   Plot three
  plane trajectories
ParametricPlot3D[sol, {t, a, b}]   Plot the space trajectory
```

■ The Lorenz Model

Define the following model:

```
vars = {x[t], y[t], z[t]};

f = p (y[t] - x[t]);
g = q x[t] - y[t] - x[t] z[t];
h = x[t] y[t] - r z[t];

eqns = {x'[t] == f, y'[t] == g, z'[t] == h};
```

This is the Lorenz model, which describes convective currents in atmosphere (see Borrelli and Coleman 1998, p. 500–514). The constants p, q, and r are positive. The variable x is the amplitude of the convective currents, y is the temperature difference between rising and falling currents, and z is the deviation from the normal temperatures. The equations result in approximating nonlinear partial differential equations of turbulent flow.

The equilibrium points are as follows:

```
equi = Solve[{f == 0, g == 0, h == 0}, vars] // Simplify
```

$\{\{y[t] \to 0, z[t] \to 0, x[t] \to 0\},$
$\{y[t] \to -\sqrt{(-1+q) r}, z[t] \to -1+q, x[t] \to -\sqrt{(-1+q) r}\},$
$\{y[t] \to \sqrt{(-1+q) r}, z[t] \to -1+q, x[t] \to \sqrt{(-1+q) r}\}\}$

Define the following numerical values for the constants:

```
p = 3; q = 268 / 10; r = 1;
```

The equilibrium points are now as follows:

```
equi // N
```

$\{\{y[t] \to 0., z[t] \to 0., x[t] \to 0.\},$
$\{y[t] \to -5.07937, z[t] \to 25.8, x[t] \to -5.07937\},$
$\{y[t] \to 5.07937, z[t] \to 25.8, x[t] \to 5.07937\}\}$

■ Sensitivity to Numerical Inaccuracies

A note for users of *Mathematica* 4: If you try the commands presented here, you will obtain quite different results. Generally, **NDSolve** in *Mathematica* 5 gives more accurate results. In addition, in the first two commands that use **NDSolve**, you should add the option **MaxSteps → 6000**.

Calculating a numerical solution of the Lorenz model may cause trouble because the model is *chaotic*. The basic property of such models is extreme sensitivity of the solution to the initial conditions and numerical inaccuracies. To demonstrate the effect of numerical inaccuracies, we first calculate one solution of the system with normal decimal numbers and ask for the value of the solution at $t = 31$:

```
inits = {x[0] == 0, y[0] == 1, z[0] == 0};

sol1 = vars /. NDSolve[Join[eqns, inits], vars, {t, 0, 35}][[1]];

sol1 /. t → 31
```

$\{-2.75379, -7.14803, 11.7918\}$

Then we calculate the solution with high-precision numbers by asking that the calculations are done to 20 decimals. We also increase the values of the precision and accuracy goals (see Section 23.4.1, p. 627):

```
sol2 = vars /. NDSolve[Join[eqns, inits], vars, {t, 0, 35},
    WorkingPrecision → 20, PrecisionGoal → 12, AccuracyGoal → 12][[1]];

sol2 /. t → 31.
```

$\{7.50216, 14.0227, 25.0762\}$

The values at $t = 31$ are wholly different in the two solutions. To see the differences more clearly, we plot the difference of the two solutions:

```
Map[Plot[#, {t, 0, 35}, PlotPoints → 300,
    PlotRange → All, DisplayFunction → Identity] &, sol1 - sol2];

Show[GraphicsArray[%]];
```

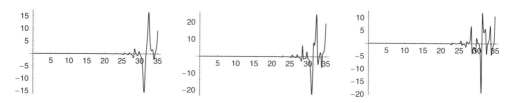

The figures show that, from about $t = 25$ on, the solutions differ greatly.

We can trust the solution **sol2** more, but, actually, how accurate it is? We can solve the equations once more and use even tighter precision and accuracy conditions:

```
sol3 = {sx, sy, sz} = vars /.
    NDSolve[Join[eqns, inits], vars, {t, 0, 35}, WorkingPrecision → 20,
        PrecisionGoal → 16, AccuracyGoal → 16, MaxSteps → 14000][[1]];
```

The difference of **sol2** and **sol3** is displayed as follows:

```
Map[Plot[#, {t, 0, 35}, PlotPoints → 300,
    PlotRange → All, DisplayFunction → Identity] &, sol2 - sol3];

Show[GraphicsArray[%]];
```

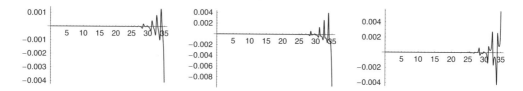

The solutions are practically the same up to about $t = 30$, so **sol2** may be quite good to this point. However, we use the more accurate **sol3** in the sequel.

To summarize, when calculating the solution of a chaotic model, it is important to use high precision in the calculations. Set **WorkingPrecision** to about 20 or higher. Increase also the precision and accuracy goals (their default value is 8). Remember to express all constants of the model with high accuracy. For example, do not set **q = 26.8** but rather **q = 268/10** or **q = 26.8`20** (the last definition sets the precision of 26.8 to 20; see Section 11.2.2, p. 289).

■ Sensitivity to Initial Conditions

To demonstrate the sensitivity of the Lorenz model to the initial conditions, we solve the equations by now setting $z(0) = 10^{-5}$ instead of $z(0) = 0$:

```
sol4 = vars /. NDSolve[Join[eqns, {x[0] == 0, y[0] == 1, z[0] == 10^-5}],
    vars, {t, 0, 30}, WorkingPrecision → 20, PrecisionGoal → 16,
    AccuracyGoal → 16, MaxSteps → 12000][[1]];
```

We compare solutions **sol3**, where $z(0) = 0$, and **sol4** by plotting $x(t)$:

```
Plot[Evaluate[{sol3[[1]], sol4[[1]]}], {t, 0, 30},
    PlotPoints → 300, PlotStyle → {AbsoluteDashing[{1.5}], {}}];
```

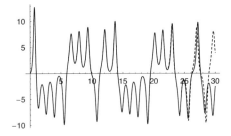

The solutions begin to differ from about $t = 25$ on, and from $t = 29$ on the solutions are wholly different. This sensitivity makes the prediction of a chaotic system impossible for a long time period, because the initial conditions are hardly known exactly.

■ Plotting the Solution

First we plot the three curves in the same figure:

```
Plot[Evaluate[sol3], {t, 0, 30}, PlotPoints → 300,
    PlotStyle → {GrayLevel[0.4], {}, AbsoluteDashing[{1}]}];
```

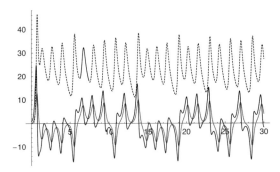

Then we plot three separate figures:

```
Map[Plot[#, {t, 0, 30}, PlotPoints → 300,
        Ticks → {{30}, Automatic}, DisplayFunction → Identity] &, sol3];

Show[GraphicsArray[%]];
```

The components seem to evolve quite unpredictably and chaotically.

The pairwise plane trajectories show an interesting behavior:

```
Map[ParametricPlot[#, {t, 0, 30}, PlotPoints → 1000,
    DisplayFunction → Identity] &, {{sx, sy}, {sx, sz}, {sy, sz}}];

Show[GraphicsArray[%]];
```

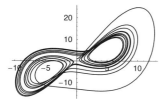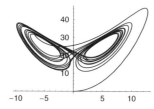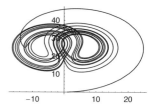

Lastly we plot the space trajectory. To celebrate this extraordinarily fine figure, we produce a two-image stereogram, which involves two versions of the figure with slightly different viewpoints (see Section 7.3.2, p. 194). If you are able to superimpose the two figures with your eyes by focusing behind the paper, you will get an amazing stereographic view. (To get a smooth curve, we gave a high value for **PlotPoints**; to prevent the figure from becoming too complex, we plotted only up to the value $t = 10.5$.)

```
fig = Map[ParametricPlot3D[sol3, {t, 0, 10.5},
    PlotPoints → 1000, DisplayFunction → Identity, ViewPoint → #] &,
    {{1.3, -2.4, 0.8}, {1.4, -2.3, 0.8}}];

Show[GraphicsArray[fig, GraphicsSpacing → 0]];
```

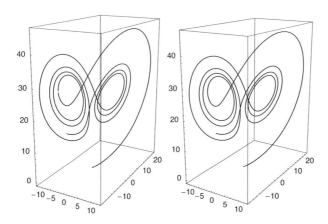

We can also plot points along the trajectory. A figure like this gives an impression of the speed of the process:

```
points = Table[sol3, {t, 0, 10.5, 0.03}];
```

```
fig = Map[Graphics3D[{AbsolutePointSize[1], Map[Point, points]},
       BoxRatios → Automatic, ViewPoint → #] &,
    {{1.3, -2.4, 0.8}, {1.4, -2.3, 0.8}}];

Show[GraphicsArray[fig, GraphicsSpacing → 0.15]];
```

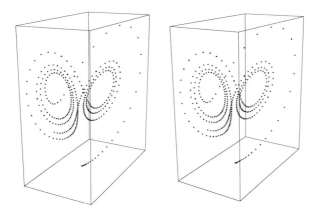

23.4 More about Numerical Solutions

23.4.1 Options

As usual, for each option, the default value is mentioned first.

Options for **NDSolve**:

WorkingPrecision   Precision used in internal computations; examples of values:
  MachinePrecision (❋5), **20**

PrecisionGoal   If the value of the option is **p**, the relative error of the solution at each
  point considered should be of the order 10^{-p}; examples of values: **Automatic**
  (usually means **8**), **10**

AccuracyGoal   If the value of the option is **a**, the absolute error of the solution at each
  point considered should be of the order 10^{-a}; examples of values: **Automatic**
  (usually means **8**), **10**

Method   Method to use; possible values: **Automatic** (means **"Adams"** for nonstiff and
  "BDF" for stiff problems), **"Adams"**, **"BDF"** (❋5), **"ExplicitRungeKutta"** (❋5),
  "ImplicitRungeKutta" (❋5), **"SymplecticPartitionedRungeKutta"** (❋5)

StartingStepSize   Initial step size used; examples of values: **Automatic**, **0.01**

MaxStepSize   Maximum size of each step; examples of values: **Automatic**, **0.01**

MaxStepFraction (❋5)   Maximum fraction of the solution interval to cover in each
  step; examples of values: **0.1**, **0.05**

MaxSteps   Maximum number of steps to take; examples of values: **Automatic** (means
  10000 (❋5)), **20000**

NormFunction (❋5)   Norm to use for error estimation; default value: **Automatic**

DependentVariables (❋5)   List of all dependent variables; default value: **Automatic**

StoppingTest Integration is stopped if the test gives **True**; examples of values: **None**, **Abs[y[x]] > 5**

SolveDelayed Whether the derivatives are solved symbolically at the beginning (**False**) or at each step (**True**); possible values: **False, True**

Compiled Whether to compile the equations; possible values: **True, False**

StepMonitor (❀5) Command to be executed after each step of the method; examples of values: **None, n++, AppendTo[steps, {t, y[t]}]**

EvaluationMonitor (❀5) Command to be executed after each evaluation of the equation; examples of values: **None, n++, AppendTo[points, t]**

The default value **MachinePrecision** of **WorkingPrecision** means that standard arithmetic is used in computations; a high value like 20 should be given for numerically sensitive problems. The default value of **PrecisionGoal** is usually 8 (6 in *Mathematica* 4) (the general value is **WorkingPrecision**/2), so **NDSolve** tries to give you an answer with a relative error of order 10^{-8} (this is only a goal; the actual relative error may be smaller or larger). The default value of **AccuracyGoal** also is usually 8 (6 in *Mathematica* 4) (the general value is **WorkingPrecision**/2), so the goal is for the result to have an absolute error of the order 10^{-8}.

For a given value of the independent variable, **NDSolve** accepts the present value of the dependent value if the precision and accuracy goals are satisfied. If you increase **Preci- sionGoal** or **AccuracyGoal**, you may also have to increase **WorkingPrecision**. See Section 11.3.1, p. 292, for more details about these options.

NDSolve uses an adaptive method in which the step size is varied as needed; this means that, if the solution seems to change rapidly, the step size is reduced so that the solution can be followed more closely. For nonstiff problems, an Adams predictor-correc- tor method of order between 1 and 12 is used. For stiff problems, a backward difference formula method (i.e., Gear's method) of order between 1 and 5 is used. **NDSolve** detects the existence of stiffness automatically and chooses the correct method as well as a suit- able order for it (the method and the order may be varied during the solution process). In a stiff system, the components vary at very different rates. An example could be the system $x'(t) = -100 x(t)$, $y'(t) = y(t) + x(t)$, $x(0) = 1$, $y(0) = 0$. Here x decreases very rapidly, whereas y increases very slowly.

With the **Method** option, we can also select the method ourselves. **"Adams"** means the predictor-corrector Adams method with orders 1 through 12, and **"BDF"** means implicit backward differentiation formulas with orders 1 through 5. The method can also be defined with controllers and submethods (see the *Help Browser*). Linear boundary value problems are solved with the Gel'fand–Lokutsiyevskii chasing method.

The default value of **MaxSteps** is **Automatic**, but this means, for ordinary differential equations, the value 10,000 (1000 in versions of *Mathematica* prior to 5). If the interval of solution is long or the solution varies widely, you may receive a message that the maxi- mum number of steps has been reached at a given point. You get the solution up to this point only. You can then increase the value of **MaxSteps** and solve anew. If the solution has a singular point, then **MaxSteps** prevents **NDSolve** from reducing the step size infinitely.

In the *Help Browser*, you can read advanced information about **NDSolve**.

■ Example 1: Precision and Accuracy

We solve an equation both symbolically and numerically:

```
eqn = y'[t] == -t y[t] + Exp[-t];
```

```
symbsol = y[t] /. DSolve[{eqn, y[0] == 1}, y[t], t][[1]]
```

$$\frac{1}{2} e^{-\frac{1}{2}-\frac{t^2}{2}} \left(2\sqrt{e} + \sqrt{2\pi} \, \text{Erfi}\left[\frac{1}{\sqrt{2}}\right] + \sqrt{2\pi} \, \text{Erfi}\left[\frac{-1+t}{\sqrt{2}}\right]\right)$$

```
numsol = y[t] /. NDSolve[{eqn, y[0] == 1}, y[t], {t, 0, 5}][[1]];
```

```
Plot[numsol, {t, 0, 5}];
```

We plot both the absolute and the relative error (❀4!):

```
Block[{$DisplayFunction = Identity},
  p1 = Plot[symbsol - numsol, {t, 0, 5}, PlotRange → All,
    PlotPoints → 300, Ticks → {Range[5], {-3. 10^-7, 3. 10^-7}}];
  p2 = Plot[1 - numsol / symbsol, {t, 0, 5}, PlotRange → All,
    PlotPoints → 300, Ticks → {Range[5], {3. 10^-6}}];]
```

```
Show[GraphicsArray[{p1, p2}]];
```

The absolute error is at most about 3×10^{-7}, and the relative error is at most about 3×10^{-6}. The goal with **NDSolve** is that both of these errors should be less than 10^{-8} in the whole interval of solution. The goal is not strictly achieved, but, nevertheless, the errors are small enough for us to be well satisfied with the numerical solution.

In the following way we can see all of the steps of the numerical solution (❀5):

```
steps = {{0, 1}}; y[t] /. NDSolve[{eqn, y[0] == 1},
  y[t], {t, 0, 5}, StepMonitor :> AppendTo[steps, {t, y[t]}]];
```

```
Show[Graphics[{Map[Line[{{#[[1]], 0}, #}] &, steps]},
  Axes → True, AspectRatio → 0.2, Ticks → {Range[5], Automatic}]];
```

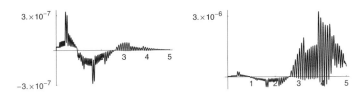

The step size varies according to the rate of change of the solution, which demonstrates the adaptivity property of **NDSolve**.

■ Example 2: A Singular Point

If the solution has a singular point in the interval of solution, then, to be able to follow the solution, the step size is reduced until it becomes effectively zero or until the maximum number of steps (10,000) is reached. In such situations, the solution process is stopped and a message is written. In the following example, we ask for the solution in the interval (0, 2) but get the solution only in the interval (0, 1), because there is a singularity at t = 1:

```
sol = y[t] /. NDSolve[{y'[t] == 1 / (t - 1), y[0] == 0}, y[t], {t, 0, 2}];
```

```
NDSolve::ndsz :
  At t == 0.9999999999974905`, step size is effectively zero;
    singularity or stiff system suspected. More…
```

```
Plot[sol, {t, 0, 1}];
```

23.4.2 Runge–Kutta Method

In Sections 15.1.1, p. 385, and 15.2.2, p. 396, we developed programs for Euler's method to solve differential equations. This method—because it is not sufficiently accurate—has mainly pedagogical value and is rarely used in practice.

The fourth-order Runge–Kutta method is a popular and accurate method. Consider the following system of differential equations:

$$y_1'(t) = f_1(t, y_1, ..., y_m),$$

$$...$$

$$y_m'(t) = f_m(t, y_1, ..., y_m).$$

We form a function **rungeKuttaStep** to perform a single step of the Runge–Kutta method. We then use **NestList** to apply the step **n** times in **rungeKuttaSolve**.

```
rungeKuttaStep[f_List, ty_List, tyi_List, h_] := Module[{k1, k2, k3, k4},
  k1 = f /. Thread[ty → tyi];
  k2 = f /. Thread[ty → tyi + h / 2 Flatten[{1, k1}]];
  k3 = f /. Thread[ty → tyi + h / 2 Flatten[{1, k2}]];
  k4 = f /. Thread[ty → tyi + h Flatten[{1, k3}]];
  tyi + h Flatten[{1, (k1 + 2 k2 + 2 k3 + k4) / 6}]]

rungeKuttaSolve[f_List, ty_List, ty0_List, t1_, n_] :=
  With[{h = N[(t1 - ty0[[1]]) / n]},
    NestList[rungeKuttaStep[f, ty, #, h] &, N[ty0], n]]
```

Here **f** is the list of righthand-side expressions of the differential equations, **ty** is the list of the independent and dependent variables, **ty0** contains their initial values, **t1** is the endpoint of the interval of solution, and **n** is the number of steps to be performed.

Remember that the result of **Thread[ty → tyi]** is a list of the form **{t → ti, y1 → y1i, ..., ym → ymi}**. **Flatten** is needed to form unnested lists. **NestList** performs the iterations in which the result of one iteration is the starting point of the next iteration (see Section 15.2.2, p. 394, for details).

■ Example 1

First we solve the equation $y'(t) = t - y(t)^2$, $y(0) = 3$, in the interval $(0, 6)$ using 30 steps:

```
sol = rungeKuttaSolve[{t - y^2}, {t, y}, {0, 3}, 6, 30];

sol // Last     {6., 2.40582}

ListPlot[sol, PlotRange → All];
```

We compare the solution with the solution given by **NDSolve** at $t = 6$:

```
y[t] /. NDSolve[{y'[t] == t - y[t]^2, y[0] == 1}, y[t], {t, 0, 6}][[1]] /. t → 6

2.40583
```

The value 2.40582 given by our program is very near to the value given by **NDSolve**.

■ Example 2

Now we solve the system of two equations $y'(t) = t - z(t)^2$, $z'(t) = y(t)$, $y(0) = 1$, $z(0) = 2$:

```
sol = rungeKuttaSolve[{t - z^2, y}, {t, y, z}, {0, 1, 2}, 6, 40];

sol // Last     {6., 0.273551, 0.744243}
```

NDSolve gives quite similar numbers:

```
{y[t], z[t]} /.
  NDSolve[{y'[t] == t - z[t]^2, z'[t] == y[t], y[0] == 1, z[0] == 2},
    {y[t], z[t]}, {t, 0, 6}][[1]] /. t → 6

{0.273997, 0.74375}
```

To plot the solution, we first extract the three components of the solution:

```
{tt, yy, zz} = Transpose[sol];
```

We plot **yy** and **zz** as well as the trajectory:

```
ttyy = Transpose[{tt, yy}];
ttzz = Transpose[{tt, zz}];
yyzz = Transpose[{yy, zz}];
```

```
Show[Graphics[{AbsolutePointSize[2], Map[Point, ttyy], Line[ttyy],
    GrayLevel[0.4], Map[Point, ttzz], Line[ttzz]}], Axes → True];
```

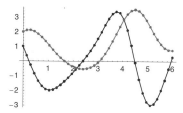

```
Show[Graphics[{AbsolutePointSize[2], Map[Point, yyzz], Line[yyzz]}],
    Axes → True];
```

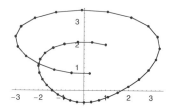

With the **NumericalMath`Butcher`** package, we can investigate various Runge–Kutta methods. The package calculates the order conditions for any pth-order s-stage Runge–Kutta method.

23.4.3 Boundary Value Problems

■ Linear Problems

Consider the following boundary value problem:

$$y''(t) = p(t)\,y'(t) + q(t)\,y(t) + r(t), \quad y(t_0) = y_0, \quad y(t_1) = y_1.$$

Here p, q, and r do not depend on y. We solve this problem by solving two initial value problems:

$$z''(t) = p(t)\,z'(t) + q(t)\,z(t) + r(t), \quad z(t_0) = y_0, \quad z'(t_0) = 0,$$
$$v''(t) = p(t)\,v'(t) + q(t)\,v(t), \qquad v(t_0) = 0, \quad v'(t_0) = 1.$$

It is easy to show that the following function is the solution of the original boundary value problem:

$$y(t) = z(t) + v(t)\,(y_1 - z(t_1))/v(t_1).$$

This requires that $v(t_1) \neq 0$. If $v(t_1) = 0$ and $z(t_1) = y_1$, then $y(t) = z(t) + c\,v(t)$ is a solution for all constants c. If $v(t_1) = 0$ and $z(t_1) \neq y_1$, the problem has no solutions. Here is a module that implements this method (without the exceptional cases).

```
linearBVP[p_, q_, r_, t_, t0_, t1_, y0_, y1_] :=
 Module[{eqn, sol1, sol2, sol},
  eqn = y''[t] == p y'[t] + q y[t] + r;
  sol1 = y[t] /. NDSolve[{eqn, y[t0] == y0, y'[t0] == 0}, y[t], {t, t0, t1}];
  sol2 =
   y[t] /. NDSolve[{eqn /. r -> 0, y[t0] == 0, y'[t0] == 1}, y[t], {t, t0, t1}];
  sol = sol1 + sol2 (y1 - sol1 /. t -> t1) / (sol2 /. t -> t1);
  FunctionInterpolation[sol, {t, t0, t1}][t]]
```

Here **FunctionInterpolation** (see Section 21.4.2) forms a single interpolating function from the solution **sol**, which contains the two interpolating functions **sol1** and **sol2**.

Let us solve the equation $y''(t) = y'(t) - t\,y(t)$ with the boundary conditions $y(0) = 2$ and $y(4) = 2$. Now $p = 1$, $q = -t$, and $r = 0$:

```
sol = linearBVP[1, -t, 0, t, 0, 4, 2, 2]
```

```
InterpolatingFunction[{{0., 4.}}, <>][t]
```

```
Plot[sol, {t, 0, 4}];
```

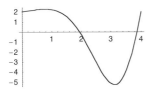

In Example 3 of Section 23.3.1, p. 613, we solved the same problem using **NDSolve**.

■ Nonlinear Problems

Consider the following nonlinear boundary value problem:

$$f(t, y, y', y'') = 0, \quad y(t_0) = y_0, \quad y(t_1) = y_1$$

(**NDSolve** does not solve such problems.) Keiper (1993, p. 49) has presented the following idea for solving this problem. First, define the equation:

```
eqn = y''[t] == y'[t] - y[t]^2 + t^2;
```

We want to solve this equation with the boundary conditions $y(0) = 1$ and $y(3) = 2$. Define $y(3)$ or **y1** as a function of $y'(0)$ or **dy0**:

```
y1[dy0_ ? NumericQ] := (sol = y[t] /.
      NDSolve[{eqn, y[0] == 1, y'[0] == dy0}, y[t], {t, 0, 3}][[1]]) /. t -> 3
```

We investigate the situation by giving some values for $y'(0)$ and calculating the corresponding values of $y(3)$:

```
tries = Table[{dy0, y1[dy0]}, {dy0, 0, 2, 0.5}]
```

```
{{0, 4.66136}, {0.5, 4.27298},
 {1., 3.21499}, {1.5, 1.40707}, {2., -2.61151}}
```

```
Show[Graphics[{AbsolutePointSize[2], Map[Point, tries], Line[tries],
    Line[{{0, 2}, {2, 2}}]}], Axes → True, AxesLabel → {"y'[0]", "y[3]"}];
```

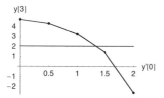

It seems that $y'(0)$ must be about 1.3 to get $y(3) = 2$. We now find the value of **dy0** that gives $y(3) = 2$ by using 1.3 and 1.4 as the starting values:

```
FindRoot[y1[dy0] == 2, {dy0, 1.3, 1.4}]
```

$\{dy0 → 1.37153\}$

Thus, if $y'(0) = 1.37153$, then $y(3) = 2$. To plot the solution, note that the solution is already in the variable **sol**:

```
Plot[sol, {t, 0, 3}];
```

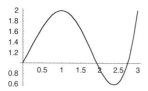

23.4.4 Estimation of Differential Equations

How to estimate unknown parameters that appear in a differential equation if we have some data? We can distinguish two situations. If the equation has a closed form symbolic solution, then use **FindFit** or **NonlinearRegress** to estimate the parameters of the solution (see Example 1). If a closed form solution is not available, then use **FindMinimum** to minimize a least-squares criterion (see Example 2).

■ **Example 1**

In Example 2 of Section 22.1.4, p. 581, we have already considered an experiment where the growth of a yeast culture was measured at time instances 0, 1, 2, …, 18 (Pearl 1927). The measurements were as follows:

```
yeast = {9.6, 18.3, 29.0, 47.2, 71.1, 119.1, 174.6, 257.3, 350.7, 441.0,
    513.3, 559.7, 594.8, 629.4, 640.8, 651.1, 655.9, 659.6, 661.8};
```

We plot the data:

```
tt = Range[0, 18]; data = Transpose[{tt, yeast}];
```

```
p1 = ListPlot[data, PlotStyle → AbsolutePointSize[2],
    Ticks → {Range[18], Automatic}, Epilog → Line[{{0, 663}, {18, 663}}]];
```

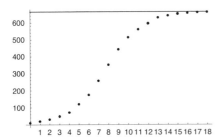

The data shows a logistic growth that can be described by the differential equation $y'(t) = r\,y(t)\,(M - y(t))$. This equation has a closed-form solution:

```
sol = y[t] /. DSolve[{y'[t] == r y[t] (M - y[t]), y[0] == y0}, y[t], t][[1]]
```

$$\frac{e^{Mrt}\,M\,y0}{M - y0 + e^{Mrt}\,y0}$$

We can easily estimate the parameters of the solution (✺5):

```
fit = FindFit[data, sol, {{M, 100}, {y0, 10}, {r, 0.1}}, t]
```

$$\{M \to 663.022,\ y0 \to 9.13552,\ r \to 0.000825002\}$$

In *Mathematica* 4, do as follows:

```
<< Statistics`NonlinearFit`
```

```
fit = BestFitParameters /.
    NonlinearRegress[data, sol, t, {{M, 100}, {y0, 10}, {r, 0.1}},
      RegressionReport → BestFitParameters, MaxIterations → 40]
```

The fit is good:

```
p2 = Plot[sol /. fit, {t, 0, 18}, DisplayFunction → Identity];
```

```
Show[p1, p2];
```

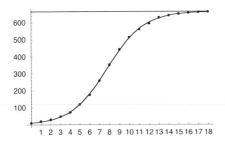

■ Example 2

From 1909 through 1934, the yearly numbers of hare and lynx pelts sold to the Hudson Bay Trading Company were as follows:

```
hare = {25, 50, 55, 75, 70, 55, 30, 20, 15, 15, 20,
    35, 60, 80, 85, 60, 30, 20, 10, 5, 5, 10, 30, 80, 100, 80};
```

```
lynx = {2, 4, 10, 14, 19, 14, 8, 9,
    2, 1, 1, 2, 4, 4, 8, 7, 9, 7, 4, 3, 2, 3, 3, 5, 7, 7};
```

(We considered an extended hare and lynx pelt data set in Section 9.1.2, p. 234.) The data is displayed as follows:

```
tt = Range[0, Length[hare] - 1];
th = Transpose[{tt, hare}];
tl = Transpose[{tt, lynx}];

p1 = Show[
    Graphics[{GrayLevel[0.4], Line[th], Line[tl], AbsolutePointSize[2],
        Map[Point, th], Map[Point, tl]}], Axes -> True, AspectRatio → 0.4];
```

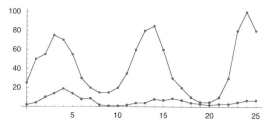

For later use, we combine the data into one list:

```
data = Flatten[Transpose[{hare, lynx}]]
```

```
{25, 2, 50, 4, 55, 10, 75, 14, 70, 19, 55, 14, 30, 8, 20,
  9, 15, 2, 15, 1, 20, 1, 35, 2, 60, 4, 80, 4, 85, 8, 60, 7, 30,
  9, 20, 7, 10, 4, 5, 3, 5, 2, 10, 3, 30, 3, 80, 5, 100, 7, 80, 7}
```

The predator–prey model of Section 23.3.2, p. 615, would possibly suit this data. Let $x(t)$ and $y(t)$ be the amounts of hare and lynx pelts, respectively. The equations are $x' = x(p - qy)$ and $y' = y(-P + Qx)$:

```
vars = {x[t], y[t]}; pars = {p, q, P, Q};

eqns[a_] := {x'[t] == x[t] (a[[1]] - a[[2]] y[t]),
    y'[t] == y[t] (-a[[3]] + a[[4]] x[t]), x[0] == 25, y[0] == 2}
```

Here **a** is a list that contains the parameters p, q, P, and Q. The initial conditions $x(0) = 25$ and $y(0) = 2$ are from the data. We also define the time interval and the time step of the data:

```
t0 = 0; t1 = 25; dt = 1;
```

To estimate the parameters of the model, define the following functions:

```
model[a_] := vars /. NDSolve[eqns[a], vars, {t, t0, t1}][[1]]
pred[a_] := Flatten[Table[Evaluate[model[a]], {t, t0, t1, dt}]]
crit[a_] := Apply[Plus, (data - pred[a])^2]
```

Here **model** gives the solution of the equations for given parameters, **pred** calculates predictions from the model (i.e., values of the solution at the same time instances as the data), and **crit** defines the expression to be minimized (i.e., the sum of the squares of the differences between the data and the model).

We will minimize `crit` using `FindMinimum`. To get reasonable starting values to the parameters, we can sample the criterion function at some random points. We guess that p, q, P, and Q may be in the intervals (0.5, 1), (0, 0.5), (0, 0.5), and (0, 0.2), respectively, and so we write the following:

```
SeedRandom[2];
s = Table[{Random[Real, {0.5, 1}], Random[Real, {0, 0.5}],
    Random[Real, {0, 0.5}], Random[Real, {0, 0.2}]}, {3000}];

(cs = Map[crit[#] &, s];) // Timing

{42. Second, Null}
```

We ask for the best value and the corresponding values of the parameters:

```
{Min[cs], s[[Position[cs, Min[cs]][[1, 1]]]]}

{5440.84, {0.875225, 0.198998, 0.446628, 0.0143937}}
```

We choose starting values near these values (we need to set the value of a system option; this is needed only in *Mathematica* 5, not in earlier versions):

```
Developer`SetSystemOptions["EvaluateNumericalFunctionArgument" → False];

{minvalue, est} = FindMinimum[crit[{p, q, P, Q}], {p, 0.87, 0.88},
  {q, 0.19, 0.20}, {P, 0.45, 0.44}, {Q, 0.015, 0.014}]

{2217.23, {p → 0.865392, q → 0.216316, P → 0.470179, Q → 0.0118829}}

Developer`SetSystemOptions["EvaluateNumericalFunctionArgument" → True];
```

The order of the initial values of the parameters may have some effect. In this example, it is advantageous to give the initial values in ascending order for p and q and in descending order for P and Q.

The estimated equations are as follows:

```
eqns[pars /. est]

{x'[t] == x[t] (0.889716 - 0.227723 y[t]),
 y'[t] == (-0.455849 + 0.011512 x[t]) y[t], x[0] == 25, y[0] == 2}
```

Here is the estimated model:

```
estModel = model[pars /. est]

{InterpolatingFunction[{{0., 25.}}, <>][t],
 InterpolatingFunction[{{0., 25.}}, <>][t]}
```

We show both the data and the model:

```
p2 = Plot[Evaluate[estModel], {t, 0, 25},
    PlotStyle → AbsoluteThickness[1], DisplayFunction → Identity];
```

Show[p1, p2];

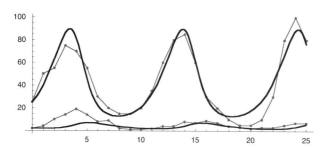

Partial Differential Equations

Introduction

> *A mathematician was given a test in which he had to produce steam starting with a block*
> *of ice which was stored in the refrigerator. He successfully described in great detail all the*
> *steps involved in the procedure, such as thawing the ice and boiling the water. Next he was*
> *asked to produce steam starting with the contents of a small pond. He replied: "Put a bucket*
> *of water from the pond into the refrigerator and apply the result of the previous problem."*

For partial differential equations (PDEs), *Mathematica* has the same two commands **DSolve** and **NDSolve** as for ordinary differential equations (ODEs).

    DSolve can in practice only find *general* solutions for quasilinear *first*-order PDEs; this means that **DSolve** cannot solve problems with initial and boundary conditions, and it can find general solutions for higher-order equations only accidentally. This is a consequence of the basic difficulty of solving PDEs; only in limited cases a general solution can be found. Many equations can be solved only if initial and boundary conditions are present, and even then the solution cannot usually be expressed in terms of standard functions but only as infinite series.

    NDSolve uses the method of lines. The number of independent variables is not restricted (❀5), and the problem can also have more than one dependent variable. In the case of two independent variables (typically one space and one time variable), the region of solution is a rectangle that has initial and boundary conditions on up to three sides; typical examples of such problems are 1D *parabolic* and *hyperbolic* problems. Elliptic problems cannot be solved with **NDSolve**.

If **DSolve** or **NDSolve** does not solve your problem, you can use other methods. The basic method is the separation of variables, which leads to series solutions; this method is considered in Section 24.2. We present several examples of how to handle such solutions. The idea in the use of this kind of series is to truncate it and use the resulting finite sum as an approximate solution. The Laplace transform can also be tried, and if *Mathematica* is not able to find the inverse transform, we can consult a table of Laplace transforms. For elliptic problems, we present a finite difference method in Section 24.3.3.

With the series solutions obtained by the method of separation of variables or by using the Laplace transform, we can obtain accurate numerical results (with up to, for example, six-digit precision). Thus, if you can find the series solution for your problem in a book or can derive such a solution by yourself, use it; otherwise, resort to numerical methods.

Using numerical methods, we can quickly get a low- or medium-precision solution (with up to, for example, three-digit precision). However, obtaining a high-precision result may either require a lot of computing time and memory or simply be impossible. You can easily give *Mathematica* or your computer a problem that is too hard to solve.

For more about PDEs with *Mathematica*, see Abell and Braselton (1997), Kythe, Puri, and Schäferkotter (1996), or Ganzha and Vorozhtsov (1996).

24.1 Symbolic Solutions

24.1.1 General Solutions

A quasilinear first-order PDE with two independent variables is of the following form:

$$p(x, y, u)\, u_x + q(x, y, u)\, u_y = r(x, y, u),$$

Here, x and y are the independent variables; u, which is a function of x and y, is the dependent variable; u_x and u_y are partial derivatives of u with respect to x and y; and p, q, and r are given functions of x, y, and u. **DSolve** is able to solve equations of this form.

> **DSolve[eqn, u[x, y], {x, y}]**   Give the general solution of the PDE

If the solution is requested with the command **DSolve[eqn, u, {x, y}]**, the solution is given as a pure function. A solution of this kind is useful if we want to calculate with it (e.g., if we want to check the solution).

As Example 4 below shows, **DSolve** can also find a general solution for some second-order equations.

■ Example 1: Constant Coefficient Equations

First we solve an equation with constant coefficients:

```
eqn = a D[u[x, y], x] + b D[u[x, y], y] + c u[x, y] == d
```
$$c\, u[x, y] + b\, u^{(0,1)}[x, y] + a\, u^{(1,0)}[x, y] == d$$

```
DSolve[eqn, u[x, y], {x, y}] // Simplify
```

$$\left\{\left\{u[x, y] \rightarrow \frac{d}{c} + e^{-\frac{cx}{a}} C[1]\left[-\frac{bx}{a} + y\right]\right\}\right\}$$

The solution contains an arbitrary function $C[1]$ with the argument $-\frac{bx}{a} + y$. We could simplify the arbitrary function to $C[1][ay - bx]$. To check the solution, we ask for it as a pure function:

```
DSolve[eqn, u, {x, y}]
```

$$\left\{\left\{u \rightarrow \text{Function}\left[\{x, y\}, \frac{e^{-\frac{cx}{a}}\left(d\,e^{\frac{cx}{a}} + c\,C[1]\left[\frac{-bx+ay}{a}\right]\right)}{c}\right]\right\}\right\}$$

```
eqn /. % // Simplify      {True}
```

Next we solve an equation with three independent variables:

```
eqn =
   a D[u[x, y, z], x] + b D[u[x, y, z], y] + c D[u[x, y, z], z] + d u[x, y, z] == e;
```

```
DSolve[eqn, u[x, y, z], {x, y, z}] // Simplify
```

$$\left\{\left\{u[x, y, z] \rightarrow \frac{e}{d} + e^{-\frac{dx}{a}} C[1]\left[-\frac{bx}{a} + y, -\frac{cx}{a} + z\right]\right\}\right\}$$

■ Example 2: A Quasilinear Equation

Here is a quasilinear equation:

```
eqn = x^2 D[u[x, y], x] - x y D[u[x, y], y] == -y u[x, y]
```

$$-xy\,u^{(0,1)}[x, y] + x^2\,u^{(1,0)}[x, y] == -y\,u[x, y]$$

```
DSolve[eqn, u[x, y], {x, y}]
```

$$\left\{\left\{u[x, y] \rightarrow e^{\frac{y}{2x}} C[1][xy]\right\}\right\}$$

■ Example 3: A Birth–Death Process

Let $X(t)$ be the size of a population at time t, and suppose that the population develops according to a birth–death process with birth rate λ and death rate μ (suppose $\lambda \neq \mu$). It can be shown that the probability-generating function $p(s, t)$ of $X(t)$ satisfies the following quasilinear first-order PDE:

```
eqn = (1 - s) (μ - λ s) D[p[s, t], s] - D[p[s, t], t] == 0
```

$$-p^{(0,1)}[s, t] + (1 - s)(-s\lambda + \mu)\,p^{(1,0)}[s, t] == 0$$

The general solution of this equation is as follows:

```
sol = p[s, t] /. DSolve[eqn, p[s, t], {s, t}][[1]]
```

$$C[1]\left[\frac{t\lambda - t\mu + \text{Log}[-1 + s] - \text{Log}[s\lambda - \mu]}{\lambda - \mu}\right]$$

If there are k individuals in the population at time 0, we know that $p(s, 0) = s^k$. We write this initial condition:

```
cond = (sol /. t → 0) == s^k
```

$$C[1]\left[\frac{\text{Log}[-1 + s] - \text{Log}[s\lambda - \mu]}{\lambda - \mu}\right] == s^k$$

From this condition, we can solve the function c[1] simply by first denoting the argument of the function by z:

```
def = cond[[1, 1]] == z
```

$$\frac{\text{Log}[-1 + s] - \text{Log}[s \lambda - \mu]}{\lambda - \mu} == z$$

We then solve for s:

```
sz = Solve[def, s][[1]]
```

$$\left\{s \to \frac{e^{z\mu} - e^{z\lambda}\mu}{e^{z\mu} - e^{z\lambda}\lambda}\right\}$$

We lastly insert this into the righthand side of the condition:

```
c1 = cond[[2]] /. sz
```

$$\left(\frac{e^{z\mu} - e^{z\lambda}\mu}{e^{z\mu} - e^{z\lambda}\lambda}\right)^k$$

Thus, c[1][z] is this expression. We can now write the final expression for the probability-generating function **sol**. We simply insert the argument of c[1] into **sol** for z in **c1** and simplify:

```
sol2 = c1 /. z → sol[[1]] // FullSimplify
```

$$\left(1 + \frac{-\lambda + \mu}{\lambda + \frac{e^{t(-\lambda+\mu)}(-s\lambda+\mu)}{-1+s}}\right)^k$$

We can now calculate, for example, the expectation of $X(t)$:

```
Limit[D[sol2, s], s → 1]
```

$$e^{t(\lambda-\mu)} k$$

We can see that, when t approaches infinity, the expected value of $X(t)$ goes to zero if $\lambda < \mu$ and to infinity if $\lambda > \mu$. The probability of zero individuals at time t (meaning extinction) is as follows:

```
sol2 /. s → 0
```

$$\left(1 + \frac{-\lambda + \mu}{\lambda - e^{t(-\lambda+\mu)}\mu}\right)^k$$

From this we can show that, when t approaches infinity, the probability of zero individuals goes to one if $\lambda < \mu$ and to $(\mu/\lambda)^k$ if $\lambda > \mu$. By also solving the case $\lambda = \mu$, we could show that the probability of zero individuals goes to one even in this case, although the mean number of individuals is the constant k for all t.

■ Example 4: The Wave Equation

Some second-order equations also have a general solution. One of these equations is the wave equation:

```
eqn = D[u[x, t], t, t] - c^2 D[u[x, t], x, x] == f
```

$$u^{(0,2)}[x, t] - c^2 u^{(2,0)}[x, t] == f$$

```
Simplify[DSolve[eqn, u[x, t], {x, t}], c > 0]
```

$$\left\{\left\{u[x, t] \to -\frac{f x^2}{2 c^2} + C[1]\left[t + \frac{x}{c}\right] + C[2]\left[t - \frac{x}{c}\right]\right\}\right\}$$

■ Example 5: A Nonlinear Equation

For most nonlinear equations, general solutions cannot be obtained; however, with a package, we can calculate a *complete integral*. This is a sufficiently representative family of particular solutions. Indeed, solutions to almost all boundary value problems can be expressed in quadratures of the complete integral. For example:

> `eqn = D[u[x, y], x] D[u[x, y], y] == c`
>
> $u^{(0,1)} [x, y] u^{(1,0)} [x, y] == c$

We load the package and ask for a complete integral:

> `<< Calculus`DSolveIntegrals``
>
> `sol = u[x, y] /. CompleteIntegral[eqn, u[x, y], {x, y}][[1]]`
>
> $B[1] + \dfrac{c\,x}{B[2]} + y\,B[2]$

Here B[1] and B[2] are arbitrary parameters. The solution satisfies the equation:

> `eqn /. u → Function[{x, y}, Evaluate[sol]]`   True

24.1.2 Using the Laplace Transform

In using the Laplace transform to solve PDEs, the critical point is to find the inverse transform. It is probable that *Mathematica* cannot find it. In particular, *Mathematica* cannot find inverse transforms that are in the form of an infinite sum.

■ Example 1: A Wave Problem

Consider the following wave equation:

$$u_{tt} - c^2 u_{xx} = 0, \quad 0 < x < 1, \quad t > 0,$$

$$u(x, 0) = d \sin(2 \pi x), \quad u_t(x, 0) = 0, \quad u(0, t) = u(1, t) = 0.$$

First, take the Laplace transform of the equation:

> `eqn = D[u[x, t], t, t] - c^2 D[u[x, t], x, x] == 0`
>
> $u^{(0,2)} [x, t] - c^2 u^{(2,0)} [x, t] == 0$
>
> `lapeqn1 = LaplaceTransform[eqn, t, s]`
>
> `s^2 LaplaceTransform[u[x, t], t, s] -`
>   `c^2 LaplaceTransform[u`$^{(2,0)}$`[x, t], t, s] - s u[x, 0] - u`$^{(0,1)}$`[x, 0] == 0`

Simplify the notation, and take the initial conditions into account:

> `lapeqn2 = lapeqn1 /. {LaplaceTransform[u[x, t], t, s] → U[x],`
>   `LaplaceTransform[D[u[x, t], x, x], t, s] → U''[x],`
>   `u[x, 0] → d Sin[2 π x], Derivative[0, 1][u][x, 0] → 0}`
>
> $-d\,s\,Sin[2 \pi x] + s^2 U[x] - c^2 U''[x] == 0$

Solve the transformed equation by using the boundary conditions:

```
lap = U[x] /. DSolve[{lapeqn2, U[0] == 0, U[1] == 0}, U[x], x][[1]] // Simplify
```

$$\frac{d\,s\,\text{Sin}[2\,\pi\,x]}{4\,c^2\,\pi^2 + s^2}$$

Mathematica succeeds in inverting this:

```
sol = InverseLaplaceTransform[lap, s, t]
```

d Cos[2 c π t] Sin[2 π x]

Plot the solution:

```
Plot3D[Evaluate[sol /. {c → 1, d → 1}], {x, 0, 1}, {t, 0, 2},
    AxesLabel → {"x", "t", ""}, Ticks → {{0, 1}, {0, 1, 2}, {-1, 0, 1}}];
```

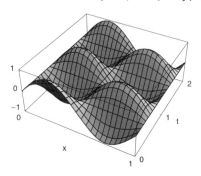

■ Example 2: A Heat Problem

Consider the following heat problem:

$$u_t - c\,u_{xx} = 0, \quad 0 < x < a, \quad t > 0,$$

$$u(x, 0) = u_0, \quad u_x(0, t) = 0, \quad u(a, t) = u_1.$$

Proceed as in Example 1:

```
eqn = D[u[x, t], t] - c D[u[x, t], x, x] == 0
```

$u^{(0,1)}[x, t] - c\,u^{(2,0)}[x, t] == 0$

```
lapeqn1 = LaplaceTransform[eqn, t, s]
```

s LaplaceTransform[u[x, t], t, s] -
    c LaplaceTransform[$u^{(2,0)}$[x, t], t, s] - u[x, 0] == 0

```
lapeqn2 = lapeqn1 /. {LaplaceTransform[u[x, t], t, s] → U[x],
    LaplaceTransform[D[u[x, t], x, x], t, s] → U''[x], u[x, 0] → u0}
```

$-u0 + s\,U[x] - c\,U''[x] == 0$

The solution of the transformed equation is as follows:

```
lap = U[x] /. DSolve[{lapeqn2, U'[0] == 0, U[a] == u1 / s}, U[x], x][[1]] //
    ExpToTrig // FullSimplify
```

$$\frac{u0 + (-u0 + u1)\,\text{Cosh}\left[\frac{\sqrt{s}\,x}{\sqrt{c}}\right]\,\text{Sech}\left[\frac{a\,\sqrt{s}}{\sqrt{c}}\right]}{s}$$

Mathematica cannot invert this, but we can consult a table of Laplace transforms. For example, formula 33.153 in Spiegel (1999) is appropriate, and we get the solution in the form of an infinite series:

$$u_0 + (u_1 - u_0)\left(1 + \frac{4}{\pi} \sum_{n=1}^{\infty} \frac{(-1)^n}{2n-1} \exp\left(-\frac{c(2n-1)^2 \pi^2 t}{4a^2}\right) \cos\left(\frac{(2n-1)\pi x}{2a}\right)\right).$$

Let us assume that $u_0 = 0$ and $u_1 = c = a = 1$. We take 60 terms from the sum and plot the resulting approximate solution:

```
uappr = 1 + 4 / π Sum[ (-1) ^n / (2 n - 1)
          Exp[- (2 n - 1) ^2 π^2 t / 4] Cos[ (2 n - 1) π x / 2], {n, 60}];

Plot3D[Evaluate[uappr], {x, 0, 1}, {t, 0, 2}, ViewPoint → {-2, -1.5, 0.8},
    AxesLabel → {"x", "t", ""}, Ticks → {{0, 1}, {0, 1, 2}, {0, 1}}];
```

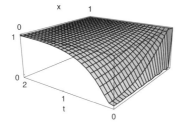

24.2 Series Solutions

24.2.1 1D Parabolic Problems

Consider the following heat problem:

$$u_t - c\, u_{xx} = F(x, t), \quad 0 < x < a, \quad t > 0,$$

$$u(x, 0) = f(x), \quad u(0, t) = u(a, t) = 0.$$

This model can be interpreted as follows. The value of $u(x, t)$ is the temperature of a bar at position x and time t. The bar is assumed to be slender and homogeneous and of uniform cross-section. The bar lies along the x axis with ends at $x = 0$ and $x = a$. The lateral surface of the bar is insulated. The value of $F(x, t)$ is the amount of heat per unit volume per unit time generated at the point x at time t. The constant c depends on the properties of the bar (c is the thermal conductivity divided by the product of the specific heat and the material density). The ends of the bar are kept at the constant temperature 0. The initial temperature of the bar at x is $f(x)$. The goal is to determine the temperature of the bar for $t > 0$.

Separation of variables is a well-known method for solving PDEs. The solution is usually in the form of an infinite sum. The solution of this problem, using the of separation of variables, is as follows (Dennemeyer 1968, p. 309):

$$u(x, t) = \sum_{n=1}^{\infty} [A_n \exp(-c\, v_n^2\, t) + H_n(t)] \sin(v_n x), \quad v_n = \frac{n\pi}{a}, \quad A_n = \frac{2}{a} \int_0^a f(x) \sin(v_n x)\, dx,$$

$$H_n(t) = \int_0^t F_n(\tau) \exp[-c\, v_n^2 (t - \tau)]\, d\tau, \quad F_n(t) = \frac{2}{a} \int_0^a F(x, t) \sin(v_n x)\, dx.$$

■ Calculating the General Term

Consider the following example:

```
a = 1; c = 1; F = 0; f = x (1 - x);
```

First calculate the coefficients:

```
vn = n π / a;
```

```
An = Simplify[2 / a Integrate[f Sin[vn x], {x, 0, a}], n ∈ Integers]
```

$$-\frac{4\left(-1+(-1)^{n}\right)}{n^{3}\pi^{3}}$$

```
Fn = 2 / a Integrate[F Sin[vn x], {x, 0, a}]        0
```

```
Hn = Integrate[(Fn /. t → τ) Exp[-c vn^2 (t - τ)], {τ, 0, t}]        0
```

Here is the nth term of the series solution:

```
term = (An Exp[-c vn^2 t] + Hn) Sin[vn x]
```

$$-\frac{4\left(-1+(-1)^{n}\right)e^{-n^{2}\pi^{2}t}\sin[n\pi x]}{n^{3}\pi^{3}}$$

If exact integration with **Integrate** does not succeed, we can use **NIntegrate**. Exact integration is handy, because we have to do only one integration for a general n, and from the result we obtain all of the coefficients we are going to use. If we have to use numerical integration, we must separately integrate each coefficient we need.

■ Choosing an Approximation

How many terms from the infinite series should we choose so that the results would be precise enough? We investigate the solution when $x = 0.5$. By making experiments with series of different lengths, we would find that, to obtain the correct value 0.25 for $u(0.5, 0)$ to 6 decimal places, we need 63 terms; here are the corresponding values of $u(0.5, t)$ for $t = 0, 0.1, 0.2, 0.3, 0.4,$ and 0.5:

```
Table[Sum[term, {n, 63}] /. x → 0.5, {t, 0, 0.5, 0.1}]
```

```
{0.25, 0.0961619, 0.0358408, 0.0133581, 0.00497868, 0.00185559}
```

If we are satisfied with 5 correct decimals for $u(0.5, 0)$, then 31 terms suffice. For 4-digit precision, 13 terms suffice.

We choose to take 15 terms and form an approximate solution (note that the terms corresponding with even n values are zero):

```
uappr = Sum[term, {n, 15}] // N
```

$0.258012\ 2.71828^{-9.8696\ t}\ \sin[3.14159\ x]\ +$
$\quad 0.00955601\ 2.71828^{-88.8264\ t}\ \sin[9.42478\ x]\ +$
$\quad 0.0020641\ 2.71828^{-246.74\ t}\ \sin[15.708\ x]\ +$
$\quad 0.000752222\ 2.71828^{-483.611\ t}\ \sin[21.9911\ x]\ +$
$\quad 0.000353926\ 2.71828^{-799.438\ t}\ \sin[28.2743\ x]\ +$
$\quad 0.000193848\ 2.71828^{-1194.22\ t}\ \sin[34.5575\ x]\ +$
$\quad 0.000117438\ 2.71828^{-1667.96\ t}\ \sin[40.8407\ x]\ +$
$\quad 0.0000764481\ 2.71828^{-2220.66\ t}\ \sin[47.1239\ x]$

■ Using the Solution

Calculate some numerical values:

```
Table[uappr /. x → 0.5, {t, 0, 0.5, 0.1}]

{0.249969, 0.0961619, 0.0358408, 0.0133581, 0.00497868, 0.00185559}
```

Note that the exact value of $u(0.5, 0)$ is 0.25 according to the initial condition, so the approximate value 0.249969 is quite accurate. Here is a plot of the approximate solution:

```
Plot3D[Evaluate[uappr], {t, 0, 0.5}, {x, 0, 1}, PlotRange → All,
    AxesLabel → {"t", "x", ""}, Ticks → {{0, 0.5}, {0, 1}, {0, 0.2}}];
```

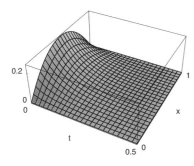

■ Animating

If we want to animate the time evolution of the solution, we first plot the solution for several values of t. With the following command, we could create 51 plots (we do not show these plots here; they can be found in the *Help Browser* version of this book):

```
Do[Plot[Evaluate[uappr /. t → ti], {x, 0, 1},
    PlotRange → {0, 0.26}, Ticks → {{1}, {0.1, 0.2}}], {ti, 0, 0.5, 0.01}]
```

We would then double-click on one of the plots, and the animation would start. See Section 5.1.3, p. 116, for more about animating. Here we only show some separate figures as a graphics array:

```
pp = Table[Plot[Evaluate[uappr /. t → ti],
    {x, 0, 1}, PlotRange → {0, 0.26}, Ticks → {{1}, {0.1, 0.2}},
    DisplayFunction → Identity], {ti, 0, 0.45, 0.05}];

Show[GraphicsArray[Partition[pp, 5]]];
```

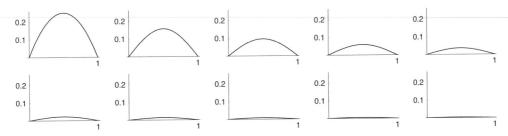

24.2.2 1D Hyperbolic Problems

Consider the following wave problem:

$$u_{tt} - c^2\,u_{xx} = F(x,\,t), \quad 0 < x < a, \quad t > 0,$$

$$u(x,\,0) = f(x), \quad u_t(x,\,0) = g(x), \quad u(0,\,t) = u(a,\,t) = 0.$$

One interpretation of the model is as follows. The value of $u(x,\,t)$ is the transverse displacement of a homogeneous thin string at time t of the point on the string with the abscissa of x. The string is assumed to be perfectly flexible and subject to uniform tension. In addition to the tension, an external force transverse $F(x,\,t)$ (force/unit mass) is acting on the string; one example is gravity. The constant c^2 depends on the properties of the string (c^2 is the tension divided by the linear density of the string). The ends of the string are fastened at $x = 0$ and $x = a$. The string is pulled aside according to the function $f(x)$ and then released at the speed $g(x)$. The problem is to determine the subsequent motion of the string.

Using the method of separation of variables (Dennemeyer 1968, pp. 170–175), we obtain the following solution:

$$u(x,\,t) = \sum_{n=1}^{\infty} [A_n \cos(c\,v_n\,t) + B_n \sin(c\,v_n\,t) + H_n(t)]\sin(v_n\,x), \quad v_n = \frac{n\,\pi}{a},$$

$$A_n = \frac{2}{a}\int_0^a f(x)\sin(v_n\,x)\,dx, \quad B_n = \frac{2}{n\,\pi\,c}\int_0^a g(x)\sin(v_n\,x)\,dx,$$

$$F_n(t) = \frac{2}{a}\int_0^a F(x,\,t)\sin(v_n\,x)\,dx, \quad H_n(t) = \frac{1}{c\,v_n}\int_0^t F_n(\tau)\sin[c\,v_n(t-\tau)]\,d\tau.$$

■ Calculating the General Term

Consider the following example:

```
a = 1; c = 1; F = -9.80665; f = 10 x^2 (1 - x)^2; g = 0;
```

(9.80665 is acceleration due to gravity.) First we calculate the coefficients:

```
vn = n π / a;
```

```
An = Simplify[2 / a Integrate[f Sin[vn x], {x, 0, a}], n ∈ Integers]
```

$$\frac{40\,(-1 + (-1)^n)\,(-12 + n^2\,\pi^2)}{n^5\,\pi^5}$$

```
Bn = 2 / (n π c) Integrate[g Sin[vn x], {x, 0, a}]      0
```

```
Fn = Simplify[2 / a Integrate[F Sin[vn x], {x, 0, a}], n ∈ Integers]
```

$$\frac{-6.24311 + 6.24311\,(-1)^n}{n}$$

```
Hn = 1 / (c vn) Integrate[(Fn /. t → τ) Sin[c vn (t - τ)], {τ, 0, t}]
```

$$\frac{-1.98724 + 1.98724\,(-1)^n + (1.98724 - 1.98724\,(-1)^n)\,\mathrm{Cos}[n\,\pi\,t]}{n^3\,\pi}$$

Here is the nth term of the series solution:

```
term = (An Cos[c vn t] + Bn Sin[c vn t] + Hn) Sin[vn x]
```

$$\left(\frac{40 \, (-1 + (-1)^n) \, (-12 + n^2 \, \pi^2) \, \text{Cos}[n \pi t]}{n^5 \, \pi^5} + \right.$$
$$\left. \frac{-1.98724 + 1.98724 \, (-1)^n + (1.98724 - 1.98724 \, (-1)^n) \, \text{Cos}[n \pi t]}{n^3 \, \pi} \right) \text{Sin}[n \pi x]$$

■ Choosing an Approximation

About 150 terms are needed for 6-digit precision; here are the corresponding values of $u(0.5, t)$ for $t = 0, 0.4, ..., 2.4$:

```
Table[Sum[term /. x → 0.5, {n, 150}], {t, 0, 2.4, 0.4}]

{0.625, -0.703533, -2.69653, -2.69653, -0.703533, 0.625, -0.703533}
```

If we want 5-digit precision, about 65 terms are needed, and about 30 terms suffice for 4-digit precision.

We choose 30 terms from the series and form an approximation to the solution:

```
uappr = Sum[term, {n, 30}] // N // Chop;

Short[uappr, 8]

(0.55693 Cos[3.14159 t] + 0.31831 (-3.97449 + 3.97449 Cos[3.14159 t]))
    Sin[3.14159 x] + ≪13≫ + (-0.000105637 Cos[91.1062 t] +
        0.0000130514 (-3.97449 + 3.97449 Cos[91.1062 t])) Sin[91.1062 x]
```

■ Using the Solution

First we calculate some numerical values:

```
Table[uappr /. x → 0.5, {t, 0, 2.4, 0.4}]

{0.624953, -0.703626, -2.69652, -2.69652, -0.703626, 0.624953, -0.703626}
```

The exact value $u(0.5, 0)$ is 0.625 according to the initial condition $u(x, 0) = 10 \, x^2 (1 - x)^2$. We plot the movement of the center point $u(0.5, t)$ when t is in the interval $(0, 10)$:

```
Plot[Evaluate[uappr /. x → 0.5], {t, 0, 10}, AspectRatio → 0.3];
```

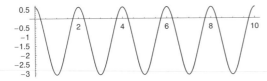

Here is a plot of the solution for t in $(0, 4)$:

```
Plot3D[Evaluate[uappr], {t, 0, 4}, {x, 0, 1},
   AxesLabel → {"t", "x", ""}, Ticks → {{0, 1, 2, 3, 4}, {0, 1}, {-3, 0}}];
```

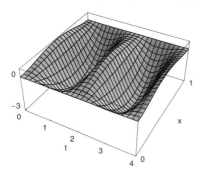

■ Animating

We could plot the solution for several values of *t* with the following command:

```
Do[Plot[Evaluate[uappr /. t → ti], {x, 0, 1},
   PlotRange → {-3.3, 0.7}, Ticks → {{1}, {-3}}], {ti, 0, 1.95, 0.05}];
```

(We do not show the 40 plots here; they can be found in the *Help Browser* version of this book.) The animation would start when one of the plots is double-clicked. Here we only show some plots in a graphics array:

```
pp =
   Table[Plot[Evaluate[uappr /. t → ti], {x, 0, 1}, PlotRange → {-3.3, 0.7},
      Ticks → None, DisplayFunction → Identity], {ti, 0, 2, 0.1}];

Show[GraphicsArray[Partition[pp, 7]], GraphicsSpacing → 0.2];
```

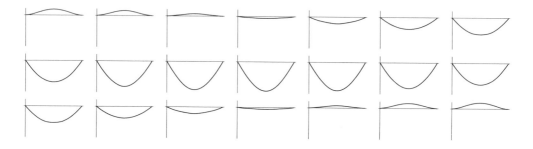

24.2.3 2D Hyperbolic Problems in Cartesian Coordinates

Consider the following wave problem:

$$u_{tt} - c^2(u_{xx} + u_{yy}) = F(x, y, t), \quad 0 < x < a, \quad 0 < y < b, \quad t > 0,$$

$$u(x, y, 0) = f(x, y), \quad u_t(x, y, 0) = g(x, y),$$

$$u(0, y, t) = u(a, y, t) = u(x, 0, t) = u(x, b, t) = 0.$$

Using the method of separation of variables (Dennemeyer 1968, pp. 191–194, 263–264), we obtain the following solution:

$$u(x, y, t) = \sum_{m=1}^{\infty} \sum_{n=1}^{\infty} [A_{mn} \cos(\omega_{mn} t) + B_{mn} \sin(\omega_{mn} t) + H_{mn}(t)] \phi_{mn}(x, y),$$

$$\phi_{mn}(x, y) = \sin(v_m x) \sin(w_n y), \quad v_m = \frac{m\pi}{a}, \quad w_n = \frac{n\pi}{b}, \quad \lambda_{mn} = v_m^2 + w_n^2, \quad \omega_{mn} = c\sqrt{\lambda_{mn}},$$

$$A_{mn} = \frac{4}{ab} \int_0^a \int_0^b f(x, y) \phi_{mn}(x, y) \, dx \, dy, \quad B_{mn} = \frac{4}{ab\omega_{mn}} \int_0^a \int_0^b g(x, y) \phi_{mn}(x, y) \, dx \, dy,$$

$$F_{mn}(t) = \frac{4}{ab} \int_0^a \int_0^b F(x, y, t) \phi_{mn}(x, y) \, dx \, dy, \quad H_{mn}(t) = \frac{1}{\omega_{mn}} \int_0^t F_{mn}(\tau) \sin[\omega_{mn}(t - \tau)] \, d\tau$$

■ Calculating the General Term

Consider the following example:

```
a = 1; b = 1; c = 1; F = 0; f = 10 x y (1 - x) (1 - y); g = 0;
```

First we calculate the integrals:

```
vm = m π / a; wn = n π / b; λmn = vm^2 + wn^2; ωmn = c Sqrt[λmn];

φmn = Sin[vm x] Sin[wn y];

Amn = 4 / (a b) Integrate[f φmn, {x, 0, a}, {y, 0, b}]
```

$$\frac{1}{m^3 n^3 \pi^6} \left(160 \left(m\pi \cos\left[\frac{m\pi}{2}\right] - 2\sin\left[\frac{m\pi}{2}\right]\right)\right)$$
$$\sin\left[\frac{m\pi}{2}\right] \left(n\pi \cos\left[\frac{n\pi}{2}\right] - 2\sin\left[\frac{n\pi}{2}\right]\right) \sin\left[\frac{n\pi}{2}\right]\right)$$

```
Bmn = 4 / (a b ωmn) Integrate[g φmn, {x, 0, a}, {y, 0, b}]      0

Fmn = 4 / (a b) Integrate[F φmn, {x, 0, a}, {y, 0, b}]      0

Hmn = 1 / ωmn Integrate[(Fmn /. t → τ) Sin[ωmn (t - τ)], {τ, 0, t}]      0
```

Here is then the (m, n)th term of the series solution:

```
term = (Amn Cos[ωmn t] + Bmn Sin[ωmn t] + Hmn) φmn // Simplify
```

$$\frac{1}{m^3 n^3 \pi^6} \left(160 \cos\left[\sqrt{m^2 + n^2}\, \pi t\right] \left(m\pi \cos\left[\frac{m\pi}{2}\right] - 2\sin\left[\frac{m\pi}{2}\right]\right) \sin\left[\frac{m\pi}{2}\right]\right.$$
$$\left.\left(n\pi \cos\left[\frac{n\pi}{2}\right] - 2\sin\left[\frac{n\pi}{2}\right]\right) \sin\left[\frac{n\pi}{2}\right] \sin[m\pi x] \sin[n\pi y]\right)$$

■ Choosing an Approximation

The values of $u(0.5, 0.5, t)$ for $t = 0, 0.3, 0.6, 0.9, 1.2$, and 1.5 are as follows if we use 55 as the upper bound for both m and n:

```
Table[Evaluate[Sum[term /. {x → 0.5, y → 0.5}, {m, 55}, {n, 55}]],
   {t, 0, 1.5, 0.3}]
   {0.624996, 0.202005, -0.650998, -0.394842, 0.355062, 0.659583}
```

(**Evaluate** speeds up the computation.) These values for the upper bounds seem to suffice for 5-digit precision (for 6-digit precision, it seems that the upper bound for m and n has to be about 130). For 4- and 3-digit precision, 25 and 15, respectively, suffice as the upper bound.

We choose 25 as the upper bound for *m* and *n* (this gives us a total of 625 terms, many of which are, however, zero) and form an approximation to the solution:

```
uappr = Sum[term, {m, 25}, {n, 25}] // N;
```

```
Short[uappr, 5]
```

$0.665703 \, \text{Cos}[4.44288 \, t] \, \text{Sin}[3.14159 \, x] \, \text{Sin}[3.14159 \, y] + \ll 167 \gg +$
$2.72672 \times 10^{-9} \, \text{Cos}[111.072 \, t] \, \text{Sin}[78.5398 \, x] \, \text{Sin}[78.5398 \, y]$

■ Using the Solution

First we calculate the solution at the point (0.5, 0.5) for some values of *t*:

```
Table[uappr /. {x → 0.5, y → 0.5}, {t, 0, 1.5, 0.3}]
```

$\{0.625036, 0.202054, -0.651003, -0.394831, 0.355097, 0.659562\}$

The exact value when *t* is 0 is 0.625 according to the initial condition. Then we plot the movement of the point *u*(0.5, 0.5, *t*) when *t* is in the interval (0, 20):

```
Plot[Evaluate[uappr /. {x → 0.5, y → 0.5}], {t, 0, 20}, PlotPoints → 50,
    AspectRatio → 0.2, Ticks → {Range[0, 30, 5], {-0.5, 0.5}}];
```

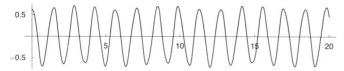

With the following command, we could plot the surface profile for a sequence of values of *t* when *y* is 0.5 (the 71 plots can be found, for animation purposes, in the *Help Browser* version of this book):

```
Do[Plot[Evaluate[uappr /. y → 0.5], {x, 0, 1}, Ticks → None,
    PlotRange → {{0, 1}, {-0.7, 0.7}}], {t, 0, 1.4, 0.02}];
```

Here we only show a collection of plots:

```
pp =
    Table[Plot[Evaluate[uappr /. y → 0.5], {x, 0, 1}, PlotRange → {-0.7, 0.7},
        Ticks → None, DisplayFunction → Identity], {t, 0, 1.4, 0.1}];
```

```
Show[GraphicsArray[Partition[pp, 5], GraphicsSpacing → 0.2]];
```

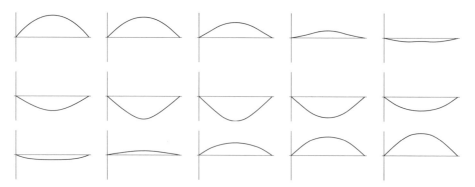

The surface is displayed as follows when t is 1.5:

```
Plot3D[Evaluate[uappr /. t → 1.5], {x, 0, 1}, {y, 0, 1},
  AxesLabel → {"x", "y", ""}, Ticks → {{0, 1}, {0, 1}, {0, 0.6}}];
```

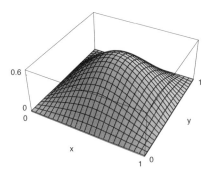

■ Animating

First we form several plots for various values of t (the plots are not shown here; they can be found in the *Help Browser* version of this book):

```
Do[Plot3D[Evaluate[uappr], {x, 0, 1}, {y, 0, 1}, Ticks → None,
  PlotRange → {-0.7, 0.7}, BoxRatios → Automatic], {t, 0, 1.48, 0.02}]
```

We could then double-click one of the plots to start the animation. Here is a graphics array of some plots:

```
pp = Table[Plot3D[Evaluate[uappr], {x, 0, 1}, {y, 0, 1},
    PlotRange → {-0.7, 0.7}, Ticks → None, BoxRatios → Automatic,
    PlotPoints → 15, DisplayFunction → Identity], {t, 0, 1.4, 0.2}];

Show[GraphicsArray[Partition[pp, 4]]];
```

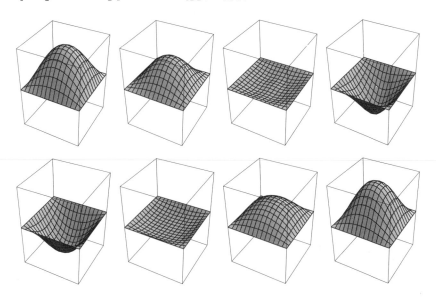

24.2.4 2D Hyperbolic Problems in Polar Coordinates

Consider the following circularly symmetric wave problem in polar coordinates:

$$u_{tt} - c^2(u_{rr} + u_r/r) = F(r), \quad 0 < r < a, \quad t > 0,$$

$$u(r, 0) = f(r), \quad u_t(r, 0) = g(r), \quad u(a, t) = 0.$$

By the method of separation of variables (Dennemeyer 1968, pp. 201–202), we obtain the following solution:

$$u(r, t) = \sum_{n=1}^{\infty} [A_n \cos(v_n t) + B_n \sin(v_n t) + H_n(t)] J_0(\xi_n r/a), \quad v_n = \frac{c\,\xi_n}{a},$$

$$A_n = \frac{2}{a^2 J_1(\xi_n)^2} \int_0^a r\,f(r)\,J_0(\xi_n r/a)\,dr, \quad B_n = \frac{2}{a^2 v_n J_1(\xi_n)^2} \int_0^a r\,g(r)\,J_0(\xi_n r/a)\,dr,$$

$$G_n = \frac{2}{a^2 J_1(\xi_n)^2} \int_0^a r\,F(r)\,J_0(\xi_n r/a)\,dr, \quad H_n(t) = \frac{2\,G_n}{v_n^2} \sin^2(v_n t/2).$$

J_0 and J_1 are Bessel functions of the first kind of order zero and one, and ξ_n is the nth zero of J_0.

■ Calculating an Approximation

Consider the following example:

```
a = 1; c = 1; F = 0; f = 1 - r^2; g = 0;
```

We use 30 terms from the infinite series. First we need 30 zeros of J_0. In a package, we have the ready-to-use command **BesselJZeros[order, n]**, which gives a list of the first n zeros of **BesselJ[order, x]**, the Bessel function of the first kind of order **order**:

```
<< NumericalMath`BesselZeros`
```

```
zeros = BesselJZeros[0, 30]
```

```
{2.40483, 5.52008, 8.65373, 11.7915, 14.9309, 18.0711, 21.2116, 24.3525,
 27.4935, 30.6346, 33.7758, 36.9171, 40.0584, 43.1998, 46.3412,
 49.4826, 52.6241, 55.7655, 58.907, 62.0485, 65.19, 68.3315, 71.473,
 74.6145, 77.756, 80.8976, 84.0391, 87.1806, 90.3222, 93.4637}
```

Then we calculate coefficients, and now we must use numerical integration because of the Bessel functions in the formulas. Each coefficient must be calculated separately:

```
v = c zeros / a;
```

```
A = Map[2 / (a^2 BesselJ[1, #] ^2)
      NIntegrate[r f BesselJ[0, # r / a], {r, 0, a}] &, zeros]
```

```
{1.10802, -0.139778, 0.0454765, -0.0209909, 0.0116362,
 -0.00722118, 0.00483787, -0.00342568, 0.00252953, -0.00193015,
 0.00151221, -0.00121077, 0.000987185, -0.000817394, 0.000685835,
 -0.000582113, 0.000499091, -0.000431745, 0.000376465, -0.000330609,
 0.000292208, -0.000259772, 0.000232161, -0.000208491, 0.000188066,
 -0.000170336, 0.000154861, -0.000141285, 0.000129319, -0.000118724}
```

Because $g(r)$ and $F(r)$ are zero, we know that all coefficients B_n, G_n, and H_n are zero. Anyway, here are the formulas for calculating them:

```
B = Map[2 / (a^2 (c # / a) BesselJ[1, #]^2) NIntegrate[r g BesselJ[0, # r / a],
     {r, 0, a}, AccuracyGoal → 10] &, zeros]
```

```
G = Map[2 / (a^2 BesselJ[1, #]^2) NIntegrate[r F BesselJ[0, # r / a],
     {r, 0, a}, AccuracyGoal → 10] &, zeros]
```

```
H = Table[2 G[[n]] / v[[n]]^2 Sin[v[[n]] t / 2]^2, {n, 30}]
```

Now we can form an approximate solution (note that all operations with lists are automatically done element by element):

```
uappr = Total[(A Cos[v t] + B Sin[v t] + H) BesselJ[0, zeros r / a]];
```

```
Short[uappr, 7]
```

```
BesselJ[0, 2.40483 r] (1.10802 Cos[2.40483 t] + 0. Sin[2.40483 t]) + ≪28≫ +
  BesselJ[0, 93.4637 r] (-0.000118724 Cos[93.4637 t] + 0. Sin[93.4637 t])
```

If we made enough experiments, it would turn out that 100 terms from the series solution does not quite suffice for 5-digit precision. Here are values of $u(0, t)$ for t = 0, 0.2, 0.4, ..., 1, using 100 terms:

```
{0.999997, 0.919997, 0.679997, 0.279996, -0.280007, -0.999731}
```

About 50 terms suffice for 4-digit precision, and 30 terms are sufficient for 3-digit precision.

■ Using the Solution

Here are some values of the solution:

```
Table[uappr /. r → 0, {t, 0, 1, 0.2}]
```

```
{0.999943, 0.919941, 0.679934, 0.279917, -0.280132, -0.998375}
```

The exact value of $u(0, 0)$ is 1 according to the initial condition. Here is the movement of the center point when t is in the interval (0, 30):

```
Plot[Evaluate[uappr /. r → 0], {t, 0, 30},
  AspectRatio → 0.2, Ticks → {Range[10, 50, 10], {-1, 1}}];
```

Next we plot the form of the surface for r in $(-1, 1)$ for various values of t (the plots can be found in the *Help Browser* version of this book):

```
pp = Table[Plot[Evaluate[uappr], {r, -1, 1}, Ticks → None,
    PlotRange → {{-1.1, 1.1}, {-1.25, 1.25}}, AspectRatio → Automatic,
    DisplayFunction → Identity], {t, 0, 1.1, 0.1}];
```

```
Show[GraphicsArray[Partition[pp, 6]]];
```

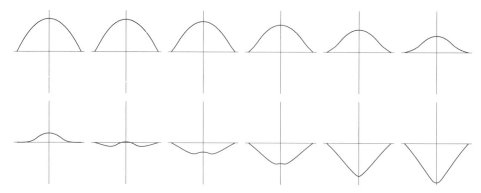

Lastly we plot the surface at times 0, 0.7, 1.2, and 2.8 (we use only the first 10 terms of **uappr**, because the plotting is time-consuming; the *Help Browser* version of this book contains more plots):

```
pp = Map[ParametricPlot3D[
    Evaluate[{r Cos[th], r Sin[th], Take[uappr, 10] /. t → #}],
    {r, 0, 1}, {th, 0, 2 π}, PlotRange → {-1.25, 1.25},
    Boxed → False, Axes → False, PlotPoints → {10, 20},
    DisplayFunction → Identity] &, {0, 0.7, 1.2, 2.8}];

Show[GraphicsArray[pp, GraphicsSpacing → -0.2]];
```

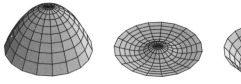  

24.2.5 2D Elliptic Problems

Consider the following elliptic problem:

$$u_{xx} + u_{yy} = 0, \quad 0 < x < a, \quad 0 < y < b,$$

$$u(0, y) = u(a, y) = u(x, b) = 0, \quad u(x, 0) = f(x).$$

One interpretation of the model is as follows. A thin rectangular homogeneous thermally conducting plate lies in the (x, y) plane and occupies the rectangle $0 \le x \le a$, $0 \le y \le b$. The value of $u(x, y)$ is the steady-state temperature of the plate at point (x, y). The faces of the plate are insulated, and no internal sources or sinks of heat are present. The edge $y = 0$ is kept at temperature $f(x)$, while the remaining edges are kept at zero temperature.

The series solution by the method of separation of variables is as follows (Dennemeyer 1968, pp. 147–148; the solution is derived below):

$$u(x, y) = \sum_{n=1}^{\infty} A_n \sin(v_n x) \sinh(v_n(b - y)), \quad v_n = \frac{n \pi}{a}, \quad A_n = \frac{2}{a \sinh(v_n b)} \int_0^a f(x) \sin(v_n x)\, dx.$$

■ Calculating the General Term

Consider the following example:

```
a = 1; b = 1; f = 4 x (1 - x);
```

Calculate the coefficient and the nth term of the series:

```
vn = n π / a;
```

```
An =
 Simplify[2 / (a Sinh[vn b]) Integrate[f Sin[vn x], {x, 0, a}], n ∈ Integers]
```

$$-\frac{16\,(-1+(-1)^n)\,\text{Csch}[n\,\pi]}{n^3\,\pi^3}$$

```
term = An Sin[vn x] Sinh[vn (b - y)]
```

$$-\frac{16\,(-1+(-1)^n)\,\text{Csch}[n\,\pi]\,\text{Sin}[n\,\pi\,x]\,\text{Sinh}[n\,\pi\,(1-y)]}{n^3\,\pi^3}$$

■ Choosing an Approximation

To get 6-digit precision, about 65 suffices for the upper bound of n. Here are the corresponding values of $u(0.5, y)$ for y = 0, 0.1, ..., 0.5:

```
Table[Sum[term, {n, 65}] /. x → 0.5, {y, 0, 0.5, 0.1}]
```

```
{1., 0.739132, 0.542517, 0.395755, 0.28663, 0.205315}
```

For 5- and 4-digit precision, about 20 and 10 terms suffice, respectively.

We choose 10 terms from the series and form an approximate solution:

```
uappr = Sum[term, {n, 1, 10}] // N
```

```
0.0893647 Sin[3.14159 x] Sinh[3.14159 (1. - 1. y)] +
  6.16932 × 10⁻⁶ Sin[9.42478 x] Sinh[9.42478 (1. - 1. y)] +
  2.48851 × 10⁻⁹ Sin[15.708 x] Sinh[15.708 (1. - 1. y)] +
  1.69356 × 10⁻¹² Sin[21.9911 x] Sinh[21.9911 (1. - 1. y)] +
  1.48804 × 10⁻¹⁵ Sin[28.2743 x] Sinh[28.2743 (1. - 1. y)]
```

■ Using the Solution

Some numerical values are as follows:

```
Table[uappr /. x → 0.5, {y, 0, 0.5, 0.1}]
```

```
{1.00049, 0.73915, 0.542518, 0.395755, 0.28663, 0.205315}
```

The exact value of $u(0.5, 0)$ is 1 (see the boundary condition). Next we plot the solution:

```
Plot3D[Evaluate[uappr], {x, 0, 1}, {y, 0, 1}, ViewPoint → {2.0, -2.4, 0.9},
  AxesLabel → {"x", "y", ""}, Ticks → {{0, 1}, {0, 1}, {0, 1}}];
```

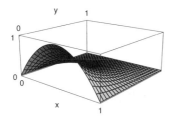

■ Derivation of the Separable Solution

As an example of the method of separation of variables, we solve the elliptic problem considered at the beginning of this section. We try to find the solution in the form $u(x, y) = X(x) Y(y)$ (i.e., in a form where the variables x and y are separated). For the homogeneous boundary conditions to be satisfied, we set $X(0) = X(a) = 0$ and $Y(b) = 0$. For the PDE to be satisfied, we must have $X'' Y + X Y'' = 0$. This can also be written as $X''/X = -Y''/Y$. Because this must hold for all x and y, both X''/X and $-Y''/Y$ must be the same constant; let us denote it $-\lambda$. We get two ODEs $X''/X = -\lambda$ and $Y''/Y = \lambda$. The general solution of the first equation is as follows:

```
Remove["Global`*"]
```

```
solx = DSolve[{X''[x] + λ X[x] == 0}, X, x]
```

$\{\{X \to \text{Function}[\{x\}, C[1] \cos[x \sqrt{\lambda}] + C[2] \sin[x \sqrt{\lambda}]]\}\}$

From the boundary conditions for X, we get the following conditions:

```
{X[0] == 0, X[a] == 0} /. solx
```

$\{\{C[1] == 0, C[1] \cos[a \sqrt{\lambda}] + C[2] \sin[a \sqrt{\lambda}] == 0\}\}$

Thus, $C[1]$ is 0. To get a nontrivial solution, we do not choose $C[2]$ to be 0 but rather $\sqrt{\lambda}$ to be $n\pi/a$; we denote $v_n = n\pi/a$. So, we get $X(x) = d_1 \sin(v_n x)$. For Y, we get the following solution:

```
soly =
  Y[y] /. DSolve[{Y''[y] - λ Y[y] == 0, Y[b] == 0}, Y[y], y][[1]] // ExpToTrig //
    FullSimplify
```

$2 e^{-b\sqrt{\lambda}} C[2] \sinh[(b - y) \sqrt{\lambda}]$

So, we can write $Y(y) = d_2 \sinh[v_n (b - y)]$. We denote $u_n(x, y) = A_n \sin(v_n x) \sinh[v_n(b - y)]$ and take an infinite sum of these terms to get the superposition $u(x, y) = \sum_{n=1}^{\infty} u_n(x, y)$. This satisfies all other conditions but still not the condition $u(x, 0) = f(x)$. We form the equation $u(x, 0) = f(x)$, multiply this equation by $\sin(v_m x)$, and then integrate both sides for x from 0 to a. From the resulting infinite sum on the lefthand side, only the mth term is nonzero, and it is $a/2$. This can be seen as follows:

```
Simplify[Integrate[Sin[m π x / a] Sin[n π x / a], {x, 0, a}], {m, n} ∈ Integers]
```

0

```
Simplify[Integrate[Sin[m π x / a] ^2, {x, 0, a}], m ∈ Integers]
```
$\dfrac{a}{2}$

We then know that $A_m \sinh(v_m b) a/2 = \int_0^a f(x) \sin(v_m x) dx$. From this equation, we can solve A_m, and so we get the solution mentioned earlier.

24.2.6 3D Elliptic Problems

Consider the following elliptic problem:

$$u_{xx} + u_{yy} + u_{zz} = 0, \quad 0 < x < a, \quad 0 < y < b, \quad 0 < z < c,$$

$$u(0, y, z) = u(a, y, z) = u(x, 0, z) = u(x, b, z) = u(x, y, c) = 0, \quad u(x, y, 0) = f(x, y).$$

Here we find the steady-state temperature in a solid, the bottom of which (at $z = 0$) is kept at a temperature $f(x, y)$ and the other sides at a temperature 0. Using the method of separation of variables, we obtain the following solution (Dennemeyer 1968, pp. 150–151):

$$u(x, y, z) = \sum_{m=1}^{\infty} \sum_{n=1}^{\infty} A_{mn} \sin(v_m x) \sin(w_n y) \sinh(\omega_{mn}(c - z)), \quad v_m = \frac{m\pi}{a}, \quad w_n = \frac{n\pi}{b},$$

$$A_n = \frac{4}{a b \sinh(\omega_{mn} c)} \int_0^a \int_0^b f(x, y) \sin(v_m x) \sin(w_n y) \, dx \, dy, \quad \omega_{mn} = \sqrt{v_m^2 + w_n^2}.$$

■ Calculating the General Term

Consider the following example:

```
a = b = c = 1; f = 20 x y (1 - x) (1 - y);
```

The highest temperature, 1.25, is at the center of the bottom. Calculate the general term:

```
vm = m π / a; wn = n π / b; ωmn = Sqrt[vm^2 + wn^2];
```

```
Amn = 4 / (a b Sinh[ωmn c])
    Integrate[f Sin[vm x] Sin[wn y], {x, 0, a}, {y, 0, b}] // Simplify
```

$$\frac{1}{m^3 n^3 \pi^6} \left(320 \, \text{Csch}\left[\sqrt{m^2 + n^2} \, \pi\right] \left(m \pi \cos\left[\frac{m\pi}{2}\right] - 2 \sin\left[\frac{m\pi}{2}\right]\right)\right.$$
$$\left.\sin\left[\frac{m\pi}{2}\right] \left(n \pi \cos\left[\frac{n\pi}{2}\right] - 2 \sin\left[\frac{n\pi}{2}\right]\right) \sin\left[\frac{n\pi}{2}\right]\right)$$

```
term = Amn Sin[vm x] Sin[wn y] Sinh[ωmn (c - z)] // Simplify
```

$$-\frac{1}{m^3 n^3 \pi^6} \left(320 \, \text{Csch}\left[\sqrt{m^2 + n^2} \, \pi\right]\right.$$
$$\left(m \pi \cos\left[\frac{m\pi}{2}\right] - 2 \sin\left[\frac{m\pi}{2}\right]\right) \sin\left[\frac{m\pi}{2}\right] \left(n \pi \cos\left[\frac{n\pi}{2}\right] - 2 \sin\left[\frac{n\pi}{2}\right]\right)$$
$$\left.\sin\left[\frac{n\pi}{2}\right] \sin[m\pi x] \sin[n\pi y] \sinh\left[\sqrt{m^2 + n^2} \, \pi (-1 + z)\right]\right)$$

■ Choosing an Approximation

To get an answer at 6-digit precision, the upper bound for m and n must be about 65:

```
uappr = Sum[term, {m, 65}, {n, 65}];
```

```
Table[Evaluate[uappr /. {x → 0.5, y → 0.5}], {z, 0, 1, 0.2}]
```

```
{1.25, 0.534528, 0.222296, 0.0897094, 0.0316101, 0.}
```

For 5- and 4-digit precision the upper bound should be about 35 and 15, respectively. We take 15 as the upper value for both m and n:

```
uappr = Sum[term, {m, 1, 15}, {n, 1, 15}] // N;
```

```
{uappr // First, uappr // Last}
```

```
{-0.0313243 Sin[3.14159 x] Sin[3.14159 y] Sinh[4.44288 (-1. + z)],
 -2.66686 × 10^-36 Sin[47.1239 x] Sin[47.1239 y] Sinh[66.6432 (-1. + z)]}
```

■ Using the Solution

The steady-state temperature at some points along a vertical line inside the solid is as follows:

```
Table[uappr /. {x → 0.5, y → 0.5}, {z, 0, 1, 0.2}]
```

{1.24969, 0.534528, 0.222296, 0.0897094, 0.0316101, 0.}

(The exact value at the bottom of the solid is 1.25.) We plot the temperature along this line:

```
Plot[Evaluate[uappr /. {x → 0.5, y → 0.5}],
  {z, 0, 1}, AxesLabel → {"z", "temp"}];
```

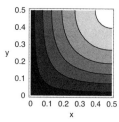

We also show a contour plot at the height $z = 0.02$ (we plot the contours only in the region $0 < x < 0.5, 0 < y < 0.5$):

```
ContourPlot[Evaluate[uappr /. z → 0.02],
  {x, 0, 0.5}, {y, 0, 0.5}, Contours → Range[0.1, 1.1, 0.2],
  FrameLabel → {"x", "y"}, RotateLabel → False];
```

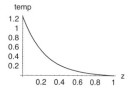

Then we plot surfaces of constant value (again in the region $0 < x < 0.5$, $0 < y < 0.5$) (for **ContourPlot3D**, see Section 7.4.2, p. 199):

```
<< Graphics`ContourPlot3D`
```

```
ContourPlot3D[Evaluate[uappr], {x, 0, 0.5}, {y, 0, 0.5},
  {z, 0, 1}, Contours → Range[0.1, 1.1, 0.2], PlotPoints -> {4, 5},
  PlotRange → {0, 0.605}, Axes → True, AxesLabel → {"x", "y", "z"},
  AxesEdge → {{-1, -1}, {1, -1}, {1, 1}}, ViewPoint → {3.1, -1.3, -0.2},
  Ticks → {{0, 0.2, 0.4}, Range[0, 0.5, 0.1], Range[0, 0.6, 0.1]}];
```

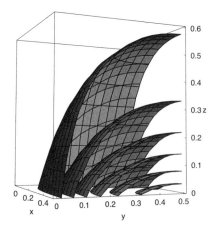

Here we can see the surfaces where the temperature is 0.1 (the highest surface), 0.3, 0.5, 0.7, 0.9, and 1.1 (the lowest surface).

24.3 Numerical Solutions

24.3.1 Parabolic and Hyperbolic Problems

NDSolve uses the method of lines and is typically suitable for solving 1D problems of parabolic or hyperbolic type. However, the problem may also have more than one space variable, and it may consist of several equations with several dependent variables. The command is not suitable for elliptic problems or for problems with singularities. For 2D elliptic problems, we present a finite difference method in Section 24.3.3, p. 667.

In problems with one space and one time variable, initial and boundary conditions can be given on three sides in a rectangular region of the space–time plane. The boundary conditions may contain derivatives, and they may be time dependent.

Here are typical commands for problems with one equation of parabolic or hyperbolic type:

> `sol = u[x,t] /. NDSolve[eqns, u[x,t], {x,a,b}, {t,c,d}][[1]]` Solve the problem
>
> `Plot3D[sol, {x,a,b}, {t,c,d}]` Plot the solution

■ Example 1: A Heat Problem

We solve the same problem that we solved in Section 24.2.1, p. 646:

> `eqns = {D[u[x, t], t] - D[u[x, t], x, x] == 0,`
> `  u[x, 0] == x (1 - x), u[0, t] == 0, u[1, t] == 0}`
>
> $\{u^{(0,1)}[x, t] - u^{(2,0)}[x, t] == 0, u[x, 0] == (1 - x) x, u[0, t] == 0, u[1, t] == 0\}$
>
> `sol = u[x, t] /. NDSolve[eqns, u[x, t], {x, 0, 1}, {t, 0, 0.5}][[1]]`
>
> `InterpolatingFunction[{{0., 1.}, {0., 0.5}}, <>][x, t]`

The result of **NDSolve** is a 2D interpolating function (see Section 21.2.2, p. 558). Here is the solution:

> `Plot3D[sol, {t, 0, 0.5}, {x, 0, 1}, PlotRange → All,`
> `  AxesLabel → {"t", "x", ""}, Ticks → {{0, 0.5}, {0, 1}, {0.1, 0.2}}];`

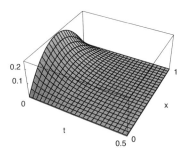

In Section 24.2.1, we obtained the following values by the method of separation of variables, using 63 terms:

    {0.25, 0.0961619, 0.0358408, 0.0133581, 0.00497868, 0.00185559}

These numbers were of approximately 6-digit precision. Here are the corresponding values from the solution given by **NDSolve**:

    Table[sol /. x → 0.5, {t, 0, 0.5, 0.1}]

    {0.25, 0.096167, 0.035922, 0.0133288, 0.00494671, 0.00183095}

The numbers are of approximately 2-digit precision. To get a more accurate result, we can set the goals for precision and accuracy:

    sol2 = u[x, t] /. NDSolve[eqns, u[x, t], {x, 0, 1},
    {t, 0, 0.5}, PrecisionGoal → 6, AccuracyGoal → 6][[1]];

Now we get numbers of about 4-digit precision:

    Table[sol2 /. x → 0.5, {t, 0, 0.5, 0.1}]

    {0.25, 0.0961618, 0.0358408, 0.0133588, 0.00497946, 0.00185615}

■ Example 2: A Wave Problem

We solve the same problem that we solved in Section 24.2.2, p. 648:

    eqns =
    {D[u[x, t], t, t] - D[u[x, t], x, x] == -9.80665, u[x, 0] == 10 x^2 (1 - x)^2,
    Derivative[0, 1][u][x, 0] == 0, u[0, t] == 0, u[1, t] == 0}

    {u$^{(0,2)}$[x, t] - u$^{(2,0)}$[x, t] == -9.80665,
    u[x, 0] == 10 (1 - x)^2 x^2, u$^{(0,1)}$[x, 0] == 0, u[0, t] == 0, u[1, t] == 0}

Note that **Derivative** is handy for specifying both the orders of the derivative and the point at which it is calculated (see Section 16.1.1, p. 418). We could also have written **(D[u[x, t], t] /. t → 0) == 0**. The solution is as follows:

    sol = u[x, t] /. NDSolve[eqns, u[x, t], {x, 0, 1}, {t, 0, 4}][[1]]

    InterpolatingFunction[{{0., 1.}, {0., 4.}}, <>][x, t]

    Plot3D[sol, {t, 0, 4}, {x, 0, 1}, PlotRange → All,
    AxesLabel → {"t", "x", ""}, Ticks → {{0, 1, 2, 3, 4}, {0, 1}, {-3, 0}}];

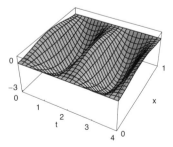

In Section 24.2.2, we obtained the following values via the method of separation of variables using 150 terms:

    {0.625, -0.703533, -2.69653, -2.69653, -0.703533, 0.625, -0.703533}

These numbers were of approximately 6-digit precision. Here are the corresponding values from the solution given by **NDSolve**:

```
Table[sol /. x → 0.5, {t, 0, 2.4, 0.4}]
```

```
{0.625, -0.702674, -2.70552, -2.6887, -0.714614, 0.62981, -0.701985}
```

The numbers have approximately 2 digits of precision.

■ Example 3: A System of Equations

In the following example, we have two dependent variables, u and v:

```
eqns = {D[u[x, t], t] - D[u[x, t], x, x] == v[x, t],
    D[v[x, t], t, t] - D[u[x, t], x, x] == 0, u[x, 0] == x (1 - x),
    u[0, t] == 0, u[1, t] == 0, v[x, 0] == 10 x^2 (1 - x)^2,
    Derivative[0, 1][v][x, 0] == 0, v[0, t] == 0, v[1, t] == 0};
```

```
sol = {u[x, t], v[x, t]} /.
    NDSolve[eqns, {u[x, t], v[x, t]}, {x, 0, 1}, {t, 0, 4}][[1]]
```

```
{InterpolatingFunction[{{0., 1.}, {0., 4.}}, <>][x, t],
 InterpolatingFunction[{{0., 1.}, {0., 4.}}, <>][x, t]}
```

```
Show[GraphicsArray[
    Map[Plot3D[#, {t, 0, 4}, {x, 0, 1}, PlotRange → All, PlotPoints → 20,
        AxesLabel → {"t", "x", ""}, DisplayFunction → Identity] &, sol]]];
```

24.3.2 Options and a Program

■ Options

The options of **NDSolve** were mentioned in Section 23.4.1 when we considered ODEs. The same options (except **StoppingTest**) can also be used for PDEs. The meaning of some options may, however, be somewhat new. In particular, some options accept a several-component list as a value; this reflects the number of independent variables. Which value belongs to which variable is inferred from the order of the independent variables in **NDSolve**. As usual, for each option in the following table, the default value is mentioned first.

Options for **NDSolve** *when solving PDEs:*

WorkingPrecision Precision used in internal computations; examples of values:
   MachinePrecision (⌘5), **20**

PrecisionGoal   If the value of the option is **p**, the relative error of the solution at each
   point considered should be of the order 10^{-p}; examples of values: **Automatic**
   (usually means **4**), **6**, **{6, 8}**

AccuracyGoal If the value of the option is **a**, the absolute error of the solution at each
   point considered should be of the order 10^{-a}; examples of values: **Automatic**
   (usually means **4**), **6**, **{6, 8}**

Method Method to use; possible values: **Automatic** (means **"Adams"** for nonstiff and
   "BDF" for stiff problems), **"Adams"**, **"BDF"** (⌘5), **"ExplicitRungeKutta"** (⌘5),
   "ImplicitRungeKutta" (⌘5), **"SymplecticPartitionedRungeKutta"** (⌘5)

StartingStepSize Initial step size used; examples of values: **Automatic**, **0.01**,
   {0.01, 0.02}

MaxStepSize Maximum size of each step; examples of values: **Automatic**, **0.01**,
   {0.01, 0.03}

MaxStepFraction (⌘5) Maximum fraction of the solution interval to cover in each
   step; examples of values: **0.1**, **0.05**

MaxSteps Maximum number of steps to take; examples of values: **Automatic** (means
   10000 (⌘5)), **20000**, **{100, 5000}**

NormFunction (⌘5) Norm to use for error estimation; default value: **Automatic**

DependentVariables (⌘5) List of all dependent variables; default value: **Automatic**

SolveDelayed Whether the derivatives are solved symbolically at the beginning
   (**False**) or at each step (**True**); possible values: **False**, **True**

Compiled Whether to compile the equations; possible values: **True**, **False**

StepMonitor (⌘5) Command to be executed after each step of the method; examples
   of values: **None**, **n++**, **AppendTo[steps, {t, u[x,t]}]**

EvaluationMonitor (⌘5) Command to be executed after each evaluation of the
   equation; examples of values: **None**, **n++**, **AppendTo[points, {t, u[x,t]}]**

NDSolve uses the method of lines. A typical situation is as follows:

```
Show[Graphics[{Table[Line[{{i, 1}, {i, 0}}], {i, 5}],
    AbsoluteThickness[1.5], Line[{{0, 1}, {0, 0}, {6, 0}, {6, 1}}],
    Text["u(x,0) = a₀(x)", {3, -0.05}, {0, 1}],
    Text["u(0,t) = b₀(t)", {-0.2, 0.5}, {1, 0}],
    Text["u(a,t) = b₁(t)", {6.2, 0.5}, {-1, 0}]}], PlotRange → All];
```

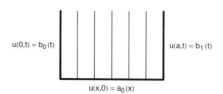

The x axis goes from left to right and the t axis from bottom to top. At $t = 0$, we have an initial condition $u(x, 0) = a_0(x)$. At $x = 0$ and $x = a$, we have boundary conditions $u(0, t) = b_1(t)$ and $u(a, t) = b_2(t)$; periodic boundary conditions of the form $u(0, t) = u(a, t)$ can also be given.

In this situation, the method of lines first discretizes the equation with respect to the space variable x by choosing some lines $x = x_i$. Along these lines, space derivatives such as u_x and u_{xx} are approximated by differences. In this way we get a system of ODEs that have only time derivatives. Initial conditions are obtained from $u(x, 0) = a_0(x)$, and the boundary conditions determine the solution along the first and last lines. This system is then solved with **NDSolve**.

The **Method** option now determines the method used to solve the space-discretized system. With **StartingStepSize**, **MaximumStepSize**, and **MaxSteps**, we can specify properties of the steps separately for each independent variable. If one value is given, it is used for all variables.

In the *Help Browser,* you can read advanced information about **NDSolve**.

■ A Program for the Method of Lines

Consider the following second-order PDE:

$$F(x, t, u, u_x, u_{xx}, u_t, u_{tt}) = 0, \quad x_0 < x < x_1, \quad t_0 < t < t_1,$$

$$u(x, t_0) = a_0(x), \quad u_t(x, t_0) = a_1(x), \quad u(x_0, t) = b_0(t), \quad u(x_1, t) = b_1(t)$$

Here we have two initial conditions: the values of u and u_t are given at t_0. The boundary conditions give the values of u at x_0 and x_1.

To derive the method of lines, we first discretize the x-interval to n_x subintervals of length h_x. Then we form $n_x - 1$ ordinary differential equations along the lines defined by these mesh points. Let $u_i(t)$ be the solution along the ith line. In the differential equations for the variables $u_i(t)$, the derivatives u_x and u_{xx} are approximated by differences. In this way, we obtain a system of $n_x - 1$ simultaneous differential equations:

$$F\left(x_i, t, u_i(t), \frac{u_{i+1}(t) - u_{i-1}(t)}{2 h_x}, \frac{u_{i+1}(t) - 2 u_i(t) + u_{i-1}(t)}{h_x{}^2}, u_i{}'(t), u_i{}''(t)\right) = 0, \quad i = 1, \ldots, n_x - 1.$$

The solutions along the lines x_0 and x_1 are known from the boundary conditions, and from the initial conditions we get initial conditions for the ordinary differential equations. The simultaneous system can then be solved using **NDSolve**. The following module implements the method of lines.

```
methodOfLines[{eqn_, a0_, a1_, b0_, b1_}, {x_, x0_, x1_, nx_},
  {t_, t0_, t1_, nt_}, {u_, ux_, uxx_, ut_, utt_}, opts___] :=
 Module[{nx1 = nx - 1, hx = N[(x1 - x0) / nx], i, xx, uu,
   icon1, icon2, uxappr, uxxappr, eqns, sol, tt, vals},
  Do[xx[i] = x0 + i hx, {i, 0, nx}];
  icon1 = Table[uu[i][0] == N[a0 /. x -> xx[i]], {i, nx1}];
  If[FreeQ[eqn, utt], icon2 = {},
   icon2 = Table[uu[i]'[0] == N[a1 /. x -> xx[i]], {i, nx1}]];
  uu[0] = Function[{t}, b0];
  uu[nx] = Function[{t}, b1];
  uxappr = (uu[i + 1][t] - uu[i - 1][t]) / (2 hx);
```

```
uxxappr = (uu[i + 1][t] - 2 uu[i][t] + uu[i - 1][t]) / hx^2;
eqns = Table[eqn /. {u → uu[i][t], ut → uu[i]'[t], utt → uu[i]''[t],
    x → xx[i], ux → uxxappr, uxx → uxxappr}, {i, nx1}];
sol = NDSolve[Join[eqns, icon1, icon2], Array[uu, {nx1}],
    {t, t0, t1}, opts][[1]];
Do[uu[i] = uu[i] /. sol, {i, nx1}];
tt = Range[t0, t1, N[(t1 - t0) / nt]];
vals = Table[Map[uu[i][#] &, tt], {i, 0, nx}];
ListInterpolation[vals, {{x0, x1}, {t0, t1}}]]
```

The argument **eqn** is the PDE to be solved. It is inputted in a special form in which u, u_x, u_{xx}, u_t, and u_{tt} are written as **u**, **ux**, **uxx**, **ut**, and **utt**, respectively. For example, a wave equation $u_{tt} - u_{xx} = -9.80665$ is written as **utt - uxx == -9.80665**. If your problem does not contain initial values for u_t, just write nothing in place of **a1** (the program assumes that initial conditions for u_t are given only if the equation contains the second derivative u_{tt}). The program accepts options for **NDSolve**. The solution given by **NDSolve** is sampled by dividing (t_0, t_1) into n_t subintervals. An interpolating surface is then constructed from the sample.

As examples, we solve the same heat and wave problems we solved with **NDSolve** in Section 24.3.1, p. 661. First we look at the heat problem:

```
sol = methodOfLines[{ut - uxx == 0, x (1 - x), , 0, 0},
    {x, 0, 1, 10}, {t, 0, 0.5, 10}, {u, ux, uxx, ut, utt}]
```

```
InterpolatingFunction[{{0., 1.}, {0., 0.5}}, <>]
```

```
Plot3D[sol[x, t], {t, 0, 0.5}, {x, 0, 1},
    PlotRange → All, AxesLabel → {"t", "x", ""},
    PlotPoints → 15, Ticks → {{0, 1, 2, 3, 4}, {0, 1}, {0, 0.2}}];
```

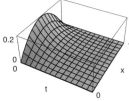

Then we take the wave problem:

```
sol = methodOfLines[{utt - uxx == -9.80665, 10 x^2 (1 - x)^2, 0, 0, 0},
    {x, 0, 1, 10}, {t, 0, 4, 10}, {u, ux, uxx, ut, utt}]
```

```
InterpolatingFunction[{{0., 1.}, {0., 4.}}, <>]
```

```
Plot3D[sol[x, t], {t, 0, 4}, {x, 0, 1}, PlotRange → All, PlotPoints → 15,
    AxesLabel → {"t", "x", ""}, Ticks → {{0, 1, 2, 3, 4}, {0, 1}, {-3, 0}}];
```

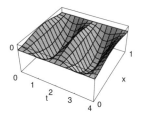

24.3.3 2D Elliptic Problems

Consider the following elliptic problem:

$$u_{xx} + u_{yy} = F(x, y), \quad x_0 < x < x_1, \quad y_0 < y < y_1,$$

$$u(x, y_1) = a_1(x),$$

$$u(x_0, y) = b_0(y), \qquad\qquad u(x_1, y) = b_1(y),$$

$$u(x, y_0) = a_0(x).$$

To derive a finite difference method for this problem, we divide the interval (x_0, x_1) into n_x subintervals of length h_x, the interval (y_0, y_1) into n_y subintervals of length h_y, and then obtain a mesh of $(n_x + 1)(n_y + 1)$ points. Denote the value of u at a mesh point with $u_{i,j}$. Approximate the partial derivatives with differences at the mesh points:

$$\frac{1}{h_x^2}(u_{i+1,j} - 2u_{i,j} + u_{i-1,j}) + \frac{1}{h_y^2}(u_{i,j+1} - 2u_{i,j} + u_{i,j-1}) = F_{i,j}$$

In this way, we obtain a system of $(n_x - 1)(n_y - 1)$ linear equations. The boundary conditions can be inserted into these equations or used as further equations. The solution of the linear system is an approximate solution of the original problem. An implementation of the finite difference method for elliptic problems is as follows:

```
ellipticFDM[{F_, a0_, a1_, b0_, b1_},
  {x_, x0_, x1_, nx_}, {y_, y0_, y1_, ny_}] :=
 Module[{hx = N[(x1 - x0) / nx], hy = N[(y1 - y0) / ny], i, xx,
    yy, vars, u, uxxappr, uyyappr, eqns, c1, c2, c3, c4, sol},
  Do[xx[i] = x0 + i hx, {i, 0, nx}];
  Do[yy[j] = y0 + j hy, {j, 0, ny}];
  vars = Array[u, {nx + 1, ny + 1}, 0];
  uxxappr = (u[i + 1, j] - 2 u[i, j] + u[i - 1, j]) / hx^2;
  uyyappr = (u[i, j + 1] - 2 u[i, j] + u[i, j - 1]) / hy^2;
  eqns = Flatten[Table[uxxappr + uyyappr == F /. {x -> xx[i], y -> yy[j]},
     {i, nx - 1}, {j, ny - 1}]];
  c1 = Table[u[i, 0] == N[a0 /. x -> xx[i]], {i, 0, nx}];
  c2 = Table[u[i, ny] == N[a1 /. x -> xx[i]], {i, 0, nx}];
  c3 = Table[u[0, j] == N[b0 /. y -> yy[j]], {j, 1, ny - 1}];
  c4 = Table[u[nx, j] == N[b1 /. y -> yy[j]], {j, 1, ny - 1}];
  sol = vars /. Solve[Join[eqns, c1, c2, c3, c4], Flatten[vars]][[1]];
  ListInterpolation[sol, {{x0, x1}, {y0, y1}}]]
```

In the program, we have written the boundary conditions as equations so that there is a total of $(n_x + 1)(n_y + 1)$ equations. The variable **sol** contains the approximations $u_{i,j}$. A neat way to represent the solution is to form an interpolating function from the solution with **ListInterpolation** (see Section 21.2.2, p. 558).

■ Example

As an example, we solve the following elliptic problem:

$$u_{xx} + u_{yy} = 0, \quad 0 < x < 1, \quad 0 < y < 1,$$

$$u(x, 0) = 4x(1 - x), \quad u(x, 1) = u(0, y) = u(1, y) = 0.$$

This is the same problem for which we considered the series solution in Section 24.2.5, p. 656. We use 10 subintervals in both directions:

```
uappr = ellipticFDM[{0, 4 x (1 - x), 0, 0, 0}, {x, 0, 1, 10}, {y, 0, 1, 10}]

InterpolatingFunction[{{0., 1.}, {0., 1.}}, <>]
```

We tabulate some values and plot the solution:

```
Table[uappr[0.5, y], {y, 0, 0.5, 0.1}]

{1., 0.740352, 0.54461, 0.39826, 0.289157, 0.207606}

Plot3D[Evaluate[uappr[x, y]],
  {x, 0, 1}, {y, 0, 1}, ViewPoint → {2.0, -2.4, 0.9},
  AxesLabel → {"x", "y", ""}, Ticks → {{0, 1}, {0, 1}, {0, 1}}];
```

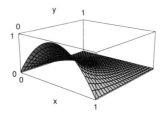

To investigate the precision of the numerical solution, we show the most accurate values of $u(0.5, y)$, $y = 0, 0.1, …, 0.5$ that we obtained by the series solution in Section 24.2.5:

```
{1., 0.739132, 0.542517, 0.395755, 0.28663, 0.205315}
```

The values we obtained using 10 and 10 subintervals (121 variables and equations) seem to be of approximately 2-digit precision. It turns out that, using 30 and 30 subintervals (961 equations), we get about 3-digit precision, and if we use 50 and 50 subintervals (2601 equations), we get nearly 4-digit precision.

Difference Equations

Introduction

*Don't let pessimistic statistics about the future to worry you. Remember in 1850
it was predicted that if the traffic kept increasing at the same rate, the entire
surface of the earth would be covered in six feet of horse manure by 1970.*

For solving difference or recurrence equations, we have **RSolve**. It can solve all constant coefficient linear equations, many variable coefficient linear equations, and also quite a few nonlinear equations. Note, however, that nonlinear difference equations are much more difficult to solve than nonlinear differential equations. For example, a solution for the logistic difference equation is only known for two values of the parameter of the model. We can study nonlinear equations by other means, and we will encounter interesting features like bifurcation, cycles, and chaos. Fractals also relate to nonlinear difference equations.

A good short introduction to linear difference equations is in Spiegel (1971). More comprehensive treatments of difference equations can be found in Sandefur (1990), Kelley and Peterson (2001), and Martelli (1999). For difference equations with *Mathematica*, see Kulenovic and Merino (2002).

25.1 Solving Difference Equations

25.1.1 One Linear Equation

RSolve solves difference equations or recurrence equations. It is used in the same way as **DSolve** is used for differential equations. In the following box, we have some typical examples of the use of **RSolve**:

RSolve[eqn, y[n], n] (✿5) Give the general solution
RSolve[{eqn, inits}, y[n], n] (✿5) Solve an initial value problem

Note that users of *Mathematica* with versions earlier than 5 have to load the **Discrete**͏ **Math`RSolve`** package before using **RSolve**. However, the **RSolve** of the package is not as powerful as the one in version 5. For example, the latter can also solve some nonlinear equations.

An example of a difference equation is $F_n = F_{n-1} + F_{n-2}$ with initial conditions $F_1 = 1$ and $F_2 = 1$; this equation defines the Fibonacci numbers. In *Mathematica*, the equation is written as **F[n] == F[n-1] + F[n-2]** and the initial conditions as **F[1] == 1** and **F[2] == 1** (remember that all equations have to be defined with **==**).

RSolve can solve all linear constant coefficient difference equations and systems of such equations (the solution is sought with matrix powers). **RSolve** can also solve many linear variable coefficient difference equations in which the coefficients are polynomial or rational. Furthermore, **RSolve** can solve many nonlinear difference equations, and it can solve q-difference equations—with terms like $y(q\,x)$ and $y(q^2\,x)$—and some partial difference equations.

We have tried to use the notation (✿4!) in places where users of *Mathematica* 4 obtain a somewhat different result. With **RSolve**, the differences between the results obtained with *Mathematica* 4 and 5 are so frequent that in Section 25.1 we do not use the notation.

■ A First-Order Constant Coefficient Equation

We ask for a general solution to a first-order constant coefficient difference equation:

    eqn = y[n + 1] == 4 / 5 y[n] + 1 / 5;

    RSolve[eqn, y[n], n]

$$\left\{\left\{y[n] \to 1 - \left(\frac{4}{5}\right)^n + \left(\frac{5}{4}\right)^{1-n} C[1]\right\}\right\}$$

Here **C[1]** is an arbitrary constant. Now we give an initial value:

    RSolve[{eqn, y[0] == 0}, y[n], n]

$$\left\{\left\{y[n] \to 1 - \left(\frac{4}{5}\right)^n\right\}\right\}$$

The indices can be written in various forms in the equation. The equation of the preceding example can also be written this way:

    eqn2 = y[n] == 4 / 5 y[n - 1] + 1 / 5;

```
RSolve[eqn2, y[n], n]
```

$$\left\{\left\{y[n] \rightarrow 1 - \left(\frac{4}{5}\right)^{n} + \left(\frac{5}{4}\right)^{1-n} C[1]\right\}\right\}$$

As with **DSolve**, the solution can also be requested as a pure function:

```
RSolve[eqn, y, n]
```

$$\left\{\left\{y \rightarrow \text{Function}\left[\{n\}, 1 - \left(\frac{4}{5}\right)^{n} + \left(\frac{5}{4}\right)^{1-n} C[1]\right]\right\}\right\}$$

This solution has the advantage that it can easily be inserted into the equation to check the correctness of the solution:

```
eqn /. % // Simplify    {True}
```

■ **Calculating Values**

If we calculate with the solution, it may be useful to ask for the value of y_n (as was the case with **DSolve**). We continue with the preceding example:

```
sol = y[n] /. RSolve[{eqn, y[0] == 2}, y[n], n][[1]] // FullSimplify
```

$$1 + 5^{-n} \left(2^{1+2n} - 4^{n}\right)$$

Once we have the solution, values y_n can be calculated and plotted:

```
vals = Table[{n, sol}, {n, 0, 20}];
```

```
ListPlot[vals, PlotJoined → True,
    Epilog → {AbsolutePointSize[2], Map[Point, vals]}];
```

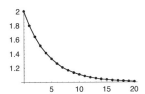

Note, however, that if you calculate a large number of values, it is often more efficient to calculate values directly from the recursive relation. One way to implement this is the following:

```
y = 2.; Prepend[Table[y = 4 / 5 y + 1 / 5, {6}], 2.]
```

```
{2., 1.8, 1.64, 1.512, 1.4096, 1.32768, 1.26214}
```

```
y = .
```

Another way is to define a recursive function (see Section 14.1.3, p. 356):

```
z[0] = 2.;
z[n_] := z[n] = 4 / 5 z[n - 1] + 1 / 5
```

Then use it:

```
Table[z[n], {n, 0, 6}]
```

```
{2., 1.8, 1.64, 1.512, 1.4096, 1.32768, 1.26214}
```

```
Remove[z]
```

However, the most compact and fastest way to calculate the values of a difference equation is the use of **NestList** (see Section 15.2.2):

```
NestList[4 / 5 # + 1 / 5 &, 2., 6]

{2., 1.8, 1.64, 1.512, 1.4096, 1.32768, 1.26214}
```

If the values of n are needed, you can get them in one of the following ways:

```
Transpose[{Range[0, 6], NestList[4 / 5 # + 1 / 5 &, 2., 6]}]

{{0, 2.}, {1, 1.8}, {2, 1.64}, {3, 1.512},
 {4, 1.4096}, {5, 1.32768}, {6, 1.26214}}

NestList[{#[[1]] + 1, 4 / 5 #[[2]] + 1 / 5} &, {0, 2.}, 6]

{{0, 2.}, {1, 1.8}, {2, 1.64}, {3, 1.512},
 {4, 1.4096}, {5, 1.32768}, {6, 1.26214}}
```

■ A Set of Trajectories

A direction field may be informative; it shows the direction of movement at several points. First, write the equation in the form $y_{n+1} - y_n = f(n, y_n)$, and then plot a direction field of $(1, f)$. In our example, the equation $y_{n+1} = \frac{4}{5} y_n + \frac{1}{5}$ can be written as $y_{n+1} - y_n = -\frac{1}{5} y_n + \frac{1}{5}$:

```
<< Graphics`PlotField`

PlotVectorField[{1, -1 / 5 y + 1 / 5}, {n, 0, 20}, {y, 0, 1.5},
   PlotPoints → 11, Axes → True, AspectRatio → 1 / GoldenRatio];
```

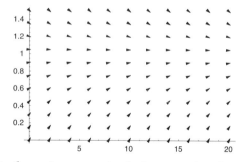

We then plot a set of solutions starting from various points. The solution of the equation with a general starting value **y0** is as follows:

```
sol = y[n] /. RSolve[{eqn, y[0] == y0}, y[n], n][[1]]
```

$$1 - \left(\frac{4}{5}\right)^n + \left(\frac{4}{5}\right)^n y0$$

Compute a set of solutions using various starting points:

```
solset = Table[Table[{n, sol}, {n, 0, 20}], {y0, 0, 1.4, 0.2}];
```

Show the solutions:

```
Show[Graphics[{AbsolutePointSize[2], Map[Point, Flatten[solset, 1]],
    Map[Line, solset]}], Axes → True, PlotRange → All];
```

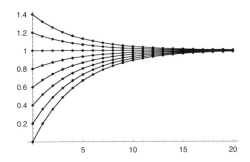

■ Some First-Order Variable Coefficient Equations

We try some variable coefficient equations (*Mathematica* 4 does not solve the third example):

```
RSolve[{y[n + 1] == (n + 1) y[n], y[0] == 1}, y[n], n]
```

$\{\{y[n] \to \text{Gamma}[1 + n]\}\}$

```
RSolve[{y[n + 1] == a y[n] + b n, y[0] == c}, y[n], n] // FullSimplify
```

$\left\{\left\{y[n] \to \dfrac{a^n \left(b + (-1 + a)^2 c - \left(\frac{1}{a}\right)^n b \left(1 + (-1 + a) n\right)\right)}{(-1 + a)^2}\right\}\right\}$

```
RSolve[{y[n + 1] == y[n] / (n + 1) + 1, y[0] == 1}, y[n], n]
```

$\{\{y[n] \to (e - \text{ExpIntegralE}[2, -1] + \text{ExpIntegralE}[2 + n, -1] \text{ Gamma}[2 + n]) / (e \text{ Gamma}[1 + n])\}\}$

■ Fibonacci Numbers

The first few Fibonacci numbers are as follows:

```
F[1] = 1; F[2] = 1;
F[n_] := F[n] = F[n - 1] + F[n - 2]
```

```
Table[F[n], {n, 15}]
```

$\{1, 1, 2, 3, 5, 8, 13, 21, 34, 55, 89, 144, 233, 377, 610\}$

```
Remove[F]
```

The *n*th Fibonacci number is given in this way:

```
RSolve[{F[n] == F[n - 1] + F[n - 2], F[1] == F[2] == 1}, F[n], n] // FullSimplify
```

$\left\{\left\{F[n] \to \dfrac{-\left(\frac{1}{2} \left(1 - \sqrt{5}\right)\right)^n + \left(\frac{1}{2} \left(1 + \sqrt{5}\right)\right)^n}{\sqrt{5}}\right\}\right\}$

Note that actually we do not need to calculate Fibonacci numbers ourselves, because *Mathematica* has the built-in **Fibonacci[n]**.

■ Chebyshev Polynomials

Consider the following second-order equation:

```
eqn = T[n + 2] - 2 x T[n + 1] + T[n] == 0;
```

With initial conditions $T_0(x) = 1$ and $T_1(x) = x$, it defines Chebyshev polynomials. The solution is as follows:

```
sol = T[n] /. RSolve[{eqn, T[0] == 1, T[1] == x}, T[n], n][[1]]
```

$$\frac{1}{2} \left(\left(x - \sqrt{-1 + x^2} \right)^n + \left(x + \sqrt{-1 + x^2} \right)^n \right)$$

This is not of the familiar form $T_n(x) = \cos(n \arccos(x))$, but for given values of n we can verify that the expressions agree:

```
Table[sol == Cos[n ArcCos[x]], {n, 0, 5}] // Simplify
```

```
{True, True, True, True, True, True}
```

Note that *Mathematica* has the built-in **ChebyshevT[n,x]**.

```
Remove["Global`*"]
```

25.1.2 Two Linear Equations

> **RSolve[eqns, {x[n], y[n]}, n]**   Solve two difference equations

■ A Constant Coefficient System

Consider the following system:

```
eqns = {x[n + 1] == 7 / 10 x[n] + 4 / 10 y[n],
    y[n + 1] == -3 / 10 x[n] + 6 / 10 y[n]};
```

Here is the coefficient matrix:

```
A = 1 / 10 {{7, 4}, {-3, 6}};
```

Its eigenvalues and their absolute values are as follows:

```
Eigenvalues[A]
```
$\left\{ \frac{1}{20} (13 + i \sqrt{47}), \frac{1}{20} (13 - i \sqrt{47}) \right\}$

```
Abs[%] // N
```
$\{0.734847, 0.734847\}$

Because the common absolute value is smaller than one, the trajectories are spiral approaching the origin (Kelley and Peterson 2001, p. 148).

Solve the system with general starting values **x0** and **y0**:

```
vars = {x[n], y[n]};
inits = {x[0] == α, y[0] == β};
```

```
sol[{α_, β_}] =
 vars /. RSolve[Join[eqns, inits], vars, n][[1]] // ComplexExpand //
  FullSimplify
```

$$\left\{ \frac{1}{47} \, 2^{-n/2} \, 3^{3n/2} \, 5^{-n} \right.$$
$$\left(47 \, \alpha \, \text{Cos}\left[n \, \text{ArcCot}\left[\frac{13}{\sqrt{47}} \right] \right] + \sqrt{47} \, (\alpha + 8 \beta) \, \text{Sin}\left[n \, \text{ArcCot}\left[\frac{13}{\sqrt{47}} \right] \right] \right), \frac{1}{47} \, 2^{-n/2}$$
$$\left. 3^{3n/2} \, 5^{-n} \left(47 \, \beta \, \text{Cos}\left[n \, \text{ArcCot}\left[\frac{13}{\sqrt{47}} \right] \right] - \sqrt{47} \, (6 \, \alpha + \beta) \, \text{Sin}\left[n \, \text{ArcCot}\left[\frac{13}{\sqrt{47}} \right] \right] \right) \right\}$$

Calculate the solution from the starting point $(60, 60)$:

```
xy = Table[sol[{60., 60.}], {n, 0, 20}];
```

Plot the trajectory in the (x, y) plane:

```
Show[Graphics[{AbsolutePointSize[2], Map[Point, xy], Line[xy]}],
    Axes → True, AspectRatio → Automatic, PlotRange → All];
```

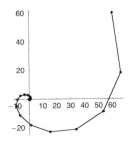

To plot the components as functions of n, write the following:

```
{xx, yy} = Transpose[xy]; tt = Range[0, 20];
xt = Transpose[{tt, xx}]; yt = Transpose[{tt, yy}];
```

```
Show[Graphics[{AbsolutePointSize[2],
    GrayLevel[0.4], Map[Point, yt], Line[yt], GrayLevel[0],
    Map[Point, xt], Line[xt]}], Axes → True, PlotRange → All];
```

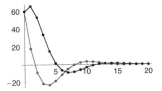

■ A Set of Trajectories

We continue the preceding example by plotting a direction field. To this end, we write the equations in the form $x_{n+1} - x_n = f(n, x_n, y_n)$, $y_{n+1} - y_n = g(n, x_n, y_n)$ and then plot a direction field of (f, g):

```
<< Graphics`PlotField`
```

```
PlotVectorField[{-3 / 10 x + 4 / 10 y, -3 / 10 x - 4 / 10 y},
    {x, -60, 60}, {y, -60, 60}, PlotPoints → 11, Axes → True];
```

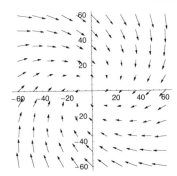

Then we calculate some trajectories from a set of starting points (here we use the function `sol[{α_,β_}]` we defined above):

```
solset = Map[Table[sol[#], {n, 0, 20}] &, {{60, 60}, {30, 60}, {-60, 60},
    {-60, 30}, {-60, -60}, {-30, -60}, {60, -60}, {60, -30}}];
```

The trajectories are displayed as follows:

```
Show[Graphics[{AbsolutePointSize[2],
    Map[Point, Flatten[solset, 1]], Map[Line, solset]}],
  Axes → True, PlotRange → All, AspectRatio → Automatic];
```

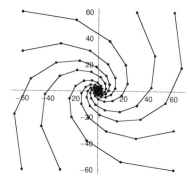

25.1.3 Some Techniques

■ Using Generating Functions

One method of solving difference equations is the use of generating functions. Consider, for example, the equation $y_{n+1} = a\,y_n + b$. Define the generating function $G(t) = \sum_{n=0}^{\infty} y_n\,t^n$. Multiply both sides of the equation by t^n and sum from 0 to ∞:

$$\sum_{n=0}^{\infty} y_{n+1}\,t^n = a \sum_{n=0}^{\infty} y_n\,t^n + b \sum_{n=0}^{\infty} t^n.$$

Because the lefthand side can be written as $\frac{1}{t} \sum_{n=1}^{\infty} y_n\,t^n$, we get the equation

$$\frac{1}{t}\,(G(t) - y_0) = a\,G(t) + b\,\frac{1}{1-t}.$$

The solution of this equation is as follows:

```
eqn = 1 / t (G - y0) == a G + b / (1 - t);
```

```
Gsol = G /. Solve[eqn, G][[1]] // Factor
```

$$\frac{b\,t + y0 - t\,y0}{(-1 + t)\,(-1 + a\,t)}$$

The coefficient of the term t^n in the series expansion of this expression is y_n, which is also the solution of the difference equation:

```
<< DiscreteMath`RSolve`
```

```
SeriesTerm[Gsol, {t, 0, n}]
```

$$\frac{(-1 + a^n)\,b + (-1 + a)\,a^n\,y0}{-1 + a}$$

(We considered `SeriesTerm` in Section 16.2.2, p. 427.) This expression agrees with the solution given by `RSolve`:

```
eqn = {y[n + 1] == a y[n] + b, y[0] == y0};
```

```
RSolve[eqn, y[n], n] // Simplify
```

$$\left\{\left\{ y[n] \rightarrow \frac{(-1 + a^n)\, b + (-1 + a)\, a^n\, y0}{-1 + a} \right\}\right\}$$

In the `DiscreteMath`RSolve`` *package:*

`GeneratingFunction[eqn, y[n], n, t]` Generating function of a difference equation

`ExponentialGeneratingFunction[eqn, y[n], n, t]` Exponential generating function

With these commands, we can directly calculate the generating function of a difference equation. For example:

```
GeneratingFunction[eqn, y[n], n, t]
```

$$\left\{\left\{ \frac{b t + y0 - t\, y0}{(-1 + t)\,(-1 + a\, t)} \right\}\right\}$$

We obtained the same generating function as above. With the exponential generating function ($H(t) = \sum_{n=0}^{\infty} \frac{y_n}{n!}\, t^n$), we can also get the solution:

```
ExponentialGeneratingFunction[eqn, y[n], n, t] // FullSimplify
```

$$\left\{\left\{ \frac{-b\, e^t + e^{a\, t}\,(b + (-1 + a)\, y0)}{-1 + a} \right\}\right\}$$

```
n! SeriesTerm[%, {t, 0, n}] // Simplify
```

$$\left\{\left\{ \frac{(-1 + a^n)\, b + (-1 + a)\, a^n\, y0}{-1 + a} \right\}\right\}$$

■ Using the Z Transform

Difference equations can also be solved with the Z transform (see Section 17.4.4, p. 462), in a way that is similar to how differential equations can be solved with the Laplace transform. To solve the difference equation of the Fibonacci numbers, we rewrite the equation and the initial values as follows:

```
eqn = F[n + 2] == F[n + 1] + F[n];
```

```
inits = {F[0] → 0, F[1] → 1};
```

Take the Z transform of the equation:

```
ZTransform[eqn, n, z]
```

```
-z² F[0] - z F[1] + z² ZTransform[F[n], n, z] ==
  -z F[0] + ZTransform[F[n], n, z] + z ZTransform[F[n], n, z]
```

Use the initial conditions:

```
zeqn = % /. inits
```

```
-z + z² ZTransform[F[n], n, z] ==
  ZTransform[F[n], n, z] + z ZTransform[F[n], n, z]
```

Solve the Z transform:

```
Solve[zeqn, ZTransform[F[n], n, z]]
```

$$\left\{\left\{\text{ZTransform}[F[n], n, z] \rightarrow \frac{z}{-1 - z + z^2}\right\}\right\}$$

Find the inverse Z transform:

```
sol = F[n] /. InverseZTransform[%, z, n][[1]]
```

$$\frac{-\left(\frac{1}{2}\left(1 - \sqrt{5}\right)\right)^n + \left(\frac{1}{2}\left(1 + \sqrt{5}\right)\right)^n}{\sqrt{5}}$$

■ Cobweb

The *cobweb* is an interesting way to illustrate how the values computed from a difference equation proceed. Consider the equation $y_{n+1} = -0.8\,y_n + 0.2$, $y_0 = 0.6$. Calculate and plot some values:

```
vals = Transpose[{Range[0, 20], NestList[-0.8 # + 0.2 &, 0.6, 20]}];

Show[Graphics[{AbsolutePointSize[2], Map[Point, vals], Line[vals]}],
    Axes → True];
```

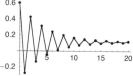

We then introduce the program **cobwebPlot**:

```
cobwebPlot[f_, y_, y0_, n_, a_, b_, opts___] := Module[{p1, p2, x = y0},
    p1 = Plot[{y, f}, {y, a, b}, DisplayFunction → Identity];
    p2 = Graphics[{Line[Prepend[
        Flatten[Table[{{x, x = f /. y → x}, {x, x}}, {n}], 1], {y0, 0}]]}];
    Show[p1, p2, DisplayFunction → $DisplayFunction, opts]]
```

Here **f** is the righthand-side function of the difference equation $y_{n+1} = f(y_n)$; in the example above, **f** is **-0.8 y + 0.2**. In addition, **y0** is the starting value; **n** is the number of values to be calculated (in addition to **y0**); (**a**, **b**) is the interval in which the figure is plotted; and **opts** is a set of graphics options. For our example, the figure is displayed like this:

```
cobwebPlot[-0.8 y + 0.2, y, 0.6, 20, -0.4, 0.7, PlotRange → {-0.35, 0.5}];
```

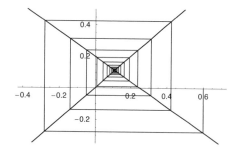

In the figure we have plotted the functions y and $f(y)$. The starting point for the cobweb is $(0.6, 0)$. From this point on, follow the broken line. Each time you meet a vertical line, the point where this line (or its extension) intersects the x axis represents the next value of the solution sequence. The horizontal lines lead you to the consecutive vertical lines. The figure nicely shows the convergence of the sequence to a point.

25.1.4 Nonlinear Equations

■ Example 1

Several nonlinear difference equations can also be solved. An example is an equation of the form $\text{expr}_1 = \text{expr}_2$ in which the two expressions contain products, quotients, and powers (not sums). For example (note that *Mathematica* 4 does not solve this, nor other nonlinear examples in this section):

```
eqns = {y[n + 2] y[n] == y[n + 1], y[0] == 1, y[1] == 2};
```

```
sol = RSolve[eqns, y[n], n]
```

```
Solve::ifun :
   Inverse functions are being used by Solve, so some solutions may not
      be found; use Reduce for complete solution information. More…
```

$$\left\{\left\{y[n] \to 2^{\frac{2 \sin\left[\frac{n\pi}{3}\right]}{\sqrt{3}}}\right\}\right\}$$

The equation could have been reduced, with a logarithmic transform, to a linear one. The solution has period six:

```
Table[y[n] /. sol[[1]], {n, 0, 17}]
```

$$\left\{1, 2, 2, 1, \frac{1}{2}, \frac{1}{2}, 1, 2, 2, 1, \frac{1}{2}, \frac{1}{2}, 1, 2, 2, 1, \frac{1}{2}, \frac{1}{2}\right\}$$

■ Example 2

This example is a homogeneous constant coefficient Riccati equation:

```
eqns = {y[n + 1] y[n] + p y[n + 1] + q y[n] == 0, y[0] == y0};
```

```
sol = RSolve[eqns, y[n], n]
```

$$\left\{\left\{y[n] \to \frac{\left(-\frac{p}{q}\right)^{-n} (p + q) \, y0}{p + q + y0 - \left(-\frac{q}{p}\right)^{n} y0}\right\}\right\}$$

The equation could have been reduced, with the transformation $y_n = \frac{1}{z_n}$, to a linear one.

■ Example 3

Now we have an equation that contains a convolution:

```
eqns = {y[n + 1] == Sum[y[i] y[n - i], {i, 0, n}], y[0] == 1}
```

$$\left\{y[1 + n] == \sum_{i=0}^{n} y[i] \, y[-i + n], \, y[0] == 1\right\}$$

Although **RSolve** (in *Mathematica* 5) is not able to solve this equation, we can use generating functions:

```
<< DiscreteMath`RSolve`
```

$$\texttt{GeneratingFunction[eqns, y[n], n, t]} \qquad \{\{\frac{1-\sqrt{1-4\,t}}{2\,t}\}\}$$

The solution for $y[n]$ is then as follows:

$$\texttt{SeriesTerm[\%, \{t, 0, n\}]} \qquad \{\{\frac{\texttt{Binomial[2\,n, n]}}{1+n}\}\}$$

■ Example 4: The Logistic Equation

One of the most famous nonlinear difference equations is the logistic equation:

```
eqns = {y[n + 1] == a y[n] (1 - y[n]), y[0] == y0};
```

The solution of this equation is unknown, except for two values of the parameter a, $a = 2$ and $a = 4$ (see Kelley and Peterson 2001, p. 173):

```
Off[Solve::ifun]

sol2 = y[n] /. RSolve[eqns /. a → 2, y[n], n][[1]]
```

$$\frac{1}{2}\left(1-(1-2\,\texttt{y0})^{2^n}\right)$$

```
sol4 = y[n] /. RSolve[eqns /. a → 4, y[n], n][[1]]
```

$$\frac{1}{2}\left(1-\texttt{Cos}[2^n\,\texttt{ArcCos}[1-2\,\texttt{y0}]]\right)$$

A realization of the solution for $a = 2$ is as follows:

```
vals2 = Table[{n, sol2} /. y0 → 0.001, {n, 0, 15}];

ListPlot[vals2, PlotJoined → True,
   Epilog → {AbsolutePointSize[2], Map[Point, vals2]}];
```

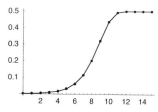

The solution for $a = 4$ behaves completely differently. To calculate the values accurately, we start with 40 digits of precision:

```
vals4 = Table[{n, sol4} /. y0 → 0.001`40, {n, 0, 100}];

ListPlot[vals4, PlotJoined → True, AspectRatio → 0.2,
   Epilog → {AbsolutePointSize[2], Map[Point, vals4]}];
```

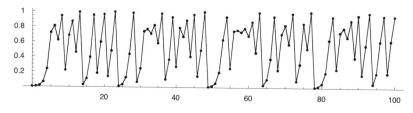

The values seem to develop rather chaotically. Next we study the logistic equation in more detail.

```
Remove["Global`*"]
```

25.2 The Logistic Equation

25.2.1 Trajectories

■ The Logistic Model

Most nonlinear difference equations cannot be solved to a closed-form expression. We have to resort to other means of investigating them. As an example of nonlinear equations, we consider in this section the famous logistic model $y_{n+1} = a\, y_n\, (1 - y_n)$. The methods presented for this model may also be used for other nonlinear models.

A logistic equation is often written as follows:

```
eqn = z[n + 1] - z[n] == k z[n] (K - z[n])
```

$$-z[n] + z[1 + n] == k \ (K - z[n]) \ z[n]$$

This can be reduced to the standard form by making the change of variable $z_n = \frac{1+kK}{k}\, y_n$:

```
eqn /. z[n_] → (1 + k K) y[n] / k // FullSimplify
```

$$\frac{(1 + k\,K)\ ((1 + k\,K)\ (-1 + y[n])\ y[n] + y[1 + n])}{k} == 0$$

Now denote $1 + kK = a$.

■ Trajectories

From a direction field, we get an impression of the solution. Write the equation in the form $y_{n+1} - y_n = a\, y_n(1 - y_n) - y_n$, and assume that $a = 1.5$:

```
<< Graphics`PlotField`

PlotVectorField[{1, 1.5 y (1 - y) - y}, {n, 0, 20}, {y, 0, 0.5},
    PlotPoints → 11, Axes → True, AspectRatio → 1 / GoldenRatio];
```

To plot sets of trajectories, we write the following program:

```
logisticPlot[a_, n_, y01_, y02_, dy0_, opts___] :=
  With[{points = Table[NestList[{#[[1]] + 1, a #[[2]] (1 - #[[2]])} &,
      {0, y0}, n], {y0, y01, y02, dy0}]},
    Show[Graphics[{AbsolutePointSize[2], Map[Point, Flatten[points, 1]],
      Map[Line, points]}], Axes → True, opts]]
```

In the program, we first calculate a solution set by starting from various points and iterating the equation **n** times. The starting points are chosen between **y01** and **y02** in steps of **dy0**. When $a = 1.5$, we get the following trajectories:

logisticPlot[1.5, 15, 0.01, 0.41, 0.025, PlotRange → All];

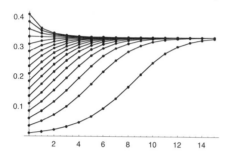

All trajectories in this figure seem to approach a certain value. When $a = 3.3$, the trajectories seem to approach a cycle of two points:

logisticPlot[3.3, 20, 0.01, 0.31, 0.05, AspectRatio → 0.4];

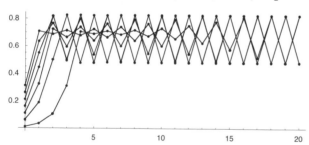

For $a = 3.52$, we get a cycle of four points:

logisticPlot[3.52, 30, 0.01, 0.31, 0.05, AspectRatio → 0.3];

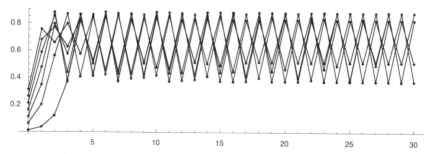

When $a = 3.7$, the trajectories appear to be chaotic (to calculate the values accurately, we start with 40 digits of precision):

```
logisticPlot[3.7`40, 20, 0.01`40, 0.31`40, 0.05`40, AspectRatio → 0.3];
```

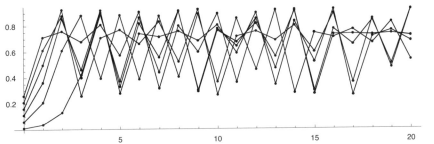

■ Cobwebs

```
cobwebPlot[f_, y_, y0_, n_, a_, b_, opts___] := Module[{p1, p2, x = y0},
    p1 = Plot[{y, f}, {y, a, b}, DisplayFunction → Identity];
    p2 = Graphics[{Line[Prepend[
        Flatten[Table[{{x, x = f /. y → x}, {x, x}}, {n}], 1], {y0, 0}]]}];
    Show[p1, p2, DisplayFunction → $DisplayFunction, opts]]
```

We have already plotted a cobweb in Section 25.1.3, p. 678. Now we investigate the logistic model by using the same values of *a* as above (again, to calculate the values accurately for *a* = 3.7, we start with 40 digits of precision):

```
Block[{$DisplayFunction = Identity},
    p1 = cobwebPlot[1.5 y (1 - y), y, 0.05, 20, 0, 0.4];
    p2 = cobwebPlot[3.3 y (1 - y), y, 0.03, 20, 0, 0.9];
    p3 = cobwebPlot[3.52 y (1 - y), y, 0.08, 30, 0, 0.95];
    p4 = cobwebPlot[3.7`40 y (1 - y), y, 0.02`40, 40, 0, 1];]
```

```
Show[GraphicsArray[{{p1, p2}, {p3, p4}}]];
```

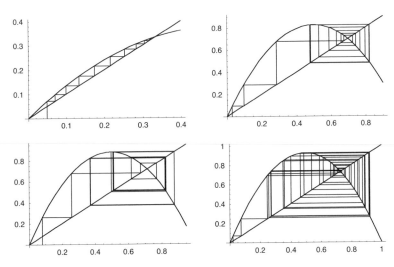

From these figures we can see the same things as we did with the trajectories: convergence to a point, a two-point cycle, a four-point cycle, and chaos.

■ Sensitivity to Numerical Inaccuracies

Chaotic models are very sensitive to numerical inaccuracies. To see this, first compute 107 values from the logistic model with $a = 3.7$:

```
Nest[3.7 # (1 - #) &, 0.02, 107]     0.70686
```

We compare this value to the value we get if we use high-precision numbers:

```
Nest[3.7`70 # (1 - #) &, 0.02`70, 107]     0.2684980945
```

Here we started with numbers that have 70 correct digits. *Mathematica* takes care that all digits in the result 0.2684980945 are correct. So, we see that the first result 0.70686 is totally incorrect.

To further illustrate the numerical instability, we first calculate 120 values by using the normal decimal numbers and then plot the last 40 values:

```
vals1 = Take[NestList[3.7 # (1 - #) &, 0.02, 120], -40];
```

```
p1 = ListPlot[vals1, PlotJoined → True, PlotStyle → GrayLevel[0.5],
    AspectRatio → 0.3, DisplayFunction → Identity];
```

We do the same thing again, but now we use the high-precision numbers:

```
vals2 = Take[NestList[3.7`80 # (1 - #) &, 0.02`80, 120], -40];
```

```
p2 = ListPlot[vals2, PlotJoined → True, DisplayFunction → Identity];
```

Now we combine the plots:

```
Show[p1, p2, DisplayFunction → $DisplayFunction, PlotRange → {0, 1},
    Ticks → {Transpose[{{10, 20, 30, 40}, {90, 100, 110, 120}}], Automatic}];
```

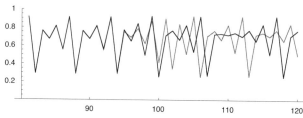

The correct values are black. From about iteration 95 on, the values begin to differ. It is thus important to use high-precision numbers when calculating long sequences from chaotic models (for high-precision numbers, see Section 11.2.2, p. 289).

■ Sensitivity to Initial Values

Chaotic models are also very sensitive to the initial value. To show this, compute, with $a = 3.7$, 20 iterations using 30 starting points: $0.02 + 10^{-i}$, $i = 1, ..., 30$. Plot then the last point of each of the 30 different iterations:

```
finals = Table[
    NestList[3.7`70 # (1 - #) &, 0.02`70 + 10^-i, 20] // Last, {i, 1, 30}];
```

```
ListPlot[finals, PlotStyle → AbsolutePointSize[2], PlotRange → {0, 1}];
```

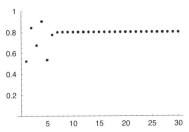

We see that, even if the starting point differs from 0.02 by a tiny 10^{-6} or more (see the first 6 points in the plot), the value of y_{20} significantly differs from the value that results when starting from 0.02. We do the same calculations but now with 100 iterations:

```
finals = Table[
    NestList[3.7`70 # (1 - #) &, 0.02`70 + 10^-i, 100] // Last, {i, 1, 30}];
```

```
ListPlot[finals, PlotStyle → AbsolutePointSize[2], PlotRange → {0, 1}];
```

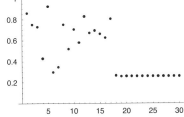

If the starting point now differs from 0.02 by a very tiny 10^{-17} or more, the value of y_{100} differs significantly from the value that results when starting from 0.02.

25.2.2 Bifurcation Diagrams

The *bifurcation diagram* or *final-state diagram* shows the long-run values calculated from the difference equation when a parameter of the model gets a range of values. The trajectories calculated above showed that, for different values of the parameter a, the logistic model behaves very differently. We say that a bifurcation occurs for a certain value of the parameter if the behavior of the final values undergoes a qualitative change at this point.

■ Limit Values

To prepare a bifurcation diagram, first choose a set of values of the parameter, then directly calculate a long sequence from the difference equation for these values, and lastly plot the limit values. We write a function to calculate limit values (Dickau 1997):

```
limits = Compile[{a},
    Map[{a, #} &, Union[Drop[NestList[a # (1 - #) &, 0.5, 1000], 301]]]];
```

The function **limits** calculates the limit values for a single value of a. **NestList** does most of the work: we iterate the recursion function 1000 times starting from 0.5. Of these values, the first 301—considered to be transient—are dropped. Of the remaining 700 values we drop duplicates with **Union**; lastly, with **Map**, we attach the value of a to each limit value of y. We have compiled the function **limits** to speed up this process (compiling was explained in Section 14.2.2, p. 365). (If your computer is not fast and does not have much RAM, you may prefer to replace 1000 with a smaller value such as 600.)

Try the function for $a = 1.3$:

```
limits[1.3]      {{1.3, 0.230769}}
```

The approximate limiting value is 0.230769. Try then $a = 3.3$:

```
limits[3.3]
```

```
{{3.3, 0.479427}, {3.3, 0.479427}, {3.3, 0.479427},
 {3.3, 0.823603}, {3.3, 0.823603}, {3.3, 0.823603}}
```

Here we see the two points forming a cycle. Each point appears three times: **Union** has found that all 16 digits are not the same. Consider then $a = 3.7$:

```
Length[limits[3.7]]      700
```

All 700 values are different. Indeed, for this value of a, the system behaves chaotically.

Normal decimal numbers are used in **limit**. If we remember the numerical problems of a chaotic logistic model, we may wonder whether the limiting points are correct at all. Contrary to expectation, the bifurcation diagram obtained with **limit** correctly displays all of the essential things. We could write a function like the following:

```
limits2[a_] :=
  Map[{a, #} &, Union[Drop[NestList[a # (1 - #) &, 0.5`400, 1000], 301]]];
```

High-precision numbers are here used (the starting value 0.5 has 400 digits of precision!). However, the bifurcation diagrams obtained with **limit** and **limit2** differ very little: only close examination of the chaotic region reveals some slight differences. So, we can be satisfied with **limit**, in particular because it is many times faster than **limit2** and uses much less memory.

■ **Bifurcation Diagram**

Now we form a function to calculate the limit values for a range of values for a:

```
bifurcation[a0_, a1_, n_] :=
  Flatten[Table[limits[a], {a, a0, a1, (a1 - a0) / n}], 1]
```

The function **bifurcation** constructs, for $n + 1$ values of a between a_0 and a_1, the corresponding set of limit values of y for the logistic model. As an example, we take 801 values of a in the interval $(0, 4)$ and plot the corresponding limit points of the recursion formula (if your computer has limited resources, replace 800 with a smaller value such as 400):

```
points = bifurcation[0, 4, 800];
```

```
ListPlot[points, PlotStyle → AbsolutePointSize[0.2],
  AxesOrigin → {0, -0.02}, AxesLabel → {"a", "y"}];
```

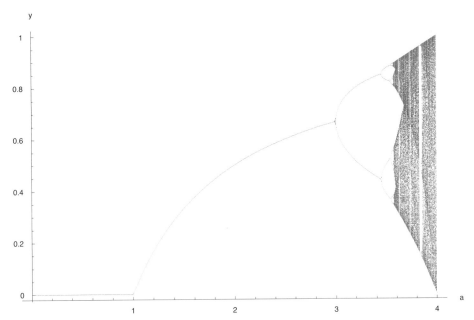

The figure shows that, for $0 < a < 1$, the limit value is 0, and, for $1 < a < 3$, it is another value (depending on a). Then we have cycles of 2, 4, 8, ... points, and for about $3.6 < a < 4$, the model behaves chaotically. For $a > 4$, the limit value is $-\infty$. Next we take a closer look at the interesting interval (3.5, 4) (the computation takes about 47 megabytes of RAM):

```
points = bifurcation[3.5, 4, 800];
```

```
ListPlot[points,
    PlotStyle → AbsolutePointSize[0.2], AxesOrigin → {3.5, 0}];
```

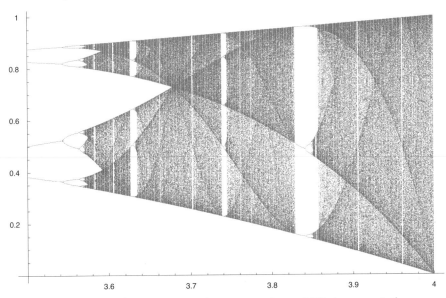

We see that, after the chaotic region begins at about 3.57, here and there we suddenly again have small intervals of periodic behavior.

25.2.3 Equilibrium and Periodic Points

The bifurcation diagram shows limit values, but we also want to know their mathematical expressions. So, here we calculate equilibrium points and points that make up periods of various lengths. With them, we can plot a better bifurcation diagram for $a < 3.57$.

■ Equilibrium Points

Consider the difference equation $y_{n+1} = f(y_n)$. If $f(y_n) = y_n$, then $y_{n+1} = y_n$ and the state remains the same. Such a y_n is an *equilibrium point*. For the logistic equation, the equilibrium points are as follows:

```
f[y_] := a y (1 - y)
```

```
sol1 = Solve[f[y] == y, y]        {{y → 0}, {y → -1 + a/a}}
```

An equilibrium point y^* is asymptotically stable if $|f'(y^*)| < 1$. For the logistic model, we have the following:

```
f'[y] /. sol1 // Simplify        {a, 2 - a}
```

Assume that $a > 0$. Points $y^* = 0$ and $y^* = 1 - \frac{1}{a}$ are asymptotically stable if $|a| < 1$ and $|2 - a| < 1$, respectively, that is, if $0 < a < 1$ and $1 < a < 3$, respectively. We plot these equilibrium points in their regions of stability (to save space, we do not yet show the plots):

```
p1 = Plot[Evaluate[y /. sol1[[1]]], {a, 0, 1}, DisplayFunction → Identity];
```

```
p2 = Plot[Evaluate[y /. sol1[[2]]], {a, 1, 3}, DisplayFunction → Identity];
```

■ 2-Periodic Points

Consider again the difference equation $y_{n+1} = f(y_n)$, and calculate $y_{n+2} = f(f(y_n))$. If $f(f(y_n)) = y_n$, we have $y_{n+2} = y_n$, which means that the same point appears by doing two iterations. Such a point y_n is called a *2-periodic point*, and the points y_n and $f(y_n)$ form a *cycle*. To find the 2-periodic points of the logistic model, first calculate $f(f(y))$:

```
f2[y_] = f[f[y]]        a² (1 - y) y (1 - a (1 - y) y)
```

Then solve the equation $f(f(y)) = y$:

```
sol2 = Solve[f2[y] == y, y] // FullSimplify
```

$$\left\{ \{y \to 0\}, \left\{ y \to \frac{-1 + a}{a} \right\}, \right.$$

$$\left. \left\{ y \to \frac{1 + a - \sqrt{(-3 + a)(1 + a)}}{2a} \right\}, \left\{ y \to \frac{1 + a + \sqrt{(-3 + a)(1 + a)}}{2a} \right\} \right\}$$

The first two points are the equilibrium points (they are, of course, also 2-periodic points). The last two points are genuine 2-periodic points; they exist if $a > 3$. We pick these points:

```
sol2a = Take[sol2, {3, 4}];
```

A 2-periodic point is asymptotically stable if the absolute value of the derivative of $f(f(y))$ is less than one at the 2-periodic points. The derivatives are as follows:

```
df2 = f2'[y] /. sol2a // Simplify
```

$$\{4 + 2a - a^2, 4 + 2a - a^2\}$$

Thus, the 2-periodic points are asymptotically stable if $|4 + 2a - a^2| < 1$. We solve this inequality, also taking into account the requirement $a > 3$ (users of *Mathematica* 4 have to load the package **Algebra`InequalitySolve`** and the replace **Reduce** with **Inequality‹Solve**):

> **Reduce[Abs[df2[[1]]] < 1 && a > 3, a]** $3 < a < 1 + \sqrt{6}$

> **% // N**   3. < a < 3.44949

For example, when $a = 3.3$, the equilibrium and 2-periodic points are as follows:

> **y2 = sol2 /. a → 3.3**

> $\{\{y \to 0\}, \{y \to 0.69697\}, \{y \to 0.479427\}, \{y \to 0.823603\}\}$

From a cobweb plot (see Section 25.2.1, p. 683), we can see the 2-periodic points and how the sequence $y_{n+2} = f(f(y_n))$ approaches one of the 2-periodic points:

> **cobwebPlot[f2[y] /. a → 3.3, y, 0.09,**
> **40, 0, 1, Epilog → {Hue[0], AbsolutePointSize[3],**
> **Point[{y, y}] /. y2[[3]], Point[{y, y}] /. y2[[4]]}];**

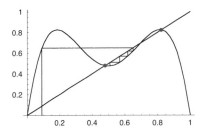

We plot the points that form the cycle:

> **p3 = Plot[Evaluate[y /. sol2a],**
> **{a, 3, 1 + Sqrt[6]}, DisplayFunction → Identity];**

■ 4-Periodic Points

To calculate 4-periodic points, define the 4-times nested function:

> **f4[y_] = f[f[f[f[y]]]]**

> $a^4 (1 - y) y (1 - a (1 - y) y) (1 - a^2 (1 - y) y (1 - a (1 - y) y))$
> $(1 - a^3 (1 - y) y (1 - a (1 - y) y) (1 - a^2 (1 - y) y (1 - a (1 - y) y)))$

We can solve the condition of 4-periodic points:

> **sol4 = Solve[f4[y] == y, y];**

However, the solution is not (and cannot be) expressed as closed-form formulas but rather as **Root**-objects. Here is one of the roots:

> **sol4[[5]]**

> $\{y \to \text{Root} [$
> $1 + a^2 - a^2 \, \#1 - a^3 \, \#1 - a^4 \, \#1 - a^5 \, \#1 + 2 a^3 \, \#1^2 + a^4 \, \#1^2 + 4 a^5 \, \#1^2 + a^6 \, \#1^2 + 2 a^7 \, \#1^2 - a^3 \, \#1^3 - 5 a^5 \, \#1^3 -$
> $4 a^6 \, \#1^3 - 5 a^7 \, \#1^3 - 4 a^8 \, \#1^3 - a^9 \, \#1^3 + 2 a^5 \, \#1^4 + 6 a^6 \, \#1^4 + 4 a^7 \, \#1^4 + 14 a^8 \, \#1^4 + 5 a^9 \, \#1^4 + 3 a^{10} \, \#1^4 -$
> $4 a^6 \, \#1^5 - a^7 \, \#1^5 - 18 a^8 \, \#1^5 - 12 a^9 \, \#1^5 - 12 a^{10} \, \#1^5 - 3 a^{11} \, \#1^5 + a^6 \, \#1^6 + 10 a^8 \, \#1^6 + 17 a^9 \, \#1^6 +$
> $18 a^{10} \, \#1^6 + 15 a^{11} \, \#1^6 + a^{12} \, \#1^6 - 2 a^8 \, \#1^7 - 14 a^9 \, \#1^7 - 12 a^{10} \, \#1^7 - 30 a^{11} \, \#1^7 -$
> $6 a^{12} \, \#1^7 + 6 a^9 \, \#1^8 + 3 a^{10} \, \#1^8 + 30 a^{11} \, \#1^8 + 15 a^{12} \, \#1^8 - a^9 \, \#1^9 -$
> $15 a^{11} \, \#1^9 - 20 a^{12} \, \#1^9 + 3 a^{11} \, \#1^{10} + 15 a^{12} \, \#1^{10} - 6 a^{12} \, \#1^{11} + a^{12} \, \#1^{12} \, \&, \, 1] \}$

For a given value of *a*, we can ask for the solution:

```
sol4 /. a → 3.52
```

```
{{y → 0}, {y → 0.715909}, {y → 0.424275}, {y → 0.859816},
 {y → 0.373084}, {y → 0.512076}, {y → 0.823301}, {y → 0.879487},
 {y → 0.0489016 - 0.0232193 i}, {y → 0.0489016 + 0.0232193 i},
 {y → 0.165614 - 0.0737384 i}, {y → 0.165614 + 0.0737384 i},
 {y → 0.505554 - 0.173586 i}, {y → 0.505554 + 0.173586 i},
 {y → 0.985956 - 0.00678703 i}, {y → 0.985956 + 0.00678703 i}}
```

Here are the equilibrium and 2-periodic points:

```
sol2 /. a → 3.52
```

```
{{y → 0}, {y → 0.715909}, {y → 0.424275}, {y → 0.859816}}
```

Thus it follows that the 4-periodic points are the next 4 points: 0.373084, 0.512076, 0.823301, and 0.879487. We select these solutions from **sol4**:

```
sol4a = Take[sol4, {5, 8}];
```

To find the values of *a* for which the 4-periodic points are asymptotically stable, we have to find the values of *a* for which the absolute value of the derivative of $f(f(f(f(y))))$ at the 4-periodic points is less than 1. At $a = 1 + \sqrt{6}$, the derivatives are 1:

```
df4 = f4 '[y];
```

```
(df4 /. a → 1. + Sqrt[6]) /. (sol4a /. a → 1. + Sqrt[6])
```

```
{1., 1., 1., 1.}
```

At *a* = 3.5, the derivatives are −0.0305:

```
(df4 /. a → 3.5) /. (sol4a /. a → 3.5)
```

```
{-0.0305, -0.0305, -0.0305, -0.0305}
```

At *a* = 3.544, the derivatives are almost −1:

```
(df4 /. a → 3.544) /. (sol4a /. a → 3.544)
```

```
{-0.997943, -0.997943, -0.997943, -0.997943}
```

To find the value for which the derivatives are exactly −1, we solve an equation:

```
a4 = a /. FindRoot[(df4 /. sol4a[[1]]) == -1, {a, 3.544, 3.545}]
```

```
3.54409
```

Thus, the 4-periodic points are asymptotically stable for $1 + \sqrt{6} < a < 3.54409$. We plot the 4-periodic points in this interval:

```
p4 = Plot[Evaluate[y /. sol4a],
    {a, 1 + Sqrt[6], a4}, DisplayFunction → Identity];
```

■ 8-Periodic Points

To find 8-periodic points, the method we have used does no more work; the algebra is too demanding. We proceed in a different way. Define first the 8-times nested function and its derivative:

```
f8[a_, y_] = Nest[f, y, 8];
```

```
df8[a_, y_] = D[f8[a, y], y];
```

Define then a function that gives, for a given a, the 8-periodic point near y_0:

```
period8[a_, y0_] := y /. FindRoot[f8[a, y] == y, {y, y0}][[1]]
```

For example, for $a = 3.55$, the 8-periodic point that is near 0.35 is as follows:

```
period8[3.55, 0.35]        0.3548
```

To find the last value of a for which the 8-periodic points are stable, we find the value of a for which the derivative of **f8** at an 8-periodic point is -1 (in this computation, we need to set a system option; the option is not needed in *Mathematica* 4):

```
Developer`SetSystemOptions["EvaluateNumericalFunctionArgument" → False];

a8 = a /. FindRoot[df8[a, period8[a, 0.35]] == -1, {a, 3.55, 3.56}]

3.56441

Developer`SetSystemOptions["EvaluateNumericalFunctionArgument" → True];
```

So, the 8-periodic points are stable in $3.54409 < a < 3.56441$.

From the second bifurcation diagram in Section 25.2.2, p. 687, we can see that 8-period points are near the points 0.35, 0.37, 0.50, 0.55, 0.81, 0.83, 0.88, and 0.89. To plot the 8-periodic points as functions of a, we produce 8 figures that correspond with these 8 starting values (the plotting takes some time):

```
p5 = Map[Plot[period8[a, #], {a, a4, a8}, DisplayFunction → Identity] &,
    {0.35, 0.37, 0.50, 0.55, 0.81, 0.83, 0.88, 0.89}];
```

■ A Bifurcation Diagram

Now we can combine all of the plots to produce a better bifurcation diagram for $0 < a < 3.56441$:

```
Show[p1, p2, p3, p4, p5, PlotRange → All,
    AxesOrigin → {0, -0.02}, DisplayFunction → $DisplayFunction];
```

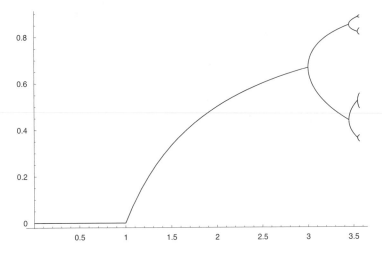

25.2.4 Lyapunov Exponents

The *Lyapunov exponent* gives still another view of a difference equation. It shows stability and instability and measures the sensitive dependence on initial conditions. It is defined, for a given a, by $\lambda = \lim_{n\to\infty} \frac{1}{n} \sum_{i=0}^{n-1} \log|f'(y_i)|$, where $\{y_i\}$ is the sequence calculated from the iteration formula $y_{i+1} = f(y_i)$. In numerical calculations, we approximate the limit by calculating only a finite sequence. Because the derivative of the righthand-side function of the logistic model is $a(1-2y)$, we can write the following function to calculate the Lyapunov exponent for a given a:

```
lyapunovExponent = Compile[{a, {n, _Integer}},
    With[{p = NestList[a # (1 - #) &, 0.6, n - 1]},
        Apply[Plus, Log[Abs[a (1 - 2 p)]]] / n]];
```

Note that, because **p** is a list, **a (1 - 2 p)** is the list of values $\{f'(y_0), ..., f'(y_{n-1})\}$.

■ Numerical Questions

When calculating the Lyapunov exponent, we again face the numerical problems if the system is chaotic. To examine the effect of loss of precision for an a for which the system is chaotic, we define another function in which we use high-precision numbers:

```
lyapunovExponent2[a_, n_] :=
    With[{p = NestList[a # (1 - #) &, 0.6`1150, n - 1]},
        Apply[Plus, Log[Abs[a (1 - 2 p)]] / n]];
```

We calculate the same exponent with both functions by using 2000 calculated values:

```
lyapunovExponent[3.7, 2000]      0.357374
```

```
lyapunovExponent2[3.7`1150, 2000]      0.3457317224630509
```

The latter result has no rounding errors, and all of decimals are correct (the result, of course, contains a truncation error due to the finiteness of the sum). The difference of the results is about 0.01; this cannot be regarded as very small, but at least if we are interested in plotting the exponent, using normal decimal numbers may give acceptable results.

To examine the convergence of the Lyapunov exponent, we calculate estimates of the exponent for $a = 3.3$ and $a = 3.7$ using sequences of length 1000, 2000, ..., 100,000:

```
Show[GraphicsArray[
    Map[ListPlot[Table[lyapunovExponent[#, n], {n, 1000, 100000, 1000}],
        PlotRange → All, Ticks → {{{50, 50000}, {100, 100000}}, Automatic},
        DisplayFunction → Identity] &, {3.3, 3.7}]]];
```

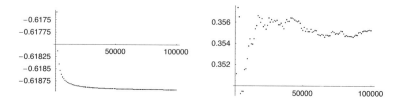

It seems that, for an *a* for which the system is not chaotic, the estimate of the exponent converges rapidly and that, for plotting purposes, where about two correct decimals suffice, a few thousand values gives satisfactory results. For an *a* for which the system is chaotic, the estimate does not converge as rapidly, but, for plotting purposes, perhaps 20,000 values suffice. Next, when we plot the Lyapunov exponent, we use 30,000 values.

■ Plotting the Lyapunov Exponent

We plot, for the logistic model, the Lyapunov exponent as a function of *a* when *a* is in (0, 4):

```
Plot[lyapunovExponent[a, 30000], {a, 0, 4}, PlotPoints → 200,
    AspectRatio → 0.35, PlotRange → {{-0.05, 4.05}, {-6.1, 1.1}}]; // Timing
```

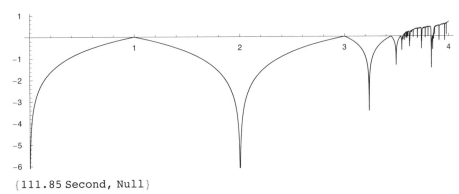

{111.85 Second, Null}

For values of *a* for which the system has a stable equilibrium point or a stable cycle, the Lyapunov exponent λ is negative. For these *a* values, the trajectories are not sensitive to initial conditions. At bifurcation points, we have $\lambda = 0$. For values of *a* for which the system is chaotic, the exponent is positive. For these *a* values, the trajectories are sensitive to initial conditions.

As noted above, for $\lambda > 0$, the numeric behavior of the sequence is problematic, and the plot is not accurate. The plot does, however, describe the overall behavior of the Lyapunov exponent. As with the bifurcation diagram, we take a closer look at the exponent when $3.5 < a < 4$:

```
Plot[lyapunovExponent[a, 30000], {a, 3.5, 4}, PlotPoints → 200,
    AspectRatio → 0.35, PlotRange → All, AxesOrigin → {3.5, 0}]; // Timing
```

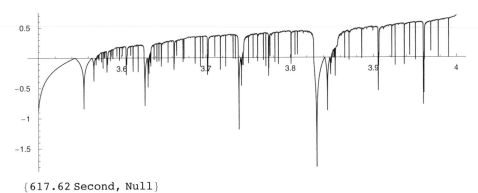

{617.62 Second, Null}

For advanced function iteration and chaos with *Mathematica*, see Maeder (1995b) (from http://www.mathematica-journal.com/issue/v5i2/ you can load the programs of this article). Knapp and Sofroniou (1997) use external C programs with *MathLink* to perform the heavy calculations needed to produce the bifurcation diagrams and plots of the Lyapunov exponent. Kaplan and Glass (1995) contains an introduction to nonlinear time series analysis.

25.3 More about Nonlinear Equations

25.3.1 A Predator–Prey Model

In Section 23.3.2, p. 615, we analyzed a predator–prey model governed by the differential equations $x'(t) = x(t)(p - q\,y(t))$ and $y'(t) = y(t)(-P + Q\,y(t))$. By approximating the derivatives by divided differences, we could introduce the discrete-time model $x_{n+1} - x_n = x_n(p - q\,y_n)$, $y_{n+1} - y_n = y_n(-P + Q\,y_n)$. However, it turns out that this model does not have realistic properties. Indeed, predators and prey fluctuate in a growing fashion, and the trajectories are outgoing spirals.

Other discrete-time predator–prey models have been introduced, one of which is the following: $x_{n+1} = a\,x_n(1 - x_n - y_n)$, $y_{n+1} = b\,x_n\,y_n$ (see Kelley and Peterson 2001, p. 184, or Martelli 1999, p. 246). We assume that $a > 0$ and $b > 0$ and begin to analyze the model.

■ **Equilibrium Points**

Define the righthand-side expressions:

```
f = a x (1 - x - y);
g = b x y;
```

Calculate the equilibrium points:

```
equi = Solve[{x == f, y == g}, {x, y}]
```

$$\left\{\{x \to 0,\ y \to 0\},\ \left\{x \to 1 - \frac{1}{a},\ y \to 0\right\},\ \left\{y \to 1 - \frac{1}{a} - \frac{1}{b},\ x \to \frac{1}{b}\right\}\right\}$$

We denote these points with e_1, e_2, and e_3. The Jacobian is as follows:

```
jac = {{D[f, x], D[f, y]}, {D[g, x], D[g, y]}}
```

$$\{\{-a\,x + a\,(1 - x - y),\ -a\,x\},\ \{b\,y,\ b\,x\}\}$$

Calculate the eigenvalues of the Jacobian at e_1:

```
eig1 = Eigenvalues[jac /. equi[[1]]]          {0, a}
```

An equilibrium point is asymptotically stable if the spectral radius (the largest of the absolute values of the eigenvalues) is less than one. So, e_1 is asymptotically stable if $0 < a < 1$.

Calculate then the eigenvalues of the Jacobian at e_2:

```
eig2 = Eigenvalues[jac /. equi[[2]]] // Simplify
```

$$\left\{2 - a,\ \frac{(-1 + a)\,b}{a}\right\}$$

The requirement that the absolute values of the eigenvalues are less than one is in simplified form as follows (users of *Mathematica* 4 have to load the package **Algebra`Inequality Solve`** and the replace **Reduce** with **InequalitySolve**):

```
Reduce[Thread[Abs[eig2] < 1], {a, b}, Reals]
```

$$1 < a < 3 \text{ \&\& } -\frac{a}{-1+a} < b < \frac{a}{-1+a}$$

Thus, the point e_2 is asymptotically stable if $1 < a < 3$ and $0 < b < \frac{a}{a-1}$. In this region, the x coordinate of e_2 or $\frac{a-1}{a}$ is positive (the y coordinate is 0).

Consider then the point e_3:

```
eig3 = Eigenvalues[jac /. equi[[3]]]
```

$$\left\{ \frac{-a + 2b - \sqrt{a^2 + 4ab + 4b^2 - 4ab^2}}{2b}, \quad \frac{-a + 2b + \sqrt{a^2 + 4ab + 4b^2 - 4ab^2}}{2b} \right\}$$

Compute the conditions under which the absolute values of these are less than one:

```
Reduce[Thread[Abs[eig3] < 1], {a, b}, Reals]
```

$$1 < a \le 3 \text{ \&\& } \frac{a}{-1+a} < b \le \frac{a}{2(-1+a)} + \frac{1}{2}\sqrt{\frac{a^3}{(-1+a)^2}} \ \ ||$$

$$3 < a < 9 \text{ \&\& } \frac{3a}{3+a} < b \le \frac{a}{2(-1+a)} + \frac{1}{2}\sqrt{\frac{a^3}{(-1+a)^2}}$$

Because the upper bound for b can be written as $\frac{a(1+\sqrt{a})}{2(a-1)}$ or $\frac{a}{2(\sqrt{a}-1)}$, the requirement is thus that either $\{1 < a < 3$ and $\frac{a}{a-1} < b < \frac{a}{2(\sqrt{a}-1)}\}$ or $\{3 < a < 9$ and $\frac{3a}{3+a} < b < \frac{a}{2(\sqrt{a}-1)}\}$.

In summary, the situation is as follows:

```
Block[{$DisplayFunction = Identity},
  p1 = Plot[{a / (a - 1), a / (2 (Sqrt[a] - 1))}, {a, 1, 3}];
  p2 = Plot[{3 a / (3 + a), a / (2 (Sqrt[a] - 1))}, {a, 3, 9}];]

p3 = Show[p1, p2, DisplayFunction → $DisplayFunction,
    Ticks → {{1, 3, 9}, {1, 2, 3}}, PlotRange → {0, 3.6},
    AxesLabel → {"a", "b"}, PlotRange → {0, 4.1},
    Epilog → {Line[{{1, 0}, {1, 3.6}}], Line[{{3, 0}, {3, 3 / 2}}],
      Text["e₁ stable", {0.5, 1.0}, {0, -1}, {0, 1}],
      Text["e₂ stable", {2.0, 0.4}, {0, -1}, {0, 1}],
      Text["e₃ stable", {2.5, 1.8}, {-1, 0}]}];
```

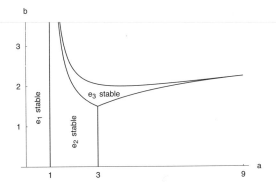

■ Bifurcations

Let us investigate how the trajectories behave for various values of a and b. First we define the programs **limits** and **bifurcation** anew so that they compute both x and y values:

```
limits = Compile[{a, b}, Transpose[
    Drop[NestList[{a #[[1]] (1 - #[[1]] - #[[2]]), b #[[1]] #[[2]]} &,
    {0.4, 0.4}, 600], 301]]];

limits2[a_, b_] := With[{lims = limits[a, b]},
    {Map[{a, #} &, Union[lims[[1]]]], Map[{a, #} &, Union[lims[[2]]]]}]

bifurcation[a0_, a1_, n_, b_] := Map[Flatten[#, 1] &,
    Transpose[Table[limits2[a, b], {a, a0, a1, (a1 - a0) / n}]]]
```

For example, when $b = 1.6$, we expect, from the figure above, that, when a grows from 0, the trajectory approaches first e_1, then e_2, then e_3, and then after that it will behave in some other way. The bifurcation diagrams for x and y confirm this:

```
{xx, yy} = bifurcation[0., 4., 400, 1.6];

Block[{$DisplayFunction = Identity},
    p1 = ListPlot[xx, PlotRange → All, AxesOrigin → {0, -0.02},
        PlotStyle → AbsolutePointSize[0.4], AxesLabel → {"a", "x"}];
    p2 = ListPlot[yy, PlotRange → All, AxesOrigin → {0, -0.002},
        PlotStyle → AbsolutePointSize[0.4], AxesLabel → {"a", "y"}]];

Show[GraphicsArray[{p1, p2}], GraphicsSpacing → 0];
```

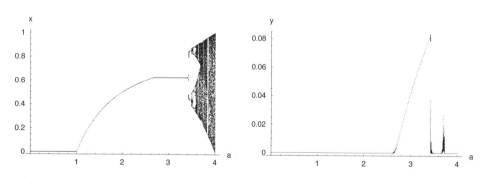

In the case where $b = 3.1$, we have interesting behavior near $a = 4$:

```
{xx, yy} = bifurcation[0., 4.2, 400, 3.1];

Block[{$DisplayFunction = Identity},
    p1 = ListPlot[xx, PlotRange → All, AxesOrigin → {0, -0.02},
        PlotStyle → AbsolutePointSize[0.4], AxesLabel → {"a", "x"}];
    p2 = ListPlot[yy, PlotRange → All, AxesOrigin → {0, -0.02},
        PlotStyle → AbsolutePointSize[0.4], AxesLabel → {"a", "y"}]];
```

```
Show[GraphicsArray[{p1, p2}], GraphicsSpacing → 0];
```

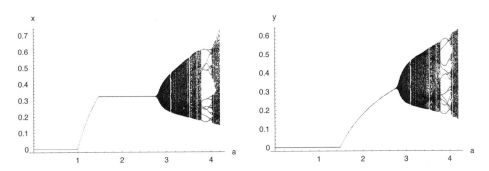

■ Trajectories

```
predatorPreyPlot[a_, b_, n_, x0_, y0_, opts___] :=
  With[{sol = NestList[{a #[[1]] (1 - #[[1]] - #[[2]]), b #[[1]] #[[2]]} &,
    {x0, y0}, n]},
   Show[Graphics[{AbsolutePointSize[2], Map[Point, sol], Line[sol]}],
    Axes → True, opts]]
```

With this program, we can plot trajectories of the predator–prey model. For example, when $a = 3$ and $b = 1.6$, the equilibrium points are as follows:

```
equi /. {a → 3., b → 1.6}
```

$\{\{x \to 0, y \to 0\}, \{x \to 0.666667, y \to 0\}, \{y \to 0.0416667, x \to 0.625\}\}$

The trajectory approaches the last point, e_3:

```
predatorPreyPlot[3, 1.6, 50, 0.6,
  0.02, AxesLabel → {"x", "y"}, PlotRange → All];
```

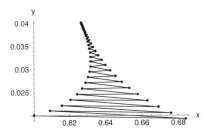

We then consider some cases in which $b = 3.1$. First, here are plots for $a = 1.4$ and $a = 2.5$:

```
Block[{$DisplayFunction = Identity},
  p1 = predatorPreyPlot[1.4, 3.1, 30, 0.3, 0.1, PlotRange → All];
  p2 = predatorPreyPlot[2.5, 3.1, 60, 0.4, 0.25, PlotRange → All];]
```

```
Show[GraphicsArray[{p1, p2}]];
```

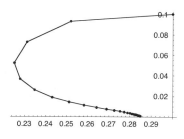 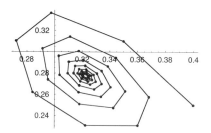

In the first plot, the trajectory approaches e_2, and in the second plot, the trajectories approach e_3 as a spiral. Next we use the value $a = 3$ (b still has the value 3.1):

```
Block[{$DisplayFunction = Identity},
 p1 = predatorPreyPlot[3, 3.1, 200, 0.32, 0.33];
 p2 = ListPlot[
   Take[NestList[{3 #[[1]] (1 - #[[1]] - #[[2]]), 3.1 #[[1]] #[[2]]} &,
     {0.32, 0.33}, 600], -400]];]
```

```
Show[GraphicsArray[{p1, p2}]];
```

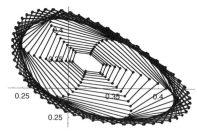 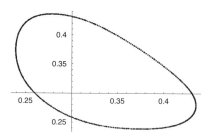

In the first plot, the trajectories approach a cycle whose center is e_3. The cycle can be seen more clearly from the second plot, where we have dropped some early points and plotted only the points, not the connecting lines. The limiting points seem to be on a closed curve. A more detailed study would reveal that the points evolve around the cycle with an approximate period of six. Lastly we use the values $a = 3.1$ and $a = 4.2$:

```
Block[{$DisplayFunction = Identity},
 p1 = predatorPreyPlot[3.1, 3.1, 100, 0.32, 0.34];
 p2 = predatorPreyPlot[4.2`150, 3.2`150, 200, 0.32`150, 0.43`150];]
```

```
Show[GraphicsArray[{p1, p2}]];
```

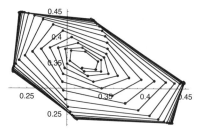 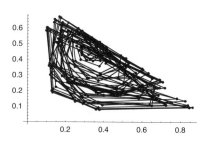

In the first plot, we have a 6-cycle, whereas the second plot shows chaotic behavior.

25.3.2 Estimation of Difference Equations

■ Return Plot

A useful way to explore a data set is to plot a *return plot*. If we have data $y_1, \ldots, y_n$, then the return plot simply contains points at (y_i, y_{i+1}), $i = 1, \ldots, n-1$. If we assume that the data is generated by a difference equation model $y_{i+1} = f(y_i)$, then the return plot contains the points $(y_i, f(y_i))$, which means that the points are on the curve $f(y)$. For example, if the model is $y_{i+1} = 3.7 \, y_i (1 - y_i)$ (a logistic model), then the points in the return plot are on the curve $3.7 \, y(1 - y)$.

Let us compare three series. The first series contains white noise (i.e., random numbers from a normal distribution; the distribution has mean 1 and standard deviation 0.5). The second series is obtained from the linear difference equation $y_{n+1} = 1.1 \, y_n$, $y_0 = 1$, and the third from the logistic difference equation $y_{n+1} = 3.7 \, y_n (1 - y_n)$, $y_0 = 0.02$. The series are as follows:

```
<< Statistics`NormalDistribution`

s1 = Table[Random[NormalDistribution[1, 0.5]], {50}];
s2 = NestList[1.1 # &, 1, 50];
s3 = NestList[3.7 # (1 - #) &, 0.02, 50];

Show[GraphicsArray[
    Map[ListPlot[#, PlotJoined → True, DisplayFunction → Identity] &,
    {s1, s2, s3}]]];
```

For each sequence, we produce the return plot:

```
pairs1 = Transpose[{Drop[s1, -1], Drop[s1, 1]}];
pairs2 = Transpose[{Drop[s2, -1], Drop[s2, 1]}];
pairs3 = Transpose[{Drop[s3, -1], Drop[s3, 1]}];

Show[GraphicsArray[Map[ListPlot[#, PlotRange → All, AxesOrigin → {0, 0},
        DisplayFunction → Identity] &, {pairs1, pairs2, pairs3}]]];
```

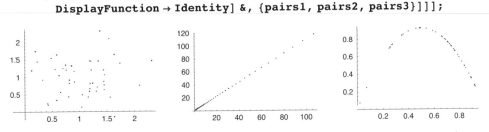

The points in the first figure do not have a clear form, but they are concentrated around the point (1, 1). The points in the second figure are on the line $1.1 \, y$, whereas the points in the third figure are on the curve $3.7 \, y(1 - y)$.

If we have data and we are searching a difference equation model for it, a return plot may give us information about a suitable difference equation model. Indeed, one possibility to use to estimate a first-order difference equation is to fit a function to the paired observations. If a fit is $f(y)$, then an approximate difference equation model is $y_{n+1} = f(y_n)$. This model may be good enough, or it can be used as a starting point for nonlinear optimization to get an even better model. Two examples follow.

■ **Drug in the Blood**

A drug dosage of 1 mg was injected into the blood, and the amount of the drug still in the blood was measured after 1, 2, ..., 12 days. The measurements were as follows (these are not real data):

```
drug = {1, 0.739, 0.537, 0.394, 0.298,
     0.211, 0.161, 0.112, 0.088, 0.060, 0.048, 0.032, 0.023};

p1 = ListPlot[Transpose[{Range[0, 12], drug}],
     PlotStyle → AbsolutePointSize[2],
     PlotRange → {{-0.2, 12.2}, Automatic}, Ticks → {Range[12], Automatic}];
```

Produce the return plot:

```
pairs = Transpose[{Drop[drug, -1], Drop[drug, 1]}];

p2 = ListPlot[pairs, PlotStyle → AbsolutePointSize[2],
     PlotRange → {{0, 1.1}, {0, 0.75}}];
```

Fit a linear function to the pairs:

```
f = Fit[pairs, {y}, y]        0.735232 y
```

A possible difference equation model is thus $y_{n+1} = 0.735232\, y_n$.

To validate the model, calculate predictions from the model:

```
g = Function[{y}, Evaluate[f]]

Function[{y}, 0.735232 y]
```

```
pred = NestList[g, 1, 12]
```

```
{1, 0.735232, 0.540565, 0.397441, 0.292211, 0.214843, 0.157959,
  0.116137, 0.0853872, 0.0627794, 0.0461574, 0.0339364, 0.0249511}
```

The differences between the data and the predictions are negligible:

```
drug - pred
```

```
{0, 0.00376849, -0.00356538, -0.0034407, 0.00578907,
  -0.00384269, 0.00304089, -0.00413652, 0.00261277,
  -0.00277938, 0.00184262, -0.00193636, -0.00195108}
```

We can be satisfied with the model.

In this example, a slightly better model can be obtained using nonlinear optimization. A linear difference equation model is nice in that the equation can be solved, and then it is straightforward to use **FindFit**. In the drug example, the solution of the difference equation is as follows (remember that if you use *Mathematica* 4, you have to load the **Discrete⌐ Math`RSolve`** package before using **RSolve**):

```
sol = y[n] /. RSolve[{y[n + 1] == a y[n], y[0] == 1}, y[n], n][[1]]
```

a^n

We estimate the parameter a (for nonlinear fitting, see Section 22.1.4, p. 580) (✻5):

```
FindFit[Transpose[{Range[0, 12], drug}], sol, a, n]
```

```
{a → 0.734947}
```

So, a slightly better model would be $y_{n+1} = 0.734947 \, y_n$. Users of *Mathematica* 4 have to do as follows:

```
<< Statistics`NonlinearFit`
```

```
NonlinearFit[Transpose[{Range[0, 12], drug}], sol, n, a]
```

0.734947^n

■ Yeast Culture: First Estimation

Consider the example of yeast culture (Pearl 1927) that we looked at in Section 23.4.4, p. 634. The measurements are as follows:

```
yeast = {9.6, 18.3, 29.0, 47.2, 71.1, 119.1, 174.6, 257.3, 350.7, 441.0,
  513.3, 559.7, 594.8, 629.4, 640.8, 651.1, 655.9, 659.6, 661.8};
```

```
p1 = ListPlot[yeast, PlotStyle → {AbsolutePointSize[2], GrayLevel[0.5]},
  Ticks → {{5, 10, 15}, {200, 400, 600}}];
```

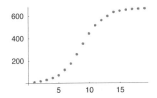

Look at the return plot:

```
pairs = Transpose[{Drop[yeast, -1], Drop[yeast, 1]}];
```

```
ListPlot[pairs, PlotStyle → AbsolutePointSize[2],
    Ticks → {{200, 400, 600}, {200, 400, 600}}];
```

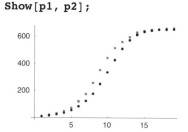

A quadratic fit may be adequate for the pairs:

```
f = Fit[pairs, {y, y^2}, y]     1.55769 y - 0.000852666 y²
```

Thus, a possible difference equation model would be $y_{n+1} = 1.55769\, y_n - 0.000852666\, y_n^2$; this is the same as the logistic model $y_{n+1} = y_n + 0.000852666\, y_n(654.049 - y_n)$.

To validate the model, calculate predictions from it:

```
pred = NestList[Function[{y}, Evaluate[f]], 9.6, 18];
```

Plot the data and the predictions:

```
p2 = ListPlot[pred,
    PlotStyle → AbsolutePointSize[2], DisplayFunction → Identity];

Show[p1, p2];
```

The black points of the predictions differ too much from the grey data points, and so we have to conclude that our model is not adequate.

```
pred =.
```

■ **Yeast Culture: Second Estimation**

We use the preliminary model obtained above as a starting point to nonlinear optimization to find a better logistic model $y_{n+1} = y_n + k\, y_n(K - y_n)$. First, define the righthand side of the equation:

```
eqn = # + k # (K - #) &;
```

Then define the following functions:

```
pred[eqn_, data_] := NestList[eqn, data[[1]], Length[data] - 1]
crit[eqn_, data_] := Apply[Plus, (data - pred[eqn, data])^2]
```

Here **pred** calculates predictions from the equation, and **crit** defines a least-squares criterion to be minimized. Now use **FindMinimum**:

```
est = FindMinimum[crit[eqn, yeast],
    {k, 0.000852666, 0.000852667}, {K, 654.049, 654.050}]
```

$\{3756.96, \{k \to 0.000999256, K \to 643.887\}\}$

This gives the model $y_{n+1} = y_n + 0.000999256\, y_n(643.887 - y_n)$.

To validate the model, calculate predictions from it:

```
a = pred[eqn /. est[[2]], yeast];

p5 = ListPlot[a,
    PlotStyle → AbsolutePointSize[2], DisplayFunction → Identity];

Show[p1, p5];
```

The model seems to be acceptable. (Remember from Section 23.4.4, p. 634, that a continuous logistic model was very good.)

25.3.3 Fractals and Cellular Automata

■ Fractals

Fractal images may be familiar to you. Mandelbrot figures are the best known of them, and they are based on the nonlinear difference equation $z_{i+1} = z_i^2 + c$ for different complex numbers c. The values of z_i may tend toward infinity, but the numbers c for which the values do not tend toward infinity constitute the Mandelbrot set. Here is a program for investigating whether a point belongs to the Mandelbrot set.

```
mandelbrot = Compile[{{c, _Complex}},
    Module[{z = 0 + 0. I, n = 0},
        While[Abs[z] < 2 && n < 50, z = z^2 + c; n++]; n]];
```

We have compiled the function **mandelbrot** to speed up the execution (compiling was explained in Section 14.2.2, p. 365). We have given the complex number $0 + 0.i$ as the starting value for z so that the compiler understands that z is a complex variable. The program returns, for a given number c, the number of iterations (n) needed for the absolute value of z to exceed 2; at most 50 iterations are done. The points c for which the full 50 iterations can be done are considered to belong to the Mandelbrot set. For example, the point $0.2 + i$ does not belong to the Mandelbrot set, but the point $0.2 + 0.2\,i$ does:

```
mandelbrot[0.2 + I]      4
```

```
mandelbrot[0.2 + 0.2 I]      50
```

Giving c complex values $x + iy$ for many x and y, we get the corresponding numbers of iterations n. These numbers can then be plotted with **DensityPlot** to get a Mandelbrot image:

```
DensityPlot[mandelbrot[x + I y], {x, -2, 1}, {y, -1.5, 1.5},
  PlotPoints → 200, Mesh → False, ColorFunction → Hue,
  FrameTicks → {{-2, -1, 0, 1}, {-1, 0, 1}, None, None}];
```

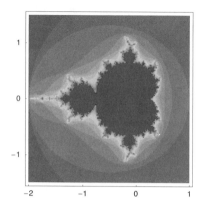

To get other colorings of the plot, try for example, instead of **Hue**, perhaps **(Hue[0.7#]&)** or **(Hue[0.8(1-#)]&)**. One of the **ExtendGraphics** packages of Wickham-Jones (1994) (see Section 8.2.2, p. 225) is **ExtendGraphics`Mandelbrot`**; it plots Mandelbrot images very quickly.

■ Cellular Automata

Cellular automata (see Wolfram 2002) do not use a recurrence equation in the usual sense, but, they are nevertheless discrete models that are based on recursive rules: the next step depends on the previous step.

CellularAutomaton[rule, init, n] Generate a list that represents the evolution of the cellular automaton rule **rule** from initial condition **init** for **n** steps

---

To generate 3 steps of cellular automaton rule 30, starting from a single 1 surrounded by 0s, write

```
CellularAutomaton[30, {{1}, 0}, 3] // MatrixForm
```

$$\begin{pmatrix} 0 & 0 & 0 & 1 & 0 & 0 & 0 \\ 0 & 0 & 1 & 1 & 1 & 0 & 0 \\ 0 & 1 & 1 & 0 & 0 & 1 & 0 \\ 1 & 1 & 0 & 1 & 1 & 1 & 1 \end{pmatrix}$$

Now we generate 40 steps:

```
ca = CellularAutomaton[30, {{1}, 0}, 40];
```

```
Show[Graphics[Raster[1 - Reverse[ca]]]];
```

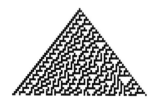

26

Probability

Introduction

> *The generation of random numbers is too important to be left to chance.* — *Robert R. Coveyou*

Mathematica has, in various packages, a wealth of information about most standard probability distributions. We can calculate probabilities with cumulative distribution functions and probability density functions, and we can also ask for means, variances, random numbers, and so on for the distributions. As examples of discrete, continuous, and multivariate distributions, we consider in more detail the binomial, multinomial, normal, and bivariate normal distributions in Section 26.1.

One very useful property of *Mathematica* is the ease with which random numbers can be generated; this makes it very convenient to perform simulations. We will give several examples of simulation, particularly when we consider stochastic processes. A variety of processes is simulated, from a simple random walk to a queuing process (in the simulations, we will mainly use the functional programming style explained in Section 15.2). From the simulations, we can actually see how realizations of the processes are displayed.

26.1 Probability Distributions

26.1.1 Uniform Random Numbers

With **Random**, we can generate uniformly distributed random numbers, or, more exactly, pseudorandom numbers. The random numbers can be real, integer, or complex valued.

For a uniform random number of type **Real**, **Integer**, *or* **Complex**:

Random[type]   A uniform random number from [0, 1]

Random[type, b]   A uniform random number from [0, b]

Random[type, {a, b}]   A uniform random number from [a, b]

Random[type, {a, b}, n]   A uniform random number with n-digit precision

---

Of special interest are random numbers of the form **Random[type]**, which are uniform random numbers from the interval [0, 1]. The default type is **Real**, so if we generate real numbers, **Random[]** with no argument suffices (however, **Random[Real]** also works). Here are commands to generate random numbers from [0, 1]:

Random[]   A real random number uniformly distributed over [0, 1]

Random[Integer]   An integer random number being 0 or 1 with probability $\frac{1}{2}$

Random[Complex]   A complex random number when the real and imaginary parts are uniform over [0, 1]

Random uses a Marsaglia–Zaman subtract-with-borrow generator for real and complex random numbers and the Wolfram rule 30 cellular automaton generator for random integers.

■ Real Random Numbers

A real uniform distribution over [a, b] is sometimes denoted as U(a, b). The probability density function of this distribution is $1/(b - a)$ for $a \le x \le b$ and zero otherwise. Here are random numbers from the distribution U(0, 1):

```
Table[Random[], {6}]
```
```
{0.314188, 0.757011, 0.528689, 0.963698, 0.0362114, 0.131348}
```

We also generate random numbers from the distributions U(0, 10) and U(−1, 1):

```
Table[Random[Real, 10], {6}]
```
```
{2.94531, 0.279744, 9.66775, 4.39286, 6.70844, 1.2432}
```

```
Table[Random[Real, {-1, 1}], {6}]
```
```
{-0.924222, -0.96668, 0.0355658, 0.914223, 0.530541, -0.384461}
```

Next we ask for real random numbers from U(0, 1) with precision of 20:

```
Table[Random[Real, {0, 1}, 20], {3}]
```
```
{0.93895647987551140912, 0.58804750808090640199, 0.62493605996520523407}
```

■ Integer Random Numbers

Let DU(*a*, *b*) be the integer-valued uniform distribution among the integers *a*, *a* + 1, ..., *b*. The probability function of this distribution is $P(X = k) = 1/(b - a + 1)$ when $k = a, a + 1, ...,$ *b*. First we generate 20 random numbers from DU(0, 1), that is, random numbers being 0 or 1 with probability 1/2:

    Table[Random[Integer], {20}]

    {1, 0, 1, 0, 1, 1, 1, 0, 0, 0, 1, 0, 1, 1, 0, 0, 1, 1, 1, 1}

This sequence can be interpreted as being a realization of 20 coin tossings (0 could mean a head and 1 a tail). Next we generate 20 random numbers from DU(1, 6), that is, $P(X = k) = 1/6$, $k = 1, 2, ..., 6$. An interpretation of the simulation is that we throw a die 20 times:

    Table[Random[Integer, {1, 6}], {20}]

    {1, 2, 3, 6, 6, 6, 4, 3, 2, 2, 1, 6, 4, 5, 3, 3, 6, 3, 2, 2}

■ Complex Random Numbers

Uniform complex random numbers are uniformly distributed in a rectangle in the complex plane. Here are uniform complex random numbers from the unit square, which are complex random numbers with real and imaginary parts distributed as U(0, 1):

    Table[Random[Complex], {3}]

    {0.230498 + 0.630346 i, 0.532115 + 0.872901 i, 0.21775 + 0.265483 i}

Next we generate complex random numbers with real parts distributed as U(5, 6) and imaginary parts as U(3, 4):

    Table[Random[Complex, {5 + 3 I, 6 + 4 I}], {3}]

    {5.21793 + 3.11589 i, 5.68906 + 3.30178 i, 5.18172 + 3.98454 i}

■ Controlling the Random Numbers

SeedRandom[ ]	Reseed the random number generator with the time of day
SeedRandom[n]	Reseed the random number generator with the integer *n*
$RandomState	Current state of the random number generator

Each time we generate a sequence of random numbers, we get a different sequence, because *Mathematica* uses a different seed for the numbers each time (the seed is the time of day measured in small fractions of a second):

    Table[Random[], {6}]

    {0.39453, 0.27381, 0.214941, 0.545256, 0.723687, 0.14949}

    Table[Random[], {6}]

    {0.177052, 0.528596, 0.205904, 0.192378, 0.411781, 0.220826}

If we want the same sequence several times, we can set the seed with **SeedRandom[n]**:

```
SeedRandom[15]; Table[Random[], {6}]
```

{0.585799, 0.12565, 0.987909, 0.255315, 0.907106, 0.428568}

```
SeedRandom[15]; Table[Random[], {6}]
```

{0.585799, 0.12565, 0.987909, 0.255315, 0.907106, 0.428568}

If, while generating random numbers, we want to perform some subsidiary generations, we can use **Block** to localize the value of **$RandomState** during the subsidiary calculations. The subsidiary generations will not then affect the main generations.

■ Random Permutations, Partitions, and Subsets

In the **DiscreteMath`Combinatorica`** *package:*

RandomPermutation[list] Elements of **list** in random order
RandomKSetPartition[list, k] Elements of **list** in **k** random blocks
RandomKSubset[list, k] Random subset of **list** with **k** elements

In place of **list**, the commands accept a given positive integer **n**, and then the operations are done for the set of integers {1, 2, ..., **n**}. Note that the package has even more commands that relate to randomness (after loading the package, type **?*Random***). For example:

```
<< DiscreteMath`Combinatorica`

RandomPermutation[10]      {10, 6, 7, 4, 3, 1, 9, 2, 5, 8}

RandomKSetPartition[10, 4]      {{1, 3, 7}, {2, 5, 8}, {4}, {6, 9, 10}}

RandomKSubset[10, 4]      {3, 4, 8, 9}
```

■ Testing the Equality of Two Expressions

Random numbers can be used—in addition to simulation—in the testing of the equality of two complex expressions. As an example, consider the following integral:

```
math = Integrate[Sin[a x^n], {x, 0, ∞}, Assumptions → n > 1 && a > 0]
```

$$a^{-1/n} \, \text{Gamma}\left[1 + \frac{1}{n}\right] \text{Sin}\left[\frac{\pi}{2\,n}\right]$$

Spiegel (1999, formula 18.51) gives the following value for the integral:

```
spiegel = 1 / (n a ^ (1 / n)) Gamma[1 / n] Sin[π / (2 n)];
```

Mathematica is able to show that the two expressions are equal:

```
FullSimplify[math - spiegel]      0
```

Another way to test the equality of the two expressions is to insert random numbers in place of *a* and *n* and check whether the difference of the expressions is practically zero:

```
math - spiegel /. {a → Random[], n → Random[Integer, {1, 10}]}
```

-2.77556×10^{-17}

By repeating this command several times, we could observe that the result is always about zero, and this confirms the equality of the two expressions.

26.1.2 Univariate Discrete Distributions

Discrete and continuous probability distributions are defined in several packages:

```
Statistics`DiscreteDistributions`
Statistics`MultivariateDiscreteDistributions`

Statistics`ContinuousDistributions`
Statistics`NormalDistribution`
Statistics`MultinormalDistribution`
```

The distributions defined in the **Statistics`NormalDistribution`** package are actually a subset of the continuous distributions defined in the **Statistics`ContinuousDistribu﹂tions`** package, the former package defining the normal distribution and distributions related to it. The **Statistics`MultinormalDistribution`** package defines the multivariate normal distribution and other related statistical distributions.

■ Properties of the Univariate Distributions

For each univariate distribution (both discrete and continuous), we can ask for the following information. In place of **dist**, simply write a particular distribution (such as **Binomial﹂Distribution[6, 1/6]**).

Domain[dist]   Range of values of the random variable
PDF[dist,x]   Value of the probability density function at **x**
CDF[dist,x]   Value of the cumulative distribution function at **x**
Quantile[dist, q]   The **q** quantile
CharacteristicFunction[dist, t]   Value of the characteristic function at **t**

Mean[dist], **Variance[dist]**, **StandardDeviation[dist]**,
Skewness[dist], **Kurtosis[dist]**, **KurtosisExcess[dist]**

ExpectedValue[f, dist]   Expected value of a pure function **f**
ExpectedValue[f, dist, x]   Expected value of a function **f** of **x**

Random[dist]   Random number from the distribution
RandomArray[dist, dims]   Random array with dimensionality **dims**

Note that the probability density function (PDF) in *Mathematica* means, for continuous random variables, the derivative of the cumulative distribution function (CDF) and, for discrete variables, the function often called the probability function. In mathematical terms, let $F(x) = P(X \leq x)$ be the CDF of a random variable X. The PDF in *Mathematica* is $f(x) = F'(x)$ if X is continuous and $f(x) = P(X = x)$ if X is discrete. Standard deviation is the square root of the variance.

   RandomArray[dist, n] gives a list of n random numbers. **RandomArray[dist, {m, n}]** gives an $(m \times n)$ matrix of random numbers.

■ Example: The Binomial Distribution

Let X be the number of sixes when a die is tossed six times; X has the binomial distribution with parameters 6 and $\frac{1}{6}$:

```
<< Statistics`DiscreteDistributions`
```

```
dist = BinomialDistribution[6, 1 / 6];
```

Ask for the domain, mean, and variance:

```
{Domain[dist], Mean[dist], Variance[dist]}
```

$$\left\{\{0, 1, 2, 3, 4, 5, 6\}, 1, \frac{5}{6}\right\}$$

The probability density function at k is as follows:

```
PDF[dist, k]
```
$$\frac{5^{6-k}\ \text{Binomial}[6, k]}{46656}$$

Calculate the numerical values of the density function:

```
t1 = Table[{i, PDF[dist, i]}, {i, 0, 6}] // N
```

```
{{0., 0.334898}, {1., 0.401878}, {2., 0.200939}, {3., 0.0535837},
 {4., 0.00803755}, {5., 0.000643004}, {6., 0.0000214335}}
```

```
ListPlot[t1, PlotStyle → AbsolutePointSize[3]];
```

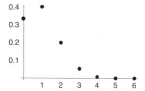

■ Discrete Distributions

When listing univariate distributions, we use the following function, which shows the name of the distribution, the PDF, the CDF, the mean, and the variance:

```
<< Statistics`DiscreteDistributions`
```

```
listDistributions[distributions_List] :=
 Map[{Apply[StringDrop[ToString[#[[0]]], -12], #],
    PDF[#, x], CDF[#, x], Mean[#], Variance[#]} &, distributions]
```

All names of distributions end with **Distribution**. The function above shortens the names by dropping the "**Distribution**," to make room in the table. The table is produced with the following function:

```
tabulateDistributions[list_, opts___] :=
 TableForm[list, TableAlignments → Center, TableHeadings →
   {None, Map[StyleForm[#, FontWeight → "Bold"] &, {"Distribution",
     "PDF", "CDF", "Mean", "Variance"}]}, opts] // TraditionalForm
```

In the **Statistics`DiscreteDistributions`**, package we have the following univariate discrete probability distributions:

```
discreteDistributions = {DiscreteUniformDistribution[n],
    BernoulliDistribution[p], BinomialDistribution[n, p],
    GeometricDistribution[p], NegativeBinomialDistribution[n, p],
    HypergeometricDistribution[n, K, N],
    PoissonDistribution[μ], LogSeriesDistribution[θ]};
```

We list these distributions and some of their properties:

```
list1 = ReplacePart[listDistributions[discreteDistributions] // Together,
    "*", {{6, 3}, {8, 3}}] /. {(1 - p) → q, (p - 1) → -q} // FullSimplify;
```

Here we have replaced the CDF of the hypergeometric and log series distribution with an asterisk (*), because these CDFs are long expressions. Also, we have used the notation $q = 1 - p$, which is quite common in probability texts. The list of distributions could be presented as a table by giving the following command:

```
tabulateDistributions[list1, TableSpacing → {1.5, 1.5}]
```

However, we have somewhat arranged the terms in the expressions, and we present the edited table below.

Distribution	PDF	CDF	Mean	Variance
DiscreteUniform[n]	$\frac{1}{n}$	$\frac{\lfloor x \rfloor}{n}$	$\frac{n+1}{2}$	$\frac{n^2-1}{12}$
Bernoulli[p]	If[x == 0, q, p]	q	p	$p\,q$
Binomial[n, p]	$\binom{n}{x} p^x q^{n-x}$	$I_q(n - \lfloor x \rfloor, \lfloor x \rfloor + 1)$	$n\,p$	$n\,p\,q$
Geometric[p]	$p\,q^x$	$1 - q^{\lfloor x \rfloor + 1}$	$\frac{q}{p}$	$\frac{q}{p^2}$
NegativeBinomial[n, p]	$\binom{n+x-1}{n-1} p^n q^x$	$I_p(n, \lfloor x \rfloor + 1)$	$n\,\frac{q}{p}$	$n\,\frac{q}{p^2}$
Hypergeometric[n, K, N]	$\frac{\binom{K}{x}\binom{N-K}{n-x}}{\binom{N}{n}}$	*	$n\,\frac{K}{N}$	$n\,\frac{K}{N}\left(1 - \frac{K}{N}\right)\frac{N-n}{N-1}$
Poisson[μ]	$e^{-\mu}\,\frac{\mu^x}{x!}$	$Q(\lfloor x \rfloor + 1, \mu)$	μ	μ
LogSeries[θ]	$\frac{-1}{\log(1-\theta)}\,\frac{\theta^x}{x}$	*	$\frac{-\theta}{(1-\theta)\log(1-\theta)}$	$\frac{-\theta\,(\theta+\log(1-\theta))}{(1-\theta)^2\,\log^2(1-\theta)}$

As we stated earlier, here the names of the distributions are shortened by dropping "Distribution". Also, $1 - p$ is replaced with q. The CDFs of the hypergeometric and log series distributions, which take up a lot of space, are not shown. Above $\lfloor x \rfloor$ is Floor[k], I_p the BetaRegularized, and Q the GammaRegularized function.

The domain (or range of values) of the discrete uniform distribution is $\{1, 2, \ldots, n\}$; of the Bernoulli distribution $\{0, 1\}$; of the binomial distribution $\{0, 1, \ldots, n\}$; of the geometric, negative binomial, and Poisson distributions $\{0, 1, 2, \ldots\}$; of the hypergeometric distribution $\{\max(0, K + n - N), \ldots, \min(K, n)\}$; and of the logarithmic series distribution $\{1, 2, \ldots\}$.

Many discrete distributions can be interpreted in terms of some experiments. In the discrete uniform distribution, the experiment has n equally probable outcomes $1, 2, \ldots, n$ (an example is die tossing with $n = 6$).

With the Bernoulli distribution, we have two outcomes: 1 (success) and 0 (failure), which occur with probabilities p and $1 - p$ (an example is coin tossing with $p = \frac{1}{2}$).

The binomial distribution emerges as the distribution of the successes in n trials, with each trial being a success or failure with probabilities p and $1 - p$ (an example is $p = \frac{1}{6}$ corresponding with tossing a die n times and counting the sixes).

In the geometric distribution, we repeat an experiment and count the number of failures before the first success. In the negative binomial distribution, we count the number of failures before the nth success.

The hypergeometric distribution may be interpreted as follows: from an urn containing N balls—K of them being black and $N - K$ being white—we pick, without replacement, n balls and count the number of black balls.

Next we continue with the preceding example and consider in more detail the binomial distribution.

26.1.3 The Binomial Distribution

■ Probabilities

In the example of Section 26.1.2, we plotted the PDF of the binomial distribution as a point plot. We can also plot a bar chart (see Section 9.3.3, p. 250). First, we recalculate the probabilities and add a third element 1 to each result. The 1 is the width of the bars:

```
<< Statistics`DiscreteDistributions`

dist = BinomialDistribution[6, 1 / 6.];

t2 = Table[{i, PDF[dist, i], 1}, {i, 0, 6}]
{{0, 0.334898, 1}, {1, 0.401878, 1}, {2, 0.200939, 1}, {3, 0.0535837, 1},
 {4, 0.00803755, 1}, {5, 0.000643004, 1}, {6, 0.0000214335, 1}}

<< Graphics`Graphics`

GeneralizedBarChart[t2,
  BarStyle → {GrayLevel[0.8]}, AxesOrigin → {-0.5, 0}];
```

We also plot the CDF; the value of this function at a point x is the probability that we will obtain at most x sixes:

```
Plot[CDF[dist, x], {x, -2, 8},
  Ticks → {Range[-2, 8], Automatic}, PlotPoints → 50];
```

What is the probability that we will obtain at most two sixes?

> **CDF[dist, 2]**   0.937714

What is the probability that we obtain two to six sixes? We can use the CDF or the PDF:

> **CDF[dist, 6] – CDF[dist, 1]**   0.263224

> **Sum[PDF[dist, k], {k, 2, 6}]**   0.263224

The sum of the probabilities of a distribution is one:

> **Apply[Plus, Table[PDF[dist, k], {k, 0, 6}]]**   1.

■ Quantiles

A q quantile is the smallest point at which the CDF has at least the value q (quantiles form the inverse function of the CDF). What is the smallest k that we obtain at most k sixes with a probability of at least 0.95?

> **Quantile[dist, 0.95]**   3

We can check that this is the right value by noting that two sixes at most do not suffice but that three sixes at most do suffice to give a probability of at least 0.95:

> **{CDF[dist, 2], CDF[dist, 3]}**

> {0.937714, 0.991298}

Here is the quantile function:

> **Plot[Quantile[dist, x], {x, 0, 1}, PlotRange → All];**

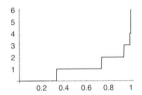

■ Random Numbers

To simulate a sequence of 20 tries of tossing a die 6 times and counting sixes, we could write the following:

> **Table[Random[dist], {20}]**

> {1, 1, 1, 2, 0, 0, 1, 0, 0, 2, 1, 1, 4, 1, 2, 1, 2, 0, 2, 0}

However, we have another command designed to produce arrays of random numbers:

> **RandomArray[dist, 20]**

> {2, 1, 0, 0, 3, 1, 0, 1, 1, 3, 2, 1, 3, 1, 4, 0, 2, 1, 3, 1}

In this latter sequence we obtained two sixes in the first try of six tosses, one six in the second try, zero sixes in the third try, and so on.

Next we do 50,000 tries of tossing a die 6 times:

> **SeedRandom[1];**
> **t3 = RandomArray[dist, 50000];**

The frequencies of the results 1, 2, ..., 6 are

```
freq = Table[Count[t3, k], {k, 0, 6}]
```

{16950, 20023, 10023, 2557, 416, 30, 1}

We succeeded in getting one result where all six throws were sixes. Five sixes were obtained 30 times. The remarkable result of 6 sixes occurred at the 45,735th try:

```
Position[t3, 6]      {{45735}}
```

Calculate relative frequencies by dividing the frequencies by 50,000:

```
freq / 50000.
```

{0.339, 0.40046, 0.20046, 0.05114, 0.00832, 0.0006, 0.00002}

These numbers are quite close to the exact probabilities:

```
Table[PDF[dist, i], {i, 0, 6}]
```

{0.334898, 0.401878, 0.200939,
0.0535837, 0.00803755, 0.000643004, 0.0000214335}

By looking at the discrete distributions package, we can see that it calculates random numbers of the binomial distribution using the following compiled function:

```
binomial = Compile[{{n, _Real}, {p, _Real}},
  Apply[Plus, Table[If[Random[] > p, 0, 1], {n}]]]
```

■ Expectations

Calculate the first and second moments of the binomial distribution we have considered:

```
m1 = ExpectedValue[k, dist, k]      1.
```

```
m2 = ExpectedValue[k^2, dist, k]      1.83333
```

The variance can be calculated in two ways:

```
ExpectedValue[(k - m1)^2, dist, k]      0.833333
```

```
m2 - m1^2      0.833333
```

Of course, we can also use **Variance**:

```
Variance[dist]      0.833333
```

■ The Characteristic Function

Calculate the characteristic function of the binomial distribution:

```
char = CharacteristicFunction[BinomialDistribution[n, p], t]
```

$(1 - p + e^{i t} p)^n$

From this we can get the moment-generating function:

```
mom = char /. t → -I t      (1 - p + e^t p)^n
```

We can also get the probability-generating function:

```
prob = char /. t → -I Log[z]      (1 - p + p z)^n
```

From the probability-generating function, we can calculate, for example, the mean and the variance:

```
m1 = D[prob, z] /. z → 1    n p
```

```
(D[prob, z, z] /. z → 1) + m1 - m1^2 // Simplify    - n (-1 + p) p
```

The coefficients in the series expansion are the probabilities of the distribution:

```
Series[prob /. {n → 6, p → 1. / 6}, {z, 0, 10}]
```

$0.334898 + 0.401878 z + 0.200939 z^2 + 0.0535837 z^3 +$
$0.00803755 z^4 + 0.000643004 z^5 + 0.0000214335 z^6 + O[z]^{11}$

26.1.4 Univariate Continuous Distributions

For univariate continuous distributions, we can use the same commands **Domain**, **PDF**, **CDF**, and so on that were mentioned in Section 26.1.2, p. 709. Here they all are:

> **Domain, PDF, CDF, Quantile, CharacteristicFunction, Mean, Variance, Standard⸱**
> **Deviation, Skewness, Kurtosis, KurtosisExcess, ExpectedValue, Random,**
> **RandomArray**

■ Continuous Distributions with a Finite or Half-infinite Domain

In the **Statistics`ContinuousDistributions`** package, we have continuous probability distributions. The domain of the following distributions is either a finite or a half-infinite interval:

```
<< Statistics`ContinuousDistributions`
```

```
continuousDistributions1 = {BetaDistribution[α, β], ChiDistribution[n],
    ChiSquareDistribution[n], ExponentialDistribution[λ],
    FRatioDistribution[m, n], GammaDistribution[α, λ],
    HalfNormalDistribution[θ], LogNormalDistribution[μ, σ],
    ParetoDistribution[k, α], RayleighDistribution[σ],
    UniformDistribution[a, b], WeibullDistribution[α, λ]};
```

We list some properties of these distributions with the program we presented in Section 26.1.2:

```
list2 = MapAt[Simplify[#, a < x < b] &,
    MapAt[FullSimplify, listDistributions[continuousDistributions1],
    {2, 5}], {{11, 2}, {11, 3}}];
```

Here we have simplified the PDF and CDF of the uniform distribution and the variance of the chi distribution. The list of distributions could be presented as a table by giving the following command:

```
tabulateDistributions[list2, TableSpacing → {1.5, 0.7}]
```

However, we have again somewhat edited the expressions, and we present here the edited table:

Distribution	PDF	CDF	Mean	Variance
Beta[α, β]	$\frac{1}{B(\alpha,\beta)}\, x^{\alpha-1}(1-x)^{\beta-1}$	$I_x(\alpha,\beta)$	$\frac{\alpha}{\alpha+\beta}$	$\frac{\alpha\beta}{(\alpha+\beta)^2\,(\alpha+\beta+1)}$
Chi[n]	$\frac{\sqrt{2}}{\Gamma(\frac{n}{2})}\left(\frac{x}{\sqrt{2}}\right)^{n-1} e^{-\left(\frac{x}{\sqrt{2}}\right)^2}$	$Q\left(\frac{n}{2},0,\frac{x^2}{2}\right)$	$\sqrt{2}\,\frac{\Gamma\left(\frac{n+1}{2}\right)}{\Gamma\left(\frac{n}{2}\right)}$	$n-2\left(\frac{\Gamma\left(\frac{n+1}{2}\right)}{\Gamma\left(\frac{n}{2}\right)}\right)^2$
ChiSquare[n]	$\frac{1}{2\Gamma(\frac{n}{2})}\left(\frac{x}{2}\right)^{\frac{n}{2}-1} e^{-\frac{x}{2}}$	$Q\left(\frac{n}{2},0,\frac{x}{2}\right)$	n	$2n$
Exponential[λ]	$\lambda\, e^{-\lambda x}$	$1-e^{-\lambda x}$	$\frac{1}{\lambda}$	$\frac{1}{\lambda^2}$
FRatio[m, n]	$\frac{m^{\frac{m}{2}}\, n^{\frac{n}{2}}}{B(\frac{m}{2},\frac{n}{2})}\,\frac{x^{\frac{m}{2}-1}}{(n+mx)^{\frac{m+n}{2}}}$	$I_{\frac{mx}{n+mx}}\left(\frac{m}{2},\frac{n}{2}\right)$	$\frac{n}{n-2}$	$\frac{2n^2\,(m+n-2)}{m\,(n-2)^2\,(n-4)}$
Gamma[α, λ]	$\frac{1}{\lambda\,\Gamma(\alpha)}\left(\frac{x}{\lambda}\right)^{\alpha-1} e^{-\frac{x}{\lambda}}$	$Q\left(\alpha,0,\frac{x}{\lambda}\right)$	$\alpha\lambda$	$\alpha\lambda^2$
HalfNormal[θ]	$\frac{2\theta}{\pi}\, e^{-\left(\frac{\theta x}{\sqrt{\pi}}\right)^2}$	$\mathrm{erf}\left(\frac{\theta x}{\sqrt{\pi}}\right)$	$\frac{1}{\theta}$	$\frac{\pi-2}{2\theta^2}$
LogNormal[μ, σ]	$\frac{1}{\sqrt{2\pi}\,\sigma x}\, e^{-\left(\frac{\log(x)-\mu}{\sqrt{2}\,\sigma}\right)^2}$	$\frac{1}{2}+\frac{1}{2}\,\mathrm{erf}\left(\frac{\log(x)-\mu}{\sqrt{2}\,\sigma}\right)$	$e^{\mu+\frac{1}{2}\sigma^2}$	$e^{2\mu+\sigma^2}\left(e^{\sigma^2}-1\right)$
Pareto[k, α]	$\frac{\alpha}{k}\left(\frac{k}{x}\right)^{\alpha+1}$	$1-\left(\frac{k}{x}\right)^{\alpha}$	$\frac{\alpha k}{\alpha-1}$	$\frac{\alpha k^2}{(\alpha-1)^2\,(\alpha-2)}$
Rayleigh[σ]	$\frac{x}{\sigma^2}\, e^{-\left(\frac{x}{\sqrt{2}\,\sigma}\right)^2}$	$1-e^{-\left(\frac{x}{\sqrt{2}\,\sigma}\right)^2}$	$\sqrt{\frac{\pi}{2}}\,\sigma$	$\frac{4-\pi}{2}\,\sigma^2$
Uniform[a, b]	$\frac{1}{b-a}$	$\frac{x-a}{b-a}$	$\frac{a+b}{2}$	$\frac{(b-a)^2}{12}$
Weibull[α, λ]	$\frac{\alpha}{\lambda}\left(\frac{x}{\lambda}\right)^{\alpha-1} e^{-(\frac{x}{\lambda})^{\alpha}}$	$1-e^{-(\frac{x}{\lambda})^{\alpha}}$	$\lambda\,\Gamma\left(\frac{\alpha+1}{\alpha}\right)$	$\lambda^2\left[\Gamma\left(\frac{\alpha+2}{\alpha}\right)-\Gamma\left(\frac{\alpha+1}{\alpha}\right)^2\right]$

Note again that here the names of the distributions are shortened by dropping "`Distri`-`bution`". Remember also that B is the `Beta`, I the `BetaRegularized`, and Q the `Gamma`-`Regularized` function. The domain of the uniform distribution is [a, b]; that of the beta-distribution [0, 1]; that of the Pareto distribution [k, ∞); and that of all of the other distributions [0, ∞).

■ **Continuous Distributions with an Infinite Domain**

Here are continuous distributions for which the domain is the interval $(-\infty, \infty)$:

```
<< Statistics`ContinuousDistributions`
```

```
continuousDistributions2 =
  {CauchyDistribution[a, b], ExtremeValueDistribution[α, β],
   LaplaceDistribution[μ, β], LogisticDistribution[μ, β],
   NormalDistribution[μ, σ], StudentTDistribution[n]};
```

To simplify some `Sign`-expressions to `Abs`-expressions, we load a package:

```
<< ProgrammingInMathematica`Abs`
```

Next we list the distributions:

```
list3 = listDistributions[continuousDistributions2] /. Indeterminate → _;
```

Here we replaced indeterminate values of the mean and variance of the Cauchy distribution with _. The list of distributions could be presented as a table by giving the following command:

```
tabulateDistributions[list3, TableSpacing → {1.5, 1.5}]
```

However, we present an edited version of the table:

Distribution	PDF	CDF	Mean	Variance
Cauchy[a, b]	$\left\{\pi b\left[1+(\frac{x-a}{b})^2\right]\right\}^{-1}$	$\frac{1}{2}+\frac{1}{\pi}\tan^{-1}(\frac{x-a}{b})$	–	–
ExtremeValue[α, β]	$\frac{1}{\beta}e^{-\frac{x-\alpha}{\beta}-e^{-\frac{x-\alpha}{\beta}}}$	$e^{-e^{-\frac{x-\alpha}{\beta}}}$	$\alpha+\gamma\beta$	$\frac{\pi^2}{6}\beta^2$
Laplace[μ, β]	$\frac{1}{2\beta}e^{-\frac{|x-\mu|}{\beta}}$	$\frac{1}{2}+\frac{\mathrm{sgn}(x-\mu)}{2}\left(1-e^{-\frac{|x-\mu|}{\beta}}\right)$	μ	$2\beta^2$
Logistic[μ, β]	$\frac{1}{\beta}e^{-\frac{x-\mu}{\beta}}\left(1+e^{-\frac{x-\mu}{\beta}}\right)^{-2}$	$\left(1+e^{-\frac{x-\mu}{\beta}}\right)^{-1}$	μ	$\frac{\pi^2}{3}\beta^2$
Normal[μ, σ]	$\frac{1}{\sqrt{2\pi}\,\sigma}e^{-\frac{1}{2}(\frac{x-\mu}{\sigma})^2}$	$\frac{1}{2}+\frac{1}{2}\mathrm{erf}(\frac{x-\mu}{\sqrt{2}\,\sigma})$	μ	σ^2
StudentT[n]	$\frac{1}{\sqrt{n}\,\mathrm{B}(\frac{n}{2},\frac{1}{2})}(\frac{n}{n+x^2})^{\frac{n+1}{2}}$	$\frac{1}{2}+\frac{\mathrm{sgn}(x)}{2}I_{(\frac{n}{n+x^2},1)}(\frac{n}{2},\frac{1}{2})$	0	$\frac{n}{n-2}$

Note that the second parameter in the normal distribution is the standard deviation and not the variance. The γ in the mean of the extreme value distribution is **EulerGamma**.

■ Normal and Related Distributions

> In the **Statistics`NormalDistribution`** *package:*
>
> **NormalDistribution[μ, σ]**
> **ChiSquareDistribution[n]**
> **FRatioDistribution[m, n]**
> **StudentTDistribution[n]**

Note that these four distributions are also in the **Statistics`ContinuousDistributions`** package.

■ Noncentral Distributions

> In the **Statistics`ContinuousDistributions`** *package:*
>
> **NoncentralChiSquareDistribution[n, λ]**
> **NoncentralFRatioDistribution[m, n, λ]**
> **NoncentralStudentTDistribution[n, λ]**

26.1.5 The Normal Distribution

■ Probabilities

Note that the parameters of the normal distribution in *Mathematica* are the mean and the standard deviation of the distribution (in some books, the second parameter is the variance). Here are plots of the PDF, CDF, and quantile function of a normal distribution with mean 2 and standard deviation 1.5:

```
<< Statistics`NormalDistribution`
```

```
dist = NormalDistribution[2, 1.5];

Block[{$DisplayFunction = Identity},
  p1 = Plot[PDF[dist, x], {x, -3, 7}];
  p2 = Plot[CDF[dist, x], {x, -3, 7}];
  p3 = Plot[Quantile[dist, q], {q, 0, 1}, PlotRange → {-3, 7}]];

Show[GraphicsArray[{p1, p2, p3}]];
```

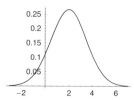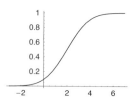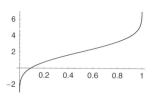

What is the probability of getting a value in the interval $(-1, 3)$? The most convenient way to answer this question is to use the distribution function, but we can also integrate the density function over the interval:

```
CDF[dist, 3] - CDF[dist, -1]      0.724757
```

```
Integrate[PDF[dist, x], {x, -1, 3}]      0.724757
```

What is a value of x such that we obtain, with probability 0.95, at most the value x? A quantile provides the solution to this problem:

```
Quantile[dist, 0.95]      4.46728
```

We check that this is the correct value:

```
CDF[dist, %]      0.95
```

■ Tabulating Probabilities

For situations in which we do not have access to *Mathematica*, we can prepare a table of probabilities (for **Grid**, see Section 13.1.3, p. 330):

```
t = Table[NumberForm[CDF[NormalDistribution[0, 1], y + x], 4],
  {y, 0.0, 3.9, 0.1}, {x, 0, 0.09, 0.01}];

rows = Map[StyleForm[#, FontWeight → "Bold"] &, Range[0.0, 3.9, 0.1]];

cols = Map[StyleForm[#, FontWeight → "Bold"] &, Range[0, 9]];

<< LinearAlgebra`MatrixManipulation`

t2 = BlockMatrix[
  {{{{""}}, {cols}, {{""}}}, {Transpose[{rows}], t, Transpose[{rows}]},
   {{{""}}, {cols}, {{""}}}}];

StyleForm[Grid[t2, RowSpacings → 0.3, ColumnSpacings → 1.5,
    RowAlignments → Bottom, ColumnAlignments → Decimal, RowLines →
     Join[Flatten[Table[{{True}, Table[False, {9}]}, {4}]], {True}],
    ColumnLines → Join[{True}, Table[False, {9}], {True}]],
  FontFamily → "Times", FontSize → 5]
```

	0	1	2	3	4	5	6	7	8	9	
0.	0.5	0.504	0.508	0.512	0.516	0.5199	0.5239	0.5279	0.5319	0.5359	**0.**
0.1	0.5398	0.5438	0.5478	0.5517	0.5557	0.5596	0.5636	0.5675	0.5714	0.5753	**0.1**
0.2	0.5793	0.5832	0.5871	0.591	0.5948	0.5987	0.6026	0.6064	0.6103	0.6141	**0.2**
0.3	0.6179	0.6217	0.6255	0.6293	0.6331	0.6368	0.6406	0.6443	0.648	0.6517	**0.3**
0.4	0.6554	0.6591	0.6628	0.6664	0.67	0.6736	0.6772	0.6808	0.6844	0.6879	**0.4**
0.5	0.6915	0.695	0.6985	0.7019	0.7054	0.7088	0.7123	0.7157	0.719	0.7224	**0.5**
0.6	0.7257	0.7291	0.7324	0.7357	0.7389	0.7422	0.7454	0.7486	0.7517	0.7549	**0.6**
0.7	0.758	0.7611	0.7642	0.7673	0.7704	0.7734	0.7764	0.7794	0.7823	0.7852	**0.7**
0.8	0.7881	0.791	0.7939	0.7967	0.7995	0.8023	0.8051	0.8078	0.8106	0.8133	**0.8**
0.9	0.8159	0.8186	0.8212	0.8238	0.8264	0.8289	0.8315	0.834	0.8365	0.8389	**0.9**
1.	0.8413	0.8438	0.8461	0.8485	0.8508	0.8531	0.8554	0.8577	0.8599	0.8621	**1.**
1.1	0.8643	0.8665	0.8686	0.8708	0.8729	0.8749	0.877	0.879	0.881	0.883	**1.1**
1.2	0.8849	0.8869	0.8888	0.8907	0.8925	0.8944	0.8962	0.898	0.8997	0.9015	**1.2**
1.3	0.9032	0.9049	0.9066	0.9082	0.9099	0.9115	0.9131	0.9147	0.9162	0.9177	**1.3**
1.4	0.9192	0.9207	0.9222	0.9236	0.9251	0.9265	0.9279	0.9292	0.9306	0.9319	**1.4**
1.5	0.9332	0.9345	0.9357	0.937	0.9382	0.9394	0.9406	0.9418	0.9429	0.9441	**1.5**
1.6	0.9452	0.9463	0.9474	0.9484	0.9495	0.9505	0.9515	0.9525	0.9535	0.9545	**1.6**
1.7	0.9554	0.9564	0.9573	0.9582	0.9591	0.9599	0.9608	0.9616	0.9625	0.9633	**1.7**
1.8	0.9641	0.9649	0.9656	0.9664	0.9671	0.9678	0.9686	0.9693	0.9699	0.9706	**1.8**
1.9	0.9713	0.9719	0.9726	0.9732	0.9738	0.9744	0.975	0.9756	0.9761	0.9767	**1.9**
2.	0.9772	0.9778	0.9783	0.9788	0.9793	0.9798	0.9803	0.9808	0.9812	0.9817	**2.**
2.1	0.9821	0.9826	0.983	0.9834	0.9838	0.9842	0.9846	0.985	0.9854	0.9857	**2.1**
2.2	0.9861	0.9864	0.9868	0.9871	0.9875	0.9878	0.9881	0.9884	0.9887	0.989	**2.2**
2.3	0.9893	0.9896	0.9898	0.9901	0.9904	0.9906	0.9909	0.9911	0.9913	0.9916	**2.3**
2.4	0.9918	0.992	0.9922	0.9925	0.9927	0.9929	0.9931	0.9932	0.9934	0.9936	**2.4**
2.5	0.9938	0.994	0.9941	0.9943	0.9945	0.9946	0.9948	0.9949	0.9951	0.9952	**2.5**
2.6	0.9953	0.9955	0.9956	0.9957	0.9959	0.996	0.9961	0.9962	0.9963	0.9964	**2.6**
2.7	0.9965	0.9966	0.9967	0.9968	0.9969	0.997	0.9971	0.9972	0.9973	0.9974	**2.7**
2.8	0.9974	0.9975	0.9976	0.9977	0.9977	0.9978	0.9979	0.9979	0.998	0.9981	**2.8**
2.9	0.9981	0.9982	0.9982	0.9983	0.9984	0.9984	0.9985	0.9985	0.9986	0.9986	**2.9**
3.	0.9987	0.9987	0.9987	0.9988	0.9988	0.9989	0.9989	0.9989	0.999	0.999	**3.**
3.1	0.999	0.9991	0.9991	0.9991	0.9992	0.9992	0.9992	0.9992	0.9993	0.9993	**3.1**
3.2	0.9993	0.9993	0.9994	0.9994	0.9994	0.9994	0.9994	0.9995	0.9995	0.9995	**3.2**
3.3	0.9995	0.9995	0.9995	0.9996	0.9996	0.9996	0.9996	0.9996	0.9996	0.9997	**3.3**
3.4	0.9997	0.9997	0.9997	0.9997	0.9997	0.9997	0.9997	0.9997	0.9997	0.9998	**3.4**
3.5	0.9998	0.9998	0.9998	0.9998	0.9998	0.9998	0.9998	0.9998	0.9998	0.9998	**3.5**
3.6	0.9998	0.9998	0.9999	0.9999	0.9999	0.9999	0.9999	0.9999	0.9999	0.9999	**3.6**
3.7	0.9999	0.9999	0.9999	0.9999	0.9999	0.9999	0.9999	0.9999	0.9999	0.9999	**3.7**
3.8	0.9999	0.9999	0.9999	0.9999	0.9999	0.9999	0.9999	0.9999	0.9999	0.9999	**3.8**
3.9	1.	1.	1.	1.	1.	1.	1.	1.	1.	1.	**3.9**
	0	1	2	3	4	5	6	7	8	9	

■ Confidence Intervals

What is an interval $(\mu - a\,\sigma,\ \mu + a\,\sigma)$ such that a normal random variable with mean μ and standard deviation σ is in that interval with probability 0.95? The constant a is 1.95996:

```
CDF[NormalDistribution[μ, σ], μ + 1.95996 σ] –
   CDF[NormalDistribution[μ, σ], μ – 1.95996 σ]

0.95
```

How can we derive the constant a? This constant is seen from the $(1 + 0.95)/2$ quantile:

```
Quantile[NormalDistribution[μ, σ], (1 + 0.95) / 2]

μ + 1.95996 σ
```

■ Random Numbers

Now we generate 100 observations from a normal distribution with mean 2 and standard deviation 1.5. We plot both the original observations and the sorted observations. Most observations can be seen to be close to the mean 2:

```
SeedRandom[1];
t1 = RandomArray[dist, 100];

Block[{$DisplayFunction = Identity},
  p1 = ListPlot[t1, PlotRange → All];
  p2 = ListPlot[Sort[t1], PlotRange → All]];
```

```
Show[GraphicsArray[{p1, p2}]];
```

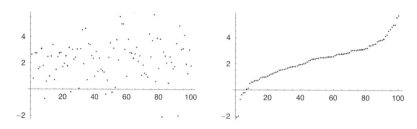

By the way, at

http://support.wolfram.com/mathematica/graphics/decorations/probabilityplot.html,

you can find a program that plots sorted data on a *probability graph paper* where the *y* axis is scaled according to a given distribution (normal distribution is the default). Normally distributed data are near to a straight line on such a paper:

```
NormalProbabilityPlot[t1];
```

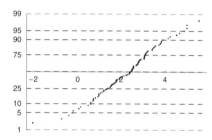

We also plot a histogram in which the categories are $[-3, -2)$, ..., $[6, 7)$:

```
<< Graphics`Graphics`
```

```
Histogram[t1, HistogramCategories → Range[-3, 7, 1],
   BarStyle → {GrayLevel[0.8]}, Ticks → {Range[-2, 7, 1], Automatic}];
```

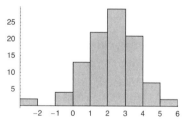

By looking at the normal distribution package, we can see that *Mathematica* calculates an array of normally distributed random numbers with the following compiled function:

```
normalpair =
  Compile[{{mu, _Real}, {sigma, _Real}, {q1, _Real}, {q2, _Real}},
    mu + sigma Sqrt[-2 Log[q1]] {Cos[2 Pi q2], Sin[2 Pi q2]}]
```

Here **q1** and **q2** are random numbers obtained with **Random[]**. The result is a pair of random variables from the normal distribution.

■ Characteristic Function

Calculate the characteristic function of the normal distribution:

```
char = CharacteristicFunction[NormalDistribution[μ, σ], t]
```

$$e^{i\,t\,\mu - \frac{t^2\,\sigma^2}{2}}$$

From this, we can get the moment-generating function:

```
mom = char /. t → -I t
```

$$e^{t\,\mu + \frac{t^2\,\sigma^2}{2}}$$

Here are the first five moments:

```
Table[D[mom, {t, n}] /. t → 0, {n, 1, 5}]
```

$$\{\mu,\ \mu^2 + \sigma^2,\ \mu^3 + 3\,\mu\,\sigma^2,\ \mu^4 + 6\,\mu^2\,\sigma^2 + 3\,\sigma^4,\ \mu^5 + 10\,\mu^3\,\sigma^2 + 15\,\mu\,\sigma^4\}$$

■ Normal Approximation

A binomial distribution with parameters n and p can be approximated with a normal distribution with mean np and variance npq, if n is large. Let us toss a die 100 times and count sixes. Here are the probabilities of getting 0, 1, ..., 30 sixes (the probability of getting more than 30 sixes is practically zero):

```
<< Statistics`DiscreteDistributions`

dist = BinomialDistribution[100, 1/6];

t1 = Table[{i, N[PDF[dist, i]], 1}, {i, 0, 30}];
```

The 1 as the third element of the sublists is the width of the bars in a bar chart to be plotted next. We also plot the PDF of the approximating normal distribution:

```
<< Statistics`NormalDistribution`

apprdist = NormalDistribution[100 * 1/6, Sqrt[100 * 1/6 * 5/6]];

<< Graphics`Graphics`

Block[{$DisplayFunction = Identity},
  p1 = GeneralizedBarChart[t1, BarStyle → {GrayLevel[1]}];
  p2 = Plot[PDF[apprdist, x], {x, 0, 30}]];

Show[p1, p2];
```

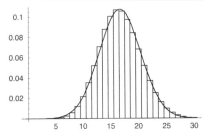

As is seen, the binomial distribution is close to the normal distribution.

26.1.6 Multivariate Distributions

■ Properties of Multivariate Distributions

The following information exists for multivariate discrete and continuous distributions:

```
Domain, PDF, CDF, CharacteristicFunction, Mean, Variance, StandardDeviation,
  Skewness, Kurtosis, KurtosisExcess, ExpectedValue, Random, RandomArray
```

These are otherwise the same commands mentioned for univariate distributions in Section 26.1.2, p. 709, but **Quantile** is lacking (in place of **Quantile** we can use, for multivariate normal and *t* distributions, **EllipsoidQuantile**; see below). In addition, for multivariate distributions, we have the following:

```
CovarianceMatrix, CorrelationMatrix, MultivariateSkewness, Multivariate\
  Kurtosis, MultivariateKurtosisExcess
```

■ Multivariate Discrete Distributions

In the **Statistics`MultiDiscreteDistributions`** *package:*

```
MultinomialDistribution[n, p]
NegativeMultinomialDistribution[n, p]
MultiPoissonDistribution[μ0 , μ]
```

■ The Multinomial Distribution

We toss a die three times. The numbers of the results 1, 2, ..., 6 form a multinomial distribution:

```
<< Statistics`MultiDiscreteDistributions`

dist = MultinomialDistribution[3, Table[1 / 6, {6}]]
```

MultinomialDistribution$\left[3, \left\{\frac{1}{6}, \frac{1}{6}, \frac{1}{6}, \frac{1}{6}, \frac{1}{6}, \frac{1}{6}\right\}\right]$

The PDF is as follows:

```
PDF[dist, Table[k_i, {i, 1, 6}]]
```

$\text{If}[k_1 + k_2 + k_3 + k_4 + k_5 + k_6 == 3,$
$\quad 6^{-k_1-k_2-k_3-k_4-k_5-k_6}\ \text{Multinomial}[k_1, k_2, k_3, k_4, k_5, k_6], 0]$

For example, the probability of getting 0, 1, 0, 0, 2, 0 times the result 1, 2, 3, 4, 5, 6, respectively, is as follows:

```
PDF[dist, {0, 1, 0, 0, 2, 0}]
```

$\frac{1}{72}$

We repeat four times the experiment of tossing a die three times. Here are the results:

```
RandomArray[dist, {4}] // ColumnForm
```

{1, 1, 0, 0, 1, 0}
{0, 0, 0, 2, 0, 1}
{1, 1, 1, 0, 0, 0}
{1, 0, 0, 0, 0, 2}

All possible different results are

```
Domain[dist]
```

{{3, 0, 0, 0, 0, 0}, {0, 3, 0, 0, 0, 0}, {0, 0, 3, 0, 0, 0}, {0, 0, 0, 3, 0, 0}, {0, 0, 0, 0, 3, 0},
{0, 0, 0, 0, 0, 3}, {2, 1, 0, 0, 0, 0}, {2, 0, 1, 0, 0, 0}, {2, 0, 0, 1, 0, 0}, {2, 0, 0, 0, 1, 0},
{2, 0, 0, 0, 0, 1}, {1, 2, 0, 0, 0, 0}, {1, 0, 2, 0, 0, 0}, {1, 0, 0, 2, 0, 0}, {1, 0, 0, 0, 2, 0},
{1, 0, 0, 0, 0, 2}, {0, 2, 1, 0, 0, 0}, {0, 2, 0, 1, 0, 0}, {0, 2, 0, 0, 1, 0}, {0, 2, 0, 0, 0, 1},
{0, 1, 2, 0, 0, 0}, {0, 1, 0, 2, 0, 0}, {0, 1, 0, 0, 2, 0}, {0, 1, 0, 0, 0, 2}, {0, 0, 2, 1, 0, 0},
{0, 0, 2, 0, 1, 0}, {0, 0, 2, 0, 0, 1}, {0, 0, 1, 2, 0, 0}, {0, 0, 1, 0, 2, 0}, {0, 0, 1, 0, 0, 2},
{0, 0, 0, 2, 1, 0}, {0, 0, 0, 2, 0, 1}, {0, 0, 0, 1, 2, 0}, {0, 0, 0, 1, 0, 2}, {0, 0, 0, 0, 2, 1},
{0, 0, 0, 0, 1, 2}, {1, 1, 1, 0, 0, 0}, {1, 1, 0, 1, 0, 0}, {1, 1, 0, 0, 1, 0}, {1, 1, 0, 0, 0, 1},
{1, 0, 1, 1, 0, 0}, {1, 0, 1, 0, 1, 0}, {1, 0, 1, 0, 0, 1}, {1, 0, 0, 1, 1, 0},
{1, 0, 0, 1, 0, 1}, {1, 0, 0, 0, 1, 1}, {0, 1, 1, 1, 0, 0}, {0, 1, 1, 0, 1, 0},
{0, 1, 1, 0, 0, 1}, {0, 1, 0, 1, 1, 0}, {0, 1, 0, 1, 0, 1}, {0, 1, 0, 0, 1, 1},
{0, 0, 1, 1, 1, 0}, {0, 0, 1, 1, 0, 1}, {0, 0, 1, 0, 1, 1}, {0, 0, 0, 1, 1, 1}}

Here are their probabilities:

```
Map[PDF[dist, #] &, %]
```

$$\Big\{ \frac{1}{216}, \frac{1}{216}, \frac{1}{216}, \frac{1}{216}, \frac{1}{216}, \frac{1}{216}, \frac{1}{72}, \frac{1}{72}, \frac{1}{72}, \frac{1}{72}, \frac{1}{72}, \frac{1}{72}, \frac{1}{72}, \frac{1}{72}, \frac{1}{72}, \frac{1}{72}, \frac{1}{72},$$
$$\frac{1}{72}, \frac{1}{72}, \frac{1}{72}, \frac{1}{72}, \frac{1}{72}, \frac{1}{72}, \frac{1}{72}, \frac{1}{72}, \frac{1}{72}, \frac{1}{72}, \frac{1}{72}, \frac{1}{72}, \frac{1}{72}, \frac{1}{72}, \frac{1}{72}, \frac{1}{72}, \frac{1}{72}, \frac{1}{36},$$
$$\frac{1}{36}, \frac{1}{36}, \frac{1}{36}, \frac{1}{36}, \frac{1}{36}, \frac{1}{36}, \frac{1}{36}, \frac{1}{36}, \frac{1}{36}, \frac{1}{36}, \frac{1}{36}, \frac{1}{36}, \frac{1}{36}, \frac{1}{36}, \frac{1}{36}, \frac{1}{36}, \frac{1}{36}, \frac{1}{36} \Big\}$$

The mean and variance vectors are as follows:

```
{Mean[dist], Variance[dist]}
```

$$\Big\{ \Big\{ \frac{1}{2}, \frac{1}{2}, \frac{1}{2}, \frac{1}{2}, \frac{1}{2}, \frac{1}{2} \Big\}, \Big\{ \frac{5}{12}, \frac{5}{12}, \frac{5}{12}, \frac{5}{12}, \frac{5}{12}, \frac{5}{12} \Big\} \Big\}$$

Here is the correlation matrix:

```
CorrelationMatrix[dist] // MatrixForm
```

$$\begin{pmatrix} 1 & -\frac{1}{5} & -\frac{1}{5} & -\frac{1}{5} & -\frac{1}{5} & -\frac{1}{5} \\ -\frac{1}{5} & 1 & -\frac{1}{5} & -\frac{1}{5} & -\frac{1}{5} & -\frac{1}{5} \\ -\frac{1}{5} & -\frac{1}{5} & 1 & -\frac{1}{5} & -\frac{1}{5} & -\frac{1}{5} \\ -\frac{1}{5} & -\frac{1}{5} & -\frac{1}{5} & 1 & -\frac{1}{5} & -\frac{1}{5} \\ -\frac{1}{5} & -\frac{1}{5} & -\frac{1}{5} & -\frac{1}{5} & 1 & -\frac{1}{5} \\ -\frac{1}{5} & -\frac{1}{5} & -\frac{1}{5} & -\frac{1}{5} & -\frac{1}{5} & 1 \end{pmatrix}$$

■ Multivariate Continuous Distributions

> *In the* `Statistics`MultinormalDistribution`` *package:*
>
> `MultinormalDistribution[`μ`, `Σ`]` μ = mean vector, Σ = covariance matrix
> `MultivariateTDistribution[R, m]` R = correlation matrix, m = dof
> `WishartDistribution[`Σ`, m]` Σ = scale matrix, m = dof
> `HotellingTSquareDistribution[p, m]` p = dimensionality parameter, m = dof
> `QuadraticFormDistribution[{A, b, c}, {`μ`, `Σ`}]`   Distribution of $z^t A z + b^t z + c$,
>   where z has the multivariate normal distribution with parameters μ and Σ

Here dof means degrees of freedom. The multinormal and the multivariate t distribution are vector-valued, the Wishart distribution is matrix-valued, and the Hotelling T^2 and the quadratic form distribution are scalar-valued distributions. For the two scalar-valued distributions we can use `Quantile`. For the multinormal and multivariate t distributions, we have, in place of `Quantile`, `EllipsoidQuantile` (and its inverse `RegionProbability`):

> `EllipsoidQuantile[dist, q]`   Ellipsoid containing $100\,q\%$ of the probability
> `RegionProbability[dist, ellipsoid]`   Probability of the ellipsoid

■ The Bivariate Normal Distribution

Ask for the PDF of a bivariate normal distribution with means 1 and 2, variances 2 and 1, and covariance 1/2:

```
<< Statistics`MultinormalDistribution`

dist = MultinormalDistribution[{1, 2}, {{2, 1 / 2}, {1 / 2, 1}}];

f = PDF[dist, {x, y}] // Simplify
```
$$\frac{e^{-\frac{2}{7}\,(7+x^2-7\,y-x\,y+2\,y^2)}}{\sqrt{7}\;\pi}$$

Plot the PDF both as a surface and as contours:

```
Block[{$DisplayFunction = Identity},
  p1 = Plot3D[f, {x, -2.5, 4.5}, {y, -0.5, 4.5}, BoxRatios -> {7, 5, 4}];
  p2 = ContourPlot[f, {x, -2.5, 4.5},
    {y, -0.5, 4.5}, PlotPoints -> 40, AspectRatio -> Automatic]];

Show[GraphicsArray[{p1, p2}]];
```

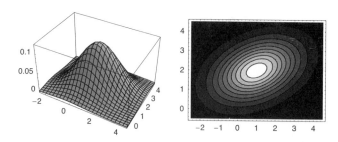

With the CDF, we can calculate the probabilities of the form $P(X \le x, Y \le y)$. For example, here is the probability that $X \le 2$ and $Y \le 3$:

```
CDF[dist, {2, 3}]      0.669716
```

We plot the CDF:

```
Plot3D[CDF[dist, {x, y}], {x, -2, 4},
   {y, 0, 4}, PlotPoints → 15, BoxRatios → {7, 5, 4}];
```

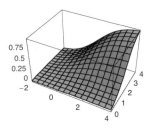

An ellipsoid centered on the mean that encompass 99% of the probability is given as follows:

```
e99 = EllipsoidQuantile[dist, 0.99]

Ellipsoid[{1, 2}, {4.50868, 2.70237},
   {{0.92388, 0.382683}, {-0.382683, 0.92388}}]

p1 = Show[
   Graphics[{e99, Point[{1, 2}]}, Axes → True, AspectRatio → Automatic]];
```

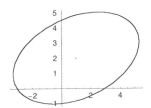

The ellipsoid really does contain 99% of the probability:

```
RegionProbability[dist, e99]      0.99
```

We plot the ellipsoids that encompass $100\,p\%$ of the probability, for $p = 0.04, 0.09, 0.14, 0.19, ..., 0.94, 0.99$:

```
e = Table[EllipsoidQuantile[dist, p], {p, 0.04, 0.99, 0.05}];

Show[Graphics[e, Axes → True, AspectRatio → Automatic]];
```

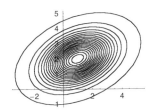

Lastly we generate 1000 random two-component vectors and plot them together with the 99% ellipsoid:

```
SeedRandom[7];
p2 = ListPlot[RandomArray[dist, 1000],
    PlotRange → All, DisplayFunction → Identity];

Show[p1, p2];
```

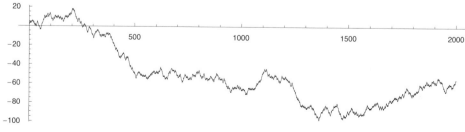

Of the points, 12 are outside the 99% region. Theoretically, the region should contain about 990 points out of 1000.

26.2 Stochastic Processes

26.2.1 Random Walks and Brownian Motion

■ Random Walk in One Dimension

We have already simulated one- and two-dimensional random walks and coin tossing in Sections 15.2.2, p. 396, and 15.2.4, p. 402. Here we repeat the random walk function presented in Section 15.2.2, but it is slightly modified:

```
randomWalk[n_] := NestList[# + (-1) ^ Random[Integer] &, 0, n]

showRandomWalk[n_, opts___] :=
 ListPlot[Transpose[{Range[0, n], randomWalk[n]}], PlotJoined → True,
   PlotRange → All, PlotStyle → AbsoluteThickness[0.1], opts]
```

We simulate 2000 steps:

```
showRandomWalk[2000, AspectRatio → 0.25];
```

Next we do 20 simulations of 2000 steps:

```
rw = Table[showRandomWalk[2000, DisplayFunction → Identity], {20}];
```

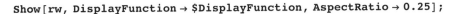

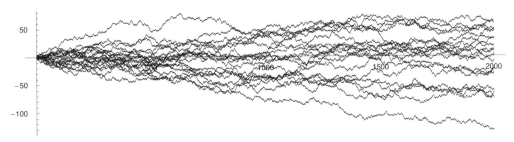

■ Random Walk in Two Dimensions

Now we generate a random walk in two dimensions. The point moves from the present
point a step of unit length in a random direction; the starting point is the origin.

```
randomWalk2D[n_] := FoldList[Plus, {0., 0.},
  With[{a = N[2 π]}, Map[{Cos[#], Sin[#]} &, Table[Random[Real, a], {n}]]]]

showRandomWalk2D[n_, opts___] := With[{rw = randomWalk2D[n]},
  ListPlot[rw, PlotJoined → True, PlotRange → All,
    AspectRatio → Automatic, PlotStyle → AbsoluteThickness[0.1], opts,
    Epilog → {AbsolutePointSize[4], RGBColor[1, 0, 0], Point[Last[rw]]}]]
```

We used **With** to define the local constant **a** so that **N[2 π]** need not be calculated **n**
times but only once. Here is a realization of this random walk with 5000 steps. The last
point is shown in red:

```
showRandomWalk2D[5000];
```

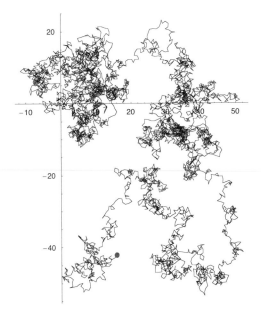

■ Gambler's Ruin

Suppose you have \$$k$ and your friend has \$$(K - k)$. You play the following game. You toss a coin. If the result is heads, your friend gives you \$1, but if the result is tails, you have to give your friend \$1. The game stops when either of the players has lost the last dollar. Let us simulate this game.

```
gamblersRuin[k_, K_] := FixedPointList[
  # + (-1) ^ Random[Integer] &, k, 1000, SameTest → (#2 == 0 || #2 == K &)]

showGamblersRuin[k_, K_, opts___] := With[{gr = gamblersRuin[k, K]},
  ListPlot[Transpose[{Range[0, Length[gr] - 1], gr}],
    PlotJoined → True, PlotRange → {0, Automatic}, opts]]
```

The function **gamblersRuin** calculates a game. The path is a random walk, but it has a stopping criterion. The walk stops if 0 or K is reached or when 1000 iterations have been done. As an example, suppose both you and your friend have \$10 initially:

```
SeedRandom[3];
showGamblersRuin[10, 20, AspectRatio → 0.2];
```

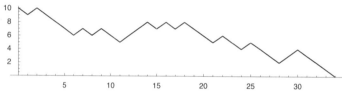

Sorry, you lost. Next we show 20 paths:

```
SeedRandom[4];
tt = Table[showGamblersRuin[10, 20, DisplayFunction → Identity], {20}];

Show[tt, DisplayFunction → $DisplayFunction, AspectRatio → 0.2];
```

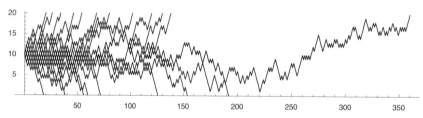

You lost 8 times and won 12 times. One game lasted for about 360 iterations.

Now we simulate the game 1000 times and only show the duration of the games:

```
frequencies[a_List] := Map[{#, Count[a, #]} &, Union[a]]

durationOfGamblersRuin[k_, K_, repl_] := With[
  {t = Table[gamblersRuin[k, K], {repl}]}, frequencies[Map[Length, t] - 1]]

showDurationOfGamblersRuin[durations_, opts___] :=
  ListPlot[durations, PlotRange → {0, Automatic}, PlotJoined → True,
    opts, Epilog → {AbsolutePointSize[2], Map[Point, durations]}]
```

Here is a simulation:

```
SeedRandom[1]; durations = durationOfGamblersRuin[10, 20, 1000];

showDurationOfGamblersRuin[durations, AspectRatio → 0.2];
```

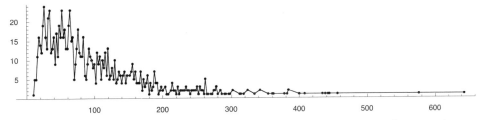

We see that most games last at most for about 200 iterations, but, in this simulation, one game lasted for about 640 iterations.

■ Brownian Motion

A Brownian motion or a Wiener process is a continuous-time–continuous-state process. Realizations of a Brownian motion are continuous, but they are nowhere differentiable. An approximating discretization has to be done for simulation purposes. The following approximation is from Cox and Miller (1965, p. 205).

```
δp[μ_, σ_, τ_] := {σ Sqrt[τ], 0.5 (1 + μ Sqrt[τ] / σ) }

brownianMotion[μ_, σ_, τ_, n_] := Module[{δ, p},
   {δ, p} = δp[μ, σ, τ];
   NestList[# + If[Random[] ≤ p, δ, -δ] &, 0, n]]

showBrownianMotion[μ_, σ_, τ_, n_, opts___] :=
  ListPlot[Transpose[{Range[0, n τ, τ], brownianMotion[μ, σ, τ, n]}],
    PlotJoined → True, AxesOrigin → {0, 0}, PlotRange → All,
    PlotStyle → AbsoluteThickness[0.1], opts]
```

The program **brownianMotion** simulates a Brownian motion where the state $X(t)$ at time t has the normal distribution with mean μt and variance $\sigma^2 t$. The approximate process moves n times a small step τ. At each step, the next value of the process is the previous value plus either δ or $-\delta$, with probabilities p and $1 - p$, respectively. Here $\delta = \sigma \sqrt{\tau}$ and $p = 0.5\left(1 + \frac{\mu}{\sigma} \sqrt{\tau}\right)$. The time step τ should be small so that δ is considerably larger than τ. The probability p should not differ much from 0.5. The Brownian motion is, mathematically, the result of the discretization if τ approaches zero.

First we show a path where the drift μ is zero:

```
δp[0, 0.5, 0.01]     {0.05, 0.5}
```

```
showBrownianMotion[0, 0.5, 0.01, 2000, AspectRatio → 0.25];
```

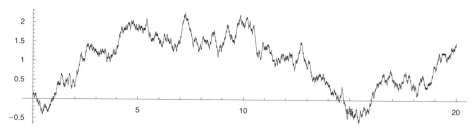

Next we draw 20 paths with a drift of 0.3:

```
δp[0.3, 0.5, 0.01]      {0.05, 0.53}
```

```
Table[showBrownianMotion[0.3, 0.5,
    0.01, 2000, DisplayFunction → Identity], {20}];
```

```
Show[%, DisplayFunction → $DisplayFunction, AspectRatio → 0.25];
```

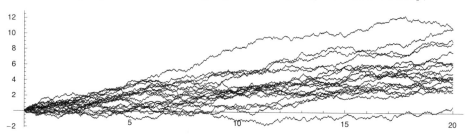

26.2.2 Discrete-Time Markov Chains

■ Precipitation Data

There is a precipitation monitor at the Snoqualmie Falls in western Washington. It defines a day as wet if the precipitation is as least 0.01 inches. Such days are denoted by 0. Other days are denoted by 1; let us call these days dry. The file **precipitation** (available on the CD-ROM of this book) contains codes for all days of January for the 36 years from 1948 to 1983. The data are reprinted with permission from Guttorp (1995, p. 17, Figure 2.1); copyright CRC Press, Boca Raton, Florida.

I have saved the file in the **MNata** folder of my **Documents** folder. First we read the file (see Section 4.2.1, p. 90):

```
prec =
    Import["/Users/ruskeepaa/Documents/MNData/precipitation", "Table"];
```

```
<< LinearAlgebra`MatrixManipulation`
```

```
prec2 = BlockMatrix[
    {{{{""}}, {Range[31]}}, {Transpose[{Range[1948, 1983]}], prec}}];
```

```
StyleForm[Grid[prec2, RowSpacings → {1, 0}, ColumnSpacings → {0.5, -0.5},
    ColumnsEqual → True, RowLines → {True, False},
    ColumnLines → {True, False}], FontFamily → "Times", FontSize → 7]
```

	1	2	3	4	5	6	7	8	9	10	11	12	13	14	15	16	17	18	19	20	21	22	23	24	25	26	27	28	29	30	31
1948	0	0	0	0	0	0	0	0	0	0	0	0	0	1	1	1	1	1	1	1	1	0	1	0	0	0	0	1	1	1	0
1949	0	0	1	1	0	0	0	1	1	1	1	1	0	1	1	0	1	0	1	0	1	1	1	1	0	0	1	1	0	1	0
1950	0	0	0	0	0	0	0	0	0	0	0	0	0	1	0	0	0	1	0	0	0	0	0	0	0	0	0	1	1	1	0
1951	0	0	0	0	0	1	0	0	1	0	0	0	0	0	0	0	0	0	0	0	0	0	0	0	0	0	0	1	1	1	1
1952	1	1	0	0	0	0	0	0	0	0	0	0	1	0	1	1	1	0	0	0	0	0	0	0	0	0	0	1	0	0	0
1953	0	0	0	0	1	0	0	0	0	0	0	0	0	0	0	0	0	0	0	0	0	0	0	0	0	0	0	0	0	0	0
1954	0	0	0	0	0	0	0	0	0	1	1	1	0	0	0	0	0	0	0	0	0	0	0	0	0	0	0	0	1	1	0
1955	0	0	1	0	0	1	1	1	1	1	1	0	0	0	0	0	0	1	1	0	1	0	0	0	0	1	1	0	1	0	0
1956	0	0	0	0	0	0	0	0	0	0	0	1	0	0	0	0	0	0	0	0	0	0	0	0	1	1	1	0	0	1	1
1957	0	0	0	0	1	1	0	0	1	0	0	1	1	0	1	1	0	1	0	0	0	0	1	1	1	1	1	1	0	0	0
1958	1	0	0	1	1	1	1	0	0	0	0	0	0	0	0	0	0	0	1	0	0	0	0	0	1	0	0	0	0	0	0
1959	0	0	1	1	0	0	0	0	0	0	0	0	0	0	0	0	0	0	0	0	0	1	0	0	0	0	0	0	0	0	1
1960	1	0	1	1	0	0	0	0	1	0	0	0	1	1	0	0	1	1	1	1	0	0	0	0	0	0	0	0	0	0	0
1961	0	1	1	0	0	0	0	0	0	0	0	0	0	0	0	0	1	1	1	1	0	0	1	1	1	0	0	0	0	0	0
1962	1	0	0	1	0	0	0	0	1	1	1	0	0	0	0	0	0	1	1	1	1	1	0	0	0	0	0	1	1	1	0
1963	0	0	0	1	0	0	0	0	0	1	0	0	0	1	0	1	1	1	1	1	1	0	1	1	1	1	1	0	0	0	0
1964	0	0	0	0	0	0	0	0	1	1	0	0	0	0	0	0	0	0	1	0	0	1	0	0	0	0	0	0	0	0	0
1965	0	0	0	0	0	0	0	1	1	1	1	0	1	1	0	0	0	0	0	0	0	0	0	0	0	0	0	0	0	0	1
1966	0	0	0	0	0	0	0	0	0	0	0	0	1	0	0	0	1	1	1	1	0	1	0	1	0	0	0	0	0	0	0
1967	0	0	0	0	0	0	1	1	0	0	0	0	0	0	0	0	0	1	0	0	0	0	0	0	0	0	0	0	1	1	0
1968	0	1	0	0	0	1	0	0	0	0	1	0	0	0	0	0	0	0	0	1	1	0	0	1	1	0	0	0	0	0	0
1969	0	1	0	0	0	0	0	0	0	0	0	0	0	0	0	0	1	0	0	1	1	0	1	0	1	0	0	0	0	0	0
1970	1	1	0	1	1	1	1	1	0	0	0	0	0	0	0	0	0	0	0	0	0	0	0	1	1	1	0	0	0	0	0
1971	1	1	1	1	1	1	0	0	0	0	1	0	0	0	0	0	0	0	0	0	0	0	1	1	0	0	1	0	0	0	1
1972	0	0	1	0	0	0	0	0	0	0	0	0	1	0	0	0	0	0	0	0	0	0	1	1	1	1	1	0	0	1	1
1973	0	0	1	0	0	1	1	1	1	0	0	0	0	0	0	0	0	1	1	0	0	1	1	1	0	0	1	0	0	0	1
1974	1	1	1	1	1	1	1	1	0	0	0	0	0	0	0	0	0	0	0	0	0	0	0	0	0	0	0	0	0	0	0
1975	0	0	0	0	0	0	0	0	0	0	1	1	0	0	0	0	0	0	0	0	0	0	1	1	1	0	0	0	0	0	0
1976	1	1	0	0	0	0	0	0	0	0	0	1	1	1	0	0	1	1	0	1	1	0	1	1	1	1	1	1	1	1	1
1977	0	0	0	1	1	1	1	1	1	0	0	0	0	0	0	0	0	1	1	1	1	1	1	1	0	0	0	0	0	0	0
1978	1	0	0	0	0	0	0	1	0	1	0	1	0	0	0	1	1	0	0	0	1	1	0	0	0	1	1	1	1	1	1
1979	1	1	1	1	1	1	1	0	0	0	1	0	0	1	0	0	1	1	0	1	0	0	1	1	0	1	1	1	1	1	0
1980	0	0	1	0	0	1	0	0	0	0	0	0	1	1	1	0	1	1	1	0	1	1	1	0	1	1	1	1	1	1	0
1981	1	1	1	0	0	1	0	1	1	1	1	1	1	1	0	0	1	0	0	0	0	0	0	0	0	0	0	0	0	0	1
1982	0	0	0	1	1	1	1	1	0	0	0	1	1	0	0	0	0	0	0	0	1	1	0	0	0	0	0	0	0	0	0
1983	0	0	0	0	0	0	0	0	0	0	1	0	0	0	0	0	0	0	0	0	1	0	0	0	0	0	0	1	0	1	1

We see that the first 12 days of January 1948 were wet, the next 7 days dry, and so on. Here is a plot of the data. Each black square denotes a wet day:

```
ListDensityPlot[prec, MeshRange → {{0.5, 31.5}, {1947.5, 1983.5}},
    MeshStyle → {GrayLevel[0.4]},
    FrameLabel → {"Day", "Year"}, AspectRatio → Automatic];
```

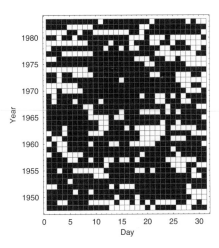

■ Runs

With `Split`, we can find runs, which are sequences of like values. For example, here are the seven runs of January 1948:

    Split[prec[[1]]]

    {{0, 0, 0, 0, 0, 0, 0, 0, 0, 0, 0, 0},
     {1, 1, 1, 1, 1, 1, 1}, {0}, {1}, {0, 0, 0, 0}, {1, 1, 1}, {0, 0, 0}}

Next we calculate the numbers of runs for all 36 Januaries:

    Map[Length, Map[Split, prec]]

    {7, 16, 6, 6, 8, 3, 5, 13, 6, 13, 8, 5, 11, 7, 9, 11, 7,
     6, 11, 6, 11, 11, 6, 5, 6, 10, 2, 4, 7, 5, 12, 11, 11, 9, 5, 8}

Calculate the mean number of runs per month:

    Apply[Plus, %] / 36 // N   7.97222

What would the expected number of runs be if the weather were to change randomly from day to day? If the probability of a dry day is p and there are n days, the expected number of runs is as follows:

    expRuns[n_, p_] := 1 + 2 (n - 1) p (1 - p)

We have a total of 1116 days, out of which 325 are dry days:

    {ndays, ndry} = {Length[Flatten[prec]], Apply[Plus, Flatten[prec]]}

    {1116, 325}

We estimate the probability of a dry day:

    probDry = ndry / ndays // N   0.291219

The expected number of runs in a month for random weather is thus about the following:

    expRuns[31, probDry]   13.3846

The actual mean number was only about eight. This is an indication that the weather does not change randomly. The weather on a certain day has a tendency to remain stable for a few days. Next we estimate the transition probabilities.

■ Estimating the Transition Probabilities

The data for January 1948 are as follows:

    d = prec[[1]]

    {0, 0, 0, 0, 0, 0, 0, 0, 0, 0, 0, 0,
     1, 1, 1, 1, 1, 1, 1, 0, 1, 0, 0, 0, 0, 1, 1, 1, 0, 0, 0}

With `ReplaceList`, we can easily obtain information about all kinds of transitions. For example, we form a list consisting of as many 1s as there are pairs {1,1} in the list:

    ReplaceList[d, {___, 1, 1, ___} → 1]

    {1, 1, 1, 1, 1, 1, 1, 1}

The three underscores (`___`) mean zero or more elements (see Section 15.3.3, p. 408). A pair {1, 1} means a transition from a dry day to another dry day. The sum of the 1s is the number of such pairs of days:

```
Apply[Plus, ReplaceList[d, {___, 1, 1, ___} → 1]]     8
```

Similarly we can find the numbers of other kinds of transitions. We write a function for doing this:

```
transitions[d_List] := Map[Apply[Plus, ReplaceList[d, # → 1]] &,
    {{___, 0, 0, ___}, {___, 0, 1, ___}, {___, 1, 0, ___}, {___, 1, 1, ___}}]
```

For example:

```
transitions[d]     {16, 3, 3, 8}
```

To get the numbers of transitions for all years, we use **Map**:

```
Map[transitions[#] &, prec]
```

```
{{16, 3, 3, 8}, {5, 8, 7, 10}, {23, 3, 2, 2}, {22, 3, 2, 3}, {20, 3, 4, 3},
 {28, 1, 1, 0}, {23, 2, 2, 3}, {11, 6, 6, 7}, {22, 3, 2, 3}, {10, 6, 6, 8},
 {20, 3, 4, 3}, {25, 2, 2, 1}, {14, 5, 5, 6}, {18, 3, 3, 6}, {14, 4, 4, 8},
 {10, 5, 5, 10}, {23, 3, 3, 1}, {21, 3, 2, 4}, {17, 5, 5, 3}, {23, 3, 2, 2},
 {18, 5, 5, 2}, {19, 5, 5, 1}, {19, 2, 3, 6}, {20, 2, 2, 6}, {20, 3, 2, 5}, {15,
 {22, 0, 1, 7}, {23, 2, 1, 4}, {16, 3, 3, 8}, {14, 2, 2, 12}, {16, 5, 6, 3},
 {9, 5, 5, 11}, {11, 5, 5, 9}, {13, 4, 4, 9}, {21, 2, 2, 5}, {22, 4, 3, 1}}
```

With **Transpose**, we can form separate lists for the four kinds of transitions:

```
{t00, t01, t10, t11} = Transpose[%];
```

For example:

```
t00
```

```
{16, 5, 23, 22, 20, 28, 23, 11, 22, 10, 20, 25, 14, 18, 14, 10, 23, 21,
 17, 23, 18, 19, 19, 20, 20, 15, 22, 23, 16, 14, 16, 9, 11, 13, 21, 22}
```

We calculate the sum of each kind of transition:

```
Map[Apply[Plus, #] &, %%]
```

```
{643, 128, 123, 186}
```

So we know that there were 643 transitions $0 \to 0$ (from a wet day to another wet day), 128 transitions $0 \to 1$, 123 transitions $1 \to 0$, and 186 transitions $1 \to 1$. Overall, there were 643 + 128 = 771 transitions from a wet day and 123 + 186 = 309 transitions from a dry day. It is thus natural to estimate the probability of $0 \to 0$ as $643/771 = 0.834$, of $0 \to 1$ as $128/771 = 0.166$, of $1 \to 0$ as $123/309 = 0.398$, and of $1 \to 1$ as $186/309 = 0.602$. (The same estimates could also be obtained by the method of maximum likelihood.) So we arrive at the following transition matrix:

```
P = {{0.834, 0.166}, {0.398, 0.602}};
```

■ Predicting

Suppose Monday is dry; the initial distribution is thus {0, 1}. We predict the weather for Tuesday, Wednesday, Thursday, and Friday:

```
TableForm[Table[{0, 1}.MatrixPower[P, i], {i, 0, 4}],
  TableHeadings → {{"Monday", "Tuesday", "Wednesday",
    "Thursday", "Friday"}, {"P(wet)", "P(dry)"}}]
```

	P(wet)	P(dry)
Monday	0	1
Tuesday	0.398	0.602
Wednesday	0.571528	0.428472
Thursday	0.647186	0.352814
Friday	0.680173	0.319827

The predictions converge to a limit, which is achieved to 6-digit precision in 18 steps:

```
{0, 1}.MatrixPower[P, 18]
```

```
{0.705674, 0.294326}
```

■ Stationary Distribution

The stationary distribution is obtained with the aid of the following linear equations:

```
eqns = Thread[{r1, r2}.P == {r1, r2}]
```

```
{0.834 r1 + 0.398 r2 == r1, 0.166 r1 + 0.602 r2 == r2}
```

One of these equations can be dropped, and the equation **r1 + r2 == 1** has to be taken into account:

```
Solve[Prepend[Drop[eqns, -1], r1 + r2 == 1]]
```

```
{{r1 → 0.705674, r2 → 0.294326}}
```

This solution is the same as the limit we obtained above. A general program to calculate the stationary distribution is as follows:

```
stationaryDistribution[P_?MatrixQ, s0_, s1_] :=
  Module[{st = Array[v, {s1 - s0 + 1}], eqn},
    eqn = Prepend[Drop[Thread[st.P == st], -1], Apply[Plus, st] == 1];
    Transpose[{Range[s0, s1], st /. Solve[eqn, st][[1]]}]]
```

Here **P** is the transition matrix and **s0** and **s1** the minimum and maximum state. For example:

```
stationaryDistribution[P, 0, 1]
```

```
{{0, 0.705674}, {1, 0.294326}}
```

■ Simulating the Process

By using the following functions, we can simulate a discrete-time Markov chain:

```
dtMarkovChainStep[dist_?VectorQ] :=
  With[{r = Random[]}, Position[dist, x_ /; x ≥ r, {1}, 1]][[1, 1]]

dtMarkovChain[x0_, P_?MatrixQ, n_] := Module[{cumP, mc},
  cumP = Map[Rest[FoldList[Plus, 0, #]] &, P];
  mc = NestList[dtMarkovChainStep[cumP[[#]]] &, x0 + 1, n] - 1;
  Transpose[{Range[0, n], mc}]]
```

```
showdtMarkovChain[x0_, P_?MatrixQ, n_, opts___] :=
 With[{mc = dtMarkovChain[x0, P, n]},
  ListPlot[mc, PlotJoined → True,
   AxesOrigin → {0, -0.2}, PlotRange → {-0.2, Length[P] - 0.8},
   Epilog → {AbsolutePointSize[1.5], Map[Point, mc]}, opts]]
```

The program **dtMarkovChainStep** calculates the next state of the process. The states are numbered 1, 2, ... The input for this function is a cumulative distribution **dist** for the next state. For example, if **dist** is {0.3, 0.8, 1}, the next state is 1, 2, or 3 with probabilities 0.3, 0.5, and 0.2, respectively. With the help of **Position**, the program searches for the positions of the cumulative distribution in which the cumulative probability is at least a given uniform random number. The **{1}** in **Position** means that the search is done at the first level of **dist**. This specification is unnecessary, but because we use a fourth argument in **Position**, we have to write something as the third argument. The fourth argument **1** specifies that we, in fact, need only the first position that satisfies **x ≥ r**. The result of this **Position** could be, for example, **{{2}}**. Finally, taking the part **[[1, 1]]** gives the position as a number, for example, 2.

The program **dtMarkovChain** generates **n** steps for a chain with initial state **x0** and transition matrix **P**. The function first calculates the cumulative distribution **cumP** for the rows of the transition matrix. In **NestList**, the current state is denoted by **#**. From this state, we go on to the next state according to a cumulative distribution, which is the **#**th row of the cumulative transition matrix: **cumP[[#]]**. So the next state is given by **dtMarkovChainStep[cumP[[#]]]**. **NestList** does this iteration **n** times. The result is a list of states, and we subtract 1 from all of the states, because we want them to be numbered 0, 1, 2, ... Lastly, we add the ordinal numbers of the steps to the states.

For example, if the initial day is wet, the initial state is 0. Here are 10 simulated steps:

```
dtMarkovChain[0, P, 10]
```

```
{{0, 0}, {1, 0}, {2, 0}, {3, 0}, {4, 1},
 {5, 0}, {6, 0}, {7, 0}, {8, 0}, {9, 0}, {10, 1}}
```

We see that the next three days are wet, followed by one dry day and so on. Next we simulate 100 steps, assuming that the starting day is wet:

```
showdtMarkovChain[0, P, 100,
  AspectRatio → 0.1, Ticks → {Automatic, {0, 1}}];
```

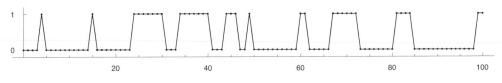

```
P = .
```

■ An Example of Diffusion

We have *n* black and *n* white balls. They are put into urns *A* and *B* so that there are *n* balls in both urns. Then we take, at times 1, 2, …, one ball at random from both urns, and then they are put back into the urns: the ball taken from urn *A* is put into urn *B*, and the ball taken from urn *B* is put into urn *A*. This is a simple model of diffusion.

We define the state of the system to be the number of black balls in urn A. The transition probabilities are $p_{i,i-1} = (\frac{i}{n})^2$, $p_{i,i+1} = (\frac{n-i}{n})^2$, and $p_{i,i} = 2\frac{i(n-i)}{n^2}$. The transition matrix is as follows:

```
P[n_] := Table[Which[j == i - 1, (i / n) ^ 2, j == i, 2 i (n - i) / n^2,
    j == i + 1, ((n - i) / n) ^ 2, True, 0], {i, 0, n}, {j, 0, n}]
```

As an example, suppose we have 50 black and 50 white balls. Calculate and plot the stationary distribution:

```
st = stationaryDistribution[P[50] // N, 0, 50];
```

```
ListPlot[st, PlotRange → All];
```

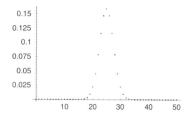

We see that, with high probability, there are, in the long run, about 18 to 32 balls in urn A. Calculate the exact probability:

```
Sum[st[[i, 2]], {i, 19, 33}]        0.997476
```

We simulate 200 steps of the diffusion, assuming that there are initially 0 black balls in urn A:

```
showdtMarkovChain[0, P[50] // N, 200, AspectRatio -> 0.3];
```

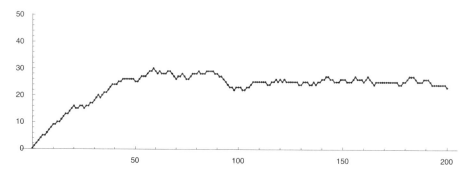

26.2.3 Continuous-Time Markov Chains

■ The Poisson Process

In a Poisson process, some events happen as time goes on. For example, calls arrive at a database, customers arrive at a service point, or particles arrive at a particle detector. When we look at the realization of a Poisson process, we should observe that the events are on the time axes randomly and uniformly. The state of the process at time t is the number of events that have occurred up to that time. A Poisson process is a special case of a continuous-time discrete-state Markov chain.

Let λ be the expected number of events in a time unit. Then the number of events in a time interval of length t has a Poisson distribution with mean λt, and the interarrival times have an exponential distribution with parameter λ (i.e., with mean $1/\lambda$); the interarrival times are independent.

A Poisson process can be simulated by generating the interarrival times from an exponential distribution:

```
poissonProcess[λ_, n_] :=
  FoldList[Plus, 0, Table[-1 / λ Log[1 - Random[]], {n}]]

showPoissonProcess[λ_, n_, opts___] := With[{pp = poissonProcess[λ, n]},
    Show[Graphics[{AbsoluteThickness[0.2],
        Table[Line[{{pp[[i]], i - 1}, {pp[[i + 1]], i - 1}}], {i, 1, n}]}],
      Axes → True, opts]];
```

The function **poissonProcess** first generates the interarrival times of the Poisson process (remember that, if X has the uniform distribution on $(0, 1)$, then $-\frac{1}{\lambda} \log(1 - X)$ has the exponential distribution with a mean of $\frac{1}{\lambda}$) and then calculates their cumulative sums, resulting in a list of the instants of the events. The function **showPoissonProcess** draws, at levels 0, 1, 2, and so on, lines that have lengths that are the interarrival times.

Let us assume that, on average, 4 calls arrive at a database per minute. Below then is a simulated sequence of 100 arrivals; 100 events occurred in about 25 minutes.

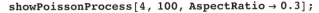

```
showPoissonProcess[4, 100, AspectRatio → 0.3];
```

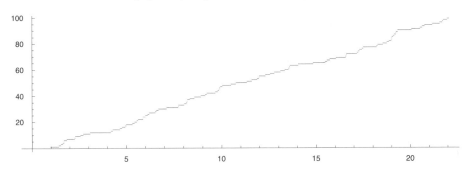

◼ General Continuous-Time Markov Chains

Consider a continuous-time discrete-state process. Let T_{ij} be the time the system stays at state i if the system then goes to state j. Assume that the random variable T_{ij} has an exponential distribution with parameter q_{ij} (or with mean $1/q_{ij}$) and that these random variables are independent of each other and also independent of the history of the process before arriving at state i (define $q_{ii} = -\sum_{j \neq i} q_{ij}$). Collect the parameters q_{ij} into a matrix Q, which is called the rate matrix or the generator of the process. It can be shown that the process is a continuous-time Markov chain. In addition, the system stays at state i for an exponential time with parameter $q_i = \sum_{k \neq i} q_{ik}$ and then goes to state j with probability $p_{ij} = q_{ij} / q_i$ (if $q_i = 0$, then define $p_{ii} = 1$) (Kulkarni 1995, p. 245).

The following programs simulate **n** steps for a continuous-time Markov chain. The initial state is the scalar **x0**, and the generator matrix is **Q**.

```
ctMarkovChain[x0_, Q_?MatrixQ, n_] :=
 Module[{R = Q, m = Length[Q], q, P, cumP, t = 0, x = x0, tt = {0}, xx = {x0}},
  Do[R[[i, i]] = 0, {i, 1, m}];
  q = Map[Apply[Plus, #] &, R];
  P = Table[If[q[[i]] ≠ 0, R[[i]] / q[[i]],
     Table[If[j ≠ i, 0, 1], {j, 1, m}]], {i, 1, m}];
  cumP = Map[Rest[FoldList[Plus, 0, #]] &, P];
  Do[If[q[[x + 1]] ≠ 0, t = t - 1 / q[[x + 1]] Log[1 - Random[]],
    Print["Absorption at ", x]; Break[]];
   x = With[{r = Random[]},
     Position[cumP[[x + 1]], z_ /; z ≥ r, {1}, 1][[1, 1]] - 1];
   tt = {tt, t}; xx = {xx, x}, {n}];
  {Flatten[tt], Flatten[xx]}]

showctMarkovChain[x0_, Q_?MatrixQ, n_, opts___] := Module[{tt, xx},
 {tt, xx} = ctMarkovChain[x0, Q, n];
 Show[Graphics[{AbsoluteThickness[0.2],
    Table[Line[{{tt[[i]], xx[[i]]}, {tt[[i + 1]], xx[[i]]}}],
     {i, 1, Length[tt] - 1}]}], Axes → True,
  AxesOrigin → {0, -0.2}, PlotRange → {-0.2, Max[xx] + 0.1}, opts]]
```

The states are numbered 0, 1, and so on. The program first sets the diagonal elements of Q to zero and then calculates the q_i and p_{ij} numbers into vector **q** and matrix **P**. The rows of **cumP** are the cumulative sums of elements of the rows of **P** (they are needed to generate the next state). The vector **q** is used to generate the exponential visiting times. The instances of events are gathered in the list **tt**, and the states are gathered in the list **xx**.

As examples of continuous-time Markov chains, we consider the Poisson process, a birth–death process, and the M/M/1 queue.

■ The Poisson Process Revisited

In a Poisson process, the state goes from i to $i + 1$ when the next event happens so that $T_{i,i+1}$ is an exponential random variable with parameter λ. So, if we generate at most m events, the generator can be written as follows:

```
Q[λ_, m_] :=
 Table[Which[j == i, -λ, j == i + 1, λ, True, 0], {i, 0, m}, {j, 0, m}]
```

Here is a small example of the generator:

```
Q[λ, 4] // MatrixForm
```

$$\begin{pmatrix} -\lambda & \lambda & 0 & 0 & 0 \\ 0 & -\lambda & \lambda & 0 & 0 \\ 0 & 0 & -\lambda & \lambda & 0 \\ 0 & 0 & 0 & -\lambda & \lambda \\ 0 & 0 & 0 & 0 & -\lambda \end{pmatrix}$$

Here is a simulation:

```
showctMarkovChain[0, Q[4, 100], 100, AspectRatio -> 0.3];
```

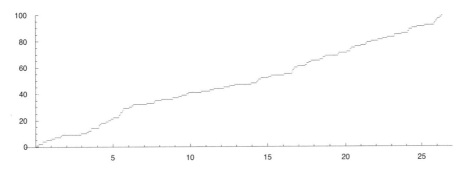

■ A Birth–Death Process

Consider a population that initially consists of x_0 individuals. The length of the life of each individual has the exponential distribution with mean $1/\mu$. Each individual produces, during its whole lifetime, descendants in such a way that the time between successive births has the exponential distribution with mean $1/\lambda$. Such a process is a simple birth–death process with a state that is the size of the population. This process was also considered in Section 24.1.1, p. 641.

To derive the generator matrix, note that the state goes from i to $i+1$ if one of the i individuals gives birth to a child so that $T_{i,i+1}$ is the minimum of i exponential random variables with parameter λ, that is, $T_{i,i+1}$ has an exponential distribution with parameter $i\lambda$. Similarly, $T_{i,i-1}$ is exponentially distributed with parameter $i\mu$. So, if we suspect that the population will not exceed the value m, the generator is as follows:

```
Q[λ_, μ_, m_] := Table[Which[j == i - 1, i μ, j == i,
    -i (λ + μ), j == i + 1, i λ, True, 0], {i, 0, m}, {j, 0, m}]
```

Here is a small example:

```
Q[λ, μ, 4] // MatrixForm
```

$$\begin{pmatrix} 0 & 0 & 0 & 0 & 0 \\ \mu & -\lambda - \mu & \lambda & 0 & 0 \\ 0 & 2\mu & -2(\lambda + \mu) & 2\lambda & 0 \\ 0 & 0 & 3\mu & -3(\lambda + \mu) & 3\lambda \\ 0 & 0 & 0 & 4\mu & -4(\lambda + \mu) \end{pmatrix}$$

As an example, let the average lifetime be 2 time units and the average time between births, for a given individual, 3 time units. Then $1/\mu = 2$ and $1/\lambda = 3$ so that $\mu = \frac{1}{2}$ and $\lambda = \frac{1}{3}$. Suppose we initially have 20 individuals. We generate 6 steps:

```
SeedRandom[1];
ctMarkovChain[20, Q[1 / 3, 1 / 2, 50], 6]
```

```
{{0, 0.0662825, 0.153276, 0.316852, 0.398011, 0.405338, 0.485722},
 {20, 21, 20, 21, 22, 21, 22}}
```

So, at time 0 we have 20 individuals, at time 0.0662825 one of these individual gives birth to a child, at time 0.153276 one individual dies, and so on. Here is a longer simulation:

```
SeedRandom[1];
showctMarkovChain[20, Q[1 / 3, 1 / 2, 50], 200, AspectRatio -> 0.3];
```

Absorption at 0

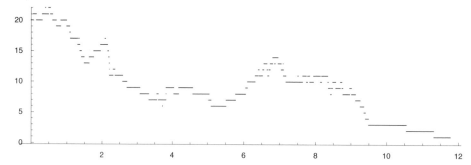

The population died out at about $t = 12$.

■ The M/M/1 Queue

In an M/M/1 queuing model, customers arrive at a service place, and a single person does the serving. Customers are served in the order of their arrival. There is room for all arriving customers to queue if the server is busy, and all customers wait until they get service (i.e., customers cannot leave the place). Customers arrive as a Poisson process with a mean of λ customers per time unit. The service time for each customer has an exponential distribution with mean $1/\mu$. The state of the system is the number of customers in the service place (customers in the system consist of the one receiving service and the others standing in the queue).

The state goes from i to $i + 1$ if a new customer arrives so that $T_{i,i+1}$ has an exponential distribution with parameter λ. Similarly, the state goes from i to $i - 1$ if a customer is served so that $T_{i,i-1}$ has an exponential distribution with parameter μ. So, if we suspect that the population will not exceed the value m, the generator is as follows:

```
Q[λ_, μ_, m_] := Table[Which[i == 0 && j == 0, -λ, j == i - 1,
    μ, j == i, -λ - μ, j == i + 1, λ, True, 0], {i, 0, m}, {j, 0, m}]
```

Here is a small example:

```
Q[λ, μ, 4] // MatrixForm
```

$$
\begin{pmatrix}
-\lambda & \lambda & 0 & 0 & 0 \\
\mu & -\lambda - \mu & \lambda & 0 & 0 \\
0 & \mu & -\lambda - \mu & \lambda & 0 \\
0 & 0 & \mu & -\lambda - \mu & \lambda \\
0 & 0 & 0 & \mu & -\lambda - \mu
\end{pmatrix}
$$

We will simulate a queuing system in which customers arrive at the service point at the mean rate of 4.0 arrivals per hour (one customer every 15 minutes) and in which the server has a mean service rate of 4.55 customers per hour. This means that $\lambda = 4$ and $\mu = 4.55$. One customer is then served in an average of $1/4.55$ hour = 13.2 minutes. We simulate 400 events (arrivals and departures), when there are initially 0 customers:

```
SeedRandom[5];
showctMarkovChain[0, Q[4., 4.55, 50], 400, AspectRatio -> 0.3];
```

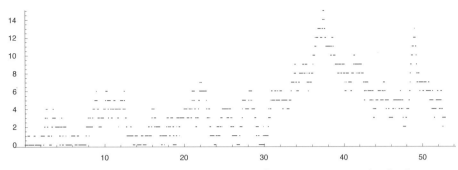

In the simulation, 400 events took about 53 hours to occur. The highest amount of customers in the service place was 15. Overall, it seems that one server is too few for this system: the queue is too long for too large a portion of the time.

With the following module, we can calculate various average values if the system has reached a steady state:

```
steadyStateAverages[λ_, μ_] := Module[{ρ = λ / μ, L, W, Wq, Lq},
  L = ρ / (1 - ρ); W = L / λ; Wq = W - 1 / μ; Lq = λ Wq;
  Print["L  = ", L, " (steady state mean number of customers)\n",
   "Lq = ", Lq, " (steady state mean length of the queue)\n",
   "W  = ", W, " (steady state mean time in the system)\n",
   "Wq = ", Wq, " (steady state mean time of queueing)\n",
   "ρ  = ", ρ, " (steady state server utilization)"]]
```

For our example, the averages are as follows:

```
steadyStateAverages[4.0, 4.55]
```

```
L  = 7.27273 (steady state mean number of customers)
Lq = 6.39361 (steady state mean length of the queue)
W  = 1.81818 (steady state mean time in the system)
Wq = 1.5984 (steady state mean time of queueing)
ρ  = 0.879121 (steady state server utilization)
```

So, if our system is in steady state, the mean number of customers in the system in a long time interval should be near 7.3. About 6.4 customers are queuing, and they each spend about 1.8 hours in the system (this time consists of queuing time and service time). Queuing takes about 1.6 hours. The server is busy a fraction of 0.88 of the time. In steady state, the number of customers in the system has the geometric distribution with parameter ρ: the probability of n customers being in the system is $(1 - \rho)\, \rho^n$, $n = 0, 1, 2$, and so on.

Statistics

Introduction

> *A statistician was about to undergo a serious operation and asked the surgeon what his chances of survival were. "Your chances are excellent," said the surgeon, "Nine people out of ten die from this operation, and the last nine patients I've operated on have died."*

Almost all statistical functions of *Mathematica* are contained in packages. Here are the packages in the order we consider them:

```
Statistics`DescriptiveStatistics`
Statistics`MultiDescriptiveStatistics`

Statistics`DataManipulation`
Statistics`DataSmoothing`

Statistics`ConfidenceIntervals`
Statistics`HypothesisTests`
Statistics`ANOVA`

Statistics`LinearRegression`
Statistics`NonlinearFit`
```

In addition to these topics, we consider, in Section 27.7, Bayesian statistics. The power of *Mathematica* for integration, interpolation, and random number generation helps with the solving of statistical problems that are related to Bayesian models. Two of the methods we consider are Gibbs sampling and Markov chain Monte Carlo.

As for fitting and regression, note that we have five commands. **Fit** (built-in) and **Regress** (in **Statistics`LinearRegression`**) fit linear models, and the latter also gives statistical information about the fit. **FindFit** (❀5), **NonlinearFit** (used in versions of *Mathematica* earlier than 5), and **NonlinearRegress** (the last two in **Statistics`** **NonlinearFit`**) fit nonlinear models, and the last one also gives statistical information. **Fit** and **FindFit** have been considered in Sections 22.1.1, p. 570, and 22.1.4, p. 580. Estimation of differential and difference equation models have been considered in Sections 23.4.4, p. 634, and 25.3.2, p. 699.

Probability distributions were considered in Chapter 26. In particular, the **Statistics`** **NormalDistribution`** package contains the normal, Student t, chi-square, and F-ratio distributions. The **Statistics`ContinuousDistributions`** package contains, in addition to these four distributions, also the noncentral t, noncentral chi-square, and noncentral F-ratio distributions (together with many other continuous distributions). The **Statistics`MultinormalDistribution`** package contains the multinormal distribution and other related distributions.

Note that the plotting of data is considered in Chapters 9 and 10. Some of the plots are especially useful in statistical reasoning. These plots include dot and multiway dot plots, box-and-whisker plots, pairwise scatter plots, labeled plots, and quantile-quantile plots; these are considered in Sections 9.4 and 9.5.

For time series analysis, Wolfram Research has published a separate product called **Time Series**. Its topics include ARIMA models, Akaike's information criterion, innovations algorithm, maximum likelihood method, best linear predictor, and spectrum estimation. For more about statistics with *Mathematica*, see Abell, Braselton, and Rafter (1999) and Rose and Smith (2002).

27.1 Descriptive Statistics

27.1.1 Univariate Descriptive Statistics

■ Basic Descriptive Statistics

> *Location statistics:*
> **Mean[data]** ($m = \frac{1}{n} \sum x_i$), **Median[data]**, **Quantile[data, q]** (❀5)
>
> *Dispersion statistics:*
> **Variance[data]** ($s^2 = \frac{1}{n-1} \sum (x_i - m)^2$), **StandardDeviation[data]** (s) (❀5)

These commands can be used without loading a package (they are new in version 5; users of *Mathematica* with version prior to 5 have to first load the **Statistics`Descrip-tiveStatistics`** package). The sample mean m is an unbiased estimate of the population mean μ. The median is the observation in the center of the sorted observations (or the average of the two most central observations if there is an even number of observations). The q-quantile gives a value such that $100\,q\%$ of the observations are at most this value ($0 < q < 1$). The sample variance s^2 is an unbiased estimate of the population variance σ^2. For example:

```
d = {1, 1, 2, 2, 3, 4, 4, 4, 5, 5};

{Mean[d], Median[d], Quantile[d, 0.9], Variance[d]} // N

{3.1, 3.5, 5., 2.32222}
```

With the **Statistics`DescriptiveStatistics`** package, we can calculate various other univariate descriptive numbers for a set of observations. This package is automatically loaded when most other statistical packages are used. Next we consider the descriptive statistics defined in this package in some groups. Note that, for the sake of brevity, we have mostly not shown the argument of the commands. The argument, if it is not shown, is a list of observations.

■ Location Statistics

> **GeometricMean** ($\prod x_i^{1/n}$) **HarmonicMean** $\left(n / \sum \frac{1}{x_i} \right)$, **RootMeanSquare** $\left(\sqrt{\frac{1}{n} \sum x_i^2} \right)$
> **Mode, TrimmedMean[data, f], TrimmedMean[data, {f1, f2}]**
> **InterpolatedQuantile[data, q], Quartiles[data]**
> **LocationReport** (mean, harmonic mean, and median)

The mode is the most frequently occurring observation. Interpolated quantile uses linear interpolation. Quartiles gives the 25%, 50%, and 75% quantiles. For example:

```
<< Statistics`DescriptiveStatistics`

LocationReport[d] // N

{Mean → 3.1, HarmonicMean → 2.23048, Median → 3.5}
```

■ **Dispersion Statistics**

VarianceMLE $\left(\frac{1}{n} \sum (x_i - m)^2\right)$, **StandardDeviationMLE** $\left(\sqrt{\frac{1}{n} \sum (x_i - m)^2}\right)$

VarianceOfSampleMean $(s_m{}^2 = s^2/n)$, **StandardErrorOfSampleMean** (s_m)

CoefficientOfVariation (s/m)

MeanDeviation $\left(\frac{1}{n} \sum |x_i - m|\right)$, **MedianDeviation**, **QuartileDeviation**

SampleRange, **InterquartileRange**

DispersionReport (variance, standard deviation, sample range, mean deviation, median deviation, and quartile deviation)

The maximum likelihood estimate of the variance has the divisor n instead of $n - 1$ and is not unbiased. The theoretical variance of the sample mean m is σ^2/n, and an unbiased estimate of this is the variance of sample mean $s_m{}^2 = s^2/n$.

> **{VarianceMLE[d], SampleRange[d]} // N**   {2.09, 4.}

Mean and variance are easy to program:

```
ownMean[data_] := Apply[Plus, data] / Length[data]
ownVariance[data_] :=
  Apply[Plus, (data - ownMean[data]) ^ 2] / (Length[data] - 1)
```

> **{ownMean[d], ownVariance[d]} // N**   {3.1, 2.32222}

■ **Shape Statistics**

Skewness, **PearsonSkewness1**, **PearsonSkewness2**, **QuartileSkewness**

Kurtosis, **KurtosisExcess**

CentralMoment[data, r] $\left(\frac{1}{n} \sum (x_i - m)^r\right)$

ShapeReport (skewness, quartile skewness, and kurtosis excess)

> **{Skewness[d], Kurtosis[d]} // N**   {-0.174749, 1.59925}

■ **Other Commands**

ZeroMean[data], **Standardize[data]**

ExpectedValue[f, data, x]

ZeroMean transforms the data so that the result has zero as its mean. **Standardize** transforms the data so that the mean is zero and the unbiased estimate of variance is unity. **ExpectedValue** calculates the expected value of a function **f** of **x**.

> **Standardize[d] // N**
>
> {-1.37806, -1.37806, -0.72184, -0.72184, -0.0656218, 0.590596, 0.590596, 0.590596, 1.24681, 1.24681}
>
> **ExpectedValue[Abs[x - Mean[d]], d, x] // N**   1.3

■ Autocorrelation

Autocorrelation is important in time series analysis. Let ρ_k be the autocorrelation at lag k. An estimate of ρ_k is $r_k = \sum_{t=1}^{n-k} (x_t - m_x)(x_{t+k} - m_x) / \sum_{t=1}^{n} (x_t - m_x)^2$. The following program calculates all autocorrelations up to lag k:

```
ownAutocorrelation[data_, k_] := Module[{diff, denom},
  diff = data - ownMean[data];
  denom = Apply[Plus, diff^2];
  Table[{i, Drop[diff, -i].Drop[diff, i] / denom}, {i, 0, k}]]
```

As an example, we consider the same data as we did in Section 9.5.1, p. 261 (the data file **environmental** is on the CD-ROM that comes with this book):

```
data = Rest[Import["/Users/ruskeepaa/
      Documents/MNData/visdata/environmental", "Table"]];
```

```
Short[data, 2]
```

```
{{1, 41, 190, 67, 7.4}, {2, 36, 118, 72, 8.},
  ≪107≫, {110, 18, 131, 76, 8.}, {111, 20, 223, 68, 11.5}}
```

The file contains 111 observations of ozone, radiation, temperature, and wind. Extract the components of the data:

```
{no, ozone, radiation, temperature, wind} = Transpose[data];
```

Consider the temperature:

```
ListPlot[temperature, PlotJoined → True,
  AspectRatio → 0.15, Epilog → {AbsolutePointSize[2],
    Map[Point, Transpose[{Range[111], temperature}]]}];
```

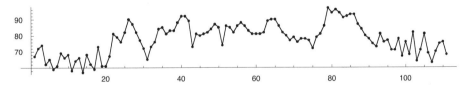

The estimated autocorrelation function is as follows:

```
ac = ownAutocorrelation[temperature // N, 25]
```

```
{{0, 1.}, {1, 0.777313}, {2, 0.694894}, {3, 0.632474}, {4, 0.546984},
  {5, 0.487101}, {6, 0.39759}, {7, 0.401579}, {8, 0.345652}, {9, 0.248692},
  {10, 0.241805}, {11, 0.199781}, {12, 0.220929}, {13, 0.208489},
  {14, 0.205918}, {15, 0.193053}, {16, 0.148181}, {17, 0.0833282},
  {18, 0.0378245}, {19, 0.0199929}, {20, -0.029604}, {21, -0.0628506},
  {22, -0.0468014}, {23, -0.0179397}, {24, -0.0147237}, {25, -0.0562379}}
```

```
Show[
  Graphics[{AbsoluteThickness[1.2], Map[Line[{{#[[1]], 0}, #}] &, ac]}],
  Axes → True, PlotRange → All, Ticks → {{5, 10, 15, 20, 25}, Automatic}];
```

27.1.2 Multivariate Descriptive Statistics

■ Basic Descriptive Statistics

Assume that the data are in the form of a matrix that contains as many rows as there are observations and as many columns as there are variables. The rows are treated as independent identically distributed multivariate observations. The following commands that we considered in Section 27.1.1 for univariate data can also be used for multivariate data (loading the `Statistics`MultiDescriptiveStatistics`` package is required in *Mathematica 4*): `Mean`, `Median`, `Quantile`, `Variance`, and `StandardDeviation`. The statistics are calculated for each column of the data separately. As an example, consider the same data for which we calculated the autocorrelation function in Section 27.1.1:

```
data = Rest[Import["/Users/ruskeepaa/
        Documents/MNData/visdata/environmental", "Table"]];
```

```
{no, ozone, radiation, temperature, wind} = Transpose[data];
```

Now we are interested in wind and ozone:

```
d2 = Transpose[{wind, ozone}] // N;
```

```
m = Mean[d2]      {9.93874, 42.0991}
```

With the `Statistics`MultiDescriptiveStatistics`` package, we can calculate various other multivariate descriptive statistics.

First, loading the package generalizes most univariate descriptive statistics so that they are applied to each column of a multivariate data. Of the commands listed in Section 27.1.1, only `LocationReport`, `CoefficientOfVariation`, and `DispersionReport` cannot be used for multivariate data (this may be a bug).

Second, the package defines several new statistics that have been especially designed for multivariate data. Next we study these statistics.

■ Location Statistics

```
MultivariateTrimmedMean[data, f]
SpatialMedian[data], SimplexMedian[data], ConvexHullMedian[data]
MultivariateMode[data]

EllipsoidQuantile[data, q], EllipsoidQuartiles[data]
PolytopeQuantile[data, q], PolytopeQuartiles[data]
```

`ConvexHullMedian`, `PolytopeQuantile`, and `PolytopeQuartiles` can only be used for two-variate data.

`EllipsoidQuantile[data, q]` gives an ellipsoid that is centered at `Mean[data]` and has $100\,q\%$ of the points inside of it. `EllipsoidQuartiles` gives the 25%, 50%, and 75% quantiles. We plot the data and the ellipsoidal quartiles:

```
p1 = ListPlot[d2,
    PlotStyle → AbsolutePointSize[1], DisplayFunction → Identity];
```

```
<< Statistics`MultiDescriptiveStatistics`
```

```
elli = EllipsoidQuartiles[d2]
```

```
{Ellipsoid[{9.93874, 42.0991}, {26.4495, 2.22566},
   {{-0.0658877, 0.997827}, {0.997827, 0.0658877}}],
 Ellipsoid[{9.93874, 42.0991}, {38.6654, 3.2536},
   {{-0.0658877, 0.997827}, {0.997827, 0.0658877}}],
 Ellipsoid[{9.93874, 42.0991}, {51.5526, 4.33802},
   {{-0.0658877, 0.997827}, {0.997827, 0.0658877}}]}
```

```
Show[p1, Graphics[elli], Epilog → {AbsolutePointSize[3], Point[m]},
    PlotRange → {-10, Automatic}, AxesLabel → {"wind", "ozone"},
    DisplayFunction → $DisplayFunction];
```

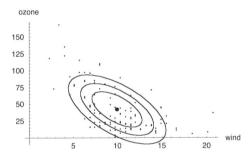

The ellipsoidal quartiles do not suit our data very well. In fact, ellipsoidal quartiles and quantiles suit data that have an elliptically contoured distribution.

`PolytopeQuantile[data, q]` gives a convex hull that is centered at `ConvexHull` `Median[data]` and has $100\,q\%$ of the points inside of it. We calculate the convex hull median and the polytope quartiles (we do not show some messages):

```
md = ConvexHullMedian[d2]      {10.3, 29.5}
```

```
poly = PolytopeQuartiles[d2];
```

For example, the 25% polytope quantile is as follows:

```
poly[[1]]
```

```
Polytope[{{12.1852, 22.2919}, {12.0209, 34.0552},
    {11.5343, 39.2713}, {11.2977, 41.555}, {9.87241, 54.7992},
    {9.1399, 60.6117}, {8.81707, 57.5468}, {8.4693, 51.279},
    {7.99981, 38.6214}, {7.96401, 36.1017}, {9.17284, 22.8395},
    {9.43336, 22.0154}, {10.7115, 18.8703}, {11.1869, 18.0443},
    {11.5533, 17.4891}, {12.1764, 22.0592}}, -Connectivity-]
```

The quartiles are displayed as follows:

```
Show[p1, Graphics[poly], Epilog → {AbsolutePointSize[3], Point[md]},
    PlotRange → {-10, Automatic}, DisplayFunction → $DisplayFunction];
```

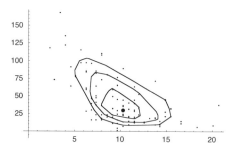

The polytope quartiles are better suited for our data. Note also that the mean is sensitive to outliers, whereas the median, being robust, does not have this drawback.

■ Dispersion Statistics

```
Covariance[xdata, ydata], CovarianceMLE[xdata, ydata]

CovarianceMatrix[xdata], CovarianceMatrixMLE[xdata],
CovarianceMatrix[xdata, ydata], CovarianceMatrixMLE[xdata, ydata],
CovarianceMatrixOfSampelMean[xdata], DispersionMatrix[xdata]

GeneralizedVariance[data], TotalVariation[data], ConvexHullArea[data]
MultivariateMeanDeviation[data], MultivariateMedianDeviation[data]
```

The formula $\mathrm{cov}(X, Y) = \frac{1}{n-1} \sum (x_i - m_x)(y_i - m_y)$ gives the unbiased covariance of two variables:

```
Covariance[ozone, wind]      - 72.5957
```

CovarianceMatrix gives all covariances:

```
CovarianceMatrix[d2]

{{12.668, -72.5957}, {-72.5957, 1107.29}}
```

Note that **Covariance** and **CovarianceMatrix** have the option **ScaleMethod** that can have the values **MeanDeviation**, **MedianDeviation**, and **QuartileDeviation** (see Section 27.1.1, p. 746). With this option, we get robust measures of covariance:

```
Map[Covariance[ozone, wind, ScaleMethod → #] &,
    {MeanDeviation, MedianDeviation, QuartileDeviation}]

{-68.8891, -39.3306, -52.7174}
```

■ Association Statistics

```
Correlation[xdata, ydata], SpearmanRankCorrelation[xdata, ydata],
   KendallRankCorrelation[xdata, ydata]
CorrelationMatrix[xdata], CorrelationMatrix[xdata, ydata],
   AssociationMatrix[xdata]
```

`Correlation` and `CorrelationMatrix` have the option `ScaleMethod`. Pearson's correlation coefficient is $\text{cov}(X, Y) / \sqrt{\text{var}(X)\,\text{var}(Y)}$:

    `Correlation[ozone, wind]` -0.612951

    `CorrelationMatrix[d2]` $\{\{1., -0.612951\}, \{-0.612951, 1.\}\}$

Correlation can be computed as follows (here we use **ownMean** and **ownVariance** which were defined in Section 27.1.1, p. 746):

```
ownCorrelation[xdata_, ydata_] :=
  (xdata - ownMean[xdata]) . (ydata - ownMean[ydata]) /
   ((Length[xdata] - 1) Sqrt[ownVariance[xdata] ownVariance[ydata]])
```

    `ownCorrelation[ozone, wind]` -0.612951

■ Shape Statistics

```
MultivariateSkewness[data], MultivariatePearsonSkewness1[data],
  MultivariatePearsonSkewness2[data]
MultivariateKurtosis[data], MultivariateKurtosisExcess[data]
CentralMoment[data, {r1, …, rp}]
```

■ Other Commands

```
Standardize[data, Decorrelate → True]
PrincipalComponents[data]
```

27.2 Frequencies and Data Manipulation

27.2.1 Frequencies

The `Statistics`DataManipulation`` package contains many commands for data manipulation, particularly for calculating frequencies. For integer observations, we typically calculate frequencies for each value that occurs in the data. For continuous observations, we typically count frequencies of observations falling within some given intervals.

■ Frequencies of Integer Data

    `Frequencies[list]`   Frequencies of the observations

Throw a die 20 times:

    `SeedRandom[2]; data = Table[Random[Integer, {1, 6}], {20}]`

    $\{5, 5, 3, 6, 1, 4, 4, 5, 3, 3, 5, 5, 6, 6, 1, 5, 1, 1, 6, 1\}$

Calculate the frequencies:

    `<< Statistics`DataManipulation``

```
fr1 = Frequencies[data]
```

```
{{5, 1}, {3, 3}, {2, 4}, {6, 5}, {4, 6}}
```

Show a bar chart for the frequencies (for bar charts, see Section 9.3):

```
<< Graphics`Graphics`
```

```
BarChart[fr1, BarGroupSpacing → 0, BarStyle → GrayLevel[0.8]];
```

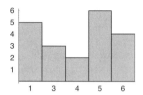

This is not a satisfactory plot, because the bar for $x = 2$ having zero frequency is lacking. It is better to calculate the frequencies as follows:

```
fr2 = Map[{Count[data, #], #} &, Range[1, 6]]
```

```
{{5, 1}, {0, 2}, {3, 3}, {2, 4}, {6, 5}, {4, 6}}
```

This list of frequencies has the advantage that it contains the frequencies of all *possible* results 1, 2, ..., 6. If only frequencies of all *occurred* results are wanted, then replace **Range[1,6]** with **Union[data]** (the result is then identical to the result given by **Frequencies**). Now **BarChart** gives a good plot:

```
BarChart[fr2, BarGroupSpacing → 0, BarStyle → GrayLevel[0.8]];
```

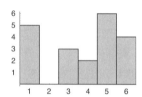

If we want to plot frequencies, it is easiest to use **Histogram**. It can be used in two forms. First, if the original data are input, then **Histogram** both calculates the frequencies and plots them:

```
<< Graphics`Graphics`
```

```
Histogram[data, HistogramCategories → Range[0.5, 6.5],
    BarStyle → GrayLevel[0.8]];
```

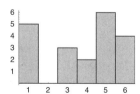

Second, if we have calculated the frequencies by **Count**:

```
fr3 = Map[Count[data, #] &, Range[1, 6]]
```

```
{5, 0, 3, 2, 6, 4}
```

or by `CategoryCounts` (see below):

    fr3 = CategoryCounts[data, Range[6]]

    {5, 0, 3, 2, 6, 4}

then we can input them to `Histogram` by giving the option `FrequencyData → True`:

    Histogram[fr3, FrequencyData → True,
      HistogramCategories → Range[0.5, 6.5], BarStyle → GrayLevel[0.8]];

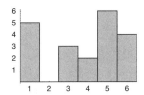

■ More General Frequencies

> `BinCounts[list, {min, max, dx}]`   Numbers of elements in intervals (min, min + dx],
>   (min + dx, min + 2 dx], …, (max − dx, max]
> `RangeCounts[list, {c1, …, cn}]`   Numbers of elements in intervals (−∞, c1), [c1, c2),
>   [c2, c3), …, [cn, ∞)
> `CategoryCounts[list, {e1, e2, … }]`   Numbers of elements matching each of the `ei`
> `CategoryCounts[list, {{e11, e12, … }, {e21, e22, … }, … }]`   Numbers of elements
>   matching any of the elements in each list `{ei1, ei2, … }`

Note that these commands generalize to data of several dimensions (see the manual of the packages).

If we want to plot the numbers of the results of our data in the sets (or categories) {1, 2}, {3, 4}, and {5, 6}, we can use `Histogram`:

    Histogram[data, HistogramCategories → Range[0.5, 6.5, 2],
      BarStyle → GrayLevel[0.8]];

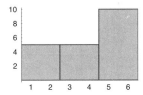

We can also first calculate the frequencies of the categories. This can be done in three different ways:

    << Statistics`DataManipulation`

    cc = BinCounts[data, {0.5, 6.5, 2}]   {5, 5, 10}

    cc = RangeCounts[data, {2.5, 4.5}]   {5, 5, 10}

    cc = CategoryCounts[data, {{1, 2}, {3, 4}, {5, 6}}]   {5, 5, 10}

Then we can plot these frequencies:

```
Histogram[cc, FrequencyData → True,
   HistogramCategories → Range[0.5, 6.5, 2], BarStyle → GrayLevel[0.8]];
```

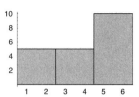

The commands mentioned above calculate the numbers of elements in some sets. The following commands show all of the corresponding elements.

> **BinLists[list, {min, max, dx}]** Elements in the intervals mentioned for **BinCounts**
> **RangeLists[list, {c1, …, cn}]** Elements in the intervals mentioned for **RangeCounts**
> **CategoryLists[list, {e1, e2, … }]** Elements matching each of the **ei**
> **CategoryLists[list, {{e11, e12, … }, {e21, e22, … }, … }]** Elements matching any
>   of the elements in each list **{ei1, ei2, … }**

Here are the individual results in the intervals:

```
BinLists[data, {0.5, 6.5, 2}]
```

```
{{1, 1, 1, 1, 1}, {3, 4, 4, 3, 3}, {5, 5, 6, 5, 5, 5, 6, 6, 5, 6}}
```

■ Frequencies of Real Data

Real-valued observations are often all distinct, and, for this reason, to calculate frequencies, the observations are first grouped. This can be done nicely with **BinCounts** or **Range Counts**. Generate a set of 1000 observations:

```
<< Statistics`NormalDistribution`
```

```
SeedRandom[1];
data = RandomArray[NormalDistribution[0, 1], 1000];
```

Calculate the number of these observations falling in the intervals $(-4, -3.75]$, $(-3.75, -3.5]$, …, $(3.75, 4]$:

```
<< Statistics`DataManipulation`
```

```
counts = BinCounts[data, {-4, 4, 0.25}]
```

```
{0, 1, 0, 0, 1, 6, 6, 7, 16, 25, 33, 51, 67, 80, 92,
 84, 100, 100, 83, 89, 59, 35, 28, 20, 9, 4, 0, 2, 1, 0, 1, 0}
```

Check that all observations have been taken into account:

```
Apply[Plus, counts]      1000
```

Plot the frequencies:

```
<< Graphics`Graphics`
```

```
Histogram[counts, FrequencyData → True,
    HistogramCategories → Range[-4, 4, 0.25], BarStyle → GrayLevel[0.8]];
```

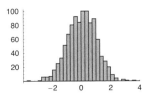

■ Relative Frequencies

The relative frequencies are as follows:

```
p1 = Histogram[counts / (1000 0.25), FrequencyData → True,
    HistogramCategories → Range[-4, 4, 0.25], BarStyle → GrayLevel[0.8]];
```

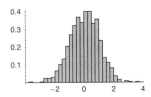

Next we show both the relative frequencies and the probability density function of the normal distribution from which the observations were sampled:

```
p2 = Plot[PDF[NormalDistribution[0, 1], x],
    {x, -4, 4}, DisplayFunction → Identity];

Show[p1, p2];
```

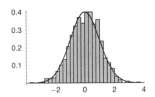

■ Cumulative Frequencies

`CumulativeSums[list]`	Cumulative sums of `list`

The cumulative sums or cumulative frequencies of **counts** are as follows:

```
<< Statistics`DataManipulation`
```

```
cumcounts = CumulativeSums[counts]
```

{0, 1, 1, 1, 2, 8, 14, 21, 37, 62, 95, 146, 213, 293, 385, 469, 569, 669, 752,
    841, 900, 935, 963, 983, 992, 996, 996, 998, 999, 999, 1000, 1000}

We plot the relative cumulative frequencies and the cumulative distribution function of the normal distribution:

```
p3 = Histogram[cumcounts / 1000 ,
    FrequencyData → True, HistogramCategories → Range[-4, 4, 0.25],
    BarStyle → GrayLevel[0.8], DisplayFunction → Identity];

p4 = Plot[CDF[NormalDistribution[0, 1], x],
    {x, -4, 4}, DisplayFunction → Identity];

Show[p3, p4, DisplayFunction → $DisplayFunction];
```

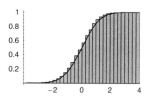

Cumulative sums can also be calculated using **Rest[FoldList[Plus, 0, list]]**.

27.2.2 Data Manipulation

Manipulating rows and columns is frequently needed when working with data. This topic is considered in Section 18.2.3, p. 477; here we recall some techniques.

---

data[[n]]   Give row number **n**

data[[All, n]] or **Transpose[data][[n]]**   Give column number **n**

xy = Transpose[{x, y}]   Pair the corresponding elements of **x** and **y**

{x, y} = Transpose[xy]   Extract the first and second components of **xy**

Suppose we have separate lists for the independent and dependent variables:

```
x = {1, 2, 3, 4, 5};
y = {14, 12, 15, 16, 13};
```

We want to pair the corresponding elements of **x** and **y**:

```
xy = Transpose[{x, y}]
```

```
{{1, 14}, {2, 12}, {3, 15}, {4, 16}, {5, 13}}
```

Then we extract the **x** and **y** values:

```
{x, y} = Transpose[xy]
```

```
{{1, 2, 3, 4, 5}, {14, 12, 15, 16, 13}}
```

Now the value of **x** is {1, 2, 3, 4, 5, 6} and the value of **y** is {14, 12, 15, 16, 13}.

```
x =.; y =.;
```

The **LinearAlgebra`MatrixManipulation`** package has commands to manage columns and combine matrices: **TakeColumns**, **TakeRows**, **TakeMatrix**, **AppendColumns**, **AppendRows**, and **BlockMatrix** (see Section 18.2.3). The **Statistics`DataManipulation`** package also contains similar commands: **Column**, **ColumnTake**, **ColumnDrop**, **ColumnJoin** and **RowJoin** and other commands like **DropNonNumeric** and **DropNonNumericColumn** (to drop nonnumeric elements, rows, and columns) and **LengthWhile**, **TakeWhile**, and **BooleanSelect**.

27.3 Smoothing

27.3.1 Smoothing with a Kernel

Suppose we have data $\{x_1, \ldots, x_n\}$ and we want to smooth the data with a *kernel* $\{k_1, \ldots, k_m\}$ ($m < n$) by forming the sums $\sum_{j=1}^{m} k_j x_{i+j}$, $i = 0, 1, \ldots, n - m$. For example, if $n = 5$ and $m = 3$, the result is the following smoothed values:

```
Map[{k₁, k₂, k₃}.# &, {{x₁, x₂, x₃}, {x₂, x₃, x₄}, {x₃, x₄, x₅}}]
```

$\{k_1\, x_1 + k_2\, x_2 + k_3\, x_3,\ k_1\, x_2 + k_2\, x_3 + k_3\, x_4,\ k_1\, x_3 + k_2\, x_4 + k_3\, x_5\}$

This is easy to program:

```
listCor[kernel_, list_] :=
  Map[kernel.# &, Partition[list, Length[kernel], 1]]
```

We can also use **ListCorrelate**:

ListCorrelate[kernel, list]	Form the correlation of **kernel** with **list**
ListConvolve[kernel, list]	Form the convolution of **kernel** with **list**

We verify that we get with **ListCorrelate** the same result that we obtained above:

```
ListCorrelate[{k₁, k₂, k₃}, {x₁, x₂, x₃, x₄, x₅}]
```

$\{k_1\, x_1 + k_2\, x_2 + k_3\, x_3,\ k_1\, x_2 + k_2\, x_3 + k_3\, x_4,\ k_1\, x_3 + k_2\, x_4 + k_3\, x_5\}$

ListConvolve gives the following result:

```
ListConvolve[{k₁, k₂, k₃}, {x₁, x₂, x₃, x₄, x₅}]
```

$\{k_3\, x_1 + k_2\, x_2 + k_1\, x_3,\ k_3\, x_2 + k_2\, x_3 + k_1\, x_4,\ k_3\, x_3 + k_2\, x_4 + k_1\, x_5\}$

So, **ListConvolve[kernel, list]** means **ListCorrelate[Reverse[kernel], list]**. Note that both commands have more general forms and that they also apply to multi-dimensional kernels and data.

■ Examples of Kernels

Moving averages are obtained by giving a constant kernel:

```
ListCorrelate[{1, 1, 1} / 3, {x₁, x₂, x₃, x₄, x₅}] // Simplify
```

$\left\{\frac{1}{3}\ (x_1 + x_2 + x_3),\ \frac{1}{3}\ (x_2 + x_3 + x_4),\ \frac{1}{3}\ (x_3 + x_4 + x_5)\right\}$

The kernel $\{-1, 1\}$ gives successive differences:

```
ListCorrelate[{-1, 1}, {x₁, x₂, x₃, x₄, x₅}]
```

$\{-x_1 + x_2,\ -x_2 + x_3,\ -x_3 + x_4,\ -x_4 + x_5\}$

A Gaussian kernel is of the following form:

```
gaussianKernel[denom_, max_] :=
  With[{t = Table[Exp[-n^2 / denom] // N, {n, -max, max}]}, t / Apply[Plus, t]]
```

For example:

```
gk = gaussianKernel[2, 3]
```

{0.00443305, 0.0540056, 0.242036,
 0.39905, 0.242036, 0.0540056, 0.00443305}

```
ListPlot[gk, AxesOrigin → {1, 0}, PlotStyle → AbsolutePointSize[2]];
```

In the example below, we use the Gaussian kernel.

■ An Example of Smoothing

We try to smooth noisy data. Our signal is as follows:

```
s = Sin[5 2 Pi x] - 0.8 Cos[9 2 Pi x];
```

```
signalPlot =
  Plot[s, {x, 0, 1}, AspectRatio → 0.2, PlotStyle → GrayLevel[0.5]];
```

To generate noisy observations from this signal, we first sample the signal:

```
xx = Range[0, 1, 0.01];
```

```
yy = s /. x → xx;
```

Then we generate noise from a normal distribution with mean of 0 and standard deviation of 0.4:

```
<< Statistics`NormalDistribution`
```

```
SeedRandom[1];
noise = Table[Random[NormalDistribution[0, 0.4]], {101}];
```

We add the noise to the signal:

```
ydata = yy + noise;
```

Then we plot our data:

```
data = Transpose[{xx, ydata}];
```

```
dataPlot = ListPlot[data, PlotRange → All, AspectRatio → 0.2];
```

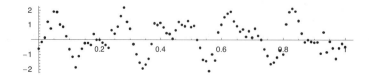

We smooth the data using the Gaussian kernel we calculated above:

```
smooth = ListCorrelate[gk, ydata];
```

When plotting the smoothed values, note that data is lost at both ends:

```
smoothPlot = ListPlot[Transpose[{Range[0.03, 0.97, 0.01], smooth}],
    PlotJoined → True, AspectRatio → 0.2];
```

We can compare the smoothed values with the data and with the signal:

```
Show[dataPlot, smoothPlot];
```

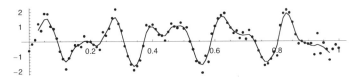

```
Show[signalPlot, smoothPlot];
```

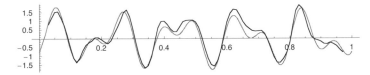

27.3.2 Other Methods of Smoothing

■ A Package

> *In the* **Statistics`DataSmoothing`** *package:*
>
> **MovingAverage[data, r]** Simple r-term moving average
>
> **MovingMedian[data, r]** Moving median of span r
>
> **LinearFilter[data, {c₀ , …, cᵣ₋₁ }]** Linear filter $\sum_{j=0}^{r-1} c_j\, x_{t-j}$, $t = r, r+1, \ldots, n$
>
> **ExponentialSmoothing[data, a]** Exponential smoothing with smoothing constant a

Consider data $x_1, \ldots, x_n$. A simple r-term moving average smoother calculates the average of all r successive terms. Similarly, a moving median smoother of span r calculates the median of all r successive terms. **MovingMedian** accepts, for odd r, the option **RepeatedSmoothing → True**, and then the smoother is applied repeatedly until convergence. A linear filter is a generalization of a moving average: now the weights of the data are not necessarily equal. The result of these three commands is a series of $n - r + 1$ terms.

If $\{x_t\}$ is the original set of data, then an exponential smoother calculates the values $y_{t+1} = y_t + a(x_{t+1} - y_t)$ for a smoothing constant a, $0 < a < 1$. The smaller a is, the stronger the smoothing is. The default is that the starting value y_0 is x_1, but y_0 can also be given as a third argument to **ExponentialSmoothing**.

The smoothing commands work for both univariate and multivariate data. The data contain only the dependent variable(s), not the independent variable (such as time).

Note that, for univariate data, **MovingAverage[data, r]** is equivalent to **List⸴ Correlate[Table[1/r, {r}], data]**, and **LinearFilter[data, kernel]** is equivalent to **ListConvolve[kernel, data]**.

Here is an example of exponential smoothing:

```
<< Statistics`DataSmoothing`

es = ExponentialSmoothing[ydata, 0.6];

ListPlot[Transpose[{xx, es}], PlotJoined → True,
   AspectRatio → 0.2, DisplayFunction → Identity];
Show[dataPlot, %];
```

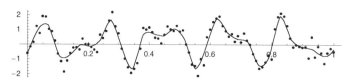

■ Local Regression

In Section 27.6.3, p. 781, we present a program to calculate a local regression curve. The method finds a smooth curve through the points by fitting a sequence of low-order polynomials. We try this method with our data:

```
fit = showLocalRegress[data, 30, 0.05, 1, AspectRatio → 0.2];
```

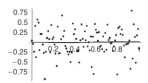

```
showLocalResiduals[data, fit, AxesOrigin → {0, 0}];
```

```
Sum of squared residuals is 10.854
```

■ Discrete Fourier Transform

The discrete Fourier transform (see Section 17.4.3, p. 461) can be used to smooth or filter data. First we calculate the Fourier transform of the y-values and plot the absolute values of the transform:

```
fou = Fourier[ydata];
```

```
ListPlot[Transpose[{Range[0, 100], Abs[fou]}],
    PlotRange → All, AspectRatio → 0.2];
```

We see two peaks at the frequencies 5 and 9 (and the corresponding symmetric peaks on the righthand side), which correspond with the frequencies 5 and 9 of the signal. All other frequencies can be considered to be caused by the noise.

We try to filter the data by simply replacing with zeros all of the values of the Fourier transform except for the four peaks. We do this by replacing with zero all frequencies whose absolute value is less than, in this case, 2:

```
filt = Chop[fou, 2]
```

```
{0, 0, 0, 0, 0, 5.12415 i, 0, 0, 0, -4.11842, 0, 0, 0, 0, 0, 0,
 0, 0, 0, 0, 0, 0, 0, 0, 0, 0, 0, 0, 0, 0, 0, 0, 0, 0, 0, 0,
 0, 0, 0, 0, 0, 0, 0, 0, 0, 0, 0, 0, 0, 0, 0, 0, 0, 0, 0, 0, 0,
 0, 0, 0, 0, 0, 0, 0, 0, 0, 0, 0, 0, 0, 0, 0, 0, 0, 0, 0, 0, 0,
 0, 0, 0, 0, 0, 0, 0, 0, -4.11842, 0, 0, 0, -5.12415 i, 0, 0, 0, 0}
```

Then we find the inverse transform:

```
filteredYData = Chop[InverseFourier[filt]];
```

```
filteredData = Transpose[{xx, filteredYData}];
```

We compare the filtered values with the data and the signal:

```
p = ListPlot[filteredData, PlotJoined → True, DisplayFunction → Identity];
```

```
Show[dataPlot, p];
```

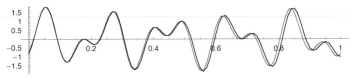

```
Show[signalPlot, p];
```

The fits seem to be good.

27.4 Confidence Intervals

27.4.1 Confidence Intervals for a Mean

With the **Statistics`ConfidenceIntervals`** package, we can compute confidence intervals for a mean, for the difference of two means, for a variance, and for the ratio of two variances. Recall that a confidence interval, for example, for the population mean, gives an interval within which the population mean lies with a given probability, say, 0.95. We assume that the observations follow a normal distribution. We will also present a confidence interval for the probability of success of independent trials.

Use the following terminology and notation: μ = population mean; m = sample **Mean**; σ^2 = population variance; s^2 = sample **Variance**; σ = population standard deviation; s = sample **StandardDeviation**; $\sigma_m = \sigma/\sqrt{n}$ = standard deviation of sample mean; $s_m = s/\sqrt{n}$ = **StandardErrorOfSampleMean**; and $s_m^2 = s^2/n$ = **VarianceOfSampleMean**.

■ **Confidence Intervals for a Mean**

```
MeanCI[data]
StudentTCI[m, s_m, n-1]
```

For the confidence interval of the population mean, we have two main commands. **MeanCI** uses the original data, while **StudentTCI** uses only computed values for the sample mean m, standard error of sample mean s_m, and degrees of freedom $n-1$ (n is the size of the sample). These two commands use the Student t distribution to calculate the confidence interval.

```
MeanCI[data, KnownVariance → σ²]
NormalCI[m, σ_m]
```

Sometimes the population variance σ^2 is known. With **MeanCI**, we can add the option **KnownVariance** (or **KnownStandardDeviation**), whereas **NormalCI** is used if we input the sample mean and the standard deviation of sample mean. These commands use the normal distribution to calculate the confidence interval.

The confidence level is 0.95 by default, but the level can be set with the option **Confi**-**denceLevel** (this holds true not only for the commands mentioned here but for all commands used to calculate confidence intervals).

■ **An Example**

To illustrate these commands, we first generate data from a normal distribution with a mean of 50 and a standard deviation of 3:

```
<< Statistics`NormalDistribution`

SeedRandom[2];
data = RandomArray[NormalDistribution[50, 3], 100];
```

```
ListPlot[data, PlotRange → All];
```

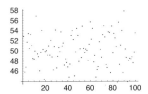

So, in this demonstration example, we know that the population mean is 50 and the standard deviation is 3. Now we proceed as if we did not know these values, and we calculate a 95% confidence interval for the population mean:

```
<< Statistics`ConfidenceIntervals`
```

```
MeanCI[data]      {49.4862, 50.7101}
```

So we know that, with a probability of 0.95, the population mean is within this interval (from the simulated data, we have the knowledge that the population mean really is in this interval). We could also first calculate the sample mean and the standard error of sample mean (note that the **Statistics`DescriptiveStatistics`** package is automatically loaded when most other statistical packages are used):

```
{m, sm} = {Mean[data], StandardErrorOfSampleMean[data]}
```

```
{50.0982, 0.308417}
```

We then use **StudentTCI**:

```
StudentTCI[m, sm, 99]     {49.4862, 50.7101}
```

For other intervals than 95%, we add the confidence level as an option:

```
MeanCI[data, ConfidenceLevel → 0.99]     {49.2882, 50.9082}
```

If we know that the population standard deviation is 3 or the variance 9, the 95% confidence interval is as follows:

```
MeanCI[data, KnownVariance → 9]     {49.5102, 50.6862}
```

We could also use **NormalCI**:

```
NormalCI[m, Sqrt[9 / 100]]     {49.5102, 50.6862}
```

With **NormalCI**, we can easily calculate the well-known confidence intervals for the normal distribution:

```
Map[NormalCI[0, 1, ConfidenceLevel → #] &, {0.95, 0.99, 0.999}]
```

```
{{-1.95996, 1.95996}, {-2.57583, 2.57583}, {-3.29053, 3.29053}}
```

■ The Meaning of a Confidence Interval

If we have a 95% confidence interval, we know that there is a 5% probability that the population mean is not in the interval. Thus, if we take 100 samples and calculate the corresponding 95% confidence intervals, we can expect that about 5% of the intervals will not contain the population mean. To illustrate this, we generate 100 samples of 100 observations, calculate for each sample the confidence interval, and investigate how many of these intervals contain the population mean 50:

```
SeedRandom[1];
samples = Table[RandomArray[NormalDistribution[50, 3], 100], {100}];

cis = Map[MeanCI[#] &, samples];

Length[Select[cis, #[[1]] < 50 < #[[2]] &]]        93
```

From the 100 samples, 93 generated a 95% confidence interval that actually contained the true population mean 50. Here are all of the 100 confidence intervals:

```
showConfidenceIntervals[cis_, mu_, n_, opts___] :=
  Show[Graphics[{Line[{{1, mu}, {n, mu}}],
     Table[Line[{{i, cis[[i, 1]]}, {i, cis[[i, 2]]}}], {i, n}]},
    Axes → {False, True}, AspectRatio → 0.2, PlotRange → All, opts]];
```

```
showConfidenceIntervals[cis, 50, 100];
```

Here are the confidence intervals in ascending order according to the mean of the samples (the means are shown with points):

```
me = Transpose[{Range[100], Sort[Map[Mean, samples]]}];
```

```
showConfidenceIntervals[Sort[cis, Mean[#1] < Mean[#2] &],
  50, 100, Epilog → {AbsolutePointSize[2], Map[Point, me]}];
```

27.4.2 Other Confidence Intervals

■ Confidence Intervals for the Difference between Two Means

```
MeanDifferenceCI[data1, data2]
```

If the populations are known to have equal variances, we can add the option **EqualVariances → True**. If we know this common value, we can simply add the option **KnownVariance → σ^2**. If the known variances are different, we can add the option **Known: Variance → {σ_1^2, σ_2^2}**.

■ Confidence Intervals for a Variance

```
VarianceCI[data]
ChiSquareCI[s², n − 1]
```

The first command uses the original data, while the other uses only the sample variance s^2 and degrees of freedom. The default confidence level is 0.95, but a different level can be set with **ConfidenceLevel**. These two commands use the chi-square distribution.

As an example, we use the simulated data that was calculated earlier:

```
<< Statistics`ConfidenceIntervals`
```

```
VarianceCI[data]    {7.33284, 12.8365}
```

Thus, with a probability of 0.95, the population variance is within this interval (the true variance 9 is in this interval). We could also first calculate the sample variance and then use the other command:

```
var = Variance[data]    9.5121
```

```
ChiSquareCI[var, 99]    {7.33284, 12.8365}
```

■ Confidence Intervals for the Ratio of Two Variances

```
VarianceRatioCI[data1, data2]
FRatioCI[s²₁ / s²₂, n₁ − 1, n₂ − 1]
```

The first command uses the two data sets, while the other uses only the ratio of the sample variances s_1^2 / s_2^2 and degrees of freedom (n_1 and n_2 are the sizes of the samples from the two populations). These two commands use the F-ratio distribution.

■ Confidence Intervals for a Probability

Suppose we have made n independent trials of which k have succeeded. The estimate of the probability of success is k/n. The following module gives an approximate $100\,\alpha\,\%$ confidence interval for the true probability of success (Johnson, Kotz, and Kemp 1992, p. 130). The probability of the true p being within the computed interval is at least α.

```
<< Statistics`NormalDistribution`
```

```
probabilityCI[succ_, total_, α_] :=
  Module[{n1 = 2 succ, n2 = 2 (total − succ + 1),
    n3 = 2 (succ + 1), n4 = 2 (total − succ), q1, q2},
    q1 = Quantile[FRatioDistribution[n1, n2], (1 − α) / 2];
    q2 = Quantile[FRatioDistribution[n3, n4], (1 + α) / 2];
    {n1 q1 / (n2 + n1 q1), n3 q2 / (n4 + n3 q2)}]
```

Generate a sequence of successes and failures by assuming that each trial succeeds with a probability of 0.3:

```
<< Statistics`DiscreteDistributions`
```

```
SeedRandom[1];
data2 = RandomArray[BernoulliDistribution[0.3], 100]
```

```
{0, 0, 0, 1, 0, 0, 0, 0, 1, 0, 0, 0, 0, 1, 0, 0, 0, 1, 0, 0, 0, 1, 0, 1,
 1, 0, 0, 1, 0, 0, 1, 1, 1, 1, 0, 0, 0, 1, 0, 0, 0, 0, 0, 0, 0, 1, 0, 0,
 1, 1, 0, 1, 0, 1, 1, 1, 1, 0, 0, 0, 1, 0, 1, 0, 1, 0, 1, 0, 0, 0, 0, 1, 0, 0,
 1, 1, 0, 0, 0, 1, 0, 0, 1, 0, 0, 0, 1, 0, 0, 1, 0, 1, 1, 1, 1, 0, 0, 0, 0}
```

Calculate the number of successes:

```
k = Apply[Plus, data2]      36
```

Then calculate an approximate 95% confidence interval:

```
probabilityCI[k, 100, 0.95]      {0.266408, 0.462122}
```

27.5 Hypothesis Testing

27.5.1 Tests for a Mean

With the **Statistics`HypothesisTests`** package, we can test the mean, the difference of two means, the variance, and the ratio of two variances. Recall that when testing, for example, the population mean, we want to infer, on the basis of a sample from the population, whether the population mean is a certain value μ or if it is different from that value. We assume that the data follow a normal distribution. We also present a test for the probability of success of independent trials and a test for goodness of fit. The material here in Section 27.5 is closely analogous to that of Section 27.4. First we consider the testing of the mean.

MeanTest[data, μ]
StudentTPValue[$(m - \mu)/s_m$, $n - 1$]

To test whether the population mean could be μ, use one of the above commands. **MeanTest** uses the original data and the hypothetical value μ of the population mean, while **StudentTPValue** uses only the value of the test statistic $t = (m - \mu)/s_m$ and degrees of freedom $n - 1$. The two commands use the Student t distribution to perform the testing.

The result of the testing is a p value. This is the probability that, if the hypothetical mean value μ is true, the test statistic t (treated as a random variable) has a value at least as extreme as its computed value. If the p value is sufficiently small—smaller than, for example, 0.05 (a significance level)—then the hypothetical mean value μ can be rejected: the observations do not give sufficient support for this mean value.

MeanTest[data, μ, KnownVariance $\to \sigma^2$]
NormalPValue[$(m - \mu)/\sigma_m$]

These commands that use the normal distribution can be used if the population variance is known. In **MeanTest** we can add the option **KnownVariance**, whereas **Normal‹ PValue** is used if we input the value of the test statistic.

■ **Options**

> *Options for hypothesis testings:*
>
> **SignificanceLevel** Significance level of the test; examples of values: **None, 0.05, 0.01**
> **TwoSided** Whether to perform a two-sided test; possible values: **False, True**
> **FullReport** Whether to include additional information; possible values: **False, True**

These options can be used for most hypothesis test commands. We can add a significance level (e.g., **SignificanceLevel → 0.05**), in which case the result of the test contains the conclusion whether the hypothesis is accepted or rejected at this significance level.

The default is a one-sided test. This means that we test whether the population mean is μ against the alternative that the population mean is greater than μ or against the alternative that the population mean is smaller than μ. If we want to test that the population mean is μ against the alternative that the population mean is different from μ, then we add the option **TwoSided → True**.

If we add the option **FullReport → True**, we get, besides the p value, also the sample mean, the value of the test statistic, degrees of freedom, and the distribution used in calculating the p value. (This option can only be used in commands that use the original data.)

■ **An Example**

To illustrate the testing of a mean, we use the same simulated data we used in Section 27.4.1:

```
<< Statistics`NormalDistribution`

SeedRandom[2];
data = RandomArray[NormalDistribution[50, 3], 100];

{m, sm} = {Mean[data], StandardErrorOfSampleMean[data]}

{50.0982, 0.308417}
```

The sample mean is close to 50. We test whether the population mean could be $\mu = 50$ against the alternative that the population mean is larger than 50:

```
<< Statistics`HypothesisTests`

MeanTest[data, 50]     OneSidedPValue → 0.375452
```

This probability is not small (not smaller than, for example, 0.05), and so we cannot reject the hypothesis that the population mean is 50 (from the simulated data, we have the knowledge that the population mean really is 50). We could also first calculate the value of the test statistic t from the values of the sample mean m and the standard error of sample mean s_m:

```
t = (m - 50) / sm     0.31833

StudentTPValue[t, 99]     OneSidedPValue → 0.375452
```

We can add a significance level as an option:

MeanTest[data, 50, SignificanceLevel → 0.05]

{OneSidedPValue → 0.375452,
  Fail to reject null hypothesis at significance level → 0.05}

To test whether the population mean is 50 against the alternative that the population mean is other than 50, we write the following:

MeanTest[data, 50, TwoSided → True]   TwoSidedPValue → 0.750905

This probability is again not small, and so we cannot reject the hypothesis that the population mean is 50. Then we ask for a full report:

MeanTest[data, 50, FullReport → True]

$$\left\{ \text{FullReport} \to \begin{array}{ccc} \text{Mean} & \text{TestStat} & \text{Distribution} \\ 50.0982 & 0.31833 & \text{StudentTDistribution[99]} \end{array} \right.,$$

OneSidedPValue → 0.375452}

If we know that the population standard deviation is 3 or the variance 9, we get the following p value:

MeanTest[data, 50, KnownVariance → 9]   OneSidedPValue → 0.371735

We could also use **NormalCI**:

NormalPValue[(m - 50) / Sqrt[9 / 100]]   OneSidedPValue → 0.371735

■ Type I Error

If we accept a correct hypothesis or reject a wrong one, we make a correct decision. However, if we reject a correct hypothesis or accept a wrong one, we make a wrong decision.

If we perform several tests and always use significance level 0.05 (i.e., we always reject the hypotheses if the p value is smaller than 0.05), then we know that, in about 5% of the tests, we reject a correct hypothesis. This is a *type I error*. To illustrate this error, we generate 100 samples of 100 observations from a normal distribution with mean 50, perform for each sample the test that tells us whether the population mean is 50 against the alternative that the mean is other than 50, and investigate the number of tests in which we draw the correct conclusion that the population mean is 50 when we use the significance level 0.05:

```
<< Statistics`NormalDistribution`

SeedRandom[1];
samples = Table[RandomArray[NormalDistribution[50, 3], 100], {100}];

pvalues1 = TwoSidedPValue /.
    Partition[Map[MeanTest[#, 50, TwoSided → True] &, samples], 1];

Length[Select[pvalues1, # ≥ 0.05 &]]     93
```

From the 100 samples, in 93 we draw the true conclusion that the population mean is 50. To plot the p values, we write the following function:

```
showPValues[pvalues_, α_, n_, opts___] :=
  Show[Graphics[{AbsolutePointSize[2], Map[Point,
     Transpose[{Range[n], pvalues}]], Line[{{1, α}, {n, α}}]}], Axes → True,
    AxesOrigin → {0, -0.05}, AspectRatio → 0.2, PlotRange → All, opts]]
```

Here are the p values for all 100 samples:

```
showPValues[pvalues1, 0.05, 100];
```

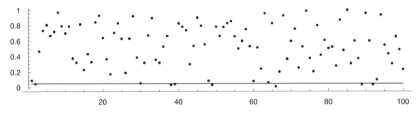

We see that seven *p* values are below the significance level 0.05. In these cases, we made the type I error: we rejected the correct hypothesis that the population mean is 50.

■ Type II Error

A *type II error* is made if the hypothesis is not true but we still accept it. To illustrate this error, we use the 100 samples of 100 observations generated in the preceding example and test whether the population mean is 51 against the alternative that the mean is not 51. We know that, in this case, the hypothesis is wrong and should be rejected:

```
pvalues2 = TwoSidedPValue /.
   Partition[Map[MeanTest[#, 51, TwoSided → True] &, samples], 1];

Length[Select[pvalues2, # < 0.05 &]]        91
```

We rightly rejected the wrong hypothesis 91 times. Here are all of the p values:

```
showPValues[pvalues2, 0.05, 100];
```

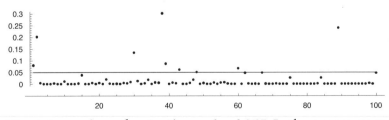

We see that 9 points are above the significance level 0.05. In these cases we made the type II error: we accepted the wrong hypothesis, which was that the population mean is 51.

27.5.2 Other Tests

■ Testing the Difference between Two Means

> MeanDifferenceTest[data1, data2, d]

Here we test whether the difference between two population means could be d. In addition to the options **EqualVariances → True** and **KnownVariance → σ^2** or **Known Variance → $\{\sigma_1^2, \sigma_2^2\}$**, we can also use the three options mentioned in Section 27.5.1: **SignificanceLevel → α, TwoSided → True**, and **FullReport → True**.

■ Testing a Variance

> VarianceTest[data, σ^2]
> ChiSquarePValue[$(n-1)s^2/\sigma^2$, $n-1$]

To test whether the population variance could be σ^2, use one of these commands. As an example, we use the simulated data that was calculated in Section 27.5.1, p. 767, and test the hypothesis that the population variance is 9 against the alternative that the variance is other than 9:

> << Statistics`HypothesisTests`

> VarianceTest[data, 9, TwoSided → True]

> TwoSidedPValue → 0.659823

We could also use the test statistic:

> ChiSquarePValue[99 Variance[data] / 9, 99, TwoSided → True]

> TwoSidedPValue → 0.659823

This probability is not small, so we cannot reject the hypothesis that the population variance is 9 (the true variance is 9 in this example).

■ Testing the Ratio of Two Variances

> VarianceRatioTest[data1, data2, r]
> FRatioPValue[$(s_1^2/s_2^2)/r$, n_1-1, n_2-1]

Use these commands to test whether the ratio of two population variances could be r.

■ Testing a Probability

Suppose we have made n independent trials of which k have succeeded. The estimate of the probability of success is k/n. The following module finds the p value to test the probability of success (Allen 1990, p. 508).

> << Statistics`DiscreteDistributions`

```
probabilityTest[succ_, total_, p0_, type_] := Module[{p1, p2},
  p1 = 1 - CDF[BinomialDistribution[total, p0], succ - 1];
  p2 = CDF[BinomialDistribution[total, p0], succ];
  Which[type === g, p1, type === l, p2,
    type === d, If[succ > total p0, 2 p1, 2 p2], True, Null]]
```

Here **succ** is the number of successes, **total** is the number of trials, **p0** is the hypothetical value of the probability of success, and **type** is the type of the alternative hypothesis. The type is **g**, **l**, or **d**, depending on whether the alternative hypothesis claims that the probability is greater than, less than, or different from **p0**. As an example, we generate a sequence of successes and failures. Each trial succeeds with a probability of 0.3:

> **SeedRandom[1]; data2 = RandomArray[BernoulliDistribution[0.3], 100];**

Calculate the number of successes:

> **k = Apply[Plus, data2]**   36

We test whether the true probability of success is 0.3 against the alternative that the probability is greater than 0.3:

> **probabilityTest[k, 100, 0.3, g]**   0.116079

This probability is not small, so we cannot reject the hypothesis that the probability of success is 0.3 (from the simulated observations, we know that the true probability is 0.3).

■ Goodness-of-Fit Test

The chi-square distribution can be used to test whether a given set of observations may have arisen from a certain distribution. Note that plotting the data on probability graph paper may help in determining whether the data follows a given distribution (see Section 26.1.5, p. 720).

When investigating days of absence at a firm during a period of 50 days, a statistician obtained the following results (these are not real data):

> **obs = {{8, 0}, {12, 1}, {14, 2}, {8, 3}, {3, 4}, {4, 5}, {1, 6}};**

This means that there were 8 days with no absences, 12 days with one absence, and so on. Do these observations follow a Poisson distribution? To find out, first separate the observed frequencies and the values:

> **{obsfreq, obsval} = Transpose[obs]**

> {{8, 12, 14, 8, 3, 4, 1}, {0, 1, 2, 3, 4, 5, 6}}

Check that the number of days is 50:

> **n = Apply[Plus, obsfreq]**   50

Then calculate the mean number of absences per day:

> **lambda = Apply[Plus, obsfreq obsval] / n // N**   2.04

Thus, in the mean, there were 2.04 absences per day. Then we calculate the first six Poisson probabilities with this parameter and multiply them by the number of days to obtain the expected frequencies:

> **<< Statistics`DiscreteDistributions`**

```
expfreq = n Table[PDF[PoissonDistribution[lambda], i], {i, 0, 5}]
```
{6.50144, 13.2629, 13.5282, 9.19917, 4.69158, 1.91416}

To these frequencies we add the expected frequency of at least six absences:

```
AppendTo[expfreq, n (1 - CDF[PoissonDistribution[lambda], 5])]
```
{6.50144, 13.2629, 13.5282, 9.19917, 4.69158, 1.91416, 0.902544}

All of the expected frequencies should be at least 5 for the test to be sufficiently accurate, so we combine the last three classes to form one class:

```
expfreq = Append[Take[expfreq, 4], Apply[Plus, Take[expfreq, -3]]]
```
{6.50144, 13.2629, 13.5282, 9.19917, 7.50828}

The sum of the expected frequencies is 50, as it should be:

```
Apply[Plus, %]     50.
```

Similarly we combine the last three observed frequencies:

```
obsfreq = Append[Take[obsfreq, 4], Apply[Plus, Take[obsfreq, -3]]]
```
{8, 12, 14, 8, 8}

Now we calculate the chi-square statistic:

```
chi2 = Apply[Plus, (obsfreq - expfreq)^2 / expfreq]     0.670651
```

Lastly we calculate the p value, which shows the probability of getting a chi-square value at least as extreme as **chi2**. The parameter of the chi-square distribution is the number of classes minus the number of estimated parameters (in this example, we estimated one parameter, namely the parameter of the Poisson distribution) minus one:

```
<< Statistics`NormalDistribution`
```

```
1 - CDF[ChiSquareDistribution[5 - 1 - 1], chi2]     0.880084
```

This probability is not small, so we cannot reject the hypothesis that the observations follow a Poisson distribution.

```
n =.
```

27.5.3 Analysis of Variance (ANOVA)

In the **Statistics`ANOVA`** *package:*

ANOVA[data]   Perform a one-way ANOVA
ANOVA[data, model, factors]   Perform a general ANOVA

With the **PostTests** option, we can tell what tests we want to apply to find significant differences. Possibilities are **Bonferroni**, **Duncan**, **StudentNewmanKeuls**, **Tukey**, and **Dunnett**. **SignificanceLevel** has the default value 0.05.

ANOVA is a way to investigate whether several populations—having normal distributions with equal variances—have equal means.

Consider the following example (Rohatgi 1984, p. 811). When a farmer investigated four fertilizers for soybeans, he received the following yields for plots of equal size:

```
fertilizer[1] = {47, 42, 43, 46, 44, 42};
fertilizer[2] = {51, 58, 62, 49, 53, 51, 50, 59};
fertilizer[3] = {37, 39, 41, 38, 39, 37, 42, 36, 40};
fertilizer[4] = {42, 43, 42, 45, 47, 50, 48};
```

We define the number of fertilizers and the number of plots for each fertilizer:

```
k = 4; n[1] = 6; n[2] = 8; n[3] = 9; n[4] = 7;
```

To do the one-way ANOVA, write the data as follows:

```
data =
Flatten[Table[Transpose[{Table[i, {n[i]}], fertilizer[i]}], {i, k}], 1]
```

```
{{1, 47}, {1, 42}, {1, 43}, {1, 46}, {1, 44}, {1, 42}, {2, 51}, {2, 58},
 {2, 62}, {2, 49}, {2, 53}, {2, 51}, {2, 50}, {2, 59}, {3, 37}, {3, 39},
 {3, 41}, {3, 38}, {3, 39}, {3, 37}, {3, 42}, {3, 36}, {3, 40},
 {4, 42}, {4, 43}, {4, 42}, {4, 45}, {4, 47}, {4, 50}, {4, 48}}
```

The results of ANOVA are as follows:

```
<< Statistics`ANOVA`
```

```
ANOVA[data, PostTests → {Bonferroni, Tukey}]
```

		DF	SumOfSq	MeanSq	FRatio	PValue
	Model	3	1015.51	338.503	31.6746	7.77406×10^{-9}
{ANOVA →	Error	26	277.859	10.6869		
	Total	29	1293.37			

	All	45.4333	
	Model[1]	44.	
CellMeans →	Model[2]	54.125	,
	Model[3]	38.7778	
	Model[4]	45.2857	

	Bonferroni	{{1, 2}, {1, 3}, {2, 3}, {2, 4}, {3, 4}}
PostTests → {Model →		}}
	Tukey	{{1, 2}, {1, 3}, {2, 3}, {2, 4}, {3, 4}}

The p-value is very small, which indicates that there are significant differences between the fertilizers. Both the Tukey and the Bonferroni test arrived at the conclusion that fertilizers 1 and 2, 1 and 3, 2 and 3, 2 and 4, and 3 and 4 differ significantly. The tests did not find a significant difference between fertilizers 1 and 4.

27.6 Regression

27.6.1 Linear Regression

The command **Fit[data, basis, vars]**, which was considered in Section 22.1.1, calculates linear least-squares fits to data: it finds the linear combination of the functions in the basis that gives the least squared error. The command **Regress** in the **Statistics`Linear** **Regression`** package does the same but, in addition, can print statistical information about the fit. **Regress** is used in the same way as **Fit**. For parameters that appear non-linearly, use **NonlinearRegress** (see Section 27.6.2, p. 779). For local regression, see Section 27.6.3, p. 781.

In the `Statistics`LinearRegression`` *package:*

`Regress[data, basis, vars]` Fit `data` by a linear combination of functions of `vars` in `basis`

Options:

`IncludeConstant` Whether a constant term is included in the model; possible values: `True`, `False`

`Weights` List of weights for each data point or a pure function; default value: `Automatic`

`ConfidenceLevel` Used for confidence intervals; examples of values: `0.95`, `0.99`

`RegressionReport` Statistics to be included in output; default value: `SummaryReport`

`BasisNames` Names of basis elements for table headings; default value: `Automatic`

`Tolerance` Numerical tolerance under which the calculation is to be effected; default value: `Automatic`

Data are normally given in the form `{{x1, f1}, {x2, f2}, ... }` (see Sections 22.1.1, p. 570, and 22.1.3, p. 577). An example of the basis is `{1, x, x^2}`.

The default value of `IncludeConstant` is `True`, which means that the constant term is included automatically even if it is not mentioned in the list of basis functions (`Fit` uses only the functions in the list of basis functions). You have to write `IncludeConstant →` `False` if you do not want the constant term.

Next we explain the use of the option `RegressionReport`.

■ Obtainable Information

`RegressionReport` controls the amount of information that is printed. The default value `SummaryReport` means the list `{ParameterTable, RSquared, AdjustedRSquared,` `EstimatedVariance, ANOVATable}`. All possible items can be seen by giving the command `RegressionReportValues[Regress]`. Here are the items classified into groups (many of these items are explained in the examples below):

- To get the fit and information about the estimated parameters: `BestFit`, `BestFit` `Parameters`, `ParameterTable`, `ParameterCITable`, `ParameterConfidenceRegion`, `CovarianceMatrix`, `CorrelationMatrix`

- To analyze variances: `ANOVATable`, `EstimatedVariance`, `CoefficientOfVariation`, `RSquared`, `AdjustedRSquared`

- To analyze predictions: `FitResiduals`, `PredictedResponse`, `SinglePrediction` `CITable`, `MeanPredictionCITable`

- To detect correlated errors: `DurbinWatsonD`

- To evaluate basis functions and detect collinearity: `PartialSumOfSquares`, `Sequen` `tialSumOfSquares`, `VarianceInflation`, `EigenstructureTable`

- To detect outliers: `HatDiagonal`, `JackknifedVariance`, `StandardizedResiduals`, `StudentizedResiduals`, `CookD`, `PredictedResponseDelta`, `BestFitParameters` `Delta`, `CovarianceMatrixDetRatio`

- To get the catcher matrix: `CatcherMatrix`

■ An Example

Consider the example of Section 22.1.1, p. 571:

```
<< Statistics`NormalDistribution`
```

```
SeedRandom[2];
data = Table[
    {x, 2 + x - 0.004 x^2 + 2 Random[NormalDistribution[0, 1]]}, {x, 0, 50}];
```

```
p1 = ListPlot[data];
```

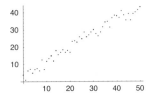

We could use **Fit** to calculate the least-squares fit. Try a second-order polynomial:

```
Fit[data, {1, x, x^2}, x]
```

$1.88055 + 1.02404\, x - 0.00445111\, x^2$

We can also use **Regress** and get more information about the fit.

```
<< Statistics`LinearRegression`
```

```
result = Regress[data, {1, x, x^2}, x, RegressionReport →
    {BestFit, BestFitParameters, RSquared, EstimatedVariance,
     ParameterTable, ParameterCITable, ANOVATable, DurbinWatsonD}]
```

$\{$BestFit $\to 1.88055 + 1.02404\, x - 0.00445111\, x^2$,

BestFitParameters $\to \{1.88055, 1.02404, -0.00445111\}$,

RSquared $\to 0.972548$, EstimatedVariance $\to 4.19652$, ParameterTable $\to$

	Estimate	SE	TStat	PValue
1	1.88055	0.827882	2.27152	0.0276376
x	1.02404	0.0765744	13.3732	0.
x^2	-0.00445111	0.00148106	-3.00535	0.0042091

ParameterCITable $\to$

	Estimate	SE	CI
1	1.88055	0.827882	$\{0.215983, 3.54512\}$
x	1.02404	0.0765744	$\{0.870079, 1.17801\}$
x^2	-0.00445111	0.00148106	$\{-0.00742898, -0.00147323\}$

ANOVATable $\to$

	DF	SumOfSq	MeanSq	FRatio	PValue
Model	2	7136.22	3568.11	850.254	0.
Error	48	201.433	4.19652		
Total	50	7337.65			

DurbinWatsonD $\to 2.00781\}$

BestFit gives the fit in the same form as **Fit**, while **BestFitParameters** gives only the parameters. **RSquared**—the square of the multiple correlation coefficient—is also called the coefficient of determination and is in the interval [0, 1]. It tells us how much the full, fitted model improves a reduced model that contains only a constant term. **Estimated Variance** is the estimated error variance or the residual mean square.

ParameterTable contains information to test whether a specific parameter is zero against the alternative that the parameter is not zero. The test can be done with the t statistic and the corresponding p value. A small p value (not larger than, say, 0.05) indicates that the observations do not support the hypothesis that the parameter is zero. In our example, all p values are small, which tells us that all coefficients are statistically significantly different from zero. **ParameterCITable** contains 95% confidence intervals for the parameters.

With **ANOVATable**, we can test the null hypothesis that the data could be described by a model containing only the constant term. A large F ratio and a small p value indicate that we can reject the null hypothesis.

The Durbin–Watson d statistic is between 0 and 4. Values near 2 mean uncorrelated errors (this is the assumption in regression analysis). Values that are less than 2 mean positive correlation, and values greater than 2 indicate negative correlation. In our model, the d statistic is near 2, so the errors are not correlated.

We show the data and the fit:

```
p2 = Plot[BestFit /. result, {x, 0, 50}, DisplayFunction → Identity];

Show[p1, p2];
```

Next we continue the analysis of our example by studying residuals, confidence region of the data, confidence region of the fitted curve, and confidence regions of the parameters.

■ Residuals

The residuals are as follows:

```
res = FitResiduals /.
    Regress[data, {1, x, x^2}, x, RegressionReport → FitResiduals];

xx = Map[First, data]; xres = Transpose[{xx, res}];
```

```
ListPlot[xres, PlotJoined → True, PlotRange → All,
    Epilog → {AbsolutePointSize[2], Map[Point, xres]}];
```

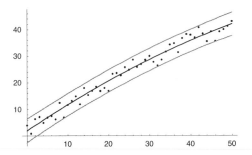

■ Confidence Region of the Data

With **SinglePredictionCITable**, we can ask for the confidence interval for *a single observed response* at each of the values of the independent variables. In this way we get a region that is likely to contain all possible observations. First we extract the components of the result: the observed values, the predicted values, the standard errors of the predicted response, and the confidence intervals:

```
{observed, predicted, se, ci} =
    Transpose[(SinglePredictionCITable /. Regress[data, {1, x, x^2},
        x, RegressionReport → SinglePredictionCITable])[[1]]];
```

Then we plot the data, the predicted values, and the lower and upper values of the 95% confidence intervals:

```
pred = Transpose[{xx, predicted}];
lowerCI = Transpose[{xx, Map[First, ci]}];
upperCI = Transpose[{xx, Map[Last, ci]}];

Show[Graphics[{AbsolutePointSize[1.5], Map[Point, data], Line[pred],
    GrayLevel[0.5], Line[lowerCI], Line[upperCI]}], Axes → True];
```

■ Confidence Region of the Curve

With **MeanPredictionCITable**, we can ask for the confidence interval for *the mean response* at each of the values of the independent variables. In this way we get a region that is likely to contain the regression curve. We do as we did above:

```
{observed, predicted, se, ci} =
    Transpose[(MeanPredictionCITable /. Regress[data, {1, x, x^2},
        x, RegressionReport → MeanPredictionCITable])[[1]]];
```

```
lowerCI = Transpose[{xx, Map[First, ci]}];
upperCI = Transpose[{xx, Map[Last, ci]}];

Show[Graphics[{AbsolutePointSize[1.5], Map[Point, data], Line[pred],
    GrayLevel[0.5], Line[lowerCI], Line[upperCI]}], Axes → True];
```

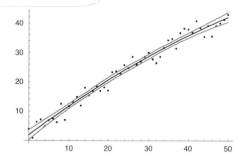

■ Confidence Regions of the Parameters

The correlation matrix gives the correlations between the parameters:

```
Regress[data, {1, x, x^2}, x, RegressionReport → CorrelationMatrix]
```

$$\{CorrelationMatrix \rightarrow \begin{pmatrix} 1. & -0.856214 & 0.7305 \\ -0.856214 & 1. & -0.967074 \\ 0.7305 & -0.967074 & 1. \end{pmatrix}\}$$

For example, the coefficients of 1 and x have a negative correlation -0.86 whereas 1 and x^2 have a positive correlation 0.73. The sign of the correlation can also be seen from the confidence regions of the parameters. We plot the 95% confidence region of the coefficients of 1 and x:

```
cr = ParameterConfidenceRegion[{1, x}] /. Regress[data, {1, x, x^2},
    x, RegressionReport → {ParameterConfidenceRegion[{1, x}]}]
```

```
Ellipsoid[{1.88055, 1.02404}, {2.09792, 0.099622},
  {{0.996865, -0.0791262}, {0.0791262, 0.996865}}]
```

```
Show[Graphics[cr], Axes → True];
```

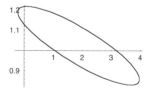

■ Designed Regress

In the **Statistics`LinearRegression`** *package:*

DesignedRegress[designmatrix, response] Fit the model represented by **designmatrix** given the vector **response** of response data
DesignMatrix[data, basis, vars] Give the design matrix

27.6.2 Nonlinear Regression

Recall from Section 22.1.4, p. 580, that we have **FindFit** (✲5) for nonlinear fitting (in versions of *Mathematica* prior to 5, we have **NonlinearFit** in the **Statistics`Nonlinear⌐Fit`** package). However, if you want statistical information about the fit, then use **NonlinearRegress**.

In the **Statistics`NonlinearFit`** *package:*

NonlinearRegress[data, funct, var, params]   When **funct** is an expression of the variable **var** and contains the parameters **params**, find values for the parameters such that the function fits the observations in **data** in the best way (in the sense of least squares)

Options:

WorkingPrecision   Precision used in internal computations; examples of values: **MachinePrecision** (✲5), **20**

PrecisionGoal   If the value of the option is **p**, the relative error of the χ^2 merit function should be of the order 10^{-p}; examples of values: **Automatic** (usually means **8**), **10**

AccuracyGoal   If the value of the option is **a**, the absolute error of the χ^2 merit function should be of the order 10^{-a}; examples of values: **Automatic** (usually means **8**), **10**

Method   Method used; possible values: **Automatic** (usually means **Levenberg⌐Marquardt**), **LevenbergMarquardt**, **Gradient**, **ConjugateGradient** (✲5), **Newton**, **QuasiNewton**

MaxIterations   Maximum number of iterations; examples of values: **100**, **200**

ShowProgress   Whether the result of each iteration is printed; possible values: **False**, **True**

Weights   List of weights for each data point or a pure function; default value: **Automatic**

Tolerance   Numerical tolerance for certain matrix operations; default value: **Automatic**

Gradient   Gradient used; examples of values: **Automatic**, **Symbolic** (✲5), **Finite⌐Difference** (✲5)

ConfidenceLevel   Used for confidence intervals; examples of values: **0.95**, **0.99**

RegressionReport   Statistics to be included in output; default value: **SummaryReport**

The parameter specification **params** is of the form **{aspec, bspec, cspec, … }** where all specifications are of the same form, which is one of the following:

a   Starting value is **1.0** for parameter **a**

{a, a0}   Starting value is **a0**

{a, amin, amax}   Starting value is a special point from the given interval; stop iteration if it goes outside of the interval

{a, a0, amin, amax}   Starting value is **a0**; stop iteration if it goes outside of the interval

The method of finding the best parameters is based on the so called χ^2 merit function, which is the sum of the squares of the residuals.

■ Obtainable Information

RegressionReport controls the amount of information printed. The default value **Summary Report** means the list **{BestFitParameters, ParameterCITable, EstimatedVariance, ANOVATable, AsymptoticCorrelationMatrix, FitCurvatureTable}**. All possible items can be seen by giving the command **RegressionReportValues[NonlinearRegress]**. Many of the items are the same for **NonlinearRegress** as they are for **Regress**, but the former also has five new items:

- To get the fit and information about the estimated parameters: **BestFit, BestFit⁚ Parameters, ParameterTable, ParameterCITable, ParameterConfidenceRegion, AsymptoticCovarianceMatrix** (new), **AsymptoticCorrelationMatrix** (new)

- To analyze variances: **ANOVATable, EstimatedVariance**

- To analyze predictions: **FitResiduals, PredictedResponse, SinglePrediction⁚ CITable, MeanPredictionCITable**

- To detect outliers: **HatDiagonal, StandardizedResiduals**

- To get other information: **FitCurvatureTable** (new), **ParameterBias** (new), **Start⁚ ingParameters** (new)

■ An Example

We consider the same model of exponential growth as in Section 22.1.4, p. 581:

```
<< Statistics`NormalDistribution`
```

```
SeedRandom[2];
data = Table[{x, Exp[0.3 + 0.2 x] + 0.5 Random[NormalDistribution[0, 1]]},
    {x, 0, 10, 0.2}];
```

```
p1 = ListPlot[data, AxesOrigin → {0, 0}];
```

```
<< Statistics`NonlinearFit`
```

```
result = NonlinearRegress[data, Exp[a + b x], x, {{a, 0}, {b, 0}},
    ShowProgress → True, RegressionReport → {BestFit, BestFitParameters,
    EstimatedVariance, ParameterTable, ParameterCITable}]
```

```
Iteration:1 ChiSquared:548.379 Parameters:{0.29844, 0.0653427}

Iteration:2 ChiSquared:109.963 Parameters:{0.607848, 0.123449}

Iteration:3 ChiSquared:16.167 Parameters:{0.439136, 0.188757}

Iteration:4 ChiSquared:12.5859 Parameters:{0.329881, 0.196837}

Iteration:5 ChiSquared:12.583 Parameters:{0.325602, 0.197223}

Iteration:6 ChiSquared:12.583 Parameters:{0.325638, 0.197218}
```

Iteration:7 ChiSquared:12.583 Parameters:{0.325637, 0.197218}

Iteration:8 ChiSquared:12.583 Parameters:{0.325637, 0.197218}

Iteration:9 ChiSquared:12.583 Parameters:{0.325637, 0.197218}

$\{$BestFit $\to e^{0.325637+0.197218\,x}$, BestFitParameters $\to \{$a $\to 0.325637$, b $\to 0.197218\}$,

EstimatedVariance $\to 0.256795$, ParameterTable $\to$

	Estimate	Asymp. SE	TStat	PValue
a	0.325637	0.0536284	6.0721	1.81201×10^{-7},
b	0.197218	0.00667518	29.545	0.

ParameterCITable $\to$

	Estimate	Asymptotic SE	CI
a	0.325637	0.0536284	$\{0.217867, 0.433408\}\}$
b	0.197218	0.00667518	$\{0.183804, 0.210632\}$

We plot the fit:

```
p2 = Plot[BestFit /. result, {x, 0, 10}, DisplayFunction → Identity];

Show[p1, p2, PlotRange → All];
```

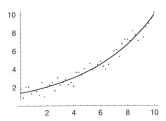

The residuals are as follows:

```
res = FitResiduals /. NonlinearRegress[data, Exp[a + b x],
    x, {{a, 0}, {b, 0}}, RegressionReport → FitResiduals];

xx = Map[First, data]; xres = Transpose[{xx, res}];

ListPlot[xres, PlotJoined → True, PlotRange → All,
    Epilog → {AbsolutePointSize[2], Map[Point, xres]}];
```

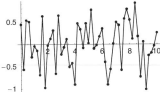

27.6.3 Local Regression

Sometimes the form of the data is so complex or obscure that it does not easily suggest a form for the approximating function, that is, a parametric family of functions. In such situations a *local regression* may be suitable (local regression or locally weighted regression falls in the category of *nonparametric regression*).

In local regression we choose a set of points from the range of the independent variable and fit a set of low-order polynomials, each polynomial describing the behavior of the data only near one of the chosen points (this is achieved by appropriately weighing the observations). Each polynomial is evaluated at the corresponding point, and so we obtain smoothed values. When these points are connected, the result is a local regression curve. Each part of the curve describes the average behavior of the data near that part.

In the following we apply the local regression method described in Cleveland (1993, p. 91–101); the method is also called *loess*. According to Cleveland, the method has some desirable statistical properties, is easy to compute (though computing intensive) and easy to use. We already presented an example of local regression in Section 27.3.2, p. 760, where we considered smoothing.

■ Explaining Ozone by Wind

To illustrate the method, we read the **environmental** data that comes on the CD-ROM of this book (we have already considered this data set in Section 9.5.1, p. 261; the same data is analyzed also in Cleveland (1994, p. 172–175)):

```
env = Rest[Import["/Users/ruskeepaa/
       Documents/MNData/visdata/environmental", "Table"]];
```

The data has 111 observations, each of which contains the number of the observation and the value of ozone, radiation, temperature, and wind. First we separate the components:

```
{no, ozone, radiation, temperature, wind} = Transpose[env];
```

We consider only wind and ozone:

```
data = Transpose[{wind, ozone}];

pdata =
   ListPlot[data, PlotStyle → AbsolutePointSize[1.5], AspectRatio → 1];
```

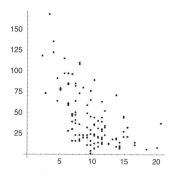

A descending pattern is clear, but otherwise the form of the data is somewhat obscure. A third-order polynomial fit is quite good:

```
fit = Fit[data, {1, x, x^2, x^3}, x]
```

$201.663 - 31.8806 x + 1.84631 x^2 - 0.0350707 x^3$

```
pfit = Plot[fit, {x, 2, 21}, DisplayFunction → Identity];
```

```
Show[pdata, pfit, PlotRange → {{0, 21}, All}];
```

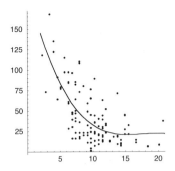

■ Local Regression

```
<< Statistics`LinearRegression`

T = Compile[{u}, If[Abs[u] < 1, (1 - Abs[u]^3)^3, 0]];

ww = Compile[{{i, _Integer}, {x, _Real}, {xx, _Real, 1}, {q, _Integer}},
   T[Abs[xx[[i]] - x] / Sort[Abs[xx - x]][[q]]]];

localRegress[data_, localpols_, α_, λ_] :=
 Module[{xx, ff, a, b, x, q, xwei, y, locfit},
   {xx, ff} = Transpose[data];
   a = Min[xx]; b = Max[xx];
   x = Range[a, b, (b - a) / (localpols - 1)];
   q = Floor[α Length[xx]];
   xwei = Map[{#, Table[ww[i, #, xx, q], {i, Length[xx]}]} &, x];
   locfit = Map[{#[[1]], BestFit /.
         Regress[data, Table[y^k, {k, 0, λ}], y, Weights → #[[2]] + 10^-15,
         RegressionReport → BestFit] /. y → #[[1]]} &, xwei];
   Interpolation[locfit]]
```

In the box above, we have programs to use to calculate a local regression curve. Now we try to explain the programs (you may want to first move to **Using the Programs** below and then come back here later on). The package **Statistics`LinearRegression`** has to be loaded, because we use **Regress** from that package (we cannot use **Fit**, because **Fit** does not have an option to weigh the data). The function **T** is the key to use to weigh the data; it is $(1 - |u|^3)^3$ for $|u| < 1$ and 0 otherwise. It looks like this:

```
Plot[T[u], {u, -1, 1}];
```

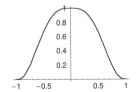

Most weight is given if the argument is near zero, and the weight is zero outside $(-1, 1)$. The function **ww** then computes the weights for the data. The wind varies between the following values:

```
{Min[wind], Max[wind]}     {2.3, 20.7}
```

Consider, for example, the wind observations with ordering numbers 1, 5, 9, 3, and 7:

```
Map[wind[[#]] &, {1, 5, 9, 3, 7}]

{7.4, 8.6, 9.2, 12.6, 20.1}
```

We plot the weights of these observations as functions of the point where the local low-order polynomial will be fitted:

```
pp =
  Map[Plot[ww[#, x, wind, 88], {x, 2.3, 20.7}, PlotRange → {{0, 21}, {0, 1}}
     Epilog → {AbsolutePointSize[3], Point[{wind[[#]], 0}]},
     Ticks → {Range[5, 20, 5], {1}}, PlotPoints → 100,
     DisplayFunction → Identity] &, {1, 5, 9, 3, 7}];

Show[GraphicsArray[pp],
   DisplayFunction → $DisplayFunction, GraphicsSpacing → -0.1];
```

For example, the first observation with wind value 7.4 receives most weight when a low-order polynomial is fitted at 7.4. For polynomials fitted at points far from 7.4, the first observation receives less weight.

Assume that we have n data points. Let $\Delta_i(x) = |x_i - x|$ be the distance between x_i and x, and let $\Delta_{(i)}(x)$ be the ith smallest of these distances. Let $\alpha \le 1$ be given, and let q be the product αn truncated to an integer. The function **ww** is $T(\Delta_i(x)/\Delta_{(q)}(x))$. If α is near 1, the smoothing of the data is strong. Lower values of α smooth less.

The program **localRegress** calculates the weight for each data point **xx[[i]]** and for each point **x[[k]]** where a local polynomial is calculated (this is quite a computing-intensive task, and to speed up the computations we have compiled the functions **T** and **ww**). Local polynomials are calculated a total of **localpols** times. For **localRegress**, we input also α and λ; λ is the degree of the local polynomials (either 1 or 2). The calculated points are connected by calculating a piecewise third-order interpolating function through the points (see Section 21.2.1, p. 555).

■ Using the Programs

We compute a local regression curve by computing 20 first-order polynomials ($\lambda = 1$) and using the value 0.9 for α:

```
fit = localRegress[data, 20, 0.9, 1]

InterpolatingFunction[{{2.3, 20.7}}, <>]
```

The result is an interpolating function. We plot it and show the curve together with the data:

```
pfit = Plot[fit[x], {x, 2.3, 20.7},
    PlotRange → All, DisplayFunction → Identity];

Show[pdata, pfit];
```

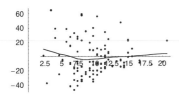

We use the following program to compute the residuals and a local regression curve for them (by calculating 20 local first-order polynomials with $\alpha = 0.8$).

```
showLocalResiduals[data_, fit_, opts___] :=
 Module[{xx, ff, resf, res, pres, resfit, presfit},
  {xx, ff} = Transpose[data];
  resf = ff - Map[fit[#] &, xx];
  res = Transpose[{xx, resf}];
  pres = ListPlot[res, DisplayFunction → Identity,
    Epilog → {AbsolutePointSize[1.5], Map[Point, res]}];
  resfit = localRegress[res, 20, 0.8, 1];
  presfit =
   Plot[resfit[x], {x, Min[xx], Max[xx]}, DisplayFunction → Identity];
  Show[pres, presfit, DisplayFunction → $DisplayFunction,
   PlotRange → All, opts];
  Print["Sum of squared residuals is ", resf.resf]]
```

```
showLocalResiduals[data, fit, AxesOrigin → {0, 0}];
```

```
Sum of squared residuals is 59829.9
```

The local fit of the residuals should be near zero. However, the local fit in the figure does not satisfy this property. Accordingly, we calculate a new curve—now using a lower value for α—to adapt the curve more closely to the data. We use the following program, which calculates a local regression curve and shows it together with the data.

```
showLocalRegress[data_, localpols_, α_, λ_, opts___] :=
 Module[{xx, ff, pdata, fit, pfit},
  {xx, ff} = Transpose[data];
  pdata = ListPlot[data, DisplayFunction → Identity,
    Epilog → {AbsolutePointSize[1.5], Map[Point, data]}];
  fit = localRegress[data, localpols, α, λ];
  pfit = Plot[fit[x], {x, Min[xx], Max[xx]}, DisplayFunction → Identity];
  Show[pdata, pfit, DisplayFunction → $DisplayFunction, opts];
  fit]
```

We use the value $\alpha = 0.6$:

```
fit = showLocalRegress[data, 20, 0.6, 1, AspectRatio → 1];
```

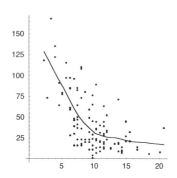

```
showLocalResiduals[data, fit, AxesOrigin → {0, 0}];
```

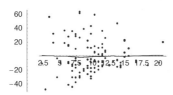

```
Sum of squared residuals is 58054.1
```

The fit seems good, and the local fit to the residuals is now practically zero.

27.7 Bayesian Statistics

27.7.1 Introduction

■ Posterior Joint Density

Suppose we have data $Y = (y_1, y_2, \ldots)$ and we want to describe the data with a model that contains unknown parameters $\theta = (\alpha, \beta, \gamma, \ldots)$. We have some prior information about the parameters in the form a probability density function $f(\theta)$, which is called the *prior (joint) density*. In addition, we know the conditional density $f(y_i \mid \theta)$.

We want to derive statistical information about θ based on the data Y. The solution is sought in the form of the density $f(\theta \mid Y)$, which is called the *posterior (joint) density*. From this density we can then calculate the mean, variance, and so on of θ. According to Bayes' theorem, $f(\theta \mid Y) = f(\theta) \, f(Y \mid \theta) / f(Y)$, or, in terms of proportionality, $f(\theta \mid Y) \propto f(\theta) \, f(Y \mid \theta)$. Assuming independent data (conditionally on θ), we have $f(Y \mid \theta) = \prod f(y_i \mid \theta)$, which is often called the *likelihood function*. So, we arrive at the formula $f(\theta \mid Y) \propto f(\theta) \prod f(y_i \mid \theta)$, which means that *posterior* $\propto$ *prior* $\times$ *likelihood*.

■ Example

Green (2001, p. 5) considers the following model. Let the data $Y = (y_1, \ldots, y_n)$ come from a normal distribution: $(y_i \mid \mu, \sigma) \sim N(\mu, \sigma)$, where $\mu \sim N(\xi, \frac{1}{\kappa})$, $\frac{1}{\sigma^2} \sim \Gamma(\alpha, \frac{1}{\beta})$; assume that μ and σ^{-2} are independent and that ξ, κ, α, and β are known. The parameters of interest are μ and σ^{-2}. The posterior joint density of these parameters is as follows:

$$f(\mu, \sigma^{-2} \mid Y) \propto f(\mu, \sigma^{-2}) \, f(Y \mid \mu, \sigma^{-2}) = f(\mu) \, f(\sigma^{-2}) \prod_{i=1}^{n} f(y_i \mid \mu, \sigma).$$

To write down this expression, we denote the argument of the density of σ^{-2} by `σ2i` (i for inverse) so that the prior joint density is as follows:

```
<< Statistics`ContinuousDistributions`
```

```
prior = PDF[NormalDistribution[ξ, 1 / κ], μ]
    PDF[GammaDistribution[α, 1 / β], σ2i] // PowerExpand
```

$$\frac{e^{-\frac{1}{2} \kappa^2 \, (\mu - \xi)^2 - \beta \, \sigma2i} \, \beta^\alpha \, \kappa \, \sigma2i^{-1+\alpha}}{\sqrt{2 \pi} \, \text{Gamma}[\alpha]}$$

To form the likelihood function $f(Y \mid \mu, \sigma^{-2})$, we look at the density function $f(y_i \mid \mu, \sigma)$:

```
PDF[NormalDistribution[μ, σ], yi]
```
$$\frac{e^{-\frac{(yi-\mu)^2}{2\sigma^2}}}{\sqrt{2 \pi} \, \sigma}$$

If we denote $\sum y_i$ with `sum` and $\sum y_i^2$ with `sum2`, we see that the likelihood function is as follows:

```
likelihood = (2 Pi) ^ - (n / 2) σ2i ^ (n / 2) Exp[-σ2i / 2 (sum2 - 2 μ sum + n μ^2)]
```

$$e^{-\frac{1}{2} (\text{sum2} - 2 \, \text{sum} \, \mu + n \, \mu^2) \, \sigma2i} \, (2 \pi)^{-n/2} \, \sigma2i^{n/2}$$

The posterior joint density $f(\mu, \sigma^{-2} \mid Y)$ is now proportional to the following:

```
prior likelihood // Simplify
```

$$\frac{1}{\text{Gamma}[\alpha]} \left(e^{-\frac{1}{2} (\kappa^2 \, (\mu-\xi)^2 + (\text{sum2}+2\beta-2\,\text{sum}\,\mu+n\,\mu^2) \, \sigma2i)} \, (2\pi)^{-\frac{1}{2}(1+n)} \, \beta^\alpha \, \kappa \, \sigma2i^{-1+\frac{n}{2}+\alpha} \right)$$

We can simplify this further by picking up only the terms that contain μ or σ^{-2}:

```
pμσ2i = Select[%, ! FreeQ[#, μ] || ! FreeQ[#, σ2i] &]
```

$$e^{-\frac{1}{2} (\kappa^2 \, (\mu-\xi)^2 + (\text{sum2}+2\beta-2\,\text{sum}\,\mu+n\,\mu^2) \, \sigma2i)} \, \sigma2i^{-1+\frac{n}{2}+\alpha}$$

Thus, the posterior joint density is proportional to this expression.

■ Posterior Marginal Densities

Posterior *marginal* densities $f(\alpha \mid Y)$, $f(\beta \mid Y)$, and so on can be obtained by integrating out all other parameters from the posterior joint density. For example, $f(\alpha \mid Y) = \int_{-\infty}^{\infty} d\beta \int_{-\infty}^{\infty} d\gamma \ldots f(\theta \mid Y)$. We consider four methods to calculate the posterior marginal densities: using integration, using interpolation, Gibbs sampling, and Markov chain Monte Carlo.

Using integration (see Section 27.7.2, p. 789). The integrals needed to calculate the marginal densities are often tedious or difficult (or both) to handle with pencil and paper, but mathematical programs like *Mathematica* should be tried for them. A program may be able to calculate difficult integrals, thereby reducing the need to resort to approximative methods. However, even *Mathematica* may not be able to calculate some of the integrals you need. Consider then the interpolation method.

Using interpolation (see Section 27.7.3, p. 792). This method relies on *Mathematica*'s ability to form a representation of a complicated expression by interpolation. Indeed, **Function**, **Interpolation** samples the expression at many points and forms an interpolating function that passes through these points (see Section 21.4.2, p. 566). With this method, we can get numerical but accurate representations of the posterior marginal densities. If this method fails, try Gibbs sampling.

Gibbs sampling (see Section 27.7.4, p. 793). This method uses the full conditional distributions of the parameters and requires that random numbers from these distributions be generated. The posterior marginal densities are approximated by using a random sample from the full conditional distributions. What if we do not have a random number generator for these distributions? Consider the next method.

Markov chain Monte Carlo (see Section 27.7.5, p. 797). This method is often shortened as "MCMC". The method has two significant advantages over Gibbs sampling. First, MCMC only requires the expressions of the full conditional densities up to proportionality. Second, random numbers are not required from these distributions. The posterior marginal densities are approximated by using random numbers from a suitably selected distribution for which we have a random number generator.

Note that the advanced methods of integration and interpolation of *Mathematica* may somewhat reduce the need to resort to Gibbs sampling or MCMC.

■ Data

In Sections 27.7.2 through 27.7.5, we continue with the above example. The starting point is always $p\mu\sigma 2\mathbf{i}$, which is the posterior joint density (up to proportionality) of the parameters μ and σ^{-2}. In all sections, we need the following package:

```
<< Statistics`ContinuousDistributions`
```

The data to be used are generated from $N(15, 2)$:

```
SeedRandom[1]; data = RandomArray[NormalDistribution[15, 2], {20}]
```

```
{15.8762, 13.4342, 15.9945, 15.99, 14.4053, 14.5665,
 16.4596, 14.7129, 11.3782, 13.1363, 15.6506, 13.6493, 15.686,
 16.8855, 13.2946, 14.3303, 15.6365, 16.0104, 16.0565, 14.6229}
```

Calculate their sum and sum of squares:

```
{datasum = Apply[Plus, data], datasum2 = Apply[Plus, data^2]}
```

```
{297.776, 4470.13}
```

(The symbolic values of these sums were denoted by **sum** and **sum2** in the example.) Use the following values of constants (which lead to [improper] noninformative priors):

```
vals = {ξ → 0, κ → 0, α → 0, β → 0, n → 20, sum → datasum, sum2 → datasum2};
```

Non-Bayesian estimates of μ and σ are as follows:

```
{Mean[data], StandardDeviation[data]}
```

```
{14.8888, 1.38794}
```

■ Posterior Joint Density of μ and σ^{-2}

Given the data and the constants, we get the following expression for the posterior joint density of μ and σ^{-2} in our example:

```
pμσ2i /. vals
```
$$e^{-\frac{1}{2}(4470.13 - 595.552\,\mu + 20\,\mu^2)\,\sigma 2i}\,\sigma 2i^9$$

Although the scaling constant of this density is lacking, we can still plot its contours of constant value, because the forms of the contours do not depend on scaling:

```
ContourPlot[%, {μ, 14.2, 15.6}, {σ2i, 0.2, 0.9}, PlotPoints → 40,
  FrameTicks → {{14.5, 15, 15.5}, {0.2, 0.4, 0.6, 0.8}, None, None},
  FrameLabel → {"μ", "σ⁻²"}, RotateLabel → False];
```

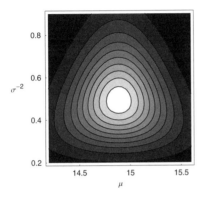

Later on, we do not need warnings about possible spelling errors, so we turn off that option:

```
Off[General::spell]
Off[General::spell1]
```

27.7.2 Using Integration

We continue the example of Section 27.7.1 and try to calculate the posterior marginal densities by direct integration. The marginal density of μ, which is $f(\mu \mid Y)$, is proportional to the following:

```
Integrate[pμσ2i, {σ2i, 0, ∞},
  Assumptions → {α, β, μ, κ, ξ, σ2i, n, sum, sum2} ∈ Reals]
```

```
Integrate::gener : Unable to check convergence. More…
```

$$\text{If}\left[n + 2\,\alpha > 0 \;\&\&\; 2\,\text{sum}\,\mu < \text{sum2} + 2\,\beta + n\,\mu^2,\right.$$
$$2^{\frac{n}{2}+\alpha}\,e^{-\frac{1}{2}\kappa^2\,(\mu-\xi)^2}\,(\text{sum2} + 2\,\beta + \mu\,(-2\,\text{sum} + n\,\mu))^{-\frac{n}{2}-\alpha}\,\text{Gamma}\left[\frac{n}{2}+\alpha\right],$$
$$\text{Integrate}\left[e^{-\frac{1}{2}(\kappa^2\,(\mu-\xi)^2 + (\text{sum2}+2\,\beta-2\,\text{sum}\,\mu+n\,\mu^2)\,\sigma 2i)}\,\sigma 2i^{-1+\frac{n}{2}+\alpha},\;\{\sigma 2i, 0, \infty\},\right.$$
$$\text{Assumptions} \to (\alpha \mid \beta \mid \mu \mid \kappa \mid \xi \mid \sigma 2i \mid n \mid \text{sum} \mid \text{sum2}) \in \text{Reals} \;\&\&$$
$$\left.\left.(n + 2\,\alpha \le 0 \;||\; \text{sum2} + 2\,\beta + n\,\mu^2 \le 2\,\text{sum}\,\mu)\right]\right]$$

To simplify the expression, we make the assumptions given in the result and, at the same time, select only the terms that contain μ:

```
pμ = Select[Integrate[pμσ2i, {σ2i, 0, ∞},
    Assumptions → n + 2 α > 0 && 2 sum μ < sum2 + 2 β + n μ^2], ! FreeQ[#, μ] &]
```

Integrate::gener : Unable to check convergence. More…

$$e^{-\frac{1}{2} \kappa^2 (\mu - \xi)^2} (\text{sum2} + 2 \beta + \mu (-2 \, \text{sum} + n \, \mu))^{-\frac{n}{2} - \alpha}$$

In the same way we get that the marginal density of σ^{-2}, which is $f(\sigma^{-2} \mid Y)$, is proportional to the following:

```
pσ2i = Select[Integrate[pμσ2i, {μ, -∞, ∞},
    Assumptions → κ^2 + n σ2i > 0], ! FreeQ[#, σ2i] &]
```

$$\frac{1}{\sqrt{\kappa^2 + n \, \sigma2i}} \left(e^{-\frac{\sigma2i \, (-2 \, \text{sum} \, \kappa^2 \, \xi + n \, \kappa^2 \, \xi^2 - \text{sum}^2 \, \sigma2i + \text{sum2} \, (\kappa^2 + n \, \sigma2i) + 2 \, \beta \, (\kappa^2 + n \, \sigma2i))}{2 \, (\kappa^2 + n \, \sigma2i)}} \, \sigma2i^{-1 + \frac{n}{2} + \alpha} \right)$$

Next we study these densities in more detail.

■ Posterior Density of μ

Calculate the scaling constant of $f(\mu \mid Y)$:

```
cμ = 1 / NIntegrate[pμ /. vals, {μ, 0, 30}]    5.47408 × 10^15
```

The posterior density of μ is then the following:

```
fμ = cμ pμ /. vals
```
$$\frac{5.47408 \times 10^{15}}{(4470.13 + \mu (-595.552 + 20 \, \mu))^{10}}$$

We plot the density:

```
plotfμ = Plot[fμ, {μ, 13, 17}];
```

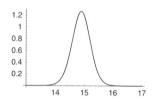

We then calculate its mean, median, and mode (note that from this on we do not show some warning messages about valid limits of integration):

```
NIntegrate[μ fμ, {μ, 10, 20}]    14.8888
```

```
FindRoot[NIntegrate[fμ, {μ, 10, medianμ}] == 0.5, {medianμ, 14, 15}]
```
$\{medianμ → 14.8888\}$

```
FindMinimum[-fμ /. μ -> modeμ, {modeμ, 15}][[2]]
```
$\{modeμ → 14.8888\}$

Calculate a 95% confidence interval:

```
q /. {FindRoot[NIntegrate[fμ, {μ, 10, q}] == 0.025, {q, 13, 14}],
    FindRoot[NIntegrate[fμ, {μ, q, 20}] == 0.025, {q, 15, 16}]}
```
$\{14.2392, 15.5384\}$

■ Posterior Densities of σ^{-2}, σ^2, and σ

Calculate the scaling constant of $f(\sigma^{-2} \mid Y)$:

co2i = 1 / NIntegrate[po2i /. vals, {o2i, 0, ∞}] 3.69265×10^7

The posterior density of σ^{-2} is then the following:

fo2i = co2i po2i /. vals $8.25701 \times 10^6 \; e^{-18.3007 \, o2i} \; o2i^{17/2}$

(This is a gamma distribution.) We plot the density:

Plot[fo2i, {o2i, 0, 1.2}];

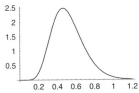

Actually we are interested in the posterior density of σ^2 or σ. Because the cumulative distribution function of σ^2 is $F_{\sigma^2}(s) = P(\sigma^2 \le s) = P(\frac{1}{\sigma^2} \ge \frac{1}{s}) = 1 - F_{\sigma^{-2}}(\frac{1}{s})$, then the density function of σ^2 is $f_{\sigma^2}(s) = \frac{1}{s^2} f_{\sigma^{-2}}(\frac{1}{s})$:

fo2 = (fo2i /. o2i → (1 / o2)) (1 / o2^2);

Plot[fo2, {o2, 0, 5}, PlotRange → All];

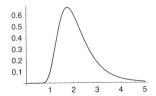

Calculate the posterior density of σ:

fo = (fo2 /. o2 → o^2) 2 o;

plotfo = Plot[fo, {o, 0.7, 2.5}, AxesOrigin → {0.7, 0}];

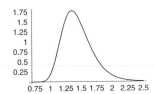

Its mean, median, and mode are as follows:

NIntegrate[o fo, {o, 0, 10}] 1.44591

```
FindRoot[
  NIntegrate[fo, {o, 0, mediano}, AccuracyGoal → 8] == 0.5, {mediano, 1, 2}]
```

$\{mediano \to 1.41279\}$

```
FindMinimum[-fσ /. σ → modeσ, {modeσ, 1.5}][[2]]
```

{modeσ → 1.3528}

Calculate a 95% confidence interval:

```
q /. {FindRoot[NIntegrate[fσ, {σ, 0, q}] == 0.025, {q, 1, 2}],
   FindRoot[NIntegrate[fσ, {σ, q, 10}, AccuracyGoal → 4] == 0.025, {q, 2, 3}]]}
```

{1.05552, 2.02719}

27.7.3 Using Interpolation

In Section 27.7.2, we succeeded in obtaining the posterior marginal densities of μ and σ^{-2} by integrating the posterior joint density. If the integration fails, one possibility is to use interpolation to get a close approximation of the marginal densities. In our example, we define anew the posterior joint density that we calculated in Section 27.7.1:

```
p[μ_, σ2i_] = pμσ2i /. vals
```

$$e^{-\frac{1}{2}(4470.13 - 595.552\,\mu + 20\,\mu^2)\,\sigma2i}\ \sigma2i^9$$

■ Posterior Density of μ

Suppose that (contrary to reality) *Mathematica* is not able to integrate the posterior joint density with respect to μ or σ^{-2}. We move to numerical integration. To calculate the posterior density of μ, we first define the numerical integral as a function of μ:

```
pμa[μ_?NumberQ] := NIntegrate[10^7 p[μ, σ2i], {σ2i, 0.001, 1}]
```

To help the numerical calculations, we multiplied the function by 10^7 and replaced the upper bound ∞ with 1. Now we can find a representation of the posterior density of μ as an interpolating function (see Section 21.4.2, p. 566):

```
pμ = FunctionInterpolation[pμa[μ], {μ, 10, 20}, MaxRecursion → 7]
```

```
InterpolatingFunction[{{10., 20.}}, <>]
```

Calculate the scaling constant:

```
cμ = 1 / NIntegrate[pμ[μ], {μ, 10, 20}]        1.48637
```

The density and its mean are as follows:

```
fμ = cμ pμ[μ];
```

```
Plot[fμ, {μ, 13, 17}];
```

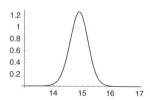

```
NIntegrate[μ fμ, {μ, 10, 20}]        14.8888
```

■ Posterior Densities of σ^{-2} and σ

Similarly we can calculate the posterior density of σ^{-2}:

```
pσ2ia[σ2i_?NumberQ] := NIntegrate[10^7 p[μ, σ2i], {μ, 10, 20}]

pσ2i = FunctionInterpolation[pσ2ia[σ2i], {σ2i, 0.001, 2.2}]

InterpolatingFunction[{{0.001, 2.2}}, <>]
```

Calculate the scaling constant:

```
cσ2i = 1 / NIntegrate[pσ2i[σ2i], {σ2i, 0.001, 2.2}]     1.47315
```

The density is then as follows:

```
fσ2i = cσ2i pσ2i[σ2i];

Plot[fσ2i, {σ2i, 0.001, 1.2}];
```

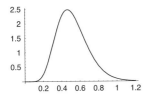

The posterior density of σ and its mean are as follows:

```
fσ = (fσ2i /. σ2i → (1 / σ^2)) 2 / σ^3;

Plot[fσ, {σ, 0.7, 2.5}, AxesOrigin → {0.7, 0}];
```

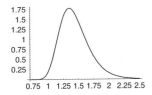

```
NIntegrate[σ fσ, {σ, 0.7, 5}]     1.44591
```

27.7.4 Gibbs Sampling

Gibbs sampling is based on the *full conditional distributions*. They are the conditional distributions of each parameter given the other parameters and the data. Thus, if the parameters are α, β, γ, ... and we have data Y, then the full conditional distributions are the distributions of $(\alpha \mid \beta, \gamma, \delta, ..., Y)$, $(\beta \mid \alpha, \gamma, \delta, ..., Y)$, $(\gamma \mid \alpha, \beta, \delta, ..., Y)$, ...

Because $f(\alpha \mid \beta, \gamma, ..., Y) = f(\alpha, \beta, \gamma, ... \mid Y) / f(\beta, \gamma, ... \mid Y)$ where the denominator does not depend on α, then $f(\alpha \mid \beta, \gamma, ..., Y) \propto f(\alpha, \beta, \gamma, ... \mid Y)$, which means that $f(\alpha \mid \beta, \gamma, ..., Y)$ is proportional to the terms of the posterior joint density that contain α.

■ Full Conditional Distribution of μ

In Section 27.7.1, p. 787, $f(\mu, \sigma^{-2} \mid Y)$ was proportional to

```
pμσ2i     e^{-\frac{1}{2} (κ² (μ-ξ)² + (sum2+2 β-2 sum μ+n μ²) σ2i)} σ2i^{-1+\frac{n}{2}+α}
```

Now we want to find full conditional densities $f(\mu \,|\, \sigma^{-2}, Y)$ and $f(\sigma^{-2} \,|\, \mu, Y)$. First, $f(\mu \,|\, \sigma^{-2}, Y)$ is proportional to the terms of **pμσ2i** that contain μ, which makes it proportional to the following:

    pμ = Select[pμσ2i, ! FreeQ[#, μ] &]

$e^{-\frac{1}{2}\,(\kappa^2\,(\mu-\xi)^2 + (\text{sum2} + 2\,\beta - 2\,\text{sum}\,\mu + n\,\mu^2)\,\sigma2i)}$

The exponent is a quadratic polynomial of μ:

    Collect[pμ[[2]], μ, Simplify]

$\frac{1}{2}\,\mu^2\,(-\kappa^2 - n\,\sigma2i) + \mu\,(\kappa^2\,\xi + \text{sum}\,\sigma2i) + \frac{1}{2}\,(-\kappa^2\,\xi^2 - \text{sum2}\,\sigma2i - 2\,\beta\,\sigma2i)$

Pick the coefficients of μ^2 and μ:

    {a = Coefficient[%, μ^2], b = Coefficient[%, μ]}

$\left\{ \frac{1}{2}\,(-\kappa^2 - n\,\sigma2i),\ \kappa^2\,\xi + \text{sum}\,\sigma2i \right\}$

More generally, consider a probability density function $f(x)$ that is proportional to an expression of the form $e^{a\,x^2 + b\,x + c}$ with $a < 0$. Because $a\,x^2 + b\,x + c = \frac{1}{2}\left(x + \frac{b}{2a}\right)^2 / \left(\frac{1}{2a}\right) + d$ for a constant d that does not depend on x, then the random variable has a normal distribution with mean $-\frac{b}{2a}$ and variance $-\frac{1}{2a}$. Accordingly, the full conditional distribution of μ is the following normal distribution:

    fμ = NormalDistribution[-b / (2 a), Sqrt[-1 / (2 a)]] // Simplify

$\text{NormalDistribution}\left[\dfrac{\kappa^2\,\xi + \text{sum}\,\sigma2i}{\kappa^2 + n\,\sigma2i},\ \sqrt{\dfrac{1}{\kappa^2 + n\,\sigma2i}}\ \right]$

■ **Full Conditional Distribution of** σ^{-2}

Recall $f(\mu, \sigma^{-2} \,|\, Y)$:

    pμσ2i $e^{-\frac{1}{2}\,(\kappa^2\,(\mu-\xi)^2 + (\text{sum2} + 2\,\beta - 2\,\text{sum}\,\mu + n\,\mu^2)\,\sigma2i)}\ \sigma2i^{-1 + \frac{n}{2} + \alpha}$

The density function $f(\sigma^{-2} \,|\, \mu, Y)$ is proportional to the terms of this expression that contain σ^{-2}. The density function of a general gamma distribution is proportional to the following:

    Select[PDF[GammaDistribution[γ, λ], x], ! FreeQ[#, x] &]

$e^{-\frac{x}{\lambda}}\,x^{-1 + \gamma}$

We now see that the full conditional distribution of σ^{-2} is a gamma distribution with the following parameters:

    {λ = -1 / Coefficient[pμσ2i[[1, 2]], σ2i],
      γ = 1 + Exponent[pμσ2i, σ2i]} // Simplify

$\left\{ \dfrac{2}{\text{sum2} + 2\,\beta + \mu\,(-2\,\text{sum} + n\,\mu)},\ \dfrac{n}{2} + \alpha \right\}$

So, the full conditional distribution of σ^{-2} is as follows:

    fσ2i = GammaDistribution[γ, λ]

$\text{GammaDistribution}\left[\dfrac{n}{2} + \alpha,\ -\dfrac{2}{-\text{sum2} - 2\,\beta + 2\,\text{sum}\,\mu - n\,\mu^2} \right]$

■ Gibbs Sampling

With Gibbs sampling, we need to generate random numbers (or sample or draw) from all of the full conditional distributions. This is easy if we can infer that each of these distributions is one of the well-known distributions (as above, in our example) and we have a random number generator for it.

Gibbs sampling is a method to sample the posterior joint distribution. The sample is then used to infer the form of the marginal densities. First select initial values α_0, β_0, ... for the parameters. In the ith step, draw from each full conditional distribution in turn, using the values of the previous step and the values already drawn in the present step. So, draw from $f(\alpha \mid \beta_{i-1}, \gamma_{i-1}, \delta_{i-1}, ..., Y)$, $f(\beta \mid \alpha_i, \gamma_{i-1}, \delta_{i-1}, ..., Y)$, $f(\gamma \mid \alpha_i, \beta_i, \delta_{i-1}, ..., Y)$, ... in turn until you have a random number from all full conditional distributions. Then go to the next step.

The values of the parameters from steps 0, 1, 2, and so on in Gibbs sampling, form a Markov chain. It can be shown that the chain converges to the posterior joint distribution and that the iterative sampling scheme draws—in the limit—a value from this distribution. In practice, a sample from the posterior joint distribution can be obtained by deleting some early draws and then considering the remaining draws to be—approximately—draws from the posterior joint distribution. For Gibbs sampling, see, Gamerman (1997) or Green (2001).

■ An Example

In our example, we first insert the numerical values of the constants into the full conditional densities:

```
gμ = fμ /. vals // PowerExpand
```

$$\text{NormalDistribution}\left[14.8888, \ \frac{1}{2\sqrt{5}\ \sqrt{\sigma 2 i}}\right]$$

```
gσ2i = fσ2i /. vals
```

$$\text{GammaDistribution}\left[10, \ -\frac{2}{-4470.13 + 595.552\,\mu - 20\,\mu^2}\right]$$

We initialize Gibbs sampling with the sample values:

```
rμ = Mean[data]; rσ2i = 1 / Variance[data];
```

Then we do 11,000 iterations of Gibbs sampling and drop the first 1000 points:

```
SeedRandom[3];
gibbs = Drop[Table[{rμ = Random[gμ /. σ2i → rσ2i],
    rσ2i = Random[gσ2i /. μ → rμ]}, {11000}], 1000];
```

The points are as follows:

```
ListPlot[gibbs, PlotRange → All, AxesOrigin → {13, 0}];
```

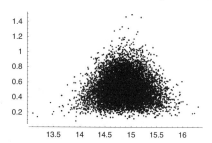

■ An Illustration

To illustrate the method, we plot 10 first points (the starting point is large in the plot):

```
{gibbsμ, gibbsσ2i} = Transpose[gibbs];

μμ = Flatten[Partition[Take[gibbsμ, 11], 2, 1]];

σσ = Drop[Prepend[
    Flatten[Partition[Take[gibbsσ2i, 11], 2, 1]], First[gibbsσ2i]], -1];

μμσσ = Transpose[{μμ, σσ}];

Show[Graphics[{AbsolutePointSize[2], Map[Point, Take[μμσσ, {1, 20, 2}]],
    Line[Drop[μμσσ, -1]], AbsolutePointSize[3.5], Point[First[μμσσ]]}],
    Axes → True, AxesLabel → {"μ", "σ⁻²"}, AxesOrigin → {14.3, 0.2}];
```

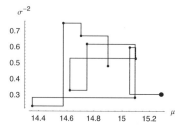

■ Posterior Histograms of μ and σ

The draws for μ and σ are as follows:

```
gibbsσ = 1 / Sqrt[gibbsσ2i];

ppp = ListPlot[Transpose[{gibbsμ, gibbsσ}],
    PlotRange → All, AxesOrigin → {13, 0.7}];
```

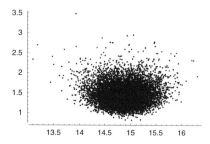

The histograms for μ and σ are as follows:

```
<< Graphics`Graphics`

hμ = Histogram[gibbsμ, HistogramCategories → Range[12.5, 17.5, 0.1],
    BarStyle → GrayLevel[0.7], HistogramScale → 1,
    DisplayFunction → Identity];

hσ = Histogram[gibbsσ, HistogramCategories → Range[0, 5, 0.1], BarStyle →
    GrayLevel[0.7], HistogramScale → 1, DisplayFunction → Identity];

Show[GraphicsArray[{hμ, hσ}]];
```

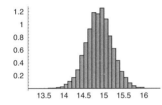 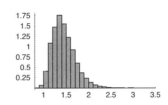

If we compare the histograms with the exact densities that we calculated in Section 27.7.2, p. 789, we can see a close agreement, which confirms that Gibbs sampling is a working method:

```
Show[GraphicsArray[{Show[hμ, plotfμ, DisplayFunction → Identity],
    Show[hσ, plotfσ, DisplayFunction → Identity]}]];
```

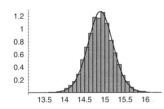 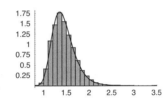

The means and variances of the samples are

```
{Mean[gibbsμ], Variance[gibbsμ]}      {14.8817, 0.11105}

{Mean[gibbsσ], Variance[gibbsσ]}      {1.44947, 0.0655214}
```

27.7.5 Markov Chain Monte Carlo (MCMC)

The great advantage of MCMC is that we do not need to *draw* from the full conditional distributions as in Gibbs sampling, but we need only to *evaluate* the full conditional densities at given points, and even the scaling constants are not needed. For MCMC, see Gilks, Richardson, and Spiegelhalter (1996), Gamerman (1997), or Green (2001).

MCMC has some variants. The basic method is called the *Metropolis method*, and a generalization is called the *Metropolis–Hastings method*. Actually, Gibbs sampling is also a variant of MCMC. We describe here the *random walk Metropolis method*.

Suppose the parameters of interest are α, β, γ, ..., and let $\theta = (\alpha, \beta, \gamma, ...)$, which is a d-vector. We have a posterior joint distribution $f(\theta \mid Y)$ from which we want to draw in order to estimate the posterior marginal distributions of the parameters (in literature,

$f(\theta \mid Y)$ is often denoted by $\pi(\theta)$ and called the target distribution). Let the successive draws be denoted as θ_1, θ_2,.... If we already have draws θ_1, ..., θ_i, then the next draw θ_{i+1} is obtained as follows.

Choose an arbitrary (really!) d-variate distribution that is symmetric around zero (a d-variate normal distribution with mean zero, for example), and draw a random number w_i from this density. Calculate $\phi_i = \theta_i + w_i$, which is called the *proposal*. The proposal is accepted with probability min $\{1, f(\phi_i \mid Y)/f(\theta_i \mid Y)\}$ (this can by implemented with drawing a random number that is uniform on $(0, 1)$). If the proposal is accepted, we set $\theta_{i+1} = \phi_i$, but if the proposal is rejected, we set $\theta_{i+1} = \theta_i$.

It can be shown that this Metropolis method generates a Markov chain that converges to the posterior joint distribution, and the iterative sampling scheme draws—in the limit—a value from this distribution. In practice, a sample from the posterior joint distribution can be obtained by deleting some early draws and then considering the remaining draws to be—approximately—draws from the posterior joint distribution.

■ Programming MCMC

We continue studying the model from Section 27.7.1, p. 786. The posterior joint density $p(\mu, \sigma^{-2} \mid Y)$ was shown to be proportional to the following:

```
p[μ_, σ2i_] = pμσ2i /. vals
```

$$e^{-\frac{1}{2} (4470.13 - 595.552\,\mu + 20\,\mu^2)\, \sigma2i}\ \sigma2i^9$$

We calculate the ratio needed in the acceptance probability of the Metropolis method:

```
ratio[{μ1_, σ2i1_}, {μ2_, σ2i2_}] = p[μ2, σ2i2] / p[μ1, σ2i1] // Simplify
```

$$\frac{1}{\sigma2i1^9}\left(e^{(2235.07 - 297.776\,\mu1 + 10.\,\mu1^2)\,\sigma2i1 + (-2235.07 + 297.776\,\mu2 - 10.\,\mu2^2)\,\sigma2i2}\ \sigma2i2^9\right)$$

Let $\theta = (\mu, \sigma^{-2})$. We generate the proposals with a two-variate normal distribution that has independent components with means of zero and known standard deviations:

```
proposal[θ_, std_] := θ + Map[Random[NormalDistribution[0, #]] &, std]
```

The proposals may need to satisfy some restrictions. For example, a draw for σ^{-2} has to be positive, so in our example we write the following test:

```
test[φ_] := If[φ[[2]] > 0, True, False]
```

The Metropolis method can then be written as follows:

```
metropolisStep[θ_, std_] := With[{φ = proposal[θ, std], r = Random[]},
  If[test[φ], If[r ≤ ratio[θ, φ], φ, θ], θ]]

metropolis[initialState_, std_, steps_] :=
  NestList[metropolisStep[#, std] &, initialState, steps]
```

Here, **metropolisStep** calculates one step of the method, and **metropolis** repeats the step a total of **steps** times. In this way, we get a sequence of random vectors.

■ **Adjusting the Standard Deviations**

The working of MCMC is greatly affected by the standard deviations of the distribution that gives the proposals. In our example, we first try the standard deviations 2 (used to draw μ) and 0.5 (used to draw σ^{-2}). We start from the sample estimates and generate 500 draws:

```
m = Mean[data]; s2i = 1 / Variance[data];

SeedRandom[3];
{metroμ, metroσ2i} = Transpose[metropolis[{m, s2i}, {2, 0.5}, 500]];
```

The draws for μ are as follows:

```
ListPlot[metroμ, AspectRatio → 0.2, PlotStyle → AbsolutePointSize[0.5]];
```

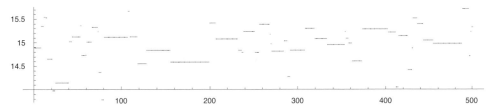

We see that the chain stays for long amounts of time in the same state before going to a new state. Calculate the different values of the chain:

```
Length[Union[metroμ]]      57
```

The small number of different values is an indication that the standard deviations are too large: a large standard deviation allows for large values for the proposal ϕ_i, and then the ratio $p(\phi_i \mid Y)/p(\theta_i \mid Y)$ is often small, which results in small acceptance probabilities and frequent rejection of the proposal.

We try smaller standard deviations 0.5 and 0.1:

```
SeedRandom[3];
{metroμ, metroσ2i} = Transpose[metropolis[{m, s2i}, {0.5, 0.1}, 500]];

ListPlot[metroμ, AspectRatio → 0.2, PlotStyle → AbsolutePointSize[0.5]];
```

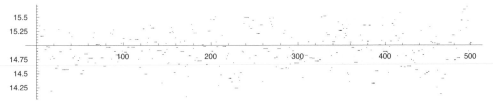

```
Length[Union[metroμ]]      274
```

Now the values vary more, and about half of the proposals are accepted. We then use still smaller standard deviations 0.1 and 0.02:

```
SeedRandom[3];
{metroμ, metroσ2i} = Transpose[metropolis[{m, s2i}, {0.1, 0.02}, 500]];
```

```
ListPlot[metroμ, AspectRatio → 0.2, PlotStyle → AbsolutePointSize[0.5]];
```

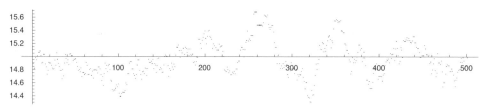

```
Length[Union[metroμ]]    449
```

Now about 90% of proposals are accepted, but the sequence has a high autocorrelation, and it moves too slowly to various parts of the domain of the posterior distribution of μ, which means that a long sequence will be needed to get a satisfactory estimate of the posterior density.

In conclusion, experimenting is often needed to adjust the standard deviations of the proposal distribution. The standard deviations should be small enough so that enough proposals will be accepted but large enough so that the autocorrelation is not too large.

Next we use the standard deviations 0.5 and 0.1 and study MCMC in more detail.

■ Using the Results

We calculate 21,000 steps and reject the first 1000, which are considered to be transient:

```
SeedRandom[3];
metro = Drop[metropolis[{m, s2i}, {0.5, 0.1}, 21000], 1000];
```

Calculate the number of different points:

```
Length[Union[metro]]    10103
```

The points are displayed as follows:

```
ListPlot[metro, PlotRange → All,
    AxesLabel → {"μ", "σ⁻²"}, AxesOrigin → {13.5, 0}];
```

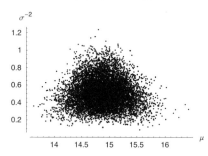

The draws for μ and σ^{-2} are as follows:

```
{metroμ, metroσ2i} = Transpose[metro];
```

```
ListPlot[metroμ, AspectRatio → 0.3, PlotRange → All,
   PlotStyle → AbsolutePointSize[0.5], AxesOrigin → {0, 13.5}];
```

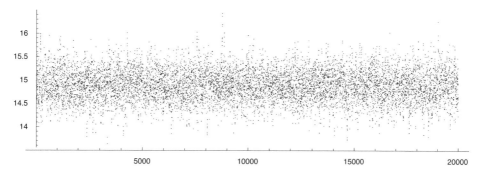

```
ListPlot[metroσ2i, AspectRatio → 0.3, PlotRange → All,
   PlotStyle → AbsolutePointSize[0.5], AxesOrigin → {0, 0}];
```

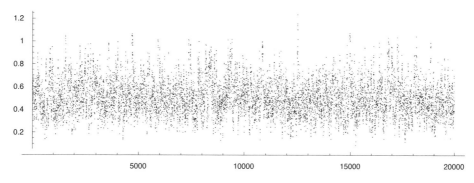

```
{Mean[metroμ], Mean[metroσ2i]}
```

```
{14.8865, 0.510997}
```

The draws for μ and σ are as follows:

```
metroσ = 1 / Sqrt[metroσ2i];
```

```
ppp = ListPlot[Transpose[{metroμ, metroσ}], PlotRange → All,
   AxesLabel → {"μ", "σ"}, AxesOrigin → {13.5, 0.8}];
```

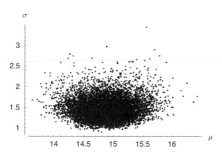

The histograms for μ and σ are as follows:

```
<< Graphics`Graphics`
```

```
hμ = Histogram[metroμ, HistogramCategories → Range[12.5, 17.5, 0.1];
    BarStyle → GrayLevel[0.7], HistogramScale → 1,
    DisplayFunction → Identity];

hσ = Histogram[metroσ, HistogramCategories → Range[0, 5, 0.1], BarStyle →
    GrayLevel[0.7], HistogramScale → 1, DisplayFunction → Identity];

Show[GraphicsArray[{hμ, hσ}]];
```

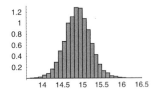 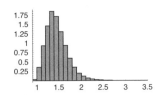

If we compare the histograms with the exact densities that we calculated in Section 27.7.2, p. 789, we can see a close agreement, which confirms that MCMC is a working method:

```
Show[GraphicsArray[{Show[hμ, plotfμ, DisplayFunction → Identity],
    Show[hσ, plotfσ, DisplayFunction → Identity]}]];
```

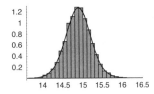 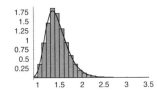

References

The references can also be found from the Index.

Abell, M. L., and J. P. Braselton (1997): *Differential Equations with Mathematica,* Second Edition. Academic Press; Boston.

Abell, M. L., J. P. Braselton, and J. A. Rafter (1999): *Statistics with Mathematica.* Adacemic Press; Boston.

Allen, A. O. (1990): *Probability, Statistics, and Queueing Theory: With Computer Science Applications,* Second Edition. Academic Press; Boston.

Bhatti, M. A. (2000): *Practical Optimization Methods with Mathematica Applications.* Springer; New York.

Borrelli, R. L., and C. S. Coleman (1998): *Differential Equations: A Modeling Perspective.* Wiley; New York.

Bulmer, M., and M. Carter (1996): Integer Programming with Mathematica. *The Mathematica Journal* 6(3), 28–36.

Burghes, D. N., and M. S. Borrie (1981): *Modelling with Differential Equations.* Ellis Horwood; Chichester.

Cheng, A. H-D., P. Sidauruk, and Y. Abousleiman (1994): Approximate Inversion of the Laplace Transform. *The Mathematica Journal* 4(2), 76–82.

Cleveland, W. S. (1993): *Visualizing Data.* Hobart Press; Summit, New Jersey.

Cleveland, W. S. (1994): *The Elements of Graphing Data.* Hobart Press; Summit, New Jersey.

Cox, D. R., and H. D. Miller (1965): *The Theory of Stochastic Processes.* Methuen; London.

Dennemeyer, R. (1968): *Introduction to Partial Differential Equations and Boundary Value Problems.* McGraw-Hill; New York.

Dickau, R. M. (1997): Compilation of Iterative and List Operations. *The Mathematica Journal* 7(1), 14–15.

Gamerman, D. (1997): *Markov Chain Monte Carlo: Stochastic Simulation for Bayesian Inference.* Chapman & Hall; London.

Ganzha, V. G., and E. V. Vorozhtsov (1996): *Numerical Solutions for Partial Differential Equations: Problem Solving Using Mathematica.* CRC Press; Boca Raton, Florida.

Gilks, W. R., S. Richardson, and D. J. Spiegelhalter (Eds.) (1996): *Markov Chain Monte Carlo in Practice.* Chapman & Hall; London.

Gray, J. W. (1997): *Mastering Mathematica: Programming Methods and Applications,* Second Edition. Academic Press; Boston.

Green, P. J. (2001): A Primer on Markov Chain Monte Carlo. In: O. E. Barndorff-Nielsen, D. R. Cox, and C. Klüpperberg (Eds.): *Complex Stochastic Systems.* Chapman & Hall; Boca Raton.

Guttorp, P. (1995): *Stochastic Modeling of Scientific Data.* Chapman & Hall; London.

Johnson, N. L., S. Kotz, and A. W. Kemp (1992): *Univariate Discrete Distributions,* Second Edition. Wiley; New York.

Kaplan, D., and L. Glass (1995): *Understanding Nonlinear Dynamics.* Springer; New York.

Keiper, J. (1993): Numerical Computation I. In: Wolfram Research: *Mathematica: Selected Tutorial Notes.* Wolfram Research; Champaign, Illinois.

Kelley, W. G., and A. C. Peterson (2001): *Difference Equations: An Introduction with Applications,* Second Edition. Academic Press; San Diego.

Knapp, R., and M. Sofroniou (1997): Difference Equations and Chaos in Mathematica: Symbolic and Numerical Mathematics at Work. *Dr. Dobb's Journal* 22(11), 84–90, 95–99. (A version of the article can be found as item number 0209-012 in *MathSource*.)

Kulenovic, M. R. S., and O. Merino (2002): *Discrete Dynamical Systems and Difference Equations with Mathematica.* Chapman & Hall/CRC; Boca Raton.

Kulkarni, V. G. (1995): *Modeling and Analysis of Stochastic Systems.* Chapman & Hall; London.

Kythe, P. K., P. Puri, and M. R. Schäferkotter (1996): *Partial Differential Equations and Mathematica.* CRC Press; Boca Raton, Florida.

MacHale, D. (1993): *Comic Sections: The Book of Mathematical Jokes, Humour, Wit and Wisdom.* Boole Press; Dublin.

Maeder, R. E. (1994): *The Mathematica Programmer.* Academic Press; Boston.

Maeder, R. E. (1995a): Single-Image Stereograms. *The Mathematica Journal* 5(1), 50–61.

Maeder, R. E. (1995b): Function Iteration and Chaos. *The Mathematica Journal* 5(2), 28–40.

Maeder, R. E. (1997): *Programming in Mathematica,* Third Edition. Addison-Wesley; Reading, Massachusetts.

Martelli, M. (1999): *Introduction to Discrete Dynamical Systems and Chaos.* Wiley; New York.

Mesterton-Gibbons, M. (1989): *A Concrete Approach to Mathematical Modelling.* Addison-Wesley; Redwood City, California.

Murrell, H. (1994): Planar Phase Plots and Bifurcation Animations. *The Mathematica Journal* 4(3), 76–81.

Nachbar, R. B. (1995): Genetic Programming. *The Mathematica Journal* 5(3), 36–47.

Pearl, R. (1927): The Growth of Populations. *Quarterly Review of Biology* 2(4), 532–548.

Rohatgi, V. K. (1984): *Statistical Inference.* Wiley; New York.

Rose, C., and M. D. Smith (2002): *Mathematical Statistics with Mathematica*. Springer, New York.

Sandefur, J. T. (1990): *Discrete Dynamical Systems: Theory and Applications*. Clarendon Press; Oxford.

Shaw, W. T., and J. Tigg (1994): *Applied Mathematica: Getting Started, Getting It Done*. Addison-Wesley; Reading, Massachusetts.

Skeel, R. D., and J. B. Keiper (1993): *Elementary Numerical Computing with Mathematica*. McGraw-Hill; New York.

Smith, C., and N. Blachman (1995): *The Mathematica Graphics Guidebook*. Addison-Wesley; Reading, Massachusetts.

Spiegel, M. R. (1963): *Schaum's Outline of Theory and Problems of Advanced Calculus*. McGraw-Hill; New York.

Spiegel, M. R. (1971): *Schaum's Outline of Theory and Problems of Calculus of Finite Differences and Difference Equations*. McGraw-Hill; New York.

Spiegel, M. R. (1999): *Mathematical Handbook of Formulas and Tables*, Second Edition. McGraw-Hill; New York.

Wagner, D. B. (1995): Dynamic Programming. *The Mathematica Journal* 5(4), 42–51.

Wagner, D. B. (1996): *Power Programming with Mathematica: The Kernel*. McGraw-Hill; New York.

Wickham-Jones, T. (1994): *Mathematica Graphics*. Springer; New York.

Wolfram, S. (2002): *A New Kind of Science*. Wolfram Media; Champaign, Illinois.

Wolfram, S. (2003): *The Mathematica Book,* Fifth Edition. Wolfram Media/Cambridge University Press; Champaign, Illinois.

Wolfram Research (2003): *Mathematica 5 Standard Add-on Packages*. Wolfram Media/Cambridge University Press; Champaign, Illinois.

Index

Mathematica names are written in this style: **Factorial**; menu commands are shown like this: **Abort Evaluation**. Names that begin with a lowercase letter are either names of programs, like **bifurcation**, or names of data sets, like **barley**. Names like **Algebra`Root`Isolation`** are packages. The adjectives one-, two-, three-, and four-dimensional are written as 1D, 2D, 3D, and 4D, respectively. *Mathematica* names beginning with **$** are listed at the end of the index.

Note that many of the names listed below are in packages. Such names are denoted by a bullet after the name, like **Animate•**. To use such a command, you first have to load the correct package, which is explained in Section 4.1.1.

Note also that if a cell in a *Mathematica* document is too long to fit at the bottom of a page, *Mathematica* may divide the cell into two parts, and the parts are printed on consecutive pages. However, index entries are associated with whole cells, and *Mathematica* has adopted the convention that the page number of an index entry will correspond with the page where the cell *ends*. Thus, the page numbers of the index entries that are associated with a divided cell and are located at the bottom of the first page are actually one less than their page number in the index.

Some topics listed below may deserve independent study. Such topics may include the following: animating, examples of; data sets; models, examples of; plots for data, advanced examples of; plots for functions, advanced examples of; programming, examples of graphics; programming, examples of numerical; programming, examples of symbolic; and two-image stereograms. Note that the CD-ROM contains all of the animations and all of the data sets that are used in the book.

TERM

This Agreement will remain in effect until terminated pursuant to the terms of this Agreement. You may terminate this Agreement at any time by removing from Your system and destroying the CD-ROM Product. Unauthorized copying of the CD-ROM Product, including without limitation, the Proprietary Material and documentation, or otherwise failing to comply with the terms and conditions of this Agreement shall result in automatic termination of this license and will make available to Elsevier Science legal remedies. Upon termination of this Agreement, the license granted herein will terminate and You must immediately destroy the CD-ROM Product and accompanying documentation. All provisions relating to proprietary rights shall survive termination of this Agreement.

LIMITED WARRANTY AND LIMITATION OF LIABILITY

NEITHER ELSEVIER SCIENCE NOR ITS LICENSORS REPRESENT OR WARRANT THAT THE INFORMATION CONTAINED IN THE PROPRIETARY MATERIALS IS COMPLETE OR FREE FROM ERROR, AND NEITHER ASSUMES, AND BOTH EXPRESSLY DISCLAIM, ANY LIABILITY TO ANY PERSON FOR ANY LOSS OR DAMAGE CAUSED BY ERRORS OR OMISSIONS IN THE PROPRIETARY MATERIAL, WHETHER SUCH ERRORS OR OMISSIONS RESULT FROM NEGLIGENCE, ACCIDENT, OR ANY OTHER CAUSE. IN ADDISTION, NEITHER ELSEVIER SCIENCE NOR ITS LICENSORS MAKE ANY REPRESENTATIONS OR WARRANTIES, EITHER EXPRESS OR IMPLIED, REGARDING THE PERFORMANCE OF YOUR NETWORK OR COMPUTER SYSTEM WHEN USED IN CONJUNCTION WITH THE CD-ROM PRODUCT.

If this CD-ROM Product is defective, Elsevier Science will replace it at no charge if the defective CD-ROM Product is returned to Elsevier Science within sixty (60) days (or the greatest period allowable by applicable law) from the date of shipment.

Elsevier Science warrants that the software embodied in this CD-ROM Product will perform in substantial compliance with the documentation supplied in this CD-ROM Product. If You report significant defect in performance in writing to Elsevier Science, and Elsevier Science is not able to correct same within sixty (60) days after its receipt of Your notification, You may return this CD-ROM Product, including all copies and documentation, to Elsevier Science and Elsevier Science will refund Your money.

YOU UNDERSTAND THAT, EXCEPT FOR THE 60-DAY LIMITED WARRANTY RECITED ABOVE, ELSEVIER SCIENCE, ITS AFFILIATES, LICENSORS, SUPPLIERS AND AGENTS, MAKE NO WARRANTIES, EXPRESSED OR IMPLIED, WITH RESPECT TO THE CD-ROM PRODUCT, INCLUDING, WITHOUT LIMITATION THE PROPRIETARY MATERIAL, AN SPECIFICALLY DISCLAIM ANY WARRANTLY OF MERCHANTABILITY OR FITNESS FOR A PARTICULAR PURPOSE.

If the information provided on this CD-ROM contains medical or health sciences information, it is intended for professional use within the medical field. Information about medical treatment or drug dosages is intended strictly for professional use, and because of rapid advances in the medical sciences, independent verification f diagnosis and drug dosages should be made.

IN NO EVENT WILL ELSEVIER SCIENCE, ITS AFFILIATES, LICENSORS, SUPPLIERS OR AGENTS, BE LIABLE TO YOU FOR ANY DAMAGES, INCLUDING, WITHOUT LIMITATION, ANY LOST PROFITS, LOST SAVINGS OR OTHER INCIDENTAL OR CONSEQUENTIAL DAMAGES, ARISING OUT OF YOUR USE OR INABILITY TO USE THE CD-ROM PRODUCT REGARDLESS OF WHETHER SUCH DAMAGES ARE FORESEEABLE OR WHETHER SUCH DAMAGES ARE DEEMED TO RESULT FROM THE FAILURE OR INADEQUACY OF ANY EXCLUSIVE OR OTHER REMEDY.

U.S. GOVERNMENT RESTRICTED RIGHTS

The CD-ROM Product and documentation are provided with restricted rights. Use, duplication or disclosure by the U.S. Government is subject to restrictions as set forth in subparagraphs (a) through (d) of the Commercial Computer Restricted Rights clause at FAR 52.22719 or in subparagraph (c)(1)(ii) of the Rights in Technical Data and Computer Software clause at DFARS 252.2277013, or at 252.2117015, as applicable. Contractor/Manufacturer is Elsevier Science Inc., 655 Avenue of the Americas, New York, NY 10010-5107 USA.

GOVERNING LAW

This Agreement shall be governed by the laws of the State of New York, USA. In any dispute arising out of this Agreement, you and Elsevier Science each consent to the exclusive personal jurisdiction and venue in the state and federal courts within New York County, New York, USA.

ELSEVIER SCIENCE CD-ROM LICENSE AGREEMENT

About the CD-ROM

The CD-ROM of *Mathematica Navigator* contains

- **the entire book** from Preface to Index, easily installable into the *Help Browser*;
- **all data sets** discussed in the book and several other data sets;
- **some packages** containing most programs developed in the book;
- the style sheet used in the book; and
- a document explaining the use of the AuthorTools package.

The CD-ROM also contains detailed instructions for installation and usage. When you have installed the book into the *Help Browser* of *Mathematica*,

- the entire book is accessible from within *Mathematica* so that you can easily read sections of the book, experiment with the examples, run animations, see the figures in color, and copy material from the book;

- the index entries of the book appear in the Master Index of the *Help Browser* so that you can easily find information about various topics, whether they are from the built-in material or from *Mathematica Navigator*; and

- the hyperlinks in the text and in the Index directly lead you to the appropriate points of the book.

The CD-ROM can be read using Windows, Macintosh (Mac OS X), and Unix computers.